Advances in Telephone
Survey Methodology

The Wiley Bicentennial–Knowledge for Generations

Each generation has its unique needs and aspirations. When Charles Wiley first opened his small printing shop in lower Manhattan in 1807, it was a generation of boundless potential searching for an identity. And we were there, helping to define a new American literary tradition. Over half a century later, in the midst of the Second Industrial Revolution, it was a generation focused on building the future. Once again, we were there, supplying the critical scientific, technical, and engineering knowledge that helped frame the world. Throughout the 20th Century, and into the new millennium, nations began to reach out beyond their own borders and a new international community was born. Wiley was there, expanding its operations around the world to enable a global exchange of ideas, opinions, and know-how.

For 200 years, Wiley has been an integral part of each generation's journey, enabling the flow of information and understanding necessary to meet their needs and fulfill their aspirations. Today, bold new technologies are changing the way we live and learn. Wiley will be there, providing you the must-have knowledge you need to imagine new worlds, new possibilities, and new opportunities.

Generations come and go, but you can always count on Wiley to provide you the knowledge you need, when and where you need it!

William J. Pesce
President and Chief Executive Officer

Peter Booth Wiley
Chairman of the Board

Advances in Telephone Survey Methodology

JAMES M. LEPKOWSKI

Institute for Social Research
University of Michigan
Ann Arbor, MI

CLYDE TUCKER

Bureau of Labor Statistics
U.S. Department of Labor
Washington, DC

J. MICHAEL BRICK

Westat
Rockville, MD

EDITH DE LEEUW

Department of Methodology and Statistics
Utrecht University
The Netherlands

LILLI JAPEC

Department of Research and Development
University of Stockholm
Stockholm, Sweden

PAUL J. LAVRAKAS

Nielsen Media Research
New York, NY

MICHAEL W. LINK

Centers for Disease Control and Prevention
Atlanta, GA

ROBERTA L. SANGSTER

Bureau of Labor Statistics
U.S. Department of Labor
Washington, DC

A JOHN WILEY & SONS, INC., PUBLICATION

Published by John Wiley & Sons, Inc., Hoboken, New Jersey
Published simultaneously in Canada

Wiley Bicentennial Logo: Richard J. Pacifico

Library of Congress Cataloging in Publication Data:

International Conference on Telephone Survey Methodology (2nd : 2006 : Miami, Florida)
Advances in telephone survey methodology / James M. Lepkowski … [et al.].
p. cm.
Includes bibliographical references and index.
ISBN 978-0-471-74531-0 (pbk.)
1. Surveys–Methodology–Technological innovations–Congresses. 2. Telephone surveys–Methodology–Technological innovations–Congresses. I. Lepkowski, James M. II. Title.
HA31.2.I563 2006
001.4′33–dc22 2007005999

Printed in the United States of America

10 9 8 7 6 5 4 3 2 1

This monograph is dedicated to the memory of Joseph Waksberg and Warren Mitofsky, who both passed away in 2006. The Mitofsky–Waksberg method of RDD sampling that they invented and developed was the cornerstone for the spectacular growth in telephone sampling in the United States. The two were close lifelong friends and collaborators, inspirations to those who had the good fortune to interact with them, and ambassadors for the use of sound statistical methods. While they had different styles and personalities, both were totally engaging enthusiastic colleagues who will be dearly missed by all.

Joe Waksberg was born in Poland in 1915 and immigrated to the United States in 1921. After graduating from City University of New York, he worked at the Census Bureau for 33 years, retiring in 1973 as the Associate Director for Statistical Methods, Research, and Standards. He then joined Westat, where he worked and became Chairman of the Board from 1990 until his death. He also consulted with CBS news and other networks on election nights from 1967 to 2004. Among many honors, the journal *Survey Methodology* invited paper issue honored his contributions to survey methodology.

Warren Mitofsky was born in the Bronx, New York, in 1934 and moved to Monticello, New York as a teenager. After graduating from Guilford College, he studied at the University of North Carolina and the University of Minnesota. He began his career at the U.S. Bureau of the Census before going to CBS News in 1967 to work on election coverage. He started the CBS News/New York Times Poll, conducting the first election-day polls. In 1990, he formed the Voter News Service, and in 1994 he began Mitofsky International. Among many honors, he received the American Association of Public Opinion Research lifetime achievement award.

Contents

Contributors xi

PART I PERSPECTIVES ON TELEPHONE SURVEY METHODOLOGY

1 **Telephone Survey Methods: Adapting to Change** 3
Clyde Tucker and James M. Lepkowski

PART II SAMPLING AND ESTIMATION

2 **Sampling and Weighting in Household Telephone Surveys** 29
William D. Kalsbeek and Robert P. Agans

3 **Recent Trends in Household Telephone Coverage in the United States** 56
Stephen J. Blumberg, Julian V. Luke, Marcie L. Cynamon, and Martin R. Frankel

4 **The Influence of Mobile Telephones on Telephone Surveys** 87
Vesa Kuusela, Mario Callegaro, and Vasja Vehovar

5 **Methods for Sampling Rare Populations in Telephone Surveys** 113
Ismael Flores Cervantes and Graham Kalton

6 **Multiplicity-Based Sampling for the Mobile Telephone Population: Coverage, Nonresponse, and Measurement Issues** 133
Robert Tortora, Robert M. Groves, and Emilia Peytcheva

7 **Multiple Mode and Frame Telephone Surveys** 149
J. Michael Brick and James M. Lepkowski

8 **Weighting Telephone Samples Using Propensity Scores** **170**
Sunghee Lee and Richard Valliant

PART III DATA COLLECTION

9 **Interviewer Error and Interviewer Burden** **187**
Lilli Japec

10 **Cues of Communication Difficulty in Telephone Interviews** **212**
Frederick G. Conrad, Michael F. Schober, and Wil Dijkstra

11 **Oral Translation in Telephone Surveys** **231**
Janet Harkness, Nicole Schoebi, Dominique Joye, Peter Mohler, Timo Faass, and Dorothée Behr

12 **The Effects of Mode and Format on Answers to Scalar Questions in Telephone and Web Surveys** **250**
Leah Melani Christian, Don A. Dillman, and Jolene D. Smyth

13 **Visual Elements of Questionnaire Design: Experiments with a CATI Establishment Survey** **276**
Brad Edwards, Sid Schneider, and Pat Dean Brick

14 **Mode Effects in the Canadian Community Health Survey: A Comparison of CATI and CAPI** **297**
Yves Béland and Martin St-Pierre

PART IV OPERATIONS

15 **Establishing a New Survey Research Call Center** **317**
Jenny Kelly, Michael W. Link, Judi Petty, Kate Hobson, and Patrick Cagney

16 **CATI Sample Management Systems** **340**
Sue Ellen Hansen

17 **Measuring and Improving Telephone Interviewer Performance and Productivity** **359**
John Tarnai and Danna L. Moore

18 **Telephone Interviewer Voice Characteristics and the Survey Participation Decision** **385**
Robert M. Groves, Barbara C. O'Hare, Dottye Gould-Smith, José Benkí, and Patty Maher

19 Monitoring Telephone Interviewer Performance 401
Kenneth W. Steve, Anh Thu Burks, Paul J. Lavrakas, Kimberly D. Brown, and J. Brooke Hoover

20 Accommodating New Technologies: Mobile and VoIP Communication 423
Charlotte Steeh and Linda Piekarski

PART V NONRESPONSE

21 Privacy, Confidentiality, and Respondent Burden as Factors in Telephone Survey Nonresponse 449
Eleanor Singer and Stanley Presser

22 The Use of Monetary Incentives to Reduce Nonresponse in Random Digit Dial Telephone Surveys 471
David Cantor, Barbara C. O'Hare, and Kathleen S. O'Connor

23 The Causes and Consequences of Response Rates in Surveys by the News Media and Government Contractor Survey Research Firms 499
Allyson L. Holbrook, Jon A. Krosnick, and Alison Pfent

24 Response Rates: How have they Changed and Where are they Headed? 529
Michael P. Battaglia, Meena Khare, Martin R. Frankel, Mary Cay Murray, Paul Buckley, and Saralyn Peritz

25 Aspects of Nonresponse Bias in RDD Telephone Surveys 561
Jill M. Montaquila, J. Michael Brick, Mary C. Hagedorn, Courtney Kennedy, and Scott Keeter

26 Evaluating and Modeling Early Cooperator Effects in RDD Surveys 587
Paul P. Biemer and Michael W. Link

References 619

INDEX 679

Contributors

The 2nd International Conference on Telephone Survey Methodology, which stimulated the chapters in this volume, could not have been held without financial support from government agencies, private firms, and academic research centers in the field of survey methodology. The contributors to the conference are listed below. The editorial team thanks each of these contributors for the financial and material assistances provided to the conference. This support is a sign of the growing intellectual and professional maturity of this field. The field is indebted to each of these contributors.

Sponsoring Organizations

Survey Research Methods Section, American Statistical Association (ASA)
American Association for Public Opinion Research (AAPOR)
Marketing Research Association (MRA)
Council of American Survey Research Organizations (CASRO)
International Association of Survey Statisticians (IASS)

Contributing Organizations

Abt Associates, Inc.
Arbitron, Inc.
Australian Bureau of Statistics
Bureau of the Census, U.S. Department of Commerce
Bureau of Labor Statistics, U.S. Department of Labor
Burke Marketing Research
Edison Media Research
Gallup Organization
Market Strategies, Inc.

Marketing Systems Group
Mathematica Policy Research
National Agricultural Statistics Service, U.S. Department of Agriculture
National Opinion Research Corporation
Neilsen Media Research
Opinion Research Corporation/MACRO
Pew Research Center for the People and the Press
Princeton Survey Research Associates
Research Triangle Institute
Scarborough Research
Schulman, Rona, and Bucuvalas
Statistical Research Service, U.S. National Science Foundation
Statistics Canada
Statistics Sweden
Survey Research Center, University of Michigan
Survey Research Group, RAND Corporation
Survey Sampling, International
Westat, Inc.
World Association for Public Opinion Research (WAPOR)

PART I

Perspectives on Telephone Survey Methodology

CHAPTER 1

Telephone Survey Methods: Adapting to Change

Clyde Tucker
U.S. Bureau of Labor Statistics, USA

James M. Lepkowski
University of Michigan, and Joint Program in Survey Methodology, University of Maryland, USA

1.1 INTRODUCTION

In 1987, the First International Conference on Telephone Survey Methodology was held in Charlotte, NC. The conference generated a widely read book on telephone survey methodology (Groves et al., 1988). Although that book continues to be a standard reference for many professionals, the rapid changes in telecommunications and in telephone survey methodology over the past 15 years make the volume increasingly dated. Considerable research has occurred since 1987, including myriad advances in telephone sampling in response to changes in the telecommunication system.

The goal of the Second International Conference on Telephone Survey Methodology was to once again bring together survey researchers and practitioners concerned with telephone survey methodology and practice in order to stimulate research papers that (1) contribute to the science of measuring and/or reducing errors attributable to telephone survey design, (2) provide documentation of current practices, and (3) stimulate new ideas for further research and development. This volume presents invited papers from the conference.

This chapter provides a brief introduction to the field, where it is today, and where it might be going. It begins by reviewing where the field stood at the time of

Advances in Telephone Survey Methodology, Edited by James M. Lepkowski, Clyde Tucker, J. Michael Brick, Edith de Leeuw, Lilli Japec, Paul J. Lavrakas, Michael W. Link, and Roberta L. Sangster

the 1987 conference and goes on to detail changes that have taken place since that time. Besides discussing the rapid changes in telecommunications and the social and political environments over the past two decades, the chapter considers ways telephone survey methodologists have adapted to these changes and what further adaptations may be needed in the future. The final section provides a brief overview of the contents of the volume.

1.2 THE CHANGING ENVIRONMENT

1.2.1 The Picture in 1987

Survey research began in the 1930s with the use of quota samples (Elinson, 1992). The controversy over the use of quota samples versus probability sampling lasted until the early 1950s. By that time, academicians and government statisticians had convinced most in the survey industry that probability sampling was a necessary ingredient in the proper conduct of surveys (Frankovic, 1992). But during all those years and through most of the 1960s, much of the survey research was conducted either by mail or through personal visits to households. The telephone was used largely for follow-up purposes. Certainly, the most important national surveys in the United States (e.g., the Gallup Poll, the Current Population Survey, and the National Election Study) were conducted face-to-face.

By the late 1960s, however, the costs of personal visits were escalating while, at the same time, the proportion of households with telephones had grown to close to 90 percent, both in North America and Europe. Furthermore, the decline in response rates in face-to-face surveys in especially the commercial sector made the possibility of using the telephone as a collection mode more attractive (Nathan, 2001). Concerns about the methodological shortcomings of telephone surveys were satisfied by the results of several studies conducted in the 1960s and 1970s (Hochstim, 1967; Sudman and Bradburn, 1974; Rogers, 1976; Groves and Kahn, 1979). Survey organizations began relying more and more on telephones for conducting surveys once random digit dialing (RDD) was introduced (Cooper, 1964; Nathan, 2001), even though the problem of locating residential numbers among the universe of possible numbers was daunting. That problem was solved by Warren Mitofsky and Joseph Waksberg with the invention of the two-stage Mitofsky–Waksberg telephone sampling methodology that took advantage of the fact that residential numbers tended to be clustered in 100-banks (Mitofsky, 1970; Waksberg, 1978), an approach also suggested but not fully developed by Glasser and Metzger (1972) and Danbury (1975). Thus, by the 1980s, telephone surveys were a part of standard survey practice; however, along with the growing reliance on the telephone survey came a number of methodological problems that had to be addressed.

It is in this environment that the first International Conference on Telephone Survey Methodology was held. One focus of that conference was telephone coverage, both in the United States (Thornberry and Massey, 1988) and in other countries (Steel and Boal, 1988; Trewin and Lee, 1988), and the potential for biased estimates as a result of ignoring nontelephone households, defined only as those without landline service. At the time, this was largely assumed to be a fixed state—a household always had service or it never did. A great deal of interest about within household sampling also existed (Whitmore et al., 1988; Oldendick et al., 1988; Maklan and Waksberg, 1988).

As today, there also was the focus on sample designs for telephone surveys, but the range of discussion was much more restricted. Several papers at the conference discussed refinements of the Mitofsky–Waksberg method (Burkheimer and Levinsohn, 1988; Alexander, 1988; Mason and Immerman, 1988). Treatment of list-assisted designs was mostly limited to methods for accessing individual-listed phone numbers directly, although there was some mention of the type of list-assisted design we know today (Groves and Lepkowski, 1986; Lepkowski, 1988). Dual-frame and mixed-mode designs were covered (Nathan and Eliav, 1988; Lepkowski, 1988; Sirken and Casady, 1988), but these studies dealt only with combining telephone and address frames in the context of telephone or personal visit surveys. Issues surrounding variance estimation, survey costs, and weighting were also addressed (Massey and Botman, 1988; Sirken and Casady, 1988; Mason and Immerman, 1988; Mohadjer, 1988).

One of the most important topics at the 1987 conference was computer-assisted telephone interviewing (CATI), which was relatively new at the time. As might be expected, one study compared the results from CATI and paper surveys (Catlin and Ingram, 1988). The construction of CATI questionnaires was the focus of several papers (Futterman, 1988; House and Nicholls, 1988). Designing and using CATI systems were also addressed (Sharp and Palit, 1988; Baker and Lefes, 1988; Weeks, 1988).

The administration of telephone surveys, a topic still of interest today, was also important in 1987. Of particular interest were best practices in the administration of centralized CATI centers (Whitmore et al., 1988; Berry and O'Rourke, 1988; Bass and Tortora, 1988). Other topics in this area included the performance of telephone interviewers (Pannekoek, 1988; Oksenberg and Cannell, 1988) and the optimal calling strategies (Kulka and Weeks, 1988; Alexander, 1988).

Finally, two areas of research that have grown greatly in importance over the past two decades were covered in 1987—nonresponse and measurement error. You will note that much of this research revolved around the comparison of telephone and face-to-face surveys. Certainly, this research was sparked by the growing concern over the rising costs of in-person visits. Could telephone surveys be a viable alternative?

Unit nonresponse rates were already considered a problem in 1987, and Groves and Lyberg (1988) made it clear that the situation was likely to get only worse. Refusal rates in the United States, Canada, and Britain were reported in two studies that examined the differences between face-to-face surveys and telephone surveys (Collins et al., 1988; Drew et at., 1988). A rather limited amount of research examined possible reasons for nonresponse in telephone surveys, but few definitive results were obtained. Only the effects of interviewers on nonresponse were notable. Collins and his colleagues found differences in interviewer response rates, and Oksenberg and Cannell (1988) found that response rates differed according to interviewer vocal characteristics. Interestingly, both studies found that the interviewer's ability to project confidence and competence coincided with higher response rates.

As for measurement error, a number of papers investigated indicators of data quality. Most of them looked at differences in estimates by mode, usually face-to-face and telephone. One paper was a meta-analysis of a number of comparisons between face-to-face and telephone surveys (de Leeuw and van der Zouwen, 1988). Nathan and Eliav (1988) looked at the consistency of reporting in a panel survey depending on the mode (telephone or face-to-face), and Kormendi (1988) examined the quality of income reporting in the two same modes. Sykes and Collins (1988) reported differences in estimates, again for telephone and face-to-face surveys, for a sensitive topic—close-ended, open-ended, and scale questions concerning alcohol consumption. Bishop et al. (1988) examined differences in data quality (the effects of question order and wording) in a telephone survey compared to a self-administered one. Just looking at data quality within a telephone survey, Stokes and Yeh (1988) evaluated the effects of interviewers on survey estimates.

1.2.2 Changes in Technology

Clearly, numerous challenges to conducting telephone surveys existed in 1987; however, those challenges may seem relatively small compared to the problems faced today. In 1987, the growth in the number of telephone service providers was yet to occur. The expansion in the number of area codes, leading to a dilution in the concentration of residential numbers among all available telephone numbers, was just in the planning stages. New technologies that coincided with the economic growth during the 1990s, such as answering machines, caller ID, and mobile phones, were not on the market. Computers were not yet in enough households to be considered as a vehicle for the administration of surveys. The public's concerns about privacy and confidentiality, while certainly present, had not reached a critical level. Important changes in the demographics of the U.S. population, such as the increased immigration of Hispanics in the 1990s, had not happened.

One of the most important developments since 1987 has been the rapid changes in telephony (Tucker et al., 2002). The number of area codes, and, thus, the total number of telephone numbers in the North American system, has almost doubled. The number of valid prefixes increased by 75 percent between 1988 and 1997, and today there are 90 percent more available telephone numbers. In contrast, the number of households has increased only a bit over 10 percent. As a result, the proportion of all telephone numbers assigned to a residential unit has dropped from about 0.25 to not more than 0.12. Figure 1.1 shows the relative change in the proportion of "active banks," those with one or more listed telephone numbers, since the late 1980s. Active banks have increased, but they now are a smaller percentage of all banks. A further complication is that it became more difficult to determine if numbers were residential because telephone business offices were less forthcoming with information and often inaccurate in their determinations (Shapiro et al., 1995). Screening numbers based on tritones was also problematic particularly in the west (Rizzo et al., 1995).

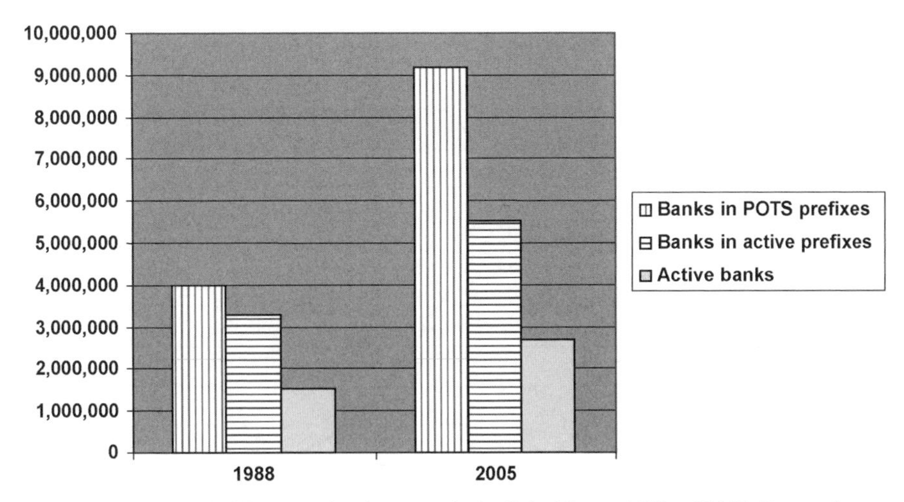

Figure 1.1. Number of 100-banks of various types in the United States, 1988 and 2005. (*Source*: Survey Sampling, Inc.)

There was also evidence that telephone companies appeared to be less systematic in the assignment of residential numbers across 100-banks. While the number of residences grew by just over 10 percent, the number of 100-banks with residential numbers has increased by over 50 percent. This increase in residential banks has resulted in a decline in the proportion of the numbers in a listed 100-bank that are residences. Figure 1.2 illustrates this change just for listed numbers. The decline has been from percentages in the low to middle 50s in 1990 to percentages in the upper 30s today. While the proportion of unlisted numbers is now approaching 30 percent in the United States and even higher in the United Kingdom (Collins, 1999), and largest in urban areas, this change cannot explain that much of the decline in the density of residential numbers in 100-banks.

In addition, there has been substantial growth in the number of households with multiple lines. Second lines dedicated to computers, fax machines, and home businesses have made it more difficult to distinguish noncontacts from nonworking numbers. Finally, there has been an increase in the assignment of whole prefixes to a single business customer. The identification of business numbers and the separation of those numbers from residential ones have become more problematic.

Accompanying these massive changes has been the amazing growth in telephone technology, and this has been a worldwide phenomenon. Besides fax machines and computers, call-screening devices have become commonplace in most homes. The first development in this area was the answering machine, and its presence in households grew dramatically during the 1990s (Nathan, 2001). In a recent Pew Research Center study (2006), almost 80 percent of U.S. households reported having either voice mail or an answering machine. Answering machines do have the advantage of potentially identifying residential numbers and allowing the interviewer to leave a message

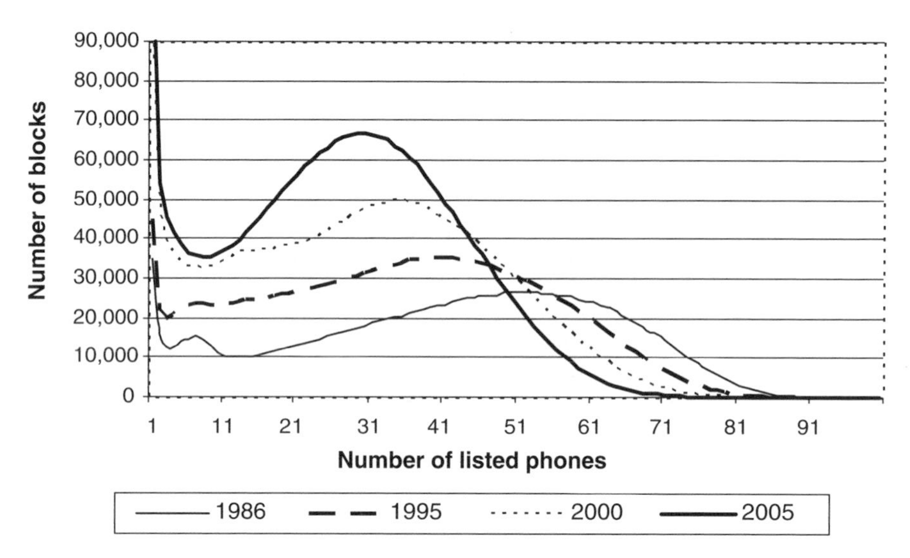

Figure 1.2. Frequency of 100-bank blocks by number of listed telephone numbers in the block. (*Source*: Survey Sampling, Inc.)

that distinguishes them from telemarketers, but they can also be used to screen out unwanted calls. With the advent of caller identification, or caller ID (one half of U.S. households now have it, according to the Pew Research Center (2006)), potential respondents could screen calls without waiting for a message to be left. Of course, households in countries such as Finland, which have heavy mobile-phone penetration (98 percent of adults have them (Kuusela et al., 2007, Chapter 4 in this volume), will have voice mail and caller ID. In fact, in Italy, the prevalence of answering machines has decreased due to the switch to mobile phones (Callegaro et al., 2007), and the proportion of households with answering machines has not changed over the past 6 years in France (Nathan, 2001; Vanheuverzwyn and Dudoignon, 2006).

While answering machines could be viewed as a device to increase the ability to communicate with the outside world, it is difficult to see how caller ID would do that. Seveal studies (Piazza, 1993; Xu et al., 1993; Oldendick and Link, 1994; Tuckel and O'Neill, 1996; Link and Oldendick, 1999) actually found that these new technologies were not having an appreciable effect on respondent cooperation in the mid-1990s, but, of course, they were unable to ascertain the effects from those potential respondents who do not answer the telephone. Now, according to the Pew study (2006), over 40 percent of U.S. households use caller ID, voice mail, or answering machines to screen calls.

Perhaps, the invention that will prove most disruptive to traditional telephone survey methodology is the mobile telephone. Until now, most telephone survey samples have been drawn only from banks of landline numbers. However, the widespread use of mobile technology, particularly the growing number of households with only

Table 1.1. Percent Distribution of Households by Telephone Status in the United States, 2004 (Current Population Survey)

Telephone status	Percent
Mobile and landline	46.4
Landline only	42.2
Mobile only	6.0
No telephone	5.4

mobile service, makes this methodology problematic. Table 1.1 gives the estimate of mobile-phone households in the United States in 2004 based on questions asked in a special supplement of the current population survey (Tucker et al., 2007). Figure 1.3 contains these numbers for the United States as well as selected European nations (Callegaro et al., 2005). Note that, while the United States had fewer than 6 percent mobile-only households, France and Italy had over 15 percent, and more than 33 percent of Finnish households were mobile-only.[1] Those individuals living

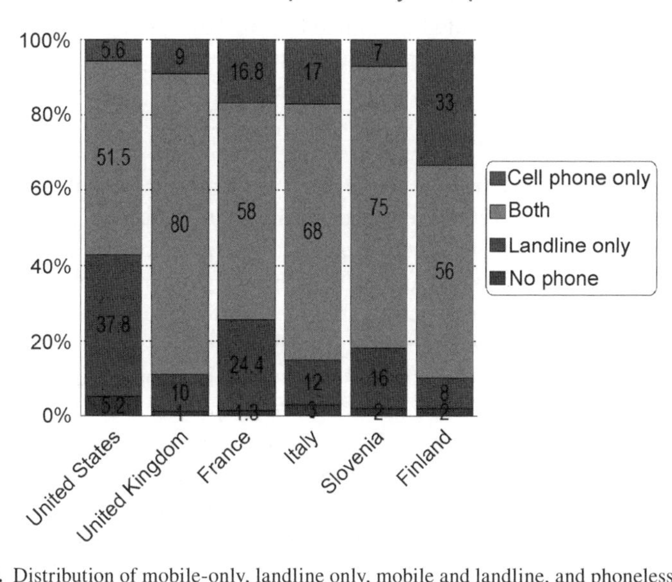

Figure 1.3. Distribution of mobile-only, landline only, mobile and landline, and phoneless households for six selected countries, 2004.

[1]Projections for the United States are, by the year 2007, the portion of mobile-only households will surpass the 10 percent level, and the percentage will be much larger for certain subpopulations, such as young, single adults.

in mobile-only households are more likely to be unemployed, have inadequate health care, and engage in risky behaviors compared to those in other households (Tucker et al., 2007; Blumberg, 2005).

Another concern is that half or more of the households in these countries had both mobile and landline phones. Those with both that rely mostly on their mobile phones might also be underrepresented in telephone surveys. Tucker et al., (2007) found that those households with both types of services receiving over half of their incoming calls on a mobile phone (a third of households with both) were more likely to be urban and younger.

Other technological developments could prove problematic to the conduct of telephone surveys (Piekarski, 2005). The Pew study (2006) reported that almost 20 percent of U.S. households have some form of electronic call blocking. Another new service, number portability, may undermine the ability to geographically target telephone samples in the years to come. Not only can a potential respondent move across the country and keep the same number, but also the number can be transferred to a wireless device without the caller being aware of that fact. Number portability has the additional troublesome effect of altering the probability of selection of a household without the respondent's knowledge. Call forwarding also presents the problem of conducting interviews on mobile phones, even when a landline has been called. Finally, with the rapid growth of home computers will come the switch to voice over Internet protocol (VoIP). Steeh and Piekarski (2007, Chapter 20 in this volume) report that it is estimated that up to 75 percent of worldwide voice traffic could be handled by VoIP by 2007. VoIP could increase the uncertainty of the location of particular telephone numbers, and VoIP service, of course, will be affected by electrical outages.

The advances in technology have had an additional effect on telephone surveys. With rapid computerization, the creation of the Internet and digital technology, new modes of survey administration have become available. As of fall 2003, the U.S. Department of Commerce (2004) reported that almost 62 percent of households had a computer and about 55 percent of households had Internet access at home (20 percent had broadband). The Harris Poll Online (2006) recently reported that over 60 percent of households now have Internet access at home. The web survey is the most widely studied alternative to the telephone survey, either as a stand-alone method or as one alternative in a multimode survey that also includes telephone administration. Many web surveys, especially marketing surveys, have employed nonprobability samples (Fischbacher et al., 1999; Poynter, 2000). Some establishment surveys early on used probability-based web designs, at least in the multimode context (Nusser and Thompson, 1998; Clayton and Werking, 1998; Tedesco et al., 1999). The same was true for some household surveys (Couper, 2000b). More recently, internet households have been recruited using probability-based methods, most notably RDD (Couper, 2000b). In some cases, the survey is administered only to those with Internet access, destroying the representativeness of the sample (Flemming and Sonner, 1999). Others attempt to remain representative by providing recruited respondents not already online with Internet access (Rivers, 2000; Couper, 2000b). Two problems arise in these latter surveys. Often the recruitment rate for what are ongoing Internet panels is quite low. In a 2003 report, Knowledge Networks (Pineau and Slotwiner, 2003) indicated that, while coverage of the U.S. population was initially 96 percent for its panels, only 37 percent of the households contacted initially agreed to participate. In addition, the recruitment process and the installation of Internet equipment can prove costly when compared to telephone surveys with response rates at the same level or higher. Amortization across the Internet panel is possible, but then there is the matter of attrition.

As summarized by Nathan (2001), other advances in electronic communication have found their way into telephone surveys. Telephone surveys can now be conducted without live interviewers (computer-assisted self-interviewing or CASI) using interactive voice recognition (IVR) and touchtone data entry (TDE). These methods, especially TDE, were tested for both establishment and household surveys at the U.S. Bureau of Labor Statistics (Werking et al., 1988; Clayton and Winter, 1992; McKay et al., 1994). Turner et al. (1998) tested the use of telephone CASI for measuring sensitive items.

1.2.3 Changes in the Sociopolitical Climate

At least in the United States, significant societal changes have accompanied the technological changes over the past two decades. Some of these changes could pose problems for those engaging in telephone survey research by increasing nonresponse and/or exacerbating measurement error. Table 1.2 provides information on changes that might be related to respondent cooperation. In some cases, actually little has changed. The total hours worked is about the same; although, given the change in household size, the hours per household member has gone up. This might be explained by the fact that the unemployment rate in 2004 was 0.7 percent lower than in 1987. Hours spent at home is a little less now, but hours spent in leisure is

Table 1.2. Change in Select Characteristics of Households and Persons in Households Effecting Respondent Cooperation, United States (American Time Use Survey)

	Annual average	
Characteristic	1987	2004
Total hours worked in household	126.32	125.10
Percent households with only 1 member	7.82	10.66
Percent households with Hispanic reference person	7.06	11.09
Mean household size	2.64	2.52
	1985	**2003**
Hours spent at home per day	6:38	6:16
Hours spent in leisure time per day	4:33	4:59
	1987	**2004**
Percent married, but only male working	21.3	16.9
Percent of families with only male head and in labor force	13.5	17.6
Percent of expenditures spent on eating out	37.2	39.5

higher. There is little increase in the amount of eating out, at least in terms of the percentage of food expenditure. What has changed is the characteristics of households themselves. The percentage of one-person households (with the average size of the household declining) has increased and so has the percentage of Hispanic households. The percentage of households where the woman stays at home has declined considerably, continuing a trend begun in the 1950s that accelerated in the 1970s and 1980s. Note also the growth in the number of households headed by only one worker. These changes may have led to increased time pressures within the American household and could explain the increases in noncontact rates as well as the decline in cooperation rates in telephone surveys over the past two decades. The larger percentage of Hispanic households may be just one indicator of the growing diversity among households making standardized survey procedures less effective. An important question is what is happening in other countries with respect to demographic change?

The attitudes of respondents (and of governments), especially in the areas of privacy, confidentiality, and respondent burden, have also changed over time. Singer and Presser (2007, Chapter 21 in this volume) present evidence in this volume of these attitudinal shifts based on a review of a number of studies involving mail, face-to-face, and telephone surveys. One note of caution is that studies of privacy, confidentiality, and respondent burden are plagued by the lack of information from nonrespondents, and, as we will see later, nonresponse in telephone surveys has been increasing. Singer and Presser found that the public's growing concerns about privacy and confidentiality coincided with the increased alienation and mistrust of government that began during the Vietnam War and Watergate (Westin, 1967). (For information on the explicit connection between political events and response rates, see Harris-Kojetin and Tucker (1999).) In more recent years, these concerns may have intensified with the advent of more sophisticated methods for monitoring the actions of the individual that came about as the result of computerization. The easy access to personal information, especially in the commercial sector, may have fueled the concern about privacy and confidentiality. Also, some technologies may result in more sensitivity to privacy concerns. Calling on a mobile phone may be considered more of an invasion of privacy than calling on a landline.

In general, privacy concerns in the United States have increased over the years. A good part of that increase, however, occurred prior to 1987. This is not surprising given the links to Vietnam and Watergate. Figure 1.4 shows the changes in trust of the Federal government from the National Election Studies since 1958. Note that a steady decline took place from 1964 to 1980. Although trust increased somewhat after that, it has never reached the levels recorded in the early 1960s, and, in fact, 1994 was a low point. That was the year the Republicans had the "Contract with America" and took control of the House of Representatives. There does seem to be more concern on the part of the U.S. public about privacy and confidentiality relative to the government than to business. In other countries, the findings on concerns about privacy and confidentiality vary widely by country. Some countries had increases in concern across time, and others had decreases. To the extent that privacy

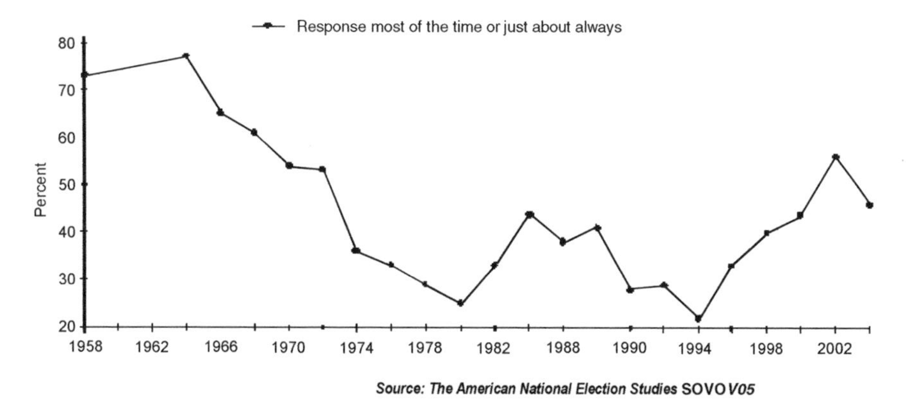

Figure 1.4. Trend in trust in U.S. federal government, American National Election Studies, 1958–2003.

concerns (and survey participation) are related to alienation and mistrust, perhaps, a more telling indicator is the decline over time in election turnout in several democracies. Figure 1.5 shows these trends, compiled largely by the International Institute for Democracy and Electoral Assistance, for selected democracies. In most of the countries, there has been a decline in turnout in general elections, with only the Netherlands and the United States countering these trends. The declines (particularly in Japan, Canada, Italy, and the United Kingdom) tend to be greater than in earlier periods.

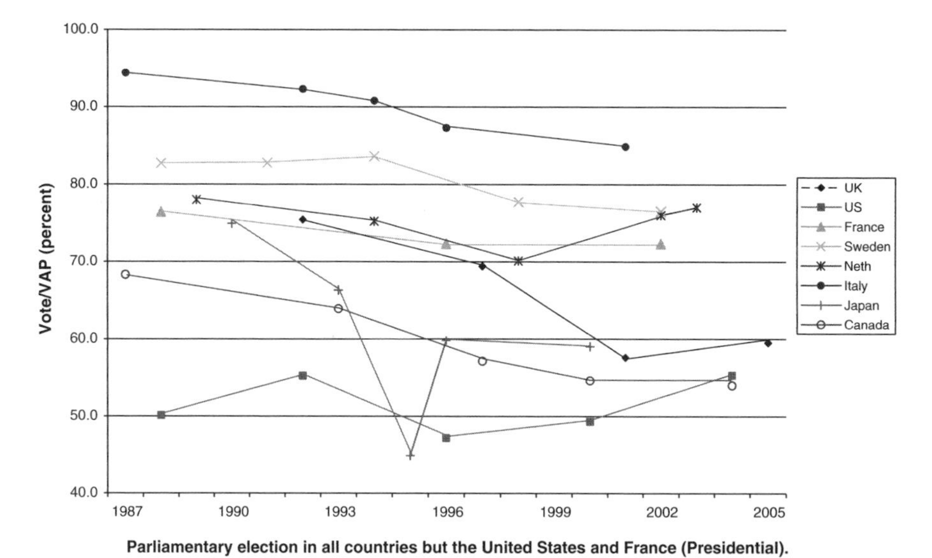

Figure 1.5. Turnout in presidential or parliamentary elections in eight selected countries, 1987–2005. Tucker, C., *Telephone Survey Methods: Adapting to Change*; © Institute for Democracy and Electoral Assistance, 2006.

Singer and Presser report that feelings about privacy and confidentiality seem to have relatively small, but detrimental effects, on a potential respondent's willingness to participate in a survey, and this effect is accentuated in the case of item nonresponse for sensitive items. But it does seem clear that the participation of those concerned about privacy and confidentiality is lower than that for those who are not. Unfortunately, assurances of confidentiality on the part of the survey organization can have either a positive or negative effect on response rates.

In terms of respondent burden, one clear finding from Singer and Presser's work is that the public either is or perceives it is receiving more survey requests today than ever before. The relationship between actual respondent burden (measured in terms of length of the interview) is less clear; perhaps, because refusals often come before the potential respondent knows the length of the survey (DeMaio, 1980). In contrast, the length of the interview could be related to the quality of the respondent's answers. Tucker et al., (1991) found this to be the case for telephone interviews collecting information about consumer expenditures. Finally, perceived burden can be much different from actual burden, and actual burden can depend on how hard a respondent is willing to work at the survey task. Tucker (1992) found that younger respondents to a consumer expenditure survey reported lower burden than older respondents, but the expenditure reports received from the older respondents were of a higher quality than those from younger respondents.

Whether or not potential survey respondents are more concerned about privacy and confidentiality today than in the past, it certainly is true that governments are. In 1991, the U.S. Congress passed the Telephone Consumer Protection Act (TCPA), and the Federal Communications Commission (FCC) issued a directive the next year restricting the use of automatic and predictive dialers (Dautch, 2005). The FCC rules apply only to telemarketers except in the case of mobile phones, where survey researchers also are prohibited from using autodialers. In 2003, the FCC established the Do Not Call (DNC) registry. Some states already had such lists. Telemarketers, but not legitimate survey researchers, were prohibited from calling the numbers placed on the list by individual consumers. The European Union (EU) passed the Directive on Data Protection in 1995 to protect the confidentiality of an individual respondent's survey data. Britain's Communications Act of 2003 provides legal recourse for those disturbed by unwanted calls, and Canada introduced similar legislation in 2004 (Singer and Presser, 2007, Chapter 21 in this volume).

Although concerns about confidentiality, privacy, and respondent burden have grown over the years (along with the decline in trust in government), there is no clearly documented correlation at the microlevel between these trends and the increases in nonresponse in telephone surveys, just as it is difficult to establish a one-to-one connection between demographic changes and changes in survey participation. In contrast, the trends do coincide over the past generation, at least in the United States, and may not bode well for the future of telephone surveys.

1.2.4 Problems Resulting from Changes

One of the more important issues resulting from the growth in the universe of telephone numbers was the increasing inefficiency of RDD designs. It became more difficult to find residential numbers using the old methods. The problem was exacerbated by the unsystematic way numbers were now being assigned. The likelihood of locating a residential number using the Mitofsky–Waksberg design declined from just over 60 percent to less than 40 percent (Steeh and Piekarski, 2007, Chapter 20 in this volume). This situation posed a problem for production telephone surveys. Either the number of 100-banks needed or the number of numbers *(k)* selected per bank would have to be increased. In either case, more would have to be spent to get a given effective sample size. If the *k* were increased, the intraclass correlation would be larger, resulting in a greater variance, and the number of calls to reach *k* households would rise. In addition, a number of survey organizations were finding the Mitofsky–Waksberg method cumbersome to manage, and because it was a clustered design, the variances were greater than desired.

With the changes in technology came the increase in the number of telephone lines into the household. Besides dedicated fax and computer lines, there were now mobile phone numbers and, with the growth of telecommuting, probably more lines in the home dedicated to business. Associated with these developments was the increase in the difficulty of determining the number of lines on which respondents could actually be reached by survey organizations. Furthermore, because of dedicated computer and fax lines as well as caller ID, it was not an easy task to identify the reasons for ring-no-answers (RNAs). At least, with answering machines or voice mail, the ability to locate a residence may have reduced the number of RNAs, but this would not necessarily lead to respondent cooperation.

The explosion in technology also gave rise to the consideration of multimode surveys. In particular, the combination of a telephone survey and a Web survey seemed attractive. Weeks (1992) discussed alternative modes that had been made possible through technological advances. One problem with multimode surveys, however, is that they open up the possibility of mode effects on estimates. Secondly, it was not clear if the design of CATI surveys could be easily transferred to the Web, or whether new design principles would have to be applied. Furthermore, the Web realistically could be used only as an alternative mode for those households with Internet capabilities.

The growth in the mobile-only population posed a coverage problem, but conducting a survey over mobile phones was problematic. How easy would it be to contact residential mobile phone owners? What about the problem of finding enough mobile-only households? Given the increasing current noncontact rates, should households with both landlines and mobile service also be candidates for a mobile phone survey? Could mobile phone respondents be convinced to do the survey, especially in the United States, where the mobile phone owner pays for the call? There were also ethical issues with which to contend (e.g., the appropriateness of conducting a survey while the respondent is driving). Perhaps, the most troublesome problem was a conceptual one. Does conducting a survey on mobile phones imply changing the sampled unit from the household to the individual?

The changes in society (both technological and social) over the past 20 years made it increasingly difficult to maintain response rates in telephone surveys with the passage of time. This situation created an even greater threat of nonresponse bias in survey estimates. Battaglia et al., (2007, Chapter 24 in this volume) illustrate the decline in response rates with examples from telephone surveys ranging from the Survey of Consumer Attitudes (SCA) (Curtin et al., 2005) to the state-based Behavioral Risk Factors Surveillance System (BRFSS). Recently, Curtin et al., (2005) showed that the overall response rate in the SCA declined considerably from 1997 to 2003 at the average annual rate of 1.5 percentage points to 48.0 percent. The National Household Education Survey (2004) reported a decline in the response rate from 72.5 percent in 1999 to 62.4 percent in 2003, an annual rate of decline of 2.5 percentage points. The U.S. Centers for Disease Control and Prevention (2003) reported that the BRFSS indicated a decline in the median response rate for the 50 states from 68.4 percent in 1995 to 53.2 percent in 2003 (an average decline of 1.9 percentage points per year). The RDD component of the National Survey of America's Families (2003) reported a decline in the overall response rate from 65.1 percent in 1997 to 62.4 percent in 1999 and to 55.1 percent in 2002 among the surveys of children, and 61.8, 59.4, and 51.9 percent among the adult surveys in 1997, 1999, and 2002, respectively. Finally (Holbrook et al., 2007, Chapter 23 in this volume), in their review of surveys from 1996 through 2003 conducted by 14 private U.S. survey organizations, found a strong negative correlation between year of the survey and response rate. They also reported that contact rates declined more than refusal rates. The same was true for face-to-face U.S. government surveys (Atrostic et al., 2001). de Leeuw and de Heer (2002) found that this trend toward increasing nonresponse rates held when looking across a number of countries.

1.3 ADAPTING TO THE CHANGING ENVIRONMENT

1.3.1 Adapting to Changing Technology

In the late 1980s, survey methodologists began searching for alternatives to the Mitofsky–Waksberg methodology. One such attempt was a dual-frame design that combined estimates from a sample of listed numbers with an RDD sample (Groves and Lepkowski, 1986). At the same time that there was growing dissatisfaction with the Mitofsky–Waksberg approach, several companies were developing sophisticated methods of processing files of residential listings, including Donnelly Marketing Information Systems. Although the listed telephone number frame itself was not suitable for direct sampling of telephone numbers because a substantial share of telephone households did not appear in the frame, by sampling numbers from 100-banks that contained listed telephone numbers, efficiencies obtained in the second stage of the Mitofsky–Waksberg could nearly be achieved. Sample selection could be simple-random, or stratified-random, selection of telephone numbers from across 100-banks containing one or more listed telephone numbers. The loss in precision due to cluster sampling was eliminated, and samples generated from list-assisted methods were less cumbersome to implement than the two-stage cluster method of Mitofsky–Waksberg. The residential hit rate for these designs was usually above 50 percent.

List-assisted methods were examined by Casady and Lepkowski (1993), who laid out the statistical theory underlying the design and presented empirical analysis on the properties of stratified list-assisted design options. Brick et al. (1995) showed that the potential bias resulting from the loss of residential telephones in 100-banks without listed numbers was small. Government agencies, academic survey organizations, and private survey firms subsequently adopted list-assisted designs.

As time went on, two commercial firms, Survey Sampling International and the Marketing Systems Group, appended information from other data sources to the records of listed numbers to make tailored list-assisted designs available along with automated sample delivery systems. Although the residential hit rate began to decline in the late 1990s, recent developments in methods to purge ineligible numbers from samples has brought the hit rates back to near the levels of the original Mitofsky–Waksberg design. The major question remaining was how many listed numbers should a 100-bank contain before being included in the sampling frame. Survey organizations today vary as to the minimum number of listed numbers required for a bank to enter the frame. This number is usually between 1 and 3, but with the better purging techniques, it may be possible in the future to actually include banks with no listed numbers without substantially adding to screening costs. To evaluate the efficiency of the list-assisted designs in the context of the growth in the size of the telephone number universe, Tucker et al., (2002) repeated Casady and Lepkowski's earlier analysis. They found that the list-assisted designs, while somewhat less efficient in an absolute sense, were relatively more efficient than the Mitofsky–Waksberg design than they had been 10 years earlier.

Another area of research has to do with selection of the respondent within the household. Binson et al. (2000) examined the effectiveness of using the Kish, Next-Birthday, or Last-Birthday method for selecting a respondent. As DeMaio (1980) found, most refusals occur before the respondent selection stage. It appears that the Kish method, requiring all household members be listed prior to respondent selection, causes the greatest problem. However, this problem seems to be an interviewer problem more than a respondent problem. In contrast, the birthday method was more prone to error (Lavrakas et al., 1992; Lavrakas et al., 2000; Lind et al., 2000). Rizzo et al. (2004a) proposed a more streamlined version of the Kish method that produced accurate results.

Coverage problems also have been a subject of study since 1987, particularly the coverage of nontelephone households. Weighting for undercoverage did not always solve the problem (Massey and Botman, 1988; Brick et al., 1992). Keeter (1995) suggested using households with intermittent telephone service to adjust for the coverage of nontelephone households. Using panel surveys, he showed that the characteristics of these "transient" households were much closer to those for nontelephone households than to those of households that always had telephone service. Brick et al. (1996b) found that collecting information on past breaks in telephone service in a cross-sectional survey could also be used to improve coverage in the same way Keeter had described. Brick et al. also used sociodemographic characteristics to improve the estimates.

As discussed in the previous section, in order to take advantage of the new technologies developed in the 1990s, survey methodologists began experimenting with multimode designs. Even with the growing problems with telephone surveys, it is likely to be retained as an important mode in multimode designs. Besides their work on combining directory listings and RDD, Groves and Lepkowski began looking at mixed-mode designs in the 1980s (Groves and Lepkowski, 1985). So did Whitmore et al. (1985). In 1997, a study of the combination of RDD and area sampling was reported by Waksberg et al. (1997) and Cunningham et al. (1997) in which the non-telephone households found in the area sample were provided with mobile phones for completing the CATI interview. Both the British Labor Force Survey (Wilson et al., 1988) and the one in the United States (current population survey) now use a mixed-mode design involving in-person interviews in the first interview and telephone interviewing in later months (U.S. Department of Labor and U.S. Department of Commerce, 2002). Pairing Internet surveys with telephone surveys is now a growing possibility (de Leeuw 2005). Biemer and Lyberg (2003) report that mixed-mode surveys are quite common today, and they have been treated in recent textbooks (Czaja and Blair, 1996; Groves et al., 2004).

There are several reasons for considering the use of a mixed-mode design. One is that coverage problems with each mode can be compensated for using the other mode, particularly when surveying special populations (Waksberg et al., 1997; Cunningham et al., 1997; Nathan, 2001; Edwards et al., 2002; Parackal, 2003; Srinath et al., 2004). As de Leeuw (2005) reports, one of the most important reasons for using a multimode design is to reduce nonresponse (Poe et al., 1990; Fowler et al., 2002). This is particularly the case when the modes are used sequentially (Jackson and Boyle, 1991; Japec, 1995; Alexander and Wetrogan, 2000; Grembowski and Philips, 2005; Dillman and Christian, 2005). The use of a Web survey in conjunction with a telephone survey is attractive because of the relatively low cost of the Web alternative (Couper, 2000; Dillman, 2000). One particular advantage that the Web survey might have over a face-to-face or telephone survey is eliciting more truthful answers regarding sensitive behaviors. Like mail surveys, the respondent is not reporting these behaviors directly to an interviewer (Aquilino and Losciuto, 1990; Presser, 1990; de Leeuw, 1992; Tourangeau et al., 1997; Fowler et al., 1998; Currivan et al., 2004). Sometimes face-to-face surveys may elicit more truthful answers than telephone surveys, but the evidence is mixed (Aquilino, 1994; Aquilino and Wright, 1996; Couper et al., 2003; Holbrook et al., 2003). de Leeuw (2005) did find a combination of the Web and telephone modes to have become more popular in market surveys.

Of course, as mentioned earlier, mixed-mode designs imply the possibility of mode effects in estimates. de Leeuw (1992) attributes mode differences to three possible sources—how a medium is used, how the information is transmitted, and the presence or absence of the interviewer. de Leeuw found that these factors appeared to have only small effects on estimates. Dillman et al., (1995) reported little in the way of primacy and recency effects when comparing mail and telephone surveys. When looking at the differences between web and telephone surveys, de Leeuw (2005) found some evidence of more extreme answers over the telephone (Oosterveld and Willems, 2003; Dillman and Christian, 2005) but with few other differences (Oosterveld and Willems, 2003). It is possible that mode effects are quite complicated and not fully understood at this point (Biemer, 2001; Voogt and Saris, 2005). Unfortunately, the truth of the matter probably lies in the interactions between the mode, the interviewer, the respondent, and the survey content and presentation.

Although not a mixed-mode design, given the growth of mobile phone ownership (especially mobile-only households), another design that must be mentioned is a dual-frame design using both landline and mobile numbers. This type of design is very new and little is known about it. After several studies, Steeh (2004a) concluded that conducting a survey with mobile phone respondents can be challenging and require a different approach than used in traditional landline surveys. Brick et al. (2006) found that producing an unbiased composite estimate using this type of design was extremely difficult.

Over the past 20 years, some work has also proceeded on weighting and estimation issues in telephone surveys. One of the more important areas has been how to produce estimates for dual-frame and mixed-mode designs. Casady et al. (1981), Groves and Lepkowski (1985), and Kalton and Anderson (1986) had done work in this area prior to the 1987 conference. Skinner and Rao (Skinner, 1991; Skinner and Rao, 1996) extended the methodology by incorporating auxiliary data, like Brick et al. (1996), to better inform the compositing of the estimates. Variance estimation for these types of composite estimates was studied by Lohr and Rao (2000).

The other aspect of weighting that has been explored to some extent since 1987 is weighting for the number of phone lines in the household. Obviously, this task has been complicated by technological innovations. To determine the correct number of lines, the researcher must eliminate dedicated fax and computer lines as well as exclude mobile phone lines (if only sampling landlines). The problem has been exacerbated by the fact more workers are now telecommuting. Fortunately, with the growth of DSL and cable, the confusion over the number of lines has been reduced to some extent (Piekarski, 2005). Lavrakas (1993) and Czaja and Blair (1996) discuss procedures for determining the correct number of lines and using the information in weighting.

The other area of development over the past 20 years, and one that will have more prominence in the next section, is the cognitive aspects of survey methodology (CASM) movement. CASM received little, if any, attention at the 1987 conference even though the first CASM seminar was held in 1983 (Jabine et al., 1984). Since the 1987 conference, the study of the role of cognition in surveys has exploded. A second CASM seminar was held in the 1990s (Sirken et al., 2000), and the subject of the last international conference on survey methodology was devoted to questionnaire design and testing (Presser et al., 2004). Several papers in that conference specifically addressed questionnaire design for different modes of administration, including CATI (Tarnai and Moore, 2004). Cognitive methods also play a role in usability testing of CATI instruments (Presser and Blair, 1994; Schaeffer and Maynard, 1996; Hansen and Couper, 2004). Of course, there is now a rich literature on questionnaire issues and pretesting for all types of surveys (Tourangeau et al., 2000; Groves et al.,

2004; Conrad and Schober, 2005). Many of these studies look specifically at questionnaire design issues in telephone surveys that sometimes consider mode effects (Dillman and Tarnai, 1991; Dovidio and Fazio, 1992; Fowler, 1992; Conrad and Schober, 2000; Dashen and Fricker, 2001).

1.3.2 Adapting to the Changing Sociopolitical Climate

Certainly sociopolitical changes have occurred, at least in the United States, over the past 20 years. The characteristics of American households have changed, fairly dramatically in some ways; and concerns about privacy and confidentiality seem to receive a lot more attention now than 20 years ago. To the extent that these changes have affected telephone surveys, they have made it more difficult to successfully conduct these surveys. This fact is most clearly seen in the greater effort needed to locate and gain the cooperation of respondents (Curtin et al., 2000; Brick et al., 2003). This is reflected in the search for new methods to gain cooperation, the increasing costs of survey projects, and the greater focus on response rates and their calculation (Groves, 1989; McCarty, 2003; AAPOR, 2004).

The changing character of the American family, especially its growing diversity in several respects, has played a part in some survey researchers call for a move away from the standardized survey advocated by Fowler and Mangione (1990) toward tailoring procedures to the respondent (Tucker, 1992; Groves and Couper, 1998). This would include altering the survey introduction or methods for refusal conversion depending on the respondent's characteristics, and these efforts have also been driven by the application of psychology to the survey situation (Dijkstra and van der Zouwen, 1987; van der Zouwen et al., 1991; Morton-Williams, 1993; Groves and McGonagle, 2001; Mayer and O'Brien, 2001; Dijkstra and Smit, 2002; O'Brien et al., 2002; Shuttles et al., 2002). Others have argued that the interview should actually have more of the characteristics of a conversation (Schaeffer, 1991; Conrad and Schober, 2000). Besides improving response rates, this approach could reduce measurement errors.

Clearly, the move away from standardization is influenced by psychological theory. This involves not only cognitive but also social psychology. For example, Groves et al. (1992) argued that the theory of reciprocity and compliance had much to offer in explaining the willingness of respondents to agree to the survey request, and the same was true with the leverage-saliency theory proposed by Groves et al. (2000). Certainly, the recent turn to the use of monetary incentives (itself, an increase in the level of effort and, of course, cost) follows from the notions of reciprocity laid out by Groves et al. On the contrary the effects of incentives are quite complicated. First of all, the promise of an incentive in RDD surveys is less effective than one delivered in person or in the mail (prepaid) (Cantor et al., 2007, Chapter 22 in this volume). Furthermore, while incentives generally will increase response rates, their effects tend to decrease with increased level of effort and the size of the incentive (Singer et al., 1999b; Brick et al., 2005c). A number of factors besides the incentive will affect the decision to participate (Groves and Couper, 1998). For instance, the sponsor of the survey can have a decided effect on the power of a monetary incentive (Rizzo et al., 2004; Curtin et al., 2005). Incentives also tend to have a greater effect when the response rate is low (Keeter et al., 2004). Finally, there is the use of the incentive just for refusal conversion. While cost effective, there is a question of ethics regarding providing incentives only to initial refusals (Singer et al., 1999a).

One area of effort in the battle to improve respondent cooperation is the use of advance letters, and these letters could be tailored to respondents if auxiliary informations were available (Goldstein and Jennings, 2002). Of course, cognitive theories of information processing could be applied to the development of advance letters too (Dillman, 1978; Dillman et al., 2002). In telephone surveys, the use of advance letters is aided by the increased ability to match telephone numbers to addresses (Brick et al., 2003b). The effects of advance letters in telephone surveys have been mixed (Traugott et al., 1987; Singer et al., 2000). Goldstein and Jennings (2002) do point out, however, that using only one version of an advance letter may limit its utility in terms of increasing response rates. Heerwegh (2005) did find that the personalization of an advance letter did improve cooperation, although Traugott et al. (1987) did not find that to be the case in an RDD survey. Link and Mokdad (2005a) recently cautioned that, while advance letters could improve response rate, this could be at the expense of increased bias.

Leaving voice messages and text messages (on mobile phones) can also be used in a manner similar to advance letters. While Xu et al. (1993) found that leaving messages increased response rates, Baumgartner (1990) saw no differences in the type of message left. In terms of the effectiveness of caller ID for reducing nonresponse, a lot depends on what information is scheduled, and that is controlled by the telephone service provider (Link and Oldendick, 1999). Yuan et al., (2005) found text messaging had no effect on nonresponse in a mobile phone survey, but it is now questionable whether or not text messaging by a survey organization is actually legal.

More operational solutions to overcoming nonresponse, especially with respect to resolving noncontacts, are call scheduling, extending the number of call attempts, and using information gathered by interviewers on previous calls to the number. Most studies of effective call scheduling find the best time for reaching a household is weekday evenings (Weeks et al., 1987). Perhaps, the best information on optimal call scheduling has been gathered by researchers working on the National Immunization Survey, and that data suggest the previous finding to be true (Dennis et al., 1999). Yuan et al. (2005) found that greater success on reaching respondents during the daytime is obtained with mobile phones as compared to landlines. Most studies on the optimal number of call attempts are old. Burke et al. (1981) reported that the optimal number of call attempts appeared to be in the range of five to seven. With the increasing problem of noncontacts, however, there has been growing concern over the residency rates among these numbers known as ring-no-answers or RNAs. Brick and Broene (1997) attempted to determine residency rates by calling telephone business offices, but, as Shapiro et al. (1995) found, business office reports are often error prone. Another method is to call a number many times (Keeter and Miller, 1998), but this is not practical in the standard survey environment. A new method for estimating residency rates was proposed recently by Brick et al. (2002). This method takes advantage of work done on survival analysis (see also Sangster and Meekins, 2004). Another line of inquiry used to reduce both noncontacts and refusals is taking advantage of information from previous calls to maximize the chance of contact and aid refusal conversion (Groves and Couper, 1998; Purdon et al., 1999; Dixon and Figueroa, 2003; Bates, 2003; Sangster and Meekins, 2004).

When all of these other methods fail, weighting adjustments for nonresponse can be made, and, as reported by Montaquila et al. (2007, Chapter 25 in this volume) and Biemer and Link (2007, Chapter 26 in this volume), this is not such a bad strategy. Unless a telephone survey is confined to local areas, where nonresponse weighting adjustments may use auxiliary information, such as exchange, that is available for all sample units, poststratification involving raking to known population totals for demographic groups is done (Brick and Kalton, 1996; Groves and Couper, 1998; Bethlehem, 2002; Groves et al., 2004). It is assumed that these groups (or weighting cells) are homogeneous with respect to the characteristics of interest (Holt and Smith, 1979). Bethlehem (2002) also discusses the use of a general regression estimator based on available data when nonresponse is present, and Gelman and Carlin (2002) illustrate the use of inverse-probability weights.

Another approach to nonresponse adjustment first proposed by Heckman (1979) shows much promise—response propensity weighting. Using information about the number of callbacks required to complete an interview, Pothoff et al. (1993) developed weights that increase as the respondent proves harder to get. Thus, those responding only after many callbacks will have greater influence on estimates than those responding with few callbacks. Groves and Couper (1998) employed a similar approach in their work on face-to-face household surveys. It is also possible to construct separate propensity models for different subgroups of respondents. The use of response propensity models can provide corrections of nonresponse bias that improve upon traditional weighting methods (Little, 1986; Kalton and Maligalig, 1991; Bethlehem, 2002).

All of the above methods were designed to reduce the potential for nonresponse bias in estimates resulting from the failure to gain cooperation from a significant portion of eligible households in an RDD survey. However, with the growing problem of nonresponse in RDD surveys, survey researchers have also turned their concern to exactly how much bias does arise from nonresponse. Recent studies by Keeter et al., (2000) and Curtin et al. (2000) suggest that level of effort does not have that much effect on estimates and that estimates remain stable across a wide range of response rates. In contrast, it is interesting that even estimates for demographic variables did not vary much across this range (Montaquila et al., 2007, Chapter 25 in this volume). Furthermore, a substantial level of nonresponse (at least 30 percent) remained. The fact is that the relationship between nonresponse and bias is likely to be a complicated one that can differ both according to the characteristic of interest as well as by subpopulation (Tucker et al., 2005).

1.4 LOOKING TO THE FUTURE

1.4.1 Technological Change

Technological change is likely to continue at a rapid pace. More and more households will move to VoIP or wireless broadband in the coming years (Steeh and Piekarski, 2007, Chapter 20 in this volume). Number portability will be an increasing problem; but, eventually, individuals could have unique communication numbers (like social security numbers), which might make telephone sampling easier in the long run (Nathan, 2001). Another possibility is that a single device will receive multiple types of communications (Ranta-aho and Leppinen, 1997; Baker, 1998). Certainly, computers and mobile phones can be used to conduct face-to-face interviews over the telephone soon. Of course, other technologies yet unimagined could have even greater ramifications for telephone surveying.

Although it is likely that the percentage of mobile-only households will reach a plateau, there is no indication that plateau is near-at-hand. It could be as large as 15 percent by the time of the 2008 Presidential election, and, as long as the billing algorithm remains the same, mobile telephone users will be reluctant to pay to do survey interviews. Furthermore, the way households use their mobile phones may change over time. It is unclear what other changes will occur in the telephone system in the coming years, and how telephone sampling might be affected. While the total size of the system should not grow as rapidly in the coming years as it has in the last decade, the allocation of numbers to different types of 100-banks shown in Table 1.2 may continue to change. Assuming change in the telephone system slows down, the listing rates (and, presumably, the unlisted numbers too) within 100-banks should increase. With the increasing densities would come an increase in efficiency for all designs. Of course, this assumes that numbers will be assigned as they have been in the past.

1.4.2 Climate Change

Technological change, of course, is not the central problem. The continued feasibility of telephone surveys depends little on how we locate respondents and what communication devices we use and much more on the respondent's willingness to cooperate. The growing concerns about privacy and confidentiality could cause response rates to deteriorate even further.

Perhaps of even greater concern might be the changing face of society. In many developed countries, household size is shrinking, making it more difficult to contact respondents. Greater mobility is also contributing to a decline in contactability. Family structure is becoming more complex, so our telephone interviewers confront a larger variety of situations once they reach a household. They must be more flexible than in the past in the ways they approach the respondent. Added to this is the growing diversity within western societies, leading to language and cultural barriers that will make surveying, in general, more problematic.

1.4.3 Research

In order to accommodate the changes that are taking place now and that will likely accelerate in the future, telephone survey methodologists must undertake a concerted research program. Given that the telephone mode is likely to be paired with other modes in the future, experimental analyses of mode effects on a large scale will be needed; these analyses should go beyond just measuring differences in mode to consider response error. Without large-scale experiments, we will not be able to uncover the interactions between mode, question, and respondent characteristics.

If future telephone surveys will be conducted on both landline and mobile phones, frequent benchmarks of the telephone service populations should be taken for use in weighting to avoid over representing mobile-only households. These are the households most likely to answer their mobile phones (see Brick et al., 2006.). The only source of these benchmarks at the present comes from the National Health Interview Survey (Blumberg et al., 2007, Chapter 3 in this volume). The growth in the use of mobile phones also presents us with an important conceptual issue. Do we now abandon household surveys in favor of personal surveys? This issue must be confronted at some point.

We do not yet know how the public will respond to all of these new telecommunication choices. Certainly, there will be an interaction between survey respondent characteristics and the use of telecommunication services. We need to understand these interactions in order to develop efficient telephone survey designs. Unfortunately, given the dynamics of an ever-changing technological world, the picture is unlikely to remain constant very long.

Between the technological changes and sociopolitical climate changes, the hope that response rates will improve or even plateau seems unrealistic at this time. First, we need to make greater efforts to estimate the costs and benefits of extra levels of effort to reduce nonresponse (see Groves, 1989; Montaquila et al., 2007, Chapter 25 in this volume). To do this, we will need more detailed information on field activities and their costs. Will incentives offer a realistic solution? How much would they need to be, and what would be an effective means of delivery? Current research on the use of incentives has provided mixed results, and promises to pay do not work as well as cash-in-hand (Singer et al., 1999b; Brick et al., 2005c; Cantor et al., 2007, Chapter 22 in this volume). Secondly, we need a better understanding of the relationship between nonresponse and bias, especially when the nonresponse rates are extremely high. This work will require investigating estimates for a variety of characteristics at subpopulation levels (see Tucker et al., 2005.) Finally, there needs to be a serious discussion about guidelines for evaluating the validity of results from telephone surveys.

1.5 OVERVIEW OF VOLUME

The 25 remaining chapters in this volume are grouped into four sections: Sampling and Estimation, Data Collection, Operations, and Nonresponse. Each section had a team of two editors who selected the papers, with advice from other editors, for the section and then worked with the authors to prepare presentations for the conference and to prepare final manuscripts for the volume. Several editors read and commented on chapters in sections besides their own. In addition, the editors requested and received editorial assistance from colleagues in the field who read and provided comments on chapters written by the editors themselves. We thank Don Dillman, Marty Frankel, Scott Keeter, and Lars Lyberg for their expert review of chapters presented in this volume.

The Sampling and Estimation section contains seven chapters, beginning with an overview of past and recent developments in sample design and estimation for telephone surveys (Kalsbeek and Agans, Chapter 2). Blumberg and colleagues present in Chapter 3 recent findings on the growing mobile-only population in the United States, and Kuusela, Callegaro, and Vehovar review the international situation in Chapter 4. Flores Cervantes and Kalton review the use of telephone sampling methods in rare population settings in Chapter 5, Tortora, Groves, and Peytcheva discuss the potential uses of multiplicity sampling methods in surveys of mobile-only households in Chapter 6, and Brick and Lepkowski examine the nature of and findings about error properties of mixed mode and multiple frame surveys with telephones as a mode or frame in Chapter 7. Lee and Valliant's, Chapter 8, concludes the section with discussion of propensity weighting methods.

Section 3 on Data Collection contains six chapters on a wide range of data collection issues in telephone surveys. Japec examines the important role interviewers play in telephone surveys in Chapter 9. Conrad, Schober, and Dijkstra in Chapter 10 discuss the kinds of indicators that arise in telephone interviews that respondents are having difficulty in understanding the questions being posed. Harkness and colleagues review in Chapter 11 another communication problem in telephone surveys, translating questions to multiple languages. Mode and format issues for questions are examined in Chapter 12 by Christian, Dillman, and Smyth; visual elements in questionnaire design for presentation to interviewers are discussed by Edwards, Schneider, and Brick in Chapter 13; and mode effects in the Canadian Community Health Survey are reviewed in Chapter 14 by Beland and St. Pierre.

Section 4 also has six chapters, opening with a review of issues in establishing a new computer-assisted telephone interviewing call center by Kelly and colleagues in Chapter 15. Hansen examines features of sample management systems in telephone surveys in Chapter 16, while Tarnai and Moore present a study of interviewer performance and productivity assessment by telephone survey organizations in Chapter 17. Groves and colleagues reexamine the relationship between interviewer voice characteristics and telephone survey response in Chapter 18. The section concludes with Steve and colleagues examination of new measures to monitor and evaluate telephone interviewer performance in Chapter 19 and Steeh and Piekarski's review of the impact of mobile and emerging voice over Internet protocol, or VOIP, technology on telephone survey design in Chapter 20.

The concluding section's six chapters examine a range of issues with respect to nonresponse. Singer and Presser examine the role that attitudes about privacy and confidentiality among the public may have on propensity to respond in Chapter 21. Cantor, O'Hare, and O'Connor conduct a review of findings about the use of incentives in improving survey response rates in Chapter 22. Holbrook, Krosnick, and Pfent present findings from a survey of research organizations that examines the relationship between procedures used in surveys and response rates in Chapter 23. Battaglia and colleagues examine trends in response rates and where they may be going in the future in Chapter 24. Montaquila and colleagues discuss models for survey nonresponse and provide a theoretical framework for the bias possibly introduced by nonresponse in Chapter 25. The final chapter by Biemer and Link examines an important area of current research, the extent to which early cooperators with telephone surveys differ from those who respond later, and whether models can be developed to adjust early cooperator data to account for late cooperators.

This volume is dedicated to the memory of Joseph Waksberg and Warren Mitofsky, two gentle men who mentored many of us as we became interested in and began working in the field of surveys and telephone survey methodology. Joseph Waksberg's death days before the conference was on the minds of all participants as they talked about topics he would have enjoyed discussing with us. Warren Mitofsky's death as the final versions of chapters were being submitted in fall 2006 left all who work in this area in sorrow at the loss of yet another colleague who had contributed so much, and friend who would be sorely missed. We all do and will miss their continued guidance and contributions to the field, and their enthusiasm and leadership.

PART II

Sampling and Estimation

CHAPTER 2

Sampling and Weighting in Household Telephone Surveys

William D. Kalsbeek and Robert P. Agans
University of North Carolina at Chapel Hill, USA

This chapter updates an earlier review by Lepkowski (1988) and provides a backdrop for in-depth treatment of sampling-related topics in the other chapters of the volume. It begins by considering features and limitations of using telephone numbers to choose probability samples of telephone households in the United States. It then tracks the evolution of sampling strategies devised to produce statistically and practically effective telephone samples. Methods and issues related to identifying respondents within households are considered next, followed by a brief review of sampling frames and related issues in other countries. The chapter then describes the process of producing sample weights needed to generate population estimates from telephone sample data. A recent national telephone survey (the National Study of Female Sexual Health) has been used to illustrate the process. The chapter concludes by discussing future issues in telephone sampling.

2.1 TELEPHONE SAMPLING FRAMES

Rapid change in the telecommunication industry during the past 18 years has revolutionized the way we communicate. Mobile telephones, pagers, faxes, and telephones that connect to the Internet all impact not only on how we communicate with each other but also on the telephone numbering system. Consequently, recent demand for more telephone numbers has changed the efficiency of calling telephone samples, and introduced new sources of error in the frames used for telephone sampling which are discussed in the next subsections.

Advances in Telephone Survey Methodology, Edited by James M. Lepkowski, Clyde Tucker, J. Michael Brick, Edith de Leeuw, Lilli Japec, Paul J. Lavrakas, Michael W. Link, and Roberta L. Sangster

2.1.1 Telephone Numbering System

The U.S. telephone numbering system is the North American Numbering Plan (NANP). It is based on 10 digits where the first three represent the area code (or, more formally, the numbering plan area (NPA)) that typically represents specific geographic areas. Digits 4 through 6 are prefixes (a.k.a., central office codes, exchanges, or *NXX* codes), and the last four digits (a.k.a., carrier identification codes or suffixes) are used to route and bill telephone traffic. Ten-digit telephone numbers may be assigned to telephones units that are accessed by both landline and mobile telephones, although the ranges of assigned numbers for these two types of telephone connections largely differ. For sampling purposes, telephone numbers with the same first eight digits define what is referred to as a "100-bank."

In 1995, the telephone industry anticipated exhausting the available area codes in the near future. The telephone numbering infrastructure was altered to allow more combinations of digits to define area codes. One outgrowth of the increased number of area codes has been that the pool of possible residential numbers has grown at a much faster rate than the actual number of residential customers. For example, the number of telephone service prefixes increased by 89 percent between 1988 and 1998, but households with telephone service only increased 11 percent (Piekarski et al., 1999). Considering the percentage of assignable numbers that are residential in 100-banks with at least one listed residential number (i.e., active 100-banks as used for list-assisted random digit dialing (RDD) frames), Piekarski (2004) found a 15 percentage point decline from 1988 to 2004 (Fig. 2.1, light diamonds), thus demonstrating the diminished calling efficiency of list-assisted frames. However, list-assisted methods still give us around a 20 percentage point increase in the percentage of called sample numbers that are residential (Fig. 2.1, light versus dark diamonds; Glasser and Metzger, 1972; Bureau of the Census, 2005b; Federal Communications Commission, 2005); and L. Piekarski, personal communication, November 11, 2005.

2.1.2 Sampling Frames

The link between telephone numbers and households or household members establishes the connection between the sampling frame and the target population. This linkage takes six forms in telephone surveys: one-to-one, one-to-none, none-to-one, many-to-one, one-to-many, and many-to-many. The *one-to-one* linkage does not contribute to survey error, but the five remaining forms of linkage do have effects on both sampling and nonsampling errors (Kish, 1965).

In a *one-to-none* linkage, where there are units on the frame (i.e., *one*) that are not in the target population (i.e., *none*), frames must be screened to identify and remove certain numbers. The cost of screening telephone samples to identify nonresidential telephone numbers, or ineligible cases, contributes to inefficiency in the calling process. Much of the innovation in telephone sampling methodology has sought to reduce this inefficiency. Sampling vendors can also identify and remove certain types of ineligible cases, such as businesses and nonworking numbers, by comparing business and residential lists against each other and then by using automated

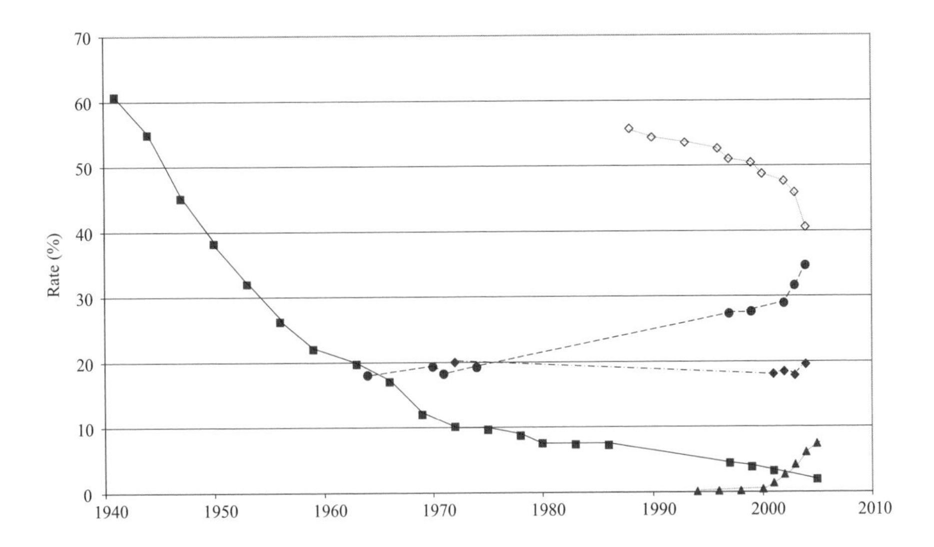

Figure 2.1. Temporal profile of statistical measures of the U.S. telephone system.

dialing equipment to resolve discrepancies. As a result, residential hit rates (number of residential numbers/number of telephone numbers on frame) between 55 and 65 percent are common.

A *none-to-one* linkage occurs in telephone sampling when the target population is the entire or a subset of the U.S. population, since households without telephones are excluded. This undercoverage problem creates a bias in estimated means, which is the product of the undercoverage rate and the differences in means between those with and without telephones (Lessler and Kalsbeek, 1992). As seen in Fig. 2.1 (squares), the rate of the telephone noncoverage in the United States has been relatively small since the 1960s (Thornberry and Massey, 1988; Bureau of the Census, 1989; Blumberg et al., 2007, Chapter 3 in this volume). Another increasingly important source of undercoverage is the households that have no landline but one or more mobile telephones. As shown in Fig. 2.1 (triangles), mobile-only households have risen to 7.4 percent in the United States, as of 2004 (Blumberg et al., 2007, Chapter 3 in this volume; and B. Meekins, personal communication, December 19, 2005). When combined with the percentage of households without telephone line service (1.9 percent), nearly 9.3 percent of the U.S. households are currently being excluded from telephone sampling frames. To the extent that members of subpopulations that differ from the remainder are excluded from the sample, bias will arise. This type of coverage bias is also seen in directory-listed frames that exclude unlisted telephone numbers. The U.S. percent of unlisted households increased from 18 percent in 1964 to 34 percent in 2004 (see Fig. 2.1, circles; Cooper, 1964; Glasser and Metzger, 1975; Piekarski, 2005); consequently, relying solely on directory-listed frames is not a valid way of picking a representative sample.

A *many-to-one* linkage (e.g., duplication or multiplicity) occurs when a single member in the target population can be selected via multiple frame entries. For example, in telephone surveys a person in a household population can be linked to two or more landline telephone numbers. Roth et al. (2001) have estimated that about 7 percent of households with telephone service can be reached by more than one landline telephone number. A sample member with multiple entries on the frame has a greater chance of inclusion than members with only one entry and is overrepresented in the samples on average. This problem can be corrected by decreasing the sampling weight attached to the record of multiple-landline households to compensate for the overrepresentation.

With *one-to-many* linkages (i.e., clustering), there is one frame entry for multiple members of the population. This is often the case in telephone samples when several people live in a household linked to only one telephone number. If we attempt to interview all eligible household respondents, the responses may be similar among residents of the same household due to a positive intraclass correlation, a factor which increases the sampling error of estimates (Kish, 1965). Contamination leading to measurement error may also occur if household members discuss the survey before all have answered. Such problems are often eliminated by selecting one member at random within a household. But this subsampling creates an overrepresentation of persons from smaller households. The sample weights are then corrected by a factor proportionate to the number of eligible persons in the household.

Finally, *many-to-many* linkages occur when multiple persons in a household are linked to multiple telephone numbers. This may occur, for example, when three household members are listed in a telephone directory, and there are two telephone lines reaching the house. The implications are the same as in the *many-to-one* scenario, and a common solution when selecting one household member at random is to multiply the sampling weight by the number of eligible respondents (3 in the above example) and divide it by the number of frame entries for the household (2 telephone lines, or a combined adjustment weight of 3/2).

There have also been developments in telephone number portability that create frame linkage problems. The Federal Communications Commission (FCC), which regulates the U.S. telecommunication industry, has mandated that service providers, where feasible, must allow customers to change providers without changing telephone numbers (a.k.a. local number portability). When telephone customers wish to carry a telephone number to a new service provider, a third number (sometimes referred to as a *shadow* or *ghost number* or a *location routing number*) is assigned, which can be selected in an RDD sample. Such numbers are used to simply manage the transfer of services from one provider to another. These numbers are technically nonworking numbers that (a) would be considered ineligible in calling through an RDD sample and (b) can increase survey costs if they remain in RDD frames. Sample vendors, however, are able to screen out most ported numbers to minimize this *one-to-none* frame problem.

2.2 SAMPLING APPROACHES

A recent historical review of telephone survey methods by Nathan (2001) traces the use of sampling in telephone surveys back to the 1930s when most telephone samples were likely to have been chosen from local telephone directories—often as a supplement to mail and face-to-face surveys. Although probability theory was well established by the end of the nineteenth century, probability sampling was relatively new in the 1930s, and the level of statistical sophistication in choosing the first telephone samples was modest by today's standards (Stephan, 1948). More rigorous sampling approaches were available for use in telephone surveys by the early 1950s (Hansen et al., 1953). By then, systematic selection and other random sampling methods had been developed for mail and face-to-face surveys, and the relatively low cost of telephone interviewing made it an attractive alternative to these other modes of data collection. Substantially greater telephone coverage (Fig. 2.1, squares), as well as the advent of computer technology for use in data collection, continued this trend in later years. Polling and marketing research organizations were among the first to routinely conduct telephone surveys and thus develop the necessary sampling approaches (Harris Poll Online, 2006). As telephone interviewing became increasingly valuable in these and other research settings, the quality of telephone sampling and associated statistical research methods improved.

The evolution of telephone sampling methods in the past 50 years has been mainly influenced by changes in telephone sampling frames and the need for statistical validity and process efficiency in calling samples of telephone numbers. Sampling frames have gone from compilations of local telephone books to electronic databases of directories and active 100-banks. There have also been important statistical issues of incomplete sample coverage because of households without any telephone access and the exclusion of unlisted numbers. The cost implications of calling inefficiency arising from the high percentage of nonresidential numbers on telephone frames have also motivated sampling innovation.

2.2.1 List Frame Sampling

Perhaps the longest standing approach to sampling in telephone surveys involves selection from existing lists of individuals, addresses, or organizations that have telephone numbers. In many cases, these lists had been developed for some business or public purposes. More recent list frames have included directories of residential telephone company customers whose telephone numbers were listed in the white pages of directories (i.e., listed telephone numbers), commercially developed community directories, and property tax record systems.

Telephone list frames exclude households without landline access, thus adding to coverage bias in surveys of the general population. These list frames have the advantage of relative availability and ease of use, while often including the name, mailing address, and other contact information to facilitate control of specific sources of survey error (e.g., sending letters in advance of data collection to minimize nonresponse). On the contrary, difficulties in working with multiple hardcopy listings for state or national samples limited the utility of list frames in the earliest telephone surveys. Telephone company directory listings had the added disadvantage of excluding households with unlisted numbers, reducing the sample coverage rate, and increasing the coverage bias of survey estimates. Thus, once the potential utility of telephone surveys became apparent, a frame that included unlisted numbers and could more easily be used for state and national surveys was needed.

2.2.2 RDD Sampling

First seen in the mid-1960s, random digit dialing methods moved toward better coverage in surveys conducted by telephone (Cooper, 1964). RDD telephone samples typically used a stratified or unstratified simple random sample of telephone numbers, each selected from an implicit list of all possible telephone numbers that could be assigned by the telephone company. One way to choose these RDD samples was to choose a random sample of all *NPA-NXX* (i.e., area code and prefix) combinations in the targeted area and then separately and independently attach a four-digit random number between 0000 and 9999 to each selected combination. For example, the selected *NPA-NXX* combination 919-782 might be combined with the selected four-digit number 7246 to form the selected RDD number, 919-782-7246. Regularly updated lists of *NPA-NXX* combinations were available from companies like Bell Communications Research (Bellcore) or Telcordia at a reasonable cost.

A variation of the RDD sampling that emerged at about the same time was "number propagation" in which the suffix of a randomly selected telephone number from a list frame (e.g., a telephone directory) was altered to obtain a sample number to call. One version of this idea, "plus one" sampling, added the integer *one* (or some other integer) to the value of the suffix of the selected telephone number. The last one or two digits of the selected telephone number could also be replaced by random digits. While this approach extended sample coverage to unlisted numbers, it suffered from low calling efficiency and failure to qualify as a true probability sample, since selection probabilities could not be determined (Landon and Banks, 1977).

In addition to increasing telephone sample coverage, RDD samples were relatively easy and inexpensive selection methods to use. Telephone coverage rates approaching 90 percent (Fig. 2.1), the advent of computer-assisted interviewing, and simplified telephone data collection led to large-scale use of RDD sampling in the 1970s. Survey research itself expanded into new areas of inquiry, with telephone interviewing the preferred mode when population information was needed quickly and resources were relatively modest.

2.2.3 Mitofsky–Waksberg RDD Sampling

Based on an idea by Mitofsky (1970), Waksberg (1978) developed an RDD sampling approach that significantly increased the percentage of residential telephone numbers among sample numbers in RDD samples at the time, thus increasing calling efficiency. This efficiency resulted from the nonrandom assignment of numbers by telephone companies to their residential customers and a clever adaptation of two-stage cluster sampling with probability proportionate to the size (PPS) selection of 100-banks of residential telephone numbers (Kish, 1965).

A sample of k clusters is sampled in the first step using a procedure that ensures that cluster i is selected with probability proportional to the number of identifiable residential telephone numbers in the cluster (M_i). The second step in Mitofsky–Waksberg RDD sampling chooses telephone numbers separately within each of the k selected 100-bank clusters. By having sampled each cluster with a probability that is proportional to M_i, the final set of residential telephone numbers were chosen from 100-banks with a higher percentage of residential telephone numbers. This increased the percentage of residential numbers when calling through the sample numbers. To avoid biasing the final sample of n residential telephone numbers in favor of 100-banks with more residential numbers, the same number of residential telephone numbers was chosen in the second stage to make sampling rates in each 100-bank inversely proportional to the number of residential telephone numbers in the 100-bank. The stage two sample was often identified by calling through a random permutation of two-digit numbers (00 through 99) serving as the ninth and tenth digits of the telephone numbers called within each 100-bank. The overall inclusion probability for all selected RDD numbers was equal in these samples, since the probability for each respondent was determined from the outcomes of the two steps that led to it being chosen:

$$\Pr\{\text{Any phone number is included in the sample}\} = \left[\frac{kM_i}{M_+}\right]\left[\frac{m}{M_i}\right] = \frac{km}{M_+} = \frac{n}{M_+} = \text{Constant} \quad (2.1)$$

where $M_+ = \sum_i^{\text{All clusters}} M_i$ is the total number of residential telephone numbers in the population and m is the sample size in each sample cluster.

While Mitofsky-Waksberg RDD samples retained the coverage advantage of RDD sampling, and they produced residential "hit" rates that were generally two to three times greater than RDD samples, calling through the second stage sample to achieve an equal probability sample can be difficult. To assure that exactly m sample telephone numbers yield a respondent in each sample cluster, it is necessary for the number of current respondents plus the number of telephone numbers still being called to equal m at any time during calling. Survey estimates from Mitofsky-Waksberg samples also tended to have slightly lower precision because of the clustering effect in their samples (Waksberg, 1978). Both of these limitations were overcome by the next major innovation in telephone sampling, the list-assisted RDD methods.

2.2.4 List-Assisted RDD Sampling

Development in the early 1990s of large national databases (e.g., MetroMail and R.H. Donnelley) containing telephone information for all households with listed telephone numbers led to the production of the RDD sampling frames that could yield higher percentages of residential telephone numbers while avoiding the statistical and operational limitations of Mitofsky-Waksberg RDD sampling. These national list frames alone were not the solution, since they missed unlisted numbers. They could be used to improve the calling efficiency of RDD samples, however. One use was to select a portion of a telephone sample from the frame of listed telephone numbers and the remainder from an RDD frame (Groves and Lepkowski, 1986). An important goal in dual-frame sampling was to use a mixture of the two frames that most effectively balanced the process efficiency of the list frame with the coverage advantage of the RDD frame.

Another important role of the national frame of listed numbers was to "assist" in producing a new type of RDD sample that is more efficient to call than the original RDD samples. As described by Casady and Lepkowski (1991, 1993), the frame for what has come to be known as *list-assisted RDD sampling* is formed by dividing the frame for basic RDD sampling into two separate sampling strata. The "high-density" stratum includes telephone numbers in those 100-banks with one or more directory-listed telephone numbers, while the "low-density" stratum includes all remaining telephone numbers in the target area.

One version of list-assisted RDD sampling calls for sampling from both strata, with much higher sampling rates in the high-density stratum. Another version "truncates" by limiting selection to the high-density stratum only. Casady and Lepkowski (1991) showed that the statistical precision of either strategy was superior to both RDD and Mitofsky-Waksberg RDD designs, although in the truncated version about 6 percent of the RDD frame was excluded, thus reducing the coverage of the truncated sample. A similar and more recent assessment of the list-assisted approach by Tucker et al. (2002) suggested that the precision gains relative to RDD sampling are even greater than in the early 1990s. Potter et al. (1991) independently developed a similar list-assisted RDD design at about the same time, but their design excluded the zero 100-banks from the frame, reasoning that the exclusion was likely to be more cost-effective.

Since RDD, Mitofsky-Waksberg RDD, and two-stratum list-assisted RDD samples are chosen from the same sampling frame, the coverage of all three designs is identical. List-assisted RDD samples produce levels of calling efficiency that are higher than those seen in RDD samples and are comparable to those expected in Mitofsky-Waksberg RDD samples. Furthermore, sampling from a list-assisted RDD frame avoids the precision loss of sampling 100-bank clusters and the process complexities of sampling telephone numbers within 100-banks associated with the Mitofsky-Waksberg RDD design.

Findings by Brick et al. (1995) established that the incremental coverage loss due to truncation has a minimal effect on the demographic composition of the sample. For this reason, selecting list-assisted RDD samples from 100-banks with one or more listed numbers (i.e., the first stratum only) soon became the standard for most list-assisted RDD samples and the basis for many survey estimates. A practical implication of list-assisted RDD sampling is that frames and selection for most RDD samples are now usually handled by commercial sampling vendors possessing the capacity to continuously update national list frames, separate the zero 100-banks from the remainder of the RDD frame, and select samples to more detailed specifications of telephone survey researchers. However, the slight decline in the calling efficiency of RDD samples (seen in Fig. 2.1) and the need for affordable samples of rare population subgroups have contributed to the increased use of multiple frames for telephone samples.

2.2.5 Other Multiple Frame Designs

Multiple frame designs are those in which an RDD frame is used in combination with other telephone frames that have higher calling efficiency but poorer coverage of the targeted population. While multiple frame telephone samples control the cost of calling, the disproportionate sampling rates that must be applied to the frames may have adverse statistical consequences due to the effect of variable sampling weights on the precision of survey estimates (Kalsbeek, 2003; Kalsbeek et al., 2007). The topic of telephone sampling from multiple frames is considered in detail in Chapter 7 in this volume (Brick and Lepkowski, 2007, Chapter 7 in this volume), while methods for sampling rare population subgroups are described in Chapter 4 in this volume (Flores-Cervantes and Kalton, 2007, Chapter 5 in this volume).

2.2.6 Boundary Problems in RDD Sampling of Local Areas

Populations targeted for study in telephone surveys have geographic limits in the definitions. Member eligibility not only establishes what type of person or household is included but also where they are located. Geographic criteria are usually tied to well-recognized geopolitical areas such as municipalities, counties, provinces, or states.

Matching the locations of households in RDD telephone samples to the area targeted for study is more difficult because of the mismatches between the boundaries of the geopolitical areas that define the target population and the boundaries of the NANP exchange areas for which telephone samples can be chosen. The reason for this boundary issue is that telephone companies provide landline services to customers in "exchange areas" corresponding largely to sets of prefixes. These exchange boundaries generally do not match common geopolitical boundaries. Figure 2.2 shows an example of a population whose target area consists of residents in four neighboring counties (i.e., Ellis, Ness, Rush, and Trego in north central Kansas). The relative location of the exchange area boundaries is denoted by lighter solid lines and county boundaries with dotted/dash lines. Particularly along the periphery of the four-county (shaded) area, parts of most exchange areas are located in the target area, but the remaining exchange area parts are in an adjacent county outside the target area.

Boundary incongruence may lead to frame under- or overcoverage that in turn impacts the statistical quality, calling efficiency, and the cost of calling in local area studies. If the set of sampled exchange areas excludes some households located inside the target area, these missed households add to the problem of sample undercoverage. If selected exchange areas include households found outside of the target area, sample households must be screened for location, and a portion otherwise willing to participate will be excluded due to geographic ineligibility. The cost of screening and the further loss of the sample due to ineligibility diminish the efficiency of the calling operation. The magnitude of these problems is greater in smaller local area

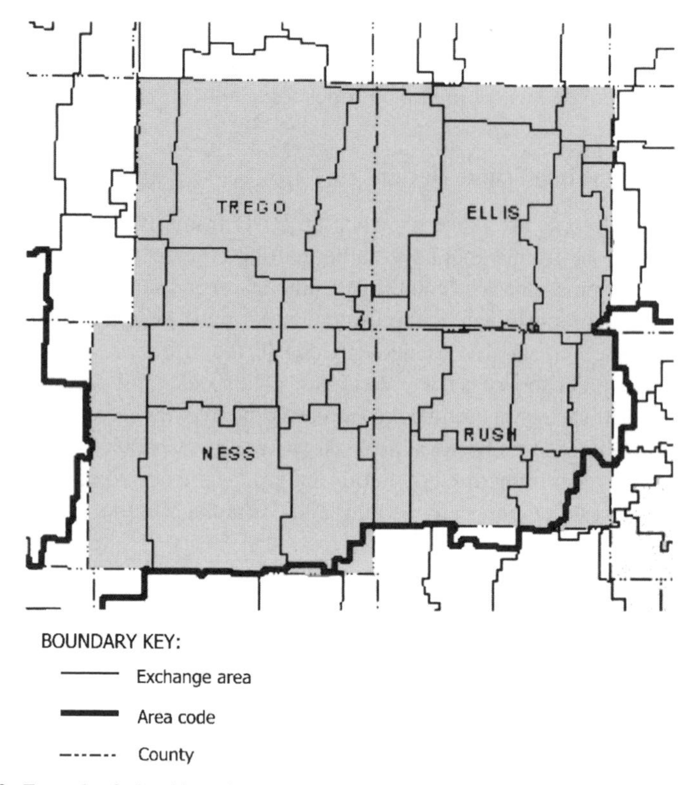

Figure 2.2. Example of a local boundary incongruency problem. (Permission to reproduce this portion of the Kansas exchange area map was received from the Kansas Corporation Commission.)

samples, since the percentage of the target population in peripheral exchange areas that affect coverage is higher

No fully satisfactory solution exists to the problem of boundary mismatches in local telephone surveys. All exchange areas whose boundaries include any part of the target area can be included to avoid increased sample undercoverage, but overcoverage and calling inefficiency are increased. Alternatively, the set of exchanges areas that best matches the target area can be chosen to provide a compromise solution to coverage and efficiency problems.

It is sometimes possible to avoid coverage problems altogether by redefining the target area to match the area actually sampled. When inference to the actual target area is necessary, and less recognizable target area boundaries (e.g., villages, towns, or townships) make screening less feasible, strategically defining the set of sampled exchanges can reduce adverse statistical effects due to boundary incongruity. For instance, one might identify the area sampled by minimizing the combined sizes of land area (or counts of households) missed inside, and included outside, the boundary of the target area. To specifically reduce estimation bias due to undercoverage, one might minimize the differences between key population indicators for those in the unintentionally included and excluded areas.[1] Effective calibration adjustment to sample weights may also help if useful sociodemographic measures are available for the target area in making inference to it (see Section 2.5).

2.3 SELECTING RESPONDENTS WITHIN HOUSEHOLDS

As noted previously, samples of telephone numbers are linked to a set of households that includes one or more members of the target population (e.g., adults, women, employed workers). Telephone numbers and households are therefore clusters in the sampling process, and sampling them can be viewed as an intermediate step in identifying the sample of the population. When population members are individuals and not all eligible individuals are to be interviewed in households, the survey designer must decide how to choose a sample within the household, and who in the household is to provide the information for the survey. Both of these design decisions can affect the error of estimates from telephone samples.

Sampling individuals within households may be done by choosing one or more members at random or by sampling residents disproportionately as a part of a strategy to oversample a rare population subgroup. Within-household sampling is thus a stage of selection in the telephone sampling of individuals. The probability of inclusion for any sample member in multistage designs is the product of selection probabilities at each stage. Therefore, the approach in selecting persons within the household determines the selection probability and consequent sample weights for respondents in telephone surveys.

Among the statistical arguments for within-household sampling of individuals is the potential to lower the effect of cluster sampling on the precision of survey estimates. Kish (1965) specified a model for a multiplicative design effect (Deff) on the variance of estimates due to cluster sampling:

$$\text{Deff} = 1 + \rho(\bar{m}-1) \tag{2.2}$$

where ρ is a measure of relative intrahousehold homogeneity (i.e., intraclass correlation) and $\bar{m}$ is the average number of population members per sample household. Typically $\rho > 0$ and thus Deff > 1, but Deff is one when $\bar{m} = 1$.

Since within-household sampling reduces $\bar{m}$ (and equals 1 when one member is chosen at random within the household), it also lowers Deff. This is the case, of course, provided the reduction in Deff is not offset by an increase in the variance of estimates due to variable sample weights from within-household sampling (Kish, 1965). Sampling one household member at random has the added advantage of avoiding measurement contamination when early interviews change response on later interviews in the same household (e.g., on knowledge or attitude items in the interview). The principal argument for not sampling within household is that taking multiple individuals per household lowers the cost per respondent (leading to a larger sample size) once the set of participating households is identified.

[1]The member level measure(s) used to define these indicators should be important predictors of key study outcome measures (e.g., socioeconomic indicators).

Several approaches have been proposed for sampling within households. Many were devised for face-to-face surveys but are adaptable to telephone administration, especially when computer technology is used for data collection. Except for a few completely nonrandom methods that can seriously skew samples and lead to biased survey estimates (e.g., choosing the first available adult, which favors those most available or willing), all approaches rely on either explicit or assumed randomization in the selection process. Some methods, however, provide no chance of selection for some household members. While the selection algorithms for these approaches differ in complexity, computer-assisted interviewing in telephone surveys has made all approaches practical to use.

A recent review of these sampling methods by Gaziano (2005) delineates 14 approaches. Some explicitly involve random selection from a constructed roster of eligible household members (such as the "Kish table" approach in Kish, 1965). Another probability method randomly chooses someone from a list of named residents or from an implied list of unique designations (not names) corresponding to a set of personal characteristics (e.g., "second oldest adult" as an entry when ordering by age). In these variations, the sample can be chosen with equal probabilities for all individuals within the household or with unequal probabilities to increase chances for those with certain characteristics (e.g., race/ethnicity or gender).

Unequal probability sampling from descriptive rosters may require that separate selection decisions be made for each roster entry, based on varying retention probabilities. For instance, for each member of a roster, a uniform random number from zero to one can be generated, say U. Then if $U \leq 0.80$ (or some other value chosen to meet target sample sizes), select the person for an interview, and if $U > 0.80$, do not select the person. Selection in this way yields a variable number of residents within a household selected, from zero to the total number of eligible residents. Choosing no one in the household may create awkwardness for the interviewer, such as having to stop contact with a willing household. Conversely, the selection scheme could select multiple household members, adding complexity to interview scheduling. These practical problems are solvable (see Nolin et al. (2004) for an example).

Sampling the household member with the last or next birthday is a "quasi-probability" form of within-household sampling that attempts to avoid the intrusiveness needed to create a household roster. Instead of explicitly using randomization to make the selection decision, it is assumed that selection occurs at some point in time that is random. Selection is thus presumed to be random.

Other methods described by Gaziano use explicit randomization as before but attempt to simplify the selection process by excluding some household members. Many are modifications of the technique first proposed by Troldahl and Carter (1964) for choosing one eligible resident per household. While the Troldahl–Carter approach retains random selection features to select a household member, its sample outcomes are limited to the oldest or youngest male or female in the household, providing no chance of selection to some household residents (e.g., the second oldest eligible male in a household of three or more eligible males). Many of the subsequent modifications to the original Troldohl–Carter approach were devised to deal with age/gender composition limitations due to nonresponse, but most still imply a selection probability of zero for some household members. Thus, neither the original Troldahl-Carter nor most of its adaptations avoid the potential for selection bias in survey estimates.

An overarching goal in the development of within-household selection methods for telephone samples has been to find statistically sound and minimally intrusive methods. The best methods are the ones that sample residents randomly but do not increase the likelihood of subsequent household nonparticipation. Methods that select the most accessible person in the household (e.g., the person who answered the telephone) to complete the interview should be avoided. Toward this end, Rizzo et al. (2004a) employ elements of several existing approaches (selection from a count of eligibles, last birthday, or Kish tables) to minimize the need to develop a roster of household members.

In some instances, the sample person cannot be interviewed (e.g., an infant) so a proxy respondent is required (e.g., a parent). The proxy may be sought because the selected person is thought to be a less reliable reporter (e.g., a young child, a cognitively impaired adult) or unable to complete the interview within the survey period (e.g., someone away until the end of data gathering). Proxy reporting is also done for all members of a cluster (household residents) by a "most knowledgeable adult" or a "primary caregiver" when, for instance, one household member reports for oneself and for all other households residents (as in the Current Population Survey; Harris-Kojetin and Mathiowetz, 1998).

The proxy respondent reduces nonresponse or improves the quality of data from sample members. It can also reduce the cost of data collection if the interviewer can collect data from an acceptable proxy when the selected household respondent is unavailable.

Moore (1988), Blair et al. (1991), Schwarz and Wellens (1997), Sudman et al. (1996), Harris-Kojetin and Mathiowetz (1998), and Tourangeau (2004) reviewed research on the merit of self- versus proxy reporting. Sudman et al. (1996) noted that early research in this area consisted mainly of empirical comparison studies establishing the degree of consistency between self- and proxy reports. More recent work examines the cognitive origins of observed self–proxy differences. Most of this research has been done in face-to-face rather than telephone mode of administration.

The cognitive bases for the hypothesis that self-reporting is superior in quality to proxy reporting are that (a) individuals remember firsthand events more vividly and in more detail and that (b) learning about these events secondhand leaves one unable to tie together basic facts into a causal sequence. Schwarz and Wellens (1997), Blair et al. (1991), and Sudman et al. (1996) suggest that respondents do not use the same strategies for self- and proxy reports, although these differences are more of degree than principle. Factors that help to explain the similarity in self- and proxy reporting strategies include (a) the degree of the respondent's participation in the events, (b) the social distance between self- and proxy respondents, and (c) the social desirability of behavior being reported. Whether or not proxy reporting should be considered, researchers have struggled to find conclusive evidence of consistent differences in the quality of self- and proxy reporting (Moore, 1988). Observing greater consistency among proxy than self-reports, for example, does not necessarily imply greater data accuracy (Schwarz and Wellens, 1997). Despite the absence of clear quality differences, proxy reporting is generally allowed, if necessary, in surveys.

2.4 TELEPHONE FRAMES AND SAMPLING IN OTHER COUNTRIES

Outside of the United States, sampling approaches vary depending on the degree of standardization of the telephone numbering system and the availability of frames. As seen in Section 2.2, the United States has a standardized telephone numbering system (*NPA-NXX-XXXX*). Standardization facilitates RDD telephone sampling procedures since RDD sampling frames are implicit rather than explicit lists of telephone numbers. Researchers in countries with nonstandardized numbering systems find RDD samples more difficult to select (Foreman and Collins, 1991). Furthermore, some countries can rely on the lists of individuals, such as population registries, that must be linked to telephone numbers to create household samples. While each country faces its own unique challenges with frames for the use in telephone sampling, countries can be grouped into sampling "scenarios" that vary by the distribution of households by telephone access. Four scenarios are considered:

Scenario A Countries such as Canada, New Zealand, Australia, and the United Kingdom that have high rates of landline telephone coverage and often a standardized telephone numbering system are in Scenario A. Researchers in many of these countries have relied on telephone directory listings as frames, but these directories have become less suitable as rates of unlisted telephone numbers increase (Noble et al., 1998; Beerten and Martin, 1999; Wilson et al., 1999; Taylor, 2003b). Australia, for example, relied on listed directories until public concern over privacy intensified and access to machine-readable directory listings became more difficult. Bennett and Steel (2000) demonstrated that list-assisted RDD methods can be useful in conducting large-scale telephone household surveys in Australia. In the United Kingdom, selecting RDD samples was problematic before the telephone numbering system was standardized. Nicolaas and Lynn (2002) showed how after the restructuring it is possible to generate efficient RDD samples using publicly available blocks of allocated landline telephone numbers. While RDD may be more appropriate for countries with high coverage rates and a standardized numbering system, there has been some reluctance to use these approaches; RDD methods have not reached the levels of use in the United States.

Though standardized telephone numbering systems facilitate RDD sampling, creative solutions for countries lacking standardized numbering have been instituted. For example, in Germany the Center for Survey Research and Methodology (ZUMA) created a frame for choosing list-assisted samples of phone numbers (see Gabler and Häder, 2002). The frame is a database consisting of all possible telephone number combinations, excluding impossible numbers given the varying number of digits in the numbering system. The database was used to create 100-banks, and those

100-banks with no directory listings were excluded. This 100-bank database was first developed in 1997 and is updated yearly.

Scenario B Scenario B countries, such as France and Italy, have somewhat lower landline coverage than countries in Scenario A, and the numbering system may be standardized or nonstandardized. In France, where the numbering system is standardized, a rapidly growing mobile-only household population is the main reason for a lower coverage rate. The mobile-only rate reaches a level where coverage bias in estimates is becoming a serious threat to the quality of survey findings. Chapters 3 and 4 in this volume suggest that the mobile-only household trend is only going to rise and that the differences between mobile-only and landline households can no longer be ignored. As the percentage of mobile-only households increases, and the difference between landline and mobile-only households widens, mobile telephone lists will have to be included in telephone sampling frames. Doing so presents several important practical and statistical challenges (see Tortora et al., 2007, Chapter 6 in this volume, for an example of one such challenge).

Scenario B may also be a setting where RDD sampling is difficult to employ because of a nonstandardized telephone numbering system, such as in Italy. Many telephone surveys in Italy use sampling methods that are fast and relatively inexpensive, but not RDD. Sampling mobile telephones in Italy will be more difficult because the number of active mobile telephone numbers equaled the total population in 2002, even though 75 percent of all persons had mobile telephones (Callegaro and Poggio, 2004). Multiple prepaid mobile telephone accounts, some of which may not be active, are linked to multiple individuals. Disentangling frame multiplicities, where persons can be reached through several different residential or personal telephones, will be a predominate concern in Italy.

Scenario C Scenario C countries are those where telephone coverage is high due to substantial coverage through landline and mobile-only households. Finland, which has the highest percent of mobile-only households, may be indicative of things to come for other Scenario C countries such as Sweden, Denmark, and the Netherlands (Forsman and Danielsson, 1997; Kuusela and Simpanen, 2002; Kuusela, 2003; Van Goor and Rispens, 2004). All have nonstandardized telephone numbering systems that make list-assisted RDD methods difficult to implement. In Finland, for example, Statistics Finland draws its samples entirely from an up-to-date population registry. Telephone numbers are not given in the registry, but entries can be linked to common databases to identify the telephone number(s). Telephone coverage among all households is very high (98 percent), and household mobile telephone penetration is the highest seen in Europe (94 percent), making it potentially easier to contact selected citizens by telephone due to the availability of multiple contact numbers (see Kuusela et al., 2007, Chapter 4 in this volume). Multiplicity is not a problem in Finland because sampling is done at the person level where the respondent has been identified in advance. This approach would suffer greatly if the databases used to identify telephone numbers for sample members ever becomes inaccessible, say, because of public concern for privacy.

Scenario D Scenario D includes most countries in the developing world (e.g., in Africa, South America, Central America, and Southeast Asia) where nontelephone household rates are high and the percentage of mobile-only access is growing. In most cases, the percentage of households with landline access is linked to socioeconomic status. Telephone sampling is not out of the question in Scenario D countries, but it is often limited to urban areas or other areas where landline coverage is likely to be high.

2.5 ESTIMATION AND SAMPLE WEIGHTS

Inference from respondent sample data to the target population in telephone surveys is made by applying estimation formulas that account for features of the sample design and assume models about population or sample origins that can influence the composition of the sample of respondents. Important design features in survey inference include stratification, cluster sampling, and variation of selection probabilities across sample members. The composition of the selected sample may be altered by limitations in the selection and data gathering processes, including frames that selectively cover the target population and differential nonresponse among members of the selected sample. A useful review of these ideas in statistical inference from survey samples is given by Särndal et al. (1992).

Inference from samples based solely on features and implementation of the survey design requires two fundamental actions by the analyst: (1) computing sample weights to account for the process of sample selection and important composition-altering forces at work on the sample during the sampling and data collection processes; and (2) using statistical formulations that utilize these weights and appropriately account for stratification and cluster sampling in generating survey findings. In this section, the first of these two actions, sample weighting, is described. Readers interested in the second estimation formulations may find useful Wolter's (1985) summarization of the conceptual basis for various approximations and replication-based methods used to estimate the variance of estimates from complex samples. Reviews of computer software that uses one or more of these methods are available from the Survey Research Methods Section of the American Statistical Association through its home page on the web (http://www.amstat.org/sections/SRMS/index.html).

A *sample weight* is a numerical statistical measurement linked to a data record for any survey respondent. In general terms, it is computed as the inverse of an adjusted probability of obtaining the data for the respondent. In some cases, the sample weight is based simply on the respondent's original selection probability. The inverse of this probability, or base weight (B_i), is often adjusted to account for sample imbalance arising during the process of conducting the survey. More than one base weight adjustment may be applied, and all are multiplicative. Unless a weight is rescaled for analytic purposes (e.g., "normalized" to sum to the number of sample respondents), its value can be interpreted as an indication of the number of population members represented by the respondent. Separate sets of weights may be necessary when data are gathered for different types of data items associated with the respondent. For example, if data in a household survey are gathered for the selected households, and for one resident who is chosen at random in each of those households, separate sets of weights would be produced for the household and for the resident data.

Whereas the general statistical rationale behind the use of weights for estimation is well established (Horvitz and Thompson, 1952), no single protocol exists for computing them in telephone surveys. Some combination of the following components is typically used in producing a set of weights for the ith respondent data record from a telephone sample:

(1) Base weight (B_i)
(2) Adjustment for nonresponse ($A_i^{(nr)}$)
(3) Adjustment for incomplete telephone coverage ($A_i^{(cov)}$)
(4) Adjustment to control variation among weights ($A_i^{(trim)}$)
(5) Adjustment to calibrate the weights ($A_i^{(cal)}$)

When all four adjustment are made the final adjusted weight (W_i) is the product of weights generated in each step, say

$$W_i = B_i A_i^{(nr)} A_i^{(cov)} A_i^{(trim)} A_i^{(cal)}.$$

The logic behind the process of producing sample weights for telephone surveys considers that a population member can only provide data if all three of the following events occur:

- *Event F*: A telephone number reaching the respondent's household is included in the sampling frame.
- *Event S*: Given Event F, one of the household's telephone numbers is chosen, and the household resident for whom the respondent is reporting is selected for the survey interview.
- *Event R*: Given Event S, once the respondent's household agrees to participate in the survey, the designated respondent in the household agrees to complete the interview and provides useful data.

The probability that the member's data are used for sample estimation is the product of the probabilities of observing these three events. Steps (1), (2), and (3) require that one calculate or estimate the probabilities of Events S, R, and F, respectively. The motivation to reduce weights variation in Step (4) is to control the incremental effect of variable weights on survey estimates. Calibration in Step (5) serves to control the sample's external validity due to factors not specifically accommodated by the nonresponse and coverage adjustments in Steps (2) and (3). These factors could be random variation in the sample's demographic composition with respect to variables not used to define sampling strata, or they could be differential nonresponse and undercoverage associated with variables other than those used to define the adjustment cells in Steps (2) and (3).

More general discussion of each of these steps is found in reviews by Lessler and Kalsbeek (1992) and by Kalton and Flores-Cervantes (2003). For simplicity, assume that these five steps are completed in the order listed above, even though this need not be the case for the last four.

A recent national telephone survey conducted by the Survey Research Unit (SRU) at the University of North Carolina at Chapel Hill is used as an example. A telephone survey was conducted for the National Study of Female Sexual Health (Agans and Kalsbeek, 2005), the general goal of which was to measure female sexual dysfunction in the United States by interviewing a national telephone sample of 2207 civilian, noninstitutionalized women between the ages of 30 and 70 years (inclusive), who had been in a stable personal relationship for at least 3 months. The sample was chosen from two nonoverlapping national frames: (1) directory-listed telephone numbers for households that were likely to have a female resident in the targeted population (the targeted list frame) and (2) a list-assisted RDD frame of telephone numbers that were not on the first frame. The targeted list frame was divided into four strata by the age (in years) of the eligible woman expected to be found in the household (i.e., 30–39, 40–49, 50–59, and 60–70 years). To reduce data collection costs, sampling rates on the more efficient targeted list were made considerably higher than on the RDD frame. Disproportionate sampling was also used to control the number of respondents by age and the presence or absence of women with a rare health trait (10 percent prevalence rate) related to age. The two youngest age strata on the targeted list frame were sampled at higher rates than the two oldest age strata. The mixture of female respondents by age and by the trait status was further manipulated through a within-household screening process in which eligible women were retained for interview with probabilities that varied depending on the age and the trait status of the screened woman.

Base weight (B_i). A base weight is normally computed first since its numerical result can be precisely determined when probability sampling is used and good records of the selection process have been kept. The base weight (B_i) for the ith respondent's data is calculated simply as $W_i^{(1)} = B_i = \pi_i^{-1}$, where π_i is the probability that the respondent was selected in the sample.

In our example, three factors affect the selection probability for the ith respondent in the hth sampling stratum: (1) the sampling rate (say, $f_h = 9.8219 \times 10^{-4}$, the rate in one of the strata) for the simple random sample of telephone numbers in the stratum of which the respondent's telephone number was a part, (2) the number of telephone landlines reaching the household (say, $t_{hi} = 1$), and (3) the age-by-trait status retention probability (say, $r_{hi} = 0.0511$). These measures generate a selection probability $\pi_{hi} = f_h \, t_{hi} \, r_{hi}$ and a base weight (B_{hi}) for the ith member of the sample in the hth sampling stratum $B_{hi} = \pi_{hi}^{-1} = (f_h \, t_{hi} \, r_{hi})^{-1}$. Using these example values for the rates in the base weight yields $B_{hi} = 19{,}924.27$.

Adjustment for nonresponse ($A_i^{(\mathrm{nr})}$). Weight adjustment is a common approach used to at least partially compensate for nonparticipation. Any sample member may be either a respondent or a nonrespondent. Bias due to nonresponse is partly determined by the member-level covariance in the population (σ_{py}) between the propensity (i.e., probability in a stochastic sense) of the ith individual member to respond (p_i) and the survey measurements (y_i) for what is being estimated from the survey data (Lessler and Kalsbeek, 1992).[2]

Adjusting sample weights for nonresponse requires that each value of p_i be estimated empirically based on the nonresponse experience in the sample. The member-level multiplicative adjustment for nonresponse is then simply the reciprocal of the estimated response propensity for the respondent $A_i^{(\mathrm{nr})} = \hat{p}_i^{-1}$, and the nonresponse-adjusted weight is then $W_i^{(2)} = B_i A_i^{(\mathrm{nr})}$.

A key issue in producing the adjustment $A_i^{(\mathrm{nr})} = \hat{p}_i^{-1}$ is how to estimate individual response propensities. Weighting class response rates and predicting response propensities from a fitted logistic model are two approaches. The *weighting class approach* estimates p_i for the ith respondent as the response rate for members of the group (i.e., "class") of population members with similar characteristics and response tendencies as the respondent. The choice of characteristics to use in defining these groupings (i.e., the "weighting class") is strategically important, since bias reduction associated with this approach is directly related to the extent of correlation between the response rate and the parameter of interest for these classes (Kalton, 1983). For any respondent in the hth weighing class where r_h out of n_h eligible sample members responded, the adjustment is computed as $A_{hi}^{(\mathrm{nr})} = \left[\hat{p}_{hi}^{(\mathrm{wca})}\right]^{-1} = \sum_{i=1}^{n_h} W_{hi}^{(1)} / \sum_{i=1}^{r_h} W_{hi}^{(1)}$ which implies that p_{hi} is estimated by the weighted response rate for its weighting class. To reduce the risk of seriously imprecise propensity estimates and widely variable values of $A_i^{(\mathrm{nr})}$, and thus $W_i^{(2)}$, minimum size requirements may be set for values of n_h.

The response propensity modeling approach, or logistic regression weighting, is an alternative to the weighting class adjustment (see Iannacchione et al. (1991), Lepkowski (1989), Kalsbeek et al. (2001) for practical applications). In this approach p_i is estimated from a fitted logistic regression model. More specifically, ancillary variables for sample respondents and nonrespondents are regressed on the binary response outcome variable R_i, where (R_i) equals 1 in the event of a response and 0 for a nonresponse. Including both main and interaction effects as appropriate, a best-fit model is then used to produce $\hat{p}_i^{(\mathrm{rpm})}$, a predicted value of p_i based on the actual values of the ancillary variables for the ith respondent. From this result, one can calculate $W_i^{(2)}$ as

$$W_i^{(2)} = B_i A_i^{(\mathrm{nr})} = B_i \left[\hat{p}_i^{(\mathrm{pm})}\right]^{-1}.$$

In principle, this approach has the benefit of utilizing the prediction of all important main effects and interactions in estimating propensity, although empirical comparisons of response propensity modeling and weighting class adjustments have shown little empirical difference in bias reduction for the two approaches (Kalsbeek et al., 2001). While both approaches have the advantage of reducing nonresponse bias, any resulting reduction in the mean squared error of estimates is at least partially offset by the increased variability of the adjusted weights arising out of

[2]The term "propensity" rather than "probability" is used in connection with survey nonresponse, since presumed stochastic behavior, rather than explicit randomization, determines the outcome of the process determining whether or not a member of a survey sample responds.

the variation among estimated values of p_i, which, in turn, increases the variance of weighted estimates.

Both of these strategies are effective in controlling bias when ancillary information that strongly predicts both p_i and y_i is available for all members of the selected sample (i.e., both respondents and nonrespondents). The reality in most telephone sampling situations, however, is that the amount of information available for the entire sample is quite limited, thus diminishing the prospects of having highly predictive ancillary variables to form weighting classes or as predictors in propensity models. Area code location or ecological information connected to the exchange area or the telephone number might be available by linking to geopolitical area units for which aggregate data are available (e.g., to the ZIP code of listed telephone numbers). Geographic sampling strata might also be used, provided there is a reasonable indication of between-stratum differences in important study estimates. Thus, when predictive ancillary variables are lacking, the nonresponse adjustment step is often skipped and the calibration adjustment used instead to adjust for any sample imbalance due to nonresponse.

The female sexual dysfunction study used a weighting class adjustment for nonresponse, where the five strata used for sampling telephone numbers (i.e., one for the RDD frame and four age strata for the targeted list frame) were designated as the adjustment cells, and the weighted stratum-specific response rate RR4 (American Association of Public Opinion Research, 2004) used to estimate the response propensity for any stratum was denoted as $RR_{w,h}$. Thus, the nonresponse adjustment for any respondent selected from the hth stratum is $A_i^{(\mathrm{nr})} = [RR_{w,h}]^{-1}$, and the nonresponse-adjusted weight is $W_i^{(2)} = B_{hi}A_{hi}^{(\mathrm{nr})}$.

The weighted RR4 response rate in one sampling stratum was 39.1 percent compared to the 56.7 percent studywide rate. Then for all respondents in this stratum, the nonresponse adjustment is $A_{hi}^{(\mathrm{nr})} = 2.558$. Applying this adjustment to the ith respondent described above, the nonresponse-adjusted weight is $W_i^{(2)} =$ [19,924.27][2.558]=50,970.24.

Adjustment for incomplete telephone coverage ($A_i^{(\mathrm{cov})}$). Sample coverage bias is determined by the extent of incomplete landline telephone coverage and the size of the difference between those with and without landline access (Lessler and Kalsbeek, 1992). Unless special effort is made to obtain survey estimates from those without telephone access, the size of this bias cannot be directly measured to remove it from reported survey estimates. The next best remedy for dealing with differential telephone coverage is to adjust for it when computing weights.

Following the same logic of using estimated response propensities to adjust for nonresponse, an estimate of the propensity for telephone coverage is needed to compute an adjustment for undercoverage in telephone samples. Brick et al. (1996b) proposed an approach to accomplish this in surveys of the general population. The key assumption behind this adjustment approach is that those in households with temporary interruption in telephone service are similar to those who have no access at the time of the survey. Brick et al. (1996b) adjust sample weights to control the totals of the number of persons in telephone and nontelephone households, while

Frankel et al. (2003b) modify this approach and require control totals by the presence of interruptions in telephone service.

The computational process for this adjustment involves separately producing adjustments $A_i^{(cov)}$ for incomplete coverage within strategically formed adjustment cells that are formed by subdividing the population by characteristics that are correlated with the likelihood of service interruption. Both Brick et al. (1996b) and Frankel et al. (2003) considered the following characteristics: parental race/ethnicity, education, housing ownership status, and geographic area. Support for the use of these variables, particularly home ownership, is in the findings in Chapter 3 in this volume. The adjustment to the existing weight for the ith respondent with temporary discontinuation in the hth cell is then computed as $A_{hi}^{(cov)}=\hat{c}_h^{-1}$, where $\hat{c}_h$ is the estimated telephone coverage rate for persons in households with service discontinuation. For all other respondents from households without service discontinuation, $A_{hi}^{(cov)} = 1$. The details for computing the coverage rate are given in Brick et al. (1996b) and Frankel et al. (2003) and depend on the nature of the control totals that are available. The coverage-adjusted weight is then obtained for all respondents as $W_i^{(3)} = W_i^{(2)} A_i^{(cov)}$.

This coverage adjustment is only practical for national studies with relatively large sample sizes. This is because (a) subnational sources of data to estimate the totals of households without telephone service or the totals of households with service interruption that currently do not have telephone service are very limited and (b) the coverage adjustment affects only those households with service interruption in the last year, a small percentage of 2.3 percent for adults with an interruption of one day or more (Brick et al., 1996b). The number of sample households with service interruption also limits the number of adjustment cells used for a coverage adjustment depending on the number of respondents in the sample, especially the number of those with service interruption.

Since only 51 of 2207 survey respondents in the sexual dysfunction survey reported service interruption of more that one week in the past year, only two adjustment cells defined by race/ethnicity (i.e., white and non-white) were used to adjust the weights for noncoverage. Suppose the sample respondent was white and her household had service interruption in the past year. Using national estimates of households by telephone service from white respondents to the March 2004 CPS, the adjustment was $A_{hi}^{(cov)} = 2.0931$, which leads to the coverage-adjusted weight for the ith individual described above as $W_i^{(3)} = W_i^{(2)} A_{hi}^{(cov)} = (50{,}970.24)(2.0931) = (106{,}683.97)$.

Adjustment to reduce the variation in weights ($A_i^{(trim)}$). Kish (1965) has shown that variation in sample weights can increase the variance of survey estimates by a factor of $\text{Deff}_w = 1+s_w^2/\bar{w}^2$, where $s_w^2/\bar{w}^2$ is the relative variance of the weights in the sample, and $\bar{w}$ and s_w^2 are the sample mean and variance of the weights, respectively. Some of the factors contributing to this variation are features of the sample design intended to reduce the variance of estimates (e.g., optimum allocations and disproportionate stratum allocation to oversample subgroups), while others are the result of strategies to control the bias in these estimates (e.g., adjustments for nonresponse and coverage). Controlling this variation effectively requires walking a fine line between reducing the adverse effect of Deff_w on the precision of estimates while minimizing change in the bias reduction benefits of weight adjustment.

Some strategies to control the variation of adjusted weights impose size restrictions on weight adjustments. For instance, Kalton and Flores-Cervantes (2003) describe the process of collapsing adjustment cells to restrict the size of adjustments produced by the final set of cells. This preventive approach might be used to control the size of adjustments for nonresponse and coverage, but it might also be used to restrict the size of calibration adjustments as a middle ground between poststratification and raking (see Deville and Sarndal, 1992).

Whether or not preventive efforts are made to control the variation of adjusted weights, their distribution among sample members often remains positively skewed, as the result of a small percentage of relatively large weights. Thus, another strategy to deal with extraordinary variation in sample weights is truncating or "trimming" the size of the largest weights, and redistributing the trimmed portions of the largest weights to the remainder of the sample to retain the overall sum of weights.

The general goal of trimming is to reduce Deff_w to a more reasonable level, while controlling the difference between weighted estimates computed before and after they are trimmed (i.e., to leave the bias reduction results of prior adjustments relatively undisturbed). Potter (1988, 1990, 1993) suggests that effective approaches are those that create multiple sets of trimmed weights based on various trimming levels and then choose the set of trimmed weights with the lowest estimated mean squared error for key survey estimates. The disadvantage of this approach is that trimmed weights producing the lowest mean squared error for one survey estimate may not produce the same result for another estimate. If trimming is applied to $W_i^{(3)}$ to yield the multiplicative trimming adjustment ($A_i^{(\text{trim})}$) from which we compute $W_i^{(4)} = W_i^{(3)} A_i^{(\text{trim})}$ then trimming generally implies that $A_i^{(\text{trim})} \leq 1$ when $W_i^{(3)}$ exceeds some value C, above which all weights are reduced, and $A_i^{(\text{trim})} \geq 1$ when $W_i^{(3)} < C$, such that $\sum_i W_i^{(4)} = \sum_i W_i^{(3)}$.

Primarily because of disproportionate sampling rates for strata involving the two frames, $\text{Deff}_w = 5.655$ for the nonresponse-and coverage-adjusted weights ($W_i^{(3)}$) in the sexual dysfunction study. To reduce this multiplicative effect on the variance of study estimates while minimizing change in the subsequent estimates (to control bias), the iterative National Assessment of Educational Progress (NAEP) trimming procedure described by Potter (1990) was used. In each iteration of this procedure, neither the modified weight nor the square of the weight was allowed to exceed specified values for that iteration, such that the modified weighted sums equal the sum of the pretrimming weights. Several sets of trimmed weights were produced and the final compromise set of trimmed weights that were ultimately used yielded a more acceptable $\text{Deff}_w = 3.285$, with absolute differences relative to comparable pretrimming estimates that varied from 0.2, 0.3, and 7.0 percent for three health indicators to 10.5 percent for a prevalence rate. For the pretrimming weight $W_i^{(3)} =$ 106,683.97, the trimming factor was $A_i^{(\text{trim})} = 1.3610$, yielding a posttrimmed weight of $W_i^{(4)} = W_i^{(3)} A_i^{(\text{trim})} = (106{,}683.97)(1.3610) = 145{,}201.40$. That is, this weight did not itself need to be trimmed since it was not among the largest. Since trimming in this way requires that sums of pre- and posttrimming weights be equal, relatively small pretrimming weights such as this one will increase.

Adjustment to calibrate the final weights$(A_i^{(\text{cal})})$. While Deville and Sarndal (1992) were the first to coin the term "calibration" in conjunction with sample weighting, approaches that in effect constrain the behavior of weights have existed for more than 60 years. In principle, a calibration weight adjustment brings weighted sums of the sample data into line with the corresponding counts in the target population. Poststratification and raking were important early forms of weight calibration, and they are special cases of the generalized calibration framework.

The role of calibration depends on which, if any, of the other adjustment steps are used, and the order in which they are made. For instance, when the order of the only three adjustments is nonresponse, incomplete sample coverage, and then calibration, the calibration adjustment corrects for any sample imbalances not specifically addressed by the first two adjustments. On the contrary, if only a calibration adjustment is used, it becomes the sole accommodation for all sources of sample imbalance.

The final set of weights may be calibrated to the population distribution based on population data from an external source (e.g., the most recent decennial census, recent findings from the CPS, the American Community Survey, and other large national surveys). This step essentially involves adjusting the weighted sample (based on prior weighting steps) to the population distribution of a set of categorical calibration variables in either of the two following ways: (1) *poststratification* (or cell weighting) to the joint population distribution of these variables or (2) *raking* (or iterative proportional fitting) to the margins of the joint population distributions of these variables in the population. As with other adjustments, calibration is most effective when the variables used to define the control distributions are highly correlated with key study variables. Although the best set of predictors often varies among study variables, common demographic indicators such as age, race/ethnicity, education, and gender are often included in the best-predictor sets, and thus are a good compromise choice. Ultimately, the final analysis weight (W_i) for the ith respondent data record is obtained as $W_i = B_i A_i^{(\text{nr})} A_i^{(\text{cov})} A_i^{(\text{trim})} A_i^{(\text{cal})}$, where $A_i^{(\text{cal})}$ is computed by some calibration strategy. ($A_i = 1$ for any adjustment step that is not included in the weighting process.)

Poststratification to produce an *adjustment to calibrate the final weights* ($A_i^{(\text{cal})}$) may be implemented by defining the cross-classification of the categorical (or categorized) calibration variables. To control variation in adjustments due to respondent cell counts that are too small (e.g., less than 30), some initial cross-class cells may be combined or "collapsed" (see Kish, 1965). The poststratification adjustment ($A_h^{(\text{cal})}$) is then computed for the gth of G adjustment cells as $A_g^{(\text{cal})} = N_g / \sum_{i=1}^{r_g} W_{gi}^{(4)}$ or $A_g^{(\text{cal})} = N_g / N \Big/ \sum_{i=1}^{r_g} W_{gi}^{(4)} / \sum_{g=1}^{G} \sum_{i=1}^{r_g} W_{gi}^{(4)}$, where N_g is the population count according to the external calibration source, such that $N = \sum_{g=1}^{G} N_g$ is the total population size, and r_g is the respondent sample size in the cell.[3] By multiplying the existing weight for any sample member times the adjustment for its cell ($A_g^{(cal)}$), the weighted relative frequency distribution of the sample using the final adjusted weight for the ith member of the gth adjustment cell (W_{gi}) now matches the corresponding population distribution with respect to the calibration variables. That is, $\sum_{i=1}^{r_g} W_{gi} / \sum_{g=1}^{G}\sum_{i=1}^{r_g} W_{gi} = N_g / N$; $\forall g$. This equality can be used as a check of the calculations for this adjustment.

Raking to force sums of weights to match marginal categorical totals for a set of calibration variables (Deming and Stephan, 1940) starts with the same type of multiway configuration of categorical calibration variables as poststratification. Each iteration of the raking involves separately forcing category sums for each variable to equal corresponding population categorical totals (see Kalton and Flores-Cervantes (2003) for a computational example with two calibration variables). Berry et al. (1996) illustrates the use of generalized raking in a large telephone survey. While marginal constraints are generally met in fewer than five iterations, it is possible for convergence to be slow or to fail. One advantage of raking is that the variation of adjusted weights is generally less than that comparably seen in poststratification, since the constraints on weights are less restrictive in forcing weighted totals to margins than they are for multiway adjustment cells. Raking is also more adaptable to smaller samples and to the use of more calibration variables than poststratification, since the sample sizes of the margins—rather than the multiway cells of the calibration variables—are important.

Respondent weights for the sexual dysfunction survey were calibrated by poststratifying the weighted distribution of the set of trimmed weights ($W_i^{(4)}$) to match the U.S. population distribution of the March 2004 CPS, which was the closest available source to the September 2004 through March 2005 data collection period for the sexual dysfunction survey. Thirty-two adjustment cells for this step were initially defined by the cross-classification of three respondent characteristics: (1) eight 5-year age categories, (2) two marital status categories (married and unmarried), and (3) two education categories (high school or less, and more than high school). Five of these cells that failed to meet the requirement of $r_g \geq 10$ were collapsed by education and, if necessary, by marital status, since they were the least and next least predictive of key study outcome measures among the three poststratification variables. The second raking formulation for $A_g^{(cal)}$ was used since the CPS asked questions about women's ages but not their relationships, so that N_g was not available but N_g/N could be conjectured from the relative frequencies of CPS findings for all females aged 30–70 years. For the weight illustrated above, the relative frequenies in the corresponding cell from the CPS and the sample were $N_g/N = 0.0926$ and $\sum_{i=1}^{r_h} W_{gi}^{(4)} \big/ \sum_{g=1}^{G}\sum_{i=1}^{r_g} W_{gi}^{(4)} = 0.1234$, respectively, meaning that respondents in this adjustment cell were overrepresented based on the adjusted weights. Thus, the trimmed weight ($W_i^{(4)} = 145{,}201.40$) was calibrated by a factor of $A_g^{(cal)} = (0.0926)/(0.1234) = 0.7504$ to produce a final adjusted weight: $W_i = (145{,}201.40)(0.7504) = 108{,}958.88$.

[3]When the external cell population counts are of lower statistical quality, the relative frequency of the population in the cell (i.e., the cell count divided by the total population count) may be a better choice for the calibration source (numerator) and sample source (denominator) of the poststratification adjustment.

2.6 FUTURE NEEDS FOR TELEPHONE SAMPLING AND WEIGHTING

Recent advances in communication technology as well as greater public resistance to survey participation have significantly impacted the current state of telephone interviewing. Given the public's growing appetite for the new communication technology and its selective wish for privacy, these recent trends are expected to continue raising important issues for telephone survey researchers. Telephone sampling will obviously be affected as a part of this trend.

2.6.1 Multiplicity in Landline Plus Mobile Frames

Declining reliance on landline telephones, increasing Internet telephone use, and growing use of mobile telephone and mobile-only access are recent trends that have important consequences for telephone sampling. If the current mix of the types of household telephone access found in some European countries like Finland is a reasonable predictor of the future for the United States, the general shift for households will be away from landline telephones and toward mobile only. This shift will reduce sample coverage for landline frames and add pressure to include mobile telephones in telephone samples. Indeed, adding mobile telephones to a landline-based sample today would improve the coverage rate, particularly of younger, transient populations as reported in Chapter 3 in this volume.

Selecting telephone samples from frames of both landline and mobile numbers adds to the complexity of the sampling and survey data collection. Among the reasons for this difficulty is that landline telephones are accessible to all household residents, while mobile telephones are far more often used by individuals who do not share their telephone with others in the household (see Kuusela et al., 2007, Chapter 4 in this volume). The current policy among U.S. telephone companies of charging both the caller and the one being called for mobile telephone use is also important. Assuming that selected mobile telephone users can serve as entry points to their household's participation, household selection in a sample chosen from a landline-plus-mobile frame would depend on the set of landlines accessing the household and the group of mobile telephones accessing the individual household members. This unique many-to-one linkage pattern between telephone number entries on the frame and households creates a new multiplicity problem establishing the need to find new, creative ways for household sampling and for within-household sampling. For a discussion of multiplicity sampling methods for mobile telephone samples see Chapter 6 in this volume.

2.6.2 Optimal Frames Mixtures in Multiple Frame Sampling

Since most existing telephone frames have declining or currently inadequate coverage rates when sampled individually, Nathan (2001) suggested that the only plausible single-frame samples in the future may come from national registers similar to those found in some European countries. Until national registries are established elsewhere, sampling from multiple frames may be the only sound short-term option for telephone surveys. Among the possible frames to combine would be those for various options of landline (RDD) and mobile telephones. These frames are likely to differ markedly in population coverage, calling efficiency, and their degree of overlap, thus affecting the precision, coverage, and calling efficiency in the combined samples.

In addition to improving sample coverage, multiple frame telephone samples have been used to improve response rates (through multimode applications) and to sample rare population subgroups. A growing body of research on the design and analysis of multiple frame and/or mode samples is summarized by Nathan (2001) and discussed in detail in Brick and Lepkowski (2007), Chapter 7 in this volume. To our knowledge, none of this work has been on the use of multiple frame telephone sampling to improve frame coverage. As noted previously, multiple frames have also been used to target sample population subgroups, with variations among frames in their subgroup coverage and calling efficiency. Individual frames are usually strong on one, but not both, of these traits.

Whether improving coverage or more effectively sampling subgroups, multiple frame sampling is expected to become increasingly popular for telephone surveys. Thus, in addition to finding solutions to multiplicity in landline plus mobile frames, research is needed to determine which frames to combine for these purposes and to determine the optimum mixture of the samples chosen from the designated frames. Cost-error optimization ideas previously devoted to dual-frame and mixed-mode applications might be adapted to this setting, although the complexity of the models driving this work may preclude closed-form solutions and require the identification of other approaches to finding practical optimum mixtures (see Boyle and Kalsbeek, 2005, for an example of this problem).

2.6.3 Evaluating Calibration Methods

If bias effects continue to increase due to nonresponse and undercoverage issues, there will be a greater need for effective calibration methods. Calibration (and nonresponse) weighting adjustment methods rely on correlated ancillary data with an assumed relationship to study variables. The ancillary variables must also be correlated with the coverage and outcome variables to be effective in reducing bias (Kalton, 1983). The difficulty with calibration is that the latter relationship between ancillary and outcome variables is not material to the method itself. While constraining the size and variation of the adjustments, and thus the impact on the variance of estimates is inherent to some applications, there is no internal mechanism in any existing calibration method to determine how effective the resulting adjustment actually is in reducing the biases intended to be controlled. Thus, the analyst can directly neither gauge the bias reduction capability of any particular calibration strategy nor compare alternative calibration strategies in this way.

Given the increasing need for effective calibration methods, it will be important to find ways to evaluate their bias reduction effectiveness. For any given strategy, one possibility would be to empirically assess bias reduction for existing surveys where study data are available to directly estimate bias (e.g., in multiround cohort designs to estimate nonresponse bias). Another possibility is to apply alternative calibration strategies to real or contrived data in which stochastic frame coverage or nonresponse outcomes in samples have been simulated, based on other research on their origins—for nonresponse bias, where work on the cognitive and process origins of nonresponse is used (see Groves and Couper, 1998). Eltinge and Yansaneh (1997) have suggested ways to reduce nonresponse bias by producing statistically effective adjustment cells.

2.6.4 Whither the Telephone Survey Sample?

As telephone frames, and thus sampling, become more complex, and the public grows less tolerant to intrusion by strangers calling them on the telephone, some might wonder about the future of telephone survey taking, and thus of telephone sampling. Arguments pointing to its demise emphasize the eventual futility as the above-stated problems get progressively worse, and there is increasingly greater penetration of technology into our lives—implying a greater utility of other modes—where information can be obtained at the convenience of the respondent.

The telephone may still be the best survey option, particularly when the survey budget is low or moderate. Mail survey costs are modest, but their response rates are relatively low and they rely on list frames that present problems for surveys of the general population. Web or online surveys may become the wave of the future if web access rates increase substantially, and statistically useful frames become available. But with current web access rates still low and web survey response rates on a par with mail surveys, it may be some time before serious general population web surveys can be done on a sample drawn from a statistically sufficient frame. In-person surveys will always be the gold standard among choices, but their much higher unit cost will never meet the sample size needs in low-to-moderate budget settings. While the telephone survey method may survive, its viability will depend on its ability to adapt to changes in technology, in the populations it examines, and on our willingness to continually search for effective and practical ways to use it alone, or in combination with, other modes of survey data gathering.

CHAPTER 3

Recent Trends in Household Telephone Coverage in the United States

Stephen J. Blumberg, Julian V. Luke, and Marcie L. Cynamon
Centers for Disease Control and Prevention,
National Center for Health Statistics, USA

Martin R. Frankel
Baruch College, City University of New York, and Abt Associates, Inc., USA

The telephone environment today looks different than it did in the 1980s when Thornberry and Massey (1988) published their chapter on trends in telephone coverage across time and subgroups in the United States. In 1985–1986, 7.2 percent of civilian adults living in households did not have telephones in their households, a decrease from 11.2 percent in 1970 and 9.2 percent in 1975. Since 1986, the percentage of adults living in households without telephones has continued to decrease, dropping below 2.0 percent in 2003. Such a low prevalence would ordinarily render moot concerns about undercoverage bias resulting from the omission of these households from telephone survey sampling frames.

However, the telephone environment has changed radically in the past 5 years. Today, an increasing proportion of adults with telephones have chosen mobile telephones (i.e., wireless telephones and cell phones) instead of landline telephones. In 2003, 3.1 percent of civilian adults living in households had at least one mobile telephone but no landline telephones in their households, and since then the size of this population has grown rapidly.

Because of legal, ethical, and technical concerns, mobile telephones are generally excluded from random digit dialing (RDD) surveys (Lavrakas and Shuttles, 2005b; see also Chapter 20 in this volume, Steeh and Piekarski, 2007). The result is that

Advances in Telephone Survey Methodology, Edited by James M. Lepkowski, Clyde Tucker, J. Michael Brick, Edith de Leeuw, Lilli Japec, Paul J. Lavrakas, Michael W. Link, and Roberta L. Sangster

undercoverage error cannot be ignored as a potential source of bias in RDD telephone surveys. RDD telephone surveys that do not reach mobile telephones may not reach a representative sample of the population. Sampling weights in a survey estimation can account for some differences between the sample and the population. However, RDD telephone surveys may still produce biased estimates if, for the substantive variables of interest, people in mobile-only households and households without any telephone service differ from demographically similar people in households with landline telephones.

This chapter investigates the telephone coverage of the U.S. household population and the resulting potential bias when telephone surveys exclude households without landline telephones. While this chapter serves as an update of Thornberry and Massey (1988), it goes beyond in a detailed consideration of the impact of mobile-only households on undercoverage bias. (See Kuusela et al. (2007), Chapter 4 in this volume for international data on landline and mobile telephone coverage.)

3.1 PRIMARY DATA SOURCE

This chapter is primarily based on the empirical analyses of data from the National Health Interview Survey (NHIS). The NHIS is an annual multistage probability household survey that collects comprehensive health-related information from a large sample of households drawn from the civilian noninstitutionalized household population of the United States (Botman et al., 2000). This face-to-face survey interview is sponsored by the National Center for Health Statistics (NCHS) of the Centers for Disease Control and Prevention and is administered by trained field representatives from the U.S. Census Bureau. Interviews are conducted throughout the year with approximately 40,000 households annually.

The NHIS interview consists of four components: (1) an introduction including questions about household composition and characteristics; (2) a family health section including basic indicators of health status, activity limitations, injuries, health insurance coverage, and access to and utilization of health care services by all family members; (3) a section with detailed questions on health conditions and activity limitations, health-related behaviors, health care utilization, and immunizations for one randomly selected child under 18 years of age from each family with children; and (4) a similarly detailed section for one randomly selected adult from each family. Questions about demographics and socioeconomic status are asked within each component. The household, family, and child-specific sections are answered by knowledgeable adult family members; the adult-specific section is answered by the selected adult, unless a physical or mental impairment requires a proxy response.

Since 1963, the NHIS included questions on residential telephone numbers to permit recontact of participants. Beginning in 1997, questions about interruptions in telephone service (i.e., whether "your household [has] been without telephone service for more than one week" during the past 12 months) were added, followed by, in 2003, additional questions to confirm that the household telephone number provided was both a family telephone number and a landline telephone number. Family respondents were also asked whether "anyone in your family has a working cellular telephone." Families were identified as mobile families if anyone in the family had a working cellular telephone.

Most telephone surveys draw samples of households rather than of families. Because households can be comprised of more than one family (approximately 2 percent of households in the NHIS have multiple families), it is necessary to also classify households as "mobile only." Households were identified as such if they included at least one mobile family and if there were no working landline telephones inside the household. Persons were identified as mobile only if they lived in a mobile-only household. A similar approach was used to identify "phoneless households" (i.e., households without any type of telephone service) and persons living in such households. Households that were either mobile only or phoneless were collectively referred to as "nonlandline households," and households with landline telephones were therefore "landline households."

NHIS annual household-level and family-level response rates are generally very high (86–92 percent). Overall response rates for the detailed adult and child interviews are lower (70–74 percent and 78–84 percent, respectively). Landline telephone status is ascertained for approximately 99 percent, and mobile telephone status is ascertained for 85–90 percent of all completed family-level interviews.

All survey data presented in this chapter are weighted to provide national estimates. NHIS sampling weights are adjusted for the probability of selection of each household and each person in the household, and they are adjusted for nonresponse and to match U.S. Census Bureau estimates for the sex, age, and race/ethnicity distribution of the U.S. civilian noninstitutionalized population. The heterogeneity of the weights and the complex sampling design of the NHIS have been taken into account when calculating adjusted odds ratios and performing tests of statistical significance ($\alpha = 0.05$).

3.2 TIME TRENDS IN LANDLINE TELEPHONE COVERAGE

Thornberry and Massey (1988) showed that the percentage of households without telephone service decreased from 1963 through 1986. In 1963, about one fifth of U.S. households did not have telephone service. This estimate had dropped to around 10 percent by the early 1970s and to around 7 percent by the early 1980s. The rate remained essentially unchanged from 1981 to 1986.

The decennial censuses conducted by the U.S. Census Bureau revealed a similar pattern (U.S. Federal Communications Commission [FCC], 2005). In 1960, 21.7 percent of households did not have access to telephone service. This estimate declined to 9.5 percent in 1970 and to 7.1 percent in 1980. Later censuses demonstrated that the trend continued to decline to 5.2 percent in 1990 and to 2.4 percent in 2000.

Table 3.2. Percent of Persons Living in Households without Landline Telephone Service, by Selected Demographic Characteristics and by Year: United States, 1997, 1999, 2001, 2003, and 2005

Characteristic	1997	1999	2001	2003	2005
All persons	4.5	3.5	2.8	5.1	9.4
Race/ethnicity					
Hispanic	9.4	6.4	4.7	6.7	13.7
White, non-Hispanic	2.8	2.3	2.0	4.2	8.0
Black, non-Hispanic	9.8	8.0	5.2	7.7	12.6
Other single race, non-Hispanic	4.2	3.7	3.2	5.7	7.8
Multiple race, non-Hispanic	7.5	4.4	5.1	8.5	12.4
Age (years)					
Under 6	7.6	5.7	4.2	7.4	12.8
6–16	5.0	4.1	3.0	4.2	7.4
17–24	7.0	5.3	4.5	9.5	19.4
25–34	5.9	4.6	3.8	8.5	17.6
35–44	3.8	3.2	2.7	4.5	8.1
45–54	2.9	2.6	1.9	3.5	5.8
55–64	1.9	1.8	1.6	2.6	4.1
65–74	1.6	1.4	1.1	1.5	2.3
75 and over	1.3	1.0	0.9	0.9	1.7
Sex					
Male	4.8	3.9	3.1	5.6	10.2
Female	4.1	3.2	2.6	4.6	8.7
Education					
Some high school or less	6.9	5.5	4.0	5.8	9.6
High school graduate or GED	4.1	3.6	2.9	5.7	9.9
Some posthigh school, but no degree	2.5	2.0	2.0	4.8	10.2
Four-year college degree or higher	0.5	0.5	0.6	2.7	6.7
Marital status					
Married	2.4	1.8	1.4	2.8	5.5
Widowed	2.1	2.0	1.8	1.9	2.9
Divorced	6.0	5.4	4.3	7.8	13.1
Separated	11.0	10.8	9.8	13.5	20.1
Never married	5.8	4.8	3.8	8.2	16.2
Under 17 years	5.9	4.6	3.4	5.3	9.3
Employment status last week					
Working at a job or business	—[a]	2.8	2.4	5.2	10.6
Keeping house	—	4.3	3.5	5.2	8.6
Going to school	—	2.6	1.9	7.4	14.9
Something else (including unemployed)	—	3.8	3.0	4.3	6.4
Number of persons in the sample	103,477	97,059	100,760	91,165	98,455

[a]Data not available

had 207.9 million mobile telephone subscribers at the end of 2005 (CTIA, 2006), more than one mobile telephone for every two persons in the United States. The number of mobile telephone subscribers has been growing at a 16 percent compound annual rate since 1998 when there were only 69.2 million mobile telephone subscribers.

Even greater growth has been reported in the number of call minutes used by mobile subscribers. Since 1998, the reported number of minutes used has grown by more than 30 percent annually reaching 1.5 trillion minutes in 2005. On a per minute basis, mobile service is now cheaper than landline service, particularly for long distance calls or when the subscriber is traveling (FCC, 2003). With mobile service providers now offering calling plans with an exceptionally large number of inexpensive minutes that can be shared among multiple telephones in families or that can be prepaid (eliminating the need for credit checks or annual commitments), the growth in subscribership and use is expected to continue to grow.

Because of the increased use, widespread availability, and the relatively low cost of mobile service, the FCC (2004) reported that some customers may now be indifferent as to whether they make a call on a landline or on a mobile telephone. Thus, mobile telephone users may be increasingly likely to substitute a mobile telephone for a residential landline telephone.

NHIS data permit an analysis of changes in the prevalence of mobile-only households from January 2003 through December 2005. Table 3.3 presents the percentage of households with only mobile telephone service, and the percentages of adults and children under 18 years of age who live in such households. As noted earlier and shown in the table, mobile telephone status is unknown for 10–15 percent of the households. Because missing data were not imputed, the percentages for mobile-only households and phoneless households (i.e., households without any telephone service) are underestimated by a small margin.

The percentage of mobile-only households, as well as the percentages of adults and children living in such households, doubled in 2 years. During the last 6 months of 2005, 8.4 percent of all households were mobile only, and 7.7 percent of adults and 7.6 percent of children lived in mobile-only households. An estimated 17 million adults and more than 5 million children were mobile only.

From January 2003, the prevalence of mobile-only adults increased at a 23 percent compound growth rate every 6 months. This 23 percent growth rate did not fluctuate over the 30-month period. The compound growth rate was lower for children (19 percent), reflecting slower growth in 2003 and early 2004. But from the first 6 months of 2004 to the first 6 months of 2005, the prevalence of mobile-only children grew at about the same rate as for mobile-only adults, followed by a jump of 31 percent during 2005.

Mobile-only households now compose the vast majority of nonlandline households. During the last 6 months of 2005, at least 75 percent of nonlandline households had a mobile telephone in the household. This proportion has increased over the 30-month period from 60 percent during the first 6 months of 2003. This increase is largely due to the relatively constant percentage of phoneless households while the percentage of mobile-only households increased.

Table 3.3. Percent of Households, Adults, and Children by Household Telephone Status and by Date of Interview: United States, January 2003–December 2005

Date of interview	Sample size	Landline households with a mobile telephone	Landline households without a mobile telephone	Landline households with unknown mobile telephone status	Nonlandline households with unknown mobile telephone status	Mobile-only households	Phoneless households
Households							
January 2003–June 2003	16,524	42.4	43.0	9.3	0.2	3.2	2.0
July 2003–December 2003	18,988	41.8	42.3	9.3	0.3	4.2	2.1
January 2004–June 2004	16,284	43.2	39.6	9.9	0.5	5.0	1.8
July 2004–December 2004	20,135	43.1	38.7	9.4	0.5	6.1	2.2
January 2005–June 2005	18,301	42.4	34.4	13.2	0.8	7.3	1.9
July 2005–December 2005	20,088	42.6	32.4	13.8	0.8	8.4	1.9
95% confidence interval[a]		41.6–43.6	31.5–33.4	13.2–14.4	0.7–1.0	7.9–8.9	1.7–2.2
Adults							
January 2003–June 2003	30,745	45.7	40.4	9.4	0.2	2.8	1.6
July 2003–December 2003	35,152	45.2	39.8	9.5	0.2	3.5	1.7
January 2004–June 2004	30,423	46.9	36.3	10.4	0.5	4.4	1.5
July 2004–December 2004	37,611	46.8	35.7	9.7	0.5	5.4	1.8
January 2005–June 2005	34,047	46.1	31.5	13.5	0.7	6.7	1.6
July 2005–December 2005	37,622	46.4	29.7	13.9	0.7	7.7	1.7
95% confidence interval[a]		45.3–47.5	28.7–30.6	13.2–14.6	0.6–0.9	7.2–8.2	1.5–1.9
Children							
January 2003–June 2003	11,937	49.6	35.7	9.6	0.3	2.9	1.8
July 2003–December 2003	13,331	49.1	35.7	9.7	0.3	3.2	2.0
January 2004–June 2004	11,718	49.6	31.6	12.6	0.7	3.7	1.8
July 2004–December 2004	14,368	49.4	31.4	11.6	0.5	4.9	2.3
January 2005–June 2005	12,903	49.3	27.0	15.8	0.7	5.8	1.5
July 2005–December 2005	13,883	50.5	23.9	15.2	0.9	7.6	1.8
95% confidence interval[a]		48.8–52.2	22.6–25.4	14.0–16.5	0.7–1.2	6.8–8.5	1.5–2.3

[a]Confidence intervals for the time period July–December 2005.

The increase in mobile-only households and persons has been found using other face-to-face surveys. Trends since 1994 are available from the Consumer Expenditure Survey (see Tucker and Lepkowski, 2007, Chapter 1 in this volume), and since 2001 from Mediamark Research Inc.'s Survey of the American Consumer (Arthur, 2004). When Mediamark data on mobile telephones were first collected from approximately 13,000 American consumers in Spring 2001, only 1.4 percent of households were mobile only. By Spring 2006, this estimate increased to 10.6 percent, with mobile-only households accounting for 85 percent of all nonlandline households.

A series of smaller scale surveys to explore mobile telephone usage patterns has been fielded annually by RoperASW (later, GfK NOP) since 2000. These nationwide face-to-face surveys of approximately 2000 adults each year provide data on the percentage of adults whose households have only mobile telephones. Tuckel et al. (2006) estimated that 1.7 percent of adults lived in mobile-only households in February 2003. This prevalence estimate increased fivefold by March 2006 to 8.5 percent.

Collectively, these surveys suggest that the prevalence of mobile-only households and persons began to rise around 2001. By the end of 2005, the prevalence had increased to approximately 8–10 percent of the U.S. households and 8–9 percent of the U.S. adults. Similar estimates were obtained from two other surveys in 2004. A February 2004 supplement to the Current Population Survey (CPS) estimated that 6.0 percent of the U.S. households had only a mobile telephone to receive incoming telephone calls (Tucker et al., 2007). In November 2004, the National Election Pool (which conducted exit polls for the major television networks and the Associated Press on the day of the presidential election) estimated that 7 percent of election-day voters were mobile only (Keeter, 2006).

3.4 DEMOGRAPHIC DIFFERENCES IN THE PREVALENCE OF MOBILE-ONLY ADULTS

The most recent trends in NHIS data on mobile-only adults are shown by selected demographic characteristics in Fig. 3.1. In the last 6 months of 2005, the prevalence of mobile-only adults was greatest for adults 18–24 years of age and adults renting their home. Adults living in poverty were more likely than higher income adults to be in mobile-only households; nearly one in 10 adults in households below the federal poverty threshold were mobile-only, nearly twice the rate for those with higher incomes (twice the federal poverty threshold or greater).

Other demographic groups were examined but not included in Fig. 3.1. One in three adults living with unrelated roommates were in mobile-only households (34.0 percent), the highest prevalence rate among all the population subgroups examined. Adults living alone were more likely to be mobile only than adults living with other related adults or adults living with children (12.4, 6.0, and 7.1 percent, respectively). Men were more likely than women to be in mobile-only

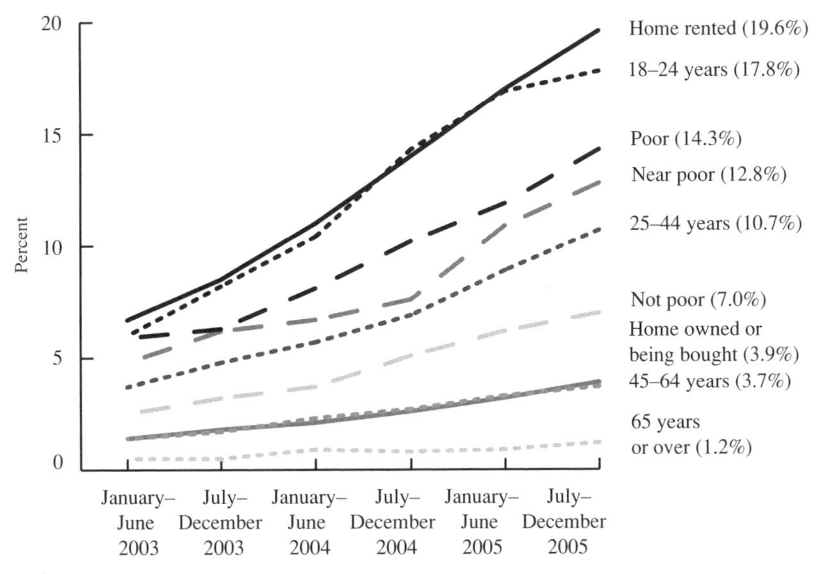

Figure 3.1. Percentage of adults living in mobile-only households, by selected characteristics and by half year: United States, 2003–2005.

households (8.6 and 7.1 percent, respectively). Hispanic adults (11.3 percent) were more likely to be in mobile-only households than were non-Hispanic white (7.0 percent) or non-Hispanic black adults (8.6 percent). Adults in the Northeast (4.7 percent) were less likely to be mobile only than were adults in the Midwest, South, or West (8.8, 9.6, and 6.3 percent, respectively). Adults in urban households were more likely than adults in rural households to be mobile-only (8.8 and 5.1 percent, respectively). Adults going to school were more likely to be mobile only than were adults working at a job or business in the week prior to the interview, keeping house, or were unemployed or doing something else (15.5, 9.2, 6.2, and 3.8 percent, respectively). Similar differences among demographic subgroups have also been shown by the Survey of the American Consumer (Arthur, 2004), the CPS supplement (Tucker et al., 2006), the National Election Pool (Keeter, 2005), the Pew Research Center for the People and the Press (2006), and Tuckel et al. (2006).

The demographic subgroups most likely to be mobile only in early 2003 are the same subgroups that were most likely to live in such households in late 2005. Still, the average rate of growth in the percentage of mobile-only adults over the past 30 months has been greater for some population subgroups than for others. The average growth rate has been greatest for adults going to school, more than tripling from the first 6 months of 2003 (4.2 percent) to the last 6 months of 2005 (15.5 percent). The growth rate was greater for Hispanic adults than for non-Hispanic adults, greater for adults living in the Northeast than for adults elsewhere, and greater for adults with a 4-year college degree than for adults with less education. For other demographic subgroups (including age, sex, poverty status, home ownership status, and urban/rural status), the rate of growth has been roughly the same as the rate of growth for the full population.

3.5 DEMOGRAPHIC COMPOSITION OF THE POPULATION OF ADULTS WITHOUT LANDLINE TELEPHONES

The data presented earlier indicate that, among adults living in nonlandline households, at least three quarters lived in households with a mobile telephone. Therefore, it is perhaps not surprising that the prevalence of mobile-only adults was greatest among many of the same subgroups that were most likely to be without landline service: young adults, males, renters, roommates, the poor, Hispanics, and those living in urban areas. However, there are differences between phoneless adults and mobile-only adults.

Table 3.4 presents the percentage of adults with selected household characteristics within each of these two groups (as well as other groups defined by household telephone status). Compared to phoneless adults, mobile-only adults were more likely to be roommates and to be living in urban areas. Phoneless adults were more likely to be poor and slightly more likely to be renters. Mobile-only adults were more likely to live in the Midwest; phoneless adults were more likely to live in the South.

Other differences between phoneless adults and mobile-only adults are presented in Table 3.5. Relative to phoneless adults, mobile-only adults were more likely to be young and more likely to be non-Hispanic white. Phoneless adults were more likely to be Hispanic or non-Hispanic black. Mobile-only adults generally had higher levels of education than phoneless adults: A majority of mobile-only adults had completed at least some post high school education, whereas more than two in five phoneless adults had not completed high school. Mobile-only adults were more likely to be working at a job or business in the week prior to the interview (although it should still be noted that a majority of phoneless adults were employed). No important sex differences were noted.

The SEARCH program (Sonquist et al., 1973) was used to investigate the extent to which demographic variables most closely related to being phoneless differ from those most closely related to being mobile-only. This program conducts a sequential series of x^2 analyses to identify the predictor variables that yield the greatest differences in the distributions of a dependent variable (i.e., maximize the likelihood-ratio x^2). We believe that the decision to have a telephone precedes and is independent from the decision to be mobile only (and the relatively constant rate of phoneless households in Table 3.3 supports this belief). Therefore, we examined two dependent variables: Phoneless or not phoneless was the dependent variable for one analysis, and mobile only or "landline telephone present" was the dependent variable for another analysis. The independent variables were the demographic variables in Tables 3.4 and 3.5. Both analyses were limited to adults 18 years of age or greater.

Table 3.4. Percentage of Adults with Selected Household Characteristics by Household Telephone Status: United States, 2005

Characteristic	Landline households with a mobile telephone	Landline households without a mobile telephone	Landline households with unknown mobile telephone status	Nonlandline households with unknown mobile telephone status	Mobile-only households	Phoneless households
Geographic region						
Northeast	19.2	20.5	20.7	14.1	11.6	12.7
Midwest	24.5	24.8	21.9	20.8	27.0	19.8
South	35.8	34.6	30.8	41.5	41.9	47.3
West	20.5	20.1	26.6	23.6	19.4	20.2
Urban/rural status						
Urban	72.1	70.4	77.2	85.8	82.9	71.9
Rural	27.9	29.6	22.8	14.2	17.1	28.1
Home ownership status						
Owned or being bought	81.7	70.2	75.2	28.1	34.8	27.2
Renting	16.9	27.3	23.2	68.2	62.6	66.7
Other arrangement	1.5	2.5	1.7	3.8	2.6	6.2
Household poverty status[a]						
Poor	4.7	15.2	9.7	24.8	15.9	48.3
Near poor	11.3	26.1	19.0	29.4	25.8	25.6
Not poor	84.1	58.7	71.3	45.8	58.3	26.1
Household structure						
Living alone	8.8	22.1	13.8	29.2	24.6	31.2
Living with roommate(s)	1.3	0.8	0.8	1.5	7.4	1.4
Living with spouse and/or related adults	46.6	46.4	41.9	28.0	34.9	34.6
Adult with children households	43.3	30.7	43.5	41.3	33.1	32.8

(continued)

Table 3.4. (*Continued*)

Characteristic	Landline households with a mobile telephone	Landline households without a mobile telephone	Landline households with unknown mobile telephone status	Nonlandline households with unknown mobile telephone status	Mobile-only households	Phoneless households
Number of persons in household						
1	8.8	22.1	13.7	29.0	24.6	31.2
2	33.1	37.8	30.7	29.5	34.7	32.2
3	20.7	15.0	20.0	20.2	18.2	14.9
4	21.3	13.2	18.7	11.8	11.9	9.7
5 or more	16.2	11.9	16.8	9.5	10.6	12.1
Number of adults in the sample	32,029	22,566	10,024	569	5,181	1,300

[a]Poverty status is based on household income and household size using the U.S. Census Bureau's poverty thresholds. "Poor" persons are defined as those below the poverty threshold. "Near poor" persons have incomes of 100% to less than 200% of the poverty threshold. "Not poor" persons have incomes of 200% of the poverty threshold or greater.

[b]Data not available.

Table 3.5. Percentage of Adults with Selected Personal Characteristics by Household Telephone Status: United States, 2005

Characteristic	Landline households with a mobile telephone	Landline households without a mobile telephone	Landline households with unknown mobile telephone status	Nonlandline households with unknown mobile telephone status	Mobile-only households	Phoneless households
Race/ethnicity						
Hispanic	10.2	13.8	15.0	24.0	17.4	22.1
White, non-Hispanic	74.7	68.9	65.4	47.0	65.4	50.1
Black, non-Hispanic	9.6	12.3	11.5	22.1	11.7	21.7
Other single race, non-Hispanic	4.6	4.0	7.4	5.7	4.3	4.3
Multiple race, non-Hispanic	0.9	1.0	0.7	0.8	1.3	1.7
Age (years)						
18–24	12.8	9.1	11.5	28.4	30.8	21.1
25–34	17.8	13.8	17.4	33.5	35.1	24.2
35–44	21.9	15.9	23.3	16.6	15.9	21.5
45–54	22.3	16.7	20.5	12.0	10.7	16.3
55–64	14.4	15.8	13.1	4.7	5.3	9.4
65–74	6.9	13.2	7.6	3.9	1.4	3.9
75 and over	4.0	15.5	6.7	0.8	0.9	3.6
Sex						
Male	48.6	45.9	47.7	54.9	53.6	54.6
Female	51.4	54.1	52.3	45.1	46.4	45.5
Education						
Some high school or less	10.3	23.6	14.3	22.3	15.8	42.0
High school graduate or GED	26.9	34.7	31.0	36.5	29.8	35.4
Some post-high school, but no degree	31.0	23.4	28.0	26.7	33.8	15.5

(continued)

Table 3.5. (*Continued*)

Characteristic	Landline households with a mobile telephone	Landline households without a mobile telephone	Landline households with unknown mobile telephone status	Nonlandline households with unknown mobile telephone status	Mobile-only households	Phoneless households
Four-year college degree or higher	31.8	18.4	26.7	14.5	20.6	7.2
Marital status						
Married	64.5	54.2	62.6	38.9	33.2	31.3
Widowed	3.3	12.2	5.5	1.4	1.4	4.3
Divorced	6.8	9.3	7.3	10.6	10.7	12.9
Separated	1.4	2.1	1.6	3.6	3.6	5.4
Never married	18.7	17.6	17.7	35.7	36.1	32.2
Living with a partner	5.2	4.7	5.5	9.8	15.0	13.9
Employment status last week						
Working at a job or business	72.6	51.5	69.3	72.6	77.0	53.3
Keeping house	6.6	8.0	6.9	7.7	5.4	10.2
Going to school	3.2	2.3	2.8	2.5	5.4	2.4
Something else (including unemployed)	17.6	38.3	21.0	17.2	12.2	34.1
Number of adults in the sample	32,029	22,566	10,024	569	5,181	1,300

[a]Data not available.

Figure 3.2 presents the results of the SEARCH analysis of phoneless adults. Home ownership status was the variable that yielded the greatest difference in prevalence rates. Relative to adults living in homes owned by someone in the household, adults living in homes that were rented or occupied by some other arrangement were more likely to be phoneless. Further differentiation was best achieved by educational attainment and household income, resulting in six final groups. The group with the greatest prevalence of phoneless adults was made up of adults with no more than a high school education living in poverty and in rented or otherwise occupied homes (11.9 percent). The group with the lowest prevalence of phoneless adults was made up of adults with at least a high school education living in households with higher incomes and in homes owned by a household member (0.4 percent).

The results of the SEARCH analysis of mobile-only adults are presented in Fig. 3.3. Similar to the results for phoneless adults, home ownership status was the variable that yielded the greatest difference in mobile-only prevalence rates. Further differentiation was best achieved by considering age, household size, and marital status, resulting in eight final groups. Among adults with some type of telephone service, the group most likely to be mobile only was made up of adults aged 18–30 years living in rented or otherwise occupied homes with no more than one other person (38.8 percent). The group least likely to have only mobile service was made up of married or widowed adults who were 37 years of age or older and living in homes owned by a household member (1.5 percent).

Thornberry and Massey (1988) found that, in 1985 and 1986, the most important variables for predicting telephone coverage were family income, age, and education. Two of these variables—income and education—were still strong predictors of being phoneless in 2005. Income and education, however, were not strongly predictive of being mobile only. Rather, young adults renting their home—"the young and the restless," as described by Steeh (2004b)—were most likely to be mobile only. Perhaps the most significant finding from these analyses is the importance of home ownership status (a variable not considered by Thornberry and Massey) in predicting phoneless or mobile-only status, and the relative insignificance of variables such as race/ethnicity, sex, employment status, and geographic region (cf. Tucker et al., 2006).

3.6 HEALTH-RELATED CHARACTERISTICS OF PERSONS IN NONLANDLINE HOUSEHOLDS

The percentage of persons living in nonlandline households is one factor that determines the potential for undercoverage bias in an RDD telephone survey. Another factor is the extent of the differences between persons with and without landline telephones on the substantive variables of interest (Groves, 1989, pp. 84 and 85). In this section, we will examine these differences, with a focus on health and health-related variables.

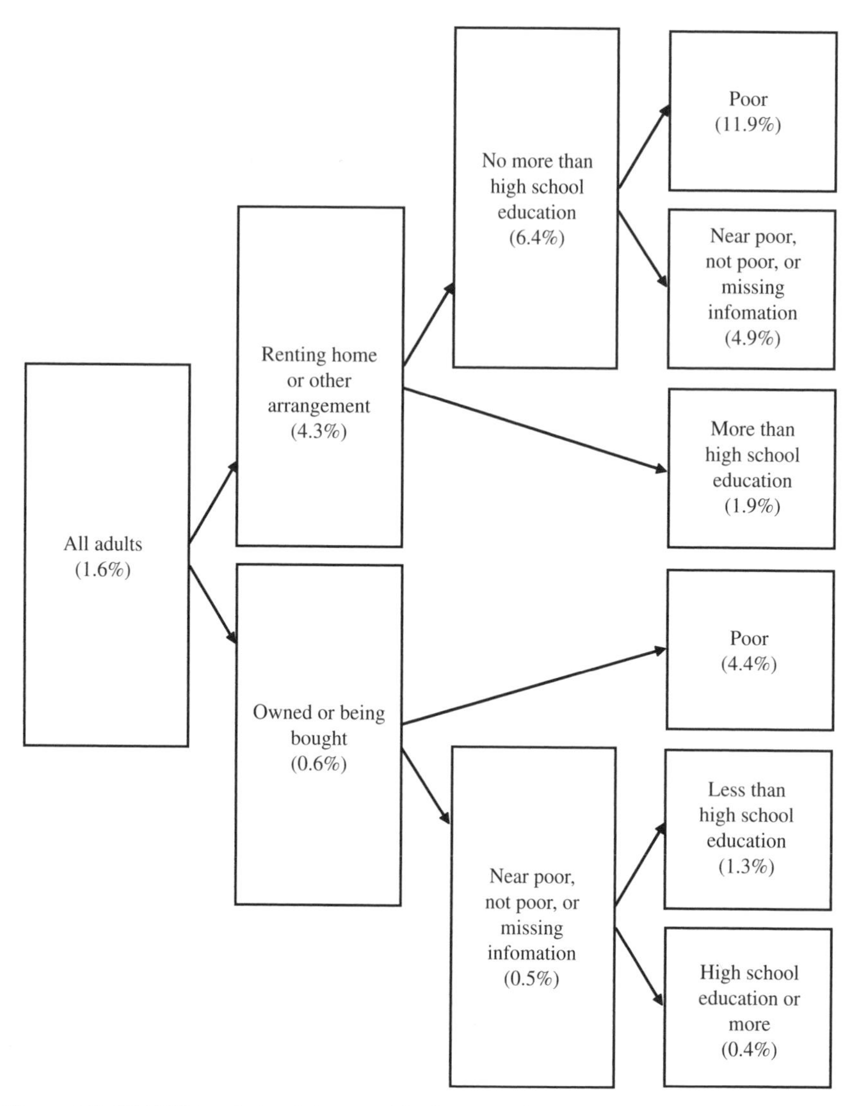

Figure 3.2. SEARCH analysis of demographic characteristics associated with living in phoneless households in 2005. Percentages represent the percentage of adults in that demographic group who are living in phoneless households. The denominator is the weighted estimate of the number of adults in that demographic group in the population.

Our approach uses a common dataset to look at differences among persons living in landline households, persons living in mobile-only households, and persons living in phoneless households. This approach is therefore similar to the previous analyses conducted with NHIS data (Thornberry and Massey, 1988; Anderson et al., 1998; Blumberg et al., 2006), National Health and Nutrition Examination Survey data (Ford, 1998), and National Household Survey on Drug Abuse data (McAuliffe et al.,

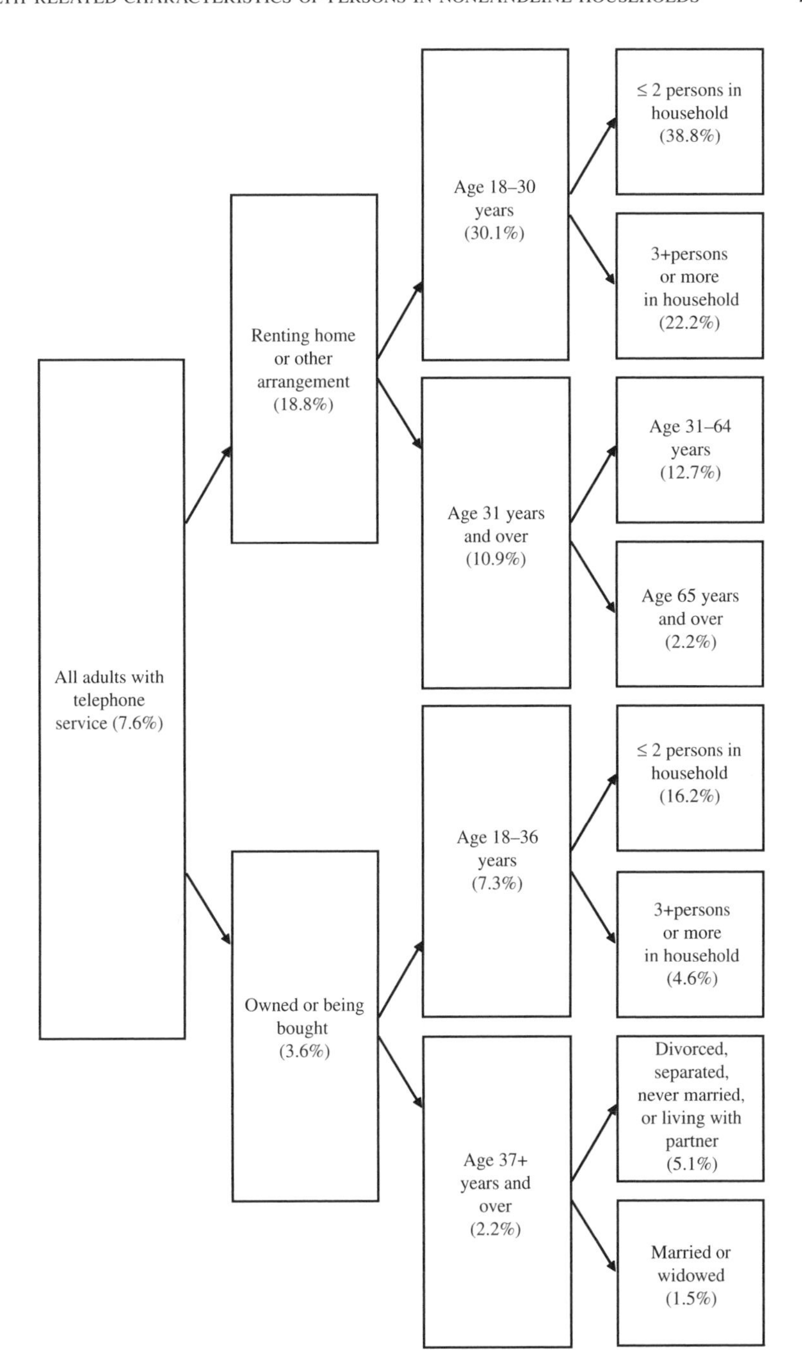

Figure 3.3. SEARCH analysis of demographic characteristics associated with living in mobile-only households in 2005. Percentages represent the percentage of adults in that demographic group living in mobile-only households. The denominator is the weighted estimate of the number of adults in that demographic group who are living in households with some type of telephone service (landline or mobile).

2002). It is, however, different from other published analyses on undercoverage bias that rely on comparisons of telephone survey estimates to face-to-face survey estimates (e.g., Anda et al., 1989; Aquilino, 1992; Nebot et al., 1994; Arday et al., 1997; Greenfield et al., 2000; Bartlett et al., 2001; Escobedo et al., 2002; Nelson et al., 2003). Comparing telephone survey estimates to face-to-face survey estimates can provide misleading estimates of undercoverage bias because differences observed could also be due to response biases and/or mode effects.

The health and health-related variables selected for this analysis are drawn from two sources: the NHIS Early Release Program (Schiller et al., 2006) and the NCHS *Summary Health Statistics for U.S. Children* report series (Bloom and Dey, 2006). The Early Release Program identified 15 key health indicators that were deemed sufficiently important to report every 3 months, before the microdata files from which they were calculated are released to the public. Data for all but one of these measures for adults are analyzed here. ("Personal care needs" was not included because the indicator is limited to adults aged 65 years and older.) For children, 10 key indicators were selected from the annual report on children's health published by NCHS. Specifications for all 24 variables are available elsewhere (Bloom and Dey, 2006; Schiller et al., 2006).

The first three columns of Table 3.6 present the percentages of adults and of children with selected health-related conditions and behaviors, as well as other selected measures of health status. Among adults living in mobile-only households, the prevalence of binge drinking (i.e., having five or more alcoholic drinks in one day during the past year) was twice as high as the prevalence among adults living in landline households. Mobile-only adults were also more likely to be current smokers, to report that their health status was excellent or very good, and to have engaged in regular leisure-time physical activity. They were also less likely to have been diagnosed with diabetes. When comparing phoneless adults to landline adults, the results were very similar. Notable exceptions were phoneless adults who were substantially less likely to have engaged in regular leisure-time physical activity or to have described their health as excellent or very good and more likely to have experienced serious psychological distress in the 30 days prior to the interview.

The health status of children in landline households was similar to the health status of children in mobile-only households and in phoneless households, with two exceptions. Relative to children in landline households, children in phoneless households were less likely to have their health status described as excellent or very good. Surprisingly, children in phoneless households were less likely to miss school due to illness or injury.

Table 3.7 presents the percentages of adults and of children with selected characteristics related to health care access and utilization. Among adults living in mobile-only households, the percentage without insurance at the time of the interview was twice as high as the percentage among adults living in landline households. Perhaps as a result, mobile-only adults were also more likely to have experienced financial barriers to obtaining needed health care, and they were less likely to have a usual place to go for medical care or to have received vaccinations for influenza

Table 3.6. Prevalence Rates and Relative Odds of Selected Measures of Health Status, Conditions, and Behaviors by Household Telephone Status: United States

	Percentage			Adjusted odds ratio[b]	
	In landline households[a]	In mobile-only households	In phoneless households	For mobile-only persons relative to landline persons	For phoneless persons relative to landline persons
Percentage of adults 18 years of age or older (January – December 2005)					
Health-related behaviors					
Five or more alcoholic drinks in 1 day at least once in the past year	17.9	38.2	19.2	1.81[c]	0.98
Current smoker	19.4	33.3	36.8	1.49[c]	1.33[c]
Engaged in regular leisure-time physical activity	29.9	36.4	21.0	1.26[c]	1.00
Health status					
Health status described as excellent or very good	61.2	68.0	47.4	1.03	0.88
Experienced serious psychological distress in past 30 days	2.8	3.5	7.8	1.03	0.99
Obese (adults 20 years of age or older)	25.5	23.8	25.8	1.02	0.86
Asthma episode in the past 12 months	3.8	4.3	4.6	1.01	0.86
Ever diagnosed with diabetes	7.8	3.4	7.3	0.83	0.90
Number of adults in the sample	27,553	2,961	722		
Percentage of children under 18 years of age (July 2004 – December 2005)[d]					
Health status					
Health status described as excellent or very good	82.8	78.9	68.2	1.06	0.85

(continued)

Table 3.6. ***(Continued)***

	Percentage			Adjusted odds ratio[b]	
	In landline households[a]	In mobile-only households	In phoneless households	For mobile-only persons relative to landline persons	For phoneless persons relative to landline persons
No school days missed due to injury or illness in the past year	28.3	28.5	39.3	0.92	1.31
Asthma episode in the past the 12 months	5.2	6.5	6.5	1.23	1.23
Ever diagnosed with ADD/ADHD	6.4	7.7	3.9	1.17	0.59
Number of children in the sample	17,421	1,428	432		

[a]Includes households that also have mobile telephone service.

[b]Odds ratios for adults were adjusted for race/ethnicity, age, sex, education, employment status, household structure, household poverty status, geographic region, urban/rural status, and home ownership. Odds ratios for children were adjusted for race/ethnicity, age, sex, highest education attained by any household member, number of parents in household, household poverty status, geographic region, urban/rural status, and home ownership.

[c]Adjusted odds are significantly different, $p < 0.05$.

[d]An 18 month time frame was necessary to achieve a sufficient sample size for reliable estimates for phoneless children.

or pneumococcal infections. Mobile-only adults were, however, more likely than landline adults to have ever been tested for HIV, the virus that causes AIDS. Similar results were revealed for phoneless adults relative to landline adults, except that phoneless adults were even more likely to be uninsured.

The results observed for children were similar. Relative to children living in landline households, the prevalence of uninsured children was half again as high for children in mobile-only households and three times as high for children in phoneless households. Perhaps as a result, children in both groups were less likely to have a usual place to go for medical care and less likely to have had contact with a dentist during the prior 6 months. Children in phoneless households were also less likely than children in landline households to have had contact with a health care provider in the prior 6 months and to have used prescription medication on a regular basis during the prior 3 months.

3.7 POTENTIAL BIAS DUE TO THE EXCLUSION OF NONLANDLINE HOUSEHOLDS

As noted earlier, the potential for undercoverage bias in an RDD telephone survey is determined by two factors: the percentage of persons living in nonlandline households and the extent of the differences between persons with and without landline telephones on the substantive variables of interest (Groves, 1989, pp. 84–85). In Table 3.3, we presented data suggesting that nearly 10 percent of U.S. adults and children live in nonlandline households. In Tables 3.6 and 3.7, we presented data suggesting that these children and adults are more likely to be uninsured and to experience barriers to care, and that these adults are more likely to binge drink alcohol, smoke, and experience serious psychological distress.

To examine the potential undercoverage bias due to the exclusion of nonlandline households from RDD sampling frames, we compared the percentages of adults and of children with selected health-related characteristics from the 2005 NHIS (which is considered to be a standard for population-based estimates) with estimates from the same dataset if the dataset were restricted only to those persons living in landline households (data not shown; see Blumberg et al., 2006 for an example of this comparison using earlier NHIS data). The potential bias is quite small. For adults, potential bias of greater than one percentage point could be observed for estimates of binge drinking, smoking, having a usual place to go for medical care, receiving influenza or pneumococcal vaccine, having had an HIV test, and being uninsured. For children, potential bias of greater than one percentage point was observed only for receipt of dental care. None of these estimates would be biased by more than two percentage points. (Similar results have been found for estimates of political attitudes; Pew Research Center, 2006.)

Recently, there has been great momentum for identifying methods and best practices for extending RDD survey interviews to mobile telephones. Steeh and Piekarski (2007, Chapter 20 in this volume) on new technologies and Brick and Lepkowski

Table 3.7. Prevalence Rates and Relative Odds of Selected Measures of Health Care Access and Utilization by Household Telephone Status: United States

	Percentage			Adjusted odds ratio[b]	
	In landline households[a]	In mobile-only households	In phoneless households	For mobile-only persons relative to landline persons	For phoneless persons relative to landline persons
Percentage of adults 18 years of age or older (January – December 2005)					
Health care service use					
Received influenza vaccine during the past 12 months	22.4	9.5	16.5	0.74[c]	1.18
Ever received a pneumococcal vaccination	17.5	7.6	9.2	0.94	0.71
Ever been tested for HIV	34.2	46.2	40.0	1.25[c]	1.00
Health care access					
Has a usual place to go for medical care	86.4	66.9	67.4	0.67[c]	0.80[c]
Failed to obtain needed medical care in the past 12 months due to financial barriers	6.4	12.9	14.6	1.34[c]	1.01
Currently uninsured	14.3	31.5	40.4	1.41[c]	1.29[c]
Number of adults in the sample	27,553	2,961	722		
Percentage of children under 18 years of age (July 2004 – December 2005)[d]					
Health care service used					
Had contact with a health care professional in the past the 6 months	74.8	72.4	66.6	0.89	0.75[c]
Had contact with dentist in the past six months	56.3	38.9	30.3	0.76[c]	0.58[c]
Regularly uses prescription medication during past the 3 months	13.5	12.9	7.0	1.00	0.52[c]

Health care access					
Has a usual place to go for medical care	95.0	92.4	86.7	0.87	0.57[c]
Failed to obtain needed medical care in the past 12 months due to financial barriers	1.9	3.9	4.6	1.41	1.62
Currently uninsured	8.3	12.9	26.0	1.18	2.23[c]
Number of children in the sample	17,421	1,428	432		

[a]Includes households that also have mobile telephone service.

[b]Odds ratios for adults were adjusted for race/ethnicity, age, sex, education, employment status, household structure, household poverty status, geographic region, urban/rural status, and home ownership. Odds ratios for children were adjusted for race/ethnicity, age, sex, highest education attained by any household member, number of parents in household, household poverty status, geographic region, urban/rural status, and home ownership.

[c]Adjusted odds are significantly different, $p < .05$.

[d]An 18-month time frame was necessary to achieve a sufficient sample size for reliable estimates for phoneless children.

(2007, Chapter 7 in this volume) on multiple frame multimode surveys include discussions of research to date. If mobile telephones could be effectively included in RDD sampling frames, the potential for undercoverage bias would be reduced because only phoneless households would be unreachable by telephone surveys. Based on the 2005 NHIS data, the potential undercoverage bias solely due to the exclusion of only phoneless households from RDD sampling frames was less than one percentage point for all estimates.

For these analyses, the reported bias is only the "potential" bias. The sampling weights that were applied to the NHIS data for adults and children in landline households were the same weights as were generated for estimates of the full U.S. civilian noninstitutionalized population. In contrast, many RDD telephone survey researchers would adjust the sampling weights for landline households so that estimates of key demographic characteristics match population control totals obtained from an independent source (such as the CPS). If the differences between landline persons and nonlandline persons on the substantive variables of interest are directly related to a demographic characteristic that can be controlled in this manner, then the undercoverage bias would be greatly reduced by this weighting adjustment.

The 2004 National Election Pool results present an example of how this adjustment can effectively reduce the undercoverage bias (Keeter, 2005). Prior to the 2004 presidential election, hundreds of news stories discussed the potential bias that could result from preelection polls because mobile telephones were excluded from these polls, younger voters were more likely to live in mobile-only households, and younger voters were more likely to support John Kerry for president. Indeed, the exit polls demonstrated that mobile-only voters were more likely to vote for Kerry. However, within each age cohort, there was little difference in support for Kerry between voters with landline telephones and voters with only mobile telephones in their homes. Provided that preelection polls adjusted sampling weights so that each age cohort was appropriately represented in the sample, the exclusion of mobile-only voters from the samples produced very little undercoverage bias.

To explore whether adjustments for demographic characteristics would greatly reduce the potential undercoverage bias in health surveys, we examined whether the differences between landline and nonlandline adults would be statistically significant after adjusting for differences in race/ethnicity, age, sex, education, employment status, household structure, household poverty status, geographic region, urban/rural status, and home ownership. Similarly, the differences between children living in landline households and in nonlandline households were examined after adjusting for differences in race/ethnicity, age, sex, highest education attained by any household member, number of parents in household, household poverty status, geographic region, urban/rural status, and home ownership. These demographic variables were chosen because they had been used to adjust the sampling weights of major RDD health surveys such as the Behavioral Risk Factor Surveillance System, the National Immunization Survey, the National Survey of Children with Special Health Care Needs, and the

National Survey of America's Families. A statistically significant adjusted odds ratio is an indirect indicator that weighting adjustments would not eliminate the potential undercoverage bias.

Even with adjustments to account for these demographic covariates, logistic regression analyses revealed that, relative to landline adults, mobile-only adults had significantly greater odds of binge drinking, smoking, exercising, having had an HIV test, being uninsured, and experiencing financial barriers to health care (see Tables 3.6 and 3.7). They also had significantly lower odds of having a usual place for medical care and of receiving influenza vaccine. Relative to landline adults, phoneless adults had significantly greater odds of smoking and being uninsured, and significantly lower odds of having a usual place for medical care. Relative to children in landline households, children in mobile-only households were less likely to have had contact with a dentist in the past 6 months. Children in phoneless households were more likely to be uninsured, less likely to have a usual place for care, less likely to have had contact with health care professionals or dentists, and less likely to regularly use prescription medications. These results suggest that undercoverage bias cannot be completely eliminated from RDD health surveys simply by adjusting the sampling weights to adequately represent each demographic subgroup in the sample.

3.8 THE DYNAMIC NATURE OF TELEPHONE OWNERSHIP

Up to this point, we have treated the presence or absence of telephone service (whether landline or mobile) as a stable characteristic of a household and of the people living in that household. However, Keeter (1995) demonstrated that telephone service can be episodic for many households. Service may be obtained or lost as a result of changes in financial situations, physical address, or use of technology. Using data from a panel survey (the 1992–1993 CPS), Keeter estimated that half of the households with no telephone service at some point during a year did in fact have service at another point during that year. These households were called "transient" telephone households.

Several statisticians have proposed that the undercoverage bias in telephone survey estimates may be reduced by adjusting the sampling weights for identified transient telephone households in the sample (e.g., Brick et al., 1996b; Frankel et al., 2003b; Davern et al., 2004). These proposals are based on the following logic: If transient telephone households are similar to nonlandline households, then the transient telephone households that can be reached with a telephone survey can be used to represent the nonlandline households that cannot be reached with a telephone survey. Although several methods for incorporating transient telephone households into survey estimation procedures have been identified, they all rely on the same assumption that transient telephone households are similar to nonlandline households.

The utility of this assumption may be examined with the 2005 NHIS data. Tables 3.8 and 3.9 present the percentage of adults with selected demographic characteristics living in households from each of three telephone service groups: continuous landline service during the prior year; transient landline service during the prior year; and no landline service at the time of the interview. The transient telephone group is limited here to adults living in households that had a landline telephone at the time of the survey because this is the only subset of the transient telephone population that can be interviewed by a telephone survey. For undercoverage bias to be reduced by including this transient telephone group in survey estimation

Table 3.8. Percentage of Adults with Selected Household Characteristics by Transient Telephone Status: United States, 2005

	Landline households		
Characteristic	Continuous service in the past year	Interruption of one week or more in the past year	Nonlandline households
Geographic region			
Northeast	20.0	13.3	12.0
Midwest	24.3	20.0	25.3
South	34.3	48.9	42.8
West	21.4	17.8	19.9
Urban/rural status			
Urban	72.3	71.5	81.2
Rural	27.7	28.5	18.8
Home ownership status			
Owned or being bought	77.4	55.4	33.1
Renting	20.9	40.6	63.6
Other arrangement	1.8	4.0	3.3
Household poverty status[a]			
Poor	8.3	21.6	21.4
Near poor	16.7	25.0	25.9
Not poor	75.0	53.4	52.7
Household structure			
Living alone	14.0	13.7	26.1
Living with roommate(s)	1.1	1.6	6.0
Living with spouse and/or related adults	46.1	34.9	34.3
Adult with children households	38.8	49.8	33.7
Number of persons in household			
1	14.0	13.7	26.0
2	34.5	27.8	33.9
3	18.7	19.5	17.8
4	18.2	16.9	11.5
5 or more	14.7	22.2	10.7
Number of adults in the sample	63,042	1,577	7,050

[a]Poverty status is based on household income and household size using the U.S. Census Bureau's poverty thresholds. "Poor" persons are defined as those below the poverty threshold. "Near poor" persons have incomes of 100% to less than 200% of the poverty threshold. "Not poor" persons have incomes of 200% of the poverty threshold or greater.

Table 3.9. Percentage of Adults with Selected Personal Characteristics by Transient Telephone Status: United States, 2005

Characteristic	Landline households: Continuous service in the past year	Landline households: Interruption of one week or more in the past year	Nonlandline households
Race/ethnicity			
Hispanic	12.0	18.6	18.7
White, non-Hispanic	71.6	60.5	61.5
Black, non-Hispanic	10.6	17.5	14.1
Other single race, non-Hispanic	4.9	2.2	4.4
Multiple race, non-Hispanic	0.9	1.1	1.3
Age (years)			
18–24	11.1	20.1	29.0
25–34	16.2	24.4	33.1
35–44	20.1	18.7	16.9
45–54	20.2	15.4	11.7
55–64	14.8	11.9	5.9
65–74	9.2	5.2	2.0
75 and over	8.4	4.4	1.3
Sex			
Male	47.6	46.8	53.9
Female	52.4	53.2	46.1
Education			
Some high school or less	15.2	24.3	20.6
High school graduate or GED	30.1	29.3	31.2
Some posthigh school, but no degree	28.0	29.6	30.3
Four-year college degree or higher	26.8	16.8	18.0
Marital status			
Married	61.1	47.1	33.3
Widowed	6.7	4.7	1.9
Divorced	7.7	10.2	11.1
Separated	1.6	3.0	3.9
Never married	18.0	24.2	35.4
Living with a partner	5.0	10.8	14.4
Employment status last week			
Working at a job or business	65.0	61.3	72.7
Keeping house	7.1	8.1	6.4
Going to school	2.8	2.7	4.7
Something else (including unemployed)	25.1	27.9	16.2
Number of adults in the sample	63,042	1,577	7,050

procedures, the characteristics of adults in this transient telephone group do not need to be identical to the characteristics of nonlandline adults (Keeter, 1995). Rather, the differences between nonlandline adults and adults in transient telephone households need to be smaller than the differences between nonlandline adults and adults living in continuous landline households.

For most of the demographic variables included in Tables 3.8 and 3.9, the differences are in the desired direction, and the inclusion of the transient telephone group in survey estimation procedures should reduce the undercoverage bias. However, the bias may increase slightly for a few selected characteristics (e.g., the number of women, the number of adults working or keeping house, the number of adults living in rural areas, and the number of adults living with children) because the differences go in the opposite direction. Other weighting procedures (e.g., adjustments for demographic characteristics) may help minimize an unintended increase in bias for these groups.

3.9 MOBILE TELEPHONE SERVICE IN TRANSIENT TELEPHONE HOUSEHOLDS

When Keeter (1995) concluded that "transient telephone households are much more like nonphone households than those with continuous telephone service," he perhaps would not have predicted that the size of the population of nonlandline households would soon grow and that a majority of these households would have mobile telephones. For demographic characteristics such as sex, employment status, urban/rural status, and household structure, perhaps the reason why adults in transient telephone households are not "much more like" adults in nonlandline households is because of the changes in the nonlandline population. To explore this possibility, we examined whether the differences between phoneless adults and adults in transient telephone households are smaller than the differences between adults in continuous landline households and adults in transient telephone households. (Estimates used for this analysis are in Tables 3.4, 3.5, 3.8, and 3.9.) Even when considering the subpopulation of nonlandline adults who are phoneless adults, the conclusions are similar to those for nonlandline adults: Adults in transient telephone households are more similar to phoneless adults on estimates of race/ethnicity, age, marital status, geographic region of residence, home ownership status, and household poverty status, but not on estimates of sex, educational attainment, employment status, urban/rural status, household size, and household structure.

Perhaps the nature of transient telephone households has also changed in the past 12 years. For example, an intermittent interruption of landline telephone service may not be considered problematic by a household that has access to a mobile telephone. Therefore, the characteristics of transient telephone households with mobile service may be different from the characteristics of transient telephone households without mobile service. An investigation of this hypothesis would ideally consider mobile telephone service at the time of the interruption. Unfortunately, such data are not available from the NHIS. Instead, our investigation of this hypothesis relies on the mobile service at the time of the interview.

In 2005, more than half (59.4 percent) of all adults in the transient telephone households that had landline service at the time of the interview also had mobile telephone service. Relative to those with mobile telephone service, adults in transient telephone households without mobile telephone service were nearly three times as likely to be poor, more than twice as likely to have less than a high school education, and substantially more likely to be unemployed or keeping house, to be renting their home, and to be Hispanic or of non-Hispanic black race (data not shown). Adults in transient telephone households with mobile telephone service were five times as likely to be living with roommates and substantially more likely to have a college education. No important differences were noted by sex or urban/rural status.

These differences largely mirror the differences between mobile-only adults and phoneless adults that were highlighted earlier. One might predict that adults in transient telephone households with mobile service are more similar to mobile-only adults than to phoneless adults, and that adults in transient telephone households without mobile service are more similar to phoneless adults than to mobile-only adults. If these hypotheses are true, then perhaps undercoverage bias could be further reduced by including these two transient telephone groups separately in survey estimation procedures. That is, perhaps transient telephone households with mobile service can be used to represent mobile-only households that cannot be reached with a telephone survey, and perhaps transient telephone households without mobile service can be used to represent phoneless households that cannot be reached with a telephone survey.

There is some evidence from the NHIS to suggest that this proposed estimation strategy might be particularly useful for reducing undercoverage bias when the substantive variables of interest are related to poverty status or educational attainment. The distributions of poverty and educational attainment among adults in transient telephone households with mobile telephone service are more similar to the distributions for mobile-only adults than to the distributions for phoneless adults. Conversely, the distributions of poverty and educational attainment among adults in transient telephone households without mobile telephone service are more similar to the distributions for phoneless adults than to the distributions for mobile-only adults. A 2004 survey for the California Public Utilities Commission revealed similar results (Jay and DiCamillo, 2005). Further research is necessary to determine if this proposed strategy will reduce undercoverage bias to a greater degree than can be achieved by simply including the full transient telephone group (i.e., with or without mobile telephone service) in survey estimation procedures (e.g., Khare et al., 2006).

3.10 DISCUSSION: UNDERCOVERAGE AND NONRESPONSE

The first commercially available mobile telephone (the Motorola DynaTAC 8000X) was introduced in 1983, 5 years prior to the publication of Thornberry and Massey's (1988) chapter on telephone coverage trends. During the 1980s, few people owned mobile telephones; use was limited to the rich and powerful people in society. Since then, mobile telephones became cheaper, smaller, and more readily available and people from all walks of life bought mobile telephones because the telephones made them feel safer (e.g., because they were continuously available, especially in the event of an emergency) and were convenient (e.g., for socializing with friends, for calling home when running errands, and for maintaining lists of telephone numbers) (Rosen, 2004; Totten et al., 2005). This convenience, along with falling prices, has convinced many people to substitute mobile telephones for landline telephones.

A major focus of this chapter has been on the size of this population, its characteristics, and the potential bias resulting from its exclusion from telephone survey sampling frames. These analyses have assumed, however, that owning a landline telephone means that one is reachable on that telephone. This is in part a false assumption. Among adults living in households with both landline and mobile telephone service in 2004, nearly 9 percent received all or almost all of their calls on their mobile telephones (Tucker et al., 2006). Even among adults living in households with only landline telephone service in early 2005, nearly one in six (16.5 percent) reported that they never answered incoming voice calls (Tuckel and O'Neill, 2005). As a result, the number of households unreachable by a landline telephone survey may be greater than that estimated here based on telephone ownership alone.

The inability to contact persons who are in the sampling frame is generally considered a component of nonresponse rather than an issue related to coverage. It is useful to reflect on the relative magnitude of nonresponse and noncoverage. Both are important contributors to bias in telephone survey estimates, yet the percentage of persons in the population who do not have landline telephones (approximately 10 percent) is easily overwhelmed by the percentage of persons in landline households who will not respond to a telephone survey (typically greater than 25 percent).

Nevertheless, the percentage of persons in the population with only mobile telephones is expected to increase. As noted earlier, the population of mobile-only adults has been growing rapidly for the past 30 months. A recent survey suggests that 52 percent of adults are seriously or somewhat seriously considering disconnecting the landline telephone and using mobile telephones exclusively (Harris Interactive, 2005). Whether this seemingly inevitable growth will increase the potential for undercoverage bias depends on whether the current patterns continue, consistent with what is happening in some European countries (see Kuusela et al. 2007, Chapter 4 in this volume). Until these patterns become clear, or until methods are established for including mobile telephone numbers in RDD sampling frames, the potential for bias due to undercoverage remains a real and a growing threat to surveys conducted only on landline telephones.

CHAPTER 4

The Influence of Mobile Telephones on Telephone Surveys[1]

Vesa Kuusela
Statistics Finland, Helsinki, Finland

Mario Callegaro
University of Nebraska, Lincoln, USA

Vasja Vehovar
University of Ljubljana, Slovenia

4.1 INTRODUCTION

In surveys conducted by telephone, the telephone may have two different roles: (1) the medium of communication and (2) the sampling unit (i.e., telephone numbers). In both roles, the device itself and its characteristics impose requirements on the survey methodology.

The main problems in telephone surveys for landline telephones were solved or at least recognized a long time ago. As mobile telephones become "everyman's" devices, the essential conditions for a telephone survey change substantially. Modifications in methodology become inevitable as mobile telephones increasingly penetrate in a target population. Mobile telephone popularity has increased very rapidly worldwide, but the popularity varies considerably from country to country. Since most countries

[1]We would like to acknowledge the assistance of Aurélie Vanheuverzwyn of Médiamétrie for the data on telephone penetration in France. Kristine Clara of the International Telecommunication Union library helped us with the international data on mobile telephone subscriptions. Lastly, Jim Lepkowski, our section editor, made a great contribution in clarifying the text and keeping us on schedule.

Advances in Telephone Survey Methodology, Edited by James M. Lepkowski, Clyde Tucker, J. Michael Brick, Edith de Leeuw, Lilli Japec, Paul J. Lavrakas, Michael W. Link, and Roberta L. Sangster

do not have accurate figures with respect to telephone coverage, it is difficult to estimate mobile telephone penetration and the problems that current penetration levels introduce for telephone surveys.

The most important questions regarding mobile telephones in telephone surveys concern the total coverage of the population and which population segments have the highest nontelephone household frequency. The usual practice to describe telephone coverage in a country has been based on the service provider data on subscriptions. The estimates of telephone coverage are calculated from the estimated total number of households in the population and the known number of private telephone subscriptions. When only landline telephones were considered, resulting estimates were somewhat inaccurate, but accurate enough to show whether telephone surveys were feasible or not. However, from the service provider data it is not possible to estimate the structure of landline telephone coverage in a country, that is, how landline telephone coverage differs in different population segments.

Mobile telephones make the estimation of coverage considerably more difficult. The structure becomes even more complex as new types of households emerge: those with only mobile telephones, and those with both landline and mobile. Using service provider's figures it is possible to estimate how many households have landline telephones and how many persons have mobile telephones, but it is not possible to estimate how many households have at least one mobile telephone or how many households are without any telephone. To be able to adequately design samples in telephone surveys it is necessary to know the extent to which landline and mobile telephones are spread over households. Because of the complexity of telephone coverage and its structure in a modern society, let alone how telephones are used, coverage structure can be reliably estimated only through a survey designed for this purpose. Respondent accessibility varies considerably between population segments and depends on the telephone type. The increase in mobile telephone popularity is not uniform. In some population segments practically all may have a mobile telephone, while in another, practically none may have it. In addition, a new kind of population segment also emerges: persons in households with only mobile telephones. The proportions of these mobile-only households are increasing in all countries.

These significant problems cause both indirect and direct effects on telephone survey methods and practices. Sampling-related issues need special attention because there is a danger of obtaining biased estimates if telephone types are not dealt with adequately. Nonsampling issues also arise that become more important as the proportion of interviews conducted over mobile telephones increases. The currently applied methodology in telephone surveys and the existing literature focus mainly on using landline telephones. The methodology assumes that nearly all households have at least one landline telephone by which all the members of the household can be reached (Groves et al., 1988; Piekarski, 1996; De Leeuw and Collins, 1997). For those households that have multiple landline telephones, weights can be developed to compensate for multiple chances of selection (Massey and Botman, 1988). Random digit dialing (RDD) methods can provide a simple random sample of households or the equivalent. In this case the respondent is not directly selected, but in a second stage, using a random procedure to select a respondent (e.g., by the last/next birthday methods; see Kalsbeek and Agans (2007), Chapter 2 in this volume).

There are three different groups from whom no data are collected: nonrespondents, nontelephone households, and households who do not have the landline but only mobile telephones. The households and persons in these groups are different. The processes leading to an absence of data are different. The missingness cannot be considered the result of random processes. If telephones are the only media of data collection, then nontelephone households cannot be chosen. Nontelephone households are not covered for yielding the potential for bias. Nicolaas and Lynn (2002) point out that nontelephone households are not a random set of the general population and do constitute a source of nonignorable nonresponse.

The quality of data obtained through mobile telephones raises other important questions. For instance, do results acquired over a mobile telephone differ so much from those obtained from a landline telephone that it would cause differences in results? Do mobile telephones require different interviewing techniques? A mobile telephone brings up new situations and problems that did not exist earlier. For example, the respondent can be virtually anywhere, and the environment or the situation may have an influence on the answering process. Respondents may try to answer as quickly as possible to save battery life or to save time if they have to pay to receive an incoming call.

The essential difference between a mobile telephone and a landline telephone is that a mobile telephone is typically a personal appliance carried nearly all the time. A landline telephone is for the whole household, and it is kept in one place. A mobile telephone could be compared to a wristwatch, while a landline telephone to a clock on the wall. Consequently, mobile telephone coverage needs to be considered per person rather than per household. This difference also has some bearing on the sampling design.

Very few scientific reports or textbooks have referred to date to the potential impact that the mobile telephones may have on telephone surveys. For example, the Khursid and Sahai (1995) bibliography had no entries that dealt with mobile telephones. Even as late as 1999, Collins (1999) made only a brief comment on the growing popularity of mobile telephones in the future, even though they were fairly popular at that time. Nathan (2001) and Nicolaas and Lynn (2002) recognized the mobile telephone problem in the context of random digit dialing. Most of the discussion, though, has taken place in conferences and not in the published literature (see the references throughout the chapter).

This chapter has two parts. First, the structure of telephone coverage in several European countries is described. In these countries structural change has been fast, though it varies considerably across countries. The changes may serve as an example of the kind of modification that can take place in other countries in the future. Second, the several influences that mobile telephones can have on telephone surveys are discussed in terms of sampling and nonsampling features of telephone surveys. The potential impacts of these influences that may still be ahead are also discussed.

4.2 STRUCTURE OF TELEPHONE COVERAGE IN VARIOUS COUNTRIES

In order to be able to adequately design a telephone sample survey and assess potential biases, precise information about the structure of telephone coverage is needed. Obtaining accurate estimates about coverage is possible in surveys that have representative samples and procedures that are not dependent on respondent accessibility to telephones. The accuracy of coverage estimates based on the service provider data is questionable because the data do not reflect the actual usage of telephones. Only in very few countries are reliable surveys focused on telephone coverage available. In some cases the published data are based on methods whose accuracy may be questioned, and these results may lead to discrepancies between estimates published by different sources for the same country.

In countries where mobile telephone prepaid subscriptions are popular, estimating mobile telephone coverage may be biased using only data provided by service providers because they publish the number of subscriptions (SIM[2] cards) that have been sold to distributors. It is not possible to know what proportions of subscriptions are really in use. A share of available SIM cards has not yet been sold. Some have been discarded by the subscriber without telling the service provider but are still considered valid telephone numbers. Further, some people own more than one SIM card and use them in different occasions, such as one for work and one for private calls. Subscribers may also use different cards to call different operators to reduce the cost of calls. For example, in Italy multiple SIM card ownership was estimated to be of 20–25 percent in 2002 (Costabile and Addis, 2002). As a result, estimates using subscription data to depict mobile telephone coverage may be misleading in countries where prepaid subscriptions are popular. In some countries reported mobile telephone coverage is over 100 percent because of these problems (see Table 4.1). The best and most reliable way to obtain unbiased estimates on telephone coverage and structure is through a face-to-face probability sample of households.

4.2.1 Face-to-Face Surveys on Telephone Coverage

In some European countries, the structure of telephone coverage has been examined either by one-time surveys or in the context of ongoing surveys where probability samples or "representative" samples have been used for face-to-face interviews with respondents. U.S. telephone coverage has also been explored by similar methods (Blumberg et al., Chapter 3 in this volume).

Finland. Statistics Finland has carried out four similar surveys in 1996, 1999, 2001, and 2005 focusing mainly on the structure of telephone coverage and the usage of mobile telephones (Kuusela, 1997, 2000). In addition, telephone coverage

---

[2]SIM card (SIM = subscriber information module) is small card that contains information about the telephone, such as the telephone number, subscribed services, identification, small memory for stored telephone numbers, and others. It is about the size of a fingernail, and it can be removed from a telephone and inserted into another. All mobile telephones in Europe must have a SIM card.

Table 4.1. Trends in Mobile Telephone Subscriptions per 100 Inhabitants in Selected Countries and the Proportion of Mobile Telephones Among all Telephones (International Telecommunication Union, 2006)

Country	2000	2001	2002	2003	2004	2005	As% of total, 2003
Australia	44.7	57.4	64.0	71.9	82.6	91.4	57.0
Belgium	54.8	74.7	78.2	83.0	88.3		61.8
Canada	28.8	34.7	38.2	41.9	46.7	51.4	39.1
Czech Republic	42.3	67.9	84.9	96.5	105.3	115.2	72.8
Denmark	63.1	74.0	83.3	88.3	96.1	100.7	56.9
Finland	72.0	80.4	86.7	91.0	93.4	99.7	64.9
France	49.3	62.3	64.7	69.6	73.7	79.4	55.1
Germany	58.6	68.1	71.6	78.5	86.4	95.8	54.4
Greece	56.2	75.2	84.5	90.2	99.8	90.3	66.5
Hong Kong, China	81.7	85.9	94.2	107.9	114.5	123.5	65.9
Hungary	30.7	49.8	67.6	76.9	88.8	92.3	68.8
Iceland	76.5	86.5	90.6	96.6	66.9	103.4	59.4
Ireland	65.0	77.4	76.3	88.0	94.5	101.5	64.2
Italy	73.7	88.3	93.9	101.8	110.2	124.3	67.8
Japan	52.6	58.8	63.7	67.9	71.6	74.0	59.0
Mexico	14.2	21.9	25.8	29.5	36.6	44.3	64.9
Netherlands	67.3	76.7	74.5	76.8	91.3	97.1	55.5
New Zealand	40.0	59.0	62.2	64.8	75.9	87.6	59.1
Poland	17.5	25.9	36.0	45.1	59.8	75.7	58.6
Portugal	66.5	77.2	82.5	89.8	98.9	109.1	68.6
Russia	2.2	5.3	12.0	24.9	52.3	83.6	49.7
Slovenia	61.1	73.7	83.5	87.1	93.3	89.4	68.2
Spain	60.5	73.4	82.4	90.9	89.5	96.8	68.1
Sweden	71.8	80.5	88.9	98.0	108.5	93.3	57.2
United Kingdom	72.7	77.0	84.1	91.2	102.2		60.3
United States	38.9	45.0	48.9	54.6	62.1	67.6	46.7

in Finland has been followed on a monthly basis in the Consumer Barometer Survey. The results of these surveys and the influence of mobile telephones on survey quality have been reported by Kuusela and Notkola (1999), Kuusela and Vikki (1999), Kuusela and Simpanen (2002), and Kuusela (2003).

The rapid change of the structure of telephone coverage in Finland started when second-generation mobile telephones were introduced in the mid-1990s. New models were considerably smaller and their batteries lasted longer than the earlier models. Telephone coverage changed in the first half of the 1990s and accelerated considerably in the second half of the decade. The landline telephone coverage of households has been decreasing since 1996 (see Fig. 4.1). Already in 1999 the popularity of mobile telephones surpassed that of landline telephones. By 2006, more than 97 percent of households had one or more mobile telephones, while only 48 percent had a landline telephone. Approximately 44 percent of households had both telephones. At the end of 2006, 52 percent of households were mobile only and 4 percent landline only.

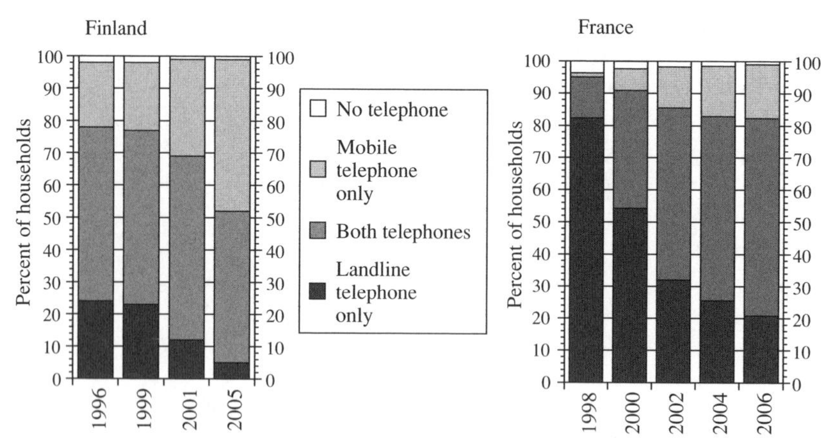

Figure 4.1. Change in the structure of telephone coverage in Finland and France.

In Finland, practically all working-age people have a mobile telephone (99 percent), and the popularity is roughly the same with teenagers. In nearly every household, each member has a mobile telephone of his or her own. Nearly all, 98 percent, of persons aged 15–34 years report that they get frequent calls on their mobile telephones, and 91 percent receive frequent SMS messages (Nurmela et al., 2004). Only elderly people, especially elderly women, have limited mobile telephone coverage (10 percent).

France. In France, Médiamétrie has carried out several surveys since 1998 concerning the structure of telephone coverage. Figure 4.1 shows trends up to 2006. The number of landline telephone only households has decreased very sharply from 82 percent to 21 percent. At the same time the frequency of mobile telephone only households has increased from 4 percent to 17 percent. The trend is consistent, and it is probable that it will continue in the immediate future.

Italy. Since 1988, the Italian Institute of Statistics (ISTAT) has collected data concerning telephone coverage in Italy by face-to-face interviews in the cross-sectional Multipurpose Household Survey (ISTAT, 2003) annually. It is a nationwide face-to-face survey with a sample size of approximately 20,000 households. A two-stage sampling design is applied: Municipalities are the primary sampling units and households are at the secondary level. Data are collected both on the ownership of landline telephones and, since 1997, on the ownership of mobile telephones within the households (Callegaro and Poggio, 2004). The trend of coverage change in Italy is close to that of France. Mobile telephones have spread very rapidly with the coverage being 75 percent of households in 2002. At the same time the number of households who have only a landline telephone has decreased sharply to 21 percent. In data not shown, households with mobile telephones have slowly increased to 13 percent in 2002.

United Kingdom. The Office of Telecommunications of the United Kingdom has published a yearly report on telecommunications issues in the United Kingdom.

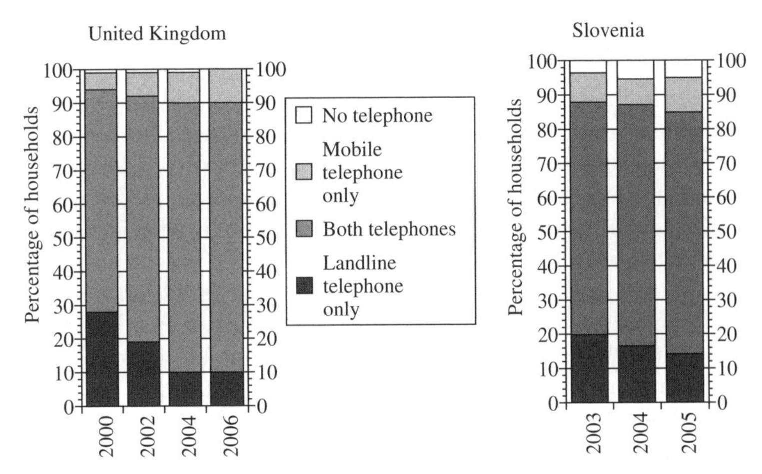

Figure 4.2. Change in the structure of the telephone coverage in the United Kingdom and Slovenia.

The data collection has been carried out by different survey organizations among a representative sample of the UK adults 15 years of age and older reflecting the UK profile of sex, age, social grade, region, and employment status. Data have also been weighted to ensure that the sample is representative of the adult population (Ofcom, 2005). Data collection was done by face-to-face interviews. Figure 4.2 shows the trends from 2000 to 2006. A striking feature is a huge drop in people living in landline only households from 28 percent in 2000 to 10 percent in 2004.

Slovenia. The structure of telephone coverage in Slovenia has been explored in the context of the Labor Force Survey, a quarterly face-to-face survey with a sample size of approximately 1500 households. Figure 4.2 shows the trends over three years. Although the time series is shorter, the trends are similar as in other countries, and they appear consistent.

Other Countries. Table 4.1 shows the trends of mobile telephone coverage from other selected countries, mainly from Europe and North America. The data come from the International Telecommunication Union (2006), which collects data from the service providers in each country. The percentages are calculated simply by dividing the number of subscriptions by the total population. Particular care should be taken when comparing these data across countries. Differences in demographics (some countries have proportionately more young people), differences in prepaid contract distributions, and differences in the percentages of people with more than one telephone number (multiple telephones or multiple SIM cards) render these comparisons problematic. Less problematic is a comparison within country across time. There is considerable variation in the frequency of mobile telephones from country to country. An important feature is that in every country the frequency of mobile telephones had increased consistently from 2001 to 2005. In some countries the increase was very fast.

4.2.2 The Nature of the Changes

The way in which mobile telephones are adopted in a society naturally depends on various device features, especially in contrast to those for a landline telephone, and on the costs of using a mobile telephone. In addition, there are also social or life situation factors that would explain the adoption process. Those for whom mobile telephones suit their life situation acquire them first, and those people who benefit little or not at all acquire mobile telephones much later, if at all. The additional factors are fashion and trends, especially among young people. There is also an important utility feature involved in the teens and children having mobile telephones: safety. Parents buy mobile telephones for their children both because children are able to call for help when needed and because parents are better able to locate and care for their children through mobile telephones.

In Finland, the structural change of telephone coverage has been followed through four similar surveys since 1996. The longitudinal data provide patterns on how mobile telephones have spread in one country. Even though installations of mobile telephone systems vary considerably between countries, there are some indications (Blumberg et al., 2007, Chapter 3 in this volume) that the pattern observable in Finland might also be realized elsewhere. The pace of adoption may show greater variation in other countries though.

Population groups show very different patterns in acquiring mobile telephones and giving up landline telephones. Figure 4.3 shows how the telephone coverage has changed in different types of household in Finland. Nontelephone households have nearly disappeared. This phenomenon has taken place in all countries that have

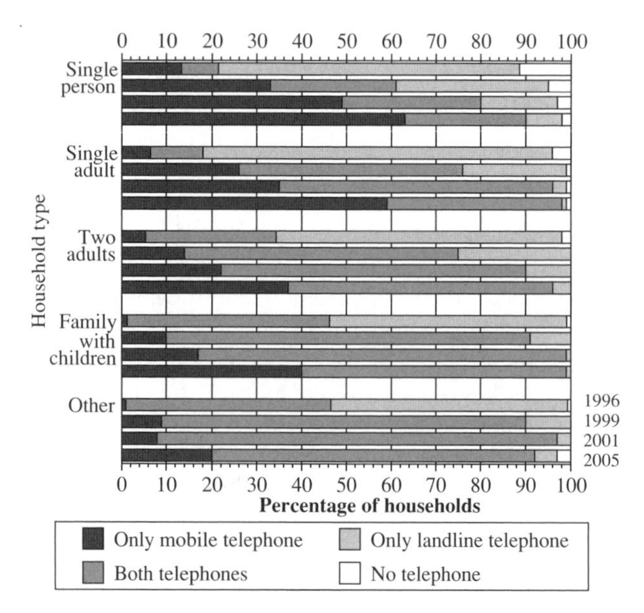

Figure 4.3. Telephone coverage from 1996 to 2005 by type of household ("other households" are family households that have more than two adults).

accurate statistics. The most dramatic change has taken place in single-person households. In 1996, 70 percent had a landline telephone only, but in 2005 90 percent had a mobile telephone and over 60 percent had only a mobile telephone. In single-adult households nearly the same pattern can be seen.

Mobile telephone coverage has also increased considerably among other types of households, but those households have not given up their landline telephones to the same extent as single-person or single-adult households have. In families with children, the proportion of mobile-only households has increased sharply.

There are two very different types of one-person households: young adults who have moved from their parents' home for studies or other reasons, and older widows who have not remarried. The lifestyles and living conditions of these two types vary considerably, and they are different from other types of households as well. Figure 4.4 shows telephone coverage of single-person households over the past 9 years. The trend actually started already before 1996 and by 2005 over 90 percent of the youngest people living alone, both males and females, had only mobile

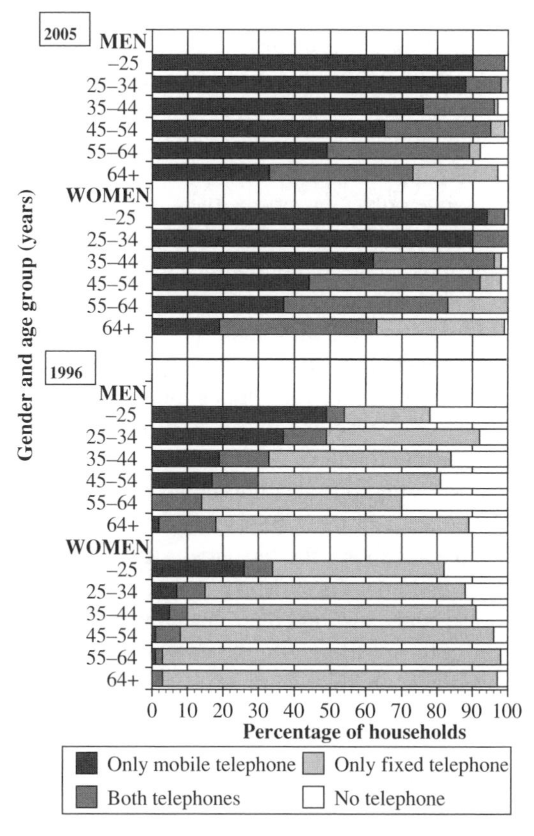

Figure 4.4. Telephone coverage in single-person households in Finland, by gender and age, in 1996 and 2005.

Table 4.2. Telephone Coverage in Finland, 2001 and 2005, by the Type of Housing Unit and Tenure

		Only mobile		Both		Only fixed		No phone	
Type	Tenure	2001	2005	2001	2005	2001	2005	2001	2005
Apartment building	Owned	20	41	62	51	18	8	0	0
	Rented	62	78	30	18	7	3	1	1
Row or semidetached house	Owned	18	40	73	56	9	4	0	0
	Rented	57	72	34	17	8	8	1	3
Single family house	Owned	11	25	79	70	10	4	1	1
	Rented	56	83	31	17	8	0	5	0

telephones. The proportion of mobile-only households decreases steadily with age and faster with women than with men. In the older age groups, landline telephone-only households are more frequent. Older females living alone more rarely have a mobile telephone, and few have only a mobile telephone.

The age structure of a household is connected with types of telephones. For example, in households where all members were aged less than 35 years of age, 76 percent had only mobile telephones and less than 1 percent had only a landline telephone. The proportion of mobile-only households decreases as the age of the oldest member of the household increases. In households where all members were over 35 years of age, 22 percent were mobile only. Conversely, the frequency of landline telephones increases in households with older members. Older people have mobile telephones too, but they are not the only means of communication.

In 2005, the structural change in Finland slowed down, but it did not stop. Over 30 percent of the households who had a landline telephone considered giving it up. The frequency of households without landline telephones has decreased steadily. There are, of course, absolute upper and lower limits to mobile and landline telephone household coverage, but it is difficult to predict how low the landline telephone coverage will go.

The stability of the life situation[3] of a household is an important factor in telephone ownership. An indicator of the stability is the type of the dwelling in which the household lives. Table 4.2 shows the telephone structure of Finnish households by the type of dwelling, and how the situation changed from 2001 to 2005. Nearly 60 percent of the households owning their dwelling had a landline telephone, irrespective of the type of the dwelling. Mobile-only households were relatively rare in owner-occupied dwellings. On the contrary, nearly 75 percent of the households living in rented dwellings were mobile only.

[3]Stable life situation of a household means that there have not been any major changes in the social or economical status recently and, more importantly, that no major changes are foreseen, expected, or hoped for. Also, dissatisfaction in the current life situation or fear may cause instability.

Generally, the proportion of mobile-only households increases as the total household income increases, but the proportion of mobile-only households was the highest among low-income households. In other words, households with lower incomes tended to be mobile only, whereas more affluent people tended to have both. Recently, there has also been a slight tendency among affluent households to give up landline telephones.

In addition to the housing type, the urban location of a household influences the prevalence of the landline and mobile telephones. In general, mobile telephones are more popular in cities; most households in rural areas have a landline telephone. It is obvious that these characteristics are not independent from each other. For example, young people may not yet have a permanent dwelling, but older people and families have lived in the same dwelling for a long time and have a landline telephone they have not given up. The type of the building where a dwelling is located is related to both the wealth of a household and its life situation. Households in single-family houses are often fairly wealthy and have no plans to move, while greater variation in household types is present in blocks of flats.

Mobile-only Households. Europe has had a substantial increase of mobile-only households. As Fig. 4.5 shows, the frequency of mobile-only households increased in all European Union (EU) countries from 2003 to 2004 (Ipsos-Inra, 2004), and the trend continues. In Finland, the proportion had increased to 52 percent in 2006. Blumberg et al. (2007, Chapter 3 in this volume), report the rates of mobile-only households in the United States.

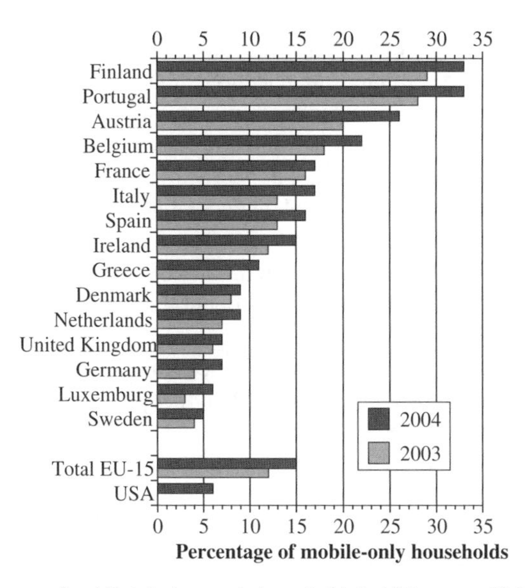

Figure 4.5. Frequency of mobile telephone only households in 15 European Union countries and in the United States (Tucker et al., 2005). *Source*: Ipsos-Inra (2004).

In four of the EU countries the proportion of mobile-only households was over 20 percent in 2004, and the total proportion in EU countries (in 2004 before the addition of new member states) was over 15 percent. The proportion of mobile-only households in the EU had already grown in 2004 to a level that in half the member countries mobile telephones could not be left out of the sampling frames.

There are three fairly different processes that are behind the mobile-only household growth. First, some households simply give up their landline telephone because they have one or more mobile telephones. The landline telephone seems less necessary, and even if it is not used, there is a cost for access.

Second, and more importantly, households moving from one dwelling to another no longer install a landline telephone in their new home. Having a landline telephone connected may be a lengthy and costly process. Mobile telephones move without any extra effort. If there is no special reason to have a landline telephone, it is not installed.

Third, newly established young persons households do not get a landline telephone at all because they already have one or more mobile telephones. A young person moving from his or her parent's household to one of his or her own does not acquire a landline telephone.

A landline telephone for many households is a luxury unless it is needed for Internet access. In other words, a mobile telephone is found to be a more versatile communication medium than a landline telephone, and the landline telephone is found unnecessary.

4.2.3 Profiles of Mobile Telephone Users

Mobile telephones users are a more heterogeneous group with respect to their ownership and usage than persons in landline telephone households. A common situation is a household with one landline telephone line that all members use freely. Most of the time the telephone is listed in the name of someone living in the household. Apart from the one landline telephone all or nearly all the members may have mobile telephones for their own use. Mobile telephones are listed infrequently, and even then, because owners of mobile telephones are often other than users, it is not certain in whose name the telephone number is listed.

Most mobile telephones are privately owned, but there are also mobile telephones that employers give to employees for business purposes. Sometimes employees use them for private purposes as well. The telephone number may be listed in the person's name or in the name of the company. Many people have both a private and a business mobile telephone.

In many countries, the so-called prepaid mobile subscriber packages are popular types of subscriptions (Table 4.3). In European and many other countries, a prepaid plan uses a SIM card for which the connection time is paid when the card is purchased. The telephone number is valid only during a period for which the prepaid minutes are available. In Europe, it is still possible to receive calls for some time after the prepaid fee has been used. In some cases, one can "reload" the card to keep the number operational, but this is not always the case. Listing of mobile telephones in

Table 4.3. Proportion of Prepaid Mobile Subscriptions Among all Mobile Phone Subscriptions in Selected Countries (OECD, 2007)

Country	2003	2004	2005
Australia	37.8	43.0	46.2
Canada	23.8	18.9	23.0
Finland	2.0	7.0	6.8
France	41.1	38.4	36.6
Germany	51.4	50.5	50.8
Greece	65.4	65.9	67.0
Hungary	77.5	73.1	68.0
Italy	91.2	91.0	90.6
Japan	3.0	3.1	2.8
Mexico	93.3	93.5	93.1
New Zealand	69.7	69.8	71.2
Poland	54.4	58.4	64.5
Portugal	79.4	79.3	81.2
Slovak Republic	62.1	57.2	
Spain	58.4	52.0	48.5
Sweden	56.8	52.7	50.9
United Kingdom	67.9	66.8	66.1
United States	7.3	8.1	11.0

directories is much less frequent than listing of landline telephones. Finland, where almost all numbers, mobile and landline are listed is an exception. Prepaid mobile telephone numbers are practically never listed.

Many people consider a mobile telephone as a personal appliance. Mobile telephones typically have one user. It is unusual that one mobile telephone has several users. Typically only the user answers a mobile telephone, and hence only its user can be reached by using the mobile telephone number.

Table 4.4 shows the age distribution of all mobile telephone users in 2005 in Finland. Slightly more than 4 percent of users were less than 10 years old, and roughly every eighth mobile telephone user was 15 years or under. The tendency has

Table 4.4. Age Distribution of Mobile Telephone Users in Finland, 2005

Age (years)	Percent	Cumulative percent
Less than 10	4.2	4.2
11–15	8.1	12.3
15–25	18.4	30.7
26–35	15.8	46.5
36–45	13.6	60.1
46–55	16.4	76.5
56–65	15.5	92.0
65+	8.0	100.0

been that younger children have a mobile telephone of their own. It is not unusual that Finnish children will have mobile telephones when they start school.

In addition to increasing mobile telephone penetration, the social context of mobile telephone usage is also important (Katz, 2003). The mobile telephone is often a private device, and its social role differs from that of landline telephones. One may expect different patterns of handling all telephones, as well as different attitudes toward use (Rheingold, 2002). A variety of mobile telephone social and usability contexts may also have an impact on the way users accept a survey request when called on their mobile telephone. Mobile telephone users compose three distinct segments: intensive pragmatic, emotional users, and less intensive users. Each type behaves differently in a mobile telephone survey process (Vehovar et al. 2004a).

4.3 MOBILE TELEPHONES' INFLUENCE ON SURVEY RESEARCH

Telephone survey methods and telephone survey practices have been developed in a landline telephone environment (Groves et al., 1988; Piekarski, 1996). Because of the growing popularity of mobile telephones, they are increasingly used as an interviewing media. But mobile telephones differ from landline telephones in many respects that force an alteration both in applied survey methodology and in the everyday practices of telephone interview centers. How extensively changes are needed depends much on local conditions. There are great differences from country to country both in the infrastructure of mobile telephone systems and in the tariffs. The mobile communication systems and mobile telephone features in Europe and in the United States, for example, are different in many respects.

In everyday life mobile telephones are practical devices for communication, much more than landline telephones. New mobile telephone features and services, practical and imaginative, are introduced all the time. Mobile telephones will grow to be even more prominent communication medium because of these changes. Mobile telephones also bring new possibilities for survey research. For example, the text message system (SMS) can be used for simple data collection and for some support functions.

Some of the problems that mobile telephones introduce to survey research can be tackled with general methodological or statistical solutions. But some problems are restricted to smaller domains, such as one or two countries or a region of a country. A primary change for telephone survey methodology will be to add households that are mobile only to telephone surveys. Exclusion of mobile telephones from the sampling frames results in undercoverage and the potential for increased bias in estimates. The noncoverage of mobile-only households at present is similar in character to nonignorable nonresponse because mobile-only households are different from the rest of the population. Conversely, if the proportion of landline telephone only households is disproportionately too high in a sample, including more mobile telephones in the sample frame can yield biased estimates unless results are properly weighted.

There is no single level of mobile phone penetration at which bias might be critical. It depends on the relative size of the noncovered population and the difference between the observed and unobserved parts of the population (Groves, 1989). Further, the proportion of mobile-only households is added to the noncoverage of nontelephone households. Excluding the mobile telephone-only households from sampling frames may cause a significant bias in results. The requirements of the survey may have an influence on the decision to include mobile telephone households. In government or official surveys the requirements for coverage are higher than in political polls. Biemer and Lyberg (2003) give an extensive treatment to the noncoverage problems and approaches to solve them.

There are also a number of other nonsampling issues that must be dealt with in order to introduce mobile phones into telephone survey practice. These nonsampling-related factors include the potential for increased survey costs to reach persons through landline or mobile telephones; the possibility of reaching mobile phone users in settings where interviews are inappropriate or dangerous, or where sensitive items may not be answered truthfully; a possible reduction in interview length; an increase in nonresponse rate or bias; the need for new operational codes to capture the full range of outcomes for mobile phones; and the availability of advance contact through messaging systems such as SMS.

In the remainder of this section, sampling and nonsampling aspects of adding mobile telephones in telephone surveys will be discussed in more detail. Mobile telephones force changes on sampling design, especially in the nature of telephone number frame[4] and samples drawn from them. Nonsampling effects are composed of the practicalities of survey work and survey costs. Both types of issues are discussed.

4.3.1 Sampling Consequences

The sampling-related issues vary between different survey designs and sampling methods. The main question is whether or not the telephone numbers are known and available before the first contact. Typically telephone numbers are known, for example, in later waves of panel surveys and in the list or register-based sampling schemes where the numbers are either known or found before the contact.

If sampling units are persons or addresses, the problem is to find a telephone number (or numbers) for each sampling unit. A critical question is how comprehensively mobile telephones are listed in directories and whether they are listed in such a manner that the sampling unit and telephone number can be matched. Generally speaking, mobile telephones facilitate such samples provided the mobile telephones are listed comprehensively in a directory.

If the sampling frame is a list of telephone numbers, or numbers are generated as in RDD methods, both sampling and estimation may become more complicated. The problems originate from coverage deficiencies of different sampling frames composed of phone numbers of different types of telephones.

[4]Telephone number frame sample means here a procedure where the sample is drawn from an explicit or implicit list of telephone numbers. The primary sampling unit is a telephone number in a telephone catalog or that obtained by random digit dialing. Other kinds of samples are drawn from lists of address or persons. Typically those samples generate telephone numbers after the sample has been drawn.

Frame Coverage. The part of the population that has only mobile telephones and the part that has only landline telephones are segregated in telephone number frames. Further, mobile telephones reach one population and landlines another. Young people in a less stable life situation tend to have only mobile telephones, while elderly people in a stable life situation have only landline telephones. Middle-aged people with families very often have both types of telephones.

An example of the uneven distribution of mobile telephones is that in Finland. The proportion of interviews completed on a mobile telephone in a computer-assisted telephone interviewing (CATI) center varies considerably across age groups. Approximately 87 percent of the interviews of respondents aged between 20 and 29 are done by mobile telephones, while only 28 percent of those respondents 65 years or older are done by the mobile telephone. Respondents' educational level is also related to the type of telephone used in an interview. For example, in the first half of 2002, 55.8 percent of interviews with respondents having a basic-level education, 48.9 percent of respondents with mid-level education, and 40.2 percent of those with high-level education were done by mobile telephone.

If either of the telephone types is left out of the sampling frame, households will be added to undercoverage for the part of the population that has no telephone (either the mobile-only or the landline-only populations). Even in panel surveys, if mobile telephones are not called, similar undercoverage would occur.

On the contrary, if both telephone types are included in the frame, overcoverage will occur. The part of the population that has both types of phones will be included in both frames. It is also possible that several persons from the same household could be included in the sample. Overcoverage hampers surveys where telephone numbers are the sampling units, but it does not hamper those surveys where the sampling units are persons or households whose telephone numbers are known before contact.

Noncoverage error is a function of two factors: the proportion of the target population that is not in the sampling frame, and the difference in the values of the statistic for those in the frame and those not in the frame. Callegaro and Poggio (2004) and Vehovar et al. (2004a) have analyzed the potential bias $Y_c = Y + N_{nc}/N\ (Y_c - Y_{nc})$, where Y_c is the value of the statistic for those covered by the sampling frame, Y is the value of the statistic for the whole target population, N_{nc} is the number in the target population not covered by the frame, N is the total size of the target population, and Y_{nc} is the value of the statistic for those not covered by the frame.

Using the data from the National Multipurpose Survey interviewed face-to-face by the Italian National Institute of Statistics, Callegaro and Poggio (2004) calculated the level of bias by comparing the results obtained from households when using the full sample to those obtained from households that could be reached on a landline line. In 2001, when the noncoverage rate in Italy was 14.3 percent (10.1 percent mobile-only households and 4.2 percent nontelephone households), a telephone survey using landline telephones would have overestimated computer ownership by 3 percent and house ownership by 3.5 percent.

Features of Telephone Number Frames. While a mobile telephone is often a personal appliance and the entire household shares a landline telephone, sampling frames composed either of mobile telephone numbers or of landline telephone numbers are different. A landline telephone number frame is composed of unequal-sized clusters (i.e., households). In the mobile telephone number frame the sampling unit is, when the device is a personal appliance, one person. Hence, sampling from a mobile telephone number frame essentially samples persons while sampling from a landline telephone number frame results in a sample of clusters of persons.

If the two number frames were completely nonoverlapping, then the differences can be readily handled. However, frames overlap creates three domains: landline telephone only, mobile telephone only, and both. Conventional dual-frame approaches may be valid. Proper sampling and weighting schemes should be based on information concerning the structure of the telephone overlap.

If all the members of a household should be interviewed, mobile telephones make implementation very difficult because the household members have different telephone numbers. There are practically no means in a survey organization to find the whereabouts of a mobile telephone. In Europe, the mobile telephone number usually does not indicate where the telephone user lives or works, and even if it did, the telephone could still be used anywhere because mobile networks are countrywide. Mobile telephone numbers begin with an operator prefix[5] that is valid within the domain of the network, in most cases the whole country. All spatially limited sampling designs using mobile telephone frames are quite difficult to implement. For example, a survey of people living in a certain city becomes so widespread geographically that it cannot be practically implemented because the telephone number does not determine location.[6]

In Europe, the number assignment for mobile telephones varies between countries and also between service providers within countries. Consequently, the optimization of RDD procedures becomes complicated. In the United States, the numbering system for mobile telephones follows the landline telephones, and local surveys using mobile telephones remain relatively feasible (Buskirk and Callegaro, 2002).

Inclusion Probabilities. If the numbers for mobile and landline telephones are homogenous, where it is impossible to distinguish the type of telephone through the number itself, estimation and weighting become more difficult.

In drawing samples from telephone number frames, mobile telephones change inclusion probabilities of households and persons. The inclusion probabilities will increase linearly with the number of telephones in household. Consequently, large households will be overrepresented in samples that use mobile telephone numbers.

In household samples the varying inclusion probabilities may cause substantial bias: rectifying the bias requires special weighting schemes. The construction of the weighting scheme is not straightforward because of the different type of telephones. Developing feasible sampling methods and producing proper weighting schemes will be a future challenge to telephone survey practitioners.

---

[5]By the telephone number, it is always known to what type telephone the call is made.

[6]Technically, it is possible to get the whereabouts of a mobile telephone very accurately. However, it is nowhere near publicly available information but, for example, in Finland, it has been under discussion whether it could be released for marketing purposes.

4.3.2 Nonsampling Considerations

Since 1998 in the Consumer Barometer Survey and in four ad hoc surveys in Finland, most interviewers have recorded the type of telephone used in the interview, allowing the study of nonsampling effects. The results of these surveys and the influence of mobile telephones on survey quality have been reported by Kuusela and Notkola (1999), Kuusela and Vikki (1999), Kuusela and Simpanen (2002), and Kuusela (2003).

An unavoidable consequence of the growing popularity of mobile telephones is an increase in the proportion of mobile telephone interviews as a greater proportion of respondents are best reached by a mobile telephone. For example, in the CATI center of Statistics Finland[7] the frequency of mobile telephone interviews has increased steadily since 1999. By July 2001, more than half of interviews were conducted on a mobile telephone. The proportion has grown steadily to 80.2 percent in September 2006. The proportion of agreed appointments over mobile telephones is slightly higher than that of completed interviews. This occurs because interviewers are instructed to try to make an appointment in many situations when calling a mobile telephone.

Survey Costs. When interviews are conducted on mobile telephones, the survey budgets have to be calculated on a different basis than interviews conducted over landline telephones. Adequate estimation of the cost of telephone calls requires knowing the number of mobile telephone interviews to be conducted and the proportion of mobile telephone-only households.

A call to a mobile telephone is usually more expensive than a call to a landline telephone in Europe. In the United States, there is no difference in calling a landline or mobile telephone. In Europe, then, the increased number of mobile telephone interviews increases the telephone expenses of surveys. For example, in Statistics Finland the telephone costs of interviewing increased by 42 percent over the 3 years from 1999 to 2001. The amount of increase depends on local tariffs, which vary substantially from country to country. Calling from a mobile telephone to another mobile telephone is usually less expensive than calling to a landline telephone. A router[8] in a telephone exchange may moderate the rise of costs since it can convert a landline call to a mobile phone call through electronic switching.

Survey Practice. In telephone surveys where calls are made to mobile telephones, interviewers need special instructions to conduct an interview. A mobile telephone is usually carried all the time and respondents can be virtually anywhere. Mobile telephones require a new "user culture" both in receiving and in making calls.

In the beginning of an interview over a mobile telephone, interviewers should always ask whether the respondent is in a place where he or she is able to speak freely and the environment is appropriate for an interview. In European countries, it is not

[7]The samples are always personal samples drawn from the population registry and telephone numbers are searched for respondents before interview.

[8]A router is a device attached to telephone exchange that converts the call to landline or mobile system based on the type of the number called.

feasible to conduct an interview when the respondent is abroad: Receiving a call is free of charge when the telephone is inside the country, but when the telephone is outside the country, the owner has to pay half the roaming charges and thus the respondent has to pay for giving the interview. Usually in such cases interviewers are instructed to cancel the interview and, when appropriate, make an appointment for a time when the respondent will be in the country. If interviewers do not know whom they are calling, they have to make sure that the respondent is eligible. Younger persons use mobile telephones increasingly. For example, in Italy 38 percent of children between 5 and 13 years of age use a mobile telephone (Doxa, 2005).

Even though a respondent agrees to be interviewed, there are situations when the interview is not appropriate to conduct. For example, if the respondent is driving a car, the interviewer should not continue with the interview but suggest an appointment at another time. When the respondent is in a noisy place where he or she might not hear the questions clearly, and he or she has to speak very loudly to be heard, an appointment should also be sought. A similar situation arises if the connection is of poor quality.

In general, it is precarious to carry on with an interview if it is suspected that the respondent would not concentrate properly. In some cases it is possible that the respondent can easily find a comfortable place nearby. Interviewers can ask a respondent to look for a quiet place to be interviewed. In cases where respondents have both landline and mobile telephone numbers, and interviewers are able to select between them, respondents should be instructed which telephone type they should call first. Often it is the landline telephone because calls to landline telephones are less expensive.

Survey Results. Analyzing whether mobile telephones as communication devices influence survey results is a difficult task. Population segments using primarily either mobile telephone or landline telephone are so different that observed differences in results may be caused by telephone type or by population differences. Till now, though, no reason has been found in Finland to indicate that results would be different because of the type of telephone used in the interview. Kuusela and Notkola (Kuusela and Notkola, 1999) first made this observation in 1999, and it has been confirmed by several later studies (see Kuusela 2003).

Apart from whether telephone type as a medium of communication has an impact on the results, the telephone type does induce indirect effects. These effects originate from structural differences in telephone coverage in different population segments. Figure 4.6 shows the distribution of answers to a question about whether the household is able to save money or is in debt, a question asked in the monthly Consumer Barometer Survey in Finland to nearly 4000 respondents. Only respondents between 30 and 49 years of age were included in order to have a subsample that was economically more homogenous. As can be seen in the figure, in every educational category respondents who were interviewed over landline telephone said that they were able to save more often than those with mobile phones. The reason for this difference is likely that households in a more stable situation and more likely to save are more often interviewed over a landline telephone because they have landline telephones more frequently, and they are more often at home. Similarly, Traugott

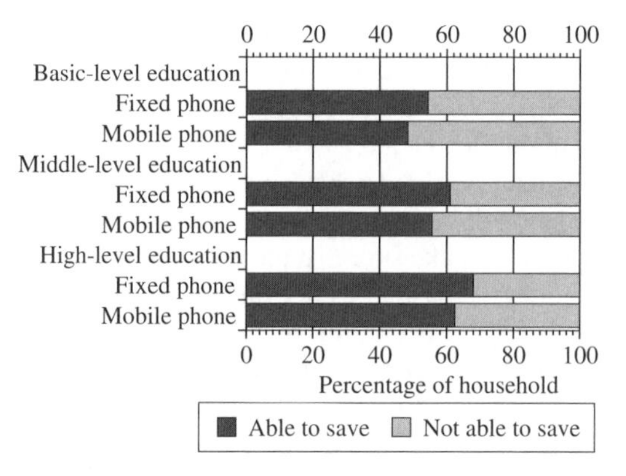

Figure 4.6. Economic situation of the household by respondent educational level and type of telephone interview among respondents aged 30–49 years.

and Joo (2003) observed that the political attitudes of mobile and landline telephone users are different. They concluded that this was caused by the strong relationship between social economic status and mobile telephone ownership.

Length of Interviews. An important question is whether mobile telephones would have an effect on interview length. The length difference between mobile and landline telephones is a difficult topic for research because the different respondent life situation may cause a very different set of questions to be asked from respondents. A few years ago, when mobile telephone devices were technically less developed than now, it was suspected that interviews over mobile telephones would become shorter than those over landline telephones. This was a real fear because the batteries of the early mobile telephone models did not last long and respondents might have to answer hastily. The batteries of modern mobile telephones last longer, but the same fear remains if respondents pay for the call, as in the United States, for example. Also, the fact that respondents could be virtually anywhere might also influence the length of interview. For example, in a situation where the conversation can be overheard, respondents may try to complete the interview as quickly as possible.

However, Kuusela and Notkola (1999) did not observe differences between telephone types in the lengths of interviews. Their early results have been confirmed in later studies. Figure 4.7 shows the interquartile ranges of interview lengths in the Consumer Barometer Survey in Finland. No consistent differences can be observed. The interview was very short, only 15–20 minutes; effects might surface only in much longer interviews. On the other hand, Steeh (2003) observed that in the United States interviews were a little longer in mobile telephone than in landline interviews.

Nonresponse. When the popularity of mobile telephones started to increase, it was feared that respondents would more frequently refuse an interview on mobile telephones than on landline telephones. There are several reasons to expect that the refusal rate would increase. Respondents might fear that batteries would not last, or

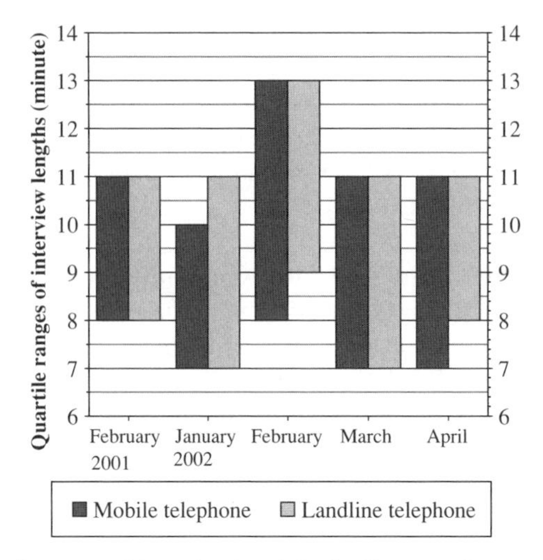

Figure 4.7. Interquartile ranges of interview length in the Consumer Barometer Survey, consecutive months, 2001–2002.

that someone might overhear their responses. In the United States respondents have to pay for incoming calls, potentially leading to even higher levels of nonresponse.

In one experiment done by Georgia State University's call center, the response rate (using American Association for Public Opinion Research, 2004) for mobile telephones was 20 percent compared to 31 percent for a comparable landline telephone sample. The refusal rate (REF 2) was 44 and 39 percent respectively (Steeh, 2005b). It was more difficult to convert a refusal for mobile telephones than for landline telephones, 8 percent versus 14 percent.

But there are contradictory experiences on whether the nonresponse and refusal rates would be higher in mobile telephone interviews. Vehovar et al. (2004) observed that the refusal rate was higher in mobile telephone interviews than in comparable landline telephone interviews. In Finland, however, there is no indication that mobile telephones increase nonresponse. Although the total refusal rate seems to remain practically unchanged, there is slight evidence of an interaction with age. When contacting young respondents with a mobile telephone, the refusal rate tends to be lower than with a landline telephone, and the opposite is true when contacting older respondents.

Apart from refusals, mobile telephones have made contacting people easier. Those respondents who were previously difficult to reach (e.g., young men, the unemployed, and students) now have mobile telephones and are more easily reached. In Finland, during holiday seasons nonresponse rates tended to be much higher than at other times of the year. The situation has become much better because of mobile telephones. In some surveys response rates now during holiday seasons are better than at other times of the year.

Fuchs (2000b) conducted an experimental study to compare nonresponse in mobile and landline telephone households. He observed that the refusal rate in mobile telephone interviews was slightly lower than in landline telephone interviews, but more call attempts were necessary to reach an adequate overall response rate. Lao carried out an experiment in Hong Kong on methodological differences between mobile and landline telephones (Lau, 2004). In an 18-minute interview, refusal rates in landline telephone interviews were much higher than in mobile telephone interviews, 47 percent versus 39 percent.

Enhanced Dispositional Coding. Survey administration when including mobile telephones poses new challenges to survey research (see Fig. 4.8). One important challenge involves the proper classification of the final disposition of each case in the sample so that response and other outcome rates can be uniformly and reliably calculated. Callegaro et al. (2007) analyzed the possible call outcomes that occur when calling a mobile telephone. There are new outcomes, and some outcomes used for landline telephone interviews have a different meaning in mobile telephone interviews. For example, a mobile telephone may be switched off, something that does not happen with a landline telephone. A "no answer" can mean that the person does not want to answer, such as when he or she has to pay to receive an incoming call.

Callegaro et al. (2007) discussed the final disposition codes sanctioned by the American Association for Public Opinion Research (2004) in light of their usefulness for surveys conducted at least partially over mobile telephones. On the basis of experiences in three surveys in which data were collected via mobile telephones in Finland, Slovenia, and the United States, the authors found that new final disposition codes would be required, and that some codes appropriate to landline telephone interviews either no longer apply or assume a different meaning. The different wireless systems in the United States and Europe affected survey outcome rates. Callegaro et al. illustrated how the development of a set of final disposition codes that apply internationally can improve survey methods by making systematic comparisons across different contexts possible.

Text Messaging (SMS). Most mobile telephones (actually the SIM cards) have text message system (SMS) installed by default. In Europe, sending and receiving text messages has grown to be very common in many countries. Typically, younger users send text messages more often than they call. Since the facility is nearly universally available, it has been utilized in survey research as well. In the United States, one has to pay for receiving a text message, but in Europe receiving a text message is free of charge.

In a study conducted on a panel of Italian high school seniors, a text message was sent to those respondents who had the mobile telephone off and no voice mail activated (Buzzi et al., 2000). The message alerted the user that the researchers were trying to reach them for conducting an interview. Almost 85 percent of those contacted called back within the same night. In an experiment done with Nextel customers in the United States (Steeh et al., 2007), respondents were randomly allocated to two conditions: to receive a cold call (no previous contact attempt made) on their mobile telephone with a request to answer a survey or to receive a prenotification text message before the call. The latter experimental condition did not result in a substantial

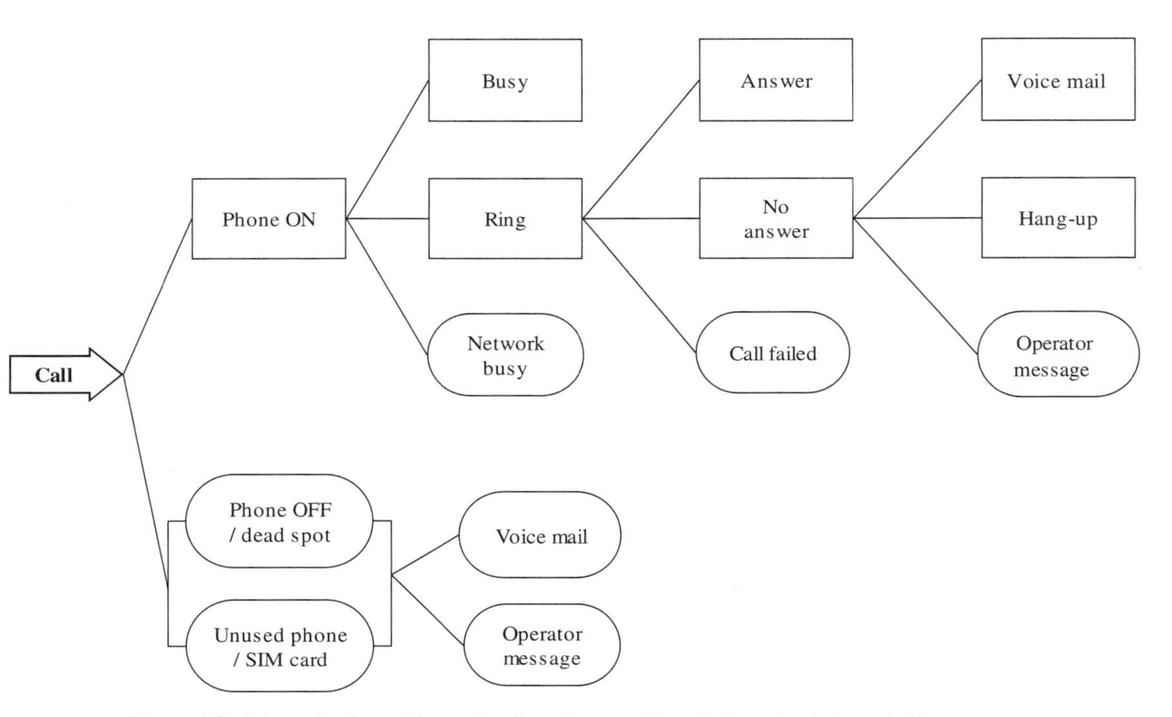

Figure 4.8. Schematic of possible results of a call to a mobile telephone (oval shapes indicate new outcomes that are specifically happening when calling a mobile telephone). From Callegaro, Steeh, Buskirk, Vehovar, Kuusela and Piekarski, 2007.

increase in response rates. It did, however, help in reducing the number of calls to resolve a case and decreased the number of unknown eligibility status outcomes. The results were expected since customers had to pay to receive text messages.

An SMS survey is one in which survey questions are asked via text messages. The technique resembles that of an interactive web survey. A problem for this mode is the size of the "screen" of the mobile telephone and the maximum length of a message (160 characters). However, this limitation can be partly evaded by randomly splitting and rotating the question set across respondents. An SMS survey is very fast to complete. Once an infrastructure has been installed, the preparation of a survey is also very fast. In a recent experience in Finland, 50 percent of responses were received in 2 hours, and nearly all responses were received in 5 hours. In the best case, 80 percent of responses had been received in 15 minutes. SMS surveys are suited well only for panels because the screen size does not allow for proper introduction, which would be necessary in a cross-sectional survey. Widman and Vogelius (2002) measured daily newspaper readership in Norway, and Down and Duke (2003) measured bank customer satisfaction and attitudes in the United Kingdom. For a methodological experiment for SMS surveys see Cooke et al. (2003).

For a survey organization, an SMS poll is a useful tool for quality control. A text message asking a respondent whether the interviewer conducted the interview and how the interviewer performed does not increase response burden much. In three split-half experiments in Finland, Virtanen et al. (2005) randomly assigned respondents to receive the reminder to complete a mail survey either as a conventional letter or as a text message (SMS) to their mobile telephones. In the group that received the SMS reminders, the response rate was significantly higher (from 6.4 percent to 9.7 percent).

In addition to SMS surveys, mobile computer self-assisted interviewing (MCASI) can also be conducted. In a study conducted in Norway, SMS and MMS[9] were used to invite an opt-in panel to answer a survey on WAP[10] enabled telephones (Tjøstein et al., 2004). The obtained response rate was 49 percent. A similar study was done in Japan by Cattell (2001) on i-mode[11] web-enabled telephones. Matthews et al. (2000) conducted a study in the United Kingdom with Nokia using WAP-enabled telephones and concluded that this technology is promising for short, simple surveys.

4.4 CONCLUDING REMARKS

Mobile telephones are an inherent part of the infrastructure in European and North American countries. Some less developed countries may leap over the landline tech-

[9]MMS Multimedia Messages, the advanced version of a text message where it is possible to embed pictures and also video.

[10]WAP Wireless Application Protocol, the capability to browse WAP-enabled Web sites with a mobile telephone.

[11]i-mode is a technology similar to WAP, mostly used in Asian countries.

nology and develop mobile telephones because the wireless infrastructure is less expensive to build. In some countries the change has been very rapid while in some others, such as the United States, it has been slower, partly because of tariff policies. Mobile telephones are becoming more popular also in the United States, and thus it is not possible anymore to disregard the fact of mobile telephones growth and the need for including them in telephone surveys.

A future scenario might be that voice communication moves completely into wireless systems. There are many reasons that point to this direction. Young people tend to have only mobile telephones and are reluctant to acquire landline telephones even when their life stabilizes. Telephone technology develops primarily within wireless technology, and new practical and imaginative services are developed all the time. Everyday communication practices change to adapt to mobile telephone features, such as having telephone numbers in a mobile telephone memory. At some point a critical mass may be reached when landline telephones become obsolete. It is difficult to foresee how long it would take for this scenario to be realized. It may happen at different speeds in different countries. In Finland, already in 2005, the greatest part of telephone calls are made by mobile telephones.

The growing popularity of mobile telephones changes the way in which telephone coverage should be assessed. Previously there were only two types of households: those with a telephone and those without a telephone. The emergence of mobile telephones has created new groups: those who have a mobile telephone, a landline telephone, or both. The sizes of these three groups vary considerably from country to country. To be able to properly design a telephone survey sample requires knowledge of the structure of telephone coverage. Assessment of the undercoverage of different frames is necessary to evaluate the potential bias caused by disregarding either of the telephone types. However, only in a few countries has this been explored with reliable methods. The current practice is to leave mobile telephones out of the sampling frame. However, in many countries mobile telephones are probably already so popular that they should be taken into account in all telephone surveys. If they are left out of the sampling frame, the resulting surveys will be biased, with the amount of bias depending on the proportion of mobile telephone-only households in the target population and, of course, on the subject of the survey.

The sampling frames of landline telephone and mobile telephone numbers are different: landline telephone numbers create a household frame but mobile telephone numbers create a person-level frame. Sampling requires a dual-frame approach that is not straightforward. Biemer and Lyberg (2003) give an illustrative account on the problems, which emerge in sampling simultaneously from two different and overlapping frames.

In addition, the nature of the number frames differs; RDD methods have to be redesigned. In the United States, mobile telephones have usually not been included in RDD samples as of yet (Lavrakas and Shuttles, 2005), but this may change in the near future. Obviously, RDD samples do not include mobile telephones elsewhere either (Nathan, 2001; Nicolaas and Lynn, 2002). In the United States, it may be impossible to tell by the number whether it is a landline or a mobile telephone (Steeh, 2004a).

It should be borne in mind that the reliability of a survey can be assessed only through applied methods. The obtained results as such cannot be proven to be correct or incorrect. Because of the emergence of mobile telephones, the sampling frames will have very decisive roles in the sampling designs. This requires knowledge of the structure on the telephone coverage in the target population. Inadequately handled sampling in a telephone survey may lead to increased bias in results.

CHAPTER 5

Methods for Sampling Rare Populations in Telephone Surveys

Ismael Flores Cervantes and Graham Kalton
Westat, USA

5.1 INTRODUCTION

Many surveys are required to produce estimates of specified precision for one or more subgroups of the overall population. Some surveys are required to produce estimates only for a single subgroup, whereas others are required to produce overall estimates as well as estimates for a number of subgroups. Thus, for example, one survey may be required to produce estimates of immunization levels only for children between the ages of 18 and 35 months, whereas another survey may be required to produce poverty estimates for all persons as well as separate estimates for race/ethnicity subgroups. Subgroups that have been of interest in different surveys are many and varied, such as persons in poverty, teenagers, women of childbearing age, the oldest old, people with disabilities, smokers, graduates in science and engineering, and men who have sex with men.

When subgroup membership cannot be identified prior to sample selection, the technique of screening a large sample of the population is widely used to produce the desired sample sizes for relatively rare subgroups. When large-scale screening is required, telephone screening is particularly attractive as compared with face-to-face screening because of the associated cost savings. For this reason, telephone survey methods are often used for surveys designed to produce estimates for rare and relatively rare subgroups.

This chapter reviews methods for improving the efficiency of the screening operation when rare subgroups of the population are sampled in telephone surveys.

Advances in Telephone Survey Methodology, Edited by James M. Lepkowski, Clyde Tucker, J. Michael Brick, Edith de Leeuw, Lilli Japec, Paul J. Lavrakas, Michael W. Link, and Roberta L. Sangster

The focus is mainly on sample designs for estimating subgroup characteristics, such as the median income of employed persons with disabilities. Estimating the prevalence or the size of a subgroup, such as the number of employed persons with disabilities in the population, represents a different objective that is not directly addressed here. However, the sampling methods described below are also relevant, particularly when the subgroup of interest is a very rare population.

There is a sizable literature on general methods for sampling rare populations or subgroups (Kish, 1965; Kalton and Anderson, 1986; Sudman and Kalton, 1986; Sudman et al., 1988; Kalton, 1993, 2003). While these methods are applicable for surveys conducted using any mode of data collection, this chapter considers their specific applicability to telephone surveys.

Section 5.2 discusses the use of screening in telephone surveys, which is the basis of most sample designs for rare subgroups. Sections 5.3 through 5.8 discuss disproportionate stratified sampling, two-phase sampling, multiple frames, multiplicity sampling, a variant of the Mitofsky–Waksberg random digit dialing (RDD) sampling method, and multipurpose surveys. Each section describes the method, assesses its applicability to telephone surveys, and provides illustrations of its application.

Before the sampling methods are discussed, two general concerns about telephone surveys deserve comment in relation to surveys involving subgroups: noncoverage and nonresponse. As discussed in Kalsbeek and Agans (2007, Chapter 2 in this volume) and Blumberg et al. (2007, Chapter 3 in this volume), noncoverage is a concern for all household surveys conducted by telephone. Under the common practice of restricting telephone surveys to households with landline telephone service, noncoverage comprises households without a telephone and households that have only mobile telephones. The exclusion of households without landline telephone service from the sampling frame raises concerns that the survey estimates may be biased because of differences in the survey variables between those on the frame and those excluded from it. There is ample evidence that these concerns are justified for a range of survey variables (see Blumberg et al., 2007, Chapter 3 in this volume; Tucker et al., 2004).

When a survey has the objective of producing one or more subgroup estimates, the noncoverage rates of interest are those for the subgroups, not the rate for the overall population. A subgroup noncoverage rate may be larger or smaller than the overall rate. For example, noncoverage rates are lower than average for middle- and upper-income households, owner-occupied households, households with more than one person, white non-Hispanics, married persons, older persons, and those with at least a 4-year college degree. In contrast, noncoverage rates are higher than average for poor households; one-person households; renters; divorced, separated, and never-married persons; persons between 17 and 35 years of age; and persons without health insurance (see Blumberg et al., 2007, Chapter 3 in this volume). When the noncoverage rate for a subgroup is high, a sizable noncoverage bias may occur, making complete reliance on telephone data collection problematic. In such a case, it may be advisable to supplement the incomplete telephone sampling frame in some way. For example, in the National Survey of America's Families (NSAF), a telephone survey was supplemented with an area sample in order to give representation to all low-income households with children, a subgroup of major interest that was not well covered by the telephone sampling frame (Judkins et al., 1999b).

Another source of noncoverage for surveys that focus on a subgroup of the overall population is that some subgroup members may not be identified in the household screening process. The failure to identify subgroup members can lead to substantial noncoverage in any survey of a subgroup population, no matter what mode of data collection is used (see, e.g., Judkins et al., 1999a; Horrigan et al., 1999). The nature of telephone interviewing for screening may, however, make it easier in some cases for respondents to deny subgroup membership.

Nonresponse is a serious concern in telephone surveys (see Battalia et al., 2007, Chapter 24 in this volume and Montaquila et al., 2007, Chapter 25 in this volume). In considering the use of telephone data collection for a subgroup, the issue becomes one of the nonresponse rate for the subgroup. For example, there are concerns that the response rate among the elderly may be low because some of them are hard of hearing. Also, a high level of nonresponse may result if subgroup membership is a sensitive topic.

In passing, it may be noted that the use of a sampling frame of all potential telephone numbers for an RDD survey of all households in effect treats the household population as a subgroup, since a large proportion of the telephone numbers on the frame are not household numbers. Several techniques have been developed to improve the efficiency of the screening sample for RDD surveys by reducing the amount of unproductive dialing to business and nonworking telephone numbers. These methods are reviewed in Kalsbeek and Agans (2007, Chapter 2 in this volume).

5.2 SCREENING

Screening large samples of the general population is the most common method used for producing desired sample sizes for small population subgroups. Screening may be the sole method used to obtain required sample sizes for subgroups, or it may be used with one of the other methods discussed in later sections. If screening is used alone, the overall sample size needed to generate a sample size of n_d elements in subclass d is $n_{0(d)} = n_d/p_d r_d$, where p_d is the proportion of elements on the sampling frame that are in subgroup d, and r_d is the response rate in subgroup d. If a survey aims to produce estimates of specified precision for a number of subgroups, with n_d being the minimum required sample size for subgroup d, then the screening sample size is given by the maximum value of $n_{0(d)}$ over the set of subgroups (d = 1,2,...).

If each of the subgroups of interest comprises a large proportion (p_d) of the frame population and has a sufficiently high response rate (r_d), then the screening sample size $n_{0(d)}$, and hence the screening costs, may be relatively modest. However, screening costs increase rapidly as the prevalence rate of a rare population decreases. For example, with a required subgroup sample size of 1000 elements and a response rate of 70 percent, a screening sample size of 2857 elements is needed for a subgroup that comprises 50 percent of the frame population. In contrast, a screening sample size of 28,570 elements is needed for a subgroup that comprises only 5 percent of the frame population.

The U.S. National Immunization Survey (NIS) provides an illustration of large-scale telephone screening. The survey is designed to produce quarterly estimates of vaccination coverage levels for children aged 19–35 months for the United States and for each of 78 Immunization Action Plan (IAP) areas. The target sample size for each IAP area is 400 completed interviews per quarter. Since fewer than 5 percent of U.S. households have children in the specified age range, more than 1 million households have to be screened each year to yield the required sample sizes (Ezzati-Rice et al., 1995b; Zell et al., 2000; Smith et al., 2001a).

As with all surveys that screen in only certain population subgroups, noncoverage is a concern for the NIS. Camburn and Wright (1996) reported a shortfall of 18 percent in observed eligibility in the 1994 NIS (4.1 percent observed eligibility versus 5 percent predicted eligibility). The survey is affected by three sources of noncoverage (Shapiro et al., 1996). The first source comprises age-eligible children who are in households that do not have landline telephones. The second source comprises eligible children in households that are excluded by the list-assisted RDD sampling method used for the NIS—eligible children in households with no listed landline telephone numbers in their 100-bank of telephone numbers. The third source, the "within-household" noncoverage that occurs when respondents do not report all of the eligible children in their households, is a major component of the overall noncoverage. Special procedures have been used to reduce the noncoverage bias in the NIS, including the use of weighting adjustments based on interruptions in telephone service to account for nontelephone households (Frankel et al., 2003b).

The NIS screener questions needed to determine the presence in the household of children aged 19–35 months are fairly straightforward and nonthreatening. Sometimes subgroup identification requires that many questions be asked, resulting in long screener interviews, increased burden on the respondent, and perhaps a high level of response error in subgroup identification. For example, in the 2002 National Transportation Availability and Use Survey, the identification of persons with disabilities was based on six disability items from the U.S. Census long form, two items based on the American Disabilities Act definition, and an item about the presence of children in the household receiving special education services (U.S. Department of Transportation, Bureau of Transportation Statistics, 2003). This survey, which used a two-phase design, is discussed in more detail in Section 5.4.

When sensitive information is required for subgroup identification, it may be necessary to defer the subgroup identification questions until rapport has been established with the respondent. Examples of such subgroups are illegal drug users, persons who have broken a law, and persons with certain sexual orientations. Nonresponse and denial of subgroup membership are generally more likely in telephone surveys than in face-to-face surveys (especially those using audio computer-assisted self-interviewing for data collection).

When many questions are needed to identify the subgroup of interest, the accumulation of response errors to the individual questions can lead to a substantial level of noncoverage, with many false negatives (i.e., subgroup members being classified as outside the subgroup) (Sudman, 1972, 1976). When subgroup membership is a sensitive issue, respondents may be inclined to avoid reporting their membership. Whether or not the questions are sensitive, some subgroup members may seek to avoid the burden of the main interview. They may deduce from the screener questions or from advance material sent before the initial contact that the subgroup is the subject of the main survey, and they may avoid the extended interview by misreporting their subgroup membership during the screener interview. In addition, it is common practice in telephone surveys for one household member to answer the screening questions on behalf of all members of the household, a feature that can also give rise to misreports of subgroup membership status. Finally, the telephone interviewer may misrecord screener responses in order to avoid the extended interview, particularly if it requires calling back another household member at a later time.

As noted above, while screener misclassification is a problem with any mode of data collection, it may be a more serious problem with telephone interviewing. The problem of false negatives is far more serious than that of false positives. The false negatives are never identified, leading to noncoverage and the associated risk of bias in the survey estimates. They also lead to a smaller sample size than planned unless an allowance has been made for them at the sample design stage. In contrast, false positives can be identified in the extended interview that follows the screener; the main concerns with false positives are that they waste resources and produce a sample size that is smaller than indicated in the screener.

5.3 DISPROPORTIONATE STRATIFIED SAMPLING

Disproportionate stratified sampling can be used to improve the efficiency of screening for a small subgroup when the population can be divided into strata that differ in subgroup prevalence. In this case, the strata with higher prevalences of the subgroup can be sampled at higher rates. In the case of a survey conducted to produce estimates for a number of subgroups, the same approach can be used to oversample strata with higher prevalence for the smaller subgroups.

Disproportionate stratification is applicable when relevant auxiliary data for creating strata are available from the sampling frame; can be obtained from some other source, such as an administrative database; and in any other situations where the cost of acquiring the auxiliary data is negligible or is an acceptable fixed cost that does not depend on the number of cases for which the data are acquired. The method is a special case of two-phase sampling in which the first-phase sample comprises the total population. There is, however, an important difference between the two methods: With two-phase sampling, the first-phase data collection can be targeted at collecting data relating to subgroup status, whereas disproportionate sampling has to rely on the population data that are available. Two-phase sampling is discussed in Section 5.4.

The simplest case of disproportionate stratification is when the survey aims to collect data for only one subgroup (Waksberg, 1973; Kalton and Anderson, 1986; Mohadjer, 1988; Kalton, 1993). Under assumptions of simple random sampling within strata, of equal subgroup means and element variances for all strata, of equal costs of the screening, and of equal costs of the main data collection for all strata, the optimum sampling fraction in stratum h is given by $f_h \propto \sqrt{P_h/[P_h(r-1)+1]}$, where P_h is the prevalence of the rare population in stratum h and r is the ratio of the cost of the extended interview to the cost of the screener interview. The value of r is often much greater than 1. However, when screening costs are relatively high, as may often be the case in telephone surveys, r may be close to 1; in this case, the sampling fraction reduces approximately to $f_h \propto \sqrt{P_h}$.

The case of $r = 1$ is of special interest because it yields the greatest gains in precision for disproportionate stratification for a given set of strata and prevalence rates. As an example, consider two strata with $r = 1$ and with a subgroup prevalence of 20 percent in one stratum and 10 percent in the other. The relative sampling fractions are in the ratio $\sqrt{2}:1$ (i.e., 1.41:1). With such a relatively minor difference in sampling fractions between the two strata, the gains in precision are modest (see the following paragraphs). The relative sampling fractions are even closer when $r > 1$; for example, when $r = 4$, the relative sampling fractions are 1.27:1.

When disproportionate sampling with optimum sampling fractions is used rather than proportionate stratification, the relative gain in precision when $r = 1$ is approximately $\text{RG} = \left[V(\bar{y}_{\text{prop}}) - V(\bar{y}_{\text{opt}})\right]/V(\bar{y}_{\text{prop}}) = 1 - \left(\sum_h \sqrt{A_h W_h}\right)^2$, where $V(\bar{y}_{\text{opt}})$ and $V(\bar{y}_{\text{prop}})$ are the variances of a stratified sample mean for the subgroup under optimum and proportional allocations, and where A_h and W_h are the proportions of the subgroup and of the total population in stratum h (Kalton, 2003). This formula clearly shows that the gains in efficiency that can be achieved from using the optimum allocation depend on the differences in the distributions of the subgroup and of the total population across the strata. The value of RG (with $r = 1$) represents the maximum possible gain in precision; the larger the cost ratio r, the smaller the gain from disproportionate stratification.

The following examples illustrate the gains achieved with optimum allocation in a variety of cases:

- Suppose that there are two equal-sized strata and that the prevalences of the subgroup in these strata are 10 and 20 percent, as in the above example, for an overall subgroup prevalence of 15 percent. Then the value of RG is 0.03, indicating only a 3 percent gain in precision as a result of the optimum allocation.
- Suppose that one stratum that comprises 5 percent of the overall population has a subgroup prevalence of 90 percent and that the other stratum has a subgroup prevalence of 15 percent, with an overall prevalence of 18.7 percent. In this case, the optimum sampling rates are in the ratio 2.45:1 and RG = 0.08, an 8 percent gain in precision.
- Finally, suppose that one stratum comprises 10 percent of the overall population and has a subgroup prevalence of 90 percent while the other stratum has a subgroup prevalence of 5 percent, with an overall prevalence of 13.5 percent. The optimum sampling rates are in the ratio 4.24:1 and RG = 0.35, a major gain in precision of 35 percent.

These examples illustrate that the gains achieved using the optimum allocation are modest unless two conditions are met. First, the prevalence of the subgroup must be much higher in one or more strata than in the total population. Second, there must be a high proportion of the subgroup in these strata. Often only one of these two conditions is met, in which case the gains from the optimum allocation are not great. However, even though major gains may not be achieved with an optimum allocation, the method is almost costless and any gains are useful.

The theory given above assumes that the proportions of the subgroup in the strata (P_h) are known. Inaccuracies in the proportions lead to non-optimum sampling rates. Estimates of these quantities are usually based on past data, such as the most recent census, and thus are likely to be inaccurate because of changes over time in the distribution of the subgroup across the strata.

Disproportionate stratification with strata defined geographically has been used in a number of area sample designs to oversample strata containing greater prevalences of small subgroups. Waksberg et al. (1997) conducted a detailed examination of the effectiveness of oversampling U.S. census blocks and block groups with higher concentrations of various race/ethnicity and poverty subgroups. Five strata were created by combining all blocks or block groups within a given range of concentrations of the subgroup of interest; for example, for Hispanics the strata were defined as less than 5, 5–10, 10–30, 30–60, and greater than 60 percent concentration. The strata were sampled at the optimum rates given above. Waksberg et al. conclude that geographic oversampling is a useful strategy for sampling blacks and Hispanics provided that the cost of the full interviews is less than 5–10 times the cost of the screener interviews. They also determined that this approach is useful for Asians/Pacific Islanders and American Indian/Eskimo/Aleut populations even with much higher cost ratios. However, geographic oversampling is not useful for low-income persons because most low-income persons do not live in high-density strata. Moreover, since income data have been collected only for the long-form sample in the census, the estimated proportions of low-income persons in block groups are subject to substantial sampling error (they are not produced for blocks). In addition, any benefits achieved by oversampling strata with high proportions of low-income persons at the time of the census are likely to disappear as the census data become outdated, and indeed the oversampling could lead to a loss in precision in the survey estimates.

In the context of RDD sampling, the only basis for forming geographic strata is the telephone numbers themselves, that is, the area codes (the first three digits) and the exchanges (the next three digits, sometimes referred to as the prefixes or the central office codes) of the 10-digit telephone numbers in the United States. Thus, with RDD sampling, this form of disproportionate stratified sampling will produce gains in efficiency only if the prevalence of the subgroup of interest varies geographically by area/exchange combinations. Since exchanges are much larger than the blocks and block groups that can be used for forming strata in area sampling, in RDD surveys lower gains in efficiency are to be expected from disproportionate

Table 5.1. Percentage Variance Reduction Achieved with Geographic-Based Oversampling Relative to Proportionate Stratification of the Same Cost for Race, Ethnicity, and Low-income Subgroups

Stratification unit	Blacks	Hispanics	Asians/Pacific Islanders	Income under \$10,000
Cost ratio: 1				
Block	48.5	45.4	46.4	NA
Block group	40.9	37.6	34.7	16.3
Exchange	29.5	31.2	27.9	7.3
Cost ratio: 2				
Block	39.2	37.0	42.8	NA
Block group	32.5	36.0	31.7	13.6
Exchange	23.6	25.0	25.2	6.2
Cost ratio: 5				
Block	24.2	23.3	34.9	NA
Block group	19.5	18.2	25.2	8.8
Exchange	14.1	15.1	19.6	4.0

NA: not available.

Source: 2000 Decennial Census SF1 and SF3 files and Genesys Sampling Systems 2005 Coverage Report for the United States.

stratification.[1] However, the ratio of the cost of a full interview to that of a screener interview (r) is likely to be lower with a telephone survey than with many face-to-face interview surveys, and the lower that ratio, the greater the gains achieved through oversampling.

Following the methods used by Waksberg et al. (1997) and using the same stratum definitions, the gains in precision achieved with optimum oversampling were computed for blocks, block groups, and telephone exchanges for various race/ethnicity subgroups and for low-income persons (persons in households with incomes under \$10,000[2]) for cost ratios of 1, 2, and 5. The results are displayed in Table 5.1. The block and block group computations used data from the 2000 Decennial Census; since census income data were not produced at the block level, stratification by blocks was not used for the low-income subgroup. The telephone exchange computations were based on 2005 population estimates rather than the 2000 Census data because 2000 Census data were not readily available for telephone exchanges. However, using the 2005 estimates rather than 2000 Census data should have had only a minimal

[1]On average, exchanges are about three times as large as block groups. In addition, they are much more variable in size than block groups.

[2]Since data were not available to compute variance reductions for persons in poverty at the exchange level, the computations were based on persons in households with incomes of less than \$10,000. For blocks and block groups, the variance reductions based on persons with incomes of less than \$10,000 are similar to those reported by Waksberg et al. (1997) based on persons in poverty.

effect on the results in the table. The table shows percentage variance reductions for an estimate of a subgroup characteristic under the assumptions that (1) the element variance is the same in each stratum and (2) the sample has been selected by simple random sampling within each stratum. The proportionate variance reductions based on the stratification of blocks and block groups are broadly similar to, although slightly less than, those computed by Waksberg et al. (1997) based on the 1990 Census. The reductions in variance achieved by oversampling using strata constructed from telephone exchanges are much lower than those achieved using strata constructed from blocks and block groups. In the case of low-income persons, the reductions in variance are modest for block groups for any cost ratio, and only half as large for exchanges. With low-cost ratios, oversampling based on exchanges is reasonably beneficial for race and ethnicity subgroups, but it gives at best only marginal gains for a low-income subgroup. Even with the race/ethnicity subgroups, it should be noted that oversampling is generally beneficial for estimates for the oversampled subgroup but may reduce the precision of estimates for other subgroups.

Some specialized commercial firms compile (and update) census data at the exchange level, which may be useful for disproportionate stratification based on telephone exchanges (see, e.g., Mohadjer, 1988). However, the gains in precision achieved using these data sources will be lower than those achieved using census data because the quality of the updated data is unlikely to match that of the census at census time.

The National Household Education Survey (NHES) program provides an illustration of the use of disproportionate stratification by exchange. There are several surveys in the NHES program, each of which is a telephone survey of a specific population of interest. Each survey requires oversampling of blacks and Hispanics to satisfy precision requirements for estimates for these two subgroups. Oversampling of exchanges with high concentrations of blacks and Hispanics has been used since the 1991 NHES Field Test when the technique was found to be useful (Mohadjer and West, 1992). For NHES 2001, a high-density stratum was defined as those exchanges in which at least 20 percent of the population was either black or Hispanic (Nolin et al., 2004).

The NHES 2001 results indicated that, overall, around 12 percent of telephone households contained youth in the given age range for each of these subgroups, with eligible youth being identified in around 25 percent of the telephone households in the high-density stratum but in only around 5 percent in the low-density stratum. The high-density stratum is estimated to contain roughly 37 percent of the total population but around 76 percent of both the black and Hispanic subgroups (based on 2005 data). Applying the formula for the maximum possible gain in precision yields a value of RG = 0.15 for each subgroup. Using a similar disproportionate stratified sample design but with a cost ratio of greater than 1, the reduction in the total cost of data collection for the NHES based on the same precision was expected at the planning stage to be only 5–6 percent (Brick et al., 2000). Although this approach produces only modest gains, it has been used repeatedly in the NHES program because it is easy and inexpensive to implement.

One of the goals of the 2003 California Health Interview Survey (CHIS) was to produce estimates for a number of small race/ethnicity subgroups in California, such as Cambodian, Japanese, Korean, and Vietnamese subgroups (California Health Interview Survey, 2005). Two approaches were used to oversample these subgroups. One was to employ disproportionate sampling with strata formed according to the density of these rare subgroups in exchanges. The other was to use a dual-frame approach that included both an RDD frame and a frame of likely subgroup members compiled from telephone directory listings based on their surnames. This section describes the use of the disproportionate stratified sampling approach for oversampling the Korean and Vietnamese subgroups. The dual-frame approach is discussed in Section 5.5.

Although the 2000 Census provided information about the Korean and Vietnamese subgroups that could be used in designing the sample for the 2003 CHIS, no information was directly available for these subgroups at the exchange level. A mapping exercise was therefore needed to allocate each exchange to either a high- or a low-density stratum. This exercise was restricted to the four largest California counties that accounted for more than 75 percent of the population of these subgroups. Since the ZIP code area was the smallest area for which information was available for these subgroups in the 2000 Census files, the subgroup densities in the exchanges were estimated from a manual mapping of the ZIP code areas to the exchanges, with each ZIP code being assigned to the exchange in which most of its population was located. The disproportionate sampling rates in the CHIS were computed to maximize the effective sample sizes for the Korean and Vietnamese subgroups while not markedly decreasing the precision of the estimates for the Japanese, Cambodian, Hispanic, and African American subgroups. Despite the limitations in the estimation of the numbers of Koreans and Vietnamese by exchange and the use of census data that were 3 years out of date, the disproportionate allocation was expected to increase the precision of estimates for the Korean and Vietnamese subgroups by 24 and 20 percent, respectively. The disproportionate allocation was also expected to increase precision by 10, 6, and 4 percent for the Chinese, Japanese, and Filipino subgroups, respectively. These increases were expected because these subgroups were also more highly represented in the oversampled stratum. In contrast, the precision of the survey estimates was expected to decrease by 5 percent for African Americans, 4 percent for Hispanics, and 2 percent for American Indians. The CHIS example illustrates the need in multipurpose surveys to consider the effect of disproportionate stratified sampling on all subgroups of interest and on estimates that cut across the subgroups.

Apart from geographical stratification, another possible source of data for forming strata is the information about household characteristics that commercial vendors have linked to individual residential telephone numbers. The value of this information for forming strata based on the likely density of a subgroup depends on the availability of information closely related to subgroup membership and on the accuracy of that information. Most of the time, the information is not closely enough related or accurate enough to make disproportionate stratification useful (Brick et al., 2000). Additionally, the information is available only for households with listed numbers. However, the information linked to individual residential telephone numbers can be useful in some cases. This approach may have greater potential in the future if the quality of the information for listed households improves.

There are a number of ways in which a high-density stratum of listed telephone numbers can be incorporated into a sample design. One is to select a sample from the high-density stratum of listed numbers, such as persons with likely Vietnamese names in telephone directories, and to sample those not in this stratum using a disproportionate stratified RDD design based on the subgroup density by exchanges, as described above. This design can be implemented either by removing telephone numbers in the high-density stratum from the RDD frame before sample selection or by treating them as blanks if they are sampled from the RDD frame. A variant is to apply the disproportionate stratified RDD component of this design to all households, including those in the high-density stratum of listed numbers. In this case, households in the high-density stratum of listed numbers have chances of being selected from both the list sample and the RDD sample. This dual-frame design is discussed and illustrated in Section 5.5.

5.4 TWO-PHASE SAMPLING

With two-phase, or double, sampling, basic data are collected from a large first-phase sample and then more detailed information is collected from a subsample of the first-phase sample (see, for example, Cochran, 1977). The screening method described in Section 5.2 is a special case of two-phase sampling: Based on the screener data, a large first-phase sample is divided into two strata, those in the subgroup and the rest. Then all members in the subgroup stratum and none of those in the other stratum are included in the second-phase data collection. This section examines an extension of the procedure in which the first-phase sample is divided into two or more strata that are sampled at different, nonzero, sampling rates. Since the approach is closely related to disproportionate stratified sampling, much of the discussion in Section 5.3 also applies here.

There are two distinct types of applications of two-phase sampling for sampling subgroups of the population. One is for estimating the characteristics of members of a single subgroup when the screener classification is imperfect. The other is for estimating the characteristics of the members of several different subgroups.

In the first type of application the screener data are used to classify the first-phase sample members into strata according to the likelihood that they are members of the subgroup of interest. Then a disproportionate stratified sample is selected for the second phase, oversampling the higher-density strata and sampling the low-density strata at lower rates, but rates greater than zero. This approach is useful when accurate determination of subgroup membership is complex and expensive, but when an inexpensive but imperfect method of subgroup identification exists. The benefits of the two-phase approach in this situation depend on the relative costs of administering the accurate and the imperfect measures: The budget for subgroup identification could either (1) be divided between administering the inexpensive measure to an initial large sample and the expensive measure to a targeted subsample or (2) allocated to administering the accurate measure directly to a sample that would be larger than the targeted subsample.

A common design for this first type of application is to divide the first-phase sample into two strata based on the first-phase responses, with the first stratum comprising those likely to be members of the subgroup of interest and the second stratum comprising those unlikely to be members of that subgroup. Then all those in the first stratum and a subsample of those in the second stratum are included in the second-phase sample. The optimum subsampling rate for the second stratum and the efficiency of this design compared with a single-phase design depend on three main factors (Kalton, 1993). Like disproportionate stratified sampling, sizable gains in precision from a two-phase design occur only when (1) the proportion of the members in the first stratum who are subgroup members is much higher than the corresponding proportion in the second stratum and (2) the proportion of the subgroup members who fall into the first stratum is large.

The third factor necessary to achieve large gains in precision from two-phase sampling is that the cost of classifying a member of the first-phase sample into one of the strata is much lower than the cost of accurately determining subgroup membership. Deming (1977) suggests that, for two-phase sampling to be effective, the cost ratio of first-phase screening to the second phase should be at least 6:1. Since the cost ratios in telephone surveys are generally lower than those in face-to-face interview surveys, two-phase sampling is likely to produce only minor gains when all data are collected by telephone. The more likely application is for surveys involving telephone data collection for the first-phase screener and face-to-face interviews for the second phase.

The second, and more common, application of two-phase sampling occurs when a survey is designed to produce estimates for a number of subgroups, and often also for the total population. As described earlier, the screener sample size that produces the required sample sizes for each subgroup can be calculated with the maximum value of these screener sample sizes then being chosen. The second-phase sample then includes all members of the subgroup requiring the maximum screener sample size and subsamples of the other subgroups. If the entire population is covered by the subgroups of interest (e.g., estimates are required for all race/ethnicity subgroups), noncoverage arising from false negatives is avoided. However, subgroup misclassification can still lead to inefficiencies in the sample design and to a failure to achieve planned subgroup sample sizes.

The National Survey of America's Families (NSAF) is an example of this second application of two-phase sampling (Judkins et al., 1999b, 2001; Brick et al., 2003a). The NSAF sample design comprised an RDD telephone sample and an area sample of nontelephone households, but this discussion is limited to the telephone component. The population was divided into 15 major strata comprising 13 individual states, one county, and the remainder of the United States, with different sampling fractions in each of these strata. The survey design required an oversample of households that had incomes below 200 percent of the poverty level (low-income households) and contained children under 18 years of age. Households containing only persons aged

65 and over were excluded from the target population. Households with no children but at least one adult under age 65 were also sampled, with a higher sampling fraction for low-income households. These requirements were met by (1) selecting the entire stratum of low-income households with children identified in the first-phase sample, (2) selecting only a subsample from the set of strata containing households with no children and households with incomes above 200 percent of poverty (high-income households), and (3) excluding all households containing only persons over age 65. The second set of strata was subsampled at rates that varied from a low of 25 percent to a high of 40 percent. The sampling rates were derived by simultaneously maximizing the effective sample sizes for low-income households with children, for all households with children, and for households without children subject to a fixed total number of interviews.

In planning the 1997 cycle of the NSAF, it was assumed that 11 percent of those classified by the screener as low-income households with children would be found not to have these characteristics at the extended interview (false positives) and that 4 percent of the households classified as high-income households with children would turn out to be low-income households with children at the extended interview (false negatives). The actual misclassification rates were 24.2 percent for false positives and 9.1 percent for false negatives, much greater than expected (Flores Cervantes et al., 1998). As a result, the sample size and effective sample size for low-income households with children were 18 and 22 percent lower than planned, respectively. However, for all households with children, the sample size and effective sample size were on target. The observed misclassification rates were used in planning the sample designs for later cycles of the NSAF. The NSAF illustrates the challenges of estimating misclassification rates at the planning stage, as well as the serious effect that underestimation can have on the precision of estimates for the oversampled subgroup.

A second example of two-phase sampling illustrates the impact of the choice of the first-phase informant on the misclassification rates. The 2002 National Transportation Availability and Use Survey had a goal of completing equal numbers of interviews with persons with and without disabilities (U.S. Department of Transportation, Bureau of Transportation Statistics, 2003). The sample was selected using a list-assisted RDD approach. In the first-phase screener interview, an adult household informant was asked to identify persons with disabilities in the household. Disability was determined through seven questions that asked if any person in the household had specific sensory and physical limitations; if anyone had long-lasting mental, self-care, or employment difficulties; and if any child was receiving special education services. In households where one or more persons were reported to have disabilities, one person with disabilities was randomly selected for an extended interview. In addition, in one third of all sampled households, one of the persons reported to be without disabilities (if any) was randomly selected for an extended interview. Under this procedure, two persons were selected in one third of the households that contained both persons with and without disabilities.

Using the second-phase responses as the gold standard, the first-phase responses incorrectly classified 11.3 percent of persons without disabilities as having disabilities (false positives) and 1.1 percent of persons with disabilities as not having disabilities (false negatives). However, the level of misclassification of disability status at the first phase varied greatly depending on whether or not the screener informant was the sampled person for the extended interview. When the screener informant was selected for the extended interview (i.e., the screener informant reported about himself or herself), only 3.6 percent of those initially classified as having disabilities were reclassified (those initially classified as not having disabilities were not reassessed in the second-phase interview). In contrast, when a person other than the screener respondent was selected for the extended interview (i.e., the screener respondent reported about somebody else in the household), the misclassification rates were much higher: 37.5 percent of the sampled persons initially classified as having disabilities were reclassified, and 6.4 percent of those initially classified as not having disabilities were reclassified. These findings indicate why the choice of the first-phase informant should be carefully considered at the survey design stage.

5.5 MULTIPLE FRAMES

Sampling of a small subgroup can be aided by the availability of separate frames on which the subgroup is more highly concentrated, even though these frames provide only partial coverage of the subgroup members. One approach with multiple frames is to uniquely identify each member of the population with a single frame, for instance, by placing the frames in a defined order and identifying members only with the first frame on which they appear (their listings on later frames are removed or treated as blanks). In this case, a disproportionate stratified design can be used, as described in Section 5.3.

An alternative approach is to select a sample from each frame without eliminating the overlaps between frames. When the frames are not disjoint, the second approach results in a multiple-frame design. A common case is a dual-frame design that combines an incomplete frame on which the subgroup is present at a high density and a general population frame that has high coverage of the subgroup but low density. With a multiple-frame design, adjustments must be made in the analysis to compensate for the fact that some population members have more than one route of sample selection. These adjustments may be made either by weighting each sampled unit by the inverse of its overall selection probability, taking into account the multiple routes of selection, or by using the multiple-frame methodology introduced by Hartley (1962, 1974). There are also numerous papers that address the optimal allocation of the sample among the frames (Kalton, 1993), as well as different approaches for estimation from multiple frames (Skinner and Rao, 1996; O'Muircheartaigh and Pedlow, 2000; Lohr and Rao, 2000, 2006).

The disproportionate stratification approach can be applied when duplicate listings can be eliminated across frames before sample selection and also when sampled units with duplicate listings can be identified from the frames after they have been selected. When duplicate listings can be identified only by contacting sampled units, it is often—but not always—preferable to employ the multiple-frame approach.

As may be expected from the similarity between the disproportionate stratification and multiple-frame approaches, they require that the same conditions be satisfied if they are to yield sizable gains in precision. For the multiple-frame approach these conditions are (1) the prevalence of the subgroup on the high-density frame (or frames) be much higher than on the other frame (or frames) and (2) a high proportion of the subgroup members be covered by the high-density frame. In practice, the high-density frame is usually a list frame that will have a high prevalence of subgroup members. However, the proportion of subgroup members on the list is often low, with the result that a multiple-frame approach will not yield major gains (unless there are great differences in data collection costs across frames). Nevertheless, a multiple-frame approach can still be beneficial even if the gains in precision are small, provided that the cost of creating the high-density list frames is low.

As noted in Section 5.3, a multiple-frame approach was used, in combination with disproportionate stratified sampling, in the 2003 CHIS to oversample Korean and Vietnamese adults (California Health Interview Survey, 2005). Two telephone list frames were constructed by identifying likely Korean and Vietnamese surnames in telephone directory listings (e.g., Lee, Kim, and Park for Koreans and Nguyen, Tran, and Pham for Vietnamese). The Korean and Vietnamese frames contained 210,000 telephone numbers and 180,000 telephone numbers, respectively, with an overlap of 82,000 telephone numbers because some surnames could not be characterized as Korean or Vietnamese surnames only. The RDD list-assisted frame contained about 29 million numbers, of which around 12 million were for households.

The sampling scheme then consisted of selecting separate telephone samples from the Korean and Vietnamese lists, and from the RDD frame using disproportionate stratified sampling as described in Section 5.3. If the household at a telephone number selected from the Korean or Vietnamese list contained a Korean or Vietnamese adult, an interview was conducted, irrespective of the list from which the household was selected. If the household did not contain a Korean or Vietnamese adult, no interview was conducted. Table 5.2 shows the prevalence and the proportion of the rare population in the surname frames used in the CHIS. Although the potential gains associated with optimum allocation would be large for a survey focusing only on Korean and Vietnamese adults (RG = 0.32 for Koreans and RG = 0.38 for Vietnamese), much smaller sampling fractions than the optima were used in the CHIS

Table 5.2. Prevalence and Proportion of the Korean and Vietnamese Population in the Korean and Vietnamese Surname Frames in the 2003 CHIS

	Prevalence rate (%)		Percentage of landline telephone population on frame	
Frame	Korean	Vietnamese	Korean	Vietnamese
Korean list	46.3	10.8	41.3	7.0
Vietnamese list	13.2	63.8	13.8	48.9
RDD	1.1	1.6	100.0	100.0

Source: UCLA Center for Health Policy Research, 2003 California Health Interview Survey.

because the sample was also designed to produce estimates for the total population by county as well as for other rare groups at the state level.

5.6 MULTIPLICITY SAMPLING

In most surveys, sampled respondents report only about themselves and sometimes about other members of their own household. With multiplicity sampling (also known as network sampling), sampled respondents are also asked to report about others linked to them in a clearly defined way (e.g., siblings) who are not living in their household. Multiplicity sampling reduces the amount of screening needed to identify members of a rare subgroup, but it results in multiple routes by which sampled members may be selected (Sirken, 1997).

As an illustration, consider a survey to collect data on women with a given health condition in which telephone households are selected with equal probability. Data are collected for all women with that condition in the sampled households and for the sisters of the women in the sampled households who have the condition. The selection probability for a sampled woman having this condition is a combination of the probability that her household will be sampled and the probabilities that the households in which her sisters are living will be sampled. This selection probability is needed to compute weights for use in analysis. If a sister lives in a household without a landline telephone, that household has no chance of selection (indeed, an added advantage of multiplicity sampling is that it provides a means for representing some who would not otherwise be covered). It matters whether two sisters live in a single household or in two different households. With an equal probability sample of households, the selection probability of a given woman is proportional to the number of telephone households in which she and her sisters reside. With an unequal probability sample of households, the selection probability of each of the households must be determined. Tortora et al. (Chapter 6, this volume) describe the use of multiplicity sampling to obtain coverage of the mobile-telephone-only population that is linked within a parent–sibling–child family network of respondents from a prior survey.

A number of requirements must be satisfied if multiplicity sampling is to be useful for sampling a rare subgroup. The first requirement is that the linkages must be clearly defined and must be operational, so that selection probabilities, and hence analysis weights, can be computed. A second requirement is that all members of the network (e.g., sisters) must be able and willing to report accurately on the subgroup status of all other members during a telephone interview. This requirement is all that is needed if the aim is simply to estimate the prevalence of the subgroup. If, however, the survey objective is to collect information about subgroup members, such as types and costs of treatment for the health condition, then other requirements must be satisfied. Either each sampled subgroup member must be able and willing to provide the information for other network members who have the condition, or each member must be able to provide valid contact information for network members with the condition so that they can be interviewed in person. A third requirement is that it must be ethically acceptable to collect information on subgroup status (and perhaps the survey responses) from network members without direct consent from the subgroup member. This issue is an especially serious concern for surveys pertaining to certain subject matter. Institutional review boards may require direct consent for the collection of information on subgroup status, thus ruling out the use of multiplicity sampling.

Multiplicity sampling has not been widely used in large-scale telephone surveys. The following examples illustrate its potential as well as its possible limitations. Multiplicity sampling was used in a 1978 telephone survey of Vietnam-era veterans conducted for the Veterans Administration at seven sites in the Midwest, the South, and the City of Los Angeles (Rothbart et al., 1982). A standard screener questionnaire was administered to identify age-eligible males in the sampled households. Interviewers also collected information about the age eligibility and veteran status of sons, brothers, and nephews of the screener respondent, regardless of whether or not they were part of the sampled household. Follow-up interviews were then conducted with those identified as possible Vietnam-era veterans. Locating the sampled veterans in the network required additional telephone calls and occasional field visits but was not otherwise problematic. The multiplicity sample produced 535 eligible Vietnam-era veterans from roughly 8700 completed screener interviews in the sampled households (i.e., an incidence rate of 6.2 percent) and an additional 476 veterans from the defined networks. Sizable savings in screening costs were achieved using multiplicity sampling, with less than an 8 percent increase in variance due to the variability in the weights.

The 1998 Field Test for the National Household Education Survey (NHES) included an evaluation of the use of multiplicity sampling for increasing the sample of 14–21-year-old youth who dropped out of school and for reducing the bias associated with telephone noncoverage (Brick and West, 1992). The data collection consisted of three components. First, a telephone screener was used to identify all 14–21-year-old youth in the sampled household, as well as the 14–21-year-old children of mothers in the household when the children lived elsewhere (out-of-household youth). Second, a Household Respondent Interview (HRI) was attempted for each of these youth. For youth living in the sampled household, the HRI could be completed by any adult household member who knew about the educational status of the youth, including self-reports by the youth. Mothers with eligible children living elsewhere were asked to complete the HRI for those youth. Third, a Youth Interview was attempted for all youth who had dropped out of school and for a 20 percent subsample of youth who had not dropped out. Out-of-household youth who lived in telephone households had two routes of selection if their mother lived in a telephone household, but only a single route of selection if their mother lived in a nontelephone household. Out-of-household youth living in nontelephone households could be selected only if their mothers were selected and would not have been covered if multiplicity sampling had not been used. The use of multiplicity sampling led to an increase of about 18 percent in the number of completed HRIs, with no real difference in response rates between in-household and out-of-household youth. However, only 57.5 percent of the out-of-household youth completed the Youth Interview, compared with 88.7 percent of the in-household youth. Important sources of nonresponse for the out-of-household youth were a failure to reach them by telephone and the inability of the HRI respondent to provide locating information.

The second objective of using multiplicity sampling in the NHES Field Test was to reduce the undercoverage of 14–21-year-olds. Using data from the Current Population Survey, it was estimated that close to 8 percent of all youth lived in nontelephone households, but approximately 5 percent of them had mothers who lived in telephone households. Among dropouts, 30 percent were estimated to live in nontelephone households, but 11 percent of their mothers lived in telephone households. The multiplicity sampling thus reduced the undercoverage of all youth from 8 percent to 3 percent and of dropouts from 30 percent to 19 percent.

5.7 MODIFIED MITOFSKY–WAKSBERG SAMPLING

Blair and Czaja (1982) proposed a variant of the standard Mitofsky–Waksberg sampling method for sampling rare populations in telephone surveys. In the standard Mitofsky–Waksberg method of RDD sampling, an equal probability sample of telephone households is selected in two stages to improve efficiency in identifying working residential numbers (Waksberg, 1978; see also Kalsbeek and Agans, 2007, Chapter 2 in this volume). In the United States, all telephone numbers are 10-digit numbers, where the first three digits are the area code, the next three digits are the exchange code, and the last four digits are the suffix. At the first stage of the Mitofsky–Waksberg method, all possible suffixes within active exchanges are grouped into clusters of 100 consecutive numbers, termed banks. A random sample of potential telephone numbers is selected within the active exchanges. Each sampled number identifies a bank that is selected. If the sampled number is a residential number, then the bank is retained and additional numbers within the bank are called until a predetermined number of households has been selected. Calling only retained banks greatly increases the likelihood of contacting a residence in the remaining numbers, while producing an equal probability sample of telephone households.

The Blair and Czaja modification of the Mitofsky–Waksberg procedure for sampling rare populations redefines the acceptance–rejection rules in terms of the rare population instead of the total population. Thus, a sampled bank is retained for further dialing if the initially sampled number is an eligible household that contains members of the rare population. Telephone numbers are sampled in a retained bank until a predetermined number of eligible households have been identified. The reduction in screening is possible in the same way as with the original Mitofsky–Waksberg procedure. The screening costs are greatly reduced when the rare population is disproportionately distributed among the banks and many banks contain no members of the rare population. The method is further improved if the first-stage sample of telephone numbers is drawn using a disproportionate stratified design that samples geographical areas with high concentrations of the rare population at higher rates.

A significant weakness in the modified design is that it is often not possible to identify the predetermined number of eligible households in the retained banks (Waksberg, 1983). The weights needed to compensate for a shortfall in a bank may lead to increases in sampling variances that make the design no more efficient than a traditional screening design. To avoid this outcome, larger banks (e.g., banks of 200 suffixes) may be used, but the banks should not be so large that the benefits of the clustering are lost.

The modified Mitofsky–Waksberg method was used in the 1980 Beliefs About Social Stratification Survey (Blair and Czaja, 1982) to oversample two subgroups using banks of 100 telephone numbers—households in which the head of the household was African American and households with an income of $30,000 or more. This method was also used in the 1984 National Black Election Survey (Inglis et al., 1987) to sample African Americans eligible to vote. A disproportionate stratified design was employed in this survey, with telephone exchanges classified into three strata based on the density of African American households, and statistical concerns about the use of this method were addressed by using banks that contained 200 telephone numbers.

5.8 MULTIPURPOSE SURVEYS

A natural way to reduce the cost of large-scale screening to identify members of a rare population is to link the rare population survey to an existing large survey. However, such a linkage is often resisted because of concerns about the increased burden on respondents in the existing survey and the possible effect on the response rate. These concerns will be greater if the data are to be collected by telephone because of a perceived need to keep telephone interviews short.

In the context of telephone surveys, the most attractive form of linkage is to conduct a single large screening survey to identify the members of several rare populations simultaneously. This multipurpose approach works best when the rare populations do not overlap because then the only additional burden on respondents is answering the screening questions required to identify the population to which they belong. If the rare populations overlap, an additional sampling step can be implemented to ensure that respondents participate in only one survey. However, this step creates the need for a larger screening sample because of both the sample size reduction and the added variability in the weights—and hence the loss in precision—associated with the additional sampling. The screening sample size must be large enough to yield the required levels of precision for the estimates for each of the rare populations.

The multipurpose survey approach is used in the U.S. State and Local Area Integrated Telephone Survey (SLAITS) (Ezzati-Rice et al., 1995a). The SLAITS builds on the large-scale telephone screening carried out for the National Immunization Survey (NIS), which is described in Section 5.2. During the course of a year, the NIS screens nearly 1 million telephone households to identify households containing children aged 19–35 months. The SLAITS takes advantage of the NIS screening to link other health care surveys for other population subgroups, with different surveys being linked at different times. The screening costs for the other surveys are greatly reduced by sharing the sampling frame and the screening efforts with the NIS. When the population of inference for the other survey overlaps with the NIS population, sample members in the overlap are included in both surveys.

Large-scale implementations of the SLAITS have included the 2000 National Survey of Early Childhood Health for children aged 4–35 months (Blumberg et al., 2002), the 2001 and 2003 National Surveys of Children with Special Health Care Needs for children under age 19 (Blumberg et al., 2003), the 2003 National Survey of Children's Health for all children under age 17 (Blumberg et al., 2005b), and the 2003 National Asthma Survey for asthma sufferers (U.S. National Center for Heath Statistics, 2006). The addition of the extra SLAITS surveys has been found to have only a small impact on the NIS response rate because the NIS screener and interview are completed before the SLAITS portion of the interview.

5.9 CONCLUDING REMARKS

Despite serious concerns about increasing nonresponse rates and increasing noncoverage rates in RDD surveys of landline telephones, telephone surveys of rare populations remain popular because of their significantly lower costs compared with face-to-face interview surveys. In assessing the suitability of telephone data collection for a survey of a given population subgroup, careful consideration should be given to the likely nonresponse and noncoverage rates and to the potential biases that may affect the survey estimates. Consideration should also be given to the suitability of the survey's subject matter for telephone data collection and to the likelihood that members of the subgroup can be accurately identified using a telephone screener.

A large-scale screening effort will generally be required in surveying most rare subgroups. The techniques described in this chapter can help to reduce the amount of screening required. Although most of the techniques result in only modest savings of time and money, they are nevertheless worthwhile because they are inexpensive and easy to implement. Although this chapter describes the methods separately, they can be applied in combination, and some examples of combinations have been given.

Most of the literature on sampling rare populations considers sample designs for surveys that focus only on a single rare population. In practice, many surveys are required to produce separate estimates for more than one rare population, for subgroups that cut across the rare populations, and for the total population. In these cases, careful attention should be given to achieving a balance in the sample design so that all of the required estimates are produced in an efficient way.

CHAPTER 6

Multiplicity-Based Sampling for the Mobile Telephone Population: Coverage, Nonresponse, and Measurement Issues

Robert Tortora
The Gallup Organization, USA

Robert M. Groves
University of Michigan and Joint Program in Survey Methodology, USA

Emilia Peytcheva
University of Michigan, USA

6.1 INTRODUCTION

The mobile or cellular telephone is both a technological development that has changed social relations throughout the world and a challenge for survey researchers. The mobile telephone poses five challenges to survey researchers: (a) adapting to a one-to-one phone-number-to-person relationship instead of a one-to-many relationship with landline telephones, (b) assessing the declining coverage of the household population by landline telephones, (c) dealing with the lack of a frame for mobile telephones in many countries, (d) researching unknown sample recruitment problems using mobile telephones, and (e) learning about the impact of diverse respondent settings during interviews. Although there is a growing literature addressing some of these issues (Cunningham et al., 1997; Kuusela and Notkola, 1999; Fuchs, 2000a; Kim and Lepkowski, 2002; Kuusela, 2003; Morganstein et al., 2004; Luke et al., 2004;

Advances in Telephone Survey Methodology, Edited by James M. Lepkowski, Clyde Tucker, J. Michael Brick, Edith de Leeuw, Lilli Japec, Paul J. Lavrakas, Michael W. Link, and Roberta L. Sangster

Steeh, 2004a; Kuusela et al., Chapter 4 in this volume), they are likely to need considerable study over the coming years.

This chapter presents the results of an experiment that assesses some of the challenges above, with a focus on frame development for mobile telephone subscribers in the United States. The experiment uses multiplicity sampling to obtain the coverage of the mobile-only population that is linked within a parents–siblings–child family network of each respondent to a prior survey. The chapter begins by reviewing the technique of multiplicity sampling as it might be applied to mobile telephones in the United States or other countries that do not construct an updated frame of mobile telephone numbers.

6.2 SAMPLING FRAME ISSUES IN MOBILE TELEPHONE SURVEYS

Despite the rapid penetration of mobile telephones in the United States, they remain a device used in diverse ways by the population. The prevalence of adults who can be reached only through a mobile telephone is small but rapidly growing (Blumberg et al., Chapter 3 in this volume). Although the United States lags behind many other countries in this prevalence, it seems clear that the future will bring ever-increasing prevalence of mobile-only households.

Currently, in many countries there is no updated sampling frame that contains listings of the mobile-only population. Further, existing telephone number frames that *do* cover the mobile telephones are increasingly integrating them with landline number series. Because of the ability of subscribers to move across service providers and simultaneously change from landline to mobile status, landline and mobile numbers are increasingly mixed together. In some countries of the world, the available frames that do contain mobile numbers are very inefficient to use. The designer would need to decide whether to exclude the mobile numbers or include them and measure the extent of multiple telephone ownership. Hence, other ways of including the mobile-only persons in a telephone survey need to be developed. One design option for mixed mobile and landline telephone surveys would be to sample the landline telephone population through one method and the mobile-only population through some other method.

Survey researchers have traditionally associated landline telephone numbers with single households, and all the members of that household became a potential focus of person-based measurement. Mobile telephones appear best linked to individuals. Thus, the existing frame of telephone numbers is growing in the complexity of its links among households and persons. A "multiplicity" is the number of different ways that a target population unit can be associated with sampling frame elements (Sirken, 1970). As the complexity of the links between persons and telephone numbers grows with increasing mobile telephones, enumerating the multiplicities of individuals becomes more complicated.

In one sense, sampling the mobile-only population is an example of a "rare population" problem in survey statistics. The traditional methods of handling rare populations (see Kalton and Anderson, 1986; Sudman et al., 1988) include locating a frame that offers large coverage of the rare population, locating a frame containing strata with varying proportions of the rare population (and oversampling the high-density strata), adaptive sampling, which sequentially selects cases based on their proximity to another eligible sampled case (Thompson and Seber, 1996), and the use of multiplicity sampling (Sirken, 1970). A more detailed discussion of some of these approaches can be found in Flores-Cervantes and Kalton (2007), Chapter 5 in this volume.

Screening for mobile-only populations can be conducted on an existing frame that has near-universal coverage (e.g, a household address frame or a telephone number frame in random digital dialing (RDD) sampling). The identification of mobile-only people can require several screening questions to ensure minimal misclassification errors, and if those errors occur they can later yield serious noncoverage problems (such as false negative reports). The screening approach to sampling mobile-only populations can be very expensive in countries where the prevalence of mobile-only population is low (e.g., 5 percent in Sweden in 2004). The use of multiple partial frames, such as different commercial lists that contain information on mobile-only telephone numbers, can be applied in the formation of a mobile-only frame. Sometimes, population members can be represented on several of the frames—in such cases compensation for overlaps has to be employed at the analysis stage (see Hartley, 1962,1974) or the frames have to be redefined (see Hansen et al., 1953).

Rejection rule sampling can be employed in countries where there are reasons to believe that mobile-only customers are clustered within groups of consecutive numbers. This procedure is analogous to the Mitofsky–Waksberg sampling approach (Waksberg, 1978), where randomly selected banks of 100 consecutive numbers are screened for working residential numbers.

Volunteer panels can also be used to create a frame for the mobile-only population. Such panels are typically mail panels and are recruited to respond periodically to surveys on a variety of topics. Because of the large number of panel members (usually, in thousands), such panels can be used to screen for mobile-only people. A serious problem with panels, however, is the low cooperation rate at the recruitment stage that can introduce unknown selection biases to later survey results (Sudman and Kalton, 1986).

Many countries in Europe have a large percentage of mobile-only persons. Data from 23 countries reported by Gallup Europe in 2004 show that the largest percentage is in Lithuania, where 52 percent of the population is mobile only (Gallup Europe, 2005). Sweden has the lowest percentage of the mobile-only population in Europe. The median percentage of mobile-only persons for these 23 countries, though, is 23 percent.

6.2.1 Multiplicity Sampling

Multiplicity, or network, sampling is frequently an option to supplement a frame that does not directly cover some members of the target population. With multiplicity designs in a base household sample, information is collected not only about the selected household members, but also about other people *linked to* the household who can be enumerated in a well-defined manner (e.g., blood relative linkage, adjacent address linkage used in surveys of ethnic minorities).

In order to determine the selection probabilities of sample members and weight inversely proportional to these probabilities in analysis, the linkage to a sampled frame element has to be clearly defined. Informants must know about the linkage and whether those linked to them have the rare trait of interest (this is the minimum required to estimate the rare trait prevalence). The informants must be willing to provide the information they possess and also provide access to the information about the network members.

In some ways, multiplicity sampling resembles other nomination sampling techniques such as snowball sampling and respondent-driven sampling (Heckathorn, 1997). However, multiplicity sampling starts the process with a probability sample of nominators; snowball and respondent-driven samplings do not necessarily. Multiplicity sampling asks the sampled person to identify the members of a well-defined network; snowball sampling and respondent-driven sampling typically ask for "friends," "acquaintances," and other relationships that can be asymmetrical and ill defined. Multiplicity sampling uses the counts of network members as a weight in estimation.

A common network used in multiplicity sampling is a family network. To apply this to the mobile-only population in the United States, an RDD sampling frame might be used, with sampled persons being asked to identify all living parents, siblings, and children (a three-generational network) who do not have residential landline telephones but do have mobile telephones. Each of those mobile-only cases could be reported by every member in that family network. Thus, the probability of the mobile-only person coming into the sample is a function of the number of different households and persons who have landline residential telephones in their network. The multiplicity approach is imperfect in practice. The multiplicity approach does not offer coverage of the mobile-only population whose networks live only in nontelephone households, who have no living family members or whose family members are outside the household population (e.g., in long-term care institutions, prisons, or are homeless).

6.2.2 Challenges of Multiplicity Sampling

Multiplicity sampling faces challenges in implementation. First, a choice must be made of the target population for the initial survey of informants about their networks. At this step the critical question is to identify an easily sampled population that is likely to be most fully linked to the rare population. For the mobile-only population, the frame used by traditional telephone surveys is likely to offer acceptable coverage of the telephone household population. If the telephone household population is well linked to the mobile-only population, a telephone list may be a suitable base frame.

Second, the researcher must choose the network definition. This decision balances two questions—what is the variation in size of the alternative network definitions and what is the impact on the measurement properties of the network reporting?

The larger the variation in size of the network, the larger the inflation of standard errors of the multiplicity estimates (other things being equal). Further, large networks require the respondent to identify and classify a larger number of persons. This can be cognitively burdensome when the network is not salient to the reporter or concerns about breaches of privacy restrict reporting (Czaja et al., 1986). Network reporters may know that the person is mobile only but may not know the telephone number. Network members may know the number but be concerned that reporting it to a survey interviewer may breach the privacy of their relative.

Third, once the mobile number is reported, the use of the number for a telephone interview faces all the existing challenges of interviewing persons on mobile telephones (Steeh, 2004a, 2005a).

Fourth, once the mobile-only respondent data record is collected, the multiplicity estimates must be constructed. It is likely that there will be a variety of forms of missing data in the counts of network members and their mobile status that complicate the construction of multiplicity weights.

6.3 QUESTIONS TO BE ADDRESSED IN MULTIPLICITY SAMPLING FOR THE MOBILE-ONLY POPULATION IN THE UNITED STATES

This chapter addresses the following questions in the use of multiplicity sampling in the mobile-only setting.

(1) Do subjects know the mobile status of family network members?

(2) Do they know the mobile telephone numbers of mobile-only network members?

(3) Do subjects express willingness to report these numbers?

(4) Do they report the numbers?

(5) Are contact and cooperation rates for mobile-only and RDD comparable?

6.4 A TRIAL OF MULTIPLICITY SAMPLING FOR THE U.S. MOBILE-ONLY POPULATION

6.4.1 Overview

This chapter reports a trial of reporting on family network's mobile-only status using a base sample survey on a recruited panel of respondents. As shown in Fig. 6.1, the design involves three sequential surveys. The first survey, Wave 1, uses an empaneled group of adults to measure family networks. Wave 2 remeasures the same panel to obtain mobile telephone numbers on those network members who do not possess landline telephones. The third survey is an experimental comparison of an RDD mobile telephone survey and a survey of nominated mobile-only network members.

The first survey is the Gallup Poll Panel, an RDD-recruited sample of about 15,000 households within the continental United States or a total of 27,000 individuals of

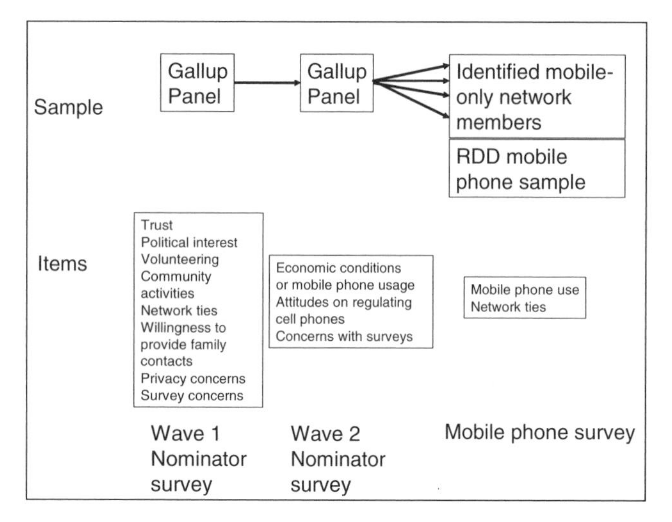

Figure 6.1. Schematic of three survey design using the Gallup Panel to generate a mobile telephone only adult sample and an RDD frame to generate a mobile phone only sample.

13 years of age or older. This panel is a discontinuous panel (Sudman and Wansink, 2002) in that members are surveyed on a variety of opinion and market research issues on average twice a month. Typically, the interviews take between 5 and 15 minutes and vary in mode of data collection—telephone, interactive voice recognition (IVR), mail, or web. Web interviews are done only with frequent web users, that is, those that typically use the Internet twice a week.

The panel is used for inference to the target population of telephone households in the United States. Response rates among the initial recruitment sample are low by traditional standards, approximately 27 percent, and further nonresponse exists for each requested survey of the panel. It is likely that panel members are more cooperative with external requests in general, and more actively interested in expressing their opinions on national and world affairs. This limitation probably produces a weaker test of the propensity to comply with the network reporting requests in the multiplicity questions. It is hoped that these subgroup comparisons are less subject to nonobservation biases. Further, a supplement sample of terminated panel members, who exhibited high nonresponse rates, is included.

Choice of informant and context. All active panel members within 14,080 sampled households in the continental United States were selected to be informants in the multiplicity sampling approach to select mobile telephone only populations. In order to determine the selection probabilities of the respondents' networks and weight inversely proportional to these probabilities in the follow-up analyses, the linkage to each network member has to be clearly defined. In addition, informants must be able to provide information on the linkage, know whether those linked to them are mobile-only persons, be willing to provide the information they possess, and thereby allow survey researchers access to the network members.

To determine the willingness to report information on network member, the context of the request was varied: in one case the requestor was the Gallup Poll and in another a survey organization. It was hypothesized that the Gallup Poll will stimulate higher reports due to familiarity and trust in the organization.

Choice of network definition. The burdensome nature of the informant's task (reporting a mobile telephone number from memory), cost of measurement, and weight variation concerns constrained the size of the network to immediate relatives: living parents, siblings, and children of 18 years of age and older who live in a mobile-only household. Eligible network members included biological and stepparents; biological, adopted, step- and half-brothers and sisters; and biological, adopted, and stepchildren.

Sample. The sample of active members consisted of 14,080 households with 20,201 active members of 18 years of age and older. The decision to sample all active panelists within a selected household was based on the expected within-household variation on network knowledge. A sample of "nonrespondents" consisting of 2635 former panelists of 18 years of age and older, who were nonrespondents in the last six Gallup Poll surveys, was also selected. To increase the response rate among inactive members, the sampled nonrespondents received a $1 cash incentive along with a mailed survey. Cost and privacy concerns determined the choice of two self-administered modes of data collection—mail and web. In order to compare the multiplicity of these mobile telephone surveys, an RDD sample of 9368 numbers was added to the research design.

6.4.2 Study Measurement Design

Wave 1 measurement. The Wave 1 measurement of knowledge and attitudes toward network reporting began in June 2005 and ended in August 2005. A total of 22,836 sampled persons were sent a Social and Family Networks Survey (12,667 active and inactive members of the panel were mailed the questionnaire, whereas 10,169 active and inactive members received an e-mail invitation to fill in a web survey). The Social and Family Networks questionnaire consisted of 13 questions in the standard mail Gallup Poll four-page format. The questions measured political and community involvement, privacy concerns, opinions about surveys in general, interest in the Social and Family Networks Survey, and difficulty of filling out the survey. The section on family networks was the focus of the questionnaire. A question about willingness to reveal mailing, e-mail address, or mobile telephone number of family network members for survey purposes was placed immediately after the grid.

In Wave 1, a split ballot design was employed, and in one version of the Social and Family Networks questionnaire, the respondents were asked about willingness to report this information to the Gallup Poll Panel, whereas in the other version they were asked to reveal willingness to report to a survey organization. The hypothesis was that respondents would express greater willingness to reveal information on mailing address, e-mail address, or mobile telephone number to the Gallup Poll Panel due to loyalty and trust. All inactive members received the Gallup Poll Panel version of the questionnaire.

About 52 percent of the panel members responded in the first wave, with negligible differences in response rates between the web requests and the mail requests. There were some differences in the backgrounds of respondents and nonrespondents. For example, respondents tend to be more active panel members (by about 20 percentage points versus the full sample), to be retired (by about 10 percentage points), to be the person in the household who makes grocery decisions (by about 9 percentage points), and to have voted in the last presidential election (by about 8 percentage points). However, most of the differences in other variables were small.

Wave 2 measurement. The Wave 2 enumeration of the network mobile-only status and telephone number began in September 2005 and ended in October 2005. The questionnaire was sent to all Gallup Panel Poll members, 50 percent of whom were respondents to Wave 1. In the period between June 2005 and September 2005, all participants in the measurement of knowledge and attitudes toward network reporting were also exposed to four other Gallup Poll surveys.

Mobile telephone survey. The purpose of Wave 2 of data collection was to collect mobile telephone numbers of family network members who live in a different household and do not have a landline telephone. In addition to the mobile telephone roster in Wave 2, there were opinion questions about mobile telephone usage while driving, questions about knowledge and attitudes toward network reporting questionnaire, and opinion questions about surveys. Approximately 7000 questionnaires were mailed, and 7000 respondents received an invitation for a web survey.

The Wave 2 enumeration of the network mobile-only status and telephone number began in September 2005 and ended in October 2005. The questionnaire was sent to all Gallup Panel Poll members, 50 percent of whom were respondents to Wave 1. In the period between June 2005 and September 2005, all participants in the measurement of knowledge and attitudes toward network reporting were also exposed to four other Gallup Poll surveys.

Data were collected from a sample of the mobile-only cases reported by Wave 2 respondents. Of the 1757 respondents to the second wave of data collection, who were also respondents in the first wave, 200 respondents provided at least one mobile telephone number of a family network member who lives in a household without a landline telephone. Of the 1285 Gallup Panel Poll respondents who did not participate in the first wave of data collection, but took part in Wave 2, 169 respondents provided at least one mobile telephone number of a family network member who lives in a household without a landline telephone. There were respondents who provided mobile phone numbers for more than one network member. In total, 524 distinct mobile telephone numbers were obtained.

The mobile phone survey took place in November 2005. Parallel to the mobile phone only sample, an RDD mobile telephone sample of 568 respondents (obtained through the assistance of Survey Sampling, Inc.) was interviewed using the same instrument. The telephone survey required approximately 5 min of interviewing time. Potential respondents were offered an incentive of $2. The interviewers noted immediately that they were aware that they were calling a mobile phone number. In compliance with current Federal government's mobile phone calling regulations, all mobile telephone numbers were dialed manually. The focus of comparison between the RDD mobile telephone and the multiplicity mobile phone survey was contact and cooperation rates.

6.5 RESEARCH RESULTS

6.5.1 Do People Know the Mobile Telephone Status of Family Members?

Even if respondents were found willing to provide information about their family members, we did not know the extent of their knowledge about mobile telephone ownership of their family members. Among persons filling out a network grid, only 2–3 percent of the family members were missing a report on whether they were mobile only. The missing data rate is highest (10 percent) for those with only one family network member in the grid.

Table 6.1 also permits a comparison to other benchmark estimates of the percentage of adults who do not have landline telephones but do have a mobile telephone. Blumberg et al. (2007), Chapter 3 in this volume, using data based on the face-to-face National Health Interview Survey, showed that the 2005 percentage of adults with no landline but a mobile telephone was between 6.8 and 7.8 percent, depending on month of the year. Table 6.1, for all network sizes, shows higher rates of mobile-only persons. There are several possible explanations for this, none of which can be directly tested: (a) the Gallup Panel overrepresents persons in mobile-only networks; (b) isolates (those persons who are not connected to a network) have much lower mobile-only prevalence; (c) Gallup Panel members overreport mobile telephone ownership of network members. We were able to examine partially this latter explanation in the mobile telephone follow-up survey (see below).

Table 6.1. Percent Distribution of Telephone Status by Number of Network Family Members in Wave 1

Network size	Mobile only	Landline telephone only	Both mobile and landline	No telephone	Missing	Total
1 Member	18.0	21.1	47.8	3.2	9.9	100.0
2 Members	15.0	22.0	57.2	3.6	2.2	100.0
3 Members	13.8	22.1	60.0	3.0	1.1	100.0
4 Members	11.3	21.5	62.1	2.5	2.6	100.0
5 Members	11.4	24.9	57.9	3.6	2.1	100.0
6 Members	9.6	25.4	56.5	4.2	4.3	100.0
7 Members	10.2	26.6	57.7	3.5	2.0	100.0
Total	12.2	23.5	58.4	3.3	2.6	100.0

Case base: Children, siblings, and parents listed by respondents, weighted by inverse of number of family members with a landline telephone or landline and mobile telephone multiplied by the Gallup Panel poststratification weight (adjusting the sample to population margins for number of adults, number of telephone landlines, census region, age and gender, and education).

6.5.2 Do People Know the Telephone Number of Mobile-Only Family Members?

For each network member reported as "mobile-only," the Gallup Panel member was asked if she/he knew the telephone number of the family members. Overall, 49.6 percent of the mobile-only nominees' telephone numbers were reported to be known from memory. In order to learn something about the factors related to this type of knowledge, knowing the mobile-only telephone number was regressed on other variables reported by the Gallup Panel member. In order to be able to retrieve the probabilities of selection and, thus, create a multiplicity weight, the model is fit only on siblings listed in the grids of all Gallup Panel respondents, with standard errors reflecting the clustering of mentions within Gallup Panel respondents within households.

In a multivariate model, respondents tend to know the mobile telephone numbers of relatives who live in the same state, whom they contact more frequently, whom they visit or call by telephone, whose mailing address they know, or whose e-mail address they know (when they have one). In summary, mobile telephone numbers are known in more integrated family networks, the ones who enjoy more frequent close contacts (Table 6.2).

Table 6.2. Logistic Regression Coefficients for Likelihood of Knowledge of Mobile Telephone Number "by Heart" Among Listed Siblings

Predictor	Coefficient (SE)
Intercept	1.80(0.149)**
Sibling does not live in the same state	−0.60 (0.0841)**
frequency of contact	
Less than once-a-year contact	−2.50 (0.420)**
At least once-a-year contact	−2.05 (0.147)**
At least monthly contact	−1.07 (0.0769)**
At least weekly	—
Most frequent method of contact	
Visit	−0.076 (0.0907)
E-mail	−0.88(0.136)**
Mail	−1.28 (0.450)**
Telephone	—
R does not know sibling's mailing address by heart	−0.78 (0.0793)**
R does not know sibling's e-mail address by heart	−1.18 (0.0864)**
Sibling does not have e-mail address	−0.63 (0.108)**
Sibling has mobile and landline telephone	−0.86 (0.118)**
Mail mode of data collection	0.062 (0.0814)

** $p < 0.01$, * $p < 0.05$.

Case base: Siblings listed, weighted by inverse of number of siblings multiplied by the Gallup Panel poststratification weight (adjusting the sample to population margins for number of adults, number of telephone landlines, census region, age and gender, and education).

6.5.3 Do People Express a Willingness to Reveal the Mobile Telephone Numbers of Family Members?

After the respondents were asked to provide information about their family network, a behavioral intention question was posed:

> If you were asked by (a survey organization/Gallup) to provide the mailing address, e-mail address, telephone number, or mobile telephone number of the family members listed in the previous questions, so they can be contacted for a survey, would you be willing to do so?

Regardless of wording, only a small fraction of persons expressed their willingness to provide such information about all their family members, less than 8 percent in both forms. Cross-tabulations indicated that lower socioeconomic level and lower concerns about self-protection predicted an increased willingness to provide contact information. There was no apparent difference in acceptance rates in mail or Web mode of data collection (data not shown). There is, however, a tendency for hypothetical requests from Gallup to garner more acceptance than from another organization (23.1 percent versus 14.7 percent willing to provide some information). We expect that this difference would be relatively large for fixed panel operations like the Gallup Panel because of the required respondent commitments to the sponsoring organization.

Two multinomial logistic regression models for the likelihood of willingness to provide family network member information (not presented) were fitted to the data. The first included a set of attitudinal predictors based on questions that preceded the question about willingness in the survey. These items included a variety of civic engagement attributes (e.g., voting, interest in politics, volunteering, and participation in community activities). The second model had an additional set of predictors that followed the willingness question in the survey instrument on concerns about privacy and confidentiality of survey data and concerns about surveys themselves. The models revealed that those willing to provide information on family members include females and those interested in politics. Those who administered the survey in the mail mode were more likely to express willingness to provide contact information. The second model suggested that those who worried about privacy protection were unlikely to provide information about family members.

Thus, attitudes about protecting information about oneself and possible abuses in surveys are relevant to reactions to request for family contact information.

6.5.4 When Asked, Do People Report the Mobile Telephone Numbers of Family Network Members?

The second wave of the survey again asked Gallup Panel members to list their parent, sibling, and adult child network, but this time only those who used mobile telephones exclusively. For those network members who had mobile telephones but did not have

A	B	C	D
First name/ nickname (please print)	Lives in same state as you	Relationship to you	Cell phone number (please include area code)
SAMPLE JANE	☒ Yes ☐ No	☒ Parent ☐ Sibling ☐ Child	(123) 456 – 7890 ☐ Don't know ☐ Not willing to provide
	☐ Yes ☐ No	☐ Parent ☐ Sibling ☐ Child	( ) – ☐ Don't know ☐ Not willing to provide

Figure 6.2. Image of network reporting grid.

residential landline numbers, the panel member was asked to report the mobile telephone number, using the format in Fig. 6.2.

Overall, 10 percent of respondents reported at least one mobile telephone number; 78 percent at least once checked "Not willing to provide number"; and 7 percent at least once checked "don't know mobile phone number." About 5 percent left all mobile telephone numbers blank.

These results support the perspective that many panel respondents are unwilling to reveal known mobile telephone numbers for network members (a fact learned in Wave 1). The 78 percent who checked at least once that they were unwilling to reveal a number could be compared to the 92 percent who said they were unwilling to provide contact information to Gallup for some or all network members. The difference reflects the gap between the behavioral intention to a hypothetical question and the behavior on the direct question itself.

Further analysis showed that some groups are more likely to refuse at least one number: those with higher incomes, those using the Internet, and those employed full time. Conversely, some groups are more likely to provide at least one number for a mobile-only relative: blacks, those with lower incomes, retirees, and those who do not use the Internet. Interestingly, there are no real differences in behaviors among more or less active panel members.

Table 6.3 shows the results of two logistic regression models predicting the likelihood of providing at least one mobile telephone number among those persons with eligible network members. The first model regresses an indicator for providing at least one number on variables known from the panel dataset and Wave 1 measures. The second adds the Wave 1 variables measuring the behavioral intention to provide contact information, if asked. To simplify the model, we used income as the key predictor to incorporate the effects of race and Internet use.

Respondents that provided answers in the mail paper questionnaire mode are more likely to provide at least one mobile telephone number. This comports with the fact that those connected to the Internet are less likely to provide at least one number. Coefficients for the various dummy variables on employment show that those most unlikely to provide a number are students and those who did not respond to the

Table 6.3. Coefficients for Logistic Propensity Models for Providing at Least One Mobile Telephone Number

Parameter	Coefficient(standard error)		Coefficient (standard error)	
Intercept	−0.050	(0.699)	−1.053	(0.742)
Mail mode	1.00	(0.285)**	1.080	(0.283)**
Income				
Less than $25,000	—		—	
$25,000–$49,999	0.15	(0.454)	−0.0055	(0.458)
$50,000–$74,999	−0.067	(0.451)	−0.091	(0.446)
$75,000–$99,999	−0.76	(0.541)	−0.81	(0.534)
$100,000 and above	−0.20	(0.4871)	−0.54	(0.497)
Don't know	0.47	(0.830)	0.69	(0.846)
Refused	−0.93	(0.813)	−0.94	(0.814)
Employment				
Employed part-time but not a full-time student	0.27	(0.510)	0.30	(0.482)
A full-time student	−15.062	(0.487)**	−14.72	(0.519)**
Retired	0.036	(0.329)	−0.087	(0.357)
Homemaker	−0.51	(0.524)	−0.43	(0.502)
Not employed	0.12	(0.694)	−0.013	(0.670)
Don't know	−15.22	(0.966)**	−14.76	(0.864)**
Voted in the last presidential election				
Yes	—		—	
No	1.036	(0.574)	1.085	(0.568)
Don't know	−13.61	(0.896)**	−14.51	(0.902)**
Privacy scale	−0.11	(0.0992)	−0.091	(0.106)
Attitudes toward surveys scale	−0.83	(0.271)**	−0.59	(0.275)*
Willingness to provide information on network members			1.27	(0.279)**

$^{**}p < 0.01$, $^{*}p < 0.05$.

Case base: Respondents to Wave 1 and Wave 2, weighted by Gallup Panel poststratification weight (adjusting the sample to population margins for number of adults, number of telephone landlines, census region, age and gender, and education).

employment question. Expressed concerns about surveys predict not providing at least one number, overwhelming any effects of privacy concerns.

When the response to the question about willingness to give contact information is added to the model, it has a large positive effect on whether at least one number was provided, as expected. However, the entry of the variable into the equation has no major effect on the strength of other predictors previously entered in the first model. Thus, in future efforts of this nature, which may be one-wave efforts to obtain mobile telephone numbers, the employment and concerns about surveys are useful predictors of the likelihood of obtaining numbers.

6.5.5 How Do Contact and Cooperation Rates of Multiplicity Sample Mobile Telephone Cases Compare to RDD Mobile Telephone Cases?

The final step in the data collection was a telephone survey of network members whose mobile telephone numbers were provided by panel members. Each of these numbers was reported as being used by mobile-only network members. As a benchmark sample, we chose a set of RDD numbers from mobile-only banks, using the Survey Sampling Incorporated (SSI) service. We note that many of the RDD numbers were expected to be used by persons who had both landline telephones and mobile telephones.

An elite group of Gallup telephone facility interviewers, using hand-dialing methods to comply with regulations of the Federal Trade Commission, dialed the sample numbers. Interviewers were aware of which numbers were RDD and which were multiplicity sample numbers. Indeed, in answer to the question "How did you get my number?" they were instructed to report the network nomination technique for the multiplicity sample. Dialing on both sets of numbers began without any prior notification to the subscribers.

The multiplicity sample generated higher response and cooperation rates using various AAPOR standard definitions than the rates generated by the RDD sample (Table 6.4). Using the most conservative AAPOR RR1 rate, the multiplicity sample generated a 36.4 percent rate versus a 9.7 percent response rate from the RDD sample. Much of that difference results from the large number of noncontacts in the RDD sample. The AAPOR Coop 1 rates are 57.9 percent rate for the multiplicity sample and 23.5 percent for the RDD.

Only 27 percent of the RDD respondents were mobile only. Although this low rate was expected, there was an unexpectedly large proportion of multiplicity sample respondents who reported that they also had a landline telephone—about 40 percent. This means that the RDD yielded 27 percent mobile-only respondents and the multiplicity sample about 60 percent. It appears that the network reports of mobile-only status overreported that status since many of the multiplicity sample cases were both mobile and landline subscribers. This false-positive reporting is a measurement error that needs further investigation if multiplicity techniques are to be used.

When we examine other reported characteristics of the respondent pool, we find results that are similar to other early findings of mobile telephone surveys. The mobile-only respondent pool on both RDD and multiplicity samples are overwhelmingly male (59 percent for both samples); young (mean age between 30 and 32 years); and

Table 6.4. Response Rates of Mobile Telephone Survey by Sample Type

AAPOR response and cooperation rates	RDD sample,[a] %	Multiplicity sample, %
RR1	9.7	36.4
RR3	10.5	37.3
Coop Rate 1	23.5	57.9
Coop Rate 3	25.1	60.6

[a]Includes both mobile only and other persons.

using the telephone for personal use only (between 60 and 66 percent). The RDD mobile-only cases report more sharing of the telephone among household members (16 percent) than the multiplicity sample respondents (7–8 percent).

6.6 SUMMARY

The research reported in this chapter was a necessary first step in addressing the issues of sampling mobile-only persons in diverse settings around the world. The work was motivated by the fact that many countries currently are not assembling telephone sampling frames that contain all possible numbers, including mobile and landline numbers. In the absence of a frame that includes mobile-only persons, the coverage of that population is declining as landline telephones decline in prevalence.

The chapter reports on a technique of frame supplementation—using an existing sample, originally drawn from an RDD frame (i.e., the Gallup Panel)—to identify family network members who are mobile only. The research came to clear conclusions.

(1) Knowledge of mobile telephone subscription status of family members is a function of cohesiveness of the network.
(2) There is large-scale unwillingness to provide a survey organization with mobile telephone numbers of family members.
(3) Those providing the numbers tend to be lower socioeconomic groups and those without concerns regarding surveys.
(4) Few respondents to the survey actually provide the mobile telephone numbers of their family members.
(5) The mobile-only numbers contain many mobile–landline persons, despite the attempt to obtain mobile-only persons.
(6) However, the cooperation rate of persons identified through multiplicity sampling techniques is higher than that from an RDD survey.

Given the limited reporting of mobile telephone numbers, it was not possible to address the estimation issues that arise in multiplicity sampling techniques. In all multiplicity samples, clusters of family networks are selected and an inflation of standard errors of estimates occurs. The same is expected to be true in this application of multiplicity sampling methods.

The findings do not suggest a package of tools that the survey researcher can successfully use to overcome the lack of sampling frames for mobile telephone numbers. The problem that motivated the work remains unsolved in many countries

There are a few steps to test in detail the multiplicity approach: using an interviewer-assisted mode may increase cooperation with the request for mobile-only numbers; reducing the network size (e.g., by using only siblings and children) could reduce burden on nominators; and probing the mobile-only reports might reduce the overreports of mobile-only status.

Commercial frames of mobile-only persons are likely to be inexpensive, but they have unclear coverage properties. The uncertainty arises because diverse sources of numbers are used to assemble most commercial mobile-only telephone lists (e.g., product warranty cards, credit applications, and mail-in entries).

Face-to-face surveys generally use sampling frames (e.g., addresses or area units) that offer complete coverage of mobile-only persons who live in households. In mixed-mode surveys (using both face-to-face and telephone modes), coverage of the mobile-only population can readily be attained. Alternatively, a face-to-face survey might be used to generate a sample of mobile-only persons for later study (see Flores Cervantes and Kalton (2007), Chapter 5 in this volume, for other ideas).

Solving the sampling frame problem for the mobile-only population is a key need for future prospects of mobile telephone surveys in many countries of the world.

CHAPTER 7

Multiple Mode and Frame Telephone Surveys

J. Michael Brick
Westat, and the Joint Program in Survey Methodology, University of Maryland, USA

James M. Lepkowski
University of Michigan, and the Joint Program in Survey Methodology, University of Maryland, USA

7.1 INTRODUCTION

As response rates and coverage rates decrease in telephone surveys, particularly random digit dialing (RDD) telephone surveys, survey researchers are searching for alternative designs that might reverse the decline in rates. An area probability sample with face-to-face interviewing, while it would likely increase these rates, is not a viable option for many of these surveys because of the data collection cost. A mail survey is an alternative established methodology, and the availability of improved address files for the United States has generated renewed interest in this approach. But many RDD surveys have complexities such as questionnaire branching and selection criteria that cannot be incorporated easily into mail surveys, and the selection of a respondent within a cluster such as a household is largely uncontrolled. The web could handle some of these technical complexities but not some practical ones such as randomly choosing a respondent. The web is also much weaker with respect to nonresponse and noncoverage bias than RDD surveys.

As a result, designs that interview the respondents in more than one mode (multiple-mode surveys) and those that select samples from more than one sampling frame

Advances in Telephone Survey Methodology, Edited by James M. Lepkowski, Clyde Tucker, J. Michael Brick, Edith de Leeuw, Lilli Japec, Paul J. Lavrakas, Michael W. Link, and Roberta L. Sangster

(multiple-frame surveys) are attracting more attention. For example, the American Community Survey (ACS) is a replacement for the long form of the U.S. Census that began scaling to full implementation in 2005. The ACS uses three modes, starting with a mailing to sampled households and following up nonrespondents by telephone and face-to-face interviews (U.S. Census Bureau, 2006). The Canadian Community Health Survey (CCHS) conducted by Statistics Canada (2003) uses both multiple frames and multiple modes. Three frames in the CCHS are an area frame, a list frame, and an RDD frame. In addition, two modes, telephone and face-to-face interviews, are used to collect data.

This chapter examines the error properties for multiple-mode and multiple-frame household surveys. Household surveys are those that collect data for the entire household or for persons residing in the household as contrasted with establishment surveys that collect data about characteristics of units such as businesses, schools, and hospitals. The respondents for household surveys may be sampled persons or proxy respondents, such as mothers of sampled children living in the household. Only household surveys that have the telephone as either a mode or a frame are considered here, but since the use of the telephone mode is very common in current multiple mode and frame designs this includes a large number of household surveys. On the other hand, there are many multiple mode or frame establishment surveys that do not use the telephone.

Since much of the development and application of multiple mode and frame household surveys has been done in the United States, many of the applications are drawn from this literature. However, we have attempted to include examples from other countries to the extent possible.

The next section describes costs of data collection and some of the compromises between costs and errors that must be examined when designing surveys. Sections 7.3 and 7.4 review the error structures and potential biases, respectively, in multiple-mode and multiple-frame surveys. The fifth section examines methods available for estimating these biases and making inferences from surveys when the estimates are biased. The final section discusses the role of the telephone in the future for multiple mode and frame surveys of households.

7.2 BALANCING SURVEY COSTS AND ERRORS

Whenever a survey is being contemplated to provide information on a topic, the initial design decisions focus on both the ability to collect data that appropriately measures the subject and the costs associated with collecting these data. If either the survey costs are too high or the measurement methods too inaccurate, designers might decide that a particular survey design, or even the survey process itself, is not appropriate. For many applications, the design process involves a balancing of the costs and measurement process. The goal is to design a survey that obtains data of sufficient accuracy to meet critical information needs while minimizing the cost of data collection. Survey designers have choices of mode, frame, sampling plan, interviewing techniques, respondent persuasion techniques, and other dimensions.

The options in each dimension have varying costs, and the choice process involves trading cost against inaccuracy or error across multiple dimensions. Survey modes and frames are major components in this balancing of costs and errors in design.

The literature on multiple mode and frame surveys explicitly considers many aspects of measurement error, even if the estimation of the errors is not very specific. In subsequent sections these error properties are reviewed and summarized. Relative costs of data collection, however, are infrequently addressed in the same literature despite their importance. This section reviews the little that is known about costs in the conduct of multiple mode and frame surveys.

One reason the telephone is a component of most multiple-mode or frame-household surveys is that the data collection by telephone generally is less expensive than any other interviewer-administered method. This is very important because the funding allocated for a survey exerts a great influence on the modes and frames choices examined in the design phase. Groves (1989, p. 49) states that "In practice many survey designs are fitted within cost constraints." If the funding level is limited, designs that use more expensive data collection modes or frames may not even be considered.

Despite its practical importance, costs are typically addressed only as a sample design problem, and simplistic cost models are usually posited. Casady and Lepkowski (1999) examine cost models for telephone sample designs but do not address dual mode or frame designs. Two examples of cost structures for dual mode or frame surveys are Lepkowski and Groves (1986), who give cost models for the National Crime Survey, and Elliot et al. (2000), who model costs in the ACS. In both of these examples, the authors look at alternatives for already existing surveys and do not report on the cost models used in the original designs. Dillman (1978, pp. 68–72) discusses costs in an examination of the advantages and disadvantages of mail, telephone, and face-to-face surveys. While Dillman wrote this nearly 30 years ago, it is still one of the most pertinent discussions of survey data collection costs available. The result of having relatively less complete cost data is that we are limited in our ability to give specifics on costs and differences in costs by mode and frame.

Rather than being driven by cost, multiple mode or frame designs are often chosen to reduce some sources of error. The main reason the ACS includes both telephone and face-to-face interviewing modes is to reduce nonresponse error. The more expensive face-to-face interviewing is restricted to a subsample to reduce overall costs. The CCHS multiple frame and mode design enables a larger sample size, decreases sampling error, and reduces nonresponse and noncoverage errors. In other instances, the design is chosen explicitly to reduce a particular source of error. The National Survey of America's Families (NSAF) chose a dual-frame design: an RDD sample from a telephone number frame to increase sample size and reduce sampling error by reducing per unit cost, and an area probability sample to include nontelephone households to reduce noncoverage error.

Multiple modes are frequently employed to address concerns associated with nonresponse errors or measurement errors. One of the most useful techniques for increasing response rates is to switch modes, starting with the least inexpensive mode, and adding more expensive modes in the follow-up of nonrespondents as in the ACS.

While switching modes in this fashion increases response rates, offering the respondent the choice of modes does not appear to increase rates (e.g., Dillman et al., 1994; Schneider et al., 2005).

Measurement errors have a more complex effect on multiple-mode surveys. Research has shown that responses may differ across modes (see Groves and Kahn 1978). For example, mode effects associated with social desirability, acquiescence, and primacy/recency have been observed (Tourangeau et al., 2000, pp. 289–312). These may be related to the presentation mode (aural or visual) or the presence of an interviewer. Thus, the decision on which modes to use in the same survey involves the evaluation of the measurement error implications.

The goals that lead to the use of multiple-frame surveys are typically related to reducing noncoverage errors and sampling errors. In the NSAF, if a single frame (telephone numbers) excludes an important component of the target population (nontelephone households), introducing another sampling frame to cover that component of the population can reduce errors. Alternatively, a standard RDD survey in the United States uses a frame that excludes mobile telephone numbers and mobile-only households (see Blumberg et al., 2007, Chapter 3 in this volume). It can be supplemented by sampling from a frame of mobile telephone numbers (see Brick et al., 2007).

To reduce sampling error, multiple frames mix less expensive data collection from one frame with more expensive data collection from another frame (introduced to reduce the noncoverage of the first frame). The overall per unit cost in data collection is reduced, and more completed interviews can be obtained for the same total cost. For example, multiple-frame surveys are used to estimate characteristics for rare populations (see Flores-Cervantes and Kalton, 2007, Chapter 5 in this volume). Edwards et al. (2002) discuss sampling to estimate health characteristics for certain ethnic groups using a dual-frame design that adds a special list containing households likely to be in the group to the regular RDD frame.

While multiple-mode and multiple-frame surveys have important benefits, there are also disadvantages. One concern is that a multiple mode and frame survey may cost more than a single mode and frame survey. There is the complexity of administering separate modes and drawing sample from different frames in the same survey. Dillman (2007, pp. 219–244) lays out the objectives and potential error consequences associated with using multiple modes in the same survey. These concerns are examined in the next section along with the biases that may result from using multiple modes.

7.3 MULTIPLE MODES

The "survey mode" is often confounded with other features of the interview process that cause differences and may be attributed to the mode inappropriately. For example, a survey conducted with face-to-face interviewing may be administered using paper and pencil, computer-assisted personal interview (CAPI), computer-assisted self-interview (CASI), or audio computer-assisted self-interview (ACASI) methods.

The error properties of the interviews conducted using these different methods vary, and it is very likely that differences for some items might be expected. Some researchers treat all of these as face-to-face interview mode, while others consider each alternative to be a distinct mode.

For simplicity, we consider just four modes: telephone, face-to-face, mail, and web. The first two of these modes are sometimes referred to as "personal interviewing modes," while the latter two are "self-administered." The telephone mode includes surveys conducted on landlines, mobile telephones, and by interactive voice response (IVR). As noted above, the effects of the mode on survey responses are highly dependent on other features of the interview process.

Tourangeau et al. (2000) list five important features of the mode that may result in different effects: (1) method of contact; (2) method of presentation (paper/electronic); (3) method of administration (self/interviewer); (4) method of questioning (aural/visual); and (5) method of responding (oral/written/electronic). Most research on mode effects deals with the last four of these features, and much less is known about the effect of the method of contact. This is true even though many surveys use one mode for contacting households and another mode for interviewing respondents. For example, case-control studies often use the telephone to identify controls (i.e., respondents with the same age and sex as the case that has the disease) and then interview the sample person face-to-face (e.g., Levine et al., 1999). In panel surveys, members are identified by one mode and interviewed using another, such as in the Dutch Telepanel described by Saris (1998), an early application of this approach. Knowledge Networks has more recently contacted panel members by RDD telephone methods and interviewed them by web (Huggins et al., 2002). Fricker et al. (2005) report a more complex example in which an RDD sample of households is contacted and those households with Internet access are randomly assigned to complete an interview by telephone or by web. They employed this design to be able to better evaluate "pure mode effects," those that are not confounded by other features of the data collection process.

7.3.1 Background on Mode Effects

The effect of the mode on the estimates has been the subject of a great deal of survey methodology research. de Leeuw (2005) provides a recent comprehensive review. This literature is summarized only briefly here.

Groves (1989, pp. 502–504) notes that mode effects are not simple to define or evaluate because surveys are bundles of methodologies and the mode is just one of the elements of the bundles. Some mode effects might be due to using a medium that is totally aural, but these effects are mediated by other factors such as nonresponse and noncoverage. The practical question of whether a survey conducted in different modes gives rise to the same responses as a given mode is different than the theoretical question of whether the mode "causes" the responses to differ. Various research methods have been used to "unbundle" mode effects.

A simple mode comparison is often motivated by the desire to reduce data collection costs, primarily those associated with face-to-face surveys, by substituting less expensive survey modes such as the telephone or the mail. The studies by Hochstim (1967) and Groves and Kahn (1979) are early examples. Hope (2005) more recently investigated the effect of replacing a face-to-face survey of crime victimization in Scotland with a telephone survey. Manfreda and Vehovar (2002) considered using the web as a newer and even less expensive mode of data collection. See Béland and St-Pierre (2007, Chapter 14 in this volume).

This type of mode research, motivated by a practical question, is less concerned with the theoretical determinants of mode differences than with whether different estimates will result if a less expensive mode is used. This type of research is especially appealing for repeated panel surveys. For example, in labor force surveys used to estimate the labor force status of the population at a particular time, the U.S. Department of Labor Statistics and U.S. Bureau of Census (2002) and the Office for National Statistics (2001) have examined whether the telephone could be used to collect data in subsequent waves of the panel after the initial waves of data are collected face-to-face. The main focus was the size difference in the cross-sectional estimates produced from the surveys rather than pure mode effects.

Researchers such as Bishop et al. (1988), Schwarz and Hippler (1991), and Tourangeau et al. (2000) have examined cognitive and psychological processes that could explain some observed mode differences. de Leeuw (1992) in her review classified the primary causes of mode effects as being due to media-related factors (norms related to the media), information transmission factors (visual or aural presentation and response methods), and interviewer effects. This avenue of research is especially useful in the planning stages of a survey because the goals of the research are to predict the conditions under which different types of mode differences are likely to occur.

Studies of specific mode effects show that the effects are less consistent than might be expected. For example, Krosnick and Alwin (1987) predicted that recency and primacy effects should differ by mode. However, Dillman et al., (1995) showed that the predicted mode effects were not observed consistently. Similarly, acquiescence and extreme response differences have been found inconsistently, despite theoretical expectations that these effects should be more pronounced in some modes. On the contrary, Christian et al. (2007, Chapter 12 in this volume) find considerable support for the expectation that telephone respondents more often affirm positive extreme categories.

Both the simple mode comparison and the more detailed determinant investigations can be used to understand the effect of mixing modes in the same survey. Dillman and Tarnai (1988) review extensively the early research in multiple-mode surveys and examine the importance of administrative issues in multiple-mode surveys.

Panel surveys frequently mix the mode of data collection over time (as noted above for the U.S. and the U.K. labor force surveys). The effect of change in modes over waves of a survey may have serious effects on estimating change across waves, even if cross-sectional estimates are not greatly affected. Dillman and Christian (2005) point out that when panel surveys mix modes, the effect of the change in mode may be confounded with changes in substantive characteristics such as labor force participation. Mixed modes may generate bias in an estimated change over time, a major problem in a panel survey.

Dillman and Christian (2005) also point out that mode effects sometimes arise because questions are designed differently for specific modes. Dillman (2007, pp. 447–461) discusses this in more detail, giving examples of mode-specific construction that may lead to mode effects. He also describes some of the strengths and weaknesses of single mode design principles in view of the need to reduce the undesirable effects in multiple-mode surveys.

7.3.2 Errors Related to Modes

The most obvious and consistent finding across the mode research literature is what de Leeuw (1992) categorizes as the interviewer effect. Respondents tend to give more socially desirable responses whenever interviews are administered by interviewers as compared to self-administered. This effect has been demonstrated repeatedly and consistently (see Sudman and Bradburn (1974), Presser (1990), Aquilino and Wright (1996), and Fowler et al. (1998)). This interviewer effect has often been interpreted as a mode effect because the telephone and face-to-face modes typically use interviewers to administer the data collection. As the use of computerized self-administration in both telephone and face-to-face modes becomes more common (e.g., see Turner et al., 1998), the effect may be viewed differently and be more appropriately linked to the role of the interviewer than the mode. Aquilino (1994) used a design that made it possible to isolate the effect of the mode from the role of the interviewer in an attempt to measure only mode effects. His findings showed some differences between face-to-face and telephone mode in response to sensitive items. He hypothesized that these differences were due to a greater social distance in the telephone than in the face-to-face mode and that it might be more difficult to assure respondents of confidentiality on the telephone. However, his findings also showed that the largest differences are between self-administered and interviewer-administered interviews, and mode differences were relatively small when the method of administration was controlled.

The intrinsic difference between visual and aural stimuli in the process of questioning and responding to the interview appears to be at the core of many mode differences. For example, Tourangeau and Smith (1996) report differences explained only by the presentation medium, even with sensitive questions. Redline and Dillman (2002) and Christian and Dillman (2004) provide persuasive evidence that visual design and layout have the power to influence responses. Efforts such as the unimode construction of Dillman (2007, pp. 229–244) may help to reduce the effects of the questioning and responding mediums, even though some differences remain unavoidable.

In summary, mode effects do exist, but they are difficult to isolate from other features of the survey process. The effects are also not always consistent with theoretical expectations, and they differ greatly by features of the interview process and by the item type. Common findings include the following:

(1) Small or negligible mode effects are typical for unambiguous, factual items that are not sensitive or socially desirable;
(2) Sensitive and socially desirable items are reported differently in self-administered and interviewer-administered interviews, and self-administered interviews typically yield more accurate answers;
(3) Ambiguously worded questions tend to be inconsistently reported in all modes, and context effects often interact with the mode for these items;
(4) Open-ended items tend to be reported in more detail in interviewer-mediated modes, although research on the Web with respect to open-ended items is still ongoing;
(5) Respondents on the telephone tend to report more positive extreme categories.

While differences may be present between modes in some circumstances, estimating the bias due to mode is very difficult. For example, sensitive items collected in self-administered interviews are likely to have less bias than those collected by an interviewer. Nevertheless, this does not mean the difference between the two is an estimate of bias since the self-administered item itself is not likely to be free of bias. Both modes may be subject to differential biases.

Further, survey estimates are subject to many sources of error, and it is difficult to isolate the effects from the multiple sources. Couper et al. (2005) describe a nonresponse telephone follow-up to a mail study that was designed to examine nonresponse bias. They were unable to disentangle the mode effects from the nonresponse bias, complicating their evaluation of nonresponse bias. Biemer (2001) examines both nonresponse and measurement errors associated with telephone and face-to-face interviewing. Both the measurement bias and the nonresponse bias are estimated from latent class models. He concludes that neither the telephone nor the face-to-face modes have consistently lower total bias, despite a difference of more than 20 percentage points between the face-to-face and telephone response rates. Voogt and Saris (2005) present similar findings for their study that attempts to balance the biases due to nonresponse and mode in multiple-mode surveys. Both of these latter studies rely heavily on model assumptions in order to partition components of survey error.

7.4 MULTIPLE FRAMES

Multiple-frame designs are used to reduce noncoverage error when no single frame includes all the units in the target population. Multiple-frame surveys are also used to reduce sampling errors by reducing per unit costs through sampling from frames that are less costly to survey, such as lists of people with rare characteristics, instead of screening the entire population. Flores-Cervantes and Kalton (2007, Chapter 5 in this volume) describe several applications of multiple-frame surveys for sampling rare populations.

The archetypical dual-frame survey is one that uses a complete frame (e.g., an area probability frame covering all households in the population) and an incomplete frame (e.g., a list of telephone numbers not covering households without telephones). The telephone frame might consist of a frame of landline numbers, mobile numbers, or both, or it might be a list of telephone numbers constructed to cover specific households or persons. The second frame could be an area probability frame of households, such as the addresses in the United States Postal Service delivery sequence file (DSF), or it might be e-mail address lists from the Web. We are unaware of any probability samples of households that have used a Web e-mail address list as a frame, but this may someday be possible.

Much of the multiple-frame literature covers statistical topics such as optimal allocation of the sample to the frames, methods of point estimation, and variance estimation. For this reason, a brief review of the theory of multiple frames is appropriate before presenting multiple-frame applications and issues that arise in these applications.

7.4.1 Multiple-Frame Theory

Two seminal papers by Hartley (1962, 1974) describe the basic statistical framework for multiple-frame surveys. In these papers, Hartley examined optimal allocation to the frames considering both cost and variance, and suggested methods for producing efficient, unbiased estimates using data collected from all the frames. The general dual-frame case is depicted in Fig. 7.1, where frames A and B overlap and some population units are contained in both A and B. The population units that are only on frame A are denoted as a, those only on B as b, and the overlap population as ab. For a characteristic Y, the population total is the sum of the totals for the three components, that is,

$$Y = Y_a + Y_b + Y_{ab}. \tag{7.1}$$

A key assumption, retained by virtually all multiple-frame research, is that the union of all the frames completely covers the population. In other words, multiple frames eliminate noncoverage. While some applications have noted that the union of the frames may not provide complete coverage (e.g., Traugott et al., 1987; Collins and Sykes, 1987), no estimation procedures or sampling methods have been proposed to account for the noncoverage in multiple-frame survey, even though weighting

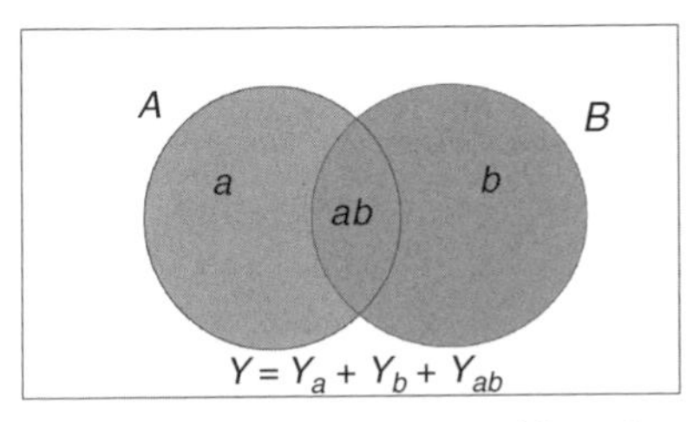

Figure 7.1. Dual-frame survey with overlap.

methods such as those discussed in Lee and Valiant (2007, Chapter 8 in this volume) do address the problem.

Following the lead of Hartley, subsequent research has studied issues associated with sample allocation or estimation. Groves and Lepkowski (1985) explore different administrative cost structures for dual-mode telephone and face-to-face dual-frame surveys, and the allocation under these structures. Biemer (1983) expands the error model by incorporating both response and coverage errors with a telephone frame. Lepkowski and Groves (1986) include three error sources (sampling error, interviewer error, and noncoverage error) and introduce a complex cost model.

Much of the research on the estimation methods for multiple frames (e.g., Lund 1968; Bankier 1986; Skinner 1991) focuses on estimating totals or means from simple random samples or stratified simple random samples. Both Bankier (1986) and Skinner (1991) examine the use of auxiliary data in raking to improve the efficiency of the estimates. Fuller and Burmeister (1972), Casady et al. (1981), and Skinner and Rao (1996) discuss estimation methods that are appropriate with complex sample designs. Variance estimation for estimates from complex multiple-frame samples, especially in combination with estimators such as raking, is examined by Skinner and Rao (1996) and Lohr and Rao (2000).

To illustrate the main ideas, we examine the case of two frames and the overlapping domains as defined in Fig. 7.1. The ideas can be extended to more than two frames (see Lohr and Rao, 2006). Suppose the samples from the two frames are independent, $\hat{y}_a$ is an unbiased estimator of Y_a, and $\hat{y}_b$ is an unbiased estimator of Y_b. Different estimates of the overlap total Y_{ab} are (1) the unbiased estimator based on the units selected from frame A that are in the overlap $\hat{y}'_{ab}$; (2) the unbiased estimator based on the units selected from frame B that are in the overlap $\hat{y}''_{ab}$; and (3) a composite of these two estimators $\hat{y}_0 = \lambda\hat{y}'_{ab} + (1-\lambda)\hat{y}''_{ab}$, where $0 < \lambda < 1$. An important but subtle feature of this estimation procedure is that there must be a mechanism that allows the identification of units in the samples from each frame that are also present in the other frame. In other words, there must be an operation that separates the sampled elements into those that are in only one of the frames and those that are in the overlap. The determination of the mixing parameter λ is also an important feature. Hartley (1962, 1974) and Fuller and Burmeister (1972) propose methods of choosing λ to give efficient estimators that require that λ be computed separately for each statistic. Lund (1968) proposes a different approach that combines all the sampled units in the overlap population to estimate the overlap total.

Other estimators have also been considered. Bankier (1986) introduces a single-frame Horvitz–Thompson type of estimator that could be used if the overlap units in the samples can be identified. Kalton and Anderson (1986) present a more general single-frame estimator that does not require information on duplicate units in the samples.

Skinner and Rao (1996) give a pseudo-maximum likelihood estimator that has the important property that a single weight can be used for all estimates. This differs from the optimal estimators of Hartley (1962, 1974) and of Fuller and Burmeister (1972) that require different weights for each estimate. Another single weight method is to use a compromise value for the compositing factor λ so as to choose the composite factor in proportion to the sample size (Burke et al., 1994). Compromise compositing factors tend to work well in practice for most estimates because in the neighborhood of the optimal value for the factor the variance is relatively flat as the value is varied above or below the optimum (e.g., Lepkowski and Groves, 1986).

7.4.2 Errors in Multiple-Frame Surveys

While multiple-frame theory is elegant, the theory does not account for issues that arise frequently and may result in substantial errors. Similar issues arise in single-frame surveys (see, e.g., Groves, 1989), but the issues tend to be more complex in multiple-frame surveys and involve errors from more sources. Further, these issues have been studied less completely in multiple-frame rather than in single-frame designs.

A comprehensive examination of all error sources in multiple-frame surveys, with telephone data collection in at least one frame, could apply a total survey error framework to each of the frames, to estimators specified for combining domain estimates from each of the frames, and to a final survey error evaluation conducted for the combined frame estimate. For example, the sampling, measurement, and nonresponse errors could be examined for each frame separately. The Hartley estimator described earlier could then be used to combine domain estimates (a, b, ab) across the multiple frames. In addition the sampling, measurement, and nonresponse errors from each frame could be combined along with an examination of remaining noncoverage error across the multiple frames. In principle, the theoretical properties of the total error could be written in a general form, as was attempted in a simple example in Lepkowski and Groves (1986), for example.

Unfortunately, the application of such a framework to multiple-frame surveys would at present be incomplete because some of these error sources are not well studied across the frames. In other words, a total survey error perspective has yet to be developed for multiple-frame (and mode) surveys. In the following, what is known from multiple frame and mode surveys about sampling error, measurement error, nonresponse, and noncoverage is reviewed. The review is limited to dual-frame designs, as illustrated in Fig. 7.1, and to those where one of the two frames involves telephone numbers. There are multiple-telephone frames (landline and mobile, for example) that are combined with area frames, address frames (such as the DSF), and other list frames. Applications have combined area with RDD landline and address with RDD landline frames, as well as RDD landline with mobile frames and RDD landline with list frames. Other combinations are possible but have not been implemented in practice yet.

Area and RDD landline frames. The Canadian Community Health Survey conducted by Statistics Canada (2003) uses area, list, and RDD frames. For simplicity consider the presumably complete area frame (A) and the incomplete RDD frame (B), a standard dual-frame survey with overlap (ab) that can be estimated by a composite estimator described previously. Béland and St-Pierre (2007, Chapter 14 in this volume) discuss the potential biases in this survey due to mode effects arising from area sample face-to-face and RDD telephone interviews. The response rates in the two frames are different as well, which could lead to nonresponse biases in the estimates. The measurement and adjustment for the biases due to both mode differences and differential nonresponse is currently limited within the framework of the theory of multiple frames.

RDD landline and RDD mobile frames. Within telephone surveys, the RDD frame of landline numbers, frame A, and the RDD frame of mobile numbers, frame B, have been combined. Households with only landlines are in domain a, those with only mobile telephones are in domain b, and those with both landlines and mobile telephones are in domain ab. Steeh (2004a) and Brick et al. (2006) describe two dual-frame surveys that use this approach. The purpose of including the mobile telephone number frame is to reduce noncoverage of households without landlines. Both of these surveys found a much higher proportion of households in b (mobile only) than expected. Brick et al. (2006) investigated the biases in the estimates and determined that they are related to differential nonresponse across the frames. None of numerous weighting schemes substantially reduced the nonresponse bias, and they concluded, remarkably, that most estimates from the regular RDD frame with its noncoverage problems are less biased than those from the dual-frame survey.

RDD landline and list frames. This combination is found in rare population surveys that use the RDD landline frame to provide more complete coverage at relatively low cost to supplement a less complete list frame for the rare population. A challenging feature of these designs is how to identify the three domains shown in Fig. 7.1 when there is no information about overlap in either the RDD landline or the list frame.

A screening design eliminates from data collection households from one frame that are in the overlap with the other frame. For example, only households that do not have a telephone are retained from the complete area probability frame A, and all households are kept from the incomplete RDD frame of telephone households. In this case, domain *a* is the set of nontelephone households, domain *b* is all telephone households, and domain *ab* is empty. The screening reduces data collection costs from the more expensive area probability sample. The National Survey of America's Families (Waksberg et al., 1998) and the Community Tracking Survey (Hall and Carlson 2002) used this design. Both surveys had a special interest in low-income households and screened the area sample to avoid noncoverage biases due to excluding nontelephone households.

In this type of screening dual-frame survey, the screening appears to increase the potential for measurement and differential nonresponse errors, especially if the target is a rare group (e.g., nontelephone households). Neither the NSAF nor the Community Tracking Survey achieved expected sample sizes of nontelephone households, possibly because of differential nonresponse. Waksberg et al. (1998) discuss biases due to noncoverage and nonresponse in the NSAF and propose weighting adjustments to reduce these biases. They do not evaluate the effectiveness of the proposed estimation procedures.

For a rare population survey, the telephone mode may be applied to a sample from a list frame of persons who are likely to be in the rare group. The principle frame is the RDD frame A that covers the entire population, while the list frame B contains a substantial proportion of the rare group members. Domain *a* consists of persons in the rare group who are not on the list B, domain *b* is empty, and domain *ab* is the rare population persons on both the list B and the RDD frame A. Srinath et al. (2004) examine such a design in the National Immunization Survey (NIS), where a commercial list of telephone numbers identifying children between 19 and 35 months of age is frame B, and A is the RDD frame. In the California Health Interview Survey (CHIS), Edwards et al. (2002) use surname lists to increase the yield of persons of specific Asian ethnicities that comprise a rare but important analytic group as frame B, and an RDD frame as A.

A common problem with this approach is that screening households for the rare group may result in differential nonresponse when households with persons in the rare group complete the interview at a higher or lower rate than households without such persons. The nonresponse may bias the estimates from one or both frames. Edwards et al. (2002) give a vivid example of this in CHIS where households were screened for the presence of persons of Cambodian origin by adding screening items to the beginning of the interview. Cambodian households were suspicious of calls because of political rivalries in Cambodia, and were more likely to refuse or answer untruthfully when the screening items were asked than when the standard CHIS questions without screening items were asked.

In a survey with a similar design, Traugott et al. (1987) attempted to reduce nonresponse in a telephone survey by using a commercial list of telephone numbers (B) added to an RDD frame (A). Advance mailings could be sent to the list sample because addresses were available for those units, but no addresses were available for the RDD sample. The authors showed that the advance mailings increased the response rates in the survey, but the lists suffered from substantial noncoverage. They did not address methods of producing unbiased estimates when one frame has noncoverage and the other has lower response rates and potentially higher nonresponse bias.

Address and RDD landline. A newer application of a dual-frame household survey combines two frames and two modes. Households are sampled from a frame of addresses (A), such as the DSF or other commercial address list. These sampled households are mailed a self-administered questionnaire and asked to complete and return it. In a second RDD frame (B), sample households and persons are contacted and interviewed by telephone. In this design, households only in the DSF (primarily nontelephone and mobile-only households) are domain *a*, and those only in the RDD (due to incompleteness of the DSF) are domain *b*. The majority of households are presumably in domain *ab*. Link and Mokdad (2006) discussed the potential for this multiple-frame design to replace standard single-frame RDD surveys, focusing particularly on the response rates obtained in the two frames. This multiple-frame design offers many challenges, both statistically and methodologically. For example, since neither the DSF nor the RDD frames are complete, the nature of the interaction of the noncoverage error is important, especially as the proportion of households without landlines increases. Procedures for selecting a probability sample of persons within a household from the DSF frame have not been resolved in a self-administered mode.

One operational aspect that causes difficulties is the inability to link the addresses from the DSF to telephone numbers from the RDD frame completely and uniquely. One reason is that both samples are essentially selected from lists, and lists over time become increasingly out-of-date. A more fundamental problem is that the DSF file is constructed for mailing purposes and does not have telephone numbers, while the RDD frame is a list of telephone numbers without addresses. The only way to link these files currently is by using commercial vendor lists that have both addresses and telephone numbers. While several firms provide this service, the degree of matching is not very high. Better linkages would provide new opportunities for more efficient designs. For example, multiple mode and frame designs specifically developed to reduce nonresponse would be more attractive.

A key statistical issue of the address-based RDD landline design is also related to the linking of numbers and addresses. The linkage is needed for the identification of units in the overlap. Since addresses are not available for all RDD numbers and telephone numbers are not available for all addresses, membership in the overlap of the DSF and RDD frames is hard to determine. This determination may not be possible for nonresponding households without the linkage. The composite estimation schemes suggested by Hartley (1962, 1974) and others, and the Horvitz–Thompson type of estimator suggested by Bankier (1986) cannot be applied without knowing the membership of the sample numbers. The single-frame estimator proposed by Kalton and Anderson (1986) and examined in more detail by Skinner and Rao (1996) may have some advantages in this case. However, methods of nonresponse adjustment have not been studied for this estimator.

One of the main reasons the DSF addresses are being explored is because response rates from RDD samples have declined substantially in the past two decades (see Battaglia et al., 2007, Chapter 24 in this volume). Link and Mokdad (2006) demonstrate that samples from the DSF may produce response rates in some circumstances that are at least equivalent to those from RDD surveys. However, their research also suggests that the nonresponse biases from samples selected from the DSF may actually be larger than those from RDD samples, even if the response rates from the DSF frame are higher. While this is problematic, research on the DSF–RDD dual frame is only just the beginning, and promising variations that address these issues are yet to be explored.

In summary, multiple-frame surveys have many attractive features, but they also have multiple error sources to contend with because multiple-frame surveys are intrinsically more complex than single-frame surveys. A problem unique to multiple-frame surveys is the handling of the overlap population, both operationally and in estimation. But other issues include differential nonresponse and measurement errors across the frames. These sources of error are often difficult to measure and may result in biases that are even more difficult to reduce by applying weighting adjustments. Despite these methodological disadvantages, the cost efficiencies and coverage improvements associated with multiple frames often dictate that they be used. Methods must be adapted to deal with the potential biases.

7.5 ESTIMATION IN THE PRESENCE OF BIAS

An important issue raised in the review of both multiple modes and multiple frames is the mixture of biases and the variable errors across the frames. If the magnitude and the direction of these errors could be evaluated when multiple mode and frame designs are used, procedures for dealing with them in making inferences could be developed. Methods for estimating the magnitude and direction of biases and the magnitude of nonsampling variable errors, and methods for compensating for the biases in inference are discussed in this section.

7.5.1 Estimating Nonsampling Errors

Producing estimates of errors or biases is one of the most difficult of all the tasks in survey methodology. Survey sampling theory supports the estimation of sampling error, but most of its theoretical foundation is not relevant for evaluating other systematic and variable nonsampling errors. Sampling theory assumes that all sampled units respond and their responses are free of all other types of errors. To deal with nonsampling biases and variability from sources other than sampling error, statistical models outside the realm of survey sampling theory must be employed.

The earlier sections identified nonresponse, noncoverage, and measurement as important sources of nonsampling errors in multiple mode and frame surveys. Nonresponse and noncoverage are both nonobservation errors where the characteristics for some units in the population are not observed. A variety of techniques have been used to estimate nonobservation errors, but estimates of bias often rely on model assumptions that are difficult to validate. For example, one technique to estimate nonresponse bias involves conducting a follow-up study of nonrespondents to determine their characteristics (e.g., Hansen and Hurwitz, 1946; Elliot et al., 2000). But the follow-up efforts themselves have substantial nonresponse, and assumptions about the nonrespondents to the follow-up must be made. Furthermore, techniques used to increase the response rate in the follow-up, such as changes in modes, may introduce other errors.

An alternative approach is to use auxiliary data from external sources, such as administrative records, to estimate nonresponse and/or noncoverage errors. The auxiliary data source may not contain information on the variables of most interest in the survey though, and administrative records typically have errors of their own that may obscure estimates of bias from the survey.

Errors are even more difficult to estimate in many multiple-mode surveys. For example, when the respondents "choose" the mode by not responding to the assigned mode (as in the ACS), a selection bias may then be confounded with mode differences involving measurement error, and the measurement errors are not estimable. As a result, differences in the ACS between the characteristics of households that complete the mail questionnaire and those that are interviewed in the telephone or face-to-face modes could be due to either the mode of the interview or to the selection bias (i.e., differential propensities to respond in one mode or the other).

An approach that avoids the selection bias problem in measurement is to conduct the survey "independently" in two different modes, say, by telephone and by face-to-face surveys (e.g., Groves and Kahn, 1979). Even in this experimental approach, it is difficult to isolate mode effects. For example, the telephone and face-to-face surveys may have different nonresponse and noncoverage rates that affect the comparisons. As mentioned earlier, Aquilino (1994) eliminated most of these problems by a careful experimental design and was able to estimate differences between telephone and face-to-face modes. He concluded that the differences were consistent with theoretical expectations and that the telephone interviews generally gave more biased estimates than the face-to-face interviews when there were differences. However, even this study is restricted in producing estimates of differences rather than biases because of the absence of known values for the variables. Both the telephone and the face-to-face interviews are subject to unknown levels of bias.

Even if a specific bias or a error source can be estimated precisely in a survey, the problem is very complex because most survey estimates are subject to multiple-error sources. Combining the contributions from different sources into one overall estimate of error is extremely difficult. While some have attempted to integrate errors from different sources in a meaningful way (e.g., Mulry and Spencer, 1990), these attempts have not been completely successful because of the complexity of possible interactions among the error sources.

Even imperfect measures of error may be very useful in the design or the inference stage. For example, findings from an interviewer variance study in the 1950 U.S. Decennial Census (Hanson and Marks, 1958) led to future censuses using self-administration to the extent possible. This change undoubtedly reduced overall error in the future censuses. In this case, the interviewer study itself had important shortcomings and could not provide definitive estimates of bias. But knowing something about an important contributor to the bias of census statistics was sufficient to help design an alternative data collection scheme with less bias.

Imperfect estimates of bias may also be useful in making better inference. One approach is to estimate bounds on the bias. For example, it may be possible to estimate the difference in the percentage of adults who report heavy use of alcohol between face-to-face and telephone interviews, along with upper and lower bounds on the difference. The bounds can be used to evaluate the sensitivity of the estimates to the potential for bias. For example, Brick et al. (1994) examine a variety of sources of error and then synthesize these to produce recommendations for making inferences from the survey. They conclude that the biases from the survey are generally small and have inconsistent directions, and recommend against adjusting the estimates for bias. Instead, they propose conservative inferential procedures such as using a smaller value of α in tests of hypotheses to reduce the chance of declaring differences to be statistically significant.

7.5.2 Minimizing Error

Since multiple-frame surveys have more sources of error, and these sources of error may well have a larger impact on the quality of the estimates, it is important to examine how to handle estimates that might be biased, assuming the bias is known.

If the bias for a statistic could be estimated precisely, then the estimates could be statistically adjusted to produce an approximately unbiased estimate. Even in the case where only upper and lower bounds are known, the bounds may be used to evaluate the sensitivity of the estimates. Since precise estimates of bias are extremely rare, this type of adjustment is not feasible. Survey estimates using unadjusted values are then used. The question is what are the properties of the unadjusted survey estimates when bias is present?

The standard approach to survey design is to choose the sample design and the estimator scheme that has the lowest cost of obtaining a specified variance of the estimate, or equivalently, of minimizing the variance of the estimate for a specified cost (e.g., Cochran 1977, pp. 96–99). This approach assumes that the estimate is unbiased or the bias is negligible.

With biased estimates, the long-standing tradition is to minimize the mean square error (MSE) of the estimate (the variance plus the bias squared) rather than the variance. For example, Lohr (1999, p. 28) notes that with biased estimates the MSE is used instead of the variance. Deming (1950, pp. 129–130) gives a more detailed discussion of the properties of MSE and why it is useful with biased estimates.

When a design and estimation scheme is selected by minimizing the MSE of the estimate, schemes that give the same MSE are treated as equivalent. For example, consider three different schemes that have the same MSE: (1) one has an unbiased estimate with a variance of 100 units, (2) a second has an estimate with a bias of 3 units and variance of 81 units, and (3) a third has an estimate with a bias of 5 units and a variance of 75 units. All three schemes have MSE = 100, and all three are equally good under the MSE criterion.

What happens to survey estimates in the presence of varying levels of bias is typically examined in the context of confidence intervals. For example, Hansen et al. (1953, pp. 56–59) compare confidence intervals $\pm z_{\alpha/2}\cdot$RMSE to those using $\pm z_{\alpha/2}\cdot\sigma$, where $z_{\alpha/2}$ is the standard normal deviate for a two-sided α-level confidence interval, RMSE is the square root of MSE of the estimate, and σ is the standard error of the estimate. They show that both of these confidence intervals have roughly the same overall coverage if the bias ratio (the bias divided by the standard error) is small. For example, when the bias ratio is 0.75, the overall coverage for ±2·RMSE is 95.9 percent instead of the nominal 95.4 percent. See Kish (1965, pp. 566–571) for an interesting graphical display of this relationship.

Figure 7.2 shows the sampling distributions of an arbitrary estimate (mean, proportion, or regression coefficient) under the three bias and variance combinations described above (zero bias and sampling variance of 100, bias of 3 units and sampling variance of 91, and bias of 5 units and sampling variance of 75). Assume the sample sizes are large enough so that the sampling distributions of the estimates are approximately normal. The figure also shows the areas in the tails of the distributions if one applies the 68 or 95 percent confidence interval, unadjusted for bias. Since the bias ratios for the three distributions are relatively small (0, $3/\sqrt{91}=0.31$, and $5/\sqrt{75}=0.58$), the overall confidence interval coverage rates for adjusted estimates are not very different from those obtained using multiples of the standard errors, consistent with the findings of Hansen et al.

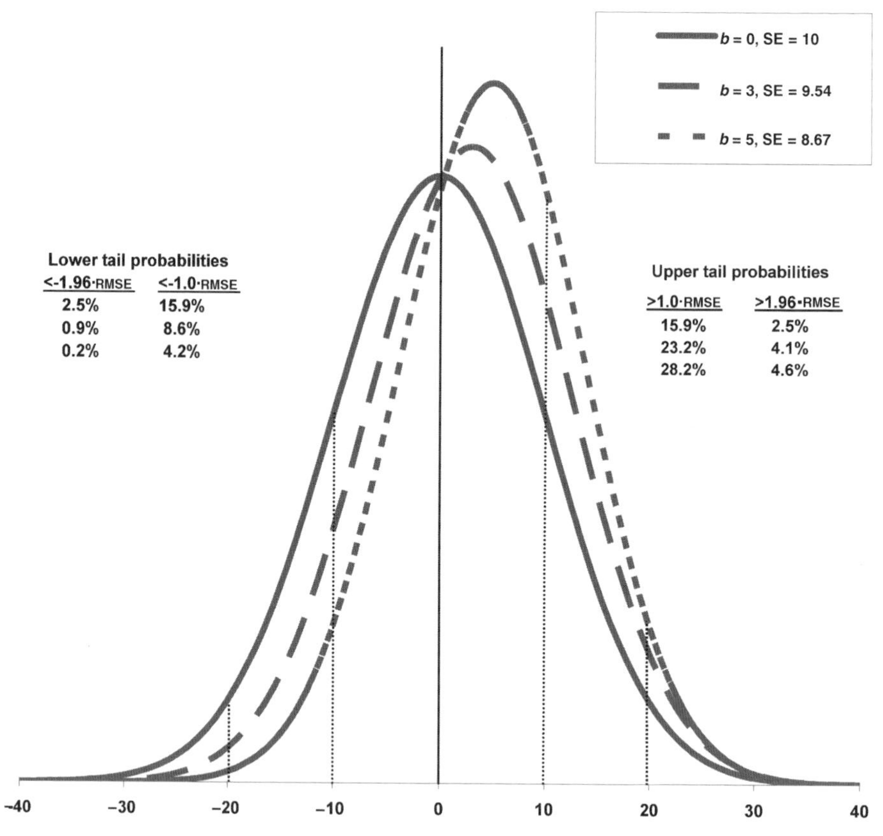

Figure 7.2. Estimators with equal mean square error.

While the overall coverage rates are reasonable with bias ratios in this range, the confidence interval coverage rates are very asymmetric. When the bias is positive, there is much less than the nominal level of coverage error in the lower tail, and more than the nominal level in the upper tail. For example, when the bias is 3 units, instead of error rates of 15.9 percent in each tail outside of 1·RMSE, the rate in the lower tail is 8.6 percent and rate in the upper tail is 23.2 percent.

Although minimizing the MSE is the standard choice, for some inferences such as one-sided tests of significance, the unequal coverages in the tails might be problematic. Many alternatives could be proposed; one is given by way of illustration. Instead of using the sum of the square of the bias and the sampling variance, as in the MSE, one could use the sum of the absolute value of the bias plus the standard error of the estimate, $|B| + \sigma$, in choosing among design alternatives, and use $\pm z_{\alpha/2} \cdot (|B| + \sigma)$ in making inferences. Figure 7.3 shows three distributions that have the same value of $|B| + \sigma = 10$ (equivalent to a MSE = 100). The unbiased distributions in Figs. 7.2 and 7.3 are identical, but the graphs appear to be different because of the different vertical scales. In Fig. 7.3, the tail probabilities are computed using $|B| + \sigma$ rather than RMSE.

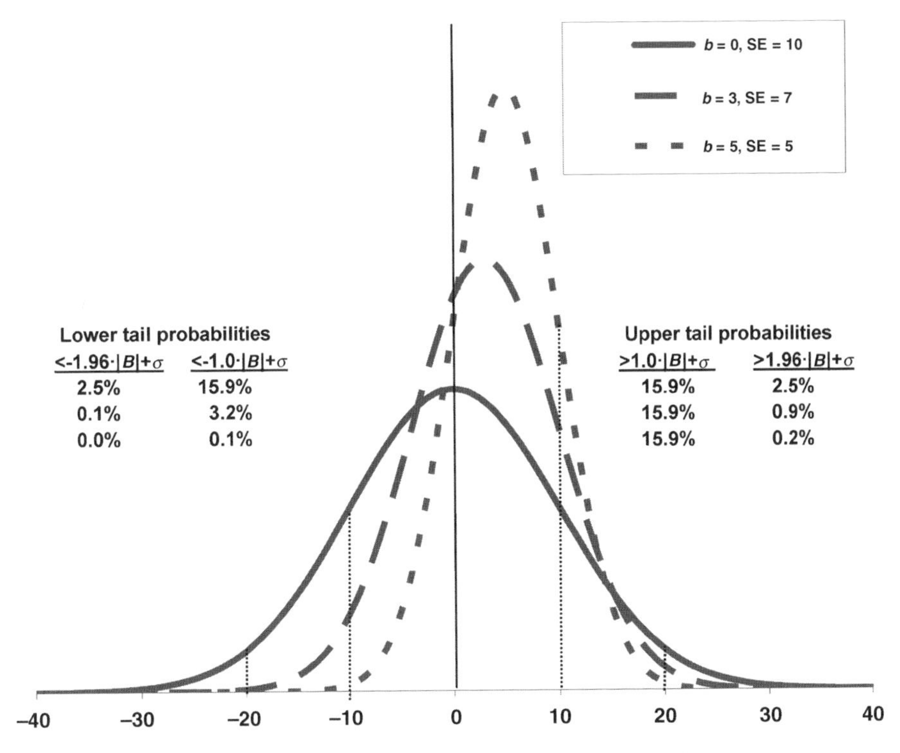

Figure 7.3. Estimators with equal absolute bias plus standard error.

Using $|B| + \sigma$ results in more conservative inferences in the sense that $|B| + \sigma$ is always larger than RMSE. As a result, coverage rates using $|B| + \sigma$ are greater than they are using RMSE. The error rates using $|B| + \sigma$ are exactly equal to the nominal level when $z_{\alpha/2} = 1$ and are less than the nominal levels for $z_{\alpha/2} > 1$. The higher coverage rate in the tails implies that fewer than expected Type I errors will be made under the nominal assumptions, even if the tests are one sided. An undesirable consequence of this conservative approach is that the overall coverage rates can be quite a bit larger. They are substantially greater than the ones using RMSE, which were close to the nominal levels. For example, when the bias is 3 units, the overall error rate of $\pm 1\cdot(|B| + \sigma)$ is 19.2 percent instead of the nominal 31.8 percent, and of $\pm 1.96\cdot(|B| + \sigma)$ it is 1.0 percent instead of the nominal 5.0 percent. For a fixed value of the bias, the distribution using $|B| + \sigma$ has a smaller variance than the corresponding distribution using RMSE. This follows because $(|B| + \sigma)^2 = B^2 + 2|B|\sigma + \sigma^2$, which is larger than the MSE by the positive term $2|B|\sigma$. When compromising between bias and variance, the $|B| + \sigma$ criterion requires a smaller variance than the MSE criterion for a fixed amount of bias. Analysts who are more concerned about bias than variance, at least for some types of analysis, might prefer the $|B| + \sigma$ criterion to the MSE criterion.

If the trade-off between bias and variance is examined more completely, loss function considerations might be beneficial. For example, it is easy to show that the loss is constant for all normally distributed estimators with the same MSE with a squared error loss function.

7.6 DISCUSSION

It is not unreasonable to expect, given the range of survey methods applied at the present time, that the telephone is likely to continue to be a component of many multiple mode and frame designs in the future. Telephone data collection continues to be relatively inexpensive compared to face-to-face, and the sampling and nonsampling errors associated with the telephone are generally competitive with other options. Technological advances may also make the telephone more attractive as the ability to switch between interviewers and self-administration on the telephone becomes more common as in IVR, which has not been heavily utilized in household telephone surveys. Further integration of IVR with interviewer-administered surveys could improve the nonsampling error properties for collecting data about sensitive topics.

However, technology developments could also counterbalance advances. For example, households in the United States began obtaining call-screening devices in large numbers beginning in the early 1990s, but the contact rate in RDD surveys was not adversely affected until around 2000. There is a constant struggle between new methods for contacting and surveying households and the household acquisition of devices and procedures that tend to affect adversely the number of calls to obtain a contact.

Many challenges remain before multiple mode and frame surveys are more widely used than single mode and frame designs. Research focusing on collecting data using more than one mode in the same survey, especially in longitudinal surveys, is an immediate need. Similarly, continued research on the errors arising from various combinations of interviewer-administered and self-administered data collection is needed.

Errors in multiple-mode surveys are generally not amenable to global statistical adjustments, such as through survey weights. The errors vary across specific items and weight adjustments affect all items for a respondent uniformly. Rather, the most promising approach seems to be to design procedures that are appropriate for all of the modes contemplated for the survey. This may mean that the design that is best overall is not optimal for any mode. The best question construction or prompt for an item in a specific mode should not be used if responses from using this structure cannot be approximated in the other survey modes. Much still needs to be learned about how and when unimode design principles should be applied.

Low response rates and high noncoverage rates in RDD surveys have encouraged the examination of multiple frames in surveys where single-frame RDD surveys were once predominant. Since area probability samples are too costly for many of these surveys, the possibility of combining RDD samples with samples from mail frames such as the DSF are attractive. Both frames can provide substantial coverage of the total household population, although each with its own coverage issues. Advances in linking addresses and telephone numbers could make using these frames in combination even more appealing. Similarly, dual-frame surveys that sample mobile telephone numbers as well as those from the traditional RDD frame could address the loss of households without landlines. But, this approach is still relatively costly and has potential nonresponse problems.

Statistical adjustment through survey weights to reduce bias in multiple-frame surveys is clearly an area of research that deserves more attention. Weighting methods are well suited to handle missing data due to nonresponse or noncoverage. The key to reducing bias is understanding the sources of the errors and obtaining auxiliary data that are highly correlated with the errors. These are not simple tasks, but the general methods that are applicable are well known.

Despite several empirical studies from RDD surveys that have not found substantial nonresponse or noncoverage biases, low response rates and increasing levels of noncoverage are eroding the acceptability of low cost surveys using only one frame or mode. These nonsampling errors force inferences from the data to be increasingly dependent on models. Multiple-frame and multiple-mode surveys may help to reduce concerns about bias from these sources. The challenges are numerous and substantial, yet the benefits associated with telephone data collection compel continued efforts to deal with them.

As we stated earlier, we believe that the telephone is likely to be an essential component for many multiple mode and frame designs of the future. Nathan (2001), in the inaugural *Waksberg Invited Paper Series*, summarizes a view of the future that is consistent with ours: "In conclusion, the advances in telesurvey methodology over the past few decades, which have made telephone surveys a viable and predominant survey instrument, will have to be continually updated to deal with the ever-changing developments in telecommunications technology and its usage. However the basic elements for these new developments are available and will continue to allow the use of advanced options to obtain high quality survey data."

CHAPTER 8

Weighting Telephone Samples Using Propensity Scores

Sunghee Lee
University of California, Los Angeles, USA

Richard Valliant
University of Michigan and Joint Program in Survey Methodology, University of Maryland, USA

8.1 INTRODUCTION

Scientific surveys often employ probability samples intended to represent target populations. In theory, these samples are expected to produce estimates that are unbiased or approximately unbiased for the population quantities. However, there is no guarantee that the theoretical unbiasedness holds in practice because of inadequate coverage by sample frames, nonresponse, response error, and other problems. While these errors may have various causes, one standard remedy to reduce the potential bias from nonresponse and poor coverage is to apply statistical adjustments to the collected survey data. The initial weights coming from the survey design are refined to compensate for imperfect field operations through postsurvey weight adjustments. This is done so that the weighted estimates from the final set of survey respondents closely match some externally available characteristics of the target population. This procedure often involves more than one stage of adjustment. Weighting is a flexible technique that can reduce biases arising from both nonresponse and inadequate coverage. The adjusted weights can be applied to most, if not all, estimation processes.

Noncoverage bias and nonresponse are major concerns in RDD telephone surveys. Telephone frames include only households with telephones and often exclude

Advances in Telephone Survey Methodology, Edited by James M. Lepkowski, Clyde Tucker, J. Michael Brick, Edith de Leeuw, Lilli Japec, Paul J. Lavrakas, Michael W. Link, and Roberta L. Sangster

households that only have mobile phones as discussed in Chapters 3 (Blumberg, et al., 2007) and 4 (Kuusela et al., 2007) in this volume. On the contrary, inferences may be desired for the general population, including groups that are not in the frame. The initial design weights for households or persons need a coverage adjustment to account for the omissions. Nonresponse is often quite high in telephone surveys (see Montaquila et al., 2007, Chapter 25 in this volume), and additional adjustments attempt to correct for potential nonresponse bias. In many surveys, nonresponse and noncoverage adjustments are made separately and sequentially, but in RDD surveys both the coverage and nonresponse adjustments are often made once and simultaneously.

When the auxiliary variables used for the adjustment are correlated with the research variables, weighting may increase precision and reduce bias (Little, 1993; Little and Vartivarian, 2005). However, weight adjustment tends to increase variances of survey estimates by adding variability to the estimates through the weights (Kalton and Flores-Cervantes, 2003). Although differential weights may be efficient for some estimates, they can be inefficient for others, especially ones where equal weighting would be more nearly optimal. Nevertheless, bias reduction is usually the top priority in weighting telephone surveys, and some degree of differential weighting is tolerated.

There are a number of ways to carry out postsurvey weighting. All methods require subjective judgment about the choice of auxiliary variables and the mode of implementation of the variables, both of which can be guided by modeling. We dichotomize existing weighting procedures into implicitly model-based weighting and explicitly model-based weighting according to the existence of specified models. Cell weighting, poststratification, ratio raking, and linear weighting belong to the first category. While it can be edifying and useful to analyze these techniques with respect to models (Little, 1993), these methods are often applied without appealing to an underlying model. Their mechanical application requires a knowledge of certain full sample or population total and can be applied without explicit formulation of a model.

Auxiliary information can sometimes be used more efficiently by explicitly specifying models to derive weights. Generalized regression (GREG) or calibration weighting (Särndal et al., 1992) and propensity score weighting are the examples of explicitly model-based weighting. These methods may also have good design-based properties in cases where there is no nonresponse or noncoverage. Explicitly model-based weighting is flexible in that many combinations of main effects, interactions, and variable transformations can be readily implemented. GREG weighting incorporates auxiliary variables through a standard linear regression. Propensity score weighting usually relies on logistic regression modeling and is sometimes termed logistic regression weighting. Discussion in this chapter focuses on propensity score weighting as an explicitly model-based weighting method.

We start this chapter with an examination of the theory for the propensity score adjustment and its use in weighting. Most of the discussion focuses on different methods for employing propensity scores in weight adjustment and the effectiveness of propensity score weighting as reported in the existing literature for telephone surveys. The chapter concludes with a summary and remarks on the limitations of propensity score weighting.

8.2 ORIGIN OF PROPENSITY SCORE ADJUSTMENT

Propensity score weighting is a variation of propensity score adjustment developed by Rosenbaum and Rubin (1983b). Propensity score adjustment aims to reduce selection bias in the estimates of treatment (or causal) effects in observational studies. D'Agostino (1998) gives a very readable account, which we paraphrase here, of the goals of the adjustment and how it is applied in observational studies. The bias arises because observational studies are not true experiments since the assignment of experimental factors cannot be controlled with randomization. The "treated" and "nontreated" groups may have large differences in their observed covariates. These differences can lead to biases in estimates of treatment effects even after applying standard covariance analysis adjustments. Nonetheless, researchers are bound to make comparisons among multiple groups of people using observational data.

As a simple example, suppose that a researcher is interested in whether diets high or low in fruits and vegetables have an effect on body mass index (BMI), a measure of whether a person is overweight or not. If data are collected in an observational study, persons are not randomly assigned to diets. They are simply classified as "high" or "low" based on their own eating habits. As a result, the distribution of persons in the high diet group on covariates such as age, race–ethnicity, and sex is likely to be markedly different from the distribution in the low diet group. For example, if the high group had a larger proportion of females than the low group and gender was a predictor of BMI, any BMI comparisons between the high and low groups would be tainted by this imbalance. This type of imbalance is referred to as "selection bias" in observational studies. A sensible step to solve this dilemma is to examine BMI after somehow balancing the covariate distribution between dietary patterns.

The selection bias in observational studies can be removed or reduced by balancing covariates $\mathbf{x}$ between the comparison groups. In other words, subsets could be created where the treated and nontreated cases had the same distribution on $\mathbf{x}$. When $\mathbf{x}$ takes a high dimension, it is not practical to obtain equivalent distributions on all covariates, although this would be theoretically desirable. An alternative is to summarize all covariates into one quantity, namely, the propensity score $e(\mathbf{x}) = \Pr(g = 1|\mathbf{x})$, where g is the indicator of treatment group membership, and to either balance or adjust all covariates based on this summary measure. The propensity score for a given case is simply the conditional probability of being treated, given all of the covariates. Ideally, when conditioned on a given propensity score, the difference in means between the treated and nontreated groups is an unbiased estimate of the treatment effect at that propensity score (Rosenbaum and Rubin, 1983b). If there are unseen or unaccounted for covariates that differ between the comparison groups for a given propensity score, the difference in means will be biased but probably less so than without the conditioning on the propensity score.

In telephone surveys the idea of propensity score adjustment can be applied in various ways. If estimates are desired for the entire population of persons in a country, noncoverage adjustment may use persons living in households with landline telephones as the "treated" group, while the "nontreated" would be everyone else, that is, persons with only mobile phones, Internet phones, or no telephone at all. If the target population were all persons in households with some type of phone, the treated group might be the persons with landline phones and the nontreated groups be the persons with some other type of phone. For nonresponse adjustment, respondents in a survey may be the treated group and nonrespondents the nontreated group. In order for propensity score adjustments to effectively remove bias in a telephone survey, some strong assumptions are required, which we describe subsequently.

8.3 WEIGHTING ADJUSTMENT USING PROPENSITY SCORES

Weighting is used to project the selected sample to the target population. Figure 8.1 illustrates the situation in many surveys. The target population can be divided into the parts that are covered (*C*) and not covered (*NC*) by the sampling frame. The frame is divided into the sample (*S*) and nonsample (*NS*), while the sample is composed of respondents (*R*) and nonrespondents (*NR*). The inferential problem is to devise the methods of estimation that allow the sample respondents *R* to be projected to either the frame or the entire target population.

Base weights are the simplest form of all weights and, in a probability sample, are the inverse of the selection probabilities. If the sample design includes oversampling of some subgroups, as is common in some telephone surveys, the base weights reflect that design element. The base weights are designed to project the sample ($R + NR$) to the frame ($S + NS$). When the sampling frame covers the entire target population and there is no nonresponse, these weights can be used to produce design-unbiased estimates of population totals and approximately design-unbiased estimates of more complicated quantities. However, as mentioned previously, telephone surveys encounter numerous operational obstacles, such as inadequate coverage and nonresponse. These are not only impediments for survey operations but also sources of biases.

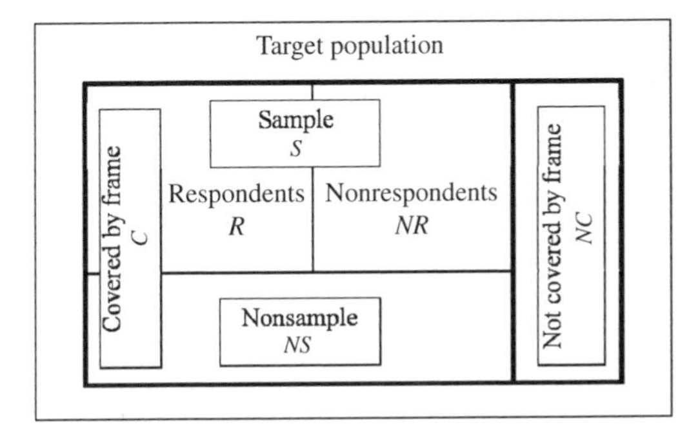

Figure 8.1. Coverage of a target population by frame and sample.

Since only the respondents are used for most estimates, the weights of sample units in R are usually adjusted to account for the nonrespondents, NR. In telephone and other types of surveys, another layer of weighting is imposed to make the respondents R represent the entire target population ($R + NR + NS + NC$). Traditionally, cell-based weighting and poststratification have been adopted as ways of reducing the potential bias from nonresponse and noncoverage (see Kalton and Flores-Cervantes, 2003, for details). These methods adjust the characteristics of one set of units to those of a larger set. For nonresponse adjustment, the sets are respondents R and all sample units $R + NR$. For coverage adjustment, the sets are the sample S and the target population $C + NC$. The basic idea of the weight adjustment is to calibrate the lower level distributions of the chosen characteristics to the higher level distributions. In these processes, adjustment factors are calculated and applied to the initial weights. If w_i is the base weight for unit i, then the weight for a unit in nonresponse adjustment cell g and poststratum c is

$$w_i^* = w_i f_g^{\mathrm{NR}} f_c^{\mathrm{PS}} \tag{8.1}$$

with f_g^{NR} being the nonresponse adjustment factor applied to all units in cell g and f_c^{PS} being the poststratification adjustment factor for all units in poststratum c. In some surveys, the final weight can be expressed with more adjustment factors as in Kalsbeek and Agans (2007, Chapter 2 in this volume). These adjustments have some theoretical support if the probability of response is uniform in each cell g and the variables measured on the units follow a model with a common mean in each poststratum c.

Characteristics used for classifying units into nonresponse adjustment cells must be available for all sample units. In RDD surveys, this is often a major constraint because little may be known about the nonrespondents. In such cases, the only adjustment that may be available is the poststratification to target population counts. For postratification, the only data needed are the poststratum classifiers for the responding units and the population counts that come from external sources. The auxiliary information used in these adjustments can be incorporated in more sophisticated ways by explicit modeling using the auxiliary variables, with propensity score adjustment being one of them.

A number of telephone studies have proposed using propensity scores in postsurvey weighting procedures to decrease biases from deficient coverage (e.g., Battaglia et al., 1995; Hoaglin and Battaglia, 1996; Duncan and Stasny, 2001; Garren and Chang, 2002). Propensity scores have also been used in surveys to address the bias due to partial response or nonresponse errors (e.g., Lepkowski et al., 1989; Göksel et al., 1991; Smith et al., 2001). Apart from telephone surveys, propensity score weighting has been used for late response adjustment (Czajka et al., 1992) and nonprobability sample selection adjustment for web surveys (Terhanian and Bremer, 2000; Terhanian et al., 2000; Varedian and Forsman, 2002; Lee, 2006). The rest of this section reviews the situations where propensity scores have been used in adjustment of design weights, assumptions underlying propensity score weighting, and the mechanics of doing the adjustments.

8.3.1 Situations where Propensity Score Weighting has been Used

Research on weight adjustment using propensity scores can be found in three main areas: errors due to nonresponse, noncoverage, and nonprobability sampling. The first application of propensity weighting in surveys is to adjust for nonresponse. The general approach is to estimate a probability of response for each sample unit based on the covariates that are available for both respondents and nonrespondents. For example, in a survey of spouses of military personnel examined in Iannacchione et al. (1991), information about the nonresponding spouses was obtained from the preceding survey on the soldiers. Given the estimated response probabilities, there are several options for constructing nonresponse adjustments, as described in Section 8.3.4. The nonresponse adjustments to the weights for the respondents R in Fig. 8.1 are intended to allow R to be projected to the frame $S + NS$.

Panel surveys produce a unique nonresponse phenomenon because respondents who completed the initial survey might drop out in the subsequent waves and return at any point. Lepkowski et al. (1989) examined the possibility of using propensities of partial nonresponse in the Survey of Income and Program Participation (SIPP), where data are collected mainly by telephone after the baseline survey is completed. All waves of SIPP included the same set of core items. This allowed adjustment using data on all respondents in the initial interview regardless of their successive response status.

Some surveys use other established survey participants as their frame. This is an example of two-phase sampling, where a rich set of information collected in the original surveys is available for the entire second-phase frame. In the National Immunization Survey (NIS), vaccination history data are collected from adults with eligible children. When verbal consent is obtained from household respondents, the history is also obtained from the children's vaccination providers. Smith et al. (2001) modeled the probability of obtaining adequate provider data for responding households as a function of various household and person-level variables. The estimated propensities were used to form cells for nonresponse adjustment. Response propensities for a telephone follow-up of the National Survey of Family Growth (NSFG) Cycle IV were estimated using the baseline information in Göksel et al. (1991). This is a special case because the NSFG itself was sampled from the National Health Interview Survey (NHIS). Data about the telephone follow-up nonrespondents were available from both the baseline NSFG and the NHIS.

There are situations when the timeliness of survey estimates is critical. Although its survey is not done by telephone, the Internal Revenue Service estimates various quantities, such as total taxable income and total Federal tax paid, based on a sample of tax returns. Czajka et al. (1992) used an advance sample of early returns for estimation. Since the outcomes are related to income level and return type (business, nonbusiness farm, and nonbusiness nonfarm), design strata were based on these variables. There was evidence that the income and tax from the early filers were different than that from the late filers. Propensities of filing a late return were calculated within each design stratum based on the previous year's data and used to decrease biases due to using only advance data for estimation.

The second type of application of propensity weighting is to adjust for imperfect coverage by a frame. This corresponds to devising a way to project the sample to the full target population $C + NC$, including the portion of the population that has no chance of being selected for the sample. Coverage is an important issue in telephone surveys because there remains a group of people without telephones of any kind who are likely to have different characteristics from people with telephones. Brick et al. (1996) and Duncan and Stasny (2001) used the people who experience telephone service interruptions to represent nontelephone households and assigned higher weights to these transient respondents. Hoaglin and Battaglia (1996) and Battaglia et al. (1995) devised model-based coverage adjustments based on propensity scores for telephone noncoverage in NIS. As described later, the NIS adjustments used coverage models developed using NHIS Immunization Supplements data on telephone and nontelephone children and applied the models to children in NIS. Models from NHIS were used because the NHIS included both telephone and nontelephone households, unlike the NIS, which was a telephone survey.

Census Public Use Microdata Samples (PUMS) may be another fertile source for telephone noncoverage bias adjustment because the PUMS contain telephone ownership status as well as a wide spectrum of sociodemographic, employment, and health data. Garren and Chang (2002) developed models for predicting propensities of telephone households using PUMS and applied the models to simulated survey data for noncoverage bias adjustment. The usefulness of this type of analysis depends on having timely data on telephone ownership. Since the proportion of mobile phone only households is steadily increasing, it would be ideal to have periodic readings on the distribution of telephone ownership from a large, well-designed survey such as the Current Population Survey or the American Community Survey.

8.3.2 Assumptions in Propensity Score Weighting

When propensity score weighting is used, biases can be reduced but a number of assumptions must be satisfied. Not all of these assumptions are equally relevant to survey applications but are listed here for completeness. These assumptions are usually phrased in terms of treatment and nontreatment in observational studies. In a survey the "treatment" corresponds to either being a respondent, being covered by the sampling frame, or responding to a particular survey over other surveys. The five assumptions as stated for observational studies are

1. Strong ignorability of treatment assignment given the value of a propensity score
2. No contamination among study units
3. Nonzero probability of treatment or nontreatment
4. Observed covariates represent unobserved covariates
5. Treatment assignment does not affect the covariates

In an observational study, strong ignorability (Rosenbaum and Rubin, 1983b; Rosenbaum, 1984a) means that an outcome variable y and whether a unit is in the treated or nontreated group are independent given the value of a propensity score. Define $e(\mathbf{x})$ to be the propensity score for a unit with a vector $\mathbf{x}$ of covariates and g to be an indicator for whether a unit is treated or not. Strong ignorability is usually symbolized as $y \perp g|e(\mathbf{x})$, where y is the research variable, $\perp$ denotes independence, and $0 < e(\mathbf{x}) < 1$. In the survey setting, g can be an indicator of whether a unit responds or not (or is covered by the frame or not). Strong ignorability means that, given a propensity score $e(\mathbf{x})$, the distribution of y is the same regardless of whether a unit is a respondent or not (or is covered by the frame or not). In this case, the inference about the entire distribution of y can be made for units that have the same score. Thus, achieving (1) is not tied to the goal of estimating a mean or some other simple, summary quantity but allows much more general inferences to be made.

Strong ignorability would hold if, for example, nonresponse or noncoverage were random within groups of units that have the same propensity score. This is analogous to the assumptions in weighting class adjustment, where the response propensity is assumed to be uniform in each nonresponse adjustment class, and the units in each poststratum are assumed to follow a model with a common mean.

When the propensity scores are not modeled correctly, this assumption could be violated even when all relevant covariates are available. If response or coverage cannot be adequately modeled with the available covariates, then strong ignorability may also be violated. This assumption is difficult or impossible to check in most surveys unless data on the nonrespondents or noncovered units are available from some source outside the survey.

Related to strong ignorability, propensity score adjustment requires another assumption—no contamination among study units. The response or coverage mechanism of one unit should not influence that of any other units. In a survey where the goal is to collect data from more than one person in a household, this assumption could be violated if the person initially contacted influences other members not to cooperate. On the contrary, in a survey where one adult is sampled at random from each sample household, within household contamination of the response mechanism would not be an issue. In general, this assumption is difficult to check but, if violated, will mean that a common propensity model cannot hold for all units.

The third assumption is that there should be nonzero probabilities of each unit having $g = 1$ or 0 for any configuration of $\mathbf{x}$. This assumption would be violated if there were certain groups of people in the target population who never responded or were never covered by the sample frame. This assumption is particularly critical if some type of propensity adjustment is used to correct for noncoverage.

Fourth, observed covariates included in propensity score models are assumed to represent unobserved covariates (Rosenbaum and Rubin, 1983a) because distributional balances obtained from observed covariates are not guaranteed for unobserved covariates. If an important covariate, for example, education were omitted from the model for $e(\mathbf{x})$ and the $g = 1$ and $g = 0$ groups had different distributions of the number of years of education, then the fourth assumption would be violated. Note that this type of problem can also occur for the weighting class adjustment.

The last assumption is that the g treatment mechanism does not affect covariates (Rosenbaum, 1984b). This assumption is unlikely to be violated in a survey, as it is hard to imagine that response status will affect, for example, education. In fact, the reverse is what propensity score attempts to address.

8.3.3 Modeling Propensity Scores

Propensity scores have to be specified in a model and estimated from the observed data. In principle, the model for a propensity score should be derived from data for the whole population, which is not possible. However, Rubin and Thomas (1992, 1996) showed that, at least in the context of matching treated and nontreated units, the estimated propensity scores from sample data reduce variances of estimated treatment effects more than using the true population propensity scores. This finding gives support for applying estimated propensity scores, although the bias correction is approximate.

There is no standard way to model propensity scores. A range of parametric models can be used: logistic regression, probit model, generalized linear and generalized additive, and classification tree. The most commonly used model is logistic regression, which models propensity scores as $\log\left[e(\mathbf{x})/(1-e(\mathbf{x}))\right]=\alpha+\beta' f(\mathbf{x})$ where $f(\mathbf{x})$ is some function of covariates. When probabilities are close to one, generalized linear regression with a log–log link may provide a better fit than logistic regression (McCullagh and Nelder, 1991). Garren and Chang (2002) advocated using generalized regression with a log–log link to estimate the propensities of telephone coverage in the form

$$\log\left[-\log\left(e(\mathbf{x})\right)\right] = \alpha + \beta' f(\mathbf{x}) \tag{8.2}$$

No matter which model is used, there has to be enough overlap between the distributions of the covariates of the comparison groups to estimate the parameters in the propensity model.

Whenever covariates are used for estimation, variable selection becomes an issue because the predictive power of the covariates in the model matters. According to Rosenbaum and Rubin (1984, p. 522), $\mathbf{x}$ is required to be related to both y and g in order to satisfy the assumption of ignorability. However, studies often examine only the covariates' significance on g but not on y, which is typically only available for respondents.

Rubin and Thomas (1996) argued that there is no distinction between highly predictive covariates and weakly predictive ones in the performance of propensity score adjustment. They recommended including all covariates, even if they are not statistically significant, unless they are clearly unrelated to the treatment outcomes or inappropriate for the model. In practice, however, some procedures are often used to select covariates: stepwise selection (e.g., Smith et al., 2001), Chi-squared Automatic Interaction Detection (CHAID) (e.g., Iannacchione, 2003), or other type of binary splitting algorithm (e.g., Lepkowski et al., 1989), or one-step covariate selection based on theoretical and/or logical relevance (e.g., Duncan and Stasny, 2001). Smith et al. (2001) used the Akaike Information Criterion (AIC) to determine the optimal set of regressors. In ongoing surveys, one may consider creating propensity score models from previous year's data, but the model should be stable across years (Czajka et al., 1992). At present, there are no universally accepted criteria for selecting variables for propensity score model building.

When it is expected that propensity scores are affected by different covariates across design strata and poststratification groups, one may consider modeling propensity scores separately within each stratum and group (Czajka et al., 1992; Battaglia et al., 1995; Hoaglin and Battaglia, 1996).

Whenever inferences use models, their specification may be an issue. It is true that there is always the possibility that model misspecification violates assumptions. Moreover, the model specification is often impossible to validate in survey practice. In the context of weighting, however, not applying propensity score weights due to potential failure from model misspecification may be more likely to produce biased estimates than applying them. Drake (1993) found that misspecifying the propensity score model, such as mistakenly adding a quadratic term or dropping a covariate, is not very serious. The misspecification of the propensity score model led to only a small bias compared to the the misspecification of the outcome response model used to simulate the response distribution.

Estimation of the parameters in a logistic regression model requires that the same covariate data $\mathbf{x}$ be available for both respondents and nonrespondents (or covered and noncovered cases). For unit nonresponse, this is a severe practical limitation since very little information may be known for the nonrespondents in telephone surveys. On the contrary, in two-phase surveys, variables collected in the first-phase can be used to model nonresponse at the second phase. In the NIS, for example, data on cooperating households was used to model response propensities of immunization providers. In some cases, imputations may be necessary for units with partially missing covariate data. As most variables used in weighting have high response rates, imputations on predictors may not lower the effectiveness of the weighting.

Parameters can be estimated using survey base weights in weighted binary regressions. This will produce approximately unbiased parameter estimates with respect to the randomization distribution. A counterpoint to this is Little and Vartivarian (2003) who argue that weighting in this type of estimation is inefficient if the response propensity model is correctly specified.

8.3.4 Mechanics

Propensity scores not only have mainly been a tool for nonresponse adjustment, but also have been used for noncoverage adjustment in cases where adequate data are available.

Nonresponse adjustment. Three alternatives for implementing propensity score adjustment were originally proposed: matching, stratification, and regression adjustment (Rosenbaum and Rubin, 1983b; D'Agostino, 1998). In surveys, two techniques are most commonly used. The first method divides the base weight of unit i by its estimated propensity score $\hat{e}(\mathbf{x}_i)$. The nonresponse-adjusted weight is then $w_i/\hat{e}(\mathbf{x}_i)$.

Lepkowski et al. (1989) is an example of directly applying this type of weight. Czajka et al. (1992) used a variation of this approach in which the individual propensity scores were used to create adjustment class, but the same weight was assigned to all units in the class. In a more recent application addressing nonignorable nonresponse, Laaksonen and Chambers (2006) imputed y data for nonrespondents before fitting the propensity model. They then computed a response propensity as $p_i = \Pr(i \text{ is respondent} | \mathbf{y}_i^*, \mathbf{x}_i)$, where $\mathbf{y}_i^*$ was a vector of analysis variables, including any imputations. When the propensity score adjustment is allowed to be different for each respondent, units with very low values of propensity scores receive large weights that can inflate the variances of estimates. This undesirable feature led to the alternative described below.

The second common technique is the response propensity stratification proposed by Little (1986). Strata (or, more properly, poststrata) are formed by sorting units based on the propensity scores and by dividing the sample into groups having the same number of units. The same weight adjustment factor is applied to all units in the same poststratum. The adjustment factor can be either (a) the sum of initial weights for the full sample divided by the sum of weights for the responding sample in the poststratum or (b) the sample average of the inverses of the probabilities of responding among units in the poststratum. For choice (a) the adjusted weight for a unit in poststratum (or adjustment class) g is $w_i f_g^{\text{NR}}$, with the adjustment factor defined as $f_g^{\text{NR}} = \sum_{i \in s_g} w_i / \sum_{i \in s_{Rg}} w_i$, where s_g and s_{Rg} are the set of all sample units and the set of responders in poststratum g. For choice (b) the adjustment for a unit in poststratum g is $f_g^{\text{NR}} = \sum_{i \in s_{Rg}} w_i / \hat{e}(\mathbf{x}_i) / \sum_{i \in s_{Rg}} w_i$. Another choice would be the unweighted mean of the $1/\hat{e}(\mathbf{x}_i)$ in a class that may have smaller variance than the weighted estimate in (b).

In either (a) or (b), f_g^{NR} is interpreted as the inverse of an estimated response probability. This interpretation assumes that the reponse propensity is nearly constant within a poststratum. In principle, this can be accomplished by creating many poststrata, but, in practice, a limited number may suffice. Rosenbaum and Rubin (1984), following Cochran (1968), recommended that five strata would be sufficient to capture almost all of the potential bias reduction from using propensity score adjustment. Based on this, many studies have used five groups, each of which has about the same number of sample responders (see Vartivarian and Little, 2002, 2003). Instead of using a predetermined number of groups, Smith et al., (2001) used a sequential method, due to Yansaneh and Eltinge (1993) and Eltinge and Yansaneh (1996), that relies on the amount of incremental bias reduction to determine the number of groups. Another option is to use propensity models to select significant and substantive predictors of nonresponse but form the adjustment cells based on the selected predictors in the order of their significance rather than on the propensity scores (e.g., Göksel et al., 1991).

Vartivarian and Little (2002) extended the idea of stratification by using the predictive mean of $y, \hat{y}(\mathbf{x})$, and forming adjustment classes based on both propensity score and predictive mean stratification. The predictive mean $\hat{y}(\mathbf{x})$ is estimated using the same set of covariates as the propensity model and must satisfy $y \perp g | \hat{y}(\mathbf{x})$. The advantage of adding predictive mean stratification is that both bias and variance can be controlled, thus improving the efficiency of estimates. The joint classification aims to provide a kind of double robustness with potential protection from misspecification of either model. However, surveys are used to estimate a vector of research variables $\mathbf{y}$ instead of a single variable y. Vartivarian and Little (2003) later used a linear combination of $\mathbf{y}$ as the canonical outcome variable, with its predictive mean stratification replacing multiple stratification from all outcome variables. They showed that the performance of the compromise joint classification is similar to that of stratification using the correct propensity model, with the added advantage of double robustness.

Noncoverage adjustment. Even though telephone samples omit nontelephone households and, often, mobile phone only households, inferences may be needed for the entire population. Adjusting for noncoverage involves projecting the initial sample S in Fig. 8.1 to the full target population C + NC. This requires the use of some data source external to the survey. Because the propensity of being covered must be estimated, this cannot be done using data only for the units that actually were covered. One option is to use another survey that, in principle, covers the entire target population and that collects covariates that are also included in your survey.

A simple approach is to poststratify a telephone sample to demographic counts for the entire population (Massey and Botman, 1988). However, if the noncovered households have different characteristics, basic poststratified estimates will be biased. In the NIS, Battaglia et al. (1995) noted that about 12 percent of age-eligible children resided in nontelephone households with the noncoverage rate ranging from 2 percent to about 25 percent, depending on the geographic area. The 1992 and 1993 Immunization Supplements to the NHIS, an area probability survey, showed that the percentage of children in nontelephone households that were up-to-date on their immunizations was less than in telephone households, even after controlling for available demographic and socioeconomic factors. Thus, standard methods of poststratification would have led to overestimates of the national immunization rates, that is, noncoverage bias would not have been corrected.

Battaglia et al. (1995) used the NHIS to estimate the probability of being in a telephone household separately for children who were vaccinated and who were not. The coverage-adjusted weight for a child i was then computed as $w_i^* = w_i f_g^{\mathrm{NR}} / \hat{\tau}(\mathbf{x}_i)$, where $\hat{\tau}(\mathbf{x}_i)$ was the estimated probability of having a telephone given the set of covariates $\mathbf{x}_i$. The logic of this adjustment does require that all units with a given $\mathbf{x}_i$ have a nonzero chance of being covered by the telephone survey frame (see assumption 3). Although this assumption is probably violated in practice, the adjustment w_i^* may still more effectively remove bias than an adjustment that takes less advantage of the covariates.

8.3.5 Use of a Reference Survey

The adjustment of the NIS for noncoverage using telephone propensity models from the NHIS is an example of using a "reference" survey. The reference survey must collect data pertinent to coverage in target survey (e.g., having a landline telephone, only a mobile telephone, both, or no telephone at all). The reference survey should also collect covariate data related to coverage (e.g., demographic and socioeconomic variables on households and individuals). The survey would then collect the same covariates found to be relevant for predicting coverage in the reference survey. The probability of being in the covered group in the survey is estimated using reference survey data. Weights in the the survey in question are adjusted as shown in w_i^*.

Having one or more well-designed, area probability surveys that have full population coverage and high response rates as reference data sets is critical for this method to work. Some organizations have attempted to build their own reference surveys to use in adjusting Web surveys (Terhanian, 2000), but this will not be feasible for most organizations. Two obvious choices for household reference surveys in the United States are the Current Population Survey (CPS) and the American Community Survey. A supplement was appended to the CPS in February 2004 to examine telephone service and usage, yielding results that made it clear that work is needed on how best to collect coverage data. Tucker et al. (2007) noted that the term "landline" was unfamiliar to elderly respondents, that the phrase "working cell phone" was ambiguous, and that there was confusion about how to treat mobile telephones used primarily for business. Research on appropriate wording of coverage questions and periodic inclusion of such questions on some major government surveys seems well worthwhile.

8.4 CONCLUSION

Propensity score weighting is a versatile way to reduce biases in survey statistics due to nonresponse and noncoverage. The propensity score modeling approach allows flexibility in choosing the form of the model and the predictor variables. The covariates in the model can be quantitative or categorical and can include higher order interactions among variables. The propensity scores give a compact, scalar summary of the effect of many covariates. The method is more flexible than weighting class adjustment since a propensity model may include main effects only or only a subset of the full set of interactions among categorical variables used in class formation. Modeling also avoids the problem of zero observations in some adjustment classes. Survey operations can also benefit from the knowledge gained through propensity score modeling. A model can identify the correlates of nonresponse and noncoverage, which can be useful in devising plans for nonresponse conversion.

For nonresponse adjustment, a model is constructed to predict the probability of a unit's responding given a set of covariates that must be measured on both the respondents and the nonrespondents. For noncoverage adjustment, a model must predict the propensity of being covered by the survey frame. This typically requires a separate reference survey that does fully cover the desired target population.

While propensity score weighting is appealing, its effectiveness depends on how closely its assumptions are met. The most critical assumption is "strong ignorability," in which, given a propensity score, the distribution of an analysis variable y is the same regardless of whether a unit is a respondent or not (or is covered by the frame or not). This assumption generally cannot be checked with the data available in most surveys. The propensity score models should include important predictors of both propensities and research variables. All units in the population should have a nonzero probability of responding (or being covered by the frame). If this assumption is violated, sound inferences are possible only if analysis variables for responders and nonresponders (or covered and noncovered) follow a common superpopulation model—another assumption that will be difficult to check.

If important predictors of response or coverage propensities are omitted from a model or are simply unavailable, the model estimates will be biased and the weight adjustments will not remove nonresponse or noncoverage bias as successfully. In addition, the relationship between the set of covariates and the research variables is likely to differ case by case. For some variables, propensity score weighting may be effective, but for others, it may not be as useful. Proponents of propensity score weighting generally recommend using as many covariates as feasible in modeling—even ones that are weak predictors.

These legitimate criticisms of the method are related mainly to the impracticality of verifying the validity of the propensity model. It should be noted, however, that similar failings apply to any method that attempts to adjust for nonresponse and noncoverage biases. In fact, making no such adjustments also relies on an assumption that survey respondents represent not only nonrespondents but also the population. On balance, adjustment based on propensity models, though not perfect, is one of the most useful techniques available for bias reduction.

PART III

Data Collection

CHAPTER 9

Interviewer Error and Interviewer Burden

Lilli Japec
Statistics Sweden, Sweden

9.1 INTRODUCTION

To reduce survey error caused by the respondent's response process, theories have been developed to help us understand *why* respondents behave the way they do (Tourangeau, 1984; Willimack and Nichols, 2001). Interviewer error has also been recognized and described in the literature (Biemer and Lyberg, 2003). This literature, however, to a lesser extent, has addressed the question of *why* interviewer error occurs and why interviewers behave in certain ways. Understanding the reasons behind certain behaviors is crucial to reducing interviewer error.

In this chapter, we address some of the mechanisms at work. We model the interview process and describe interviewers' cognitive processes. But the discussion concentrates on a multifaceted concept called interviewer burden, with components such as administration and survey design, respondent and interviewer characteristics, social environment, and tasks. We define the concept of interviewer burden and its potential effects in a model. One effect is interviewer satisficing, that is, the phenomenon that occurs when the interviewer does not expend the necessary effort to complete an interview as prescribed. Satisficing has been discussed in the survey literature but then in relation to respondents only. We use results from two interviewer studies to identify what interviewers find burdensome.

In Section 9.2, the nature of the interview process and some existing models of it are described. This provides the starting point for new models designed to integrate new and common knowledge, addressing interviewer aspects that to our knowledge

Advances in Telephone Survey Methodology, Edited by James M. Lepkowski, Clyde Tucker, J. Michael Brick, Edith de Leeuw, Lilli Japec, Paul J. Lavrakas, Michael W. Link, and Roberta L. Sangster

have not been addressed in previous models. In Section 9.3, a model is introduced that describes the cognitive steps the *interviewer* has to go through during an interview, including the interaction with the respondent. We also discuss errors that can occur in these steps. The concept of interviewer burden is described in more detail in Section 9.4. We test some of our hypotheses described in the interviewer burden model by means of two surveys conducted among Statistics Sweden's interviewers. Section 9.5 discusses methods to measure and indicators of interviewer burden. Finally, Section 9.6 discusses how interviewers handle interviewer burden and potential effects of burden on data quality and what survey designers and managers can do to reduce interviewer burden. The Appendix provides details regarding the interview organization at Statistics Sweden, and results and the data collection procedures in the two interviewer surveys.

9.2 MODELS ASSOCIATED WITH THE SURVEY INTERVIEW PROCESS

In an interview survey there are two main players, the respondent and the interviewer. Both have an influence on data quality. In this section, models and theories are reviewed that have been developed to describe the survey interview and the cognitive processes involved in various interactions.

9.2.1 Respondent

Respondents' cognitive processes have long been known to affect data quality. Tourangeau (1984) describes four cognitive steps that a respondent needs to go through to provide a response to a survey question in a response model: understand the question, retrieve relevant information, make a judgment, and select a response. Problems can occur in all four steps and jeopardize data quality. The response model is general and applies to all types of questions and data collection modes. It gives a good structure for identifying potential problems with questionnaires (Snijkers, 2002; Biemer and Lyberg, 2003; Willis, 2005) and for guiding research in questionnaire design.

Tourangeau's response model has also been expanded to fit the response process in establishment surveys (Willimack and Nichols, 2001). Beatty and Herrmann (2002) investigate more closely the second step of the response model, retrieval of information, and classify a respondent's knowledge about the requested information in one of the four states (available, accessible, generatable, and inestimable). They conclude that any of these states can result in either a response or an item nonresponse. The quality of an answer will, however, depend on the information state. They demonstrate how survey questions can be evaluated by asking respondents about the availability of the information requested.

The cognitive steps in Tourangeau's response model and the availability of information in Beatty and Hermann's model will have an effect on respondent burden. If the burden is too heavy, it is believed to have a negative effect on data quality causing nonresponse or measurement errors. The concept of respondent or response burden has been described in a number of different models (Bradburn, 1978; Fisher and Kydoniefs, 2001; Haraldsen, 2002, 2004). Respondent burden comprises not only the time it takes to participate but also other factors such as the nature of the task, topic sensitivity, topic complexity, and the setting of the interview. For instance, presence of family members or an unpleasant interview environment might add to the perceived burden.

The survey organization, the respondent, the interviewer, or all three can take measures to reduce response burden. For example, the survey organization can minimize the number of questions. An interviewer may change question wording (even though this is accepted only in studies using conversational interviewing; see Suchman and Jordan, 1990) if this will help the respondent and the interviewer is convinced that the original wording is poor. Respondents can deal with burden by deliberately choosing a strategy that minimizes it. According to satisficing theory (Krosnick, 1991, 2002), some respondents do not expend the cognitive effort needed to provide an optimal response. They take shortcuts through the cognitive steps described in Tourangeau's model and provide an answer that seems reasonable and does not require them to work hard. But it is not only the respondent that might perceive the survey request or the interaction burdensome. The interviewer might experience similar feelings. Interviewer burden will be discussed in more detail in Section 9.4. First consider other models that incorporate the interviewer.

9.2.2 Interviewer

Many studies discuss interviewer effects without explicitly referring to an interviewer model, for example, interviewer influence on response and participation (Groves and Couper, 1998), interviewer strategies and attitudes (Lehtonen, 1996; Hox and de Leeuw, 2002), interviewer variance (O'Muircheartaigh and Campanelli, 1999), race matching, interviewer voice (Oksenberg and Cannell, 1988; Groves et al., 2007, Chapter 18 in this volume), usability testing (Couper et al., 1997; Hansen et al., 1997), and interviewing techniques (Suchman and Jordan, 1990; Schober and Conrad, 1997).

There are some early models that describe the interview process. Sudman and Bradburn (1974) illustrate the interaction among the interviewer, the respondent, and the task that leads to a response. The focus is on the instrument and the effects it can have on responses. Tourangeau et al. (2000) adapt the Cannell et al. (1981) model of the response process. These models describe processes that lead to carefully considered answers and processes that lead to inadequate answers. The focus is on the respondent, and the interviewer is present as a factor that can influence a respondent's choice of response.

Schaeffer and Maynard (1996) and Schaeffer (2002) call for more detailed research on the interaction process between interviewer and respondent. They claim that such research can help identify features of survey questions that cause interviewer variability and help to improve instrument design. Hansen (2000) provides an extensive overview of earlier models and proposes together with (Couper and Hansen, 2001) a social cognition model that describes the interaction among computer, interviewer,

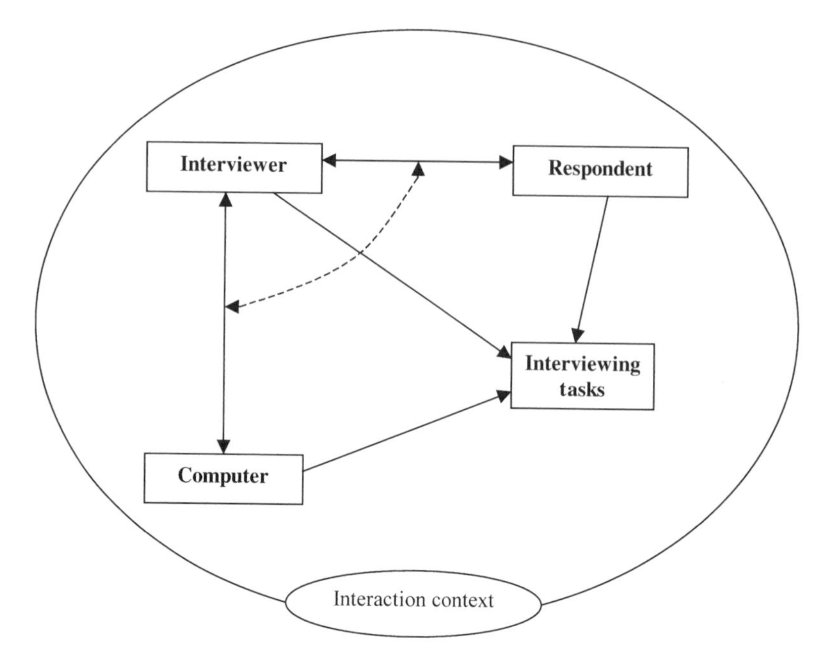

Figure 9.1. The Hansen and Couper social cognition model (Hansen, 2000; Couper and Hansen, 2001): Computer-interviewer-respondent interaction focusing on interviewing tasks within an interaction context.

respondent, and the interviewing task within an interaction context (Fig. 9.1). In a second model Hansen (2000) takes a closer look at the interaction between interviewer and respondent, and in a third model describes the interaction between the computer and interviewer. The Hansen models describe the *decisions* that the interviewer has to make and the *cognitive processes* the respondent has to go through when answering a question. For example, the interviewer has to decide, on the basis of the response, whether to clarify a question or not. The interviewer's cognitive processes underlying this decision are, however, not part of the models.

The respondent models (see Section 9.2.1) focus on respondents' cognitive processes. These models provide a framework that can guide research and theories regarding *how and why* respondents act in a specific way. These theories can help us design surveys in a way that minimizes undesirable respondent behavior. Current models that exist concerning interviewers focus on how interviewers influence respondents. These models do not help us understand *why* interviewers behave in certain ways. To reduce interviewer effects we need to identify the root causes of these effects. We need models that explicitly address interviewers' cognitive processes. Such models could guide research on *how and why* interviewers act in specific ways. An interesting component of such a model is interviewer burden, a concept analogous to respondent burden. A general model of the interview process with both associated interviewer and respondent cognitive processes is described in Section 9.3; then, a model is presented of interviewer burden aspects in Section 9.4.

9.3 THE INTERVIEW PROCESS MODEL WITH INTERVIEWER AND RESPONDENT COGNITIVE PROCESSES

9.3.1 Introducing the Interview Process Model

The model presented in Fig. 9.2 describes the question–answer process in interview surveys. The model is general and applies for both telephone and face-to-face surveys, and standardized and conversational interviewing techniques. Telephone and face-to-face interviewing constitute two different modes of data collection (de Leeuw, 2005). They both are interviews but their characteristics can differ substantially. For example, the possibility for monitoring is more limited in the face-to-face mode resulting in differing error structures. More specifically, the interviewer variance is usually larger in face-to-face surveys (Groves, 1989). Face-to-face surveys, however, are more flexible since visual aids can be used and more extensive interviewer assistance is possible. In telephone surveys, the set of available cues that interviewers pick up are different from that in face-to-face surveys (Conrad et al., Chapter 10 in this volume) and the social context in which the interview takes place also differs. There are also differences within the telephone mode depending on whether a listed sample or an RDD approach is used and whether a centralized facility or a decentralized organization where interviewers make telephone calls from their homes is used. Thus, the difference between modes and interviewing techniques stems from different sets of restrictions, and the cognitive demands on interviewers are affected by these restrictions. These cognitive demands will affect interviewer burden (see Section 9.4).

The model in Fig. 9.2 is different from earlier models in various respects. It considers both interviewer and respondent cognitive processes. The upper part of the model refers to the cognitive steps the respondent has to go through when asked to perform the survey request (Tourangeau, 1984) and the degree of availability of the requested information in memory (Beatty and Hermann, 2002). The lower part of the model reflects the interviewer's cognitive processes and the interviewer's access to concepts and definitions used. The data collection steps are marked in gray. The frame illustrates the social context in which the interview takes place. The social context in which the interview takes place affects the interaction. Issues such as whether the interview is done by telephone (landline or mobile) or face-to-face, whether the interview is carried out at the respondent's home or in a neutral setting, whether the interviewer and respondent belong to different social groups, or whether others are present or not illustrate the social context and their effect on the interviewer, the respondent, and the survey result.

9.3.2 The Interview Process—Step by Step

The model focuses on the question–answer process. Before that, however, the interviewer must find the sampled person. This task will vary depending on the type of telephone used or whether it is a face-to-face survey. In listed samples, the interviewer will have to trace some telephone numbers. Most telephone numbers, however, are in Northern European official statistics matched to individuals and based on

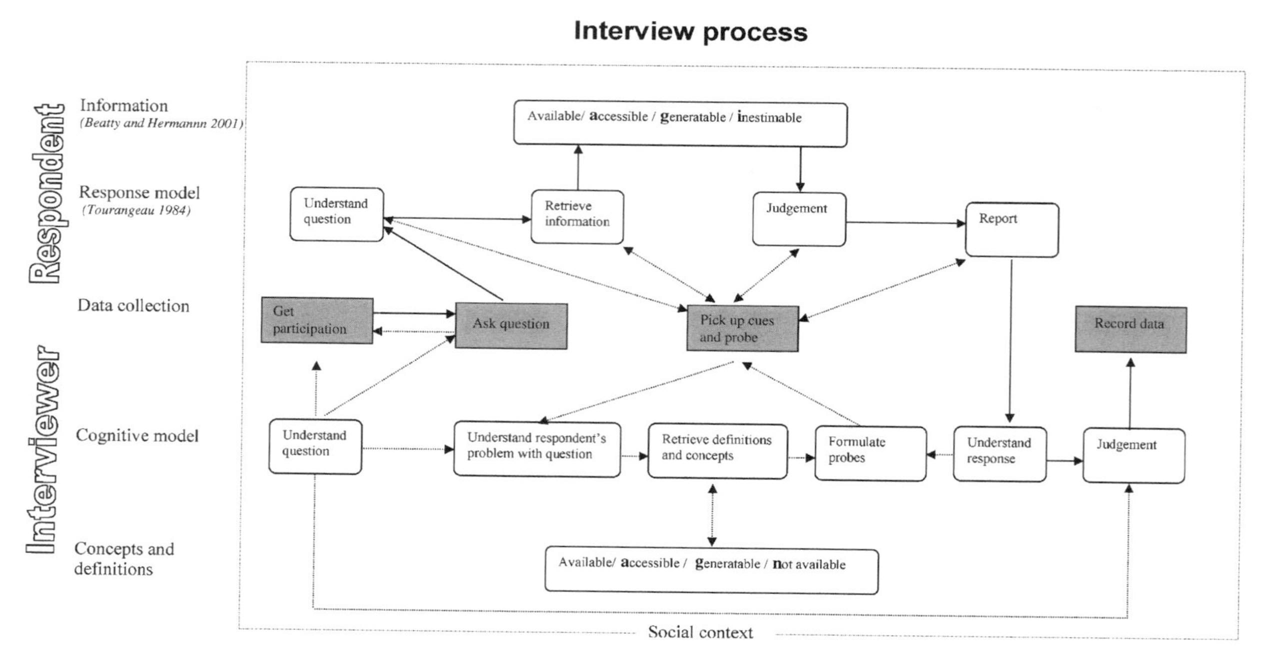

Figure 9.2. A model of the question – answer process.

the type of telephone service (landline or mobile). In RDD surveys the interviewer has to confirm eligibility, identify the household, and select the respondent. In face-to-face surveys, the identification procedure is carried out by enumerators or address lists are provided for (e.g., a census or a register) making the identification task easier for the interviewer. Once the identification and contact have been established, the interviewer might have to persuade the person to participate.

A number of issues will influence a respondent's decision as to whether or not to participate in a survey: interest in the survey topic, interviewer, attitude toward surveys, and the timing of the call attempt. All steps preceding the interview involve activities that can be rather burdensome for the interviewer, but the focus of the model in Fig. 9.2 is the interview process. In the following paragraphs, for the interviewer portion of the model, bold headings indicate interviewer tasks (illustrated as gray boxes in Fig. 9.2) and italic headings indicate the cognitive steps of the model. Data from two interviewer surveys that were carried out in order to test some of our hypotheses are used to examine the components of the model (see Appendix 9A for detail).

Get Participation

Once the interviewer has succeeded in getting a respondent to participate, the interviewer has to keep the respondent motivated to provide answers to all survey questions. If the respondent does not feel sufficiently motivated, it could result in termination of the interview, item nonresponse, or respondent satisficing (see Section 9.2.1). The interviewer also has to make the respondent feel sufficiently comfortable to give an honest answer. This is particularly important in surveys containing sensitive questions. If the respondent does not feel comfortable enough to give the correct answer, he or she might deliberately report an incorrect answer thereby causing social desirability bias. The interviewer's own motivation will have an effect on how successful he or she will be in getting respondents to answer all the questions. Interviewer motivation is further discussed in Section 9.4.

Understand Question—Comprehension

Understanding the intention of questions and how the answers will be used can help the interviewer to motivate and give correct feedback to respondents. In order to detect that a respondent has misunderstood a question, the interviewer needs to understand the question.

Variation in interviewer understanding of questions and instructions can cause interviewer variance and bias the estimates. For example, if some interviewers interpret a vague term used in a question in the same incorrect way, then this can lead to a biased estimate. If interviewers comprehend questions in individual ways, this can lead to interviewer variance. The bottom line is that interviewer comprehension of questions is crucial to survey quality but that acquiring and maintaining this knowledge can be burdensome. The obvious way for an interviewer to ease this burden is to put less emphasis on instructions, updates of instructions, and rely more heavily on standardized interviewing. Further, this step is likely to influence the amount of probing and the ability or willingness to react to respondent cues of comprehension difficulties.

Ask Question

Most survey organizations use a mix of standardized and conversational interviewing techniques, starting out with a standardized questionnaire but allowing interviewers to use probes to a certain extent.

When asked to clarify a question, the interviewer might reword the question in a leading way. Questions can contain terms that are not familiar to the respondent, and interviewers then need to use other terms to explain the meaning. Different interviewers might use different terms thus resulting in interviewer bias and variance. If the question-asking situation is such that clarification requests are met by interviewer statements such as "whatever it means to you" the interviewer burden is minimal, but as interviewing approaches a more conversational form, the burden increases.

In some situations interviewers are expected to translate questions before they can ask them (Harkness et al., 2007, Chapter 11 in this volume). Doing this on the fly triggers a new set of cognitive demands on the interviewer's part that is bound to be very burdensome and has a negative effect on data quality.

It is not uncommon for respondents to have difficulty understanding some of the questions. Therefore in the second of the interviewer surveys, we asked interviewers what they do when respondents do not understand a specific question. We stated four different strategies and asked interviewers how often they use each strategy (most of the time, sometimes, rarely, or never). The most common strategy was to repeat the question (67 percent of the interviewers said that most of the time they do this). Seventeen percent said that most of the time they try to explain the meaning of the question to the respondent in their own words. Fifteen percent said that most of the time they change a word that the respondent does not seem to understand. Finally, one percent said that most of the time they just proceed to the next question. Interviewers do not seem to distinguish between attitude questions and factual ones when choosing a strategy to make the respondent understand.

Pick Up Cues and Probes

If the interviewer does not pick up cues that a respondent might have a problem with a question and consequently does not probe, this can lead to biased estimates. This is a step in the process that is poorly understood, and it is closely linked to the subsequent three cognitive steps. What cues are interviewers picking up during the interaction with the respondent (Conrad et al., 2007, Chapter 10 in this volume)? What strategies are interviewers using for picking up cues in different interview modes? In telephone surveys, the cues are more limited compared to face-to-face surveys. How do interviewers choose the right probes? When do interviewers ignore cues and when do they address them with a probe? Different interviewers pick up different cues and the amount of probing also varies.

Understand respondent's problem with question. The interviewer should be able to understand a problem that a respondent has with a question. Sometimes the respondent might not explicitly tell the interviewer that he or she does not understand. The interviewer should then be able to pick up cues and probe what the problem is. Even if the respondent tells the interviewer what the problem is, the respondent has to express the problem in such a way that the interviewer will understand the exact problem. The amount of effort the interviewers put into understanding problems that respondents have with questions varies by question and interviewer.

Retrieve definitions and concepts. The accuracy of this step is affected by such things as the number of concepts and definitions the interviewer has to remember, the cues the instructions provide, and how long ago the interviewer last had to recall the concepts and definitions. Definitions and concepts are essential in a survey. The interviewer should be able to recall the definitions learned in training or from available instructions. The interviewer has to adopt a retrieval strategy. If an interviewer is working on many surveys at the same time, it will be difficult to remember all definitions and concepts. It could also be difficult to distinguish between concepts and definitions in different surveys if they are similar. In the second interviewer survey, we asked interviewers whether it is difficult to keep definitions apart for the surveys that they work on (see Appendix), and 47 percent said that this is the case sometimes or most of the time.

The model developed by Beatty and Herrmann (2002) regarding respondent knowledge can be adapted to fit interviewer knowledge of concepts and definitions. Interviewer knowledge can be in one of the following four cognitive states regarding a certain question:

- *Available*: the concepts and definitions can be retrieved with minimal effort
- *Accessible*: the concepts and definitions can be retrieved with some effort
- *Generatable*: the concepts and definitions are not exactly known but may be guessed using other information in memory
- *Not available*: the requested concepts and definitions are not known

The first cognitive state is the least demanding for the interviewer. The other cognitive states are likely to be increasingly more demanding and interviewers are likely to handle these states very differently. For example, if a definition is not known exactly (generatable), the interviewer might recall a similar problem from an earlier interview and adopt the same definition for the current interview. Some interviewers might choose this strategy whereas others might choose to record a "don't know" answer. In any case, burden increases from the first to the last cognitive state.

Formulate probes. The interviewer's skill in formulating probes is important for the final outcome. The probe has to be neutral and convey the intended meaning of a question. The interviewer can probe either to make the respondent understand a question or to understand a respondent's answer. In both cases the interviewer must master the survey goals, definitions and concepts, how to handle borderline cases, and other problems. Even if the interviewer is familiar with definitions and concepts, it may be difficult to describe these to the respondent. It also requires the interviewer to understand the problem a respondent might have with a question or a probe. Interviewers vary in their skill at formulating probes.

Understand response. To be able to record the data, the interviewer must understand a respondent's answer. The interviewer must grasp the essential information from what the respondent is saying. Some situations will require more effort on the part of the interviewer to understand a response, for example, when interviewing a person who does not understand or speak the language well, or who has an extreme dialect or a hearing problem. The ability of understanding a response varies among interviewers and respondents' ability to express their answers that will affect whether understanding will be perceived as burdensome by the interviewer or not.

Judgment. The process when the respondent sums up all the information, retrieved from memory, whether complete or incomplete, in order to come up with an estimate is the judgment process. The interviewer has to make different but parallel judgments. When the respondent has difficulty making a judgment, this task can then be partially transferred to the interviewer. For example, if the respondent does not remember the number of times he or she did a particular thing, the interviewer might probe and help the respondent with a strategy to fill in memory gaps to come up with an estimate.

Another type of interviewer judgment that comes to play is that when the respondent does not use the wording as stated in the response categories. Hak (2002) looks at how interviewers make such judgments and the kind of decisions they make based on those judgments. Hak and Bernts (1996) distinguish between two types of problems: when the interviewer does not understand the response given and when it is difficult to map the answer to the proper response alternative.

The interviewer might also have to make a judgment about the accuracy of the answer based on what is reasonable and what the respondent has reported earlier in the interview. The interviewer might have to probe in these cases and must find a balance between the amount of probing and the prospect of the interview continuing in a friendly atmosphere.

Record Data

The interviewer should record a respondent's answer into the proper response category. The design of the instrument will have an effect on this step—an inconsistent use of codes for response categories such as yes and no in a survey or across surveys, for example, can cause errors and add burden to the interviewer.

Respondents are believed to sometimes report or record the wrong answer in order to avoid follow-up questions. This might also be the case for an interviewer: In order not to upset the respondent, the interviewer might choose to report "no" to a question and thereby avoid further follow-up questions.

In this section, the interaction process and interviewer cognitive processes involved in the interaction have been described. The following section deals with interviewer burden.

9.4 INTERVIEWER BURDEN

9.4.1 What Interviewers Say They Find Burdensome

To illustrate the concept of interviewer burden, quotes from the interviewer surveys are valuable. One of the open-ended questions in the first survey was "Are there any particular situations in your work that you often find stressful?" About two thirds (64 percent) of the interviewers responded "Yes" and gave examples of such situations.

Many interviewers say that they find it stressful when the survey administration is poorly planned. Examples of poor planning include: the number of cases assigned to them is too large; the number of surveys they have to work on is too large; the initial fieldwork period is too short and then extended; call scheduling rules are elaborate and not taken into account in case assignments; tracing instructions are unclear; survey instructions are sent out late; workload fluctuates; and sometimes both telephone and face-to-face surveys have to be conducted during the same week. Other things that have to do with survey administration that the interviewers mention are lack of positive feedback and clearly defined expectations. Interviewers also mention matters related to the survey instrument that can be stressful. These include questionnaires that have not been tested, poor instructions, routing errors, lack of concept definitions, new instructions that are sent out in the midst of the fieldwork, telephone follow-up of mail surveys with no interviewer instructions, long introductions before questioning starts, and asking sensitive or offensive questions. Other stressful events or tasks include getting many refusals in a row, tracing sampled persons, carrying out long interviews (more than 30 minutes), persuade reluctant respondents, and dealing with respondents who are in a hurry or who are very talkative.

We also asked interviewers to rank how important the different survey aspects are to them to carry out their job on a 4-point scale: very important, important, not so important, and not at all important. Interviewers said that it is very important that questions are well-designed (86 percent), that the survey topic is important to society (62 percent), and that the survey sponsor shows interest in the survey (38 percent).

In the second survey, we asked interviewers how stressful they found certain situations. Lack of information about the purpose of a survey and not having exact definitions of concepts were considered to be either very stressful or stressful by 83 percent and 79 percent of the interviewers, respectively. Carrying out an interview with a person that does not understand Swedish well is considered to be either very stressful or stressful by 51 percent of the interviewers, and carrying out a long telephone interview (more than 30 minutes) is considered to be very stressful or stressful by 46 percent of the interviewers.

The results of the two surveys support our hypothesis that there are situations that are perceived as burdensome by the interviewers.

9.4.2 Introducing an Interviewer Burden Model

The model in Fig. 9.2 illustrates the links between the interviewer and the respondent and the cognitive processes involved. Obviously, there are many different ways to conduct an interview. A recorded answer will depend not only on the respondent but also on the interviewer and other factors. There are many steps in the interview process that are crucial to data quality. For example, many survey organizations take measures to reduce respondent burden since it is known to affect data quality. From the model it is clear that in interview surveys it is not only the respondent burden that is a concern. One also needs to consider interviewer burden. Our hypothesis is that one should avoid the interviewer burden while at the same time preserving enough varieties in the interview work to make it interesting and challenging to the

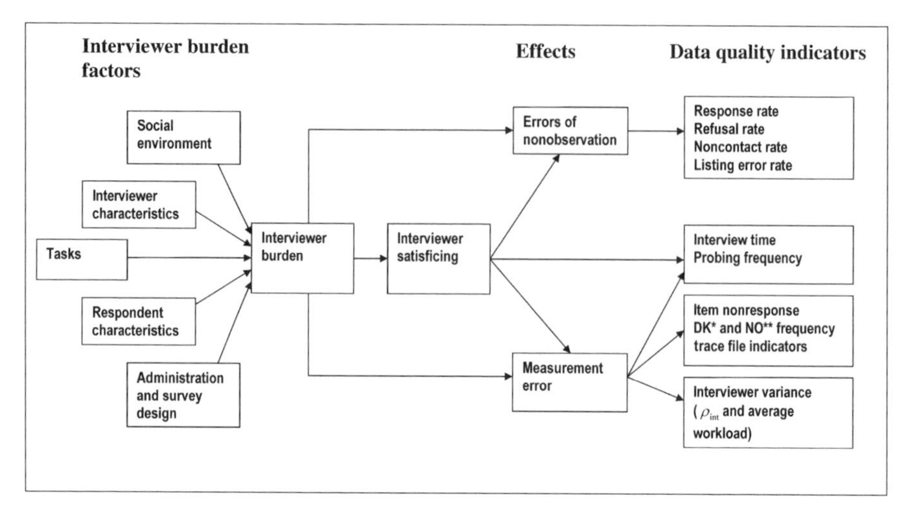

Figure 9.3. Interviewer burden factors, effects, and data quality indicators. The model contains some adaptations suggested by Geert Loosveldt, personal communication. *DK frequency is "don't know" frequency. **No frequency is "no opinion" frequency.

interviewers. Components such as complexity of definitions and concepts, number of different surveys an interviewer works on simultaneously, design of instrument and instructions, work environment, time pressure, and length of interview will all affect data quality.

We define interviewer burden as the total amount of perceived effort, both physical and cognitive, an interviewer has to put in to complete an interview according to specifications. The concept of interviewer burden is depicted in Fig. 9.3. Again in this model, respondent, interviewer, and survey characteristics are combined. We see that interviewer burden can result in interviewer satisficing and generates nonobservation errors such as nonresponse and noncoverage and measurement errors such as interviewer variance (see also Section 9.7).

In Fig. 9.3, interviewer burden is a function of the social environment, interviewer characteristics, interviewer tasks, respondent characteristics, administration, and survey design. What will be perceived as burdensome will vary between interviewers. In this section, each factor in Fig. 9.3 that can contribute to interviewer burden is discussed.

9.4.3 Interviewer Characteristics

Each interviewer has a set of attributes, skills, interests and attitudes that interact with respondent characteristics, social context, interviewer tasks, survey design, and administration affecting what the interviewer finds demanding.

Interviewer attributes. An interviewer's attributes, such as age, gender, education, experience, personal style, self-esteem, and religiosity, are all part of the interview context. These attributes will influence how an interviewer perceives an interview. For example, a religious interviewer might find it very burdensome to ask questions about sexual behavior. An anxious interviewer might be afraid that the respondent will hang-up and he or she speeds through the interview making it difficult for the respondent to fully understand the questions.

Interviewer skills. The role of the interviewer is to carry out many different tasks, such as gain cooperation from sample persons, persuade reluctant persons, motivate respondents during the interview, clarify concepts, and administer call attempts. These tasks require different skills, and interviewers' skills vary. For example, an interviewer who is very good at scheduling call attempts in an efficient way may not be very good at persuading reluctant persons. That interviewer is likely to find it more burdensome to persuade reluctant persons than to schedule call attempts. In the second interviewer survey, we asked interviewers about their views on some tasks and also how stressful they found certain situations. Many interviewers perceive persuading reluctant persons negatively (see Appendix 9A).

Interviewer tasks are such that they often require multitasking. The interviewer has to interact with the respondent and the computer, and pay attention to both while carrying out the interview (Couper and Hansen, 2001). Interviewers vary in their multitasking ability. In general, there is less multitasking involved in telephone interviews than in face-to-face interviews.

Interest and motivation. An interviewer's interests will influence interviewer motivation. His or her interests can include a personal interest in a specific topic or a specific task such as tracing sampled persons. It is likely that an interviewer who does not find a survey topic interesting will be less motivated and find it more burdensome to work on that survey than an interviewer who does find it interesting (Japec, 2004). In the interviewer survey, we asked interviewers to rank how interesting they find some of the surveys they work on. The data were analyzed together with the results from the Swedish part of the European Social Survey (ESS). Interviewers who find ESS very interesting had significantly higher response rates, lower refusal rates, recorded less do not know answers, and had longer interviews (Japec, 2005).

Attitudes. Interviewers' attitudes toward the survey client, the perceived importance of a survey, and their role as interviewers (Hox and de Leeuw, 2002; Japec, 2004) are all components that can contribute to interviewer burden. An interviewer who perceives a survey to be of little value to society might find it more burdensome to work on that survey than on others. Some of these aspects were ranked as very important by interviewers in the interviewer survey (see Appendix 9A).

9.4.4 Respondent Characteristics

In the interaction between the interviewer and the respondent, the characteristics of the respondent can add to interviewer burden. In this section, respondent characteristics and how they can contribute to interviewer burden are examined.

Respondent attributes. In Section 9.4.3, we discussed interviewer attributes such as age, personal style, and religiosity, and how these might influence interviewer burden. It is also relevant to talk about respondent attributes in similar ways. The interviewer has to adapt to different types of respondents. For example, some respondents feel embarrassed about talking about certain topics. The interviewer has to behave in such a way that respondents feel at ease when talking about sensitive topics. Respondents may not be consistent in their responses. The interviewer will have to probe and devote more cognitive effort to make sure that the respondent gives consistent answers. Again, some interviewers might perceive this to be demanding, and it can contribute to interviewer burden.

Motivation or interest in survey topic. A sampled person that is not motivated to participate or does not find the survey topic to be interesting will force the interviewer to work harder. For example, the interviewer will have to tailor persuasion efforts to gain participation, or during an interview the interviewer will have to keep the respondent motivated to expend the necessary cognitive effort to answer accurately. This can be perceived as burdensome (see Appendix 9A).

Language. Encountering respondents with language difficulties requires the interviewer to use an interpreter or to carry out interviews without any assistance. Using an interpreter takes additional time and is not straightforward. If the interviewer chooses to do an interview despite language difficulties, then he or she will have to put in additional effort to overcome the language barrier. The interview will also take longer. Irrespective of the presence of an interpreter, the situation is considered stressful by many interviewers. Fifty-one percent of the interviewers said it is stressful or very stressful to conduct an interview with a person who has language difficulty without an interpreter. Even with an interpreter, 29 percent still find the situation to be stressful or very stressful.

Attitudes toward surveys. Some respondents have a negative attitude toward surveys, government, or society and will therefore require the interviewer to expend more effort to get participation or to carry out the interview.

Behavior. Other respondent behaviors that can have an effect on interviewer burden include when the respondent seems to be reluctant, in a hurry, preoccupied, or very busy. Respondent behavior can cause interviewers to feel that they should try to end the interview as quickly as possible. This hypothesis is supported by the fact that 54 percent of the interviewers said that they feel negative or more negative than positive when faced with the task of having to persuade a reluctant respondent.

9.4.5 Administration and Survey Design

Administrative features. The general survey administration, including interviewer workload, mode, mix of face-to-face and telephone interviews, and the number of different surveys an interviewer works on, more or less simultaneously, can cause interviewer burden.

- *The workload.* The number of cases assigned to the interviewer can be large, automatically causing interviewer burden. In the interviewer surveys, we asked if interviewers had had a workload that was too heavy during the past 12 months, whether they had to skip some duties, and, if so, what duties they skipped. A worrying proportion answered in the affirmative (see the Appendix). We also asked if interviewers felt they had enough time to study instructions they received for new surveys and to brush up their knowledge about recurring surveys. Again, a worrying proportion (10 percent and 23 percent for studies and brush up, respectively) said that they rarely or never did. It is clear from what interviewers say that their strategies for handling a heavy workload will have an effect on data quality.
- *The number of surveys the interviewer is working on.* In some organizations the interviewer is working on more than one survey during the same period. In the interviewer survey 25 percent said that they were involved in too many surveys (Japec, 2005). The cognitive demands are higher if the interviewer is working on many surveys simultaneously, because there are lots of concepts and definitions to remember in multisurvey situations. Furthermore, if the concepts in some of the surveys are similar, the interviewer must remember the correct definitions for each survey.

Survey and instrument features The specifics of a survey and the instrument can be such that they require extensive effort on the part of the interviewer. Factors that can affect burden include the following.

- *Mode.* An interview can be paper-and-pencil or computer assisted. In a paper-and-pencil mode, the interviewer must follow the correct routing so that the right question is asked. This can be quite burdensome if the survey is complex. In computer-assisted interviewing, on the contrary, the routing is normally done automatically. The design features of RDD surveys are such that they require the interviewer to make many unproductive call attempts. This can be very burdensome to the interviewers (see Battaglia et al., 2007, Chapter 24 in this volume).
- *Question and questionnaire design.* If question and questionnaire design are poor, interviewers may expect negative reactions from respondents and the burden on them will increase. The interviewer must repair poorly worded questions or adjust for the effects of them. Common repair procedures include changing question wording or trying to explain the meaning of the question to the respondent.
- *Design differences and design inconsistencies across surveys.* If, for example, the code system and other design features vary across and within surveys conducted by the same interviewer organization, then the interviewers are forced to remember them all. Couper and Hansen (2001) provide examples of studies of the effects of inconsistent computer-assisted interviewing (CAI) design. For example, in an analysis of keystroke files, Couper et al (1997) found a significant amount of backups for an item due to inconsistent use of codes for "Yes" and "No" within a survey.

- *Sensitive questions.* Interviewers may feel uncomfortable asking sensitive questions and anticipate negative reactions from respondents that might influence the interaction.
- *Instructions.* Clear instructions will help the interviewer to carry out the interview. Unclear instructions, however, will have the opposite effect (see the Appendix).
- *Visual design of the instrument.* If the visual design of the instrument is such that, for example, the interviewer will have difficulty finding the correct routing or important information, this can contribute to interviewer burden.
- *Probes.* Sometimes the interviewer must probe or clarify a question for the respondent. If no scripted probes are provided, this will generate a cognitive burden that manifests itself as a need to formulate a probe on the fly. If the respondent gives an answer that is not a straightforward mapping on the response alternatives provided, the interviewer is forced to probe. Interviewer training in formulating probes is therefore important.
- *Length of the interview.* A very long interview can cause both respondent and interviewer burden. The length per se can be a problem but also the fact that the interviewer might feel stressed if he or she is expecting negative reactions from respondents as the interview proceeds (see Appendix 9A).
- *Asking the same question many times.* In household surveys the same information is sometimes requested from all household members. There are surveys where interviewers are instructed to read what they have recorded and ask the respondent for confirmation. In some surveys the same question is asked more than once during the interview to check for response consistency. Furthermore, the respondent sometimes volunteers an answer, but according to the instructions the interviewer must read all of the options. It could also be the case that prior comments from the respondent have already answered the question, such as the mentioning of a spouse or a child. In this situation, the interviewer might feel awkward reading all response options. Again, some interviewers will worry about a negative reaction from respondents to these repetitive procedures and the procedures may be perceived as burdensome.

Training, feedback, and information The results from the surveys show that the majority of the interviewers are very professional and keen to learn how to improve their skills (see Appendix 9A). Further, 83 percent of the interviewers said that it is stressful (or very stressful) to lack information about the purpose of a survey, and 79 percent of the interviewers said that it is stressful (or very stressful) to lack clear definitions of concepts in a survey.

9.4.6 Tasks

The three basic tasks that an interviewer has to perform are administering the survey, motivating sampled persons to participate, and conducting the interview according to instructions. These tasks will each require different amounts of cognitive effort, but they all contribute to interviewer burden. See also the description of interview process and interviewer tasks as discussed in Section 9.3.

9.4.7 Social Environment

The social environment within which the survey request and interview take place varies. This environment, whether the interviewer feels at ease in it or not, can affect interviewer burden. The interaction between the social environment and interviewer characteristics, respondent characteristics, and survey characteristics can cause interviewer burden.

Social group and setting. In their job interviewers meet many different people belonging to different social groups. The social distance between the interviewer and the respondent, the neighborhood, and language are aspects of the social setting that can contribute to burden.

Presence of others. Sometimes there are others present during an interview that might affect both the respondent and the interviewer. For example, in a survey on domestic violence, the interviewer might not feel comfortable asking such sensitive questions while other family members are present. In a centralized telephone facility, there are other interviewers present, and a noisy background can be disturbing when trying to interview someone.

Survey climate. The general attitude toward surveys and statistics in a country or region can contribute to interviewer burden. A negative debate in the media will make interviewers' job much harder.

9.5 INDICATORS OF INTERVIEWER BURDEN

As we have described in this chapter, we have good reason to believe that interviewer burden affects data quality. We should strive for improving our processes so that respondent burden and interviewer burden are minimized. Although interviewer burden (Fig. 9.3) consists of many different components, there are probably some components that have a greater effect on data quality than others. Reducing burden that has the largest effect on data quality in relation to costs seems like a good strategy, but we need burden indicators to be able to assess the effect.

In this chapter, clear indicators of interviewer burden, both qualitative and quantitative, were uncovered. Examples of burden indicators include heavy workload, low interviewer interest in specific survey, heavy administrative burden, poor question and questionnaire design, complex concepts, sensitive questions, poor interviewer instructions, lack of scripted probes, long interviews, and lack of training and information about the survey. Next we discuss methods on how to assess indicators of interviewer burden.

9.5.1 Interviewers' Own Reports

We have given examples in Appendix 9A of questions that can be included in an interviewer survey to illuminate problems related to administration and survey design, and to some extent also interviewer and respondent characteristics. The results can be used directly to improve procedures. Many of the indicators from the interviewer surveys can be used in evaluating the effect of interviewer burden on data quality (Japec, 2004) by linking them to interviewer results in surveys that they are working on.

Interviewers can be asked to provide notes on problems in surveys and on interactions such as the amount of probing needed and language problems in interviews that they conduct.

9.5.2 Qualitative Methods

Qualitative methods that are used to find out respondent's cognitive processes can be adapted to also learn about interviewer cognitive processes and interviewer burden. Some adaptation of methods has already been done. Methods such as interviewer focus groups and debriefings can provide information about administrative procedures that need to be improved, instructions, problems with questionnaires, complex concepts, CAI design, workload issues, design inconsistencies, training, and information needs. Also, usability tests and walkthroughs (Hansen et al., 1997) can provide information about problems that interviewers have with CAI instruments. In a walkthrough, the interviewer goes through the CAI instrument thinking aloud and commenting on potential problems in the instrument. In a usability test the interviewer performs interviews in a laboratory equipped with video- and audio-taping facilities.

Methods such as recording and behavioral coding of interactions among interviewer, respondent, and computer, and small-scale laboratory experiments can be further adapted as can think-alouds and structured probes. Combining recordings of interviews with retrospective think-alouds with interviewers can shed light on interviewer cognitive processes and problems caused in the interaction. This information can be used in order to improve instruments and interviewer training. Structured probes can be combined with retrospective think-alouds. The structured probes can focus on all stages of the interviewer's cognitive process.

9.5.3 Counts of Burdensome Situations per Interviewer or Survey and Other Process Data

Process data can be collected to control for interviewer burden. The indicators listed below are examples of process data. Other types of process data that can be collected are keystroke files for backups and errors that interviewers make. This information can be used to improve instruments.

Counts of burdensome situations can be obtained from, for example, surveys, administrative sources, and interviewer notes. Some examples are the number of call attempts, the number of refusal conversion attempts, the number of sample members with language problems per interviewer, the number of surveys at one point in time per interviewer, the number of reassignments per interviewer, the number of inconsistent codes per survey, and the number of poorly worded questions per survey.

9.5.4 Quality Control System

Many of the methods and indicators mentioned can be used in a systematic way to control for quality. For example, monitoring can provide information systematically about interaction problems due to poor questionnaire design, interviewer instructions, and lack of interviewer training in telephone surveys. Behavioral coding can be used to get information systematically about interaction problems in interviews. Finally, control charts can be used to make sure that interviewer burden is under control. For example, control charts for interviewers' workload and the number of reassignments per interviewer can be prepared. If an interviewer falls outside upper control limits, that is, has significantly larger workload and reassignments than other interviewers, the root cause can be interviewer burden (Japec, 2005).

9.6 EFFECTS OF INTERVIEWER BURDEN ON DATA QUALITY AND WHAT SURVEY MANAGERS CAN DO TO REDUCE THEM

9.6.1 Effect of Interviewer Burden on Data Quality

In most situations where the burden is heavy, interviewers will develop their own strategies to handle it. These strategies will vary among interviewers and affect data quality accordingly.

In the first interviewer survey, we asked interviewers what they usually do when they experience a workload that is too heavy. Answers included cutting down on the number of call attempts, making call attempts at times that are not optimal, not tracing all sampled persons, not trying to persuade reluctant respondents, cutting down on paperwork such as reading instructions regarding new and ongoing surveys, doing fewer face-to-face interviews if working on both telephone and face-to-face surveys, spending less time thinking about what the appropriate response alternative is for a respondent during an interview, and probing less during the interview.

Respondents sometimes take shortcuts in surveys and do not put in the cognitive effort needed to accurately answer a question. This behavior is known as satisficing (Krosnick, 1991, 2002). It should be clear that interviewers might also use various strategies to reduce the total amount of cognitive effort needed to complete an interview in a prescribed manner. This behavior might be called interviewer satisficing (Japec, 2003, 2004). For example, interviewers might short-circuit their cognitive processes by deciding not to persuade a reluctant respondent to participate, or decide not to probe, even though the respondent does not quite seem to understand a question, or they might not read question instructions.

Hansen (2000) found that some interviewers skip some questions and that interviewers sometimes do not use show cards in the U.S. National Health Interview Survey. A deliberate reduction in effort could be an indicator of interviewer burden or a sign of sloppy work and lack of motivation. It could also be a way for the interviewer to reduce respondent burden in order to be able to carry out the rest of the interview smoothly. One hypothesis is that reduction in effort that correlates with other indicators of burden calls for a system change, while when reduction of effort is a constant behavior, this calls for specific individual measures.

In our model described in Fig. 9.3, interviewer satisficing can lead to both errors of nonobservation and measurement errors. A very short interview time with little or no probing during the interview could be the result of interviewer (and respondent) satisficing (Japec, 2004).

Interviewer burden can contribute to errors of nonobservation directly or indirectly through interviewer satisficing. An example of a direct effect would be an interviewer that gets too many cases and does not have time to contact all the sampled units. This would result in a high noncontact rate for that interviewer. An example of an indirect contribution to errors of nonobservation is when an interviewer satisfices to deal with burden and does not persuade a reluctant respondent to participate in a survey.

Interviewer burden can also contribute to measurement errors directly or indirectly through interviewer satisficing. An example of a direct effect is if an interviewer finds it difficult to ask a respondent a sensitive question or chooses to tell the respondent that he or she does not have to answer the question. This strategy can result in a high "don't know," "no opinion," or item nonresponse frequencies for that interviewer on that specific question.

Interviewer variance is between-interviewer variance (Biemer and Lyberg, 2003). The problem with a large interviewer variance is that the accuracy of estimates is reduced, and this reduction cannot be measured easily. It is likely that interviewer burden contributes to interviewer variance. For example, interviewers who find it difficult to ask sensitive questions may use different strategies to handle this problem. For example, one interviewer tells respondents that they do not have to answer that question, another interviewer skips the question and records an item nonresponse, and yet another interviewer chooses to change the wording of the question so that it becomes less sensitive.

An example of measurement error directly caused by interviewer burden is the following: because of a heavy workload the interviewer does not read instructions properly and misunderstands the researcher's intent with a question. This is likely to cause bias in a situation when the interviewer should clarify a question to the respondent.

9.6.2 What Survey Designers and Managers Can Do to Reduce Interviewer Burden and Its Effects

In most, if not all, survey organizations, the concept of burden is confined to the respondent's situation or burden. At first glance it might seem as if respondent and interviewer burden have very little in common. However, many of the measures taken to reduce respondent burden also help to reduce interviewer burden. For example, formulating and testing questions so that respondents understand the meaning of the questions (thereby reducing their cognitive burden) will also reduce the amount of probing interviewers will have to do.

Interviewer burden can be a problem (Japec, 2002, 2003, 2004) and it can affect data quality indicators such as response rates, item nonresponse, and length of interview (Japec, 2004). The conclusion is that we need to pay attention to interviewer burden and find ways to reduce it. The interviewer burden components that are under the direct control of survey designers and managers are administration and survey design, and, to some extent, also tasks. Furthermore, survey managers also have the possibility of training and influencing interviewers. Respondent characteristics and social environment are components that survey managers have very little control over. Still there may be measures such as assigning interviewers to cases where their burden is likely to be low by matching respondent age and area to interviewers with certain characteristics. Not much is known about the effects of such measures; further research is needed.

Other examples of measures to reduce interviewer burden include the following.

- Develop and test questions and instructions so that they become interviewer friendly. Long or tedious interviews should be avoided both for the sake of the respondents and the interviewers.
- Identify an "optimal" workload both regarding the number of cases and the number of survey. The optimal workload will be affected by, for example, type of survey, interview length, number of call attempts required, and travel time for face-to-face interviews. The optimal workload will vary among interviewers and regions. In call centers the optimal workload should be defined in relation to interviewers' working hours.
- It is important to make sure interviewers receive survey information and instructions well in advance and that they know that it is important to read and understand the information.
- The interviewers should be kept motivated. There are many things that can decrease interviewer motivation: poorly designed instruments, sloppy survey plans, uncoordinated surveys, and confusing instructions. Feedback and interviewer incentives are measures that can help keep interviewers motivated.
- Standardized codes should be used within and across all surveys, and all concepts should be clearly defined.
- If there are sensitive questions, using audio computer-assisted self-interviewing (A-CASI) should be considered.
- Feedback and training will help interviewers to improve their skills. Sixty-two percent of the interviewers said that they would like to enhance their interviewing skills, and 55 percent of the interviewers wanted to improve their skills with motivating reluctant respondents. Thirty-six percent of the interviewers said that they only sometimes, rarely, or never have enough time to study new survey instructions, and only 44 percent said they had time to brush up on recurring surveys. Management is responsible for changing this state of affairs.

Clearly, the interviewers are very important to data quality, and everything possible should be done to improve their possibilities to do a good job. A variety of phenomena can be classified as burdensome to interviewers, and excessive burden affects data quality in a negative way. We have provided some insights into the concept of interviewer burden and conclude that interviewer burden can have as large an effect as excessive respondent burden on data quality. The good thing is that interviewer burden is easier to control and to do something about. However, the right mix of measures still needs to be identified.

APPENDIX 9A: DATA ON WHICH THE INTERVIEWER BURDEN MODEL IS BASED

This appendix gives results from the two interviewer surveys described in Section 9.3.

To obtain data for testing various hypotheses underlying the interviewer burden model, two surveys were conducted. Statistics Sweden has a centralized facility with interviewers that conduct telephone interviews and a decentralized organization with interviewers located around the country. The decentralized interviewers carry out telephone interviews from their homes and sometimes also face-to-face interviews. The vast majority of the surveys are telephone surveys. When the interviewer surveys were conducted, there were 74 interviewers in the centralized facility and 150 decentralized interviewers. Interviewers in the centralized facility worked on average with two different surveys simultaneously while the corresponding number for decentralized interviewers was six. Each decentralized interviewer receives a number of cases to interview. Interviewers in the centralized facility, however, do not have a prespecified individual workload since they pick up their cases from a common database. Decentralized interviewers have a monthly salary, and the majority of the interviewers in the centralized facility are paid by the hour. The majority of Statistics Sweden's decentralized interviewers have long experience working as interviewers: The average number of years decentralized interviewers have worked as interviewers was 8 years. Only 6 percent have worked for less than a year, and 20 percent have worked between 15 and 33 years.

The two surveys among Statistics Sweden's interviewers were conducted during the period from November 2002 to February 2003 (Japec, 2003). All interviewers received both questionnaires. The first interviewer survey (IS-1) was conducted in November 2002. Interviewers were asked about their opinions on training, feedback, working hours, particularly stressful situations, and workload. A mail questionnaire was used, and the response rate was 82 percent (183/224). Since some of the questions were deemed sensitive, the survey was anonymous. The second interviewer survey (IS-2) was mailed to the interviewers in December 2002. The response rate was 79 percent (177/224). The IS-2 contained questions on interviewers' strategies, attitudes, preferences, stress, and opinions about surveys they were working on. This survey was not anonymous. Results are presented from both surveys.

Interviewer Survey 1 (IS-1)

IS-1.14

How important are the following issues for your work?

	Very important	Important	Not so important	Not at all important
a That the survey is important to society	62%	34	3	1
b That the questions are interesting to you	13	34	43	9
c That the questions are well designed	86	13	0	1
d To get feedback regularly	19	54	22	5
e That the client shows an interest in the survey	38	44	16	2

IS-1.15a

Are you interested in enhancing your knowledge in any of the following areas? *You can choose more than one alternative.*

37% To plan the work and allocate the time between different surveys
44% To trace sampled persons
55% To persuade reluctant respondents to participate
62% To improve your interviewing skills
16% Yes, please describe

IS-1.17

In your opinion, do you have enough time to study instructions and other survey materials that are sent out together with *new* surveys?

64% Most of the time
26% Sometimes
8% Rarely
2% Never
[Item nonresponse 1/183]

IS-1.18

In your opinion, do you have enough time to brush up your knowledge about *recurring* surveys?

44% Most of the time
32% Sometimes
22% Rarely
2% Never

IS-1.23

(Decentralized interviewers only)
During the past 12 months, has it happened that you have had more sampled persons to call than the time allowed?

28% Often
43% Sometimes
20% Rarely
8% Never

IS-1.24a

(Decentralized interviewers only)
During the past 12 months, have you been forced to skip duties that you know should have been carried out?

13% Often
42% Sometimes
28% Rarely
16% Never

IS-1.24b

(Decentralized interviewers only)
If you are forced to skip some duties, what duties do you most often skip?

12% Tracing telephone numbers
58% Call attempts (visits in face-to-face surveys)
17% The "last" call attempts in telephone surveys where telephone numbers are available
14% Other

Interviewer Survey 2 (IS-2)

IS-2.1b

If you suspect that a respondent does not understand an *attitude* question, what do you do to make the respondent understand the question?

	Most of the time	Sometimes	Rarely	Never
a Change a few words that the respondent does not seem to understand	15%	43%	28%	14%
b Try to explain in your own words the meaning of the question so that the respondent understands	17	25	32	27
c Read the question again without changing any words	67	21	8	4
d Nothing, I continue to the next question	1	8	20	71
e Other	9	10	9	72

The same question was asked except that instead of attitude questions we asked about factual questions. The response distribution was more or less the same as for IS-2.1b.

IS-2.7

Is it difficult to remember and keep apart information, definitions, and instructions for surveys that you are working on?

3% Most of the time
44% Sometimes
48% Rarely
6% Never

IS-2.11

Most of the time, how do you feel about the following tasks or situations?

	Positive	More positive than negative	More negative than positive	Negative
a To persuade reluctant respondents to participate	9%	37%	38%	16%
b To trace sampled persons	39	33	20	7
c To clarify a question to the respondent	38	53	6	3
d To carry out an interview with a person that does not understand Swedish very well	21	44	29	7
e To carry out an interview with the help from an interpreter	24	41	26	9
f (Decentralized interviewers only) To carry out both face-to-face and telephone interviews during the same week	73	20	4	2

IS-2.12

How stressful do you find the following situations?
According to Norstedt's Swedish dictionary, stress is demanding conditions that cause tension often in connection with heavy workload and time pressure.

	Very stressful	Stressful	Not at all stressful
a To lack clear definitions of all concepts in a survey	10%	69%	22%
b To lack sufficient information about the purpose of a survey	21	62	18
c To carry out a telephone interview that lasts a long time (at least 30 minutes)	20	26	53
d To carry out an interview with a person that does not understand Swedish very well	5	46	49
e To carry out an interview with the help from an interpreter	4	25	70
f (Decentralized interviewers only) To carry out both face-to-face and telephone interviews during the same week	7	11	81

CHAPTER 10

Cues of Communication Difficulty in Telephone Interviews

Frederick G. Conrad
University of Michigan[1], USA

Michael F. Schober
New School for Social Research, USA

Wil Dijkstra
Vrije Universiteit of Amsterdam, The Netherlands

When people converse, they do not just send and receive messages. They also give each other ongoing visual, auditory, and textual cues about the extent to which they believe they are understanding each other—about the extent to which their utterances are "grounded," to use Clark's (1996) terminology (see also Clark and Wilkes-Gibbs, 1986; Clark and Schaefer, 1989; Clark and Brennan, 1991, among others), as well as about their emotional reactions to what their partners are saying. Cues that addressees give to speakers—nods of the head, looks of confusion, back channels like "uh-huh" or "huh?", and requests for clarification like "what do you mean?" or reactions like "ouch!"—can alter what speakers say, such that both parties can be seen as molding each other's language use simultaneously.

Much the same sort of thing goes on during survey interviews. While respondents are answering questions, they simultaneously are giving cues about their comprehension and their reactions. Respondents who simply answer a question smoothly, without delays

[1]This work was supported in part by NSF Grant No. SES-0551294. The studies described here were supported by NSF Grant Nos. SBR-9730140 and ISS-0081550, with additional support from the Bureau of Labor Statistics and the Vrije Universiteit of Amsterdam.

Advances in Telephone Survey Methodology, Edited by James M. Lepkowski, Clyde Tucker, J. Michael Brick, Edith de Leeuw, Lilli Japec, Paul J. Lavrakas, Michael W. Link, and Roberta L. Sangster

or requests for clarification, are giving evidence that they believe they have understood the question well enough. Respondents who ask for clarification explicitly, as in

I: How many hours do you usually work?
R: Does "usually" mean on average?

are indicating explicitly that the communication is in danger of going wrong without clarification of a term in the survey question. Respondents who "report" rather than directly answering a question, as in

I: How many hours do you usually work?
R: Well, some weeks I work 60 hours and other weeks I work 30.

are indicating, less explicitly, that their circumstances do not map onto the questions in a straightforward way and that they would like the interviewer to make the decision for them based on their description of their circumstances. The respondent in this example is "reporting" potentially relevant facts rather than directly answering the question (see Drew, 1984; Schaeffer and Maynard, 1996).

Respondents can also implicitly signal (intentionally or not) their need for clarification or discomfort with a question by answering in ways that indicate trouble coming up with an answer:

I: How many hours do you usually work?
R: "Well... uh... usually fifty."

As we will discuss further, paralinguistic features of language like pauses, "well," and "uh" have been shown to be potential signals (intended or not) of need for clarification, plausibly just as informative as facial cues like looks of confusion. In the arena of telephone surveys, these features may constitute "paradata" (to use Couper's term [1998a, 2000b]) that telephone interviewers, and telephone interviewing systems of the future, can exploit.

In this chapter, we explore cues of comprehension difficulty displayed by respondents in telephone interviews and how these differ from cues displayed in face-to-face and other modes of interviewing. We discuss how interviewing techniques may moderate the informativeness of these cues, presenting evidence that whether and how interviewers present clarification in response to the cues can actually change their prevalence—how likely respondents are to emit them.

The approach taken in the studies described here is laboratory-based experimentation using questions about facts and behaviors. We use relatively small samples of respondents answering survey questions in situations where we have independent evidence about the facts about which they are answering—either from fictional scenarios from which respondents are to answer or from extensive postsurvey questioning to probe further into respondents' actual circumstances. The focus has been on cues of the extent to which respondents have understood questions, as opposed to on emotional reactions or rapport. The focus has also largely been on respondents' cues for questions about nonsensitive facts and behaviors, rather than on responses to sensitive questions or questions about attitudes and opinions; obviously, these would be important areas to study further.

Our laboratory approach clearly has its strengths and weaknesses. The power of the laboratory situation is that it allows us to manipulate features of interviews, either by training interviewers in particular techniques or by simulating telephone speech interfaces, so that causal inferences can be drawn. Thus far, the evidence from one larger scale replication in a U.S. national telephone sample (Conrad and Schober, 2000) of an earlier laboratory study (Schober and Conrad, 1997) suggests that the statistically reliable findings from these laboratory studies are likely to generalize to larger populations. But obviously, without full testing it is unknown whether all the results will so generalize, and we intend our discussion here to be suggestive of areas worth exploring rather than definitive or providing immediate prescriptions for practice in larger scale telephone surveys.

10.1 CUES IN TELEPHONE CONVERSATION

To a certain extent, the grounding cues in telephone surveys reflect the constraints of grounding understanding on telephones more generally, which differ from the constraints on grounding in other media (see Clark and Brennan, 1991, for extended discussion; see also Williams, 1977; Whittaker, 2003). Telephone interlocutors are audible to each other, but they do not have access to visual cues such as eye gaze, gestures, and frowns (although this is changing with the advent of video telephony). Unlike people conversing via handwritten letters, e-mail, or instant messaging, telephone interlocutors have access to audible paralinguistic cues that can be useful for understanding what their partner means: timing cues (long delays can be a sign of trouble—see Brennan and Williams, 1995), intonational cues (rising intonation, under the right circumstances, can be an indicator of doubt [Brennan and Williams, 1995]), and other discourse features like *ums* and *uhs*, sentence restarts, and other disfluencies that can give evidence about what speakers do and do not mean (e.g., Fox Tree, 1995; Clark, 1996; Brennan and Schober, 2001). Unless the phone connection is poor, telephone interlocutors perceive each other's signals more or less instantaneously, without the kinds of delays that can occur in handwritten letters, e-mail, or even (on a different scale) "instant" messaging. Because the channel of communication is speech, it leaves no reviewable trace that would allow further inspection, again unlike e-mail or chat room discourse, or messages left on telephone answering machines. And telephone conversationalists can produce and receive messages simultaneously, allowing overlapping speech (of course, within limits—complete overlaps can lead to communication breakdown); this differs from one-way forms of communication such as voice mail messages or walkie-talkie radio communication.

Other facts about grounding understanding in telephone survey interviews arise from what is unique about standardized survey interaction. As various reviews have noted (see, e.g., Suchman and Jordan, 1990; Schober, 1999; Schaeffer, 2002; Schober and Conrad, 2002, among others), respondents in standardized surveys are in the unusual position of trying to ground their understanding with a partner (the interviewer) who is not the originator of her utterances. Typically, interviewers read questions scripted and pretested by survey designers, and the agenda of the survey interview is predetermined and instantiated entirely by that script. The agent who is genuinely responsible for what a question means (and thus the person with whom grounding by rights should happen) is entirely absent, and the interviewer, as a mere purveyor of the questions, becomes a grounding partner of uncertain status.

Respondents are also in the unusual position that their grounding cues—their requests for clarification, their sounds of hesitation—are likely to be ignored or to be responded to in ways that are quite different from what would happen in less controlled and scripted telephone conversations. So, for example, interviewers trained in the strictest form of standardization (see, e.g., Fowler and Mangione, 1990) are trained *not* to respond substantively to explicit or implicit requests for clarification, because if only some respondents receive clarification, then all respondents are not receiving the same stimuli (words). Interviewers are, instead, trained to repeat the question (thus reframing any request for clarification as if it had been a simple mishearing) or to use a neutral "whatever-it-means-to-you" response—to explicitly leave the interpretation of question meaning up to respondents. (Note that what counts as training in standardized interviewing can vary substantially from survey center to survey center [Viterna and Maynard, 2002], and interviewers all trained in the same practices can nonetheless vary substantially in how they implement that standardization [e.g., Schober et al., 2004, Study 2]. In practice, in actual telephone interviews, then, some grounding cues may function more as they do in nonsurvey telephone calls.)

The result in current standardized interviewing practice is that only some proportion of respondents' ordinary communication cues are addressed in ways that respondents are familiar with in other telephone conversations (see Schober and Conrad, 2002, for more detailed discussion and examples). As in other interactions on the phone, survey respondents who simply answer a question without further ado will be taken as having provided evidence that their comprehension was successful, and the interaction will proceed. Also, as in other telephone interactions, respondents who say "huh?" or "what was that?" when they did not hear what their partners said will be given another chance when their partner (the interviewer) repeats the question. But unlike in other telephone interactions, respondents in the most strictly standardized surveys will find that both their explicit grounding cues (requests for clarification) and their implicit grounding cues (reports, pauses, *ums* and *uhs*) are ignored or treated as if they were requests for something else.

We have demonstrated in a series of prior studies that how interviewers respond to telephone respondents' explicit and implicit communication cues can have substantial effects on the success of the communication and thus the quality of the resulting survey data. In our first laboratory study (Schober and Conrad, 1997) we contrasted two extreme ways that interviewers might handle respondents' grounding cues: strictly standardized interviewing, where requests for clarification or implicit signs of confusion are ignored or rejected, and a more collaborative form of interviewing in which participants talk about what has been said to be sure they understand each other sufficiently. (We dubbed this conversational interviewing because it relies on the conversational process of grounding [Clark and Wilkes-Gibbs, 1986; Schober and Clark, 1989; Clark and Schaefer, 1989; Clark and Brennan, 1991; Clark, 1996].)

In the study, we measured comprehension and response accuracy by having 42 telephone respondents answer 12 questions from ongoing U.S. government surveys on the basis of fictional scenarios. Because we had access to official definitions of survey terms, we knew what the survey designers hoped would be included and excluded in respondents' answers, and could judge respondents' interpretations of the questions based on their answers. For each question, we designed two alternative scenarios that a respondent might answer; in one case, there was no likely ambiguity in how the question mapped onto the respondents' (fictional) circumstances, and in the other case there was potential for a mapping ambiguity. For example, a respondent answering a question about how many bedrooms there are in his house would likely find this a straightforward question to answer when the fictional scenario is a floor plan of a house with three rooms labeled "bedroom." But that respondent is likely to have more trouble if one of those rooms is labeled "originally designed as a den, this room is being used as a bedroom." The hypothesis was that how interviewers handled grounding cues should be particularly important for these more complicated mappings.

Both in this laboratory study and in a subsequent field study in which interviewers telephoned a national sample of respondents (Conrad and Schober, 2000), the evidence showed that interviews in which interviewers were licensed to respond to explicit and implicit grounding cues substantively—conversational interviews—improved understanding and response accuracy, particularly when respondents' circumstances mapped onto questions in complicated ways. The evidence from the field study showed that these complicated mappings are frequent enough in a national sample to suggest that when grounding cues are ignored (in strictly standardized interviewing conditions) data quality may be compromised. There was, however, a cost to responding to grounding cues in both the laboratory and field study: providing clarification takes time, and so increased response accuracy is accompanied by increased interview duration.

In another laboratory study (Schober et al., 2004, Experiment 1), we made a first attempt at disentangling how interviewers' responses to explicit grounding cues (respondents' requests for clarification) and implicit grounding cues might affect data quality. In that study, we compared response accuracy under strictly standardized conditions and four versions of conversational interviewing. In all four versions, interviewers were able to clarify concepts after respondents provided explicit grounding cues (what we called respondent-initiated clarification). The grounding cues differed in whether (1) interviewers could also volunteer clarification in response to implicit cues (mixed-initiative clarification) and (2) they could use their own words to clarify the questions (paraphrased versus verbatim clarification).

As in the Schober and Conrad (1997) study, respondents answered 12 questions on the basis of fictional scenarios designed to be unambiguous (straightforward) or to include a mapping ambiguity (complicated). For example, one straightforward scenario described a prototypical nuclear family with two parents and two children living in the home. Respondents used this to answer "How many people live in this house?" The complicated counterpart described a similar family in which one child was a college student living away from home. Should this child be counted as living in the house? The definition of "living in a home" created by the survey sponsors resolved the confusion (the child should be counted as living in the home) and so an interviewer able to provide this information and ground the respondent's understanding should collect more accurate survey data.

The evidence showed that, for complicated mappings, response accuracy was greatest when interviewers responded substantively to both explicit and implicit grounding cues—and, less importantly for the current discussion, when they were licensed to do this in their own words rather than following a script. Respondents answered questions with complicated-mapping scenarios most accurately in mixed-initiative, paraphrased clarification interviews (87 percent). Accuracy was at intermediate levels when interviewers could initiate clarification in response to implicit cues or paraphrase definitions but not both (mixed initiative, verbatim: 66 percent; respondent initiated, paraphrased: 55 percent) and just as high when the only clarification that interviewers could provide was verbatim definitions requested by respondents (respondent initiated, verbatim: 59 percent). In contrast, when no clarification was available, accuracy was disturbingly low (standardized interviews: 28 percent).

The pattern of data suggests, then, that explicit and implicit grounding cues by respondents may contribute independently to the success of communication in telephone interviews. Although accuracy was high under the mixed-initiative, paraphrased approach, respondents provided explicit grounding cues—initiated clarification—for only 47 percent of the cases where clarification was given. The rest of the time interviewers initiated clarification episodes because respondents had displayed uncertainty in their answers or failed to answer the questions definitively. The most common way in which respondents showed they were uncertain was to describe their situation (reports); they did this for 58 percent (37 of 64) of the complicated cases where interviewers intervened voluntarily. Interviewers also intervened voluntarily when respondents asked them to repeat the question (9 percent of cases, 6 of 64) and when respondents explicitly said they were not sure about the answer but did not ask for clarification (6 percent of cases, 4 of 64). (See Schober and Conrad, 1997 for further details.)

Why do respondents initiate clarification sequences less than, it would seem, they might benefit from them? There are at least two possibilities. First, respondents may not realize they are misinterpreting key concepts; respondents may presume that their initial interpretation is the right one—what Clark and Schober (1991) have called "the presumption of interpretability." Second, even if they are uncertain about the meaning of the question and recognize the potential benefits of obtaining clarification, they may not be willing to invest the effort needed to articulate their uncertainty to the interviewer, or they may not be willing to acknowledge uncertainty about the meaning of an ordinary concept like *bedroom* or *living in a house*.

Either way, it is possible that respondents display implicit evidence of need for clarification that telephone interviewers could potentially exploit. Although a respondent may not be keenly aware of confusion, it could be that her processing difficulty or uncertainty (conscious or not) is reflected in her speech: in hedges like "about 50 hours a week" instead of just "50 hours a week," in paralinguistic behaviors like a doubtful-sounding tone of voice or (with video telephony) in visual indicators like a furrowed brow. A respondent who finds it too effortful or embarrassing to ask for help could nevertheless signal his need for help in similar ways. We now turn to an examination of the validity of such uncertainty cues. Do they reliably occur when respondents are in need of clarification about what a question means and how to answer?

10.2 SPOKEN CUES OF RESPONDENT NEED FOR CLARIFICATION

The studies just described show that explicit requests for clarification are reliable communication cues. They also show that telephone interviewers can successfully judge when respondents need clarification even when they have not explicitly requested it. Presumably, they do this based on respondents' implicit cues—although note that it is also possible for respondents to answer incorrectly without producing any overt indications of uncertainty. What particular cues allow interviewers to make this judgment?

A preliminary set of answers comes from Schober and Bloom's (2004) examination of respondents' paralinguistic behaviors in the first turn following a question's delivery in the mixed-initiative, paraphrased as well as standardized interviews from Schober, Conrad and Fricker (2004). The focus was on several classes of paralinguistic behavior that in ordinary (nonsurvey) discourse have been linked to speakers' planning and production difficulties (Goldman-Eisler, 1958; Fromkin, 1973, 1980; Levelt, 1989), the complexity or conceptual difficulty of what they are trying to say (Bortfeld et al., 2001; Barr, 2003), the novelty of the information they are presenting (Fox Tree and Clark, 1997), and their uncertainty or lack of confidence in what they are saying (Smith and Clark, 1993; Brennan and Williams, 1995). Among the verbal behaviors they examined were

- *Reporting.* A way of answering a question that leaves the responsibility of answering to the person who posed the question. To use the Schober and Bloom example, if person A asks person B "Do you like punk rock?" and B responds "I like The Clash," B has left it up to A to decide whether The Clash's music counts as punk rock. A survey analog of this would be answering the bedroom question with "I have two bedrooms, and we are using a room as a bedroom that was originally designed as a den."
- *Speech disfluencies.* Parts of utterances that are not words in the everyday sense and are often assumed to provide little semantic content to the utterance; these include fillers (like *ums* and *uhs*), prolonged pauses in the wrong places (like a 2 second pause at the start of a turn), and repairs (such as "Y- y- yes I did buy a fl- some furniture").
- *Discourse markers.* Words that can alert listeners that what comes next is unexpected or that the speaker is not entirely sure about the content of what is being uttered (Schiffrin, 1987). Examples include *well* (as in "Well, we have three bedrooms") and *oh* (as in "Oh, we have two bedrooms").
- *Hedges.* Words such as *about* (as in "We have about three bedrooms") and phrases like *I think* (as in "I think we have three bedrooms").

Just as the explicit requests for clarification occurred more often for complicated than for straightforward mappings, so did reporting and some speech disfluencies. In addition to producing longer pauses, respondents produced fillers, pauses, and repairs reliably more frequently for complicated than for straightforward situations. This suggests that these communication cues—whether they indicate processing trouble, uncertainty, or even intentional grounding requests—count as valid markers of need for clarification. Combinations of some cues were even more diagnostic of need for clarification. For example, fillers and repairs, and fillers and reports, appeared together more often in complicated than straightforward situations, even more than one of these cues alone. Hedges and discourse markers, in contrast, appeared no differently in answers for complicated than straightforward scenarios, which suggests that, at least for these questions, they are not diagnostic of need for clarification.

Obviously, this set of results is based on a relatively small sample (41) of interviews using particular fact-and-behavior-based questions, and so we should be cautious about overstating their generality. But it is possible that if telephone interviewers can be trained to attend to and detect the cues that are particularly reliable indicators of need for clarification, particularly in combination with one another, they might be able to volunteer clarification in particularly judicious ways, explaining the question's meaning when it is actually needed and refraining from offering help when it is not needed.

An additional set of findings is that the use of various communication cues was affected by what interviewers were licensed to respond to: Respondents used some cues differently in conversational than in standardized interviews. Not surprisingly, explicit requests for clarification were far more likely in conversational (mixed-initiative, paraphrased) than standardized interviews; respondents no doubt recognized that explicit requests in strictly standardized interviews would be unlikely to elicit substantive grounding help. And along the same lines, perhaps it is not surprising that respondents produced reports (e.g., "She bought a floor lamp" when asked "Did Dana purchase or have expenses for household furniture?") more often in conversational than standardized interviews. As with explicit requests for clarification, reporting is an effective strategy only if interviewers are able to react substantively, perhaps determining and recording the answer on the basis of the report. In a standardized interview, reporting is less likely to help the respondent to ground his understanding; the interviewer is most likely to respond with a nondirective (and nongrounding) probe like "Would that be 'yes' or 'no'?"

But some of the disfluencies were also differentially present in standardized and conversational interviews—and not always in the direction one might expect. For example, fillers (*ums* and *uhs*) were actually *less* frequent in conversational interviews than in standardized interviews. Why might this be? Perhaps, when telephone respondents are deterred from requesting clarification (e.g., as seems to be the case in standardized interviews) their speech is more likely to reflect unresolved uncertainty. There may be a trade-off between having the ability to ask for clarification and the production of certain communication cues. In any case, these findings suggest a link between how an interviewer may respond to particular grounding cues and which cues the respondent produces.

10.3 SPOKEN VERSUS VISUAL CUES OF COMPREHENSION PROBLEMS

Are the cues we have been investigating specific to telephone surveys? *Do* respondents present different cues, as grounding theory proposes, in interviews that also allow the exchange of visual information? It is possible that respondents are exactly as likely to display confusion facially, in their voices, and via explicit requests for clarification whether or not the interviewer can see them. Or are respondents sensitive to the relative richness of cues afforded by the mode of data collection? Might respondents compensate for the absence of visual cues in telephone interviews by displaying more cues of uncertainty in their speech than they do in face-to-face interviews?

At present, little is known about the particularity of communication cues in telephone surveys, despite various comparisons between telephone and face-to-face interaction in other settings (e.g., Williams, 1977; Whittaker, 2003). Large-scale comparisons between telephone and face-to-face interviewing, such as de Leeuw and van der Zouwen's (1988) meta-analysis, have found few, if any, differences in data quality that could be attributed to mode differences. Historically, the lack of difference between the modes has been good news: in the early days of telephone interviewing, this was celebrated as evidence that the new mode—and its concomitant benefits—had come of age.

We propose an alternative view of the situation: Respondents should produce mode-specific communication cues exactly to the extent that they are useful for grounding understanding or otherwise communicating useful information in the interview. In strictly standardized interviews, where interviewers are restricted from substantively responding to even explicit requests for clarification, the availability of a communication channel for grounding should make little difference. But when interviewers are licensed to use a particular communication cue—in some sort of conversational interview—the availability of a cue should be relevant because it can be exploited by the interviewer. More generally, if we open up the possibility that grounding cues are potentially useful to interviewers, then a new set of research questions emerges concerning when and how audio and visual cues might be redundant with each other and when they might complement each other in survey interviews.

As a first step toward exploring these kinds of questions, Conrad et al. (2004) carried out a laboratory study in which 42 Dutch respondents were asked about their own lives in either conversational or standardized interviews that were conducted over the telephone (four interviewers) or face-to-face (four other interviewers). After the interview, respondents self-administered a paper questionnaire that included the interview questions accompanied by definitions of the relevant concepts. Thus, if respondents changed their answers between the interview and the postinterview questionnaire, the change could be attributed to a change in their understanding brought about by reading the definition in the questionnaire, suggesting they had misinterpreted the question during the telephone interview.

The analysis was focused on one question about membership in the Dutch institution *verenigingen* or registered clubs: "I would now like to ask you some questions about your membership in clubs. Can you list all the clubs in which you are personally a member?" This type of question, which requires respondents to list their answers, is a particularly good candidate for conversational interviews because interviewers can help respondents evaluate each club they list for compliance with the definition. Indeed, answers changed more after standardized interviews than conversational ones for this question, suggesting that clarification during the (conversational) telephone interview had been beneficial to respondents' understanding and the accuracy of their answers.[2] However, there were no differences due to mode (telephone versus face-to-face). Why might this be given the extra richness in potential cues of uncertainty afforded by face-to-face interviews?

Part of the answer lies in respondents' greater disfluency over the telephone. In particular, they produced reliably more *ums* and *uhs* on the telephone (8.0 per 100 words) than face-to-face (6.1 per hundred words), as if they recognized that the interviewers could not see them on the telephone and so would need extra auditory evidence of their difficulty in answering. This was true both in standardized and conversational interviews, which suggests that interviewer responsiveness to cues is not the driving force behind the differential levels of disfluency. In the conversational interviews, telephone interviewers who provided clarification in response to disfluencies did so much sooner (after 4.2 "moves"—more or less equivalent to speaking turns) than the face-to-face interviewers (11.4 moves). Although the sample is too small for these results to be more than suggestive, it is possible that spoken cues of trouble can be taken as particularly revealing on the telephone.

What then are the visual cues available only face-to-face and for which telephone respondents may have been compensating? One such potential cue is respondents' gaze aversion, that is, their tendency to look away from the interviewer while answering. Increased gaze aversion has been associated with increased difficulty in answering questions (Glenberg, 1998) and is attributed to the respondents' attempt to avoid the distraction that is almost certainly brought about by looking at the questioner's face. (For further discussion of the communicative implications of gaze see, e.g., Doherty-Sneddon et al., 2002 and Goodwin, 2000.) The critical issue in the Conrad et al. (2004) study was whether respondents looked away more in conversational than standardized interviews when interviewers might possibly provide clarification based on these cues.

In fact, respondents did look away for larger percentages of time when answering questions posed by conversational than standardized interviewers: In cases where their answers later proved reliable, respondents looked away 15.4 percent of the time while answering in the 10 conversational, face-to-face interviews, as compared with 4.3 percent of the time in the 11 standardized, face-to-face interviews. More tellingly, in cases where their answers later proved unreliable, they looked away 28.3 percent of the time in conversational interviews (versus 0 percent of the time

[2]See Conrad and Schober (2000) for another example of how questions requiring lists as answers produce more accurate data with conversational interviewing than standardized ones.

for standardized interviews, where there was no chance they could get clarification). These data suggest that respondents were sensitive to whether the interviewers could provide clarification in response to a visual behavior. Curiously, conversational interviewers did not provide more clarification in response to this behavior, despite glancing at respondents at least once during 80 percent of their looking-away episodes. One explanation is that conversational interviewers simply had not been instructed to treat such cues as indications of respondent uncertainty and that with appropriate training they could provide more and better-timed clarification. Another possibility is that interviewers were so focused on looking at their laptop screens that they were not sufficiently aware of respondents' gaze aversion to use it as a cue of need for clarification.

In addition to verbally signaling the need for clarification, speakers may supplement these cues visually (e.g., direction of gaze). If so, understanding might suffer on current telephones because, without visual cues, interviewers may miss opportunities to provide needed clarification. Alternatively, respondents may compensate for the limits of auditory-only communication by verbalizing their comprehension problems paralinguistically. Clearly, this warrants further investigation, particularly as video telephony becomes more practical (see Anderson, 2008 and Fuchs, 2008).

10.4 INTERACTING WITH AUTOMATED TELEPHONE INTERVIEWING SYSTEMS

It is currently unknown whether all interviewers, or only the most socially sensitive interviewers, can use verbal and visual cues of respondent uncertainty as a trigger for providing clarification. The division of attention that may have limited interviewers' use of gaze aversion in the Conrad et al. (2004) study could be a serious impediment. We propose that technology may be able to help. In particular, diagnostic software could be created that could take some of the attentional burden off interviewers by monitoring for spoken or even visual cues of respondent difficulty. One could even imagine deploying such technology as part of fully automated interviews in the not-so-distant future.

We have begun studying the effectiveness of this kind of diagnosis by simulating such technology with a "Wizard-of-Oz" (WOZ) technique (e.g., Dahlbäck et al., 1993). In this approach respondents believe they are interacting with an automated system via telephone but are actually interacting with a human (wizard) who presents preexisting speech files created to sound like synthesized speech. Unlike conventional speech recognition technology (as in some of today's interactive voice response [IVR] systems), the simulated dialogue technology is not limited to utterance recognition but can take into account discourse criteria like whether a concept has already been discussed and whether respondents' speech contains the kind of markers observed by Schober and Bloom (2004).

Here we describe two experiments using the WOZ technique to simulate automated interviewing technology. Respondents answer questions asked by the simulated interviewing system on the basis of fictional scenarios, just as in our studies of human telephone surveys, so that we have independent evidence about when they have interpreted questions as the survey designers intended. Note one advantage of this sort of study: The behavior of the "interviewer" can be manipulated with algorithmic precision in a way that is far less certain in training human interviewers.

10.5 DIAGNOSING RESPONDENT'S NEED FOR CLARIFICATION FROM COMMUNICATION CUES

In the first study (Bloom, 1999; Schober et al., 2000), a Wizard-of-Oz technique was used to simulate a speech interface. Users believed they were interacting with a computer, when actually a hidden experimenter presented the questions and scripted clarification. To enhance believability, we used an artificial-sounding computer voice (Apple's "Agnes" voice); virtually all respondents were convinced they were interacting with a computerized interviewing system, and the data from the few who doubted this were removed from the study.

In the first condition, the system could not provide clarification. This was similar to one of our strictly standardized interviews in that a confused respondent could not obtain a definition; if a respondent requested clarification, the system would repeat the question. In the second condition, clarification was based on explicit respondent-initiated grounding cues—the system would provide clarification if the respondent asked for it explicitly. In the third condition, clarification was based both on explicit and implicit respondent grounding cues (the initiative was mixed)—the system would also "automatically" provide full definitions when users displayed the cues of need for clarification cataloged in Schober and Bloom (2004). These included *ums*, *uhs*, pauses, repairs, and talks other than an answer. In the fourth condition, the system always provided clarification; no matter what the user did, the system would present the full official definition for every question.

The results with this simulated system in some respects parallel those for our studies of human interaction. As in Schober et al. (2004, Study 1), respondents were almost perfectly accurate when they answered about straightforward scenarios. For complicated scenarios, respondents were substantially more accurate when they were always given clarification (80 percent) than when they were never given clarification (33 percent).

But the pattern for grounding cues was somewhat different. Unlike in Schober et al. (2004), requiring explicit grounding cues (requests) in order to provide clarification was entirely ineffective, because respondents almost never asked for clarification despite being instructed to do so if they were "at all uncertain about the meaning of a word in a question." In the respondent-initiated clarification condition, the accuracy of respondents' answers was no better (29 percent) than when they were never given clarification. Most likely it did not occur to respondents that clarification was necessary; the presumption of interpretability (Clark and Schober, 1991) probably holds in computer-administered interviews. What *was* effective was relying on respondents' implicit grounding cues; response accuracy was reliably better when the system provided clarification in response to users' disfluencies and pauses (the mixed-initiative clarification condition) (59 percent), although not as good as when clarification was given always.

When the system provided clarification in response to implicit grounding cues, respondents were actually more likely to ask explicitly for clarification: Respondents asked questions more often in the mixed-initiative condition, presumably because they were more likely to recognize that clarification might be useful. These users also spoke less fluently, producing more *ums* and *uhs*—and there is some evidence that this tendency increased over the course of the interview. We speculate that this was because these users at some level recognized that the system was sensitive to their cues of uncertainty.

Why did respondents with the computer speech interface give explicit grounding cues (ask for clarification) so rarely? Perhaps even more than the respondents in the telephone interviews in Schober, et al. (2004) they found it relatively uncomfortable to articulate their confusion or uncertainty to a computer agent. But we cannot conclude this with certainty, as there are other differences that we suspect may have been even more important: Obtaining a definition with this particular speech interface was a more daunting prospect than getting a definition from a human interviewer, because the entire definition—not just the relevant parts—would be spoken, and this was time consuming (up to 108 seconds) and impossible to shut off. In contrast, human interviewers can potentially provide just the relevant part of the definition (as in the paraphrased clarification interviews in Schober et al., 2004) and respondents can interrupt the interviewer if necessary to circumvent the full delivery of the definition. Finally, respondents in the current study could not reject a system-initiated offer to provide a definition because the system did not offer—it simply provided—the definition. In the Schober et al. (2004) interviews, it was often the case that interviewers asked respondents if they wanted clarification.

As in our studies with human interviewers, clarification took time. The more clarification a respondent received, the more time the interviews took. Sessions where clarification was always provided took more than twice as long as sessions with no clarification or when it was (rarely) respondent-initiated (12.8 versus 5.2 and 4.9 seconds per question, respectively); mixed-initiative clarification took an intermediate amount of time (9.6 seconds per question).

Respondents rated the system more positively when it was responsive (respondent or mixed-initiative conditions). When the system was not responsive (no clarification or clarification always), users wanted more control and felt that interacting with the system was unnatural. Respondents did not report finding system-initiated clarification particularly more annoying than respondent-initiated clarification—which they almost never used.

Overall, these results suggests that enhancing the collaborative repertoire and diagnostic capability of a speech-interviewing system can improve comprehension accuracy without harming user satisfaction, as long as the system provides help only when it is necessary. But these improvements come at the cost of increased task duration, which raises questions about the practicality of a system with only these characteristics in real-world survey situations.

10.6 MODELING RESPONDENTS' SPEECH TO PROVIDE MORE TAILORED CLARIFICATION

We propose that systems may be able to provide more precisely tailored clarification to respondents by attending to their grounding cues in a more nuanced way. We demonstrated this in an experiment (Ehlen, 2005; Ehlen et al., 2007) in which we modeled different groups of respondents' relevant paralinguistic behaviors. In particular, the respondent modeling techniques allowed us to distinguish behaviors more likely to signal uncertainty from those that are less likely to do so. For example, someone who regularly says *well* and *uh* as part of their daily repertoire is less likely to be signaling comprehension difficulty with well or uh than someone who rarely uses them. A listener who makes this distinction is modeling the individual speaker. The same logic can apply to groups of speakers. Older speakers have been shown to be less fluent than younger speakers (e.g., Bortfeld et al., 2001), and so the same disfluency rate for a young and old speaker may indicate different states of understanding. In other words, the same level of *um*ming might indicate problematic understanding for a younger speaker but ordinary speech for an older speaker. We applied this idea to automated interviewing by allowing the system to offer clarification on the basis of a generic model (same criteria for all respondents) and a stereotyped model (different criteria for old and young respondents).

One hundred respondents (50 older than 65 years of age and 50 under 40 years of age), answering 10 questions on the basis of fictional scenarios, participated in one of the five kinds of interviews: No Clarification, Respondent-Initiated Clarification, Required Clarification, Generic Respondent Model, and Stereotyped Respondent Model. The first two kinds of interviews, similar to their namesakes in the Bloom et al. study, generated the respondent speech that was used for the respondent models. In the Required Clarification interviews, respondents first answered each question; after this they were presented the full definition and could change their response if they chose to. These interviews served two functions. First, they provided a test bed for the models. In particular, they allowed us to ask how precisely the models predicted comprehension accuracy prior to the definition being presented. Second, they served as a benchmark of comprehension accuracy. Because definitions were presented for all questions, response accuracy under these conditions provided an upper bound on the benefits of clarification. In the Generic Respondent Model interviews, the system initiated clarification after certain speech conditions (discussed next) were met, regardless of respondent characteristics. In the Stereotyped Respondent model interviews, the conditions that triggered the system to provide clarification were different for older and younger respondents.

The respondent models were calculated with ordinary least-squares regression techniques in which the predicted behavior was response accuracy and the predictors were fillers, hedges, restarts, repeats, repairs, reports, mutters, confirmation pickups,[3] and pauses. While older respondents produced more spoken cues and longer

[3]An example of a confirmation pickup is "Usually, fifty" in response to "How many hours per week does Mindy usually work at her job?" because it picks up the term "usually" as a way of keeping it in play so that it can be confirmed or negotiated.

pauses than younger respondents, none of the cues improved the model beyond the predictive ability of pause length. If respondents answered too quickly or too slowly, they were more likely to be incorrect than if they answered within the intermediate (not too slow, not too fast) range. (We called this the "Goldilocks" range, in honor of the "just right" range of porridge temperatures and chair sizes in the "Goldilocks and the Three Bears" tale.) The Generic Goldilocks range was 2–7.2 seconds. The Goldilocks range for younger respondents ran from 4.6 to 10.2 seconds and for older respondents it ran from 2.6 to 4.35 seconds. Surprisingly, older people did not take longer to answer, in general, than did younger people. Rather, the range in which older respondents were likely to be accurate was smaller and faster than for younger respondents.

Response accuracy (again focusing on complicated scenarios) increased across the different kinds of interviews much as in the previous study (Bloom, 1999; Schober et al., 2000): poorest when no clarification was available, better when respondents could request clarification but the system could not provide it and better still when the system could also provide clarification (on the basis of models or after each question). When response accuracy prior to the required definition was used to test the models, 53 percent of the answers outside the Generic Goldilocks range were inaccurate and 83 percent of the answers outside the Stereotyped Goldilocks ranges were inaccurate.

When the system actually provided clarification, response accuracy improved reliably (linear trend) from Generic to Stereotyped Respondent Modeling to Required Clarification (after the definition had been delivered). In fact, the accuracy with Stereotyped Respondent models was as high as with Required Clarification, yet the interviews were reliably faster. It seems that by tailoring clarification to the respondent's age group, the clarification was often provided when it was needed and rarely when it was not needed, thus minimizing the temporal costs of improving clarification.

10.7 CONCLUSIONS

The data described here suggest in a preliminary way that respondents' explicit and implicit cues of their states of comprehension provide exploitable evidence that telephone interviewers and future telephone interviewing systems could use to improve survey data quality. The cues we have investigated are primarily conveyed through language (explicit requests for clarification, saying something other than a direct answer) and paralanguage (too-long and too-short delays before answers, *ums* and *uhs* in answers, etc.). But visual cues (gaze aversion, looks of confusion) are potentially exploitable in telephone interfaces that include visual information, to the extent that such cues prove nonredundant with textual and paralinguistic cues.

Of course, much more would need to be known before the findings described here are translatable into practical prescriptions for telephone survey centers; the studies described here only begin to address the larger set of theoretical and practical questions that survey researchers of the future will need answers to. And even for interpreting and applying these studies, we should be very clear about several caveats.

Most of our studies are laboratory-based, relying on small samples of respondents answering questions about nonsensitive facts and behaviors, and professional interviewers given brief training in alternate interviewing techniques. How one generalizes from experimental manipulations to actual, large-scale surveys is not at all straightforward. Experiments demonstrate that certain phenomena *can* happen but not that they necessarily do happen under all circumstances. To clearly map experimental results to large-scale surveys, one must know how often the circumstances created in the laboratory actually occur "in the wild."

We have focused on accuracy of respondents' interpretation of questions (the extent to which their answers reflect the same interpretations as the survey designers'), rather than on other important indicators of data quality in surveys, such as response rates, completion, and break-off rates. Whether the findings will extend beyond the lab to larger samples of respondents, different kinds of questions, different interviewer populations, and additional measures of survey quality remains to be seen.

Also, the effects of attending to grounding cues are apparent particularly for situations where the respondent's circumstances are ambiguous with respect to (well-pretested) questions, and so the frequency with which this occurs in real-world settings places a limit on the utility of our findings. Thus far, the evidence suggests that these sorts of "complicated-mapping" circumstances are frequent enough to worry about in broader samples and under more natural conditions—see Conrad and Schober, 2000; Suessbrick et al., 2005—and that they apply to attitude and opinion questions as well as questions about facts and behaviors, but again we recommend caution in assuming this about every sample of respondents and questions. We should also note that there are other potential sources of trouble answering questions that we have not been investigating: trouble understanding technical terms, trouble deciding on the response task (e.g., estimate or count?), trouble retrieving answers from memory, and troubles resulting from ambiguous (polysemous) terms in questions. Whether the communication cues surrounding these other kinds of trouble are the same as those investigated here is an open question.

In general, we see the findings described here as raising a set of issues that need to be explored in much greater detail, and we hope that this discussion helps to prompt further research along these lines. In particular, we think it would be important to know the following:

(1) *How diagnostic of need for clarification are communication cues across different respondents and circumstances?* While we have seen good evidence across our samples that, for example, respondents use *um* more often in their first turn after a question is asked when the answer is likely to need clarification, it is also clear that there is substantial individual variability in discourse styles, dialects, and propensity to *um*. The *um* from a respondent who never *um*s is presumably informative in a way that an *um* from a respondent who regularly *um*s is not. Gaze aversion from a steadily gazing respondent is different in meaning than gaze aversion from a respondent who never looks at the interviewer. To what extent do sensitive interviewers already attend to baseline rates of any potential communicative cue—delay in responding, reporting, gazing, *um*ming—as they decide whether to probe or clarify?

To what extent should interviewers be trained to attend to the individual variability of such cues, and to what extent should interviewing systems of the future be able to diagnose the individual variability of such cues?

It is entirely possible that what is a cue of the respondent's comprehension difficulty in one situation reflects a quite different internal state in another situation. Or, worse, the same cue might indicate different states in the same situation on different occasions. Consider the "looking away" cue. Because respondents look away longer in conversational interviews (when interviewers might react to the cue) than in standardized interviews (when they cannot), looking away would seem to be under respondents' control. But if respondents look away irrespective of interviewers' ability to react, this would more likely mean that looking away is an involuntary reflection of comprehension difficulty.

Alternatively, looking away could reflect something other than comprehension difficulty. It could indicate that the respondent is planning what to say next and does not want to be distracted by looking at the interviewer (cf. Glenberg et al., 1998). Or it could reflect a state almost diametrically opposed to needing help: Looking away could reflect respondents' desire to maintain the floor and not surrender it to the interviewer, a concern more in conversational than standardized interviews (see the discussion by Clark, 1996, of turn allocation rules, pp. 321–324). Finally, looking away (or any of the cues we have suggested reflect comprehension difficulty) could indicate ambivalence about answering truthfully. If a respondent is concerned that providing a truthful answer might somehow cause her harm, for example, by confessing to illegal conduct, she might look away or answer less fluently or pause longer before speaking than if she has no reservations about answering.

(2) *How does an interviewer's responsiveness to any communication cue affect the respondent's likelihood of using it?* Presumably, when strictly standardized interviewers ignore explicit request for clarification, they reduce the likelihood that any but the most perverse or conversationally insensitive of respondents will continue asking for clarification. Does the same go for more implicit communication cues? The preliminary evidence reported here hints that rates of disfluency may well be sensitive to interviewers' responsiveness to them. Surely this is not the sort of thing that is under respondents' conscious control, and it suggests a kind of dyadic regulation of individual process that is not part of many views of communication.

(3) *Are all interviewers equally sensitive to grounding cues?* Although reliable empirical evidence on this is rare, ordinary intuitions and clinical observation of the general population suggest that people can vary substantially in their interpersonal sensitivity: their empathy, perspective-taking ability, and ability to attend to subtle linguistic cues (see, e.g., Davis, 2005, and other chapters in Malle and Hodges, 2005; Schober and Brennan, 2003). Presumably, interviewers who are socially tone deaf do not survive long in the job; overly sensitive interviewers, for whom denying requests for clarification may be interpersonally aversive, may also not survive in a telephone survey center that requires the strictest of standardized practice. What is unknown is the extent to which sensitivity to such cues is trainable, or whether adult language users already have an ingrained repertoire of cues to which they attend that

is resistant to change. Presumably, there are individual differences in interviewers' sensitivity to grounding cues (the approach of Hall and Bernieri, 2001 might allow assessment of this kind of skill), which will constrain the effectiveness of training. In particular, for interviewers low in sensitivity to such cues, the additional task of monitoring for them may be unrealistically burdensome (see Japek, 2005, for a discussion of interviewer burden). Also, external constraints like time pressure to finish interviews may cause even the most interpersonally attuned interviewers to ignore potentially useful cues.

(4) *How nonredundant are communication cues?* Thus far, little is known—in general and in survey interviews—about the extent to which visual cues provide information distinct from that provided by tone of voice, delay, or the content of what is said. While one can find clear examples where a particular cue seems to be the only indicator of trouble, we just do not know whether a single cue is always sufficiently diagnostic to warrant intervention by an interviewer or an interviewing system. To complicate matters, it is possible that different respondents may have different discourse styles: One respondent's gaze aversion may always be accompanied by a pause and an *um*, while another's gaze aversion may provide unique nonredundant information. To the extent that cues are redundant, interviewers who already have a lot to attend to might be able to rely on the cues in the most easily available or measurable channel.

(5) *How multifunctional are communication cues?* Our approach thus far has been focused on the cognitive aspects of comprehending questions and how respondents' disfluencies and other cues provide relevant evidence. But every cue we have discussed—explicit requests for clarification, reports, hedges, and so on—is also a potential indicator of the respondent's emotional state, level of irritation, and likelihood of continuing the interview. Respondents could delay or rush responses not only because they have trouble understanding the question or have not thought hard enough, but also because they find a question intrusive, because they feel the interview has gone on too long, or because the interviewers' nonresponsiveness to a request for clarification is becoming trying. To what extent do grounding cues also provide evidence about the rapport and emotional alliance between interviewers and respondents? We suspect that although grounding cues and rapport cues are conceptually distinct, in practice they can be quite intertwined. For example, an interviewer's apology for the stiltedness of an interview ("I'm sorry, I can only repeat the question") can be a response to cues that the interview is going offtrack both on affective dimensions (the respondent's frustrated tone of voice) as well as on grounding dimensions (explicit and implicit indicators of need for clarification).

As the space of new technologies available for telephony expands, telephone interviews are beginning to share more features with face-to-face interviews (see Schober and Conrad, 2008 as well as other chapters in Conrad and Schober, 2008). Will additional visual information help improve comprehension and thus data quality? Whittaker (2003) has observed that across various (nonsurvey) domains there is little evidence in support of the *bandwidth hypothesis*: the idea that adding visual information to speech will improve the efficiency of communication. It may be that the total amount of usable information about a communicative partner's need for clarification is the same with or without video. It remains to be seen what the facts are for surveys with different populations of respondents, with individually and culturally variable communication styles, with different domains of questioning (sensitive and nonsensitive questions), and with different interviewing agents (human versus computer) with different capabilities (diagnosing and responding to requests for clarification versus leaving interpretation up to respondents). How these questions are answered will help shape future debates about how telephone interviews should be conducted.

CHAPTER 11

Oral Translation in Telephone Surveys[1]

Janet Harkness
University of Nebraska at Lincoln, USA and ZUMA, Mannheim, Germany

Nicole Schoebi and Dominique Joye
SIDOS, Neuchâtel, Switzerland

Peter Mohler, Timo Faass, and Dorothée Behr
ZUMA, Mannheim, Germany

11.1 INTRODUCTION

In the context of growing cultural diversity, telephone studies are regularly faced with the challenge of collecting data from respondents who do not speak the majority language of a sampled population. In order to avoid problems related to coverage, nonresponse, and measurement, projects need to approach these respondents in a language they can use comfortably. Three basic options are available to interview such respondents; projects can have written translations prepared, can use interpreters, or can hire bilingual interviewers to translate and interview at the same time. In this chapter, we focus on the last option, when interviewers translate orally while they are interviewing. We present and discuss transcripts of interviewers translating orally and compare these to transcripts of the same interviewers when they interview using a written translation.

Interviews for which there was a written version of the questionnaire in the language of the interview are called here *scripted interviews* and stand in contrast to the *orally translated interviews* for which interviewers did not have a written

[1]We would like to thank the Schweizer Institut für Daten in den Sozialwissenschaften (SIDOS), the Swiss Social Science Data Archive, in Neuchâtel, Switzerland, and the Zentrum für Umfragen, Methoden und Analysen (ZUMA), in Mannheim, Germany for financing the study reported here.

Advances in Telephone Survey Methodology, Edited by James M. Lepkowski, Clyde Tucker, J. Michael Brick, Edith de Leeuw, Lilli Japec, Paul J. Lavrakas, Michael W. Link, and Roberta L. Sangster

questionnaire in the language of the interview. Following terminological practice in translation studies, we call the language translated out of the *source* language and the language translated into the *target* language.

Oral translation is not restricted to interviews conducted by telephone, but telephone surveys have fostered the use of oral translation to interview respondents unable to be interviewed in the language(s) in which a written questionnaire is available. As it is not always possible to predict which language a contacted person will require, telephone interviewers can be provided with call center numbers to contact if the sampled unit reached cannot be interviewed in the language(s) the interviewer speaks (cf. Murray et al., 2004; Thorpe et al., 2006; and the Language Line Web site[2]). In cross-regional and cross-national contexts, centralized management can ensure that language proficiencies are switched to match locations called and/or sampled unit reached.

11.2 HOW WRITTEN AND ORAL TRANSLATION DIFFER

Oral translations made during telephone interviews necessarily differ in many respects from translations that are written down. The reasons outlined below have much less to do with the differences between oral language and written language (on which see, e.g., Chafe and Tannen, 1987) than with the special context of computer-assisted telephone interviewing (CATI) interviews and the numerous constraints and activities involved. We outline these below, focusing on differences of particular relevance for interview quality and comparability.

Constraints. The pressure and time constraints on interviewers who are orally translating an interview differ from those on translators producing a written translation. Translators producing written versions do not need to translate in real time, interviewers translating orally do. Translators producing written versions can focus on translating; they do not need to disguise the fact that they are translating nor carry on a conversation (interview) while they are translating.

Context. In written translation, translators can usually move back and forth in the source text and in their translation, checking meaning, words they have used, and reminding themselves of context. This helps them determine meaning and maintain consistency. In oral CATI translations, interviewers have to rely on the immediate textual context to determine meaning. There is little time and limited actual possibility to read ahead or to think about consistency. In oral translation, there is no concrete record of the text translated, and the source text on the CATI screen also soon disappears from sight. An additional complication is that the CATI screen text in the source language acts as a constant distracter from the language of performance, the target language (see *Source influence* below).

Revision. Written survey translations can be read, compared to the source, revised, and pretested. Recurring elements can be made consistent throughout.

[2]http://languageline.com/page/opi.

Oral telephone survey translations are spoken, ephemeral productions. Opportunities for considered delivery and correction are much reduced. We can expect more mistakes to be left uncorrected and translations of recurring elements to be less consistent.

Overt Revisions. The final version of a written translation normally shows no trace of earlier versions. In oral telephone interview translation, the translation revisions become part of the discourse interaction with respondents. The verbal material presented to respondents is thus more complex and potentially contradictory ("Oh, no sorry, I mean…"). This may make processing more difficult for respondents and affect their perception of the interviewer and the interview. Awareness of that, in turn, may be an added burden for the interviewer. Interviewers in the project, for example, apologize for corrections they make and other disruptions to the flow that occur. Standard revision practice for written translations includes making checks for omissions and consistency of language used.

Source Influence. Research indicates that it is especially difficult for translators translating orally to avoid source text interference (Kiraly, 1995; Gile, 1997; Agrifoglio, 2004). Interference is a term used to describe what happens when vocabulary, structures, or the sound system of the source language affect output in the target language. Time pressures, the simultaneous confrontation with the written source language and the spoken target language, as well as the cognitive processes related to this all encourage interference. In written translations, interference can also occur but can be spotted and removed.

Training and Expertise. Interviewers used as translators have been trained to interview. Even if they commonly switch languages in their everyday life, this is not the same as training on producing oral translations from written source texts. Research suggests that inexperienced translators usually stick closer to the literal meaning and surface structure of source texts than do experienced translators (Krings, 1986; Kussmaul, 1995). Kiraly (1995) reports otherwise. However, these translators were not translating into their strongest language, and performance differences are to be expected.

Simplification. Time constraints and the cognitive burden of doing multiple things at once mean that greater simplification and more omissions can be expected in oral interview translations than in written survey translations.

Translation Direction. Interviewers translating in interviews work in both directions, partly at the same time. They translate out of language A into language B, silently follow instructions in language A and also code responses presented by the respondent in language B in categories offered on the screen in language A. An added burden is automatic. Translators of written survey translations usually work in one direction at a time.

Tools. Translators producing written translations have access to tools to help with the translation that interviewers in orally translated interviews do not.

Given that written survey translations have clear advantages over oral translations, we might wonder why oral translations are used. We discuss this in the following section.

11.2.1 Arguments in Favor of Using Oral Translations

The most common reasons for using oral translations in surveys are related to practical considerations such as the cost of producing written translations, the time required to produce them, and the sometimes poor predictability of which languages respondents will need.

Cost Cutting. Nationally based projects expecting to interview in one or more minority languages often use bilingual interviewers, even if written translations are available. Bilingual interviewers may cost more than monolingual interviewers.

By having interviewers translate orally, the added expense they occasion can be offset by not paying for a written translation. If only a few respondents for a language are expected, production costs for a written translation may well be viewed as disproportionately high. In cross-national projects when multiple languages are required, the savings in outlay for written translations may appear quite considerable.

Timeliness and Flexibility. Oral translations are produced on the spot, usually with little extra preparation time for the interviewers, and they provide flexibility in the languages that can be covered. This can be especially important in studies in which languages are difficult to predict.

Practicality. In telephone surveys, language expertise can be accessed at a distance without respondents experiencing any great burden or delay.

Coverage. Coverage and unit response can be improved because people can be interviewed who might otherwise be excluded.

Formal Requirements. Sometimes a language needs to be "available" as an official option for respondents even if they are proficient in the language of the written questionnaire. Oral translation is then a low-cost option.

11.2.2 Arguments Against Using Oral Translations

Numerous arguments related to data quality and reliability can be made against using oral translations in standardized interviews, and we list these below.

Lack of Standardization. Interviewers do not have a script in the language of the interview and produce their own translated versions of questions, response categories, and instructions for each interview. There are many ways to translate and, unlike the legendary translators of the Torah, we cannot expect interviewers to arrive at the same common translation. Language is essentially open to interpretation; thus, interviewers will differ in what they understand and what they translate. Since the translated texts they produce are also ephemeral, we can expect each interviewer to vary his or her translations across and within interviews. Variation among interviewers also contributes to the correlated interviewer variance. All these factors contribute to destandardizing the interviews.

Bias. If many interviewers make the same errors, it will lead to biased estimates. If one interviewer makes repeated errors, it will increase interviewer effects.

No Pretesting. Orally translated instruments are not pretested.

No Documentation. Timeliness and costs are prime reasons for using oral translations. Transcriptions are therefore not likely to be available, since these are time-consuming and costly. As a result, we usually cannot compare the textual output or quality of orally translated interviews with scripted interviews.

Monitoring. Ongoing (simultaneous) monitoring of what is said to respondents and what is coded as having been said is more complex and expensive. Supervisors undertaking such work need to be bilingual and also versed in assessing the ephemeral translations as well as monitoring the other features of interviewer performance in the interview.

Burden. Interviewer burden is increased. Both theoretically and at face value, multiple effects can be expected.

Quality. Written survey translation is challenging enough. It is unrealistic to expect oral translations to retain the degree of faithfulness, consistency, and quality expected for written translations. Interviewers have been trained as interviewers, not as translators of written texts, and this raises further quality issues.

11.3 A TWO-COUNTRY ORAL TRANSLATION STUDY

In June and July 2005, a telephone study was conducted in Switzerland and Germany on oral translation in telephone surveys. The aim was to collect data that permitted an assessment of how orally translated interviews compare to scripted interviews, particularly in terms of quality.

One-hundred telephone interviews were planned, 50 conducted in Germany by five German interviewers and 50 conducted in Switzerland by five Swiss German interviewers. Each interviewer was to carry out five interviews translating orally from an English source questionnaire and then conduct five interviews reading off a written German/Swiss German script of the same questions. Technical difficulties with the recordings of the CATI interviews meant that a few more translated interviews were conducted (but not recorded) in Germany than planned. Forty-nine recorded interviews could be analyzed for each country.

The telephone samples drawn were for noncommercial numbers. Since the study was primarily interested in interviewer performance, quotas were used on respondent sex and age. The German sample was drawn at ZUMA and the Swiss sample was drawn by the Instititut für Markt und Meinungsforchung, M.I.S. Trend SA. This is also the survey research institute that programed the CATI applications and from whose telephone studio in Bern, Switzerland, the calls inside Switzerland as well as those to Germany were made. Respondents were asked ahead of the interview for permission to record so as to monitor interviewer performance.

Interviewers were hired on the basis of good English knowledge and telephone interviewing experience in German or Swiss German, respectively. They were informed that the study was investigating how translated interviews differ from scripted interviews. The Swiss German interviewers were telephone interviewers who worked for M.I.S. Trend SA. With one exception, the German interviewers had been trained and worked as telephone interviewers in Mannheim, Germany. When it proved difficult to find suitable German interviewers who could go to Bern on the days specified for the project, one person was hired who worked in survey research but who had not worked as a telephone interviewer before. The German interviewers were screened for the project in English and their translation abilities tested using European Social Survey questions. Staff from the ZUMA telephone unit briefed the German interviewers on general telephoning principles, and they were briefed again in Switzerland on the software at M.I.S. Trend SA. The Swiss interviewers already knew the software but, like the German interviewers, were briefed on the project by M.I.S. staff in the presence of SIDOS researchers.

The questions were selected from the European Social Survey 2002 to represent the different types of questions in that survey (opinions, values, reported behavior, sociodemographic information). The existing European Social Survey 2002 translations for Switzerland and Germany were essentially what the interviewers used in the scripted interviews. They were modified in a few respects for telephone delivery and the two German scripts harmonized slightly to reduce programing requirements for the CATI applications.

A special feature of Swiss German should be mentioned here. Written German differs considerably from spoken Swiss German in sentence structure, grammar, and vocabulary. It is thus normal practice for Swiss German interviewers to adapt their written German scripts to the norms of spoken Swiss German when interviewing. In some surveys even the written scripts are modified in the direction of spoken language in ways that are not standard for written Swiss German. In some senses, therefore, Swiss interviewers are accustomed to reading one text and speaking another, even if the text they read is written German and the text they speak is spoken Swiss German. This familiarity with doing two things at once—adapting and interviewing—was one reason for including Swiss interviewers in the project.

11.4 COMPARING THE ORALLY TRANSLATED AND SCRIPTED INTERVIEWS

This section compares the oral translations produced in the study with the read-off, scripted interviews. Space restrictions allow us to present only a few examples of some of the features that differ markedly between the two. Again, we focus on differences relevant for data collection quality.

The emphasis in the current chapter is more on the translations produced than on interviewer–respondent interaction. We have therefore kept our representation of the interview discourse as simple as possible. To increase accessibility, we have rendered everything in English, only occasionally mentioning specific German words. Later, reports on these and other questions planned from the project will also provide the German texts.

11.4.1 Simplifications and Omissions

Source questions, response categories, and instructions often provide details, explanations, and repetitions to guide respondents toward the intended meaning of the questions. Written translations are expected to retain the detail and nuance of source questions. Oral survey translations are conducted under considerable time pressure and practical constraints. One consequence of this is that elements deliberately included in a source question may be systematically dropped in oral translation. A reduction in detail and a focus on essential components is one natural outcome of oral translation, as is documented in research on interpreting and sight (oral) translation (Kiraly, 1995; Gile, 1997; Kalina, 1998).

Question B2 provides an illustration: *How often does politics seem so complicated that you can't really understand what is going on?* The words "seem" and "really" are intended to convey that the complicated state of politics is the problem and not the respondent's (in)ability to grasp complicated issues. The orally translated versions often run along the lines of *How often is politics so complicated you do not understand what is going on?* In German, this formulation could be understood to imply that respondents are unable to understand complicated issues.

Omission goes hand in hand with simplification. Multiple examples of omission and reduction of detail of potentially major and minor import can be found in the transcriptions; instructions and response categories are omitted; response categories are shortened; adverbs, modifications, and modal constructions ("should be," "might be," etc.) are simplified. As a general rule, any textual features causing the interviewers problems are dropped or glossed over as expedient.

11.4.2 Destandardization and Variation

Interviewers are expected to deliver scripted interviews in accordance with the interviewer briefing. Research has demonstrated that verbatim delivery is neither always possible nor always desirable (see, e.g., Houtkoop–Steenstra, 2000 and 2002; Maynard and Schaeffer, 2002b; Schober and Conrad, 2002; Schaeffer, 2002; Viterna and Maynard, 2002b). Nonetheless, verbatim delivery is frequently requested. The transcriptions of the orally translated interviews in the project indicate these vary far more than do interviews for which the interviewer has a script. Consequently, orally translated interviews are further removed from the intended questions in content and form than are the scripted interviews.

Table 11.1 presents one translation from each interviewer of question B33, a question that illustrates how source text challenges increase variation in translation. To save space, we have reduced "modern science" to "ms." The written English and German ESS 2002 formulations are given below the table.

The passive construction with "rely," with which the English question begins, cannot be followed in German. In order to use "rely," an active construction is necessary. The written translation actually used in the German ESS 2002 provides an example: *We can rely on modern science to solve our environmental problems.* Translation 6 in the table illustrates an interviewer beginning with a formulation that leads to problems, then beginning again in order to construct an active sentence and appropriate translation.

Having little time to process the entire question, most of the interviewers start translating the question word-to-word. Once they have translated *modern science*

Table 11.1. Sample Translation by Each Interviewer of B33

1	ms should be made responsible for our environmental problems
2	ms can be used to solve our environmental problems
3	ms can help to solve our environmental problems
4	ms can be trusted that it can solve environmental problems
5	ms should be made responsible for our environmental problems
6	ms can...one can trust ms to solve our environmental problems
7	ms can be used to solve environmental problems
8	ms... that one can rely on it to solve our environmental problems
9	ms can be used to solve environmental problems ...or should be used to solve environmental protection problems
10	one can trust ms that it can solve our environmental problems

English ESS 2002—Ms can be relied upon to solve our environmental problems.

German ESS 2002—We can rely on ms to solve our environmental problems.

can aloud and find that they cannot continue with a passive "be relied upon," they rather ingeniously find some vaguely appropriate verb, in active or passive form, that lets them finish the item. As is evident above, the outcome is often mistranslation.

Interviewers have to concentrate on producing interview text, not on remembering what they translated. As a result, it is not surprising that they vary their translations of a given phrase or question across interviews. As already noted, there are often multiple ways to translate a question. Moreover, translation research on memory and cognitive burden indicates that retention of performance as well as target language quality are particularly difficult in oral translation (Agrifoglio, 2004; Gile, 1997; Kiraly, 1995).

Respondents' "interruptions" may also increase variation. When interviewers conducting scripted interviews are interrupted, they can return to their text or tailor to accommodate the interruption. (See Houtkoop–Steenstra, 2000 and 2002 and Moore and Maynard, 2002 for general discussions of interruptions and respondent–interviewer interaction.) When an interviewer translating is interrupted, there is no written version of what she or he was saying to which to return. Table 11.2 illustrates the variance that can occur in one interviewer's translations from interview to interview.

After version 15, the respondent asked for the question to be repeated. Version 16 therefore followed only a few seconds after version 15. Either the interviewer was uncertain about how to translate, had no recollection of how she just translated, or knew but decided to change her translation. Whatever the reason, versions 15 and 16 differ considerably. Both an interviewer who translated B33 quite well the first time (see translation 10, Table 11.1) and an interviewer who translated the question wrongly throughout (see translation 1, Table 11.1) varied much less in each version they produced than the interviewer in Table 11.2. Thus the extent to which an interviewer is aware of problems and expends effort trying to deal with them may also play a role in explaining variance, not just better or poorer translation skills.

Table 11.2. Variation in Translation by One Interviewer

11	ms can be used ...or should be used to have new technol* sorry new ms should be used to solve the environmental problems
12	ms can be made responsible for solving environmental the environmental problems
13	ms can...should be made responsible for the environment... for the environmental problems...well...ms should be made responsible for solving environmental, the environmental problems
14	ms can be used to solve environmental problems...or should be used to solve environmental protection problems
15	ms should be able to be made responsible for our environmental problems
16	ms should be made responsible for solving our environmental problems

*Indicates a word is broken off.

11.4.3 Interference from English

Sticking too close to the source text structure and vocabulary can be thought of in terms of source text interference. It can reduce clarity and naturalness and can also change meaning (cf. Harkness et al., 2004). Interviewers might reasonably count as inexperienced practitioners in oral translation of written text. As indicated earlier, inexperienced translators tend to stick closer to the source text. The constraints of oral translation in CATI further encourage close translation, as do the added difficulties of oral translation contexts, even those in which translation is the only activity required.

As a result, the oral translations are at times awkward and unidiomatic, questions are harder to follow, and the translations may change question meaning. We provide only a few examples here. For instance, interviewers take English words instead of German words (*the following statement* is translated as "die folgende statement" instead of "die folgende Aussage"), or use a German word that mirrors the English word but is not the appropriate term in German. For example, Question F22 asks for the *name or title of your main job.* Interviewers use the German word "Name." However, in German, jobs have a "Bezeichnung," not a name. Sticking close to the English sometimes gets the interviewers into tight corners. The English bridge to the sociodemographic questions is *and now a few details about you.* A comparable German bridge would be more like *and now a few questions about you.* Some of the interviewers mirror the English "details" and use the German word "Details." In German this really suggests lots of detail. As a result, the interviewers follow their translation with reassurances such as "actually not many questions," "actually not very detailed," and "Don't worry, it is only a few questions."

Measurement can also be affected. The first questions of the study ask respondents about the time spent on different media *on an average weekday.* This wording is intended to guide respondents to think about working days of the week rather than all the days of the week. The word that does this well in German is like "work-day" ("Werktag"). However, the great majority of the interviewers translate "weekday" with a term that superficially matches the English, "week"+"day" ("Wochentag"). "Wochentag," however, does not identify working days of the week. Responses to these questions in the oral interviews also differ from those collected in the scripted interviews, which do refer to working days of the week.

Little research is available to date on how unidiomatic and incorrect language or unclear presentation of these kinds affects respondents' understanding of the questions or their perception of interviews or interviewers. However, when the translation output is repeatedly followed by requests from respondents to repeat the question, we can safely assume that problems have arisen (cf. Moore and Maynard, 2002). The following section looks at instances of this.

11.4.4 Translation Problems Affect Respondent Comprehension

If the source text presents a challenge to interviewers, their presentation suffers. More frequent or more dramatic repairs than in scripted interviews, false starts, long pauses, self-discourse, and apologies are examples of how interviewer output can not only be inaccurate but also be more complicated when they come up against problems in translation.

Because idioms often cannot be translated by taking one word at a time, they may pose especial problems in oral survey translation. The idiom "income levels" in B30 provides an example. This question, *The government should take measures to reduce differences in income levels*, resulted in notable interviewer hesitancy in both German and Swiss German oral translations. One part of the problem may well be that *income levels* could be translated quite well with "income differences" in German. However, because the word "differences" is used earlier in the English text, the interviewers, because they are translating closely, have also already used that word before they get to *income levels*. For them to use "differences" again (e.g., "the government should take measures to reduce differences in income differences") would be as odd in German as it is in English. At the same time, the interviewers seem to feel that something is still needed to translate "levels."

This results in the numerous problems presented below. To save space we present only Swiss translations, focusing on the core section of these, and limit our commentary to major problems. Each Swiss interviewer had one or more requests to repeat this question (see Conrad et al., 2007, Chapter 10 in this volume). We checked the ESS 2002 data in case public discourse in Switzerland and Germany meant this question was particularly difficult in some way. However, the distributions for the question in these countries do not differ greatly from those of a large group of countries in which more respondents agree with the statement than disagree with it.

Language interference may be what leads the interviewers to look for something to match *levels*. Some interviewers indeed borrow *levels*, which is an inappropriate solution here. Others use German words associated with the semantics of *level* but which are also not appropriate in this context. We indicate inappropriate translations using words associated with *level* by putting them in inverted commas.

Example 1

I A ...to reduce the difference between the different incomes... well so that well, income "level," that is simply that well, managers get much more than simple workers

R request for clarification

Interviewer A reformulates several times, reflecting hesitancy in the processing particle "well" and provides an explanation of different income levels as well as borrowing the English word.

Example 2

I B ...um take measures um to reduce the um the differences in income

R if they are going to make differences in income I agree

I B have you had the feeling that they should um take measures?

R yes

The respondent appears to misunderstand, and it is unclear in the end what it is with which he indicates agreement. The interviewer proceeds to the next question without probing further on either the degree of agreement or what precisely the respondent has understood. It seems quite possible that the respondent is more in favor of differences than against them.

Example 3

I B ...take measures to um reduce differences in the income "steps"... sorry... in the case...or yes do you um...

R (interrupts) bec* because of the school do you mean?

I B well, not here generally do you agree or not agree with that statement?

R asks to have question repeated

The interviewer does not actually finish off her presentation of the statement. As she begins to formulate "do you agree or disagree," the respondent asks for clarification. The interviewer does not provide a clear response to the respondent's first question and does not ascertain whether the respondent has understood the question or her clarification properly.

Example 4

I C ...should (pause) do more so that (pause) um the differences in the income le* income "levels," that is to say, the rich and the poor um are not so big, well, that they are reduced

R the tax differences is that correct?

I C yes, or simply the income "level" that that

R no

I C should be changed

The respondent asks for clarification before interviewer C has completed presenting an elaborated version of the item and then answers while the interviewer is still responding to his request for clarification. It is not clear whether the respondent says "no" in response to the interviewer answering "yes," that is, confirming that tax differences were the issue, or whether he took notice of the interviewer's somewhat unclear correction ("or simply the income level") and responded in the negative to that.

11.5 HOW DESIGN AFFECTS TRANSLATION

In this section, we demonstrate how the oral translations can highlight weaknesses in the source questionnaire, providing evidence for the usefulness of "advance translation," a quick translation procedure sometimes used to discover problems in draft source questionnaires (see, e.g., Harkness, 1995 and Braun and Harkness, 2005). We present only two examples here, related to quantification and frequency.

11.5.1 Regularising Quantification

Questions A8, A9, and A10 measure attitudes to other people. They are constructed similarly but, as the translations reflect, not similarly enough. A8 and A9 present statements about *most people*: *Most people can be trusted*; *Most people would try to take advantage*; *Most people would try to be fair.* Question A10, in contrast, makes statements about *most of the time* and *people*: *Most of the time people try to be helpful.* The corresponding response category label in A10, however, *people mostly try to be helpful* does not match the question formulation. This particular response formulation is ambiguous. It can be understood as "people try to be helpful most of the time" (matching the A10 statement) but also as "most people try to be helpful."

A number of the interviewers translated A10 on the pattern of A8 and A9, rendering it as something like *most people try to be helpful.* Sometimes they also changed the source response category *people mostly try to be helpful* to *most people try to be helpful.* It is worth noting that the written German translations in the ESS also regularize these questions, modeling A10 on A8 and A9. Both the oral translations and the written ESS translations thus point to the fact that the unmotivated variation in source question A10 would have been better avoided.

11.5.2 Translating and Repairing Response Scales

Languages differ in the frequencies they can readily differentiate. As a result, frequency scales are difficult to translate (e.g., Harkness, 2003; Smith, 2004). Table 11.3 records each of the translations produced for the frequency scale used in question B3: *never/seldom/occasionally/regularly/frequently.* The English scale is presented at the top of the table. It poses challenges for translation. For example, it is not certain what degree of frequency is to be associated with *regularly*, either with respect to *occasionally* or *frequently.* In English the word "regularly" is sometimes used for "often," although its semantics refers to regularity of occurrence and not frequency. If intended as "often" here, it is doubling for a synonym of the last category *frequently.* It thus raises questions about the design of the scale and

Table 11.3. Interviewer Translations of Response Scale B3

Never	Seldom	Occasionally	Regularly	Frequently
Never 1[a]	Seldom	Occasionally	Regularly	Often
Never 2	Sometimes 2[b]	Sometimes 2[b]	Often	Very often
		Sometimes 1[b]	More frequently	Frequently
		From time to time	Relatively often	Mostly
		On and off	Constantly	Constantly
		In part/some of the time		Very regularly
				Almost always
				Always
				Always or very often
				Very frequently
2 versions	2 versions	6 versions	5 versions	10 versions

[a]German has two words for *never* (*nie* and *niemals*), represented here as *never 1* and *never 2*

[b]Swiss German has two forms for *sometimes*, one the German *manchmal* and the other a Swiss German form, *mängisch*, labeled here as *sometimes 1* and *sometimes 2*.

presents challenges for translation, in particular, for oral translation. As elsewhere in the chapter, we provide English renderings of the German translations.

"Frequently" is not a difficult word to translate directly into German. Interviewers nonetheless proposed 10 different translations for it. *Frequently* is the last category the interviewers read and present to respondents, and as such it is arguably where they try to resolve issues resulting from how they have translated the problematic *regularly*. Some interviewers produce the more common scale "never/seldom/sometimes/often/always" as their version of the ESS scale. Other interviews stick to the English and translate *regularly* literally, with the word "regelmässig." However, "regelmässig" does not function particularly well as a frequency adverb. Interviewers who translate *regularly* with German "often" try to rescue their answer scale design by either strengthening the end point of *frequently* with "very" or selecting a stronger term than *frequently*, such as "constantly," "mostly," or "always." German has a word that corresponds to "often" and another that corresponds to "frequently" and some interviewers produce "never/seldom/sometimes/often/frequently," again revealing the underlying problems of the source scale.

A scale that avoided *regularly* and had a clear distinction between the fourth and fifth categories (and preferably between the second and third) would, we suggest, have caused less trouble. In the translators' hotline for the 2006 round of the ESS, a country translating this scale for the first time asked what to do about these problems. Several ESS translations in Round One and Two strengthened *frequently*, just as a number of the interviewers in our study did.

11.6 INTERVIEWER PERFORMANCE

Research into interviewer–respondent interaction has demonstrated convincingly how complex and multiple the tasks required of the interviewer may be (Japec, 2005). In orally translated telephone interviews, interviewers take on even more tasks, working and processing in two languages, translating and interviewing at the same time. Research on scripted interviews has shown that interviewers sometimes cannot "stick to their script" and are indeed sometimes better advised to tailor their script. They also sometimes abandon their script in inappropriate ways (Houtkoop-Steenstra, 2000, 2002; Conrad et al., 2007, Chapter 10 in this volume). This has raised concerns about survey error and poor data quality and about overrigid standardization. Lynch (2002) indicates that faithful recitation of the scripted text is particularly difficult in telephone interviews. This is partly because interviewers in face-to-face contexts have more means available to signal they are not finished speaking. As a result, respondents tend to "interrupt" the interviewer more in telephone surveys. By preventing the interviewer from finishing, respondents may reduce or complicate the information they receive.

The tape recordings of the translated and scripted interviews in the present study provide vivid examples of interviewer–respondent interaction problems also documented in research in which translation was not a component. In translated interviews, however, the interaction is less successful in terms of good survey practice than in scripted interviews. It seems likely that this reflects the increased burden oral translation places on the interviewer. After sequences involving comprehension problems, interviewers often fail to follow good practice requirements. They repeat responses incorrectly, present answer categories in biased fashions, and they fail to probe and negotiate suitable responses. Sometimes, as in the case of the introduction to B13–19, they seem keen to leave unresolved problems behind them. In the scripted interviews, on the contrary, the interviewers ask more often for confirmation of their coding and also probe more to ascertain which response a respondent intends. They also negotiate or coconstruct responses more frequently in scripted interviews than in translated interviews (Conrad et al., 2007, Chapter 10 in this volume).

In B30 (see Section 11.4.4) and in B13–19 (see Section 11.7), attempts at translation were sometimes so unsuccessful that the "conversation" came to a halt for several moments. Hootkoop-Steenstra (2000) points out that silences in CATI interviews can be considerably longer than the one second that Jefferson (1989) reports as the maximum standard time in ordinary conversation before a speaker will resume speaking. However, Houtkoop-Steenstra (ibid) interprets her examples as instances of the interviewer giving respondents time to elaborate on answers. In our examples, the considerable silences occurring around these translation problems seem to be motivated by interviewer uncertainty about how to continue their own contributions.

The restricted space available allows us to mention only a few of the components considered to be typical sources of survey error.

Omission. Simplifications and omissions proved to be frequent in the translated and scripted interviews. However, these are fewer and less intrusive in the scripted interviews than in the orally translated interviews. In scripted interviews, they are often preceded by a previous full presentation of information that is later not repeated or is presented in a reduced, simplified form. Thus a response scale read out for the first two or three items in a question battery might not be read out for the remaining few items in the battery or referred to in shortened form or presented only if the respondent hesitates. In translated interviews, response scales are sometimes forgotten from the start. Only when something goes wrong, do the interviewers realize they have not presented the categories available, marking their recognition with comments such as "Oh; oh yes; Sorry, I forgot to mention..." or "Sorry, I should have read out the response categories to you."

Failure to probe. When respondents offer a response not covered in the response scale, interviewers are expected to negotiate which available category fits best. Research has shown they do not always do so appropriately (cf., Schober and Conrad, 2002; van der Zouwen, 2002). A number of failures on the part of interviewers to probe can be found in examples in Section 11.4.4. The following example, in which an interviewer acknowledges receipt of an uncodable answer and moves to the next question, is by no means unique in the transcripts.

I: would you say generally speaking that people who come to live in Switzerland are good or bad for the Swiss economy? 0 means it is bad for the economy 10 means it is good for the economy.
R: well, that depends on the people
I: depends on the people?
R: yes of course!
I: OK

This is not to suggest that all translated interviews are disjointed or difficult to follow. Some of the orally translated interviews are very fluent. Flow and fluency are reflections of interviewer practice but unfortunately, in our data, they do not guarantee that the translation is appropriate. Some interviewers translate very smoothly from the start, irrespective of whether they translate correctly or not.

11.7 WHAT DO THE STATISTICS TELL US?

We had originally expected to have all of the tapes transcribed by summer 2006, but we now expect to have the complete transcripts in 2007. Thus some work remains to be done on both the textual and the statistical data, and our readings and conclusions are preliminary and cautious. Detailed analysis of possible discrepancies between what respondents answered on tape and how responses were coded may, for example, yet alter our statistics. At the same time, the textual data examined make it clear that respondents in scripted interviews were not exposed to the same stimuli as those interviewed when interviewers were translating.

The interview questions discussed here were selected on the basis of our wish to cover a variety of questionnaire components including questions, answers scales, and introductory sections and a preliminary assessment of the taped material. We also aimed to look at both translation problems and source design issues. The decision to transcribe these questions in all of the interviews and leave other material aside until later was taken before we turned to the statistical data. In other words, the statistical data did not drive our selection of questions.

We could have operated the other way round, looking only at what showed up in the data. If we had, we would have missed numerous features that, as we collect them, are beginning to form a picture of systematic processes and at least partially predictable outcomes. Such pictures are exactly what are needed to be able to inform decisions about the use of oral translation and about interviewer training as well as guide our understanding of such tasks. In addition, we are discovering how much a less-than-perfect translation can nevertheless reveal about source question design.

We could expect differences in mean or in variance to be a result of interviewer effects, a result of oral translation versus scripted interviews, a reflection of differences between Switzerland and Germany, or a combination of several or all of these. We mentioned earlier that spoken Swiss German is further removed from written German than spoken (standard) German in Germany. If, for example, the Swiss German oral translations are further removed from the scripted version than the German translations, variance could increase in the Swiss data. It might also transpire that interviewers have culturally different interviewing styles. At the same time, we could also expect cultural differences in response between the countries. The small number of cases makes it difficult to assess interviewer or cultural impact on the statistical data. Once we have the full transcriptions, however, we will at least be able to estimate the frequency of behaviors.

In the few examples presented here, differences in what respondents experienced in the interviews seem to have affected the statistical data in understandable ways. Despite the comparatively small number of cases available, the data for the questions presented differ between scripted interviews and translated interviews where one might expect them to differ and, as indicated below, do not differ where we might expect them not to differ. Thus, for example, the translation of English "average weekday" (meant as an average working day of the week) as simply "average day of the week" affects the data significantly. The multiple versions of the frequency scale (*regularly, frequently*) translated in B2, and the numerous and frequently incorrect translations of B33 (*modern science can be trusted to...*) may well be responsible for the difference in distributions between scripted and translated interviews on these questions. In contrast, the changes presented earlier in orally translated versions of questions A8–A10, in which the interviewers harmonized the formulations *most people* and *most of the time people*, would not be expected to result in differences, since the changes in the oral versions exactly matched the harmonization in the written translations. There is also no significant difference in the data.

What we do find in these last questions is a significant difference between Swiss and German data. Cultural differences can be observed in other questions not discussed here, such as question B28 about opinions on education. In focusing on the differences between scripted and translated versions, we have had to set aside considerations of culture, both in terms of country/culture and in terms of interviewing conventions. Further qualitative analysis with complete transcripts will also help determine, for example, whether interviewers change their presentations as they become more familiar with the instrument. Given the small number of cases, a statistical analysis for this is not appropriate.

The greatest obstacles for oral translation are passages where the interviewers find the source text unclear. Yet when we look at the transcripts, one of the most problematic passages for the interviewers, the introduction to questions B13–19, does not seem to affect the core question and answer interaction. There is also no significant difference in the data between scripted and translated interviews. This makes sense when we look at the larger context. The introduction to B13–19 runs as follows: *There are different ways to improve things in (country) or to help prevent things from going wrong. Have you done any of the following in the last 12 months?* Interviewers' translations of the first sentence are frequently hesitant; they mistranslate, make false starts, attempt corrections, and repeat themselves. Sometimes the exchange more or less breaks down. Here are two typical examples, by no means the most disjointed of those produced.

I: um there are different means and ways which one can use, which one can go, to prevent something in Swiss politics, that is, always related to politics

I: good, and now there are different ways one can um try to improve things in Switzerland or one can help um things that are going wrong can prevent um

The first sentence was difficult to translate orally. For a start, it is not clear whether "or" midsentence is inclusive or exclusive. If "or" is inclusive, the sentence can be likened structurally to "there are different ways to cut or mow grass," in which "to cut or mow" refers to the same general activity. If "or" is intended exclusively, the sentence can be likened to "there are different ways to prune hedges or start a compost heap." In addition, *improve* and *help prevent from going wrong* are neither synonyms (inclusive *or*) nor antonyms (exclusive *or*). They are also structurally asymmetrical; the word "improve" is set against "help prevent from going wrong." This phrase is a complex verbal construction (help + prevent from + going wrong). It also contains two negative semantic elements in "prevent" and in "going wrong," which may further complicate parsing. When translating, interviewers frequently simplify the construction by omitting *help*. A number also resolves the mismatch between *improve* and *prevent going wrong* by translating *going wrong* with an antonym of "improve" ("get worse"). It is notable that the written German and Swiss German ESS translations of this question employ a similar strategy. Clearly, the oral translations are highlighting problems, at least for translation, in the source questionnaire. However, because the respondents are not required to use the problematic "getting-better-or-worse" material to arrive at answers to the question *Have you done any of the following in the past 12 months?*, they seemed to ignore the unclear part and focus on the understandable question, producing response data that did not differ from the scripted interviews.

11.8 DISCUSSION AND OUTLOOK

Oral translation is not an established practice in Switzerland or Germany. It might therefore be argued that our interviewers are not representative of the "bilingual" interviewers used in other studies. At the same time, little is documented about the skills, experience, or performance of interviewers described as "bilingual" in other studies. It is possible that familiarity with the task does make a difference, but what this difference might be remains to be investigated. Familiarity with a task might, for example, reduce attention rather than enhance performance. Whatever the case on that point, many of the strategies adopted by the interviewers to simplify and reduce material are documented in the translation sciences as standard features of interpreting and "sight" (oral) translation. The close translation and language interference that lead our interviewers into problems have also been described as typical of oral translation even by professional translators. Furthermore, a number of the problems the interviewers encountered were also reflected in the treatment given to them in the written ESS translations.

One of the arguments for oral translation is that it allows respondents to be interviewed who speak languages not envisaged in a given study. The fact that a language is rare enough not to be envisaged may be an indicator for considerable cultural and linguistic difference between the population(s) for whom a written questionnaire is available and a population not expected. It could thus be essential to test whether the source instrument is indeed *suitable* for translation (cf. Harkness et al., 2003; Tanzer, 2005; van de Vijver and Poortinga, 2005). Moreover, if a language is rare in a given context, it could be difficult to find trained interviewers with strong spoken and written skills in the source and target language. If, however, the number of respondents who need a translation are sufficiently high to be predictable, good practice would require that the questionnaire is pretested for this population in the target language. Pretesting would imply that a written translation is also available, provided the language has a written form.

Just as some interviewers are better at interviewing than others, some interviewers in the study are better at translating than others. Nonetheless, as things stand, most of the interviewers made a reasonable or good job of conveying essentials while also carrying out the many other tasks required of survey interviewers. Our preliminary analysis of the transcripts indicates that recurring features of difference across interviews are more related to the mode of administration (oral translation instead of simply interviewing) or to the instrument design than to the individual ability of a given interviewer to translate. Thus differences across interviewers notwithstanding, the orally translated interviews deviate more on all scores from the written German texts than do the scripted interviews.

Everything else being equal, written translations provide the opportunity to ask the questions intended in the manner intended. (This is not to deny that other considerations may be important; see, e.g., Harkness et al., 2003; Harkness, 2004; Braun and Harkness, 2005.) Thus scripted interviews using written translations can provide better data quality. Orally translated interviews introduce manifold differences that work against the principles of standardization, the notion of carefully worded questions, and against good interviewing practice.

The data discussed here thus indicate that great caution is called for in using oral translations. Nonetheless, for languages that do not have a written form, oral translation and interpreting are the basic options. When oral translation and interpreting are the options, careful preparation is essential. Thus, whatever the motivation for oral translation, a greater understanding of the process and its demands needs to be pursued. Without this we cannot properly inform selection, training, or briefing of those employed to conduct orally translated and interpreted interviews.

Many questions about the viability of combining an interview with concomitant translation remain to be investigated, as do issues related to who might be best trained for the job. It might, for example, be easier to teach translators to interview than to teach interviewers to translate. Techniques developed to train oral production and on-sight (oral) translation in interpreting (see Gile, 1997; Kalina, 1998; Agrifoglio, 2004) might be usefully adapted for surveys. At the same time, process and memory issues in reading, interviewing, coding responses, writing, and translating orally all need to be addressed.

Our discussion here has focused on translation issues. The cultural gaps between cultures that do not have a written language and the cultures that determine the shape and content of many questionnaires are likely to be great. In many instances, translation will be insufficient and adaptations will also be required. Research into adaptation suggests clearly that this is not an activity that can be delegated to either translators or to interviewer-cum-translators. Different strategies and preparatory protocols will be needed (cf. Harkness, 2007).

In conclusion, oral translation, apart from changing the language, also changes the mode of question presentation and what is presented. The systematic changes noted as an integral part of oral survey translation raise basic questions about question design and data comparability. If oral translation concomitant with interviewing can convey only the essentials, we will need to reconsider how we perceive the role of carefully crafted source questions, how we define what "asking the same question" involves, and how we demonstrate whether this has indeed been achieved in orally translated interviews.

CHAPTER 12

The Effects of Mode and Format on Answers to Scalar Questions in Telephone and Web Surveys[1]

Leah Melani Christian, Don A. Dillman, and Jolene D. Smyth
Washington State University, USA

12.1 INTRODUCTION

The use of mixed-mode surveys has become increasingly popular as surveyors adapt to rapid technological advances in survey methodology and the changing lifestyles of survey respondents (Biemer and Lyberg, 2003; de Leeuw, 2005). The trend toward conducting mixed-mode surveys was only beginning at the time of the first Telephone Survey Methodology Conference in 1986 (Dillman and Tarnai, 1988) but has proliferated since then with the creation of web surveys. Utilizing multiple mode(s) to collect data from respondents allows survey designers to increase response rates and sometimes data quality by exploiting the strengths of particular survey modes while remaining within the time and cost constraints of a study.

Since the data are often combined for analysis the increased use of mixed-mode survey designs raises concern about whether respondent characteristics are being measured equivalently across modes. Previous research indicates that the mode of data collection can influence how respondents answer survey questions (de Leeuw, 1992). Generally, the various survey modes differ with respect to technical and

[1]Analysis of these data was supported by funds provided to the Washington State University Social and Economic Sciences Research Center (SESRC) under Cooperative Agreement #43-3AEU-1-80055 with the USDA-National Agricultural Statistics Service, supported by the National Science Foundation, Division of Science Resource Statistics. Data collection was financed by funds provided to the SESRC by the Gallup Organization.

Advances in Telephone Survey Methodology, Edited by James M. Lepkowski, Clyde Tucker, J. Michael Brick, Edith de Leeuw, Lilli Japec, Paul J. Lavrakas, Michael W. Link, and Roberta L. Sangster

cultural factors related to the media or mode itself, the impact of interviewer presence (or absence), and how information is transmitted or conveyed during the survey (de Leeuw, 2005). Understanding the effects of these three types of mode effects can help us evaluate the equivalency of data collected across modes. In addition to the effects of these specific survey mode factors, survey designers often introduce question format effects in mixed-mode surveys by constructing questions differently depending on the mode being used to survey respondents.

Scalar questions are one of the most commonly used types of survey questions and are frequently formatted or constructed differently across modes to maximize the effectiveness of particular questions for each mode. For example, since no visual aid is typically available for telephone interview respondents, response scales are oftentimes simplified by providing only the polar end-point labels to ease the administrative task for interviewers and the cognitive and memory burden placed on respondents. However, response scales on web surveys are often presented with all of the scale points verbally labeled, whereas for face-to-face respondents the scale might be presented visually on a show card also with all of the categories labeled. Research on scalar questions suggests that we may expect differences in responses because of the differential labeling of response categories and the overall visual presentation of the scale (Krosnick and Fabrigar, 1997; Christian and Dillman, 2004; Tourangeau et al., 2004; Dillman and Christian, 2005). Moreover, the effects of survey mode and question construction or format may both independently and/or jointly influence responses.

Our purpose in this paper is to assess the equivalency of data collected using different scalar formats within and across both telephone and web modes. More broadly, our objective is to contribute to theoretical understandings of how differences among survey modes *and* the scalar question formats influence responses to survey questions. The experimental comparisons were designed to identify ways of asking scalar questions that present equivalent stimulus to respondents across telephone and web modes so that optimal questionnaire designs can be constructed for mixed-mode surveys.

12.2 OVERVIEW OF EXPERIMENTAL COMPARISONS AND PROCEDURES

The popularity and frequent use of scalar questions means that many ways of constructing them have been developed over the years. In this paper, we analyze 70 experimental comparisons from one survey using six versions (three telephone and three web) of 13 scalar questions. We include comparisons of *similar scales across telephone and web modes* and comparisons of *different scales within these modes* to test the independent effects of both mode and scale format, and we also test for interaction effects of survey mode and scale format.

To test for mode effects, we compare telephone and web responses to 5-point fully and polar point labeled scales as well as 11-point polar point labeled scales (see Fig. 12.1). To test for scale format effects and provide insights into how various

	Telephone example	Web example
(*a*) Fully labeled scales *9 comparisons* 3 satisfied/dissatisfied 3 agree/disagree 3 construct-specific	How satisfied are you with Washington State University as a place to go to school? Would you say you are … Very Satisfied Somewhat Satisfied Neutral Somewhat Dissatisfied Very Dissatisfied	How satisfied are you with Washington State University as a place to go to school? Very Satisfied Somewhat Satisfied Neutral Somewhat Dissatisfied Very Dissatisfied
(*b*) Polar point labeled scales *13 comparisons* 6 satisfied/dissatisfied 3 agree/disagree 2 extremely/not likely 2 best/worst possible *4 comparisons* 11 category polar point w/ midpoint also labeled "average" on two comparisons best/worst possible	On a 5-point scale, where 5 means very satisfied and 1 means very dissatisfied, how satisfied are you with Washington State University as a place to go to school? You may use any of the numbers (read slowly) 5, 4, 3, 2, or 1.	How satisfied are you with Washington State University as a place to go to school? 5 Very Satisfied 4 3 2 1 Very Dissatisfied

Figure 12.1. Summary and examples of experimental comparisons testing the mode effects across telephone and web.

aspects of constructing scalar questions can influence responses, we compare different ways of formatting or constructing scales such as whether to label all of the categories or only some, whether to use verbal and numerical labels, and the overall presentation of the scale (see Fig. 12.2). The comparisons together provide insights into constructing scales optimally for mixed-mode surveys. Since we include a large number of comparisons in this paper and to facilitate connections between the theory and results, we summarize general procedures here but present our discussion of previous research, theoretical rationales, and detailed description of the manipulations with the results for each set of comparisons. We group the comparisons into two sections, mode effects and scalar format effects, and have included a summary figure (Fig. 12.3) of our experimental comparisons and key findings.

Biemer (1988) has noted that most of the literature on mode effects actually compares two (or more) "systems of data collection" where the overall survey instrument and implementation are adapted to maximize the efficiency of each mode such that the estimation of "pure mode effects" is not possible. To overcome this shortcoming, the survey implementation procedures were standardized across the web and telephone modes. We embedded the experimental comparisons in a mixed-mode survey of random samples of undergraduate students about their experiences at Washington State University's (WSU) Pullman campus conducted in the fall of 2004. We randomly assigned each student to one of the six experimental versions (three telephone and three web versions) of a 25-question survey and administered

	Telephone example	Web example
(*a*) Fully labeled versus Polar point labeled scales See Fig. 12.1 a and b	*6 telephone comparisons* *6 Web comparisons* 3 satisfied/dissatisfied 3 agree/disagree	
(*b*) Agree/disagree versus Construct-specific scale *3 telephone comparisons* *3 web comparisons*	To what extent do you agree or disagree that your instructors are accessible outside of class. Would you say you … Strongly Agree Agree Neutral Disagree Strongly Disagree How accessible are your instructors outside of class? Would you say … Very Accessible Somewhat Accessible Neutral Somewhat Inaccessible Very Inaccessible	To what extent do you agree or disagree that your instructors are accessible outside of class. Strongly Agree Agree Neutral Disagree Strongly Disagree How accessible are your instructors outside of class? Very Accessible Somewhat Accessible Neutral Somewhat Inaccessible Very Inaccessible
(*c*) Assigning the most positive category 5 versus 1 *3 telephone comparisons* *3 web comparisons* satisfied/dissatisfied	On a 5-point scale, where 5 means very satisfied and 1 means very dissatisfied, how satisfied are you with the quality of advising you have received as a WSU student? You may use any of the numbers (read slowly) 5, 4, 3, 2, or 1. On a 5-point scale, where 1 means very satisfied and 5 very dissatisfied, how satisfied are you with the quality of advising you have received as a WSU student? You may use any of the numbers (read slowly) 1, 2, 3, 4 or 5.	How satisfied are you with the quality of advising you have received as a WSU student? 5 Very Satisfied — 1 Very Satisfied 4 — 2 3 — 3 2 — 4 1 Very Dissatisfied — 5 Very Dissatisfied

Figure 12.2 Summary and examples of experimental comparisons testing the effects of various scalar formats.

	Telephone example	Web example
(*d*) Presenting the most positive versus most negative category first *2 telephone comparisons* *2 web comparisons* extremely/not at all likely	On a 5-point scale, where 5 means extremely likely and 1 not at all likely, how likely are you to continue to attend WSU until you finish your degree? You may use any of the numbers (read slowly) 5, 4, 3, 2, or 1. On a 5-point scale, where 1 means not at all likely and 5 means extremely likely, how likely are you to continue to attend WSU until you finish your degree? You may use any of the numbers (read slowly) 1, 2, 3, 4, or 5.	How likely are you to continue to attend WSU until you finish your degree? 5 Extremely likely 4 3 2 1 Not at all likely 1 Not at all likely 2 3 4 5 Extremely likely
(*e*) Telephone only		
Instruction versus No Instruction *6 comparisons* satisfied/dissatisfied	On a 5-point scale, where 5 means very satisfied and 1 means very dissatisfied, how satisfied are you with Washington State University as a place to go to school? You may use any of the numbers (read slowly) 5, 4, 3, 2, or 1.	On a 5-point scale, where 5 means very satisfied and 1 means very dissatisfied, how satisfied are you with Washington State University as a place to go to school?
Direction of instruction *2 comparisons* extremely/not at all likely	On a 5-point scale, where 5 means extremely likely and 1 not at all likely, how likely are you to continue to attend WSU until you finish your degree? You may use any of the numbers (read slowly) 5, 4, 3, 2, or 1.	On a 5-point scale, where 5 means extremely likely and 1 not at all likely, how likely are you to continue to attend WSU until you finish your degree? You may use any of the numbers (read slowly) 1, 2, 3, 4, or 5.
(*f*) Web only		
Polar point versus number box (web) *6 comparisons* satisfied/dissatisfied	How satisfied are you with Washington State University as a place to go to school? 5 Very Satisfied 4 3 2 1 Very Dissatisfied	On a 5-point scale, where 5 means very satisfied and 1 very dissatisfied, how satisfied are you with Washington State University as a place to go to school? You may use any of the numbers 5, 4, 3, 2, or 1. [ ]
5 versus 1 positive (number box) *2 comparisons* satisfied/dissatisfied	On a 5-point scale, where 5 means very satisfied and 1 very dissatisfied, how do you feel about the quality of *instruction* in the classes you have taken at WSU? You may use any of the numbers 5, 4, 3, 2, or 1. [ ]	On a 5-point scale, where 1 means very satisfied and 5 very dissatisfied, how do you feel about the quality of *instruction* in the classes you have taken at WSU? You may use any of the numbers 1, 2, 3, 4, or 5. [ ]

Figure 12.2. ***(Continued)***

SC		*t*-tests (difference in means) More positive ratings on the telephone than the web	χ^2 tests (2×2; df=1) More likely to select the most positive category on the telephone than the web
Fully labeled 5-point scales	(see Table 12.1)	8 of 9 comparisons significant	5 of 9 comparisons significant
Polar point 5-point scales	(see Table 12.2)	11 of 13 comparisons significant	7 of 13 comparisons significant
Polar point 11-point scales	(see Table 12.3)	3 of 4 comparisons significant	No significant differences (4 comparisons)

SC	
Fully labeled versus polar point (see Table 12.4)	More positive ratings to fully labeled than polar point labeled scales 6 of 6 comparisons significant on the telephone; 2 of 6 comparisons significant on the web
Agree/disagree versus construct-specific (see Table 12.5)	No consistent pattern 2 of 3 comparisons significant on both the telephone and the web
Most positive category 5 versus 1 (see Table 12.6)	More positive ratings when 5 is assigned to the most positive category versus 1 1 of 3 comparisons significant on the telephone; 0 of 3 comparisons significant on the web
Most positive versus negative first	No significant differences — 0 of 2 comparisons significant on both the telephone and the web
Instruction versus no instruction Direction of instruction (5 versus 1)	(telephone only) Slightly more negative ratings w/ instruction — 1 of 6 comparisons significant (telephone only) No significant differences (2 comparisons)
Polar point versus number box Number box (5 versus 1 positive) (see Table 12.7)	(Web only) No significant differences (6 comparisons) (Web only) More positive ratings when 5 (versus 1) is positive — 2 of 2 comparisons significant

Figure 12.3 Summary of experimental comparisons and findings.

the survey to the telephone and web samples simultaneously. To test for "pure mode effects," we needed a population with equal access to completing either a web or telephone version of the survey (i.e., we needed a population with telephone and web access so that we could randomly assign respondents to one of the six versions). Thus, students, who all have web access through the university, provided an ideal population for this experimental research. Response rates for both surveys were comparable with 59 percent of the telephone respondents completing the survey (945 completes out of 1608 sampled) and 60 percent of the web respondents completing the survey (1082 completes of 1800 sampled).

All of the students were initially contacted via postal mail letter that included a $2 incentive. Telephone respondents were then contacted by the WSU Social and Economic Sciences Research Center's telephone lab to complete the telephone survey. Up to 10 callback attempts were made. Web respondents for whom we had an e-mail address (about 2/3 of the sample) were sent an initial e-mail, which included a link to the web survey, in addition to the initial postal mailing. Subsequent contacts to web nonrespondents were sent using postal mail and e-mail.

An interactive or dynamic design was adopted for the web survey where each question appeared on a separate screen. Questions were presented in black text against a colored background with white answer spaces to provide contrast between the text, answer spaces, and background. All of the screens were constructed using HTML tables, where proportional widths were programed in order to maintain a consistent visual stimulus regardless of individual screen or window sizes. Cascading Style Sheets were used to automatically adjust font size and accommodate varying user browsers and screen resolutions. We controlled access to the web survey by assigning each student an individual identification code they had to input to gain entrance to the survey. Their unique access code was provided in each contact to the respondent.

For 10 questions, we performed a series of multinomial logistic regression models testing both the independent effects of survey mode (telephone or web) and scale format and the potential effects of the interaction of mode and format. We found several significant independent effects of survey mode and question format all consistent with the bivariate analyses; however, none of the interactions of survey mode and scale format were significant. Consequently, we present only the results from the difference of means *t*-tests and χ^2 tests of association (two-by-two, with one degrees of freedom) and their associated *p*-values for both tests of mode and scalar format effects.

12.3 MODE EFFECTS

12.3.1 How Does Survey Mode Influence Data Quality?

Since data are often combined in mixed-mode surveys, there is growing concern about whether answers collected from respondents surveyed using different modes are comparable. There are several factors that differentiate modes and can help us understand why responses to questions might differ across modes. These factors can be grouped into three general categories: media-related, interviewer impact, and information transmission (de Leeuw, 1992). Although the various mode-related factors are grouped separately into these three categories for purposes of discussion, these categories are not independent and the various types of factors influence and relate to one another.

Media-related factors include respondents' familiarity with and use of the mode itself, which can be of particular concern for web surveys since computers and the Internet are relatively new technologies now only gaining widespread use, and not everyone knows how to use or has access to either a computer or the Internet. One extremely influential media-related factor is the locus of control. In face-to-face and telephone surveys, interviewers control the delivery of the stimulus including the order in which questions are read to respondents as well as the pace and flow of the conversation (de Leeuw, 1992; Dillman, 2000). However, since mail and web surveys are self-administered, the respondent controls the survey conversation, determining what parts of the questionnaire to process when, how to answer, and how quickly to move through the survey (Dillman, 2000). Telephone interviews are often conducted at a quicker pace because the interviewer and respondent try to avoid silences or lapses in conversation over the telephone, whereas in face-to-face interviews, nonverbal communication and interaction between the respondent and interviewer make silences less of a problem (de Leeuw, 1992; Dillman et al., 1996) and in self-administered surveys, norms of silences are not an issue. Thus, telephone respondents tend to be more susceptible to various types of format effects because of this increased time pressure, a factor that may result in more "top of the head" answers (Hippler and Schwarz, 1988).

The impact of interviewer presence in face-to-face and telephone surveys and absence in mail and web surveys can have potential advantages and disadvantages. In telephone and face-to-face surveys, interviewers may facilitate communication between the researcher and the respondent and increase respondent motivation by clarifying questions and respondent answers (see Conrad et al., 2007, Chapter 10 in this volume, on the effect of paralinguistic cues in survey interviews). However, their presence may also invoke norms of social interaction where respondents tend to provide more culturally acceptable or socially desirable answers (de Leeuw, 2005; also see St-Pierre and Béland, 2007, Chapter 14, in this volume, on social desirability in computer-assisted telephone versus personal interviewing). In contrast, web and mail survey respondents have more control and privacy thus making them less likely to be influenced by interactional norms and social desirability. At the same time though, these respondents must rely solely on the questionnaire itself to infer the researcher's intentions and expectations (Schwarz et al., 1991b; de Leeuw, 1992). Recent research on the Internet, however, has focused on using the dynamic nature of web surveys to simulate the presence of an interviewer (Krysan and Couper, 2004) and provide various types of interactive feedback to help increase respondent motivation, provide clarifications or feedback to respondents, and decrease potential confusion and survey terminations (Couper et al., 2001; Crawford et al., 2001; Conrad et al., 2005).

The survey modes also differ dramatically in how information is transmitted between the researcher and the respondent and thus the cognitive stimulus the respondents receive (de Leeuw, 2005). The primary difference is whether information is transmitted aurally, visually, or both (Schwarz et al., 1991b; de Leeuw, 1992; Dillman, 2000). Aural transmission of information requires higher demands on respondents' memory capacity than visual transmission because they must remember the information rather than being able to repeatedly refer to it in the questionnaire. In addition to the presentation of stimuli aurally or visually, the various modes use different types of communication channels. Both aural and visual information transmission rely on verbal communication—the words used to convey meaning. In addition to verbal language, paralinguistic features such as voice inflection, tone or emphasis, and timing also convey meaning to respondents in face-to-face and telephone surveys (de Leeuw, 1992; Conrad et al., 2007, Chapter 10 in this volume). During face-to-face surveys, nonverbal communication transmitted through the use of gestures, facial expressions, and the body is also an important channel of communication. Similarly, for visual surveys, symbolic and graphical languages can act as a type of paralanguage where information can be emphasized, using font, bold, or italics. Additional visual features such as arrows, shapes, size, and graphical location provide other means of transmitting paralinguistic-type information to respondents (Redline and Dillman, 2002).

Face-to-face surveys, often considered the most "flexible" of modes, can use aural and visual (i.e., through the use of show cards) transmission and can convey information through verbal and nonverbal languages as well as paralinguistic communication. In comparison, telephone interviews lack visual transmission and nonverbal language cues and instead rely only on aural transmission of information through verbal and paralinguistic communication. In contrast to face-to-face and telephone surveys, mail surveys totally lack aural communication and instead rely solely on visually transmitted information communicated through verbal as well as symbolic and graphical languages to convey meaning to respondents. Finally, web surveys generally use the same visual transmission of information as mail surveys, relying mostly on verbal, symbolic, and graphical communication, but also have the potential to use aural communication or other multimedia technologies, such as pictures, to simulate facial expressions and other types of nonverbal communication.

12.3.2 Mode Effects and Scalar Questions

Previous research comparing responses to scalar questions has found a mode effect where respondents surveyed by telephone are more likely to provide extreme answers than respondents to modes where the scale is presented visually (either using a show card in a face-to-face interview or as part of the mail or web questionnaire). Specifically, two studies showed that respondents were more likely to select the extreme positive category when surveyed by telephone than by face-to-face interview, where a show card was used to visually display the scale (Groves, 1979; Jordan et al., 1980). In these two studies, the most positive category was presented first. All four categories were verbally labeled (strongly agree, agree, disagree, strongly disagree) in the Jordan et al., (1980) study, and the polar end points and midpoint were labeled in the Groves (1979) study using a 7-point satisfaction scale with the most positive category labeled "completely satisfied." de Leeuw (1992) also found that telephone respondents were more likely to select the extreme positive response category ("very satisfied") than face-to-face interview (who received a show card) and mail respondents, when all of the categories were verbally labeled and the extreme positive category was the last alternative presented. To explain these findings, de Leeuw (1992) suggests that it is more difficult for telephone respondents to keep multiple categories in memory since they do not have a visual aid displaying the response options.

Dillman and Mason (1984) found that respondents to telephone *and* face-to-face surveys (where a show card was not used) were also more likely than mail survey respondents to choose the extreme positive option "not a problem" on 7-point polar labeled scales independent of whether the category was mentioned first or last. Further, Tarnai and Dillman (1992) confirm the previous results of Dillman and Mason (1984) and also test the independent effects of "visually" presenting the response scale to respondents by experimentally comparing telephone respondents who received a copy of the questionnaire to use during the interview to telephone respondents who did not receive a copy of the questionnaire and thus did not have the visual presentation of the response scale when being interviewed. They found that providing respondents the questionnaire mitigated some of the mode effect, but differences persisted between telephone respondents who received the questionnaire and mail survey respondents (Tarnai and Dillman, 1992).

Additional research has also shown that respondents provide more extreme answers to polar point labeled scales when surveyed by telephone and IVR (interactive voice response) modes than by mail and web where the scale is provided visually (Dillman et al., 2001). Dillman et al. (2001) argue that respondents give more attention to the internal or middle categories when the scale is displayed visually. Thus, the literature suggests that the combination of increased time pressure and the task of requiring respondents to hold categories in their memory (since no visual aid is usually present) increases respondents' selection of the extreme positive end point of the response scale when they are surveyed by telephone (compared to other survey modes), regardless of whether that category is mentioned first or last or whether full or partially labeled scales are used.

12.3.3 Comparing Responses to Fully Labeled and Polar Point Labeled Scales Across Telephone and Web

To experimentally test whether telephone respondents provide more extreme answers than web respondents, particularly whether they are more likely to select the extreme positive category, we compare the results of 22 experimental comparisons of 5-point fully labeled and polar point labeled scales using various types of verbal labels (see Fig. 12.1). First, we test for mode effects across telephone and web among fully labeled scales with five categories using nine questions with three types of verbal labels: satisfied/dissatisfied, agree/disagree, and construct-specific labels. In both the telephone and web modes, all of the verbal labels are read or presented to the respondents as part of the question stimulus (see Figure 12.1(*a*)). In all nine comparisons, we find that respondents to fully labeled scales provide higher ratings when surveyed by telephone than by web (see Table 12.1). Eight of the nine mean comparisons indicate that the telephone ratings are significantly higher, and a greater percentage of telephone respondents select the extreme positive category for eight of the nine comparisons with five reaching significance.

We also test for mode effects across telephone and web using polar point labeled scales with five categories for 13 questions and with four types of verbal labels: satisfied/dissatisfied, agree/disagree, extremely likely/not at all likely, and best possible/worst possible. In both the telephone and web modes, only the positive and negative end points are labeled (see Fig. 12.1(*b*)). Again, we find that telephone respondents provide higher mean ratings than web respondents and that they are more likely to select the extreme category. In 12 of the 13 comparisons, we find that telephone respondents provide higher mean ratings than Web respondents; 11 are significant (see Table 12.2). Telephone respondents select the extreme positive category more frequently than web respondents for 10 of the 13 comparisons (seven of the 13 χ^2 tests are significant).

We also include four comparisons of 11-category polar point labeled (worst possible/best possible) scales to test for mode effects across telephone and web (see Fig. 12.1(*b*)). Similar to the above findings, we find that respondents provide more positive ratings when surveyed by telephone than by web in all four comparisons with three of the four comparisons reaching statistical significance (Table 12.3). However, responses tend to be distributed among the positive categories (6–10), with none of the χ^2 tests of the most extreme positive category significant.

Overall, telephone respondents provide more positive ratings than web respondents for five category response scales in 19 of the 22 comparisons and are more likely to select the most extreme positive category to both fully labeled and polar point labeled scales. In addition, telephone respondents to 11-category polar end-point labeled scales also provide more positive ratings than web respondents in three of the four comparisons. These findings confirm previous research that telephone respondents are more likely than mail, web, and face-to-face respondents to select the positive end-point category. Additionally, the findings appear to be quite robust as telephone respondents provide more positive ratings than web respondents regardless of whether all or only the end-point categories are labeled and across various types of substantive scales (i.e., satisfaction scales, agree/disagree, construct-specific, etc.).

Previous research comparing responses from face-to-face and telephone interviews suggests that the presence of an interviewer cannot account for these differences since telephone respondents also provided more extreme answers than face-to-face respondents (who were provided a show card). In addition, most of the questions in our survey would not be considered sensitive questions or ones where we might expect more socially desirable responses when an interviewer is present. However, we cannot rule out a self-selection bias where students with more positive attitudes toward the university were more likely to answer by telephone and students with more negative attitudes were more likely to respond by web. Further, we present the most positive category first on the telephone *and* web, so recency, where respondents

Table 12.1. Response Differences, t-Test of the Difference of Means (and Associated p Values), and χ^2 Tests of the Fifth Versus All Other Categories (and Associated p Values) for Fully Labeled 5-Point Scales Across Telephone and Web Modes

		Telephone							Web										
		Percent of R selecting category							Percent of R selecting category							Diff. means		5+	
Q	Scale type	5+[a]	4	3	2	1−	n	Mean	5+	4	3	2	1−	n	Mean	t-test	p	$\chi^2(1)$	p
2	Satisfied/dissatisfied	54.5	38.1	5.5	1.3	0.6	310	4.45	48.1	39.3	9.4	2.9	0.3	351	4.32	2.10	0.018	2.67	0.102
16		67.5	23.8	6.4	1.6	0.7	311	4.56	55.2	28.0	9.8	5.2	1.7	346	4.30	3.87	0.000	10.46	0.000
24		41.9	34.9	15.8	5.8	1.6	310	4.10	29.8	39.6	15.9	10.4	4.3	346	3.80	3.62	0.000	10.57	0.001
5	Agree/disagree	26.7	50.2	19.9	2.9	0.3	311	4.00	18.8	54.9	20.6	5.1	0.6	350	3.86	2.23	0.013	5.78	0.016
21		28.0	44.1	24.1	3.2	0.6	311	3.95	25.9	40.1	24.2	6.9	2.9	347	3.79	2.24	0.013	0.35	0.556
25		27.4	43.9	18.0	9.7	1.0	310	3.87	25.2	38.5	21.2	11.6	3.5	345	3.70	2.09	0.019	0.41	0.523
5	Construct-specific[b]	36.1	44.3	13.1	6.2	0.3	321	4.10	24.2	51.3	17.1	7.1	0.3	351	3.92	2.65	0.004	11.36	0.001
21		43.3	41.1	13.1	1.6	0.9	321	4.24	34.3	47.2	13.5	2.9	2.1	341	4.09	2.36	0.009	5.63	0.018
25		16.9	43.5	23.1	13.1	3.4	320	3.57	18.7	39.2	21.6	14.9	5.6	342	3.51	0.79	0.216	0.38	0.537
	Overall	38.0	40.9	15.8	5.3	1.0		4.09	31.1	42.5	17.4	7.7	2.4		3.92				

[a]5 is the most positive category and 1 is the most negative category.

[b]Label types for construct-specific scales are: Q5, accessible/inaccessible; Q21, central/not very central; Q25, desirable/undesirable.

Table 12.2. Response Differences, t-Test of the Difference of Means (and Associated p Values), and χ^2 Tests of the Fifth Versus All Other Categories (and Associated p Values) for Polar Point Labeled 5-Point Scales Across Telephone and Web Modes

		Telephone							Web										
		Percent of R selecting category							Percent of R selecting category							Diff. means		5+	
Q	Scale type	5+[a]	4	3	2	1−	n	Mean	5+	4	3	2	1−	n	Mean	t-test	p	$\chi^2(1)$	p
2		27.6	56.6	13.2	2.6	0.0	311	4.09	32.5	48.0	16.1	2.6	0.8	379	4.09	0.11	0.542	1.86	0.172
4		12.2	49.5	33.8	4.2	0.3	311	3.70	8.2	50.7	34.2	5.8	1.1	377	3.59	1.72	0.043	3.02	0.082
6	Satisfied/	22.3	43.5	26.1	6.1	1.9	310	3.78	14.9	43.5	29.4	8.8	3.4	377	3.58	2.83	0.002	6.27	0.012
7	dissatisfied	21.6	27.1	28.4	15.8	7.1	310	3.40	15.9	28.1	28.7	18.0	9.3	377	3.23	1.86	0.032	3.67	0.056
16		60.8	25.7	10.9	2.3	0.3	323	4.44	48.0	34.4	11.9	4.1	1.6	369	4.23	3.18	0.001	11.13	0.001
24		29.2	43.7	21.9	4.2	1.0	311	3.96	21.4	40.8	27.1	7.1	3.6	365	3.69	3.68	0.000	5.58	0.018
5		25.5	42.6	27.4	4.2	0.3	310	3.89	18.4	47.2	28.5	4.3	1.6	375	3.77	1.86	0.031	5.03	0.025
21	Agree/disagree	21.0	45.2	27.4	4.5	1.9	310	3.79	24.5	33.1	28.8	8.4	5.2	368	3.63	1.98	0.024	1.16	0.281
25		25.1	38.9	23.2	9.3	3.5	311	3.73	25.9	34.9	25.3	7.1	6.8	367	3.66	0.79	0.214	0.06	0.811
18	Extremely/not	81.9	10.7	2.6	1.6	3.2	310	4.66	70.0	15.3	9.2	3.2	2.3	347	4.48	2.65	0.004	12.60	0.000
19	likely	54.2	29.7	12.2	2.6	1.3	310	4.33	43.9	32.3	18.6	3.5	1.7	344	4.13	2.75	0.003	6.92	0.009
9	Best/worst	27.3	35.4	25.2	9.1	3.0	297	3.75	22.4	28.0	32.8	13.0	3.8	339	3.52	2.64	0.004	2.01	0.157
17	possible	20.5	56.5	21.4	1.3	0.3	322	3.96	8.4	55.7	33.0	2.9	0.0	345	3.70	4.93	0.000	19.93	0.000
Overall		33.0	39.2	21.3	5.4	1.9		3.96	39.4	55.1	36.3	10.1	4.6		3.79				

[a]5 is the most positive category and 1 is the most negative category.

Table 12.3. Response Differences, t-Test of the Difference of Means (and Associated p Values), and χ^2 Tests of the Fifth Versus All Other Categories (and Associated p Values) for 11-Point Polar Labeled Scales Across Telephone and Web Modes

		Percent of respondents selecting category …													Diff. means		5+	
Q	Format	0	1	2	3	4	5	6	7	8	9	10	n	Mean	t-test	p	$\chi^2(1)$	p
9	Telephone	1.4	1.0	2.1	3.4	4.8	15.2	11.0	15.9	22.4	13.1	9.7	290	6.82	2.33	0.010	0.03	0.869
	Web	2.2	0.6	5.5	5.7	3.3	21.0	6.8	13.1	16.7	12.6	12.6	366	6.52				
9[a]	Telephone	2.2	1.8	3.6	2.2	4.7	8.2	12.2	16.8	20.1	16.5	11.8	279	6.94	3.69	0.000	1.37	0.241
	Web	2.1	3.8	3.2	6.1	5.6	12.6	10.8	14.6	16.9	12.9	11.4	342	6.46				
17	Telephone	0.0	0.0	0.3	1.6	2.3	6.4	10.6	26.5	29.0	18.1	5.2	310	7.42	1.59	0.056	1.83	0.176
	Web	0.3	0.0	1.1	0.8	3.5	7.6	10.0	26.8	30.4	14.1	5.4	369	7.28				
17[a]	Telephone	0.0	0.3	0.0	1.0	0.3	3.9	5.1	18.1	43.1	21.2	7.1	311	7.87	1.16	0.123	0.02	0.881
	Web	0.0	0.0	0.3	1.2	3.2	6.9	8.1	23.9	33.1	18.7	4.6	347	7.47				

[a]The midpoint was also verbally labeled "average" on these comparisons.

are more likely to select later items when heard aurally, cannot explain these findings. Thus, the faster pace of the telephone interviews (mean 12.1 minutes on the phone and 21.4 minutes on the web)[2] and the lack of visual aids when answering scalar questions together seem to encourage telephone respondents to answer more extremely than respondents to other modes.

12.4 SCALAR FORMAT EFFECTS

Response scales are often constructed differently depending on the mode used to survey respondents. Thus, surveyors usually must make several decisions when constructing response scales, such as whether to label all or only some of the categories, what types of labels to choose, and how to present the scale to respondents. For example, fully labeled scales, where all of the categories are given a verbal label, are used on mail and web surveys and frequently in face-to-face interviews using show cards. Since visual communication can be used in these modes, the surveyor can visually present multiple category labels to respondents without increasing memory burden. However, to ease the interviewing task and reduce demands on respondents' memories in telephone surveys, where visual aids are not available, the same response scale is often changed to provide verbal labels for only the end points, and respondents must choose a number corresponding to the category that best represents their answer.

In addition to the across mode comparisons reported in the previous section, we also include several comparisons of different types of response scales *within* both telephone and web modes to identify how question format affects respondents answers. These within-mode comparisons include fully labeled versus polar point labeled scales, agree/disagree versus construct-specific scales, assigning 5 versus 1 to the most positive category, presenting the positive or negative end of the scale first, the use of an instruction to simulate the visual response scale for telephone respondents versus no instruction, and a number box version where the visual scale is removed for web respondents versus polar point scales. These within-mode comparisons of different scalar formats are designed to test various ways of asking scalar questions in an effort to help determine effective formats for constructing response scales that provide equivalent stimulus across modes.

12.4.1 Fully Labeled Versus Polar Point Labeled Scales

Researchers can include words and/or numbers to label categories, and they can choose to label all the categories or only some. Partially labeling the scale by only using verbal labels for some of the categories (end points and/or midpoint) may differentially attract respondents to these categories because they tend to be drawn to labeled points, and the respondent must interpret what the unlabeled categories

[2]To calculate overall response time, outliers were removed at two standard deviations away from the mean.

mean (Krosnick and Fabrigar, 1997). However, verbally labeling all of the response categories requires respondents to read, process, and interpret all of the labels. Several studies have concluded that fully labeled scales rate higher on various measures of data quality (reliability and validity as well as respondent satisfaction) than polar point scales as long as care is taken to choose verbal labels that are not overly specific or too vague (Schwarz and Hippler, 1991; Krosnick and Fabrigar, 1997). Since survey designers often construct scales with all of the labels for surveys where the scale can be presented visually (i.e., mail, web, and face-to-face surveys when a show card is used) and as polar point labeled scales for telephone surveys, we include six comparisons of fully labeled and polar point labeled scales within telephone and web modes (Fig. 12.2a).

Our telephone survey results indicate that respondents provide significantly more positive ratings, as reflected in higher mean ratings, to fully labeled than polar point labeled scales for all six comparisons (see Table 12.4). Additionally, in all of the comparisons, a greater percentage of respondents to the fully labeled scales select the most positive extreme category than respondents to the polar point labeled scales. The web survey results also indicate that respondents provide more positive ratings to the fully labeled scales than the polar point labeled scales for all six comparisons, but only two of the six difference of means tests are statistically significant (see Table 12.4). This finding appears to be linked to the scale type as the magnitude of the differences between fully labeled and polar point labeled scales is much greater for the satisfaction scales than the agree/disagree labeled scales within the web mode.

Overall, respondents seem to provide more positive ratings to fully labeled than polar point labeled scales, a finding that is particularly robust within the telephone mode where all six comparisons yield significant differences compared to only two of the six comparisons within the web. These findings, in conjunction with previous research suggesting that fully labeled scales are more reliable and rate higher in validity measures, indicate that when polar point scales are provided, respondents may answer more negatively than their actual attitude.

12.4.2 Agree–Disagree Versus Construct-Specific Scales

In addition to choosing whether to label all or only some of the response categories, researchers must choose what type of substantive labels to use. Surveyors often use agree/disagree, yes/no, and true/false response categories when designing scales because they are easy to administer and can be applied across a variety of question topics. However, research on acquiescence response bias suggests that people have a tendency to agree regardless of the content of the question and particularly in the social presence of someone else (i.e., an interviewer). Saris and Krosnick (2000) argue that scales with construct-specific response options, where verbal labels are designed using the underlying concept or construct being measured, decrease acquiescence response bias and produce less measurement error than agree/disagree scales because the process of respondents mapping their judgment to the appropriate response option should be more accurate when respondents only have to think along one dimension (the underlying

Table 12.4. Response Differences, t-Test of the Difference of Means (and Associated p Values), and χ^2 Tests (and Associated p Values) for Fully Labeled Versus Polar Point Scales for Telephone and Web Respondents

			Fully labeled							Polar point										
			Percent of R selecting category							Percent of R selecting category							Diff. means		5+	
	Q	Scale type	5+[a]	4	3	2	1−	n	Mean	5+	4	3	2	1−	n	Mean	t-test	p	$\chi^2(1)$	p
PHONE	2	Satisfied/dissatisfied	54.5	38.1	5.5	1.3	0.6	310	4.45	27.6	56.6	13.2	2.6	0.0	311	4.09	6.15	0.000	51.72	0.000
	16		67.5	23.8	6.4	1.6	0.7	311	4.56	60.8	25.7	10.9	2.3	0.3	323	4.44	1.87	0.031	5.64	0.210
	24		41.9	34.9	15.8	5.8	1.6	310	4.10	29.2	43.7	21.9	4.2	1.0	311	3.96	1.82	0.035	14.49	0.005
	5	Agree/disagree	26.7	50.2	19.9	2.9	0.3	311	4.00	25.5	42.6	27.4	4.2	0.3	310	3.89	1.73	0.042	6.42	0.131
	21		28.0	44.1	24.1	3.2	0.6	311	3.95	21.0	45.2	27.4	4.5	1.9	310	3.79	2.41	0.008	6.51	0.168
	25		27.4	43.9	18.0	9.7	1.0	310	3.87	25.1	38.9	23.2	9.3	3.5	311	3.73	1.79	0.037	7.76	0.101
WEB	2	Satisfied/dissatisfied	48.1	39.3	9.4	2.9	0.3	351	4.32	32.5	48.0	16.1	2.6	0.8	379	4.09	3.98	0.000	21.59	0.000
	16		55.2	28.0	9.8	5.2	1.7	346	4.30	48.0	34.4	11.9	4.1	1.6	369	4.23	0.95	0.171	5.37	0.251
	24		29.8	39.6	15.9	10.4	4.3	346	3.80	21.4	40.8	27.1	7.1	3.6	365	3.69	1.36	0.087	17.79	0.001
	5	Agree/disagree	18.8	54.9	20.6	5.1	0.6	350	3.86	18.4	47.2	28.5	4.3	1.6	375	3.77	1.59	0.057	8.79	0.066
	21		25.9	40.1	24.2	6.9	2.9	347	3.79	24.5	33.1	28.8	8.4	5.2	368	3.63	2.03	0.022	6.73	0.151
	25		25.2	38.5	21.2	11.6	3.5	345	3.70	25.9	34.9	25.3	7.1	6.8	367	3.66	0.54	0.294	9.72	0.045

[a]5 is the most positive category and 1 is the most negative category.

construct) instead of along two dimensions (the underlying construct *and* the agree/disagree response options provided). Experimentally testing these two scale formats, Saris and Krosnick (2000) find that construct-specific scales decrease cognitive burden and acquiescence response bias and yield data of higher quality (with higher reliability and validity ratings). To test whether respondents to agree/disagree scales are more likely to acquiesce, we include three comparisons of agree/disagree labeled scales (i.e., strongly agree, agree, neutral, etc.) and construct specific scales (i.e., very accessible, somewhat accessible, etc.) for both telephone and web respondents (see Fig. 12.2(*b*)).

For two of the three comparisons within both the web mode and the telephone mode, the mean ratings are higher for the construct-specific scales than the agree/disagree scales because respondents are more likely to select the most positive category in the construct-specific scales (see Table 12.5). Respondents seem to avoid the "strongly agree" category, and instead most respondents choose the "agree" category, whereas responses to the construct-specific scales are more evenly spread over the two positive categories (e.g., very and somewhat accessible). However, the pattern for the third comparison is different with significantly more positive ratings to the agree/disagree scales than the construct-specific scales within both telephone and web modes and with more respondents choosing the most positive category on the agree/disagree scales. This question asks students about the desirability of Pullman as a place to live while going to school, and students have tended to provide more negative ratings overall when this question was asked in previous surveys. Thus, both the topic of the question (and whether respondents are more positive or negative on that topic) and the type of response options seem to be influencing respondent answers. The across mode results reported earlier in this paper also show that telephone respondents provide more positive ratings than web respondents regardless of whether agree/disagree or construct-specific labels are used.

12.4.3 Assigning 5 Versus 1 to the Most Positive Category

Some survey designers also choose to use numbers to label response categories in addition to or instead of verbal labels. Previous research testing the effects of numeric labels suggests that respondents interpret the meaning of word labels differently when the numeric labels run from −5 to 5 than 0 to 10 (Schwarz et al., 1991a). Other research on the web suggests that labeling polar point scales from 1 to 5 does not produce significantly different answers than when the numeric labels are omitted on polar point scales (Christian, 2003). Overall, scales with numbers often take longer because respondents are required to process additional information; so unless the numeric labels are essential in helping respondents interpret the scale, they seem to unnecessarily increase respondent burden (Krosnick and Fabrigar, 1997).

Respondents tend to culturally associate higher numbers with more positive categories and lower numbers with more negative categories. Thus, it would seem that when numeric labels are chosen to be consistent with this expectation, it should ease the response task. Research using mail surveys suggests that when this a priori expectation is not met and respondents are asked to select a number from 1 to 5 where 1 is very satisfied and 5 is very dissatisfied, they often confuse which end of

Table 12.5. Response Differences, t-Test of the Difference of Means (and Associated p Values), and χ^2 Tests (and Associated p Values) for Agree/Disagree Versus Construct-Specific Scales for Telephone and Web Respondents

		Agree/disagree							Construct-specific										
		Percent of R selecting category							Percent of R selecting category							Diff. means		5+	
	Q	5+[a]	4	3	2	1−	n	Mean	5+	4	3	2	1−	n	Mean	t-test	p	$\chi^2(1)$	p
PHONE	5	26.7	50.2	19.9	2.9	0.3	311	4.00	36.1	44.3	13.1	6.2	0.3	321	4.10	1.46	0.072	13.99	0.004
	21	28.0	44.1	24.1	3.2	0.6	311	3.95	43.3	41.1	13.1	1.6	0.9	321	4.24	4.39	0.000	23.08	0.000
	25	27.4	43.9	18.0	9.7	1.0	310	3.87	16.9	43.5	23.1	13.1	3.4	320	3.57	3.78	0.000	15.86	0.003
WEB	5	18.8	54.9	20.6	5.1	0.6	350	3.86	24.2	51.3	17.1	7.1	0.3	351	3.92	0.92	0.178	5.34	0.236
	21	25.9	40.1	24.2	6.9	2.9	347	3.79	34.3	47.2	13.5	2.9	2.1	341	4.09	4.11	0.000	24.49	0.000
	25	25.2	38.5	21.2	11.6	3.5	345	3.70	18.7	39.2	21.6	14.9	5.6	342	3.51	2.37	0.009	6.41	0.170

[a]5 is the most positive category and 1 is the most negative category.

the scale is positive and which is negative and have to correct their answers accordingly (Christian and Dillman, 2004; Dillman and Christian, 2005). To test whether assigning higher versus lower numbers to more positive ratings influences respondent answers, we include three comparisons of 5-category polar point labeled scales, where the most positive category is numerically labeled 5 versus 1 on both the telephone and the web modes (see Fig. 12.2(*c*)). We find that within both the telephone and the web modes, respondents provide more positive ratings when 5 is assigned to the most positive category; however, only one of the telephone and none of the web comparisons are statistically significant (see Table 12.6). Thus, assigning 1 versus 5 to the positive end of the scale does not seem to substantially impact how respondents answer polar point scales on the telephone or the web.

12.4.4 Presenting the Most Negative or Most Positive Category First

There is a tendency for researchers to present the most positive category first on telephone surveys and the most negative category first on mail or web surveys to avoid primacy/recency effects on respondents' answers. However, respondents gain information about each category from its labels and its position in relation to other categories. In other words, they interpret additional meaning from the overall presentation of the response scale, and therefore, their responses may be different depending on whether the positive or negative end of the scale is presented first. Tourangeau et al. (2004) have suggested five heuristics respondents use to interpret meaning from the visual presentation of the response scale. Two of these heuristics, "left and top mean first" and "up means good," suggest that respondents to visual surveys expect scales to begin with the most positive category (i.e., very satisfied) and expect the successive categories to follow logically from that point (i.e., somewhat satisfied, neutral, etc.). Since these heuristics have been only tested using web surveys, it seems important to test whether they also apply to telephone surveys where the response scale is not presented visually.

To test whether presenting/mentioning the positive versus negative category first influences responses, we include two comparisons of scales presenting the most positive versus the most negative category first within both telephone and web modes (see Fig. 12.2(*d*)). Within the telephone mode the means are slightly higher when the positive (versus the negative) category is presented first, but within the web mode, the means are slightly higher when the negative (versus the positive) category is presented first. However, none of the comparisons yield significant differences within the telephone or the web modes (analysis not shown[3]). Thus, we find that presenting or mentioning the positive versus negative end of the scale first does not seem to influence how telephone or web respondents answer scalar questions; however, findings from other research on the web has found that designing scales with the positive end of the scale first facilitates the response task, suggesting that additional research is still needed.

[3]All analyses not shown are available from the authors upon request.

Table 12.6. Response Differences, t-test of the Difference of Means (and Associated p Values), and χ^2 Tests (and Associated p Values) for Polar Point Reversals for Telephone and Web Respondents

			Polar point 1+ 5−							Polar point 1+ 5−										
			Percent of R selecting category							Percent of R selecting category							Diff. means		5+	
	Q	Scale type	5+[a]	4	3	2	1−	n	Mean	1+	2	3	4	5−	n	Mean	t-test	p	$\chi^2(1)$	p
PHONE	4		12.2	49.5	33.8	4.2	0.3	311	3.70	9.6	47.3	24.8	17.0	1.3	311	3.47	3.28	0.000	31.45	0.000
	6	Satisfied/	22.3	43.5	26.1	6.1	1.9	310	3.78	24.0	40.0	23.4	11.0	1.6	308	3.74	0.56	0.287	5.59	0.228
	7	dissatisfied	21.6	27.1	28.4	15.8	7.1	310	3.40	22.7	24.3	27.5	19.7	5.8	309	3.38	0.22	0.412	2.33	0.676
WEB	7	Satisfied	15.9	28.1	28.7	18.0	9.3	377	3.23	15.4	23.7	32.9	18.6	9.4	350	3.17	0.70	0.241	2.46	0.652
	18	Extremely/	70.0	15.3	9.2	3.2	2.3	347	4.48	68.7	14.5	8.4	5.5	2.9	345	4.41	0.92	0.179	2.66	0.616
	19	not likely	43.9	32.3	18.6	3.5	1.7	344	4.13	46.2	29.5	15.5	7.3	1.5	342	4.12	0.18	0.427	6.32	0.172

[a]5 is the most positive category and 1 is the most negative category.

12.4.5 Instruction Versus No Instruction (Telephone Only)

Several researchers have found that visually manipulating the presentation of the response scale influences how respondents answer scalar questions in both mail (Schwarz et al., 1998; Smith, 1995; Christian and Dillman, 2004) and web surveys (Christian, 2003; Tourangeau et al., 2004; Dillman and Christian, 2005). The findings from these studies suggest that response differences might also occur across modes when the scale is presented visually in one mode (i.e., web) and not at all in another mode (i.e., telephone).

Dillman et al. (2000) suggest that an additional instruction such as, "you may use any of the categories 5, 4, 3, 2, or 1 to answer" might help telephone respondents visualize the response scale, thus simulating the visual display seen by web respondents and strengthening the stimulus for the internal categories when only the end-point labels are provided in the query. Additionally, having the interviewers read this instruction should help slow the pace of the interview and allow respondents more time to complete the question/answer process, perhaps reducing the number of "top of the head" answers. We discuss six comparisons where telephone respondents are asked 5-category satisfaction scales with the polar end points labeled with and without the additional instruction, "You may use any of the numbers 5, 4, 3, 2, or 1 for your answer" (see Fig. 12.2(*e*)). Interviewers were instructed to read the instruction slowly, pausing between each number. We also include the results from two comparisons in which we reverse the direction of the instruction (beginning with the most positive versus the most negative category) to determine whether this influences respondent answers (also in Fig. 12.2(*e*)).

Overall, the means are slightly lower for the version with the additional instruction, but only one of the six tests is significant (analysis not shown). Thus, including the additional instruction does not seem to influence how respondents answer polar point labeled scales on the telephone. We also find no significant differences in responses when the instruction is presented beginning with the most positive versus the most negative category first (analysis not shown). This particular instruction does not seem to significantly influence telephone responses and thus does not seem to provide a powerful enough stimulus to overcome the lack of visual presentation of the response scale in telephone surveys.

12.4.6 Polar Point Versus Number Box (Web Only)

Previous research has shown that on mail and web surveys, respondents provide more negative ratings when asked to report a number corresponding to their answer than when the scale is displayed visually with the polar end points labeled (Christian and Dillman, 2004; Dillman and Christian, 2005). In these studies, the most positive category was assigned the number 1 and the most negative category the number 5, a format that is inconsistent with respondents' a priori expectations as discussed above. We include six comparisons of polar point labeled and number box scales where the numbers are assigned consistent with respondents a priori expectations (i.e., where 5 is the most positive category and 1 the most negative category) (see Fig. 12.2(*f*)). We find that respondents provide slightly more negative ratings to the number box scales than the polar point scales, but none of the differences are significant for any of the six comparisons (results not shown). Thus, when the numeric labels attached to the scale match respondents expectations (i.e., higher numbers indicate more positive ratings), respondents appear to answer polar point labeled scales and number box entries similarly.

Since previous research indicated that respondents to number box scales confused which end of the scale was positive when the most positive category was numerically labeled 1 in the question stem, we include two experimental comparisons on the web of 5-category number box scales where the query indicates that either 5 or 1 is the most positive category. Our results show that web respondents give significantly higher ratings when 5 is labeled the most positive category (see Table 12.7). Respondents are more likely to write "4" when 1 is labeled the most positive category. These findings support previous research that respondents confuse which end of the scale is positive when the numeric labels are not consistent with the verbal labels (i.e., the highest number is assigned to the most positive category). Once the scales are constructed such that higher numbers indicate more positive ratings, no significant differences are found in how respondents answer polar point labeled and number box versions.

12.5 DISCUSSION AND CONCLUSIONS

In this paper, we included a large number of comparisons in part to integrate various findings from previous research on ways to construct scalar questions. The results from our 70 experimental comparisons indicate that survey mode and scalar format independently influence respondents' answers (see Fig. 12.3 for a summary of the experimental comparisons and the main findings). We find no significant interaction effects of mode and format. Overall, we find that telephone respondents provide more positive ratings and are more likely to use the extreme positive end point than web respondents to 5-category scales (19 of 22 difference of means tests) and to 11-category scales (three of four mean tests significant). This mode effect finding appears quite robust as telephone respondents provide more positive ratings to both 5-point and 11-point scales, regardless of whether all or only the end-point categories are labeled, and across various types of substantive scales (i.e., satisfaction scales, agree/disagree, construct-specific, etc.). These findings confirm previous research that telephone respondents are more likely than mail, web, and face-to-face respondents to select the positive end-point category. It appears that neither social desirability nor recency can explain the extremeness tendency of telephone respondents. The lack of a show card or other visual presentation to help telephone respondents remember the response categories and perhaps the faster pace of telephone interviews together seem to encourage telephone respondents to select the extreme positive category more frequently than respondents to other modes.

Our comparisons of different scalar formats within modes, particularly in combination with the findings from previous research, provide several suggestions for constructing response scales. First, it seems desirable to use fully labeled scales because they are more reliable and yield higher validity ratings than polar point labeled scales (Krosnick and Fabrigar, 1997). In addition, respondents to agree/disagree scales seem to avoid the most positive category "strongly agree" with most respondents choosing the second positive category "agree," whereas respondents to construct-specific scales are more likely to select the first two positive categories. However, since Saris and Krosnick (2000) found that construct-specific scales yield data of higher quality and decrease measurement error, construct-specific scale labels should be used when possible. If numerical labels are also used when labeling scales, it seems optimal to assign higher numbers to more positive categories particularly if no visual presentation of the scale is provided (i.e., the number box format on the web) since this is consistent with respondents' a priori expectations. Finally, our additional instruction on the telephone survey ("you may use any of the numbers 5, 4, 3, 2, or 1 for your answer") is designed to simulate the response scale and help slow down the pace of the telephone interview in an effort to provide greater equivalency across modes; however, the nonsignificant findings suggest that this stimulus is not powerful enough to visually represent the scale to telephone respondents. Further analyses of variance and covariance could also help us understand how these various formats influence respondent answers within and across modes. Also, since our experimental comparisons were tested using students, a population with typically higher levels of education and younger in age (18–25 years), it is important for future research to determine whether these same results are found in samples of the general population both in the United States and internationally.

Overall, the results from our experimental comparisons show that mode effects exist independent of format effects—telephone respondents provide more positive ratings and select the extreme positive category more often than web respondents across various types of scales. Our finding of independent mode effects suggests caution for mixed-mode surveys where data are often combined across modes as it appears that these mode effects cannot be overcome by scale format changes; it seems that none of the scalar formats tested here translate equivalently across telephone and web modes. This suggests that combining responses to scalar questions across telephone and web modes is a tenuous practice. Further, this mode effect seems rather robust historically with telephone respondents providing more extreme positive ratings than respondents to all other modes where the scale is presented visually (either on a show card or directly in the stimulus of the mail or web questionnaire), suggesting that the lack of visual presentation of the scale and perhaps the slower pace are the differentiating factors between telephone and other modes. Additionally, since neither social desirability nor recency can explain these findings, it seems urgent to develop a cohesive theoretical explanation for why telephone respondents provide more extreme positive ratings.

In addition to developing a theoretical explanation for why telephone respondents are more likely to select the positive extreme end point, one important direction for

Table 12.7. Response Differences, *t*-test of the Difference of Means (and Associated *p* Values), and χ^2 Tests (and Associated *p* Values) for Number Box Reversal for Web Respondents

		Number box 5+ 1−							Number box 1+ 5−										
		Percent of R selecting category							Percent of R selecting category							Diff. means		5+	
Q	Scale type	5+[a]	4	3	2	1−	*n*	Mean	1+	2	3	4	5−	*n*	Mean	*t*-test	*p*	$\chi^2(1)$	*p*
4	Satisfied/	7.1	48.6	35.4	8.0	0.9	350	3.53	6.6	37.3	33.2	20.6	2.3	349	3.25	4.30	0.000	27.31	0.000
6	Dissatisfied	11.8	44.7	32.1	9.7	1.7	349	3.55	9.7	36.1	30.4	19.5	4.3	349	3.28	3.80	0.000	19.20	0.000

[a]5 is the most positive category and 1 is the most negative category.

future research is to test whether other scalar constructions might mitigate the mode effects found in this paper and by other researchers. Previous research has found that unfolding or branching, where scalar questions are asked using two questions, one asking respondents to indicate the direction of their attitude (i.e., satisfied or dissatisfied) and another question asking respondents to indicate the strength or intensity of their attitude (i.e., very or somewhat ... satisfied or dissatisfied), decomposes the response task for respondents making it easier for them to provide an answer (Groves, 1979; Krosnick and Berent, 1993). In addition, branching has been shown to improve reliability and takes less time for mail and telephone respondents to answer than presenting the scale all at once in face-to-face, telephone, and mail surveys. Thus, future experimentation should compare responses to scales using branching on telephone *and* web modes to determine whether this construction decreases the extremeness found for telephone respondents.

Mixed-mode surveys will continue to proliferate as they attract the interest of surveyors attempting to balance the competing demands of survey quality, response rates, and limited budgets in today's survey world. It also seems likely that telephone and web modes will figure prominently in many of the mixed-mode surveys being conducted because these two modes together provide a relative cost-efficient combination. As a result, it is increasingly important to understand the effects of mixed-mode designs on methodological quality and to minimize the effects of various changes within different "systems of data collection" on responses to survey questions. In addition, much of the literature on mixed-mode surveys focuses on ways of reducing nonresponse error. However, that focus needs to be balanced with appropriate attention given to understanding the causes and consequences of measurement differences, like those revealed in this paper, to help identify question formats that present the equivalent stimulus to respondents across modes so that responses can be combined for analysis.

CHAPTER 13

Visual Elements of Questionnaire Design: Experiments with a CATI Establishment Survey

Brad Edwards, Sid Schneider, and Pat Dean Brick
Westat, USA

13.1 INTRODUCTION AND STATEMENT OF THE PROBLEM

Almost all telephone surveys, be they surveys of households or establishments, use computer-assisted telephone interview (CATI) systems. In establishment surveys CATI may be used, often in combination with web or mail questionnaires, to collect data about employees, business services and products, finances, and policies (de Leeuw et al., 2000). For example, the Bureau of Labor Statistics uses CATI to obtain information from businesses for the Current Employment Statistics reports.

In a CATI survey, interviewers read the questions on a computer screen and enter the responses at the appropriate places on the screen. In this chapter, we report on an experiment that explored how the design of the computer screens in a CATI establishment survey affects interviewer behavior and data entry error. We discuss the design and results of the experiment, and then suggest future research and propose screen design guidelines that may be applicable to both household and establishment CATI surveys, and may be extended to computer-assisted personal interview (CAPI) and web surveys.

Unique requirements of establishment surveys. Establishment surveys pose particular challenges for CATI screen designers. They often contain a large proportion of multipart questions. They often have questions with many qualifying clauses and statements. The design of the CATI screens could potentially facilitate the

Advances in Telephone Survey Methodology, Edited by James M. Lepkowski, Clyde Tucker, J. Michael Brick, Edith de Leeuw, Lilli Japec, Paul J. Lavrakas, Michael W. Link, and Roberta L. Sangster

interviewers' task of collecting the data efficiently and accurately. However, some fundamental questions about CATI screen design are unresolved. Should multipart questions be presented with one part per screen or with all parts on a single screen? Should complex introductions to a question be simplified? How should the labels in the data entry area be designed? Which screen designs help the interviewer address a respondent's questions?

One screen or many? Establishment surveys often contain sets of similar items. In a paper questionnaire, these sets are often presented as a form or grid, so that respondents can enter data in any order that they wish. With CATI, however, the computer largely controls the order in which the respondent provides information (Norman, 1991). A typical CATI screen contains a single question in the question text area; after the interviewer enters the response, the computer presents a new screen with the next question. For example, one screen may present "In the past year did you advertise your services in newspapers, magazines, or other publications?" with response categories of "Yes" and "No." The next screen would contain the next question in the series: "In the past year did you advertise your services on the Internet?" and so on. However, this set of questions could alternatively be displayed on a single screen in the following manner:

In the past year, did you advertise your services...

(1) In newspapers, magazines, or other publications
(2) On the Internet
(3) On the radio
(4) By direct mail
(5) On cable television

Each of these two competing formats has advantages and drawbacks. When the questions are presented on separate screens, the interviewer's attention is focused on the current question. The interviewer may ignore previous questions and may fail to recognize the general structure and context of the series. This distortion of orientation in an interview is called the segmentation effect (Groves et al., 1980; House and Nichols, 1988). The interviewer may tend to repeat each question in its entirety, even when the respondent would clearly understand the question if the interviewer stated only the last word or phrase. In addition, the interviewer may need to pause momentarily while waiting for each new screen to appear. This repetition may sound artificial or become tedious to the respondent.

When all of the questions appear on one screen, the interviewer can see all the items in the series at once and may not be as likely to pause between questions. On the contrary, the interviewer may be inclined to read only the last words of the question, even when the respondent might not remember all of the concepts in the question stem.

Simple or complex introductions? Some survey designers attempt to help respondents by including dependent clauses or additional sentences that clarify definitions or list exclusions that apply to a question. Other designers advocate parsimony and strive to create simple, short questions. Complex questions may be difficult for respondents to understand, especially when the questions contain several parts. The interviewer may have to repeat cumbersome definitions or clauses to ensure that the respondent remembers them for each part of the question. However, simple questions also have a drawback: When a question is not framed so that the respondents hear the precise time periods and other parameters that apply, the respondents may not understand the question as it was intended.

Much or little information conveyed by entry area labels? Blaise survey data collection software is widely used for government-sponsored surveys in Europe and North America. Blaise has a unique approach to question display and data entry. The screen is divided into panes; the top pane contains the question and answer category display (the "info pane"), while the bottom pane (the "form pane") contains data entry fields that correspond to the question in the top pane; the form pane also typically displays the answers recorded to previous questions and the entry fields for forthcoming questions. Thus the form pane resembles a paper form in that prior and succeeding items are visible. Blaise typically labels each data entry area with the names of the variables that are stored in the database such as "Q9F3aNL". However, the labels in the form pane can be more informative; for example, a field that recorded whether an establishment advertised on cable television might be labeled "Cable TV advertising." Moreover, the fields could be organized into sections; for example, a section might be labeled "Advertising media." These labels could conceivably reduce the segmentation effect by showing the interviewer the context of each question. But they might also distract the interviewer from the text of the questions themselves.

13.2 LITERATURE REVIEW

CATI interviewers can be viewed as both intermediaries in the research process and end users of the CATI software. As intermediaries, the interviewers strive to maintain the respondent's interest and motivation in the survey, to obtain complete and accurate answers, and to address respondent concerns. As end users of the CATI software, the interviewers require screen designs that facilitate the interview process.

Tourangeau and others have developed a cognitive model for the response process (Cannell et al., 1981; Tourangeau, 1984; Tourangeau et al., 2000). In this model, respondents first comprehend the question, then retrieve the information that they need from their memory, judge the adequacy of the information, and finally respond. Edwards and Cantor (1991) extended this model to establishment surveys for which the retrieval of information often means locating and reading documents and records. Sometimes more than one respondent provides data (Sperry et al., 1998). When the establishment survey uses CATI, the interviewer plays a central role in the response process (Cantor and Phipps, 1999). The interviewer interacts with the CATI software; this interaction is an integral part of the conversation between the interviewer and the respondent (Couper and Hansen, 2001; Japec, 2005).

Screen complexity. Designers must decide how much information to put on each Blaise CATI screen. For example, the screens can contain very little aside from the text of the current question and a means for the interviewer to record the response. Alternatively, the screens can convey the context of the question by providing information about prior and upcoming questions along with the current question.

The first alternative—presenting little more than the text that the interviewer must read—seems to have the advantage of eliminating ambiguity. Using these screens, the interviewer should immediately know where to look on the screen and what to read. This distraction-free design seems to be compatible with software usability principles, calling for screens to have an aesthetic, minimalist design (Neilsen, 1999; Krug, 2005). The purpose of simplifying screen designs is to help users perform the intended action, which in the CATI context means reading the current question and recording the response smoothly without hesitations.

The drawback of this approach, of course, is that CATI interviewers need to know more than the current question. They also need to know the content of the preceding and following questions. They need to know the time period that the question covers and any other conditions that apply to the question. When interviewers understand the context of a question, they can read the question with the appropriate inflection and are able to answer the respondents' questions or requests for clarification. When interviewers lack that understanding, they may read the questions mechanically and be unprepared for respondent's inquiries—the condition called the segmentation effect.

Segmentation effect. The segmentation effect is described in the CATI literature as the disorientation that the interviewer experiences because the questionnaire can be viewed only in very small pieces, or segments, and not in its entirety. Segmentation effects leave the interviewer with a foggy or disjointed mental map of the overall structure of the questionnaire. One result may be that the interviewer is unable to deliver the questions in a lucid, coherent fashion because the interviewer himself is experiencing mental dislocation. Another result may be simply that the interviewer becomes more error prone in general in performing data entry tasks, responding to respondent's queries, and administering the interview. Groves et al. (1980) discussed the segmentation effect and suggested ways that the instrument can be programed to give the interviewer freer movement through the instrument, thus mitigating some of the effects of segmentation. House and Nicholls (1988) took a different perspective and pointed out some of the benefits of segmentation in that the interviewer's attention is narrowly focused on the task at hand: asking the current question. Further, displaying one question at a time enforces correct question sequence. House and Nicholls then described CATI screen design principles that recognize the strengths of the CATI screen while minimizing the factors that trigger the disorientation of segmentation. In a paper on the integration of new technologies into surveys, Perron et al. (1991) mentioned that CATI establishment surveys are also subject to segmentation effects. Couper (1998a) expounded beyond segmentation to more general design issues when he noted that many CATI instruments make it difficult for the interviewer to find the next item to be read and, in general, make it difficult for the interviewer to determine the next appropriate action.

Accordingly, designers may choose to include information on each screen about prior and upcoming questions, and the conditions that apply to each question. Several items may be displayed simultaneously on one screen, along with the structure of the sets of questions. To keep this additional material from distracting the interviewer, designers can highlight the current question by presenting it in a color or size that distinguishes it from the other material on the screen (Wright, 1998). Still, the possibility exists that more complex screens can cause interviewers to hesitate or make errors.

Factors in CATI screen usability. Mockovak and Fox (2002) listed the factors relevant to screen design that make a CATI instrument difficult for interviewers: very large numbers of questions; multiple sections or modules giving rise to many paths through the instrument; complex or lengthy data entry tasks; rostering questions; inconsistencies in screen design, data entry, or user functioning; and use of table, grids, or screens that scroll. Other components of screen design that can affect the level of difficulty of the interviewer's job are font style and size, screen resolution, and color sets.

Redline et al. (2002) looked at branching errors as they relate to question complexity and found that question complexity had a tendency to increase certain errors, but not others: "The odds of making an error of commission with questions that contain a high number of answer categories is over one and a half times the odds of making an error with questions that have a low number of answer categories... However, the odds ... when the question contains a high number of words or the last category branches is significantly less than... with questions that do not possess these characteristics."

Telephone interviewers have varying levels of experience. The average tenure in many organizations is only several months. Less experienced interviewers might need simpler screen designs while more experienced and productive ones may prefer more complex designs.

Some researchers (e.g., Shneiderman, 1997) have suggested that an "inverted U" effect exists in which the optimal design is neither too simple nor too complex. When the screen design is too simple, some interviewers can lose the context of the question (the segmentation effect). Too complex screen designs may confuse and delay some interviewers and increase training costs.

Flexibility afforded by Blaise screens. The Blaise screen simultaneously presents the question, the response alternatives, and a schema of the most recent past responses. The Blaise system allows developers to manipulate these screen elements in the screen design.

Blaise has the ability to display multiquestion pages in contrast with the one-question-per-page default display conventions of most CATI systems (Pierzchala and Manners, 2001). The conventional one question per page is a recognized contributor to the segmentation effect. Furthermore, research has shown that as interviewers gain more experience with the CATI system, they tend to prefer multiquestion pages over one question per page (Westat, 2000)

Screen design standards. Very few large survey organizations actually use CATI software as it is configured out of the box. The reasons for tailoring CATI systems are manifold, including (1) to establish a configuration that can be applied uniformly across the types of surveys that the organization does and (2) to incorporate features that were used in the previous CATI system. Stanford et al. (2001) explain that the objective in developing an organization-level standard screen design is to provide the interviewers with an environment that is familiar, predictable, highly functional, efficient, and adaptable over time. They recommend standardizing features like function keys, size and type of font, labels, navigation area, and color.

Several large survey organizations generously made their screen standards available to us—the University of Michigan's Survey Research Center, Statistics Canada, the U.S. Bureau of the Census, the U.K. Office of National Statistics (ONS), and Westat. A perusal of these documents showed that most organizations have retained the default background colors and the default proportions of the panes. However, almost all organizations have changed the font and font size, and the position on the screen where the text appears. The organizations differ in their use of the navigation area, variable labels, and tags; the tabs; the control menu; help text; and function keys. Many of these differences emanate not only from the different types of surveys that each organization conducts, but also from organization-level preferences and the legacy CATI conventions practiced within each organization.

13.3 RESEARCH DESIGN

We designed a set of experiments to examine the effects of multiple items on a screen, question complexity, and informative labels on interviewer behavior. The experiments were implemented in the form of a questionnaire of workers in child care centers that was developed specifically for this study. Sixteen interviewers participated. All of these interviewers were experienced with Westat's Blaise CATI system.

13.3.1 Methods

Interviewers. Sixteen telephone interviewers served as the subjects. Five were male. The interviewers had an average of 22.7 months experience administering telephone interviews using the Blaise CATI system (range, 2–67 months). They had an average of 30.2 months experience with Westat's proprietary Cheshire CATI system (range, 0–78 months) and an average of 32.9 months with any form of telephone interviewing, CATI or paper and pencil (range, 2–78 months).

Procedure. The interviewers first received a 1-hour training session in which they learned to administer a telephone survey of child care center workers. First, the interviewers received an introduction to the survey. They then formed pairs and practiced asking the questions and entering the responses using Blaise CATI screens. The interviewers were told that they were expected to administer the surveys and record the responses accurately. The interviewers were also informed that this project was a research study and that the respondents and the CATI screens would be controlled for the purposes of the research. Each interviewer signed a consent form. After the training, the interviewers proceeded to their stations where they administered the first telephone interview.

Each of the 16 questions in the survey was in multiple choice format. Each question had five segments, in branch-and-stem format, such as "How satisfied are you with the workplace morale? The level of challenging work? The benefits? The salary or wages? The opportunities to learn new skills?" All five segments of each question had the same multiple choice alternatives.

The respondents were not actually child care center workers. They were actors or Westat employees who followed a script. Each interviewer administered eight interviews, each to a different respondent, at scheduled times over 2 or 3 days. The respondents' scripted answers differed in each of the eight interviews.

Design. The study used a fully factorial design with five factors. The first factor was called "interview number." The results for the first interview that the interviewers administered were compared with the results for the last (i.e., the eighth) interview. The results for the intermediate six interviews were not analyzed. These six interviews served to give the interviewer experience administering the survey. The results of the first interview revealed the behavior of the interviewers when they had no experience with the survey; the results from the eighth interview revealed the behavior of the interviewers when they had several hours of experience with the CATI instrument. Comparisons of the data from the first and eighth interviews revealed the effect of this experience.

The second factor was called "labeling." For half of the 16 questions in each interview, the data entry fields in the form pane were grouped together visually, question by question, separated by titles that described the questions. For example, the title "Courtesy" appeared above the data entry fields for a question about the courtesy of the agency that licenses child care centers. Also, the data entry fields for the five segments of each question had complete-word labels. For example, the data entry field for a segment within the question about courtesy was labeled "License process." These titles and labels were intended to show the interviewer the content and organization of the questions in the interview. For the other half of the questions in each interview, the labels in the form pane were coded abbreviations (e.g., "Q9F3aNL"), and no titles grouped together the data entry fields for each question. The results for the questions that had complete-word labels in the form pane were compared with the results for questions that had abbreviated labels.

The third factor was called "introduction." Half of the 16 questions in each interview had an introductory paragraph that explained the parameters of the question, such as the time frame and the concepts to be included or excluded. For example, the paragraph might explain that the question pertained to only the prior 3 months and that the respondent should consider certain events, but not others. These questions were called questions with a "complex introduction." The other half of the questions, called questions with a "simple introduction," had no introductory paragraph. These questions consisted of no more than three sentences. Questions with complex introductions contained a mean of 70.3 words (standard deviation, 12.9; range, 48–96 words), while questions with simple introductions contained a mean of 26.3 words (standard deviation, 6.4; range, 15–41 words). The results for questions that had a complex introduction were compared with the results for questions that had a simple introduction.

The fourth factor was called "screens." For half of the questions in each interview, all five segments of the question appeared on a single screen. In the earlier example, the five segments "The level of challenging work? The benefits? The salary or wages? The opportunities to learn new skills?" would all appear on a single screen. Each segment was successively highlighted on the screen, so that when the interviewer finished entering the response for one segment of the question the next segment became highlighted. For the other half of the questions, each segment of the question appeared on a separate screen, so that each question was presented on a total of five screens. The single-screen and multiple-screen formats required the same number of keystrokes to enter data. The results for the questions in which all five segments were presented on one screen were compared with the results for the questions in which each segment was presented on a separate screen.

The fifth factor was called "confusion." In each interviewer's first and eighth interviews, the respondent expressed confusion when answering half of the questions. The confusion always occurred during the fourth segment of these questions. This scripted confusion was in the form of a response like "Didn't I already answer that?" or a response that signaled to the interviewer that the respondent was experiencing some form of comprehension difficulty. During the other half of the questions in each interviewer's first and eighth interviews, the respondent did not express any confusion. In the middle six interviews, the respondent never expressed confusion. The results for the questions in which the respondent became confused during the fourth segment were compared with the results for the questions in which the respondent did not become confused.

Each interviewer's eight interviews were identical. They contained the same 16 questions in the same order. The labeling, the introductions, and the number of screens for each question remained the same across the eight interviews. The respondents for the first and eighth interviews expressed confusion during the same questions.

Construction of the questionnaires. Sixteen questions, each with five segments, were created for the questionnaire. Each question was programed in standard Blaise for Windows as commonly used in Westat's CATI centers, rather than the web version of Blaise (BlaiseIS) that allows screen designs under the control of a web style sheet. Each question had four different formats:

- Complex introduction, segments on a single screen
- Complex introduction, segments on five separate screens
- Simple introduction, segments on a single screen
- Simple introduction, segments on five separate screens

For each question, two different form panes were programed:

- Complete-word labeling
- Abbreviated labeling

Also, for each question, two kinds of scripts were written for the respondents: one in which the respondent expressed confusion while answering the fourth segment of the question, and one in which the respondent expressed no confusion.

In summary, each of the 16 questions could appear in 16 different formats, defined by four conditions: labeling (complete word or abbreviated), introduction (complex or simple), screens (multiple screen or single screen), and confusion (respondent confusion present or absent). The format used for each question was counterbalanced across the 16 interviewers using a Latin square design. The purpose of the counterbalancing was to minimize the effect of the content of the questions. Because each question appeared in each of the 16 possible formats, the Latin square design could reveal the effect of the format of the questions, irrespective of the content or the topic of the questions.

Figures 13.1 and 13.2 display two examples of the possible formats. Figure 13.1 shows a question in the format "simple introduction, segments on a single screen, complete-word labels," while Fig. 13.2 shows a screen from a question in the format "complex introduction, segments on separate screens, abbreviated labels."

Half of the interviewers administered CATI instruments that began with eight questions with complete-word labeling, followed by eight questions with abbreviated labeling. The other half of the interviewers administered CATI instruments that

Blaise Data Entry - C:\Tsm\Data\Tsm

Forms Answer Navigate Options Help

Tsm | Break Off Interview |

CASE ID: 45452

My next question is about the precautions against skin cancer that the children take at your center.

When children at your center go outside for more than one hour on a warm, sunny day, how often do they...

Wear sunscreen?

Wear long sleeves?
Wear long pants?
Wear hats?
Wear insect repellent?

1. Always	4. Rarely
2. Often	5. Never
3. Sometimes	6. Children do not go outside

Skin cancer precautions	Web site use	Job satisfaction
Sunscreen	Access child care news	Morale
Long sleeves	Participates in chat	Challenging work
Long pants	Access user directories	Benefits
Hats	Download	Salary
Insect repellent	Access messages	New skills

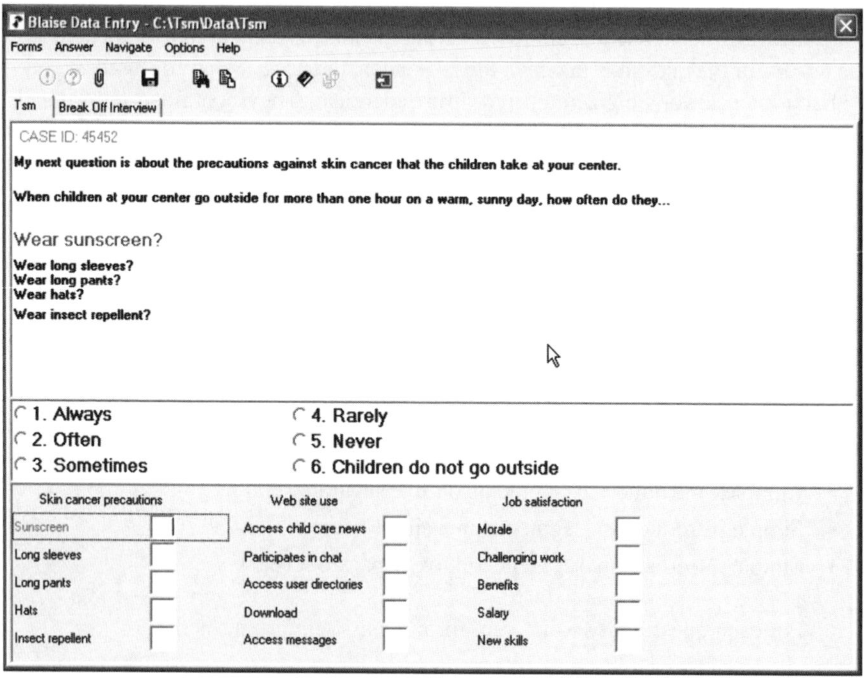

Figure 13.1. A question in the format having a simple introduction, with the question segments all on a single screen and complete-word labels in the form pane.

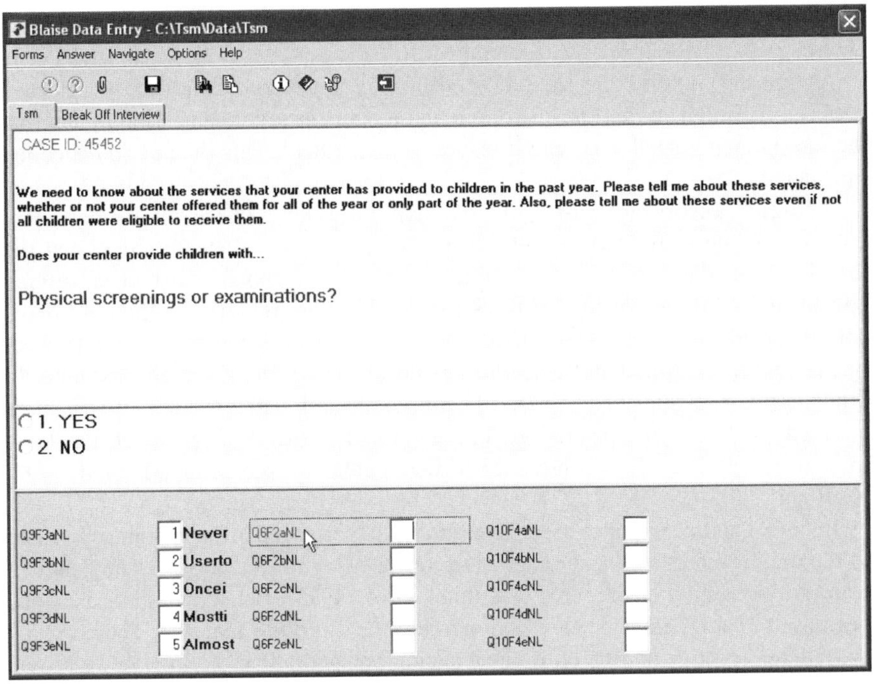

Figure 13.2. A question in the format having a complex introduction, with each question segment on separate screens and abbreviated labels in the form pane.

began with eight questions with abbreviated labeling, followed by eight questions with complete-word labeling.

In order to minimize the effect of the order of the questions, the eight questions that appeared with complete-word labels in the form pane were in a randomly determined order. The order of the eight questions that appeared with abbreviated labels in the form pane was also random.

13.3.2 Measures

Several measures of the interviewers' behavior and opinions were obtained: data entry errors, use of the backup key, latencies, hesitations, behavior codes, and debriefing responses. The interviewers' errors were identified by comparing the scripted responses with the data that the interviewers actually entered. An error was defined as a discrepancy between the entry for a question segment and the respondents' script.

While the interviewers collected data, the Blaise system automatically recorded their keystrokes in a database. All keystrokes, including the use of the Enter key and the Backup key, were recorded. These data are called an "audit trail." When interviewers pressed the Enter key, the computer screen progressed to the next question segment. When they pressed the Backup key, the computer screen returned to the prior question segment, where they could correct the response that had been entered.

As the interviewers collected data, the Blaise system recorded the time that elapsed between the moment that the cursor entered a data entry field and the moment that the interviewer started to enter the response in that field. This time period is called the "latency."

A digital recording was made of each interview. These recordings were analyzed using Bias Peak sound editing software. The software calculated the length of the silence between the moment that the respondent finished expressing confusion and the moment that the interviewer began replying to the confusion. These periods of silence were called "hesitations." In some cases, the interviewer began replying immediately, so the hesitation was scored as zero seconds. In other cases, the interviewer made hesitation sounds (like "Um"), which were counted as part of the hesitation. The length of hesitations in speech has been used in prior research as a measure of the speaker's uncertainty (Prinzo, 1998; Kjellmer, 2003).

Behavior coding was also used. A researcher experienced in the technique listened to the digital recordings and assigned scores as shown in Table 13.1. The behavior coding followed the procedures of Cannell et al. (1975). Behavior coding has long been used to study interviewers' command of the interview material. These coding procedures identify nonstandard interviewing behaviors that the interviewers were trained to avoid. Codes were assigned to the interviewers' behavior of asking the questions, reacting to the respondents' confusion, and probing unclear responses. The adequacy of probing was assessed using a procedure described by Conrad et al. (2007).

Table 13.1. Behavior Coding Scheme

Category	Possible codes
Asking the question	Made minor wording change
	Made major wording change
	Read question as worded
Delivery	Corrected an error
	Exhibited rough delivery (e.g., stumbling, poor inflection)
	No error correction behavior
Navigation	Hesitated, verbally expressed
	Hesitated, observed
	No hesitancy
Probing	Used neutral, nondirective probes
	Failed to probe when needed
	Inappropriate directive comment or probe
	No probing
Other behavior	Made extraneous comments
	Other behavior not coded elsewhere
	No extraneous behavior

At the conclusion of all the eight interviews, the interviewers completed a debriefing questionnaire that asked for their opinions about the various Blaise CATI screen formats. The questionnaire consisted of nine questions with a 5-point scale. The interviewers indicated their preferences among the CATI screen formats on this scale.

In this way, the speed and accuracy with which the interviewers enter responses, and the interviewers' satisfaction with the Blaise screens, were measured. Users' speed, accuracy, and satisfaction are three core components of usability (American National Standards Institute, 2001).

13.4 RESULTS

Data entry errors. The 16 interviewers committed a total of 38 data entry errors for a mean of 2.38 errors per interviewer. Each interviewer entered 160 responses (two interviews [the first and the eighth] times 16 questions per interview, times five segments per question). The overall error rate for the 16 interviewers was therefore 1.5 percent.

During the first interview, the interviewers committed many more data entry errors for questions in which the respondents expressed confusion than for questions in which the respondents did not express confusion (means, 0.20 versus 0.02 errors per question). During the eighth interview, however, respondent confusion was no longer associated with an increase in data entry errors (means, 0.02 versus 0.05 errors per question). This interaction effect was statistically significant ($F\ [1, 15] = 17.73, p < 0.001$).

Figure 13.3 shows that for the questions in which the respondents expressed confusion, the interviewers committed more errors on the fourth segment (the segment during which the confusion always occurred) than they did on the first or second

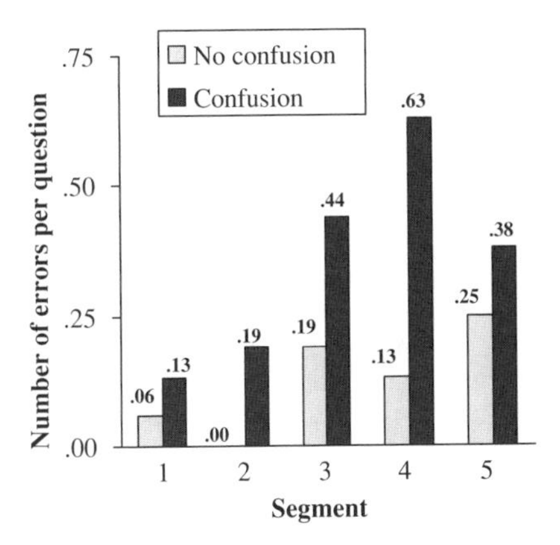

Figure 13.3. When the respondents expressed confusion, the interviewers committed more data entry errors on the fourth question segment (during which the confusion occurred) than on the first or second segment.

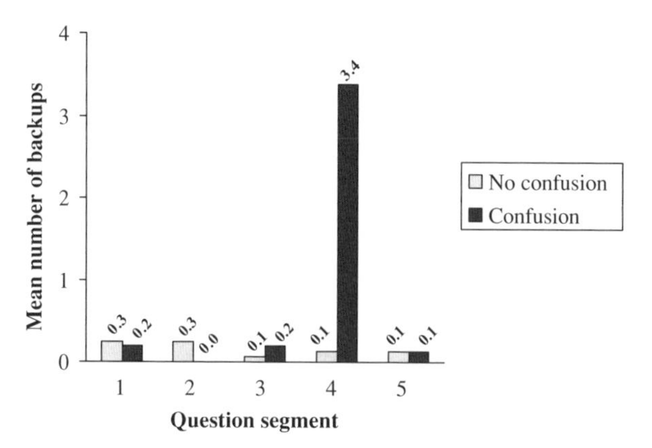

Figure 13.4. For the questions in which the respondents expressed confusion, the interviewers backed up from the fourth question segment more than they did from any of the other segments. For the questions in which the respondents did not express confusion, interviewers were equally likely to back up from all five segments.

segment (F [1, 15] = 10.00, $p < 0.01$ and F [1, 15] = 5.79, $p < 0.03$, respectively). No significant differences were detected, however, between the error rate for the fourth segment and the error rates for the third and fifth segments. No significant differences were detected among the error rates of the segments of the questions in which the respondents did not express confusion.

Backups. The interviewers were more likely to back up when the respondent expressed confusion than when the respondent did not express confusion (means, 0.24 versus 0.05 backups per question; F [1, 15] = 21.45, $p < 0.0003$). Figure 13.4 shows that for the questions in which the respondents expressed confusion, the interviewers backed up from the fourth segment—the segment in which the confusion occurred—much more frequently than they did from any other segment (3.4 versus a mean of 0.13 backups per segment). For the questions in which the respondents did not express confusion, interviewers were equally likely to back up from all five segments (mean, 0.18 backups per segment). This interaction effect was statistically significant (F [4, 60] = 23.46, $p < 0.0001$).

When the respondents expressed confusion, the interviewers were more likely to back up to the prior question segment when the segments were presented on individual screens rather than all on one screen (means, 0.32 versus 0.16 backups per question). When the respondents did not express confusion, separate screens were not associated with an increase in backups (means, 0.05 versus 0.05 backups per question). This interaction effect was statistically significant (F [1, 15] = 6.99, $p < 0.02$).

The 16 interviewers, taken together, backed up to the prior question segment a total of 48 times when the question segments were presented on five separate screens and 27 times when the question segments were all presented on a single screen. The mean duration of the backups was 24.6 seconds in the five-screen condition and 26.4 seconds in the one-screen condition. Thus, backups during the eight five-screen questions added a mean of 36.9 seconds per interview while backups during the eight single-screen questions added a mean of 22.3 seconds.

Backups were not strongly associated with data entry errors. Of the 28 questions on which the interviewers committed at least one error, the interviewers used the backup key on four.

Latency. The mean latencies (the time that elapsed from the moment that the interviewers' cursors entered a data entry field to the moment that the interviewers started to enter data in that field) were longer during the interviewers' first interviews than during their eighth interviews (means, 12.27 versus 10.56 seconds; $F[1, 15] = 32.39$, $p < 0.0001$). In other words, the interviewers tended to enter data more quickly during the eighth interview than during the first interview.

Overall, latencies were longer for questions that had complex introductions than for questions that had simple introductions (see Fig. 13.5). This finding is not surprising. It simply reflects that the interviewers needed more time to read the complex introductions than to read the simple introductions. While they read the introductions, the cursor was situated in the data entry field for the first segment of the question. A highly significant interaction effect ($F[1, 15] = 125.56$, $p < 0.0001$) showed that the latencies for the first segment of the questions with complex introductions were much longer than the latencies for the first segment of the questions with simple introductions (means, 35.88 versus 21.38 seconds).

Figure 13.5 shows that the latencies were longer when the labels in the form pane used abbreviations rather than complete words (means, 11.75 versus 11.07 seconds; $F[1, 15] = 7.04$, $p < 0.02$). This effect was apparent when the questions had complex introductions (means, 13.73 versus 12.21 seconds) but not when the questions had simple introductions (means, 9.78 versus 9.92 seconds). This interaction effect was statistically significant ($F[1, 15] = 6.88$, $p < 0.02$).

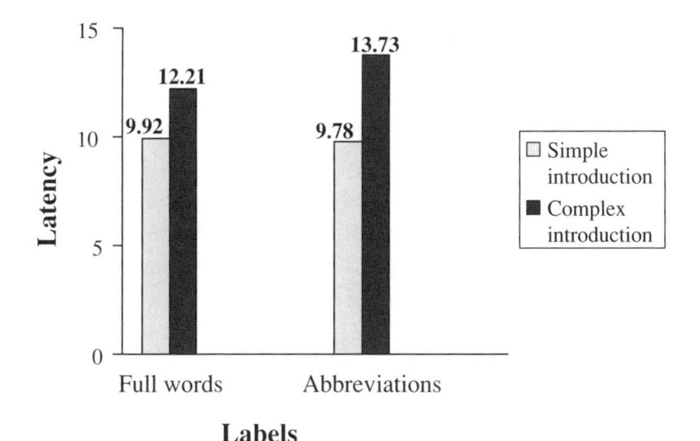

Figure 13.5. Latencies—the time until the interviewer began entering a response in the data entry field—were longer when the labels in the form pane used abbreviations rather than complete words. This effect was larger when the questions had a complex introduction than when the questions had a simple introduction.

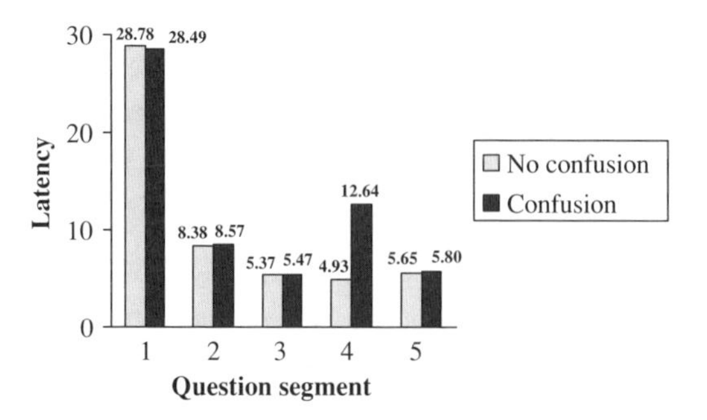

Figure 13.6. Latencies for the fourth segment of the questions—the segment during which respondents expressed confusion—were significantly longer for the questions in which the respondents expressed confusion than for the questions in which the respondents did not express confusion. Latencies were long for the first segment because interviewers read the question while the cursor was in the data entry field for that segment.

Latencies were longer for questions in which the respondent expressed confusion than for questions in which the respondent did not express confusion (means, 12.20 versus 10.62 seconds; F [1, 15] = 34.48, $p < 0.0001$). Figure 13.6 shows that latencies for the fourth segment of the questions—the segment during which respondents expressed confusion—were significantly longer for the questions in which the respondents expressed confusion than for the questions in which the respondents did not express confusion (means, 12.64 versus 4.93 seconds). This interaction effect was statistically significant (F [1, 15] = 83.40, $p < 0.0001$).

Hesitations. The analysis using Bias Peak software revealed the length of the interviewers' hesitations, defined as the elapsed time between the moment that the respondent finished expressing confusion and the moment that the interviewer began replying to that confusion. Several of these hesitations were unusually long, lasting over 4 seconds. To mitigate the impact of these outlying data, all of the hesitation times underwent a square root transformation. The results are shown in Fig. 13.7.

During the interviewers' first interviews, the complexity of the introductions of the questions had no significant effect upon the length of the hesitations (means, 0.52 versus 0.56 seconds, after the square root transformation). However, during the interviewers' eighth interviews, the hesitations were longer when the question had a complex introduction than when the question had a simple introduction (means, 0.63 versus 0.53 seconds, transformed). This interaction effect was statistically significant (F [1, 15] = 4.66, $p < 0.05$).

The hesitations were somewhat longer when the question segments were each on separate screens rather than all on one screen (means, 0.61 versus 0.51 seconds, transformed). This trend was not quite statistically significant (F [1, 15] = 3.80, $p < 0.07$).

Behavior coding. The interviewers were somewhat more likely to introduce changes to the question wording when the form pane labels were in the form of abbreviations than when they were in the form of complete words (means, 0.95

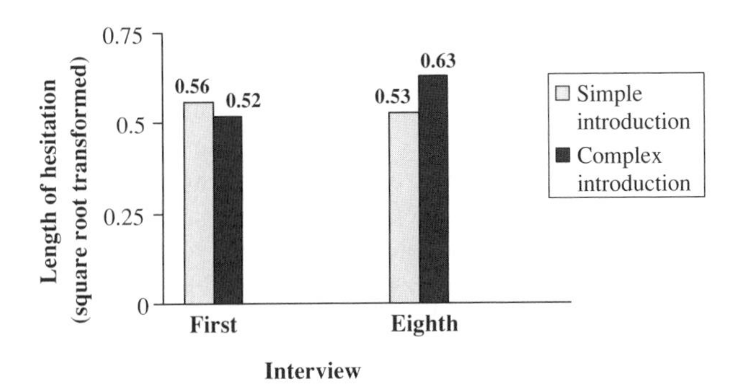

Figure 13.7. The figure shows the length of the interviewers' verbal hesitation when the respondent expressed confusion. During the interviewers' first surveys, the complexity of the introduction had no significant effect upon the length of the hesitations. However, during the interviewers' eighth surveys, the hesitations were longer when the question had a complex introduction than when the question had a simple introduction.

versus 0.77 changes per question). This effect was just below the level of statistical significance (F [1, 15] = 4.35, $p < 0.054$).

Similarly, when the respondent expressed confusion, the interviewer was more likely to ask the question with a rough delivery (making and correcting speaking errors or speaking with imperfect fluency or inflection) when the labels in the form pane were in the form of abbreviations rather than complete words (means, 0.43 versus 0.25 times per question); this effect was not apparent when the respondent did not express confusion (means, 0.26 versus 0.31 times per question). This interaction effect was statistically significant (F [1, 15] = 6.31, $p < 0.025$).

Also, when the question was divided among five pages, the interviewer was more likely to have a rough delivery when the labels were in the form of abbreviations rather than complete words (means, 0.39 versus 0.20 times per question); this effect was not apparent when the question appeared on a single page (means, 0.30 versus 0.37 times per question). This interaction effect was statistically significant (F [1, 15] = 10.25, $p < 0.01$).

The interviewers were more likely to alter the wording of a question when the question had a complex introduction rather than a simple one (means, 0.95 versus 0.77 changes per question; F [1, 15] = 4.91, $p < 0.05$). Similarly, when the question had a complex introduction, the interviewers were more likely to have a rough delivery (means, 0.36 versus 0.27 times per question; F [1, 15] = 6.92, $p < 0.02$). Of course, the longer, complex introductions afforded more opportunity for error.

The interviewers were more likely to have a rough delivery during the first interview than during the eighth interview (means, 0.45 versus 0.18 times per question; F [1, 15] = 25.65, $p < 0.0001$). Similarly, they were more likely to seem hesitant during their first interview (means, 0.34 versus 0.18 times per question; F [1, 15] = 5.86, $p < 0.03$).

The interviewers were more likely to follow up a response with a probe during the first interview than during the eighth interview (means, 0.67 versus 0.48 times per question; F [1, 15] = 6.26, $p < 0.025$). Also, as one would expect, the interviewers were more likely to probe when the respondent expressed confusion than when the respondent did not express confusion (means, 0.93 versus 0.23 times per question; F [1, 15] = 116.23, $p < 0.0001$). Moreover, the interviewers committed more probing errors (failing to probe when needed or making an inappropriately directive comment or probe) when the respondents expressed confusion than they did when the respondents did not express confusion (means, 0.22 versus 0.03 times per question; F [1, 15] = 24.48, $p < 0.0002$). Similarly, the interviewers were more likely to make extraneous comments when the respondent expressed confusion (means, 0.11 versus 0.04 times per question; F [1, 15] = 11.80, $p < 0.004$).

Interviewer debriefing questionnaire. Interviewers preferred having all question segments on a single screen rather than divided among separate screens. The interviewers' ratings used a 5-point scale in which 1 denoted a strong preference for the single-screen design, 3 denoted no preference, and 5 denoted a strong preference for the multiple-screen design. They rated the single-screen format as faster for entering data (mean = 2.23), explaining a question to a respondent (mean, 1.92), and overall (mean 2.07).

The interviewers had mixed feelings about questions' introductory paragraphs. The interviewers rated the simple-introduction format as faster for entering data (mean, 4.07). However, they believed that complex-introduction format was easier to use when explaining a question to a respondent (mean, 2.54). Overall, the interviewers did not strongly prefer either format (mean, 2.92).

The interviewers preferred complete-word labels to abbreviated labels in the form pane. They found that complete-word labels facilitated entering data (mean, 2.50) and explaining a question (mean, 2.08), and was better overall (mean, 2.17).

Correlations among the variables. The number of data entry errors committed by the interviewers was significantly correlated with the number of extraneous comments spoken by the interviewers ($r = 0.63$, $p < 0.01$). Also, the number of times the interviewers used the backup key was significantly correlated with the number of times the interviewers exhibited rough delivery of a question ($r = 0.63$, $p < 0.01$).

The number of times that the interviewers used the backup key was also significantly correlated with the interviewers' ratings on the debriefing questionnaire indicating that the multiple-screen design, in which the individual segments of each question were presented on separate screens, was helpful for explaining questions to the respondents ($r = 0.90$, $p < 0.001$).

Summary. Table 13.2 shows a summary of the results. The columns represent five features of the experimental design. The rows represent aspects of observed interviewer behavior or reported preference. Effects that were statistically significant are indicated by an arrow. The arrow indicates the direction of the association; upward-pointing arrows signify a positive association. Interaction effects are noted by text in the cells. For example, interviewers made more data entry errors than when the respondents expressed confusion than when the respondents were not confused, although this effect occurred only in the first interview, not in the eighth interview.

Table 13.2. Summary of Interviewer Behavior and Preferences by Design Feature

Interviewer behavior or preference	Respondent confusion (versus no confusion)	Question segments on separate screens (versus single screen)	First interview (versus eighth interview)	Complex introduction (versus simple introduction)	Abbreviated labels (versus complete-word labels)
Data entry error	↑ interview 1 only				
Use of backup key	↑	↑ confusion only			
Data entry latency	↑		↑	↑ first segment only	↑ complex introduction only
Hesitation at confusion	(not applicable)	↑ ($p < 0.07$)		↑ interview 8 only	
Wording alteration				↑	↑ ($p < 0.054$)
Rough delivery			↑	↑	↑ confusion on separate screens only
Hesitant delivery			↑		
Inappropriate probing	↑				
Extraneous comment	↑				
Probing, appropriately	↑		↑		
Preference		preferred single screen		mixed preference	preferred complete words

13.5 DISCUSSION

The results point to the advantages of having all parts of a multipart question on a single CATI screen, rather than having each part on its own screen. When respondents expressed confusion, the interviewers were less likely to return to earlier screens when all of the parts of the question were on a single screen. They could see the earlier parts of the question on the current screen, so they had no need to back up to earlier screens. By contrast, when each part of the question was on a separate screen, the interviewers needed to return to earlier screens to check those parts when the respondents expressed confusion. In the debriefing questionnaires, the interviewers preferred having all parts of the question on the same screen. The interviewers who backed up to earlier screens most often were most likely to have this preference.

The results suggest that the interviewers encountered some problems with the complex introductions. The interviewers were more likely to alter the wording of the question, or say the question with a rough delivery, when the question had a complex introduction. Of course, the questions with a complex introduction were longer, so the interviewers had more opportunity to make mistakes. In the debriefing questionnaires, the interviewers stated that they could enter data more easily when the question introductions were simple. However, they could explain questions to the respondents more easily when the question introductions were complex.

The results suggest that informative labels by the data entry areas are desirable. Interviewers generally preferred having informative labels by the data entry fields. If the questions had complex introductions, interviewers tended to start entering data more quickly when the labels were informative. Perhaps the labels helped interviewers feel confident that they were entering the data correctly. The behavior coding results suggest that informative labels may have helped interviewers deliver the question fluidly, even when the question was divided among many screens. Perhaps interviewers were better able to keep the meaning and the context of the question in mind with the help of the informative labels.

13.6 FUTURE RESEARCH: QUESTIONS THAT REMAIN

CATI screens ideally accomplish two distinct goals: first, they permit the interviewer to read the questions and record the responses accurately, and second, they permit the interviewer to respond to respondents' inquiries quickly and accurately. The second goal is important because interviewers may not always keep in mind all of the facts about a question; they may rely on the CATI screens to provide the information they need for this task when it is needed. Future research should focus on ways to enhance interviewers' ability to address respondent concerns.

Improvement over time. The interviewers became more proficient as they gained experience by administering the interview eight times. They committed fewer data entry errors during the eighth interview than they did during the first interview, especially when the question had a complex introduction. Latencies were shorter (i.e., data entry was quicker) during the eighth interview. The interviewers committed fewer speaking errors and hesitated less often during the eighth interview. These results suggest that interviewers became more skillful administering the questionnaire as they grew more familiar with it. This "learning curve" may be unique to each survey, but it would be useful for designers to understand it better. How many interviews must an interviewer do to become proficient? How does the quality of the first interview compare to the tenth or to the hundredth?

Segmentation effect. Some results suggest that as interviewers became more proficient over the course of the eight interviews, they also may have become less mindful of the overall structure of the interview and more focused solely on the question they were currently asking. For example, the interviewers tended to probe more often during the first interview than during the last. Since interviewers tended to probe primarily when the respondent expressed confusion (consistent with Conrad, Schober, and Dykstra, 2007, Chapter 10 in this volume), this finding suggests that the interviewers developed new ways of handling the confusion—ways that did not involve probing—as the interviewers gained experience. Indeed, two interviewers in the study were observed redirecting the respondents rather curtly when confusion was expressed (RESPONDENT: "Didn't I just answer that?" INTERVIEWER: "No, you didn't.") Developing an understanding of these skills and their impact on data quality is an important area for future research.

Moreover, when respondents expressed confusion during the eighth interview, interviewers hesitated longer when they were asking a question with a complex introduction than when they were asking a question with a simple introduction. However, during the first interviews, the complexity of the introduction had no significant effect on the interviewers' hesitation. This result suggests that as the interviewers gained experience with the CATI instrument, they may have been making less effort to keep the facts about the question in mind while asking the questions. As the interviewers acquired experience, they read the question more fluidly and relied less on remembering the details and the context of the questions. The result may be smoother respondent interactions, but less ability to address unexpected problems as interviewers acquire experience on a particular CATI survey. Future research should examine the impact of this behavior on data quality and explore ways in which the interactivity of computers might be used to accelerate the learning curve. (For example, screens might be adjusted dynamically to present definitions and suggest probes based on the interviewer's prior behavior or the respondent's prior needs.)

13.7 GUIDELINES FOR CATI USER INTERFACE DESIGN

This research allows us to propose a short set of guidelines for computer-assisted interview (CAI) user interface design that may have general applicability beyond CATI studies and establishment surveys to all CAI surveys:

Multipart Questions. Display questions with multiple parts on one screen. This enables interviewers to answer respondents' questions more readily and efficiently.

Answers to Questions. Display answers (and entry areas) to related questions on one screen. Use complete-word labels for the answer entry areas. Organization and labeling of data entry fields are important design features of the CAI user interface. Interviewers attain greater awareness of question meaning and context when the fields are grouped together, titled, and clearly identified.

Question Introductions. Use brief, simple introductions. Consider providing additional information about the question or the conditions for answering it in explanatory text set off from the question.

These guidelines result from a tradition of research on the relationship between visual survey design elements and survey error pioneered by Redline and Dillman. They provide answers based on the empirical evidence to questions about the segmentation effect posed by Groves and colleagues and by House and Nicholls in the 1980s. The guidelines amplify work by Couper and Hansen (2001), by Japec (2005), and by Cantor and Phipps (1999) in their focus on the interviewer's role as user of a CATI system in a telephone interview and in their emphasis on the importance of this role in the survey response process.

CHAPTER 14

Mode Effects in the Canadian Community Health Survey: A Comparison of CATI and CAPI

Yves Béland and Martin St-Pierre
Statistics Canada, Canada

14.1 INTRODUCTION

To ensure successful implementations of large surveys with specific objectives within reasonable costs, government statistical agencies are often forced to utilize mixed-mode design surveys. At Statistics Canada, a lot of surveys are conducted using more than one mode of data collection methods. Collection methods such as self-administered mail questionnaires, personal or telephone computer-assisted interviews including Internet surveys, electronic data reporting, and so on are often combined to optimize data collection procedures and reduce coverage and non-response errors. The subject of interest, the population under study, the interview length, the timeframe, the data collection infrastructure of the agency, and the costs are key factors that are considered to find the optimal collection procedures. However, mixing modes of collection within the same survey often comes with a price as there is an increased potential for measurement errors such as mode effects. In such cases, it is thus important to assess the magnitude of this source of error (de Leeuw, 2005).

Statistics Canada conducts the regional component of the Canadian Community Health Survey (CCHS) every 2 years (Béland et al., 2000). This CCHS component has been designed to provide health region level estimates related health determinants, health status, and utilization of the health system for more than 120 health

Advances in Telephone Survey Methodology, Edited by James M. Lepkowski, Clyde Tucker, J. Michael Brick, Edith de Leeuw, Lilli Japec, Paul J. Lavrakas, Michael W. Link, and Roberta L. Sangster

regions in Canada. The CCHS makes use of multiple sampling frames (area, list, and random digit dialing (RDD) frames of telephone numbers) and multiple data collection modes [computer-assisted telephone interview (CATI) and computer-assisted personal interview (CAPI)]. For operational and cost purposes, the mix of sampling frames and collection modes can change from one collection cycle to another complicating the comparisons of point estimates over time. The proportion of CCHS interviews completed over the phone (CATI) increased from 50 percent in CCHS 2001 to 70 percent in CCHS 2003. Therefore, with the objective of evaluating the impact of this change on survey estimates for key health-related indicators, a study on the effect of the two collection methods (CATI and CAPI) on the responses was carried on within the CCHS 2003. For the purpose of the study, 11 health regions across Canada were chosen and, for each region, two samples of similar size were selected from the same sample frame and randomly assigned to the two modes of collection. In total, a sample of 5008 individuals participated in the study (2598 CATI and 2410 CAPI).

This chapter presents the results of the various analyses performed as part of the mode study where several types of comparison between the two modes of collection were carried out. Although the study was not conceived to assess pure mode effect, it was designed to allow for valid comparisons between CATI and CAPI collection methods as conducted by Statistics Canada.

14.2 OVERVIEW OF PAST MODE COMPARISONS

An extensive literature exists on mode comparisons where several aspects such as response rates, costs, and data quality are compared. A large number of them have been studied in the meta-analysis by de Leeuw and van der Zouwen (1988). Studies included in the meta-analysis compared telephone and face-to-face surveys in terms of quantity and quality of the data. Only small differences were found between these two modes of collection. One interesting finding is that slightly less social desirability is observed for face-to-face interviews. This effect goes in the opposite direction as expected by the theoretical model used by the authors. Many reasons are suggested to try explaining why telephone interviews are not as effective in asking sensitive questions.

In another experiment by Aquilino (1994) comparing telephone and face-to-face (with or without self-administrated questionnaire (SAQ)) interviews, no significant difference was found for alcohol use between telephone and face-to-face (no SAQ) interviews. However, the difference was significant between telephone and face-to-face interviews with SAQ. For a more sensitive topic as illicit drug use, Aquilino found that telephone yielded lower prevalence rates compare to face-to-face/no SAQ. This finding support the notion, as mentioned by Aquilino, that increased social distance in the telephone mode makes it harder to assuage respondent confidentiality concerns. The telephone mode is thus more susceptible to underreporting of sensitive or socially undesirable behavior than face-to-face interviewing.

Most of the studies discussed in the meta-analysis and the one by Aquilino were conducted in the early days of telephone surveys when "paper and pencil" data collection was mainly used. Nowadays the use of computer-assisted interviewing is much more predominant, especially in large statistical agencies. Studies comparing two or more computed-assisted methods of collection are less common and not necessarily in the context of large national surveys. The following two national studies compare CATI and CAPI. First, in a recent experiment by Scherpenzeel (2001) carried out within the framework of the Swiss Household Panel, nearly no differences in terms of social desirability bias were found between CATI and CAPI in the answer distribution for sensitive topics such as alcohol consumption, self-rated health, satisfaction with life, and number of visits to a physician. However, the authors noted that it is possible that the questions chosen for the study were not as sensitive to the social desirability bias as they thought. However, another recent study comparing CATI and CAPI conducted in the United States by Holbrook et al. (2003) shows that data obtained from telephone interviews appear to be more distorted by satisficing and by a desire to appear socially desirable than are data obtained from face-to-face interviewing. This study also confirms known concerns about the sample representativeness of telephone compared to face-to-face. People who are socially disadvantaged are underrepresented in telephone surveys compared to face-to-face surveys, partly due to coverage error but mostly due to systematic nonresponse. Also, the study shows that for the elderly it is easier to obtain a response face-to-face than by telephone but that it is the opposite result for younger adults.

A study conducted by Pierre and Béland (2002) using the CCHS 2001 data indicated possible mode effects between CATI and CAPI. The study compared health indicator estimates between CATI from a phone frame sample and CAPI from an area frame sample. Among other things, mode effects were observed for activity limitation (more persons reported activity limitations in CAPI than CATI), physical inactivity (more persons reported being inactive in CAPI than CATI), and hospitalization (more persons reported being hospitalized at least once in the last year in CAPI than CATI). This study however had many limitations as some uncontrolled factors distorted the interpretation of the study results. The difference in sample frames used for each mode was one possible source of the distortion in the results, and moreover, there was no control on the mode of collection for the area frame sample; field interviewers were allowed to conduct either face-to-face or telephone interviews. An analysis of the sociocharacteristics clearly indicated that the profile of those respondents who were interviewed over the phone was highly different than those interviewed in person; field interviewers mainly conducted telephone interviews with those "hard-to-reach" individuals, for example, youths in the 12–29 age group.

These days several other forms of computerized data collection are available, such as computer-assisted self-interviewing (CASI), electronic mail surveys, and Web surveys. A taxonomy of these methods with their advantages and disadvantages are presented in de Leeuw and Nicholls (1996) and de Leeuw (2005). Only a few studies exist for comparing CATI and other computerized data collection methods (excluding CAPI). In a recent experimental study, Moskowitz (2004) has compared the effects of CATI and telephone audio computer-assisted self-interviewing (T-ACASI) on self-reports of smoking among adolescents 12–17 years of age. T-ACASI offers greater privacy than CATI, which should result in a higher reporting of socially undesirable behaviors. Indeed, the results of the study showed that the estimates of smoking for youths were significantly greater in T-ACASI than in CATI.

In summary, the results of past studies comparing CATI and other modes of collection are not consistent (either no effect or those that did find an effect found more social desirability in telephone), and several factors such as the topic analyzed and the context of the experiment could explain why the conclusions of these comparisons vary from one experiment to another.

14.3 MODE-COMPARISON STUDY IN THE CANADIAN COMMUNITY HEALTH SURVEY

The Canadian Community Health Survey consists of two alternating cross-sectional surveys conducted over a 2-year repeating cycle. The first survey (2001, 2003, 2005, etc.) collects data from over 130,000 households on a range of population health topics and aims to produce reliable estimates at the health region level. The second survey (2002, 2004, etc.), with a sample size of about 30,000 households, focuses on a particular topic that changes every cycle and aims to produce reliable estimates at the province level (mental health, nutrition, etc.).

The first survey of the first cycle was conducted in 2001 and made use of multiple sampling frames and data collection modes (Statistics Canada, 2003). The main source for selecting the sample of households for the CCHS 2001 was an area probability frame where field interviewers conducted either face-to-face or telephone interviews using a questionnaire designed for computer-assisted interviewing (CATI or CAPI). The sample was complemented by households selected from either a random digit dialing frame or a list frame of telephone numbers where call center interviewers conducted CATI interviews with the selected respondents. For operational and budgetary reasons, the ratio of area/telephone frame cases changed for the CCHS 2003 to increase the number of cases completed through CATI. Table 14.1 shows the change in the sample allocation between the two cycles. It was anticipated that such a change in the method of collection would affect the comparability of some key health indicators over the two cycles, either by artificially amplifying or by masking a real change in behaviors. The percentages in Table 14.1 reflect the fact that some area frame units and all telephone frame units are interviewed through CATI.

Table 14.1. Sample Allocation by Frame and by Mode

		2001	2003
Frame	Area	80%	50%
	Telephone	20%	50%
Mode	CAPI	50%	30%
	CATI	50%	70%

As cited in the overview of past mode comparisons, a preliminary study was already conducted using the CCHS 2001 that indicated possible mode effects between CATI and CAPI. In order to better understand the differences caused by the methods of collection (CATI and CAPI) in a large health survey, it was decided to design a special mode study and fully implement it as part of the CCHS 2003. Although it is understood that many factors could explain differences in survey estimates, it is believed that the results of this study will provide valuable indications to CCHS users on the magnitude of the differences in some key health-related estimates caused by the method of data collection.

14.3.1 Study Design

Because of operational constraints, the mode study was fully embedded in the CCHS 2003 with minimal modifications to the regular collection procedures. It is important to emphasize that it was not a true experimental design to measure pure mode effects because not all factors were controlled in the design (e.g., interviewers could not be randomized between the two modes of collection). This study however makes use of a split-plot design, a stratified multistage design where the secondary sampling units are randomly assigned to the two mode samples.

14.3.2 Sample Size and Allocation

In order to detect significant differences between point estimates at a certain α-level, a power analysis was performed to determine the minimum sample size required for each mode sample. Considering the study design effect (around 1.5), it was determined that to detect a 2 percent difference for a 10 percent prevalence and a 3 percent difference for a 25 percent prevalence at the level $\alpha = 5$ percent, a minimum sample of 2500 respondents was required for each mode sample.

To facilitate the implementation of the study design with minimal disturbance to the regular CCHS collection procedures, it was decided to conduct the study in a limited number of sites (health regions) in Canada. The 11 sites identified for this study provide a good representation of the various regions in Canada (East, Quebec, Ontario, Prairies, and British Columbia). Rural health regions with very low-density population were not considered for this study for collection cost purposes.

Each mode's sample size was allocated to the study sites proportionally to the CCHS 2003 sample sizes. Table 14.2 provides a detailed distribution of the mode study sample by site.

Extra sample was attributed to the CAPI mode in anticipation of possible telephone interviews (e.g., interviewer must finalize a case over the phone for various reasons); these cases were later excluded. For example, it is possible that an interview cannot be completed in one visit because it is time for lunch or dinner, or simply because it is too late. In such cases, some respondents would ask to complete it over the telephone. These sample sizes were inflated before data collection to take into account out-of-scope dwellings, vacant dwellings, and anticipated nonresponse.

Table 14.2. Mode Study Sample Sizes

Health Region	CAPI	CATI
St. John's, N.L.	135	100
Cape Breton, N.S.	125	100
Halifax, N.S.	200	150
Chaudière-Appalaches, Qc	230	215
Montérégie, Qc	405	390
Niagara, Ont.	235	230
Waterloo, Ont.	235	230
Winnipeg, Man.	320	320
Calgary, Alb.	350	290
Edmonton, Alb.	335	290
South Fraser, B.C.	240	240
Total	2810	2555

14.3.3 Frame, Selection, and Randomization

In the selected sites the CCHS 2003 used two overlapping sampling frames: an area frame and a list frame of telephone numbers. However, with the objective of eliminating all possible sources of noise during data analysis, it was decided to select the mode study sample from one sampling frame only. In order to keep to a minimum the changes to the regular CCHS data collection procedures, it was determined that selecting the sample from the list frame of telephone numbers and assigning the method of collection afterward would cause fewer changes in the procedures than selecting from the area frame.

The list frame of telephone numbers used by CCHS 2003 is created by linking the Canada Phone directory, a commercially available CD-ROM consisting of names, addresses, and telephone numbers from telephone directories in Canada, to Statistics Canada internal administrative conversion files to obtain postal codes. Phone numbers with complete addresses are then mapped to health regions to create list frame strata.

As mentioned earlier, the mode study makes use of a stratified two-stage design. The 11 sites represent the study design strata. The first-stage units were the Census Subdivisions (CSD) whereas the telephone numbers were the second-stage units. Within each site, the sample of telephone numbers was selected as follows:

(1) First stage: probability proportionate to size (PPS) selection of CSDs
(2) Allocation of the total sample (CATI + CAPI) of a given site to the sampled CSDs proportionally to their sizes
(3) Second stage: Random selection of telephone numbers in each CSD

Once the sample of telephone numbers was selected, those cases for which a valid address was not available were excluded from the process and added to the regular CCHS 2003 CATI sample. Those telephone numbers that represented approximately

7 percent of all numbers would have caused the implementation of severe changes to the procedures for the field interviewers (CAPI method of collection) to perform personal interviews; it was hence decided to exclude them from both mode samples.

Finally, controlling for the CSD within each study site, the telephone numbers with a valid address were assigned a method of collection (CAPI or CATI) on a random basis to constitute the two mode samples.

14.3.4 Collection Procedures

The data collection for the CCHS 2003 started in January 2003 and ended in December 2003. The sample units selected from both the area frame and the telephone frame were sent to the field or to the call centers on a monthly basis for a 2 month collection period (there was a 1 month overlap between two consecutive collection periods). Two weeks prior to a collection period, introductory letters describing the importance of participating in the survey were sent to all cases (area and telephone frames) for which a valid mailing address was available.

For the regular area frame cases, the field interviewers were instructed to find the dwelling addresses, assess the status of the dwellings (out-of-scope or in-scope), and list all household members to allow for the random selection of one individual aged 12 or older. If the selected individual was present, then the interviewer conducted a face-to-face interview. If not, then the interviewer had the choice of coming back at a later date for a face-to-face interview or completing the interview over the phone (in CCHS 2003, 40 percent of the area frame cases were completed over the phone).

For the telephone frame cases, the call center interviewers were instructed to assess the status of the phone numbers (specific questions were included in the computer application), list all household members, and conduct an interview with the selected individual at that moment or at a later date.

The data collection for the mode study took place between July and early November 2003. For the CAPI mode sample only a subset of field interviewers per site was identified to work on the study cases to facilitate the monitoring of the operations. The field project managers identified a good mix, both experienced and inexperienced interviewers, among those already working on the CCHS 2003 in the selected sites. In early July, the interviewers received the mode study cases (between 20 and 60) in a separate assignment from their CCHS assignment to clearly identify them as they were instructed to conduct only personal interviews (CAPI). To provide maximum flexibility to the interviewers, the collection period for the mode study cases was extended to 3 months. At the end of data collection, 132 field interviewers worked on the mode study and they completed 18–20 cases each on average.

The CATI mode sample cases were divided into three and simply added to the CCHS monthly CATI samples (July, August, and September) for a 2 month collection period. The CATI mode study sample cases were not known by the call center interviewers as they did not know which cases were part of the study and which ones were part of the regular CCHS sample. Only head office staff in Ottawa was able to do the distinction between these cases.

14.3.5 Questionnaire

The CCHS questionnaire contains approximately 300 questions on several population health topics related to health determinants, health status, and health system utilization. Sociodemographic information is also collected for analysis purpose. The average length of the interview is 45 minutes for both CATI and CAPI. For the study, as well as in all CCHS cycles, exactly the same version of the questionnaire was used for both CATI and CAPI modes of collection. There are no differences in the question wording and formats, and no additional visual stimuli were used for CAPI. The CCHS 2003 full questionnaire can be downloaded free of charge at http://www.statcan.ca/english/sdds/instrument/3226_Q1_V2_E.pdf.

14.3.6 Data Processing, Weighting, and Estimation

As the mode study was fully integrated with the CCHS 2003, the data collected for the study cases were processed using the CCHS processing system along with the remaining part of the CCHS sample. In addition to the main sampling weight, mode study respondents were assigned a separate and specific sampling weight just for the mode study to fully represent the target population of the 11 sites. The reader should note that the mode study cases were also part of the CCHS 2003 master data file as well.

Two weighting strategies with various adjustments were processed side by side (one for CAPI and one for CATI). Key factors determining the weighting strategy for each mode sample were

- Use of stratified, multistage design involving PPS sampling of PSUs and simple random sampling of telephone numbers
- Household-level nonresponse
- Random selection of one person based on household composition
- Person-level nonresponse

The sampling weights of each mode sample were calibrated using a one-dimensional poststratification of ten age/sex poststrata (i.e., 12–19, 20–29, 30–44, 45–64, and 65+ crossed with the two sexes).

Similar to the regular CCHS and because of the complexity of the study design, sampling error for the mode study was calculated using the bootstrap resampling technique with 500 replicates (Rust and Rao, 1996). All results presented in this chapter used the mode study sampling weights.

14.4 UNIVARIATE ANALYSIS

The main purpose of the mode study was to compare health indicators derived from data collected in person (CAPI) and those collected over the phone (CATI). This section presents univariate analyses comparing the two modes of collection. First, a comparison of the household-level and person-level nonrespondents observed in the two mode samples is presented. Then the rates of those who have given permission to share and link their responses are compared. Chi-square tests for association were used to compare the two mode samples in terms of sociodemographic characteristics and the findings are presented in Section 14.5.3. Finally, direct comparisons of several health indicators between the two modes are presented in Section 14.4.4. All comparisons were performed on weighted distributions, and the adjusted χ^2 tests for association used a 5 percent level of significance. For these comparisons, z-tests were applied to see if there was a significant difference between the estimates. Bootstrap weights were used to calculate standard deviations. As the two mode samples were not independent, the standard deviation of the difference between the estimates was calculated by measuring the dispersion of the 500 differences of estimates using the 500 bootstrap replicates.

14.4.1 Comparisons of Nonrespondents

This section presents comparison results in relation to nonresponse between CATI and CAPI, but first here is a summary of the response rates observed from each mode sample.

In total and after removing the out-of-scope units, 3317 households were selected to participate in the CAPI mode sample. Out of these selected households a response was obtained for 2788, giving a household-level response rate of 84.1 percent. Among these responding households, 2788 individuals (one per household) were selected out of which 2410 responded, giving a person-level response rate of 86.4 percent. The combined response rate observed for the CAPI mode sample was 72.7 percent.

For the CATI mode sample, 3460 in-scope households were selected to participate in the study. Out of these selected households a response was obtained for 2966, giving a household-level response rate of 85.7 percent. Among these responding households, 2966 individuals (one per household) were selected out of which a response was obtained for 2598, giving a person-level response rate of 87.6 percent. The combined response rate observed for the CATI mode sample was 75.1 percent.

As anticipated, the response rates observed in the mode study (especially for CAPI) were lower than the CCHS 2003 response rates because the extensive nonresponse follow-up procedures in place for the main survey were not fully implemented for the mode study cases for operational reasons and for better comparison. The national response rate for CCHS 2003 was 83 percent for the area frame sample and 78 percent for the telephone frame sample (*note*: response rates are available by sample frame, not by mode) for an overall response rate of 80.6 percent. Note that regular CCHS nonresponse follow-up procedures call for senior interviewers to contact nonresponding households over the telephone (Béland et al., 2001).

Within the CCHS 2003 and the mode study, total nonresponse could be divided into two categories: household-level nonresponse and person-level nonresponse. Very little information is known for the 529 CAPI and 494 CATI nonresponding households, but a comparison of the reasons for not responding shows no major differences between the two modes. For the "no one home/no contact" category, the rate was

Table 14.3. Mode Study in the CCHS 2005: Person-level Nonresponse Rate by Age Group (%)

Mode	Total	12–19	20–29	30–44	45–64	65+
CAPI	13.6	17.6	15.7	15.1	12.4	8.9
CATI	12.4	11.9	16.9	12.0	10.1	13.9

3.6 percent for CAPI and 2.1 percent for CATI. The "refusal" rates are also similar—8.7 percent for CAPI versus 10.4 percent for CATI. Person-level nonresponse is observed when interviewers successfully get through the first part (complete roster with age, sex, marital status, and highest level of education of all members) but not the second part, the actual CCHS interview with the selected respondent. Table 14.3 compares the age group distributions of the nonrespondents (person level) observed in CAPI and CATI. It is interesting to note the differences at the two ends of the distributions. A response from elderly persons (age 65 and older) is much more difficult to obtain over the telephone (13.9 percent nonresponse) than in person (8.9 percent) while the opposite is observed for the younger age group (12–19). Although the variable "age" is used in the creation of the response propensity classes for the person-level nonresponse weighting adjustment, the nonresponse bias could be nonnegligible for some characteristics. One could think that elderly persons with a physical condition might have difficulty getting to the phone. The same could be said for teenagers, where the more physically active could be home less often and hence less available for a personal interview. This would however require further research.

14.4.2 Comparisons of Share and Link Rates

The CCHS is conducted under the data sharing agreement of the Canadian Statistics Act meaning that answers of respondents who give permission can be shared with other governmental departments. For the CCHS, two specific and clear questions were asked of respondents at the end of the interview: (i) to ask permission for Statistics Canada to share their responses to federal/provincial ministries of health to avoid duplicating efforts and (ii) to ask permission to link their responses to administrative health-related databases for more in-depth analysis. For the latter, the provincial health insurance administrative number was also asked of those who had agreed to the question. As the mode study was fully embedded in the CCHS, it is thus of interest to compare the share and link agreements by mode. Table 14.4 gives share and link rates as well as rates of those who gave their provincial health insurance number by mode. Very little difference can be observed in the sharing agreement question between the two modes of collection (95.1 percent in CAPI compared to 94.8 percent in CATI). It is, however, interesting to note a larger difference between the two modes for the more intrusive "agreement to link" question (90.0 percent in CAPI compared to 85.8 percent in CATI). Finally, the data collection of the provincial health insurance administrative number, a very personal piece of information, is clearly affected by mode (84.0 percent in CAPI compared to 70.0 percent in CATI).

Table 14.4. Mode Study in the CCHS 2003: Comparison of Share and Link Agreement Rates and Rates of Those Who Gave Their Provincial Health Insurance Number (PHIN) (%)

Mode	Share	Link	PHIN
CAPI	95.1	90.0	84.0
CATI	94.8	85.8	70.0

14.4.3 Comparisons of Sociodemographic and Household Characteristics

Although both mode samples are representative of the target population and sampling weights were calibrated to age/sex groupings, differences could still be observed for other sociodemographic or household characteristics. In order to assess these possible differences, a series of χ^2 tests for association were performed.

The results of the tests can be separated in two groups: the characteristics for which no statistical differences were found between the two mode samples and those for which differences were found. No differences in the distributions were found for the following characteristics: living arrangement, household size, education of respondent, race, immigration, and job status. Statistically significant differences were however found for the following characteristics: marital status, language of interview, highest level of education in the household, and household tenure. The main differences can be summarized as follows:

- More single persons in CATI compared to CAPI (31 percent versus 29 percent), more married in CATI (53 percent versus 51 percent), but less persons living in common law in CATI (6.5 percent versus 9.5 percent) and less widowed/divorced/separated persons in CATI (10 percent versus 11 percent)
- More home owners in CATI (82.7 percent versus 79.5 percent) when comparing home ownership and home tenancy
- More CATI households where the highest level of education was a postsecondary degree (74.4 percent versus 71 percent), but less CATI households where the highest level of education was less than a secondary degree or some postsecondary degree
- More interviews were conducted in a language other than English for the CATI sample (27 percent versus 25.7 percent). Most of the interviews conducted in another language were conducted in French

For the income variables, the item nonresponse rate was too high (10 percent for CAPI and 20 percent for CATI) to allow for valid comparisons.

14.4.4 Comparisons of Health Indicators

Statistical z-tests were performed to determine if the differences in health indicators were significant. Around 70 health indicators for various age/sex domains of interest were looked at and significant differences were found for 15 indicators. Table 14.5 shows point estimates of selected indicators at the national level (11 sites) by mode.

Note that the prevalence rate for each health indicator by mode was calculated using complete cases only unless otherwise specified. For example, the prevalence rate of smoking was calculated by excluding those respondents who did not provide a response to the smoking questions. By doing so, it is assumed that item nonresponse is similarly distributed as item response that might not be totally true. It should, however, be noted that item nonresponse was very low for most of the health indicators

Table 14.5. Mode Study in the CCHS 2003: Comparison of Health Indicators Between CAPI and CATI

	CAPI		CATI		Difference
Health indicators	%	95% C.I.	%	95% C.I.	%
Obesity: BMI $\geq$ 30 (self-reported height and weight)	17.9	15.9–19.9	13.2	11.4–15.1	4.7**
Physical inactivity	42.3	39.5–45.1	34.4	31.8–36.9	7.9**
Current daily or occasional smokers—all ages	23.6	20.7–26.5	21.7	19.8–25.4	1.9
Current daily or occasional smokers—20 to 29 yrs old	37.7	31.4–44.0	28.2	21.7–34.8	9.5*
Regular alcohol drinker	80.7	78.0–82.5	78.8	76.8–80.8	1.9
At least one chronic condition	69.5	66.5–72.5	68.5	66.2–70.8	1.0
Activity limitation	25.4	22.9–27.8	26.8	24.0–29.5	−1.4
Fair or poor self-rated health	9.3	7.9–10.7	9.9	8.6–11.1	−0.6
Fair or poor self-rated mental health	4.0	2.8–5.2	3.9	2.9–4.9	0.1
Contact with medical doctors in past 12 months	83.5	81.5–85.6	78.4	76.2–80.6	5.1**
Contact with medical specialists in past 12 months	31.1	28.4–33.8	24.9	22.3–27.5	6.2**
Self-reported unmet health care needs	13.9	12.0–15.8	10.7	9.0–12.3	3.2*
Driven a motor vehicle after two drinks	13.5	11.3–15.7	7.2	5.1–9.3	6.3**
Ever had sexual intercourse	90.2	88.5–91.9	87.3	85.1–89.5	2.9*

A z-test was performed between CAPI and CATI national-level estimates (* $p < 0.05$, ** $p < 0.01$).

for both modes. For CAPI, the item nonresponse rates were always under 2 percent at the exception of the questions on sexual behaviors where nonresponse rates were between 4 and 9 percent. For CATI, the item nonresponse rates were slightly higher than for CAPI but always under 5 percent with the exception of the questions on sexual behaviors (as for CAPI) and the questions on fruit and vegetable consumption (9 percent).

The most important indicator for which significant differences were found is the obese category of the body mass index (BMI). The CCHS 2003 collected self-reported height and weight from which a BMI was derived. According to the World Health Organization, a person is considered obese if his/her BMI is 30 or higher. The obesity rate derived from mode study respondents aged 18 or older was significantly higher for CAPI (17.9 percent) than for CATI (13.2 percent). Larger differences were even observed for the 30–44 age grouping (18.1 percent CAPI and 11.4 percent CATI) and for men (20.4 percent CAPI and 14.7 percent CATI).

Another important indicator for which significant differences were found is the physical activity index. The physical activity index is an indicator that shows the amount of leisure-time physical activity done by a person during the past 3 months. It is derived from a series of questions that ask if the respondent has done any of 20 different activities, how many times, and for how long. There were significantly more inactive persons with CAPI (42.3 percent) than with CATI (34.4 percent).

For the smoking indicator (daily or occasional smokers), the rate was 2 percent higher for CAPI (23.6 percent) than for CATI (21.7 percent), but it was not statistically different at the 5 percent level of significance. However, a significant difference was observed for the 20–29 age group (37.7 percent for CAPI and 28.2 percent for CATI).

Other results show that the proportion of persons reporting contacts with medical doctors and contacts with medical specialists are higher for the sample interviewed in person. However, a more in-depth comparison for contacts with medical doctors broken down by gender shows interesting results where significant differences were found for men (80.3 percent for CAPI versus 72.5 percent for CATI) and not for women (86.7 percent for CAPI versus 84.1 percent for CATI). Also, on a related subject, significantly more unmet health care needs have been reported for CAPI (13.9 percent) than for CATI (10.7 percent).

It is worth mentioning that overall this analysis did not show any mode effect for some important health indicators and determinants such as all the chronic conditions, self-rated general and mental health, and alcohol drinking.

14.5 MULTIVARIATE ANALYSIS

To better understand the differences and to ensure that the mode effects found in the indicators comparisons are not simply due to discrepancies in the sociodemographic characteristics between the two mode samples, a series of multiple logistic regressions were performed. These analyses examine the associations between the mode of collection and each of the selected health indicators while controlling for confounding factors such as differences in sociodemographic and socioeconomic variables. These confounding factors were age, gender, marital status, personal and household income, personal and household highest level of education, immigrant status, race, job status, type of household, language of interview, month of interview, was the respondent alone during the interview, and the health region. In addition to the mode of collection, the other variables that entered into the regression models were selected using a stepwise selection technique. Also, interaction terms between the mode of collection and the confounding factors were all tested in the models.

For selected health indicators, Table 14.6 shows the odds of having the health condition or the health determinant when interviewed by telephone in comparison with when interviewed in person. In general, the results confirmed the ones found

Table 14.6. Mode Study in the CCHS 2003: Odds Ratios of the Health Condition for CATI Versus CAPI Multivariate Logistic Regression Controlling for Sociodemographics Variables

Health indicator	Factor	Odds ratio
Smoking	White 12–29	0.56**
	White 30+	1.00
	Non-white	1.49
Obesity: BMI ≥ 30 (self-reported)	Alberta	0.48**
	Elsewhere	0.79*
Physical inactivity	All	0.65**
Influenza immunization	12–15	4.48**
	16–19	1.78
	20+	1.10
Regular alcohol drinker	White nonimmigrant	0.70**
	Non-white or immigrant	1.71**
Five or more drinks on one occasion at least once a month	White and lowest or lower middle income	0.45*
	White and highest or higher middle income	0.97
	Non-white	2.45*
Self-reported unmet health care needs in past 12 months	Highest income adequacy	1.11
	Not highest income adequacy but with postsecondary	0.81
	Not highest income adequacy and no postsecondary	0.46**
Driven a motor vehicle after 2 drinks	12–19	1.23
	20–44	0.29**
	45–64	0.97
	65+	0.60
Ever had sexual intercourse	Female 15–24	0.43*
	Others	1.02

($*p < 0.05$, $**p < 0.01$)

in the univariate analysis indicating that the mode effects were not "caused" by differential nonresponse and differences in sociodemographic and socioeconomic variables. However, several interaction terms were significant between the mode of data collection and some of the confounding factors. This gives a precise indicator of the subgroups of the population for which a mode effect occurs.

The first result presented concerns the smoking indicator. As shown in Section 14.4.4, results of the univariate analysis did not show a significant mode effect at the national level for this variable. However, the multivariate analysis shows associations between the mode of collection and smoking for a subgroup of the population only. In other words, for white persons between 12 and 29 years old, being interviewed by telephone makes their odds of reporting a current daily or occasional smokers about 1.8 times (1/0.56 = 1.79) less than if interviewed in person (significantly different at the 1 percent level). For white persons, 30 years old and over, the odds are the same (1.00) for CATI and CAPI. For non-white persons, being interviewed by telephone makes their odds of reporting a current daily or occasional smoker about 1.5 times (1.49) more than if interviewed in person, but it is not significant at the 5 percent level.

As a confirmation of previous results, Table 14.6 shows that being interviewed by telephone makes the odds of being designated obese lower than if interviewed in person, and it shows that these odds vary by province. These odds are lower in Alberta (0.48), whereas elsewhere in Canada the odds are 0.79. No other interaction was found with the sociodemographic and socioeconomic variables (confounding factors).

For the physical activity index (inactive), where the univariate analysis showed a higher rate of physical activity for CATI, no interaction was found between the mode of collection and the sociodemographic variables. The odds are constant over all subgroups of the population. Overall, being interviewed by telephone makes their odds of being identified as inactive about 1.5 times (1/0.65 = 1.54) less than if interviewed in person.

For the alcohol use indicators, ethnicity, immigration status, income, and age group are characteristics for which varying mode effect is found. First, white non-immigrant persons are less likely to describe themselves as alcohol drinkers when interviewed by telephone (odds = 0.7), whereas the opposite is observed for non-white or immigrant persons (odds = 1.71). Similar results are found for the alcohol indicator related to the frequency of drinking. In other words, for non-white persons, being interviewed by telephone makes their odds of reporting to have had five or more drinks in one occasion at least once a month about 2.5 times more than if interviewed in person. However, the opposite mode effect is found for white persons in the lowest or the lower income adequacy category (odds = 0.45). For the drinking and driving indicator, a mode effect is this time found for the 20–44 age group. For these persons, being interviewed by telephone makes their odds of reporting drinking and driving about 3.4 times (1/0.29) less than if interviewed in person.

Another interesting result shows that persons not in the highest income adequacy category and without a postsecondary degree are less likely to report unmet health care needs when interviewed by telephone. A respondent had unmet health care needs when he/she felt that a particular health care was needed (in the past 12 months) and did not receive it. There are various reasons that could lead to unmet health care needs such as no doctor/specialist available, still on a waiting list, cost or transportation issues, or the respondent was simply too busy to go to see the doctor.

14.6 INTERPRETATION OF THE FINDINGS AND FUTURE DIRECTIONS

The results of the mode study are quite diverse. Nearly no differences were found between CAPI and CATI in the point estimates for the vast majority of health indicators measured by CCHS, such as tobacco use (all ages), chronic conditions, activity limitations, fruit and vegetable consumption, and others. This means that the comparability of the health indicators over the first two cycles of CCHS is not affected by the increased number of CATI in the second cycle.

Out of a total of 70 health indicators significant differences were however found between CAPI and CATI for 15 health indicators. Among others, self-reported height and weight, physical activity index, contact with medical doctors and self-reported unmet health care needs are certainly the most notable ones. Although the multivariate analysis somewhat attenuates the impact of the mode effects when sociodemographic characteristics are considered, it is believed that any comparison of the above indicators over the two cycles should take into consideration the increased number of CATI interviews in the second cycle. It is important to mention that other methodological (sample sizes, reference period, questionnaire, etc.) and contextual (changes in standards, true change, etc.) aspects should, as well, always be taken into consideration in any comparison of survey indicators over time.

As discussed earlier, extensive literature exists on comparisons between personal and telephone interview techniques, and a great deal of inconsistency in the results is certainly noticeable as these studies report varying magnitude of mode effects. Scherpenzeel (2001) suggests that the inconsistency among results is probably caused by differences in the design of the studies. The mode study conducted as part of the CCHS 2003 is no exception as any comparable studies could be found. It is, however, agreed upon that mode effects can be observed on some variables resulting in nonnegligible biases on survey estimates.

The literature on mode effects suggests that the factors that differentiate modes and that may be the origin of response differences can be grouped into three main classes: media-related factors, factors influencing the information transmission, and interviewer effects (de Leeuw, 1992), and all those factors could probably explain the differences observed in the above mode study. However, the authors of this chapter think that the differences found in the mode study of the Canadian Community Health Survey between CATI and CAPI are mainly caused by two confounding factors: social desirability and interviewer variability. The widely documented social desirability response bias is generated by people's attempts to construct favorable images of themselves in the eyes of others. It could occur at different levels and for different topics for both CATI and CAPI, and it is very difficult to quantify the magnitude of the measurement biases due to the absence of "gold standards" for many variables. Moreover, the magnitude of the bias would differ based on sociodemographic profiles, and it could even vary with time. Among all health indicators evaluated in this study, self-reported height and weight are good examples of variables for which the magnitude of the social desirability response biases differs between CATI and CAPI. Preliminary data of the 2004 Canadian Nutrition Survey conducted by Statistics Canada where exact measures of height and weight were collected on a large sample suggest that the obesity rate among Canadians of all ages is significantly higher than those calculated using the self-reported measures of the CCHS 2003 mode study (CATI and CAPI). Clearly, the measurement bias is larger in CATI than in CAPI, but they are both far from the gold standard derived from the nutrition survey.

The interviewer variability is the term used to describe the errors that are attributable to interviewers. The reasons why interviewer error occurs are not fully known. Japec (2005, 2007, Chapter 9 in this volume) discusses the interview process and addresses interviewers' cognitive processes by studying the concept of interviewer burden with components such as administration, respondent and interviewer characteristics, social environment, and tasks. Interviewer variability is inevitable in large surveys conducted by national statistical organizations. At Statistics Canada, the field interviewing staff is composed of more than 650 interviewers and 800 interviewers work in the call centers. Despite all efforts to standardize training procedures among all interviewers, some aspects of the work environments (e.g., supervision) of the two collection methods are simply so different that it is reasonable to believe that interviewers' behaviors could differ from one to the other, and hence interviewer variability biases could be introduced. For the mode study, additional information provided by the computer application systems (CATI and CAPI) such as time length of each question revealed interesting findings. The physical activity module of the CCHS questionnaire from which the physical activity index was derived took significantly less time to conduct in CAPI than in CATI suggesting that some activities (from the list of 20 activities read by the interviewers) might not have been clearly mentioned to some CAPI respondents for various reasons. In parallel, the quality control procedures implemented in the call centers have not detected such behaviors from the CATI interviewers. The authors believe that the interviewer variability explains a large part of the differences observed in the mode study for the physical activity index, but the absence of a gold standard for this variable does not allow for an assessment of the real measurement bias (CATI or CAPI).

The results of the mode study are very useful to have a better understanding of the impact of the increased CATI in CCHS 2003 compared to CCHS 2001. As well and in light of the observed results, a series of recommendations has been made for future cycles of CCHS. It was decided to implement the same 2003 sample design (area/telephone frames and CAPI/CATI ratios) for CCHS 2005. Starting in CCHS 2005, exact height and weight will be collected on a subsample of individuals to allow for national estimates of BMI categories for specific age/sex groupings. Also, interviewers' procedures will be reinforced to standardize even more collection procedures among the two collection methods. It has also been decided to update the mode study results on a regular basis by conducting a collection mode study as part of future cycles of CCHS (probably every two or three cycles). These changes should hence improve the quality of CCHS data and provide a solid basis to policy makers and health care professionals to better track changes over time and take appropriate actions to address the various issues surrounding the health of Canadians.

Mixing modes of collection is very useful for survey methodologists to reduce coverage and nonresponse errors. Large national statistical organisations with important data collection infrastructure like Statistics Canada often combine CATI and CAPI to successfully conduct their household-based surveys, especially when the questionnaire (wording, ordering of questions, etc.) can be administered in the same format in both modes. But improving response rates when using more than one mode of collection often increases the risk for measurement errors such as mode effects. It is thus of great importance to survey methodologists to determine the magnitude of this source of error to better inform the data users.

PART IV

Operations

CHAPTER 15

Establishing a New Survey Research Call Center

Jenny Kelly
National Opinion Research Center at the University of Chicago, USA

Michael W. Link
Centers for Disease Control and Prevention, Altanta, GA, USA

Judi Petty, Kate Hobson, and Patrick Cagney
National Opinion Research Center at the University of Chicago, USA

Twenty years ago, establishing a telephone survey operation was often as simple as finding unoccupied desks with telephones. Advances in computer and telecommunications technology and survey methods, however, made establishing telephone survey operations much more complex and expensive (see Tucker and Lepkowski, 2007, Chapter 1 in this volume). Call centers evolved from mere physical space where calls were placed and received to multichannel contact centers, where survey information is gathered using an array of modes in addition to voice, for example, interactive voice response (IVR), fax, e-mail, and Web chats (Link 2002a; Dawson 2004). As technology advanced and society changed, the functions of the modern survey research call center also adapted to keep pace. The concept of a call center as a centralized location is also changing: through Internet and computer/telephony integration (CTI), decentralized or virtual call centers can be set up.

Yet, practical guidance for those planning to establish a survey research call center is sparse, and the available resources tend to deal mostly with outbound call centers for commercial sales or inbound call centers for customer technical assistance. Examples of such resources are: books (e.g., Dawson, 2004 and Sharp, 2003), journals (e.g., *Customer Inter@ction Solutions* and *Call Center Magazine*),

Advances in Telephone Survey Methodology, Edited by James M. Lepkowski, Clyde Tucker, J. Michael Brick, Edith de Leeuw, Lilli Japec, Paul J. Lavrakas, Michael W. Link, and Roberta L. Sangster

Table 15.1 Comparison of Survey Research Call Centers with Other Types of Call Centers

Telephone survey research center	Other (nonresearch) telephone centers
Conducting telephone surveys (social surveys and market research surveys) is the main activity.	Telemarketing or customer service is the main activity.
Staffed with interviewers working on survey projects by asking questions.	Staffed with agents working on campaigns, marketing calls, or answering questions.
Typically outbound (i.e., the center initiates the calls), with limited inbound calls.	Often entirely inbound (i.e., the center receives the calls), or a mix of inbound and outbound.
Samples of phone numbers handled scientifically. For example, multiple callbacks might be made to a nonanswered number to increase the response rate.	Samples of phone numbers handled for maximum agent efficiency. Nonanswered numbers typically discarded.
Dialing technologies must be configured so they do not alienate respondents.	Dialing technologies configured to maximize "agent talk time." Inbound work allows callers to be kept in queues.
Typically a varying workload over the year.	Work load tends to be constant and repetitive throughout the year.
Interviewers usually paid by the hour with only a small part, if any, of their wage directly determined by output.	Agents often paid on commission based on number of calls handled or customers recruited.
Typically fewer than three interviewer classifications in a single organization, with not a great deal of difference in their work or pay.	Many different classifications of agents, allowing for more specialization and career advancement.

and a variety of websites. Although useful, these resources are of limited value to researchers, given the substantial differences between commercial call center operations and the operations required for most types of survey research (Table 15.1). This chapter is meant to fill that gap: it concentrates on the challenges of establishing a multichannel contact center for survey research.

15.1 PLANNING A NEW CALL CENTER

Establishing a survey-research call center is a complex endeavor that requires planning. The issues to be resolved are (a) the system of interviewers (centralized or a decentralized), (b) the size of the call center, (c) the location of the call center (urban, suburban, or rural; near the corporate office or near the client base), (d) the technology to use, and (e) the type of software to install.

15.1.1 New Choices, New Issues

Blankenship (1977) details the history of telephone surveys from the early 1920s through the mid-1970s. However, after the mid-1970s, call centers changed radically because of advances in telecommunication technology and changes in societal behavior:

- *Computer technology.* Personal computers revolutionized telephone survey research. Through computer-assisted telephone interviewing (CATI), researchers can conduct long, complex interviews, incorporate data edits and range checks directly into the questionnaire, and increase the speed with which they produce survey results because interviewers enter responses directly into the survey database (Lavrakas, 1993).
- *Structure of the telecommunications industry.* In the United States, the structure of the telecommunications industry changed during the 1980s (Steeh and Piekarski, 2007, Chapter 20 in this volume). Competitive pressure among suppliers of telephone equipment and long distance service forced the federal government to challenge the monopoly of AT&T. Seven local/regional phone companies and one long-distance carrier were established in 1984. Competition among carriers meant a large drop in long-distance carrier costs. Similar deregulation and reductions in long-distance pricing in Australia and other countries reduced the need for a series of dispersed centers and increased the feasibility of single large operations.
- *Dialer technology.* Telephone dialing is now often automated, saving interviewer time and reducing labor costs.
    - *Automatic dialing* via modems began in the early 1990s, thus relieving interviewers from having to dial the telephone manually. More recent versions include pacing systems that automatically start dialing of the next number within a specified amount of time of the previous call ending unless the interviewer intervenes.
    - *Automated dial outcome detection.* Time is saved when the telephony system can automatically detect busy signals, fax lines, and other nonproductive outcomes.
    - *Predictive dialing* uses statistical algorithms to predict (1) when interviewers will finish calls or (2) what ratio of numbers dialed will result in a connection to a person. Since predictive dialing allows more than one number per interviewer to be dialed at a time, the interviewers' time between contacts is reduced.
    - *Message push-out.* Many dialing systems also allow researchers to leave recorded messages on the answering machines of potential survey respondents to notify them to expect a survey call (Link and Mokdad, 2005b).
- *Voice over internet protocol (VoIP) technology.* VoIP carries voice on inexpensive Internet channels instead of telephone lines and allows voice and data to be consolidated into a single network. No longer do call centers need parallel wiring systems: one for data and another for voice. As a result, interviewers can now use *softphones* (telephone software in their computers) to make calls. The most significant effects of these are new software applications based on computer/telephony integration (CTI) and distributed operations that allow interviewing stations to be set up anywhere with a high-speed Internet connection (e.g., in an interviewer's home) (Link, 2002b).

- *Unwilling public.* Survey research companies also have to work harder these days to overcome the negative association many households make between survey research calls and telemarketing (Lavrakas and Shepard, 2003). Increasingly, response rates suffer further from the onslaught of household devices that help people screen or ignore incoming calls (e.g., caller ID, answering machines, privacy managers, and automated voice mail) (Tuckel and Feinberg, 1991; Oldendick and Link, 1994; Tuckel and O'Neill, 1996; Link and Oldendick, 1999). In addition, in the United States and in some other countries, "Do Not Call" lists restrict the type of organizations that can call households, a reflection of the public's desire for privacy.
- *Labor practices and workforce issues.* Labor practices and the available labor pool for staffing call centers also changed dramatically over the years. Because research call centers are business operations, there is always pressure to reduce costs and increase efficiency and productivity. Limited employee benefits, structured breaks, and objective productivity measures are the range of labor practices consistently reviewed. Overall, these businesses tend to make little long-term investment in these short-term employees. Moreover, as more jobs become available in other service industries, competition increases for people who might work as interviewers. Typically interviewers today are less educated and more likely to have a shorter tenure than those of several decades ago.

15.1.2 Structure

Today's research call centers can be any size and configuration. Depending on the type of survey data gathered, work load, and the organization's plans for growth, call centers may be small (fewer than 50 seats), large (more than 150 seats), or mega-sized (300 or more seats). The needs and goals of the organization will determine in large part whether a single centralized center (all work conducted under one roof) is called for or if a network of separate centers will be most useful. The purpose and function of a call center drive the physical structure and technology needs, which in turn influence how a call center functions, its organizational structure, and the resources required for maintaining an effective operation. Each call center is unique; however, Table 15.2 gives an overview of some of the key differences between small and large call centers.

Small self-contained centers. Small centers are typically a by-product of another function or need within an organization. For example, a university may establish a small call center as part of its research and teaching program or a company may set up a survey center to gather information for its core business. Such centers can be established nearly anywhere, such as an office building, a set of renovated classrooms, or a stand alone building. Universities, in particular, are creative in their use of space for survey call centers: one university set up a survey shop in the suite of a renovated hotel and another placed its center in the basement where a department's archived journals were stored.

Table 15.2 Common Differences between Small and Large Call Centers

Attribute	Small centers	Large centers
Booths	Less than 50	150 or more
Centralization	Single small site	Monolith or multiple sites
Ability to handle multichannel and inbound calls	Manual or automated	Automated
Location	Near researchers	Near labor force
Affiliation	Academia, specialized	Commercial, government
Facilities required	Almost any kind	Purpose designed
Size of contracts	Limited	Unlimited
Cost effectiveness	Low overhead	Economies of scale
Planning horizon	Can be short term	Must be long term
Technology choices	Off the shelf, home developed	Customized, controlled, integrated
Dialer sophistication	Option to be low tech	Needs to be high tech
Staffing and roles	Generalist	Specialist
Career development	On the job	Established career paths
Recruitment	Word of mouth	Systematic recruitment
Atmosphere and policies	Informal, flexible	Formal, consistent policies
Human resource office involvement	Minimal	Extensive, typically on-site
Management	Single team, personal	Complex matrix, positional

The telecommunications technologies for these small operations are typically basic and rarely involve high-end dialers or complicated networking. On the contrary, although many small shops rely on off-the-shelf software for CATI, some have complex, highly tailored sample-management and questionnaire-administration software, especially if the operation has been in place for some time and/or there is an in-house expert of long-standing who worked on the original software and has been tweaking it ever since.

The organizational structure of small survey operations is often informal, with a small core of management or supervisory staff who often are multiskilled and fill a variety of roles from desktop support through to recruitment and training to floor management. Word of mouth is often the main recruitment strategy for interviewers, and many interviewers have some other connection with the parent organization, for example, they may be students or clerks.

Small survey operations can often be established quickly (within a few months) and with considerable flexibility. Typically these shops serve a niche clientele (e.g., university professors, state and local government officials).

Large centralized operations. Large call center operations (i.e., those with 150 or more booths) are often viable entities in their own right, with facilities separate from any parent organization, in its own location, and with its own organizational structure, including human resource specialists and technology support. Economies of scale make it feasible

- To locate the operation close to the labor force (e.g., in an area of high unemployment, low rents, and local government incentives for establishing a new business)
- To invest in high-end dialers, since the expected volume of calls will provide a reasonable return on investment in the equipment
- To use computer-based training packages for staff
- To establish formal career paths for staff
- To have specialized rather than generalist roles for most staff and
- To use customized software

Large organizations often require disciplined management and structure and less "seat-of-the-pants" operations. For example,

- Human resource policies tend to be formal for efficient and consistent treatment of staff. Unlike small operations, large organizations cannot ensure consistency by having all human-resource decisions made by a single person.
- In-house software needs to be fully documented, with formal testing protocols and implementation procedures, thus avoiding dependence on a single computer expert as is sometimes the case in small call centers.

Large decentralized operations. Although a single, large operation offers considerable economies of scale and communication advantages (Mitchell, 2001), often there is a size at which a single centralized operation is no longer feasible, because of the pressure on the local labor market and the need for optimal organizational efficiency. Staffing as few as 100 booths may not be feasible in some locations if the pool of interviewer applicants is insufficient. Yet, even in locations with relatively high unemployment and an educated workforce, few operations can sustain a staff to fill more than 400 booths.

Often, organizations exceeding 300 booths will spread across several locations. However, these "smaller" locations still will have many features of the large centralized operations.

Multiple sites can also help to avoid a single point of failure. Two 100-seat telephone centers can be arguably less risky than a single 200-seat center. Unanticipated events such as fires or natural disasters (hurricanes, ice storms, flooding, and earthquakes) may bring all data collection efforts to a halt if an organization relies on a single, large centralized configuration. With a decentralized configuration, disasters such as hurricanes have less dire consequences if some centers are unaffected and remain operational. Multicenter designs come, however, with additional costs, including the need for multiple center directors, technology infrastructures, and facilities as well as added management time for coordinating all centers.

Sometimes the desire for a particular mix of people within the workforce can drive the decision to have several sites. For example, one site might be in an area with a high immigrant population if a certain language is needed, while another might be placed to take advantage of a particular regional accent.

Lastly, practical considerations or circumstances can drive this decision, such as existing building limitations or lease commitments.

Completely decentralized and distributed operation. At the far end of the decentralized spectrum, VoIP technology has made it possible to operate "virtual call centers," with interviewers, and possibly a large proportion of supervisors working out of their homes. The advantages of a fully distributed operation include

- Few facilities and low overhead
- Large face-to-face interviewer force scattered across the country
- Increased interviewer satisfaction
- Opportunities to attract qualified interviewers who are homebound, may live in remote areas, look after small children, and/or have mobility limitations

However, at the time of writing few organizations have progressed far down this path. Even those who embrace distributed operations emphasize caution and careful planning (Fadia, 2001; Kim, 2004). Disadvantages and obstacles include the following.

- Loss of control over the security and focus of the interviewing situation, ranging from problems ensuring no unauthorized persons overhear the interview to the problems of domestic distractions such as ringing doorbells, crying babies, or other distracting background noises.
- Restricted workforce supervision, including challenges in monitoring interviewers' work and providing in-person mentoring, feedback, and motivation. While remote monitoring systems usually allow both visual and audio monitoring (i.e., the supervisor can see a copy of the interviewer's screen as well as listen in on the call), as well as track idle time while the interviewer is logged in, these things do not have the same impact as they do when used on site combined with physical proximity.
- Cost of equipping the interviewer's home and the maintenance and security of the equipment so placed.
- Inefficient use of equipment. An in-office CATI booth will typically be used for at least 34 hours a week every week. In contrast, an in-home booth is used only when the at-home interviewer works, which may be as little as 15 to 20 hours per week.

15.2 FACILITY SETUP

A call center can be built in various types of space, from high-tech office space to an empty warehouse. In all cases, however, the main considerations are the same. The facility needs to have a significant telecommunications and computer network infrastructure, be a pleasant environment in which to conduct telephone surveys, and a safe and secure place of work.

15.2.1 Size and Infrastructure

Generally, call centers require 150 to 230 sqft. per interviewing booth to provide sufficient space for the interviewing booths as well as space for offices, training, meetings, breaks, and similar activities. However, with careful planning, as little as 100 sqft. per booth may suffice and still be an attractive, functional—although confined—facility. More information about space planning is available at *www.gsa.gov*, which has an overview of the topic "Facilities Standards for the Public Buildings Service."

Good acoustics and soundproofing are critical in a call center. Acoustic walls and carpeting provide adequate sound-proofing, but their effect can be diminished if the acoustic material is covered with paper or signage. Parabolic lighting is typically used to provide sufficient illumination without glare on interviewers' computer screens. As in most work environments, natural light is also a positive factor, particularly in meeting rooms or break areas. In production areas, computer monitors should be placed perpendicular to the natural light, or shades or blinds should be installed. Natural light and openness are important factors for interviewer morale.

15.2.2 General Layout

Interviewing floors can be laid out in a variety of ways, with linear stations and circular pods being the two most common layouts. Linear stations typically allow more stations to be placed in a facility and a better supervisor sightline if the supervisor stations are at the ends of each row (Fig. 15.1). However, this arrangement can also foster unproductive socializing and chit-chat among interviewers sitting adjacent to or across from each other. Although partitions may be unpopular with some call center staff, they do reduce unnecessary talk among interviewers on the production floor. Partitions of a height of about 42″ from the floor seem to work best: they dampen sound and do not completely impair visual monitoring of floor activities by supervisory staff.

In contrast, circular pods are a more open setup (Fig. 15.2). In this configuration, cabling is usually brought to a center point and fanned out to each station in the pod. Usually, this setup gives interviewers more workspace than the linear setup, but it may reduce the number of stations that fit into a space. Supervisors' line of sight to observe floor activity may also be obscured, depending on how supervisor stations are positioned on the floor, and sometimes just knowing a supervisor may be watching reduces unnecessary conversation.

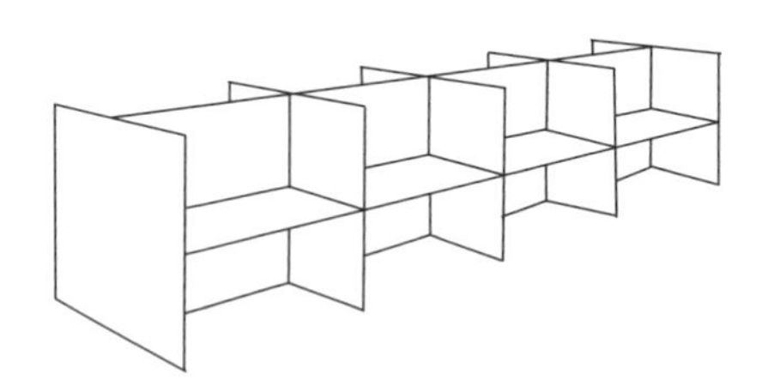

Figure 15.1. Linear workstations example.

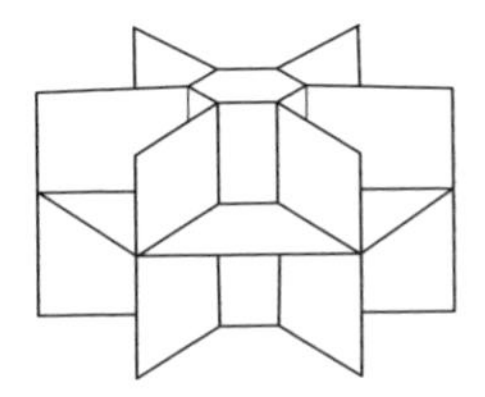

Figure 15.2. Pod workstations example.

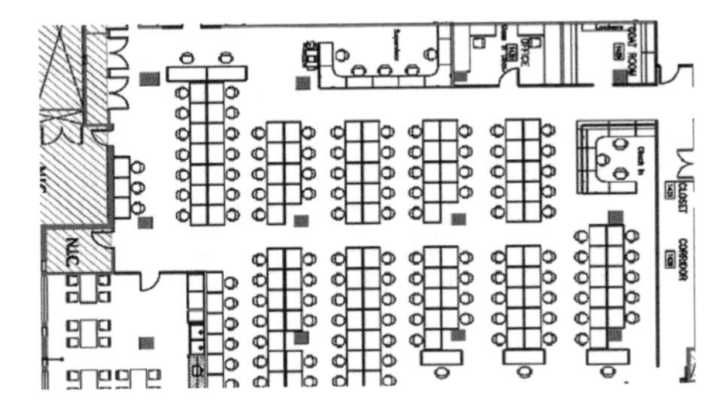

Figure 15.3. Example of production floor using a linear workstation layout.

Auto-computer-aided design (CAD) systems are useful in determining the most appropriate and efficient production floor layout for a particular organization. Examples of floor layouts with linear stations and pod layouts are in Fig. 15.3 and 15.4. Layout advice is also available from companies that specialize in call center arrangements and furniture.

It is essential to check the local building codes as they vary widely from location to location.

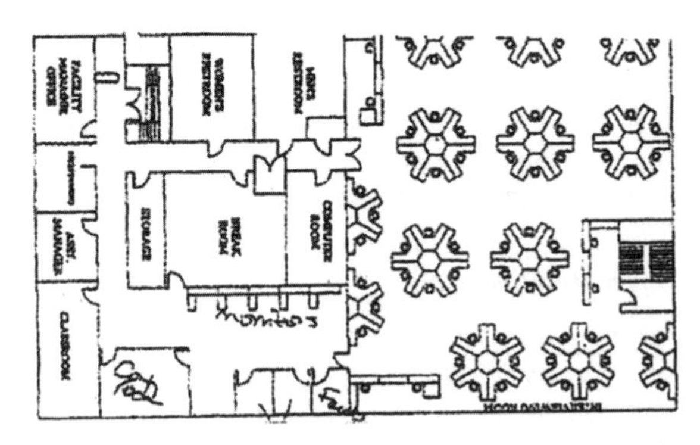

Figure 15.4. Example of production floor using a pod workstation layout.

15.2.3 Interviewing Stations

For a healthful and productive work environment, interviewing stations should be ergonomic. Fully adjustable ergonomic chairs are critical, given the amount of time interviewers will spend in them. Other required features include desks of different fixed heights or adjustable height desks and adjustable monitors to accommodate, particularly, short or tall interviewers or those with heavy legs that may not fit well between the seat and the desk, and those with back problems who prefer to work standing rather than sitting. Pull-up keyboard trays add to users' comfort and can substitute to a fair degree for adjustable height desks. Footrests also provide an inexpensive way to increase stations' comfort.

The workspace width depends partly on layout. With a pod design, a workspace with a front edge width of 5–6 ft. works well. With linear layouts, booths range from 3–5 ft. wide, with the choice depending on the desired number of stations within a space, the size of the computer monitors to be used, and the amount of paperwork interviewers must do at the stations. Booths 4 ft. in width allow only 75 percent of the stations within an area than do 3-ft. wide booths. However, narrow stations are viable only if flat screen monitors are used and little paperwork is done (i.e., little desk space is required).

Shelving or hanging racks can be used at interviewing stations to store, for example, manuals or procedures. Shelving enclosed by upholstered flipper doors provides additional sound proofing. Providing locking drawers or locker space for each employee is also important, so employees can store valuables safely.

Standardized interview work stations are recommended if stations are to be used by several people and for a uniform look throughout the center. Allowing employees to personalize their stations tends to give an untidy appearance to the entire site, causes disagreements among staffs sharing one station, and restricts seating options and practices such as overbooking (that is, scheduling more interviewers for a shift than the number of stations available in anticipation of no-shows).

15.2.4 Planning for Staff with Disabilities

By law, call centers must accommodate people with physical disabilities and ensure that there are no barriers to employing them if they otherwise are qualified. Ergonomic stations can help with the physical arrangements for some staff with disabilities. Advice on planning for staff with disabilities can often be obtained from local or state rehabilitation or outplacement centers. Sometimes organizations that promote the hiring of people with disabilities provide or fund specialized equipment to accommodate them. For example, organizations that promote hiring visually impaired staff may provide special monitors, software, and even training at their expense.

15.2.5 Other Space Requirements

In addition to production floor layouts and interviewing station requirements, a host of other space issues need to be considered, including where to place supervisory and interview monitoring stations, offices for managerial staff, private meeting rooms, training rooms, administrative areas for check-in and security control, restrooms and break areas, a place for coats, and secure storage for other personal items. In addition, call centers, particularly those that are separate from the parent company, require highly secured wire closets, or an equipment room, or both. These should have separate climate control, enhanced security, and high levels of insulation for computer and telephony systems.

Supervisor/monitoring stations. Call center managers have different philosophies about the most effective place for supervisor desks on the production floor. Some centers put them in the center of the production floor, others place them at the perimeter, and still others put them at the end of rows of interview booths. Another option is to spread the supervisor stations throughout the production floor, interspersed among the interviewer booths. Generally, supervisor areas are placed at entries to and exits from the production floor and positioned to maximize sightlines to interviewing booths. Usually clusters of supervisor booths are better than single booths throughout the center. This arrangement allows interviewers to locate a supervisor quickly if they have a question or need immediate assistance.

Supervisor stations may also be placed on a raised platform to improve visibility and lines of sight. Some call center managers believe that raised platforms provide a better method of monitoring floor activity and allow interviewers to locate supervisors more quickly when needed. Others believe, however, that raised platforms for supervisors creates an "us-versus-them" atmosphere, raising tensions and, therefore, being counterproductive.

The placement of interview-monitoring or quality control staff stations or desks calls for somewhat different considerations than that of supervisory staff. If monitors are not also supervisors, they are usually in a separate monitoring room. This arrangement is typical in large operations where interview-monitoring and quality control is handled by a specialized unit. This arrangement causes fewer visual distractions and allows monitors to focus on their tasks. It also allows the monitoring work to be standardized and well supervised. Since most interviews are conducted and monitored simultaneously via a computer or telephone, monitors do not always need to be seated with sightlines to the interviewers.

Office and private meeting space. Typically, call center managers need private office space to carry out their jobs effectively. Depending on the organization, staff in charge of hiring and staff who provide computer or network support may also require separate lockable office space. Supervisors need a quiet area to work, and although they may not need an office, they do need a quiet, soundproof area (often called coaching rooms) if they need to speak privately with an interviewer. If space permits, an unassigned office or two, off the main floor, is best for this purpose.

Training rooms. Interviewer training rooms can be configured in a variety of ways, but most centers attempt to lay them out in such a way as to provide overflow production and training space. Although classroom style (with desks arranged in rows) is a common layout, an alternative is to position stations around the perimeter of the room with tables in the center so the instructor can see what is on each screen. This allows the instructor to see how students are progressing in relation to the rest of the class. Making small work stations out of these areas allows the center to have extra interviewing stations for overflow work. If trainees have chairs with wheels, they can work or train at the computer, then turn around, and roll their chairs to the center tables for meetings or noncomputer training.

Administrative and security areas. Since telephone interviewers come and go at all times, each center needs an area where interviewers check in or a computer system that records interviewers' start time. The check-in system can be built into administrative greeting protocols that includes keeping visitors off the interviewing floor. Call center security systems and camera monitors (if used) are also usually located in the greeting area.

Wire closets and equipment rooms. Call centers require greater telecommunications and computer networking infrastructure than the typical business office. It is essential, therefore, that sufficient space be allotted to provide a secure area for on-site system components, including computer servers, telecommunications equipment, uninterrupted power supply (UPS), and other backup systems. This should be a temperature controlled, secure space with access limited to essential personnel only. A cipher lock, that is one requiring a code to unlock the door, is often used to secure this area because the code can be changed periodically.

15.2.6 Facility Features to Support Disaster Recovery

Despite the best planning, disasters do occur in call centers, and they can significantly curtail or completely stop data collection. The most common situation is power loss. When planning for such situations, most think only of a back-up system for computers such as a UPS (uninterrupted power supply) that will shut down the system gracefully (i.e., allow in-progress interviews to be completed). However, temperature control (heating/air conditioning) and on-going operations must also be considered. Depending on the size and budget of a facility, an external generator may be located on-site. Such a system should provide at least 50 percent of the temperature control and enough power for computer systems and emergency lights. A back-up power source will become even more critical as telephone centers migrate to VoIP-based infrastructures that run telecommunications through the same system as the computers. With this type of system, power loss prevents access not only to computers but to telephones as well.

Other disaster precautions could include having a system that redirects calls either to another center or to interviewers' homes during bad weather. For major disasters, a rollover system can be developed to direct calls to an unaffected area. Some companies prearrange, through a contract, to have a remote "hot site" with back-up computers. Advances in technology now make it possible to quickly regain lost call center capacity in the event of an emergency. In January 2001, a large survey operation in the United States suffered a devastating fire at one of their two call centers during several major projects. Although some work was shifted to the sister facility, the company was able to implement disaster-recovery plans and establish a temporary working call center near their destroyed facility, thereby retaining many of the already trained and experienced call center staff and also completing the projects on time. More detail on disaster planning is in Dawson (2004).

15.3 TECHNOLOGY

At the broadest level a data collection or survey center is typically a physical space where information is collected and stored through a variety of communication channels, including telephone, Web, fax, and IVR (Dawson, 2004). As concurrent multimode interviewing becomes more feasible, telephone data collection has often been paired with one or more other modes. As a result, telephone survey centers are less exclusively outbound call survey shops and more multichannel contact centers. When considering which new technologies to install in a call center, planners must decide not only which technology is best for the present but also which technology best fits the organization's near-future needs. There are a number of resources (e.g., www.acs.org.uk, Bocklund and Bengtson, 2002, Waite, 2002) available to assist in determining the best technology platforms and computer-aided interviewing (CAI) tools for an organization. These resources offer step-by-step processes and criteria to consider when selecting specific tools.

High-level understanding of current and projected capacity needs, system integration requirements, budget constraints, expectations for technological support and training availability are required. These requirements should be communicated to the design and implementation teams early in the call center design process. Business users also need to be moderately familiar with the technological options available. Consideration should also be given to whether the technology choices being made-particularly those concerning telephony and networking—are for call center operations only or for use throughout an entire enterprise.

A survey research call center has three main technological components: the base technology platform, the telephony infrastructure, and the CAI tool(s).

15.3.1 Technology Platform

The base platform is a basic operating system or environment in which both the CAI and telephony come together to facilitate the collection of data. Decisions about which environment to work in are best made by the person in charge of the organization's technology. This person also needs to understand the business's requirements for data collection. The computer platform chosen will in some cases determine or at the least affect other technology choices. Some CAI software packages will run only on one platform (e.g., Windows, DOS). In addition, some telephony choices are not compatible with certain computer platforms. Certain autodispositioning dialers, for example, work more effectively with some PBX systems than with others. If a different operating system from the one in place at the parent company or sister call center site is selected, planners must consider the difficulty, timing, and expense of migrating technologies and software or the difficulty of providing support for both the legacy and new systems simultaneously.

15.3.2 Telephony

Telephony incorporates all communication channels and equipment from the telephone used by the interviewer (desk-top phone or headset connected to the computer) to the telephone in the respondent's home. A number of sources give good introductions to basic call center telephony (e.g., Waite, 2002, Dawson, 2004, and Cleveland and Mayben, 2002). Additional issues, particularly with respect to wireless and VoIP as they relate to telephone survey work, are discussed in detail in the chapter by Steeh and Piekarski in this volume. Here we focus on some of the key aspects of telephony as they relate to survey research operations.

Voice transmission. In many areas, call center managers can choose among communication vendors to carry survey calls. Communication vendors vary widely in terms of their capabilities, quality, and costs. When choosing a vendor it is important to assess qualities such as completed call ratios, dropped line rates, and potential for service outages. Cost is obviously another driving factor in the choice of vendor. When feasible, using more than one carrier is worthwhile because a second carrier provides back up for times when the primary system fails. Also be aware that technology that is suitable in one country may not be as good in another because of issues such as the age and reliability of the telephony infrastructure or the regulation and enforcement of the use of *Standard Information Tone* (SIT).

Dialers. The dialing method is an increasingly important component of the survey telephony system, particularly in call centers with high call volumes. Dialing may be as basic as having an interviewer pick up a hand set and dial a telephone number or as complex as using an arrangement of sophisticated hardware and software that directly handles the dialing, distribution, and resolution of calls.

Dialers can automate the process in two ways: (1) at dial initiation and (2) at outcome detection. Automatic dial outcome detection relies on the SIT signals from the dialed number. These systems can recognize various tones that signal whether a number is disconnected, a fax or data line, busy, or not answered. Since dialers recognize tones and assign the appropriate outcomes to calls faster and more accurately than interviewers can, they also (depending on the system) save on costs. Some detection systems can also recognize whether an answering machine or a person picks-up the call. However, because these systems work by detecting a pause after a short greeting (as opposed to an answering machine script which is launched without waiting for a response), this feature often is apparent to the respondent. Because many people respond to this pause by immediately hanging up, which increases the survey's nonresponse, answering machine detection is typically an undesirable feature from a survey research point of view.

Automatic dialing comes at several levels.

- The most basic, the "autodialers," simply circumvent the need for the interviewer to manually punch numbers on a phone. They automatically transmit the phone number to be dialed (and often the project code to which the call should be charged) to the dialer once interviewers indicate they are ready to call.

- With the next level of dialing, the computer automatically dials a number some predetermined length of time after an interviewer completes a call. This saves interviewer time in that it eliminates the additional keystroke required to signal the computer that a new number should be dialed and reduces the pause between calls.

Predictive dialing is the most efficient level of dialing, whereby the dialer predicts either (1) the chances of an interviewer becoming free to answer a call and/or (2) the chances of a single dialing resulting in a connected call that should be passed to the interviewer for handling. Therefore, predictive dialers usually dial more numbers than there are interviewers available on the assumption that only a fraction of the numbers dialed will result in a connection that needs an interviewer's attention. The volume dialed is driven by an algorithm that constantly assesses the proportion of calls resulting in connections as well as the number of interviewers available for work.

Although predictive dialers can save on costs, they have a major drawback: the algorithm, which is based on a set of probabilities, sometimes causes too many numbers to be dialed, resulting in the respondent either having to wait for an interviewer or hanging up. This "abandonment rate" can be set to zero. However, to achieve zero, the dialer must dial only a single number at a time and only when an interviewer is available, effectively causing the system to cease functioning as a predictive dialer. Predictive dialers need a critical mass of interviewers to work effectively and rarely have any advantages with fewer than 10 concurrent interviewers.

Another problem with predictive dialers from a survey standpoint is that the number is not assigned to an interviewer until after the connection with the respondent is established, thus giving the interviewer inadequate time to check relevant call notes, especially in the case of refusal conversion. A way to ameliorate this is to use a *hybrid mode of dialing*, which is a combination of autodispositioning, autodialing, and predictive mode dialing that bases dialing mode on the entry state and history of the number being dialed so that the cost savings of predictive mode can be realized, call notes can be checked, and respondents are not disconnected.

Other dialer features: recording. Some dialers allow part or all of an interview to be recorded, which is helpful for assessing data quality and monitoring or coaching interviewers. However, recording interviews often requires additional investment in storage space (voice recordings take up considerable space), indexing applications (to assist in retrieving specific recordings by interviewer, respondent, time, or question), and compliance with legislation in some countries and states that require that permission from the respondent be obtained before a recording is made.

Other dialer features: message push-out. Some dialers can deliver a recorded message. These dialers may be useful in the survey industry as they can leave electronic messages that notify respondents about the survey, although the effectiveness of such notifications is questionable (Link and Mokdad, 2005b). In addition, legislation in some states and countries is now being introduced along with "do not call list" legislation to restrict the use of dialers with these features.

Other dialer features: sample generation and sample management. Although some dialers generate and manage samples of telephone numbers, they are often of limited use in survey situations. Because having been developed primarily for telemarketing, the methods used are often rudimentary and not of the level of sophistication required for most survey research projects.

15.3.3 Computer Assisted Interviewing (CAI)

Given that the central focus of telephone survey centers is the timely, accurate, and cost effective collection of high quality data, the choice of CAI software is often a driving technology choice. Rigorous analysis up front is required to determine the most appropriate software to fit the work being performed. For example, if a call center expects to perform a large number of studies with limited lead time for development, a CAI tool that is flexible and streamlines the implementation of studies would be of high value. If the center expects to have work that requires complex reporting, then easy access to the data is important.

The most comprehensive way to assess which is the best software is to analyze its features and compare them with the operation's needs and requirements. Business and scientific requirements need to be delineated by call center stakeholders and ranked according to importance. Software choices are then ranked against these requirements. The matrix that results helps indicate which choice is best suited to meet an organization's needs. The Association for Survey Computing (ASC) has a useful software register (http://www.asc.org.uk/Register), where software packages are listed by feature, which can assist in making a preliminary shortlist of candidate packages.

The scheduling of survey calls is driven by either the telephony or the CAI systems. Call scheduling tools that reside within the CAI or in a case management system that is directly related to the CAI are generally more sophisticated than those associated with dialers; often they can schedule calls on the basis of call history. The complexity of the call scheduling rules differs from project to project and organization to organization. Most likely some customization of the call scheduler will be required even for projects within the same call center. Be aware that the more complex the call scheduling rules and interactions with other components of the center, the higher the development, testing, and deployment costs. Hansen (2007) examines call scheduling issues in depth in Chapter 16 of this volume.

15.3.4 Quality Control Monitoring

Monitoring involves a third party (usually a quality control monitor, interviewing supervisor, project staff member, or client) seeing the interviewer's computer screen and hearing the conversation between the interviewer and respondent. Ensuring audio and visual synchronization is a necessity. For a fully comprehensive monitoring system (such as that described by Steve et al., 2007, Chapter 19 in this volume), strictly standardized tools must be in place for the monitoring staff to select which interviewers should be monitored next and to score interviews. Because most commercial monitoring systems are expensive and geared toward the needs of inbound call centers, survey research shops often must develop and maintain their own monitoring systems.

Another aspect of the quality control process is remote monitoring, which allows a fourth-party (usually a client or off-site project staff member) to link into the call and silently monitor the interview. This is of great benefit when researchers or clients wish to listen to and, in some cases, to view simultaneously the work being performed by interviewers; however, data security and confidentiality become large issues. Some CAI tools have remote monitoring capabilities. However, non-CAI remote monitoring facilities are also available. For example, Web conferencing is a cheap and easily supported way of remotely monitoring interviews without investing in a costly system as these services can be purchased on a pay-per-use basis.

15.3.5 Reporting Systems

Reports and metrics generated from the CAI and telephony system are critically important to a center not only to guarantee data reliability but also to assist in the efficient management of a workforce and projects. Planning for such reports when systems are designed is preferable to attempting to extract information after systems are in place. Most off-the-shelf CAI software comes with basic reporting capabilities and an easy means of exporting datasets and timestamps. Tarnai and Moore (2004) discuss a range of interviewer productivity measures in Chapter 17 in this volume.

15.3.6 Data Storage

All data collected must be stored securely. Many countries have laws about retaining and destroying data, and these laws are often more onerous for data collected through government studies. Usually, data need to be accessible to management for at least 2 years. An effective archiving and destruction process helps to ensure that high data volumes do not slow down or otherwise negatively affect the CAI system.

15.4 STAFFING THE CENTER

A call center may have the most sophisticated technology in place but if it is not staffed properly it likely will fail in its mission. Thus, carefully planning is needed to determine the critical staff needed. These issues need to be decided *before* deciding on the location, structure, and layout of a new call center. Ideally filling positions for a new center is top-down. That is, the top managers should be hired first. They will then be involved in hiring the next lower level. As each new level is hired, that staff will be involved in hiring the next lower level.

15.4.1 Call Center Director/Manager

Small operations usually have a manager at the helm, while large ones have a director, who is often supported by a manager at each site if the operation is multisite or an assistant director if it is a single large site. The titles differ because more levels of management are needed in a large operation than in a small one. The titles also reflect the extent to which the number of direct and indirect staff reports influences job classification and salary.

The primary leader of an operation is responsible for the functioning of the operation and facilities. This responsibility includes a host of duties such as interacting with clients; managing the operating budget and payroll for the call center; overseeing staff recruitment, training, and retention; troubleshooting problems on the production floor; working through personnel issues; ensuring that the facilities are maintained; coordinating computer and systems issues with the technical support department; and interfacing with other departments and units in the parent company. In a small organization, the primary leader is often a "jack-of-all-trades" and intimately involved in the details of running the shop, whereas in large call centers, many duties are delegated and managed by someone in a lower-level position.

Regardless of size, success in this position is usually assessed along the following lines.

- Surveys in their purview are delivered within budget at the required quality and within the agreed timeframe.
- The center generates a good return on investment (i.e., the center is not empty most of the time; jobs run cost-effectively, so they can be priced competitively).
- Staff morale is healthy.
- Actions are taken to ensure the center thrives long term.

These outcomes require different strategies and actions to achieve, and the call center manager/director's job is to ensure that appropriate plans are successfully developed and implemented to meet these goals.

15.4.2 Project Managers

Even the smallest operations usually need one or more people to meet with clients (including internal research staff) and to plan and execute projects. Duties typically include working with clients in planning call-center-related aspects of a project; defining the number and type of staff required for a project; overseeing staff training; establishing and monitoring production metrics (e.g., number of completed interviews, number of refusals, calls made per hour), ensuring data quality (through interview monitoring reports), scheduling, tracking costs, and, providing interim and final reports to clients. Having only one person responsible for these functions is risky because the project and center cannot depend on the health and presence of a single person. In small operations, supervisors can help with some of these functions, but usually a significant level of skill and expertise is required.

Dedicated project managers (who also have significant staff management responsibility if the supervisors report to them) are usually needed once the operation exceeds 30 booths. As a rule, one project manager is needed for every 5 to 10 supervisors, or for every 40 to 80 interviewers, depending on the complexity and number of projects and the number of other supporting positions.

15.4.3 Interviewer Supervisors

Production floor supervisory tasks can be as rudimentary as "walking the floor" to ensure that interviewers stay on task and answering interviewers' questions. Or the tasks can be more complex, such as scheduling interviewers, managing performance, loading surveys into the CATI system, completing shift reports, and managing the survey sample. As a rule, 5 percent to 10 percent of interviewer hours need to be scheduled for supervisory tasks, with most centers having between one and two supervisors for every 20 interviewers.

Because supervisory tasks vary in proportion to the number of interviewers, at least one third of the supervisors need to be employed on the same basis as the interviewers (i.e., intermittent and hourly). Rarely can the majority of supervisory staff be employed as regular full time employees, in part because of the small number of peak capacity hours during a week and because of the fluctuations in workload as surveys open and close. Even those employed full time need to be prepared to work part time or to endure a temporary layoff if the time between surveys is significant. That said, having some full time supervisory positions is worthwhile because it helps in retaining talented supervisors long enough to prepare them for a more senior management role.

15.4.4 Interviewer Quality Monitors/Team Leaders

The quality control monitor observes interviews to ensure that interviewers follow study protocols for obtaining cooperation from sample members, administer questions appropriately, and record respondents' answers faithfully. Most survey organizations monitor between 5 and 10 percent of interviews or interviewer time (Steve et al., 2007, Chapter 19 in this volume).

It is usually prudent to hire these staff on the same basis (e.g. intermittent, hourly) as the interviewers, since the demand for their labor is exactly proportional to the demand for interviewers' labor. However, monitors are usually hired from among the best and most reliable interviewers, so they tend to have less absenteeism and attrition, which needs to be considered when working on schedules.

15.4.5 Survey Interviewers

Although interviewers are the engine that drives a call center, little empirical research shows which factors lead to their being persistent and performing well. Anyone who has spent 10 minutes at an interviewer training session or in a monitoring room listening to an interview knows that interviewers vary greatly in their skill and ability to persuade sample members to participate in a survey. Experience is one of the most often-cited attributes linked to interviewer success (Durbin and Stuart, 1951; Groves and Fultz, 1985; Couper and Groves, 1992; Martin and Beerten, 1999), although some have disputed this finding (Singer et al., 1983). Groves and Couper (1998) caution researchers against unqualified emphasis on experience, however, noting that some interviewers may have longer tenures (i.e., be classified as more "experienced") because they came to the job with better persuasive skills and thereby persisted, rather than gained those persuasive skills as they become more experienced (see Groves et al., 2007, Chapter 18 in this volume).

Interviewers' attitudes and expectations also appear to play a role in determining their success, since interviewers with more positive expectations and confidence tend to post higher response rates (Singer et al., 1983; Groves and Couper, 1998; De Leeuw, 1999). Interest in how interviewers' attitudes contribute to survey nonresponse sparked the International Survey of Interviewer Attitudes (ISIA) at the 1998 International Workshop on Household Survey Nonresponse in Mannheim, Germany. One report generated from that study found that an interviewer's attitude, sex, and age were important predictors of job success (Japec and Lundqvist, 1999). Women, people with a positive attitude, and people aged 52 years or older had higher response rates than men, people with a negative attitude, and people younger than 52. Likewise, Martin and Beerten (1999) found that women of 52 or older, interviewers with experience, and confident interviewers had higher response rates than younger women, inexperienced interviewers, and interviewers with low confidence.

Groves and Couper (1998) posit that interviewer attributes alone might not be the most important factor that determines a person's ability to get a respondent to cooperate. Instead, the interaction of interviewer's characteristics and sample respondent's characteristics may be the key. Further, interviewer experience, sociodemographic attributes, project assignment, and the design features of the study itself affect interviewers' expectations and behaviors. These behaviors, in turn, interact with those of the sample members to drive respondents' cooperation. The extent to which these findings can be translated into recruitment strategies however is limited, since the strength of the some of these effects do not outweigh the moral, and in some cases legal, requirement not to discriminate on the basis of age, sex, and certain other characteristics.

Several other factors have also been hypothesized to relate to successful telephone interviewing. For instance, the quality of an interviewer's voice (e.g., tone, accent, pitch, speed, clarity) affects refusal rates: interviewers with "attractive" voice qualities have lower refusal rates (Oksenberg et al., 1986) than interviewers without attractive voices. Personality is another factor that may have an effect (Hyman, 1954; Axelrod and Cannell, 1959; Morton-Williams, 1993), but a strong link has yet to be demonstrated between a specific set of personality characteristics and interviewer performance.

One study of 383 newly hired telephone interviewers working on two large-scale national surveys found that interviewer success on the job results from a complex mix of individual and workplace characteristics (Link, 2006). Workplace factors such as location of the call center, shift worked, and study assignment were most relevant for predicting persistence in the job, while individual attributes, including telephone skills, previous experience as an interviewer, and a confident yet realistic attitude toward the difficulties of survey interviewing were closely linked to job performance. The study's findings have implications for recruiting and training new interviewers.

- No single factor or set of factors produces efficient, effective, long-term interviewers; rather a complex mix of an interviewer's personal characteristics and workplace characteristics tend to affect how well interviewers do their job.

- Interviewer training helps interviewers gain the confidence they need to do their job well, but it should also give them realistic expectations about what they will encounter once they are on the telephone.
- Starting interviewers on easy studies may help them gain confidence early and, therefore, persist longer on their jobs.
- Having supervisors with interviewing and mentoring experience is critical to developing a productive interviewing staff.

Planning for interviewer recruitment and attrition over time can be one of the more difficult aspects of staffing a survey call center. Interviewing is often a temporary job with a high degree of staff turnover. The job is repetitive and subject to periods when the work is insufficient to keep all of those recruited fully employed. Most survey operations have household surveys as the large part of their work mix. This restricts operating at full capacity (i.e., using every booth) to the hours when people are most likely to be at home (i.e., evenings and weekends). Call centers usually work at full capacity only 30 hours per week (3 hours each week night: 6–9 pm; and all weekend). The 30 hours can be increased to about 40 per seven days if a survey is running nationally (due to varying time zones).

Most interviewers work part-time; therefore, usually two interviewers are required to cover 40 hours of full capacity per interviewing booth. Therefore, a 60-booth facility requires at least 120 interviewers when it is operating at full capacity. If there is an appreciable load of daytime work (such as surveying businesses or institutions), then the ratio of employees-to-booths can rise to 3:1.

Absenteeism and attrition rates must also be factored in when making staffing projections, both of which will be greater the newer the workforce. Absenteeism against a schedule set about a week in advance is typically in the 10–20 percent range, caused by factors such as sickness, family emergencies, travel issues, and other work or study commitments. If a project required 1000 hours a week, and interviewers on that project were scheduled on average 20 h per week, and if absenteeism typically ran at 15 percent, then 59 interviewers would need to be scheduled as follows: *59 interviewers × 20 h per week × 85 percent attendance = 1003 hours per week.*

However, more than 59 interviewers would need to be trained for the project because of attrition. The extent of attrition depends on many factors, including

- Local labor conditions (high unemployment means a lower rate)
- Seasonal factors (e.g., student interviewers tend to leave before examinations and near the beginning of the school year)
- Attractiveness of the pay and benefits package offered
- Nature of the projects on which the interviewers will work (a project that generates fewer or less hostile refusals due to topic, brevity, advance letters, and/or incentives will be easier and more attractive to interviewers)

- The proportion of the interviewing team who were newly recruited for that project, since no matter how well interviewers are prepared for a job, there is always some proportion that will decide within the first couple of weeks that interviewing is just not for them.

So to continue the above example of the project requiring 1000 hours per week; if a large proportion of the interviewers were to be newly recruited, a 20 percent attrition rate might be planned over the first 4 weeks and so in order to be able to schedule 59 interviewers per week 74 interviewers would be trained. If the project was likely to run for more than 8 weeks, a decision would need to be made as to whether to accommodate the likely level of ongoing attrition via training more people at the start of the project (and having more hours worked initially), or "topping up" the hours with an additional mid-project training.

What is acceptable attrition? Although a low turnover is often admirable, it is of little value if it comes through an expensive recruitment process or by tolerating poor performance. At the skilled end of the labor market a high turnover of staff is expensive. However, interviewers are not highly skilled and relatively easily recruited and trained; therefore, in some situations high attrition is an acceptable tradeoff for a finer filter up front (in effect, filtering out unsatisfactory employees post hoc). There is little survey industry data on how long it takes for a new interviewer to reach full productivity. However, our experience and analysis of workforce data is consistent with that reported by Burstiner (1974) regarding telephone sales; that is, most interviewers for simple surveys reach their peak productivity between 6 and 10 weeks after being hired. Therefore, the effort, risks, and costs needed to recruit highly qualified interviewers might not be worthwhile. Instead, it might be better to simply bring on minimally acceptable applicants and see how they do. Clearly different projects may require a different approach, but the main point here is that a high short-term turnover rate is not necessarily a bad thing. Grote (2005) argues the case for constantly sloughing off the bottom 10 percent of workers, for at least the first 5 or so years it takes to develop an optimum corps of workers at which time the percentage sloughed off can be lower.

Another consideration is retaining and rewarding the best interviewers. This topic is large and complex; it includes pay and benefit packaging, performance feedback and management, incentive plans, and job structuring. Although many books and articles touch on various aspects of this topic, the best are those that discuss how these issues interrelate. Chingos (2002) has one of the most thorough discussions on these topics. Lavrakas (1993) has the most specific information in relation to interviewer pay, convincingly arguing for a payment system that contains both a base and variable component that provides an incentive to interviewers by linking their performance to their pay.

15.4.6 Other Key Call Center Roles

The size of an operation determines the remaining staffing structure, with the large operations requiring specialized staff and small ones needing "jacks-of-all-trades."

Technical support. The technical support staff (including infrastructure maintenance and survey programing) is often part of a department within the parent organization rather than a separate group at the call center. However, call centers rely heavily on technical services, so often a small staff with specialized knowledge will be assigned to deal specifically with call center issues.

The greater the amount of technical support needed, the more complex the telecommunications, networking, and CAI systems and the degree of customization required for software. For a smaller operation, technically oriented supervisors and project managers can undertake a fair amount of the technical support in terms of desktop hardware support and software maintenance. As the size of the operation grows, however, a dedicated team reporting through the technical services department is essential so that appropriate systems can be put in place to ensure consistency of hardware and software configurations in all interviewer booths and a regular program of testing and maintenance is established.

Human resources. Even the smallest operation needs access to reliable human resources support to ensure that all regulatory requirements are met. Like technical support, human resources is often a separate department within the parent organization. There are often some fundamental differences, however, between the type of staff hired for a call center and those hired for the larger organization. Because of the itinerant nature of many of the interviewers, the human resources department needs to gear up for a higher level of issues, from more frequent hirings and separations to a stunning variety of employee relations issues. However, few of these are survey-industry specific, so the issues already documented in call center literature typically apply to survey research centers as well (e.g., Dawson (2004), Butler (2004), and the various books in the series *Best of Call Center Management Review* edited by Cleveland and Hash (2004a, 2004b)).

15.5 CONCLUSION

Survey research call centers grow increasingly complex as the needs of survey research and technology evolve. No longer are these simple outbound survey call shops. Rather, they are multichannel contact centers involved in a host of data collection support activities. In this chapter, we considered the key aspects involved in establishing a modern call center for survey research projects, focusing on planning, facilities, technology, and staffing. The main themes repeated throughout are the need for thorough planning and consultation, the interdependent nature of many of the decisions that must be made, and risk minimization and preparedness for change. Regardless of the frustrations and difficulties encountered in establishing a new center, it is a rewarding and satisfying experience watching the best ideas and vision of an organization be put into practice.

CHAPTER 16

CATI Sample Management Systems

Sue Ellen Hansen
University of Michigan, Michigan, USA

16.1 INTRODUCTION

Computer-assisted telephone interviewing (CATI) sample management has the potential to improve the scheduling of interviews, monitoring of sample dispositions, and supervision of interviewers, as well as data quality. In addition, CATI allows for much more complex questionnaires than that is possible and practical with paper and pencil telephone interviewing, including the use of tailored introductions and questions, complex routing, and randomization. However, the triple constraints of time, budget, and resources—in relation to quality—need to be considered in designing instruments and sample management strategies and systems for CATI surveys.

The primary challenge in telephone surveys is finding ways to minimize the effort it takes to make initial contact and to gain compliance in order to reduce costs and increase response rates, while also minimizing bias (Groves, 1989; Sangster, 2003). Optimal strategies likely differ across specific survey populations, types of survey, sample designs, sizes of sample, data collection periods, sizes of call centers and interviewing staffs, characteristics of interviewers, and characteristics of informants and respondents. All of these factors operate in combination and influence selection of CATI sample management approaches and how they may need to vary throughout the lifecycle of a survey. Thus, CATI sample management systems need to be flexible and have features that accommodate a wide range of sample management strategies.

Brick et al. (1996) provide a useful conceptual framework for managing a CATI survey, which considers *inputs* to the system (telephone numbers and information

Advances in Telephone Survey Methodology, Edited by James M. Lepkowski, Clyde Tucker, J. Michael Brick, Edith de Leeuw, Lilli Japec, Paul J. Lavrakas, Michael W. Link, and Roberta L. Sangster

about them), *processes* that can be manipulated (scheduling protocols and procedures, interviewer staffing and sample assignments, and data collection period), and *outcomes* (completed interviews, refusals, other contacts, and noncontacts). The features of a particular CATI sample management system determine how easy it is to manage a telephone survey in relation to such inputs, processes, and outcomes. This includes the accessibility of *paradata*[1] (Couper, 1998; Couper and Lyberg, 2005), which are critical to understanding survey processes and their impact on call attempt dispositions and data quality.

This chapter provides a review of literature on CATI sample management, and the remaining sections focus on meeting CATI sample design needs, the CATI survey lifecycle, features of a CATI sample management system, optimizing call scheduling and delivery, assessing interviewer performance, assessing CATI sample management system performance, and implications for future development. Where literature that directly addresses specific topics is scant, there is discussion of what CATI sample management systems capabilities are feasible and generally in use.

16.2 MEETING SAMPLE DESIGN NEEDS

CATI sample management systems must accommodate the sample design needs of surveys, which include managing, monitoring, and reporting on different sample designs. Designs generally can be classified as random digit dialing (RDD), list, and multiple frame and/or mixed mode, each of which has distinctive sample management requirements.

16.2.1 RDD Samples

RDD samples are a means of obviating the coverage problems of telephone directories, and may be described as surveys "in which households are randomly sampled within some geopolitical area using one of several random-digit techniques that generate telephone numbers" (AAPOR, 2005).[2] There has been a variety of approaches to generating random telephone numbers (Cooper, 1964; Glasser and Metzger, 1972; Hauck and Cox, 1974; Landon and Banks, 1977; Cummings, 1979; Groves, 1989; Casady and Lepkowski, 1991; Lavrakas, 1993), focused on efficiently identifying area code and telephone exchange combinations with a higher likelihood of having working residential numbers. In the 1970s and 1980s, the dominant approach to RDD was the Mitofsky–Waksberg method of generating random numbers to dial,

[1]Paradata are data about the survey process, at the global (e.g., response and coverage rates), individual record (e.g., flag for imputed data), and item (e.g., question time stamps and keystrokes) levels (Couper and Lyberg, 2005).

[2]In the United States as of November 2004, increasing local portability of numbers (LPN), through which consumers retain their local numbers when changing local exchange carriers, location or type of service, threatens control over the areas sampled in RDD surveys (cf. Lavrakas, 2004).

which uses two-stage sampling (Mitofsky, 1970 [cited by Casady and Lepkowski, 1991]; Waksberg, 1978; Lepkowski, 1988). The first stage is to select a random list of primary numbers to call. If a primary number reaches a household, then additional numbers are generated from the primary number by randomly regenerating the last two digits. Those numbers form the second stage cluster, up to a prespecified cluster size.

Surveys that use classic Mitofsky–Waksberg designs require sample management systems that have a means of randomly generating the secondary numbers, as well as mechanisms for stopping generation of numbers when the secondary sample cluster sizes are reached, and for replacing primary numbers that do not generate additional working numbers. They also need a means of reporting on the status of sample clusters.

By the late 1980s and early 1990s, list-assisted sample designs began to supplant Mitofsky–Waksberg designs, and they are now generally used for RDD.[3] Typically these are designs that select from a sampling frame organized into banks of 100 consecutive telephone numbers with the first eight digits in common (formed by telephone number area codes, prefixes, and the first two digits of suffixes), and in which there is at least one working residential number (Casady and Lepkowski, 1991, 1999; Potter et al., 1991; Tucker et al., 2001, 2002; Brick et al., 2003b; Williams et al., 2004). Frames of 100 banks of numbers can be screened to identify business and nonworking numbers (Casady and Lepkowski, 1999; Brick et al., 2003b). They can also be checked against electronic telephone directory or postal service lists to establish whether they are listed or to identify associated mailing addresses. One-hundred banks are selected in the first phase, stratified by listed status (banks with or without one or more listings, or high and low density listings) or address status, and then subsampled, selecting at a higher rate from the high density listing or mailing address stratum (Tucker et al., 2001; Brick et al., 2003b).

The U.S. telephone system has changed substantially since the beginning of the 1990s, which has made it increasingly difficult to identify working residential numbers in telephone samples (Tucker et al., 2002). This has led to increasing efforts to purge out-of-scope numbers and to identify cell phone numbers (Battaglia et al., 2005).

Sample management of list-assisted designs is more straightforward than for Mitofsky–Waksberg designs. Lists of sample numbers are created in advance and loaded into the sample management system, which ideally allows the release of randomized subsets of the sample (replicates), over the course of the survey lifecycle.

Mitofsky–Waksberg and list-assisted frames often may have associated demographic information at the telephone exchange or prefix level, with list-assisted frames also having listed and/or mailing address status. These data can be used for modeling propensity to make contact and gain compliance and for making post-data collection adjustments to survey estimates (Williams et al., 2004). Thus, it is important for sample management systems to make it easy to associate such data

[3]Although they did not use the term "list assisted," LeBailly and Lavrakas (1981) provide an early description of a list-assisted RDD sampling approach.

with generated telephone numbers, from initial loading of sample into the system through delivery and postprocessing of survey data.

16.2.2 List Samples

A second type of telephone survey sample frame is any list of telephone numbers for a specific survey population for which there are known sample persons. Examples of list samples include a telephone directory listing, a reverse directory listing used in small area sampling, a list of panel survey participants, members of professional organizations, and employees of businesses or educational institutions. Such lists may have a variety of data associated with the telephone numbers and respondent names, which CATI sample management systems can make available to the interviewer for verification and tracking purposes, and to facilitate interaction. For example, being able to display to the interviewer the name and other information about a panel respondent makes it easier to initiate and maintain interaction than that on an RDD survey.

Data associated with list samples could include information used by the call scheduler, including best time to call based on prior contact with respondents. For longitudinal surveys, this could be day of week and time of completion in the prior wave. Important features of CATI systems for management of list samples include means for the interviewer to update respondent information and track panel respondents, as well as means of loading respondent information collected in prior waves of panel surveys.[4]

16.2.3 Multiple Frame Samples and Mixed Mode Surveys

Telephone survey designs may use multiple frames and telephone surveys may be used in conjunction with other modes (see also Brick and Lepkowski, 2007, Chapter 7 in this volume). Multiple frames and modes may be used to improve screening efficiency, response rates, and coverage, for example, combining RDD samples with telephone directory listings to allow for mailings and the use of incentives, combining RDD and list frames, or combining area probability samples and telephone samples (Casady et al., 1981; Groves and Lepkowski, 1985; Traugott et al., 1987; Waksberg et al., 1997; Srinath et al., 2004; Link et al., 2004; Mitofsky et al., 2005). Multiple frames may also be used to accommodate longitudinal designs for example, a rotating panel in which each wave has a new RDD sample and a list of respondents from an RDD sample in a prior wave.

Mixed mode surveys may involve telephone precontacts in mail or Web selfadministered surveys, in order to allow interviewer facilitation of screening and collection of contact information for surveys in other modes, or to recruit respondents (Van Liere et al., 1991; Ramirez, 1997; Manfreda et al., 2002). They may also involve telephone follow-up in mail surveys to increase response (Paxson et al., 1995;

[4]As discussed below, interviewers would not be able to review such information prior to a call if the sample management system used certain types of auto/predictive dialers.

Petrie et al., 1997). Some survey designs use multiple modes to provide response flexibility and reduce respondent burden, for example, making it possible to respond to the Current Employment Statistics survey through CATI, touchtone data entry, electronic data interchange, fax, the Web, or telephone interactive voice recognition (Rosen and O'Connell, 1997).

Increasing use of multiple frames and mixed mode surveys is driven by the need to reduce costs and nonresponse.[5] Such designs require sample management systems that facilitate cost-effective automated, manual, and self-selected assignment of respondents to modes, movement of cases across modes, monitoring and tracking of sample according to sample type and mode, and reporting on both sample type or mode and total sample progress. Mixed-mode survey designs are not new, dating back to the 1960s (Dillman and Tarnai, 1988), but it is only in the last decade through advances in and the lowering cost of technology that it has become feasible to develop fully integrated sample management systems to meet the needs of such designs.

16.3 THE CATI SURVEY LIFECYCLE

CATI RDD studies may be characterized by a survey lifecycle with three distinct phases (Buckley et al., 1998):

- An *initial phase* of data collection in which a very large proportion of fresh cases have not been dialed, and in which it is desirable to make call attempts at a very high rate in order to resolve the statuses of telephone numbers (whether working and eligible) as early in the data collection period as possible.
- A *middle phase*, the majority of the data collection period, in which there is a mix of fresh sample, worked but still not contacted sample, nonfinal contacts, and finalized sample.
- A *final phase*, with predominantly worked but nonfinal cases (partially completed interviews, initial refusals, appointments, and a variety of noncontact outcomes).

The typical survey lifecycle becomes particularly problematic on very short studies, and on two-stage Mitofsky–Waksberg RDD designs, in which replacement numbers could receive their first call attempts at the middle or end of data collection. It is generally thought that a longer study period, all else being equal, yields higher response rates.

To maximize a survey's response rates, organizations could staff differentially across these phases, with a much larger number of nonspecialized interviewers

[5]Although mixed-mode surveys have been shown to improve response rates, they may also introduce bias. In a mixed-mode experiment, Link and Mokdad (2004) found that responses to a Web survey with telephone follow-up differed from those of a baseline CATI survey, whereas the responses to a mail survey with telephone follow-up did not. They recommend testing mixed-mode approaches against an existing standard in order to detect potential bias.

at the beginning, a smaller number of more specialized refusal conversion staff at the end, and a mix of staff at a level in between during the main field period. However, the potential for increased survey costs and the typical need to have a level mix of interviewers throughout data collection often generally operate against this approach.

An alternative to changing staffing is to change call scheduling optimization strategies, and depending on strategies used, to vary them across lifecycle phases. On longer studies with very large sample sizes, research shows that it is possible to maintain even staffing and workloads by releasing sample in replicates or batches. This allows monitoring the sample and varying the size of sample releases over time, in order to keep a controlled amount of fresh sample. Buckley et al. (1998) report on a successful use of this strategy, with little detrimental effect on response rates.

In contrast, Sebold (1988) reports on a study to assess the effect of doubling a 2-week study period to 4 weeks, in an attempt to resolve the statuses of more of the remaining unanswered numbers. Although response rates increased by 3 percentage points (from 77% to 80%), refusal rates also increased by 3 percentage points (from 10% to 13%), as many of the noncontact cases resolved to refusals. In addition, the longer study period increased costs.

A later section reviews literature on optimizing call scheduling, but these examples reinforce the notion that there is no one-size-fits-all optimization strategy, and that methods to manage sample across the survey lifecycle should be tailored to the specific requirements of a survey design. Thus, CATI sample management systems need to be flexible, support the needs of different survey designs and study periods, and allow altering survey design during the survey lifecycle.

16.4 FEATURES OF A CATI SAMPLE MANAGEMENT SYSTEM

The key requirements of a CATI sample management system (Table 16.1) parallel the requirements of managing telephone survey samples in a paper and pencil environment (Lavrakas, 1993). They focus on optimal scheduling of contacts to maximize response rates and minimize costs, and managing sample in relation to sample size, staffing levels, and data collection period length. This requires efficient mechanisms for assigning number for interviewers to call, recording dispositions for each contact attempt, collecting histories of call dispositions, and monitoring and reporting on interviewer and sample performance (Ferrari et al., 1984; Sharp and Palit, 1988; Weeks, 1988; Couper et al., 1992; Edwards et al., 1998).

The typical CATI sample management system's *interviewer interface* usually allows the interviewer to select a survey to work on and request the next case to call. Because there are situations that require direct access to specific cases, typically there is an option to select a particular case for calling. Flexible call scheduling systems allow determining the level of control the interviewer has over selection of survey(s) and/or cases to call, which could vary by types of surveys, experience and attributes of interviewers, and staffing levels. The interface generally also provides a means of recording information about a call, including disposition and notes for each contact

Table 16.1 Key Features of a CATI Sample Management System

- Interviewer functions*
  - Select survey(s) to work on, if permitted
  - Review current sample assignments, e.g., soft and hard appointments
  - Select telephone number to call, if permitted
  - Review prior call history
  - Review and record tracking information
  - Record call information
    - Disposition
    - Contact notes
    - Case notes
- Supervisory and management functions
  - Set survey and interviewing parameters
    - Survey dates, restricted calling periods
    - Time slots or parts of day to call
    - Number and types of calling queues
    - Maximum number of attempts on noncontact cases
    - Interviewer attributes and calling groups
  - Review case dispositions and histories, filtered by sample type, disposition, time period, interviewer, etc.
  - Review and record tracking information
  - Reassign cases to interviewers manually
  - Set final outcome codes manually
  - Create dynamic ad hoc reports
  - Modify scheduling algorithms
- Automatic call scheduling
- Autodialing
  - Predictive
  - Nonpredictive
- Time zone and daylight savings adjustments
- Algorithm-based scheduling of specific dispositions
  - Busy signals
  - Ring no answers
  - Answering machines
  - Refusals
- Control over delivery of hard appointments
- Delivery of cases according to interviewer attributes
  - Experience
  - Refusal conversion status
  - Fluency in other languages
  - Interviewer demographics
- Automatic assignment of final dispositions
- Optional manual selection of cases
- Reporting systems
  - Study, case, and interviewer status reports
  - Shift reports
  - Cost control reports
  - Case and call record listings
  - Trend, outlier, and other diagnostic reports

*Some of these features are not possible with predictive dialers (see Section 16.4.1).

attempt, as well as case notes; a means of reviewing the prior call history and reviewing and recording respondent tracking information (for a panel survey case); and a means of reviewing current sample assignments, such as scheduled appointments.

Supervisory and management functions include the ability to set survey and interviewing parameters (e.g., survey dates, time slots or parts of days to call, number and types of calling queues, number of attempts, interviewer attributes, and calling algorithms); review case dispositions and histories; review and record tracking information; override the scheduler and manually (re)assign specific cases to interviewers; and set final disposition codes. In a multilingual environment, the system also needs to facilitate changing the language displayed. While CATI sample management systems typically provide standard reports (see "Reporting Systems" below), flexible systems also allow supervisors and managers to create as needed ad hoc reports that meet selected criteria, such as survey, call center, team leader, interviewer group, shift, survey dates, or life cycle period.

Beyond interviewer and supervisory functions, the key components of a CATI sample management system are automated call scheduling and reporting systems.

16.4.1 Automatic Call Scheduling

The basic feature of most current CATI sample management systems is that they deliver cases to interviewers to call, using an automated call scheduler, which uses protocols or rules for sequential handling of calls to cases. The scheduler determines the order in which numbers are called, assigns cases to interviewers, records the disposition of each call, and assigns final dispositions based on protocols such as maximum number of contact attempts.

Call scheduling generally includes some form of autodialing (Edwards et al., 1998). This could be simple autodialing, in which the interviewer's computer modem dials the telephone number at the interviewer's command, which is more efficient and accurate than manual ("finger") dialing. It also could involve automatic detection and disposition coding of certain telephone signals, such as busy signals, modems, fax machines, answering machines, etc. Such autodetection is extended in predictive autodialing, in which telephone numbers are dialed automatically by the system and control is transferred to the next available interviewer only after a person answers.

Predictive systems use algorithms that predict availability of interviewers to handle calls based on variables such as number of available interviewers and length of interview. If an interviewer is not available, the call is abandoned. If the abandonment rate is set too high (a lower probability of an available interviewer), a greater proportion of answered calls will be "dropped" by the dialer before an interviewer can start to interact with whoever answered. This greatly increases the likelihood of alienating a household thereby increasing the likelihood of an ultimate refusal (especially if the household has caller ID showing the survey organization name).[6]

[6]Some unpublished work Shuttles and Lavrakas (May 2002) did with Nielsen's predictive dialer call history data showed that even one abandonment raised the likelihood of an eventual refusal, and the more abandonments the greater the likelihood of an eventual refusal (Lavrakas, 2006).

In addition, such systems remove interviewer control of when a case is called and the ability to review case histories in advance of dialing. The remaining discussion of CATI sample management system features assumes a *nonpredictive dialer environment.*

The typical functions of an automated call scheduler include

- Time zone and daylight savings time adjustments
- Algorithm-based scheduling of specific dispositions, such as busy signals, ring no answers, answering machines, and refusals, designed to optimize likelihood of making contact and gaining compliance (see later section on optimizing call scheduling)
- Control over delivery of hard appointments, for example, if necessary according to the study design, the ability to deliver appointments early and/or late in order to increase the likelihood of delivering to the interviewer who set the appointment
- Ability to assign cases based on interviewer attributes or characteristics, such as experience, refusal converter status, language skills, and demographics
- Automatic assignment of final dispositions, for example, assigning final non-contact status after a maximum number of contact attempts
- Control over manual or automatic delivery of cases, across surveys and interviewers.

Flexible systems that use call scheduling algorithms may use complex rules for prioritizing calls, dynamically changing them with each attempt, based on survey design and lifecycle parameters. Edwards et al. (1998) point out, however, that "even the most sophisticated automated call scheduling system can achieve its potential only when it operates in concert with [well-managed] interviewer staffing" (p. 306). In the best of circumstances, a staff scheduling system would work in conjunction with the automated call scheduler, so that information about staff scheduled is maintained in one system interface and is available to both staff and call schedulers. This would also facilitate interviewer monitoring discussed below.

16.4.2 Reporting Systems

Most CATI sample management systems provide reports that facilitate supervision and management of staff in relation to sample (Weeks, 1988; Edwards et al., 1998). These include reports that summarize the current status of all cases in the sample, interviewer productivity and performance, and study costs in terms of interviewer time per interview. Some systems list cases of specified dispositions or states (for example, scheduled and missed appointments) to assist in supervisor intervention in case assignment. Desired features of reporting systems are the availability of both standard reports and customized reports, in order to monitor interviewer sample management and system performance (discussed below), and flexibility over when they are produced (daily, hourly, or on demand; Edwards et al., 1998).

Types of routine reports available in most systems are (Weeks, 1988 and 1992; Edwards et al., 1998)

- *Study status reports*, showing cumulative progress on the study, with average number of calls per case, and overall response, cooperation, efficiency, abandonment (if a predictive dialer environment), and other rates
- *Case-level status reports*, including number of cases in key outcome categories for final and active cases; for review of cumulative and current day results, as well as net change, in order to assess current status of sample and effect of the current day's production
- *Interviewer-level status reports*, for evaluating interviewer sample performance
- *Cost control reports*, with hours, cost, and communications charges per interview, and total cost per interview
- *Case listings*, in particular scheduling queues or specific call outcomes
- *Call record listings*, for offline review or for creating a paper coversheet
- *Trend reports*, showing key progress or cost indicators over time, such as response or efficiency rates, and hours per interview
- *Outlier reports*, such as duration and exception reports, which identify cases that have remained in a status beyond a prescribed duration and cases which have not followed a prescribed processing path, respectively
- *Other diagnostic reports*, for identifying problems with scheduling, such as missed appointments

If interviewer monitoring or other quality control systems are in place (see the following section), interviewer quality control reports are generally also available; (see also Steve et al. chapter in this book). For example, Mudryk et al. (1996) reported on a demerit-based interviewer quality control program that monitored interviewer behavior. Weekly reports included average demerit rates by interviewer, demerits by error categories, and statistical control charts for tracking errors. Sun (1996) reported on results of using an alternative approach to statistical process control for tracking interviewer errors, with different control charts for process control.

Web-based reporting systems make it possible for clients, project managers, and supervisors to review reports remotely (e.g., Dennis et al., 1998). As discussed below, interviewer and CATI system performance can be monitored and assessed more efficiently through the use of such Web-based systems, especially if they provide real-time data and reports that can be tailored to study, survey population, and survey lifecycle requirements (see Tarnai and Moore, 2007, Chapter 17 in this volume and Steve et al., 2007, Chapter 19 in this volume).

16.4.3 Additional Features

Additional features of CATI sample management systems include the ability to preload sample frame or other data into the sample management system and move data between the computer-assisted survey instrument and the sample management and reporting systems, and to monitor or record interviews for interviewer performance measurement.

Monitoring involves observation of interviewer behavior, which may be audio only, that is, listening to interviewer-respondent interaction, or audio plus video, which includes watching what the interviewer sees on the computer screen while listening to interviewer-respondent interaction. Typically this has been real-time, while the interview occurs, but recent advancements in technology make it easier to record interviews or portions of interviews and review them later (Mudryk et al., 1996; Biemer et al., 2000; Groves et al., 2007, Chapter 18 in this volume).

Most monitoring involves the use of a form of online data entry program for recording basic interviewer behaviors related to making survey introductions (voice quality and pace) and asking questions (question wording and probing behavior), and may include other characteristics of interviewers related to gaining cooperation (Cannell and Oksenberg, 1988; Chapman and Weinstein, 1990; Couper et al., 1992; Lavrakas, 1993; Groves et al., 2007, Chapter 18 in this volume; Steve et al., 2007, Chapter 19 in this volume).

Although they generally do not (Couper et al., 1992), an ideal CATI sample management system would allow for systematic random selection of interviewers and segments of the interview or activities to monitor, with differential selection rates across interviewers based on experience level and past performance, and across shifts, taking into consideration the number of interviewers working during each shift (Couper et al., 1992; Mudryk et al., 1996). This permits making generalizations about data collected and providing more systematic feedback to interviewers.

16.5 OPTIMIZING CALL SCHEDULING AND DELIVERY

As previously indicated, the key challenge in telephone surveys is finding ways to minimize the effort it takes to complete an interview, in order to reduce costs and increase response rates without introducing bias. Most research on CATI sample management and call scheduling is based on analysis of call history data from a variety of surveys (primarily RDD, but also list samples; for example see Weeks et al., 1987). It has focused on how to optimize call scheduling and delivery of sample numbers, and tends to fall into three categories: (1) best times to call; (2) refusal conversion protocols; and (3) noncontact scheduling protocols.

16.5.1 Best Times to Call

A feature of most CATI sample management systems is the ability to customize time slots or parts of day for scheduling calls, because optimal times to call vary across survey populations (Weeks, 1988). For example, for the general population, the best times to complete interviews are weekday evenings and weekends, but especially Sunday nights (Weeks et al., 1980; Vigderhous, 1981; Weeks, 1987; Weeks et al., 1987; Kulka and Weeks, 1988; Lavrakas, 1993; Ahmed and Kalsbeek, 1998). This is true not only for first call attempts, but also for second and third call attempts (Weeks et al., 1987; Kulka and Weeks, 1988; Brick et al., 1996a). In addition, contact rates may be lower if first and second calls are made during the same time slot

(Kulka and Weeks, 1988; Stokes and Greenberg, 1990), although Brick et al. (1996) did not find this to be true.

While there seems to be consensus that weekday evenings are good times to call, weekday evening calls also may be more likely to result in callbacks, refusals, and breakoffs (Brick et al., 1996a; Massey et al., 1996). Ahmed and Kalsbeek (1998) found that there were fewer refusals on Fridays, although also noted that in their study fewer calls were made on Fridays.

In contrast to surveys of the general population, weekdays would be best for calling businesses, weekdays as well as evenings and weekends would be best for samples of the elderly and others not confined to a traditional 40-hour work week, and it may be desirable in panel surveys to base best times to call on call history information or respondent preferences from prior waves. Call schedulers need to accommodate such differences and facilitate setting different time slot parameters and best times to call for different survey populations.

Some literature distinguishes between resolving eligibility, making contact, and completing interviews (e.g., Greenberg and Stokes, 1990; Massey et al., 1996; Dennis et al. 1999; Sangster and Meekins, 2004), noting differences in best times among the three outcomes. This suggests that optimal calling patterns should be developed with all three outcomes in mind; for instance, sacrificing some early efficiency in making contact in order to resolve eligibility (e.g., identifying business numbers in an RDD survey). This could be done by making at least one call on a weekday when using a three-call protocol. The inefficiency may be offset by increasing the likelihood of making contact on later calls, due to the higher percentage of eligible numbers. Noncontact scheduling protocols are discussed further below.

16.5.2 Refusal Conversion Protocols

Refusal conversion efforts can result in a significant improvement in response rates (Lind et al., 1998; Srinath et al., 2001). However, they come at a cost, and may be differentially effective depending on when follow-up calls are made. Call schedulers use refusal conversion protocols to determine how many days to wait before making an additional call attempt on a refusal, and how many attempts to make before coding a case a final refusal. Such protocols may have an impact on nonresponse bias (Ahmed and Kalsbeek, 1998; Stec et al., 1999).

Triplett et al. (2001) found that organizations varied widely in "rules of thumb" for optimal periods to hold refusals before attempting conversion. Research results on optimal waiting periods vary, which may reflect differences in survey designs. Odom and Kalsbeek (1999) reported higher likelihood of refusal conversion for each additional day the case was held back from calling, but did not specify an optimal period. In an experimental design comparing 1-, 2-, and 3-week refusal holding periods, Edwards et al. (2004) found no differences in refusal conversion rates across the three periods. Stec et al., (2004) found that delaying at least 72 hours before attempting a refusal conversion leads to better success. Triplett et al. (2001) examined data on refusal conversion rates from nine national CATI studies with approximately 5400 initial refusals, and found that while the optimal period for converting respondent refusals was 2 weeks, the optimal period for converting informant refusals was 1 week or less. They also found that converting males is easier than converting females, while it is harder to convert refusals in the Northeast, requiring on average five additional attempts. These findings suggest that refusal conversion protocols might be further optimized by utilizing respondent and telephone number characteristics.

16.5.3 Noncontact Scheduling Protocols

Noncontact cases are cases in which no previous contact with a person has been made. These include untouched or unworked sample, and cases for which all prior call attempt outcomes resulted in no contact, such as busy signals, ring no answers, and answering machines. Generally they are scheduled and delivered automatically to interviewers from the sample pool of numbers. In an effort to increase the likelihood of making contact early in the survey period, noncontact scheduling or calling protocols are rules used to determine the best days and times to call, the order in which calls are made to specific time slots, the time to wait between calls, and the maximum number of call attempts. Approaches to noncontact scheduling include

- Using one or more algorithms for scattering calls among time slots (Weeks, 1988; Edwards et al., 1998; Cunningham et al., 2003)
- Employing prespecified or static algorithms for sequencing or scattering calls across shifts or time slots (Weeks, 1988; Edwards et al., 1998)
- Conditioning delivery of a number to call at a specific time based on outcomes of prior contact attempts (Kulka and Weeks, 1988; Weeks, 1988; Groves, 1989; Brick et al., 1996a; Massey et al., 1996; Dennis et al., 1999; Stec et al., 2004)
- Assigning priority scores for all numbers based on call histories, each time the scheduler runs (Weeks, 1988; Greenberg and Stokes, 1990), for example, assigning a higher priority to answering machines over ring no answers.

Some literature reports on studies using call history and other data in multivariate analyses to develop predictive models for making contact (Greenberg and Stokes, 1990; Stokes and Greenberg, 1990; Brick et al., 1996a; Reedman and Robinson, 1997; Dennis et al., 1999; Sangster and Meekins, 2004). Stokes and Greenberg (1990) describe how a Markov decision process could use survey, call history, survey population, and shift resources data to determine actions on subsequent calls, in order to minimize number of calls required to make contact, based on transition probabilities. The approach is designed to minimize total contacts over the study period, as opposed to increasing the probability of making contact on the next call. However, it could be used in conjunction with priority scores to improve likelihood of contact on the next call (Greenberg and Stokes, 1990; Reedman and Robinson, 1997). In the literature reviewed, there was one reference to the intention of utilizing Markov decision processes in actual call scheduling protocols (Sangster and Meekins, 2004), although there was not evidence that it had actually successfully been used.

Based on multivariate analyses conducted to evaluate predictive models using call history data, information about calling protocols (such as lag between calls), workforce availability, data collection period, and demographics, Brick et al. (1996b) found that some of their results differed from those of prior research that found that contact rates for second calls were not as high as for first calls (Kulka and Weeks, 1988; Stokes and Greenberg, 1990) and that time between calls was a significant factor (Stokes and Greenberg, 1990). They concluded that such differences in findings are likely due to differences in survey designs across studies, and recommend that noncontact calling protocols take into account factors related to processes (scheduling algorithms, staff availability, data collection period, and interviewer characteristics), as well as inputs to those processes (telephone numbers and their characteristics, and call history).

Through the analysis of call history data, studies have attempted to identify optimal sequential calling patterns for increasing the likelihood of contact with fewer total attempts, generally focused on three-call patterns and conditional probabilities of making contact based on prior calls. Separating calling periods into optimal time periods (weekday evenings and weekends) and nonoptimal time periods (weekday mornings and afternoons), Kulka and Weeks (1988) found that when mixing optimal and nonoptimal time slots, it is better to call during optimal times earlier in the sequence, and that calling in an optimal time period is less productive if it is the same time period of a prior call attempt. They identified the best three-call patterns that could be used in conditional probabilities scheduling algorithms. In a similar analysis, looking at both three-call and five-call patterns, Massey et al. (1996) found that the best five calling patterns reached 90% of households, which were primarily a mix of weeknight and weekend calls. While weekdays were not considered optimal, they recommended including one weekday time slot in calling patterns to facilitate identification of businesses.

Since costs are associated with each call made, attempts have been made to determine the optimal number of calls to make on a previously noncontacted number, balancing costs and risk of nonresponse error. Harpuder and Stec (1999) used summed squared standardized z-scores for key survey statistics and demographics to estimate the potential bias at each cutoff point. In relation to costs, they found that as few as four callbacks minimized error, but recommended a more conservative six to seven callbacks, given the use of a summary estimate of bias across survey statistics. In an analysis of call history, cost, and survey data, Srinath et al. (2001) examined the impact on response rates of cutting off attempts at 5, 8, 10, 12, and 15 calls, and found 12 call attempts to be optimal, when balancing costs against mean squared error.

The optimal number of attempts may vary across different types of prior noncontact outcomes (busy numbers, answering machines, and ring no answers). For example, in analysis of the call history data for several extremely large national RDD studies, Stec et al. (2004) found very low likelihood of making contact with telephone numbers with more than five prior busies or no answers, but that it was still efficient to make up to 14 contact attempts on answering machine numbers with no more than two prior busies or no answers. This suggests that it is desirable for call schedulers to use different call attempt cutoffs for different noncontact types, based on analysis of call history data for the same or a similar survey.

In some circumstances, call scheduling might be optimized further if one were to go beyond the basic distinctions between refusals and nonrefusals, and contacts and noncontacts. For example, Murphy et al. (2003) analyzed call history data, focusing on bias introduced by following or not following types of numbers, dividing them into promising (at least one nonnegative contact, including a soft refusal), unpromising (in which a first call resulted in a hard refusal or unknown household status), and base numbers (completed on or before the seventh call). They found that (1) *unpromising* cases tended to represent younger adults, single-person households, and healthier respondents, but once contacted had higher cooperation rates; (2) cost savings outweighed bias introduced by not following *unpromising* cases beyond seven calls; (3) bias outweighed cost savings introduced by not following *promising* calls beyond seven calls; (4) only a marginal cost increase resulted from following *promising* cases to 14 calls; and (5) a seven call limit on all cases would have resulted in significant bias.

16.6 ASSESSING INTERVIEWER PERFORMANCE

Advances in technology have made it possible to easily evaluate the status of interviewers' production activity, particularly if real-time data are available to reporting systems. In addition, accessibility of reporting systems becomes easier if such systems are Web-based (see, for example, Dennis et al., 1998).

Once data are easily accessible, reporting options expand greatly, and can include trend reports, statistical process control charts, and outlier reports. Survey methods research has begun to focus on quality improvement in surveys and the use of paradata and quality control charts to assess survey quality, including interviewer efficiency and performance (Mudryk et al., 1996; Sun, 1996; Reed and Reed, 1997; Couper, 1998a; Dobbs et al., 1998; Lyberg et al., 1998; Biemer and Lyberg, 2003; Couper and Lyberg, 2005). Some measures of interviewer performance are hours per interview, hours or minutes not spent dialing or in an interview, response rates, and cooperation rates (see Tarnai and Moore, 2007, Chapter 17 in this volume). Although literature does not provide examples of systems or reports used to track such measures for process quality improvement (as opposed to monitoring interviewer behavior; see Steve et al., 2007, Chapter 19, and Groves et al., 2007, Chapter 18, in this volume), it is assumed that most CATI sample management systems make this possible. Accessible data and dynamic reporting systems also make it possible to identify outliers and track trends in interviewer performance, and to generate ad hoc reports as needed.

16.7 ASSESSING CATI SAMPLE MANAGEMENT SYSTEM PERFORMANCE

Easily accessible data also make it possible to assess overall sample management system performance. For example, Buckley et al. (1998) report on a system that produces a number of reports and charts that help monitor 78 RDD samples simultaneously. They include charts that show cumulative completed interviews; targeted versus completed interviews in each of the samples; number of active versus needed cases for each sample; summary over 2-week periods in a quarter of scheduled appointments in relation to scheduled future work (appointments and initial refusals); and number of completes across samples at the end of the quarter.

Systems that allow real time access to call history and other paradata permit close monitoring of sample status. They also facilitate adjusting scheduling protocols as needed throughout the course of the lifecycle, when combined with the ability to easily change scheduling parameters.

Some CATI sample management systems have made it possible to dynamically evaluate the current status of a study sample, in order to identify problems that need correction. For example, Fig. 16.1 shows an example of a Statistics Canada report on average calls per case to convert refusals (by region).[7] As shown by the selection criteria, the system can also produce reports on number of accesses per case, times to complete, times spent on cases, and no contacts per case. One can imagine other measures of performance such as interviewer monitoring rates and completed interviews by shift.

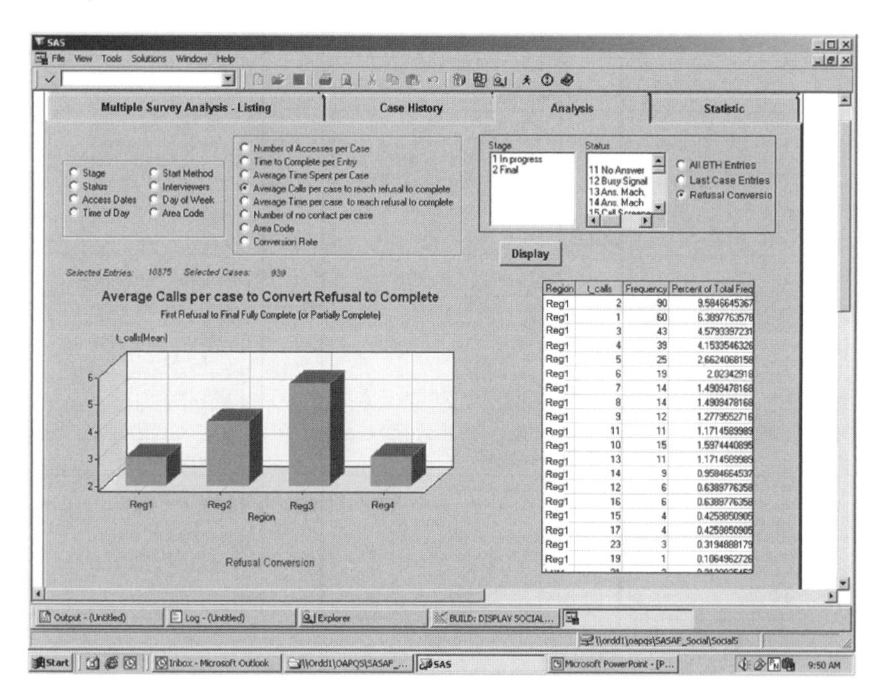

Figure 16.1 Statistics Canada report: average calls per case to convert refusals (by region). Hansen, S.E., *CATI Sample Management Systems*; © Statistics Canada, 2006.

[7]The author thanks Jacqueline Mayda for providing this example, adopted from a report produced by the Operations Research and Development Division, Statistics Canada.

16.8 IMPLICATIONS FOR FUTURE DEVELOPMENT

Considering the processes involved in conducting a CATI survey, the features of a CATI sample management system and its users, one could reframe the Brick et al. (1996b) flow of a CATI telephone survey (a series of inputs, processes, and outcomes) as inputs, actions, and outcomes (Fig. 16.2).[8] Key actors in CATI sample management are the call scheduler, interviewers, supervisors, and those who monitor interviewer performance and interview quality. Inputs to the system include study design parameters, the telephone sample and sample frame information, staffing information, and call history data. Outputs are call outcomes and indicators of study and interviewer progress, interviewer performance and interview quality, and system performance. Inputs drive or inform actions, and some outputs become new inputs to further action or influence actions directly.

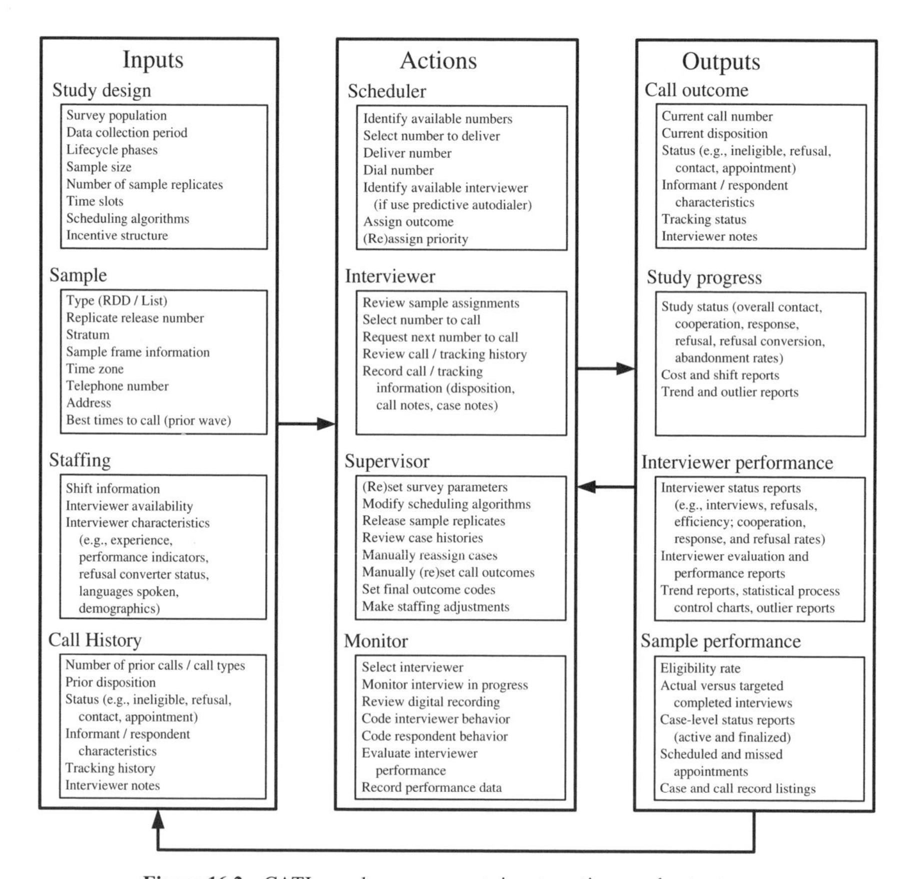

Figure 16.2 CATI sample management: inputs, actions, and outputs.

[8]The author thanks several University of Michigan colleagues—Grant Benson, Stephanie Charcoal, Gina-Qian Cheung, Patty Maher, and Vivienne Outlaw—for thoughtful input and discussion of this framework. Inputs, actions, and outputs listed are meant to be representative, not exhaustive.

Research on survey organizations (Edwards et al., 1998; Haggerty et al., 1999 [cited by Couper et al., 1992]; Steve et al., 2007, Chapter 19 in this volume) indicates that not every CATI sample management system has all the features and capabilities of systems discussed in this chapter. However, effective CATI sample management requires most if not all of the features described, ideally with systems designed to smoothly support the actions, and facilitate the inputs and outputs listed in Fig. 16.1. Thus, it requires systems very different from those in use in the early days of CATI and from many currently in use.

In addition, CATI sample management strategies and requirements are rapidly changing due to current trends in declining telephone survey response rates, increasing costs of surveys, increasing availability and use of paradata for optimizing scheduling and monitoring and managing samples, technological advances (such as web-based reporting, digital recording, voice over Internet protocol [VoIP]), and increasing use of mixed mode survey designs.

As a result of these changes, the survey industry has begun to recognize the need for CATI sample management to be tied to or part of overarching systems that manage and report on sample and survey processes across multiple modes and multiple surveys, with the ability to seamlessly move cases among modes, sites, and work groups during data collection. This has become more feasible with advances such as VoIP that make decentralized CATI more cost effective. CATI sample management systems will need to evolve to accommodate mixed-mode sample management and to use technologies that facilitate monitoring, assessing, and improving mixed-mode sample performance.

The current ease of digitally recording segments of interviews makes it possible to have cost-effective systems for CATI interview quality assurance and interviewer performance evaluation (see Tarnai and Moore, 2007, Chapter 17 and Groves et al., 2007, Chapter 18 in this volume). As new sample management systems are developed, they are increasingly likely to integrate recording of interviews with CATI and mixed-mode sample management, and to standardize approaches to recording across modes.

The availability in sample management systems of automatically collected rich paradata facilitates responsive or adaptive survey design (for example, see Heeringa and Groves 2004; Groves and Heeringa, 2006), which in turn is likely to place increasing demands on sample management systems to make paradata easily accessible, to monitor processes for quality improvement, and to make it easier to implement phased design changes at any point in the survey lifecycle. Dynamic, Web-based reporting systems such as those described have begun to meet these demands.

Finally, a focus of much of the literature reviewed has been on optimal call scheduling, although it is not always clear that results and findings have led to optimizing call scheduling in practice, or when they have, if implementation has affected CATI sample management for more than one survey or particular type of survey. More research is needed—research that uses analysis of the rich amount of available call history and other paradata to inform experimental implementation of sample management strategies for particular survey populations, survey types, sample designs, sample sizes, and data collection periods—in order to demonstrate that specific protocols do or do not lead to significantly lower costs and/or increased response rates for a specific survey design. In each case, this should include analysis of risks of nonresponse bias in relation to effort and costs.

Such research could lead to the development of improved sets of CATI call scheduling protocols and flexible sample management systems that would meet the needs of a broad range of survey designs in the coming decades.

CHAPTER 17

Measuring and Improving Telephone Interviewer Performance and Productivity[1]

John Tarnai and Danna L. Moore
Washington State University, USA

17.1 INTRODUCTION

This chapter describes how telephone interviewer performance is measured in survey research centers, and how the performance of telephone interviewers can be improved with appropriate training. We are concerned with what telephone interviewers actually do during a shift, how to assess their productivity, how this information is communicated to interviewers, and what kinds of performance can we reasonably expect from our interviewers.

Measuring interviewer productivity is important for several reasons. First, interviewer wages are often the single largest expense in a telephone survey budget. So anything that improves interviewer productivity will have an impact on the survey budget. Second, knowledge of productivity is essential to survey cost estimating. Developing a realistic estimate of survey costs requires making realistic assumptions about how many completed interviews, refusals, noncontacts, ineligibles, and callbacks are to be expected. Third, the consequences of incorrectly estimating interviewer productivity can be quite serious. Inaccurate estimates of interviewer

[1]Opinions expressed in this chapter are those of the authors and do not represent those of Washington State University. We would like to thank all of the people who participated in the Internet survey of organizations, and all those who willingly provided us with copies of their training materials. We also want to thank Jim Lepkowski and Clyde Tucker for writing the cover letter and endorsing the survey.

Advances in Telephone Survey Methodology, Edited by James M. Lepkowski, Clyde Tucker, J. Michael Brick, Edith de Leeuw, Lilli Japec, Paul J. Lavrakas, Michael W. Link, and Roberta L. Sangster

productivity can produce survey budgets that are too low or too high, creating financial problems for the survey center. They can also lead to missing project deadlines and/or survey goals for completed interviews. Fourth, interviewer productivity may be used as a basis for merit pay that is given to the most productive interviewers in an organization. Fifth, interviewer productivity is often the basis for rewarding interviewers if they are performing well, or sending them to retraining or letting them go if they are performing poorly. Interviewers are typically evaluated on their performance on a survey. Those performing below average or below some standard are then sent for further training in interviewing or persuasion techniques. Interviewers who cannot perform at some minimum standard are either reassigned to other work or their employment is terminated, and they are let go. Sixth, knowledge of interviewer productivity is helpful for training and retraining of interviewers, and for communicating performance expectations that are grounded in reality. Last, information about the productivity of telephone interviewers is essential for planning and scheduling the number of interviewers needed for fielding a telephone survey.

The first part of this chapter summarizes what the survey literature says about interviewer performance and training, and describes the kinds of productivity measurements that are used to evaluate interviewer performance, and how this information is used to train or retrain interviewers and improve their performance. The next section describes how interviewer training is used to communicate performance and productivity expectations to telephone interviewers, and describes the technique of rapid interviewer training to quickly bring interviewers in line with performance expectations. This is followed by a summary of the results of a survey of organizations that conduct telephone interviews and the kinds of productivity measures they collect to evaluate their interviewers and use in interviewer training and retraining. The chapter concludes with some recommendations for measuring the performance of telephone interviewers and improving training of telephone interviewers.

17.2 MEASURING INTERVIEWER PRODUCTIVITY

There is relatively little in the survey research literature about how to manage telephone interviewers for productivity, or about the kinds of performance measures that are collected on interviewer performance. This may reflect a prevailing view that managing telephone interviewers is more of an administrative concern rather than a research issue. However, an awareness of the importance of managing interviewer productivity is apparent in the literature. More than 25 years ago, Dillman (1978) noted the importance of interviewer productivity: "The number of interviewers needed for later sessions and the number of sessions that are needed depend on how fast interviews are completed. An alertness to the length of time it is taking to complete interviews and the proportion of refusals and no answers is essential for making adjustments in the schedule for later interviewing sessions" (p. 278). Interviewer productivity tends to be mentioned, if it is at all, in discussions of the most productive times to conduct interviews. For example, Frey (1983) suggests that "most calling takes place between 5:30 and 9:00 pm and within that the hour from

6:00 to 7:00 ranks the highest for potential completions" (p. 164). He also remarks that "when scheduling interviews ... the project director should try to have the most interviewers available at times when the probability of nonresponse is small and the likelihood of completion great" (p. 161) and "a project director would thus be wise to keep his most productive people working the longest" (p. 164). These and similar statements suggest that survey researchers are aware that interviewer productivity varies depending on a number of factors and that productivity is important to measure. Lavrakas (1993) proposes that "a basic measure of interviewer productivity is to compare the number of properly completed interviews attained per interviewing session with the number of refusals (and partials) the interviewer experiences" (p. 137). He suggests that a reasonable goal for which interviewers should be trained is to achieve at least four completions for every one refusal or partial. Lavrakas (1993) also suggests incorporating productivity measures into interviewer pay rates, such that one third of their hourly rate is based on their productivity.

17.2.1 Interviewer Productivity

In the research literature, there has been greater emphasis on evaluating the effects of interviewers on response bias and measurement error, which, while not the focus of this chapter, also suggests a need for measuring interviewer productivity. Interviewers who are more productive will have a greater influence on the survey data collected than less productive interviewers. This literature shows the quite substantial effects that interviewers have in influencing survey outcomes, and thus the importance of properly managing telephone interviewers. We do need to be concerned about the size of interviewer workloads. Groves et al. (2004) note that estimates of the standard errors in a survey are directly related to the average number of interviews per interviewer.

In discussing the management of data collection (mainly for face-to-face interviews), Weinberg (1983) suggests a supervisor-to-staff ratio of about 1:10 and says that "the interpersonal interaction between supervisor and interviewer can affect survey production, data quality, and costs." She also notes that the management of data collection should include monitoring both the quality and quantity of the work. Lavrakas (1987, 1993) similarly recommends a ratio of about 1:10 for supervisors to telephone interviewers.

Lavrakas (1987, 1993) suggests using as a measure of telephone interviewer productivity, the ratio of properly completed interviews with the number of refusals and partials obtained per interviewing session. He indicates that a good standard to strive for would be about four completions per one refusal or partial, although this may vary for different geographic regions and for different survey topics. Another element of interviewer productivity for Lavrakas is the speed at which sample call records are processed. He does not provide a standard but suggests that interviewers be told what is expected of them. For Fowler (1984) interviewer productivity is primarily the number of interviews completed.

The number of completed interviews obtained by an interviewer is frequently used as the main measure of productivity, since survey goals often emphasize achieving a certain number of completed interviews. We found one study in the literature (Thurkow et al., 2000) that compared the effects of group and individual monetary incentives on the productivity of telephone interviewers. Call completion was the main measure of productivity, although the study also looked at call attempts per hour. Individual incentives were found to be more effective at increasing productivity than either group or competitive incentives.

The literature on call centers, much of which is concerned with direct marketing, sales, and support—and not research—is concerned with the management of telephone staff, and which has applications to managing telephone interview staff. Call centers use a variety of metrics for measuring the productivity and performance of call center staff (Anton, 1997, Waite, 2002).

Lavrakas (1987, 1993) indicates that it is important to communicate productivity expectations to interviewers so that they know how to perform adequately. Weinberg (1983) also indicates that supervisors should reinforce any expectations stated in the interviewer manual. Morganstein and Marker (1997) make a case for the importance of collecting process data to identify and better control processes that produce statistical products. The processes we are interested in are those under the control of telephone interviewers.

It would be difficult to develop standards of interviewer productivity that can be applied to all telephone surveys or even to a single survey because of the differences in sample populations, interviewers, and survey requirements, and which phase of a survey is being measured. However, the International Standards Organization (ISO) is developing a set of standards for market, opinion, and social research, which will help to ensure that survey research is "undertaken to an appropriate standard and in a verifiable and consistent manner" (International Standards Organization, 2005). For example, the current draft of the ISO standards for interviewer training states a 6-hour minimum training, including the context of training (pp. 18–19).

The process for identifying staffing needs to ensure that survey projects are worked adequately and efficiently is difficult to do, although computer-assisted telephone interview (CATI) systems can help by providing basic performance and productivity data on telephone interviewers (Edwards et al., 1998). Most telephone survey research is now carried out with CATI systems, and the detailed performance and productivity data that CATI systems provide make it possible for better and more efficient management of telephone interviewers.

17.2.2 Productivity Metrics

There are several different productivity measures in a telephone survey, and modern computer-assisted telephone interview systems facilitate the collection of many of these measures. Commonly used productivity measures include the following: number of call attempts made; number of completed interviews obtained; number of refusals; number of ineligibles; total number of minutes worked; average length of completed interviews; quality of interviews completed; monitoring scores; and attendance and tardiness.

Derivatives of these measures are sometimes more useful as measures of productivity, usually calculated on a per hour or minute basis, such as number of call attempts per hour or minutes per completed interview. Some frequently used metrics include the following:

- *Call attempts per hour* is calculated as the average number of calls made by interviewers per hour, or the total number of call attempts made during a work shift, divided by the total number of hours in the work shift. This is a baseline measure of time on the telephone and is useful in ensuring that interviewers stay on task.
- *Hours per complete* is calculated as the average number of hours required to get a completed interview and is determined by dividing the total number of hours worked by the total number of completed interviews obtained during a shift. This measure generally correlates positively with days in the field, since easier to reach respondents are typically disposed of early on, and the more difficult to reach respondents that require more call attempts occur later in the field period.
- *Refusals per hour* is calculated as the average number of refusals obtained per hour, or the total number of refusals made during a work shift, divided by the total number of hours in the work shift. This measure is generally negatively correlated with days in the field.
- *Ineligibles per hour* is calculated as the average number of ineligible respondents obtained per hour, or the total number of ineligible respondents obtained during a work shift, divided by the total number of hours in the work shift. Like refusals per hour, this measure is negatively correlated with days in the field.

All of these data and more can usually be obtained from a CATI or computer-assisted interview (CAI) system used for conducting telephone surveys. In addition, the data are generally available continually and by interviewer as well. When plotted against project timelines, it is useful for assessing what adjustments if any need to be made to workloads or timelines.

Other productivity measures that researchers have suggested for comparing interviewers before and after training include the ratio of completed interviews to the number of non-immediate hang-up first refusals (CFR) and cooperation rate (contacted eligible sample units that cooperate with a survey request).

17.2.3 Using and Communicating Productivity Information

Two main uses of productivity information are (1) forecasting hours and days required to achieve project goals and (2) communicating with interviewers about performance on a survey project. To help understand how productivity metrics can be used, we have collected data from six recent telephone survey projects conducted by the authors involving both listed and random digit dialing (RDD) samples and summarized the results in Fig. 17.2, 17.3, and 17.4 showing average calls per hour, average hours per complete, and average number of refusals per hour for these six telephone survey projects. We present averaged results, rather than present individual survey results to simplify the presentation and discussion.

One of the most useful aspects of productivity information is its use in estimating the number of interviewers needed for a survey, or the number of hours/days required to reach project goals. There are three parts to the equation relating productivity to interviewer hours as shown below; however, the productivity measure (hours per complete) is generally an estimate based on the result of calls over the first few hours or days. Since hours per complete tends to increase while a project is in the field (shown in Fig. 17.2), it is useful to calculate this information daily, put it in a spreadsheet, and reestimate the number of interviewer hours needed.

$$\text{Number of interviewer hours} = \text{Hours per complete} \times \text{Number of completes}$$

Using this productivity information is invaluable in forecasting interviewing needs, since the forecast is based on actual trends instead of hypothetical or past estimates. The forecast can be further improved by linear regression, since as we show below, productivity changes as more of a sample is worked and the composition of the remaining sample changes from unworked cases to no answers, answering machine cases, and refusal conversions. With each successive day that a telephone survey is in the field, a new regression equation can be developed to predict the remaining hours/days required to complete the survey, and the prediction becomes increasingly accurate as the survey nears its end. Edwards et al., (1998) in a review of automated call scheduling features in computer-assisted interview systems suggest that "the area most sorely in need of future work is the interaction between the autoscheduling system and interviewer staffing" (p. 306) and that the ability to predict staffing needs for interviewing shifts is a particularly useful feature that is not currently available in any CAI system. Thus, survey organizations either do this manually, haphazardly, or in a limited way. Figure 17.1 displays an Excel

Day	Interviewer hours	Number CM	CM per hour	Hours per CM
1	7	4	0.57	1.75
2	3	1	0.33	3.00
3	4	5	1.25	0.80
4	25	14	0.56	1.79
5	23	25	1.10	0.91
6	15	13	0.87	1.15
7	19	14	0.76	1.32
8	16	10	0.65	1.55
9	12	11	0.90	1.11
10	18	16	0.89	1.13
11	23	13	0.57	1.77
12	18	12	0.67	1.50
13	17	6	0.35	2.83
14	14	12	0.86	1.17
TOT/AVG	213	156	0.73	1.37

Goals	
Days remaining	7
Completes	200
Forecast	
CMs to do	44
Intvr hours	60
Hours per day	9

Figure 17.1. Using productivity data to forecast interviewer hours.

sheet that uses daily productivity data to forecast the number of interviewer hours and the number of days required to meet project goals. Productivity information is also useful in communicating with interviewers about performance expectations. Tracking the number of calls per hour, hours per complete, refusals per hour, and/or ineligibles per hour that interviewers achieve over the course of a survey project and then posting this information provides interviewers, supervisors, and managers with information about what level of productivity is possible for any given survey.

There will be individual variation on all of these measures, but it is straightforward to identify control limits on these measures and determine whether some interviewers are outside the expected range of performance. However, since many of the productivity measures are interrelated, they must be used prudently so that interviewers are never penalized for good behavior. For example, interviewers who achieve a high number of completes (or ineligibles in screening interviews) will generally see their calls per hour decrease because their time is occupied in talking with respondents, which takes more time than making call attempts.

Interviewers who consistently exceed the averages for productivity measures are good candidates for becoming monitors, interviewer leads, or supervisors, but this is not always so. Interviewers who consistently perform below average, and particularly those who are outside the control limits on any of the productivity measures should be assigned to retraining sessions. Productivity measures and the expected normal range should be presented to all telephone interviewers, especially new interviewers, so that performance expectations are clearly communicated. If interviewers are rewarded for their performance, care must be taken to ensure that this is done in a way that does not encourage inappropriate behavior, such as faking completed interviews. This is another reason that monitoring of telephone interviewers is an essential part of the survey process (see Hansen, 2007, Chapter 16 in this volume).

Figure 17.2 shows the average number of calls per hour for each of the 37 days that these surveys were in the field. The overall average was about 40 call attempts per hour, and the trendline shows a slight increase in call attempts with each succeeding day in the field. This slight increase is mainly due to the increase in unproductive calling (no answers, answering machines) the longer a survey stays in the field. Figure 17.3 shows the average number of hours per completed interview for each of the 37 days in the field. The overall average was 1.48 hours per completed interview, and the trendline also shows an increase with each successive day of calling. This reflects the increasing difficulty of achieving completed interviews from the remaining active sample in the field. Figure 17.4 shows the average number of refusals per hour for each of the 37 days in the field. The overall average was 2.02 refusals per hour, and here the trendline shows a slight decrease. This also reflects the increasing difficulty of reaching respondents from the remaining active sample, with more no answers and answering machines reached than respondents.

Comparing the three figures, calls per hour shows the least variation and the trendline is horizontal. The number of hours per completed interview shows quite a bit of variability and slight increase in the trendline, indicating that each day the survey is in the field, the number of hours required to get a completed interview increases. In contrast, the average number of refusals per hour decreases each day

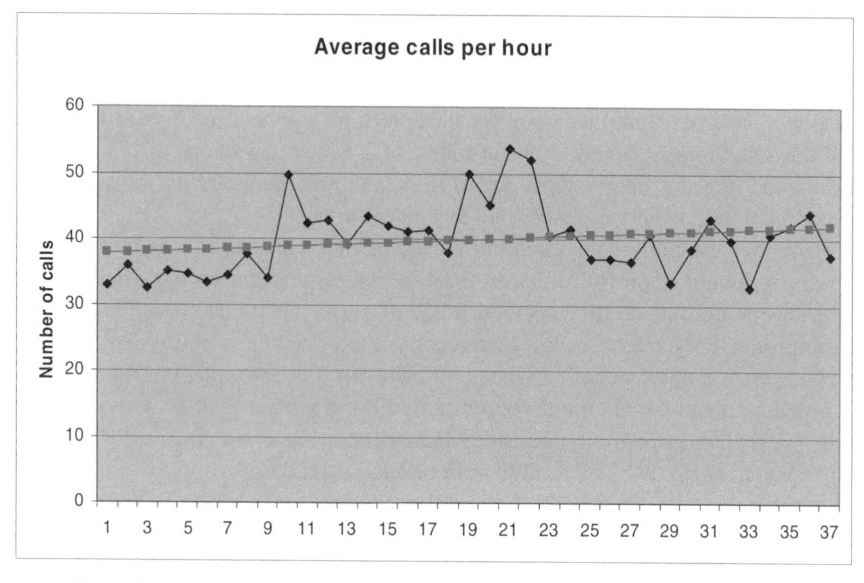

Figure 17.2. Average calls per hour. Overall average = 39.9, SD = 5.4, slope = 0.115.

that the survey is in the field. Thus, interviewer productivity as measured in these three ways is not static but changes throughout the course of a telephone survey. Some surveys may emphasize these trends more than others, but all telephone surveys probably follow a similar pattern. This makes it important to consider which phase a survey is in, when using productivity information to forecast interviewing needs and when using this to assess interviewer performance.

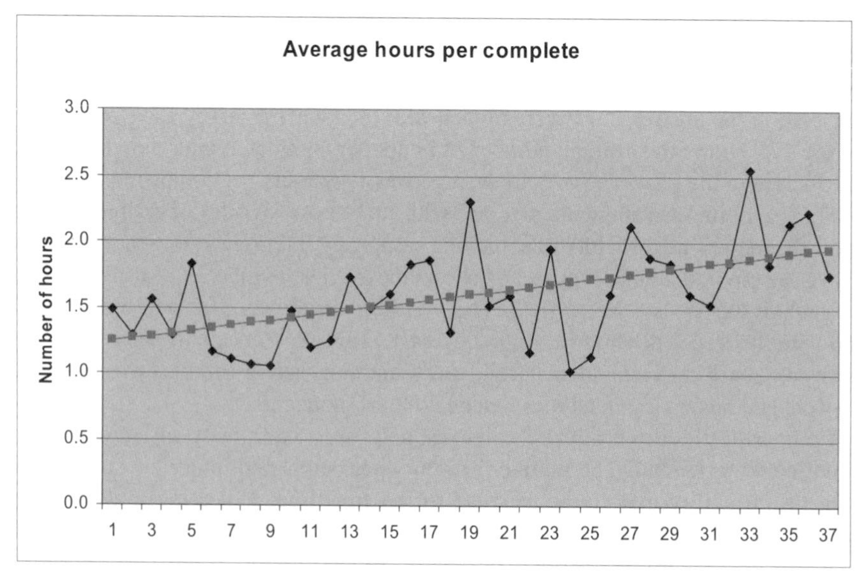

Figure 17.3. Average calls per complete. Overall average = 1.48, SD = 0.38, slope = 0.02.

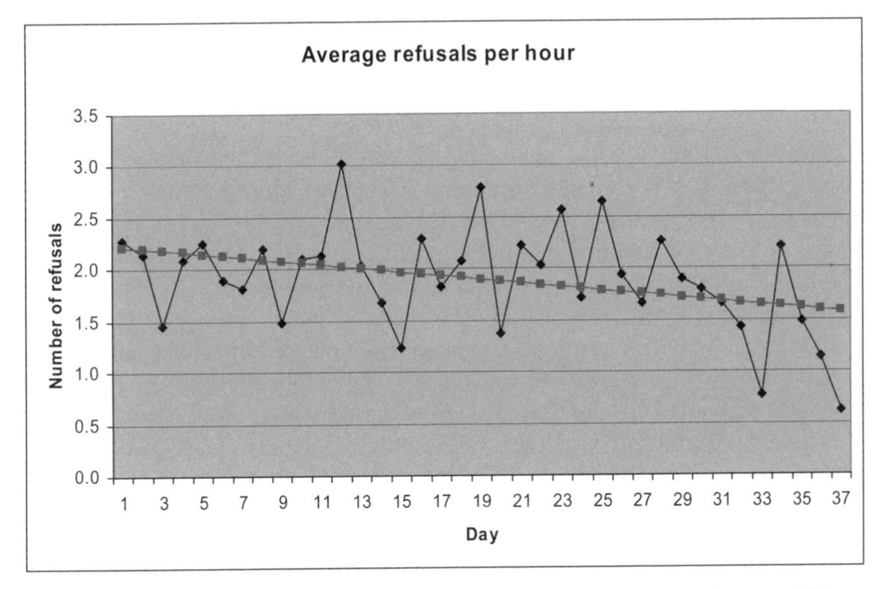

Figure 17.4. Average refusals per hour. Overall average = 2.02, SD = 0.51, slope = −0.02.

17.3 HOW INTERVIEWER TRAINING AFFECTS PRODUCTIVITY

We were unable to find any published studies of how training affects interviewer productivity per se. Most studies of interviewer training seem to focus on the effects of training on one aspect of interviewing, such as avoiding refusals (Shuttles et al., 2003) or increasing the rate of completions (Groves and McGonagle, 2001). Productivity plays an important part in interviewer training, as increasing productivity through more completions, fewer refusals, fewer missing items, more call attempts is generally the goal of training.

17.3.1 Training Protocols Designed to Improve Cooperation

Groves and McGonagle (2001) suggest that interviewer training may be under utilized as a way to influence survey participation and that most past research on interviewer trainings has primarily focused on question delivery and measurement error. In their study they show that training regimens directed at changing the behavior of interviewers, especially during survey introductions and at sample recruitment, can significantly influence sampled individuals' decisions to participate in surveys. Training regimens that taught interviewers to diagnose and provide counterstatements to respondents' concerns and to maintain interaction, in authors' experiments, outperformed and had lower rates of refusals than interviewers who did not receive this type of training. These authors point out that standardized interviewing that promotes set scripts and a "one size fits all" approach to survey introductions has failed to make significant improvements in survey cooperation over the past two decades. As a result, surveyors are seeking other avenues for improving first contact rates and are testing the requirement and "long-standing practice" in many organizations for interviewers to deliver standardized scripted introductions. In our research we are specifically interested in training protocols that enhance interviewer effectiveness as measured by their productivity.

Tailoring and maintaining interaction are two interviewer behavior constructs that have been used to formulate training protocols to address sample person participation (see Groves and Couper, 1998). Maintaining interaction is a fundamental element of tailoring, and prolonging the interaction with the respondent promotes more commitment by both actors (interviewer and respondent). As the interview and interaction progress, each actor is less likely to dismiss the other. The concept of tailoring promotes interviewers evaluating reactions of respondents and seeking cues to prolong the interaction, and using successful arguments that increase respondent receptivity toward the interview request. Inexperienced interviewers and interviewers with high refusal rates are often evaluated as creating soft refusals (respondents who mildly refuse and might be persuaded to participate with more effort) by having too few counterstatements to use and having to resort too quickly to making the interview request. We suspect that new interviewers and those that are less productive are virtually unaware of what they need to do and how to go about making connection with respondents. On the contrary, more experienced interviewers and those showing higher cooperation rates engage in more dialogue at first contact and actively communicate with respondents. The main attributes of experienced interviewers influencing survey participation are more confidence and a larger combination of behaviors proven to be effective in persuading sample persons (Groves et al., 1992).

Mayer and O'Brien (2001) conducted tests with 24 Census Bureau interviewers allocated to three experimental groups to specifically evaluate the effects of their Refusal Aversion Training (RAT) on first contact cooperation rates. Interviewers in the RAT received 8 hours of training. First contact cooperation rates were used as metrics to simplify analysis. An interesting aspect of this research was the acknowledgment that all interviewers showed a natural increase in first contact cooperation rates with interviewing experience (two data collection periods for this study) without specialized training. Overall, the outcomes of experiments showed that first contact cooperation rates increased 3–7 percent and as much as 14 percent over time for interviewers who participated in RAT. While this study focused on only two consecutive data collection periods, it showed that it took time for the effects of training to show, and authors report that interviewers admitted that it took time to become comfortable with strategies taught in the training and to put them to use.

What is not evident from any of these studies is whether training stays with interviewers over time, and whether or not they need refresher trainings to reactivate skills taught in specialized trainings. Does specialized training hold across surveys? Shuttles et al. (2003) conducted a series of three studies that randomly assigned 282 interviewers to a treatment or a control group with regard to participating in a specialized RAT. Unlike the previous two studies, these authors' results showed no significant differences between interviewers as a result of whether or not interviewers attended specialized RAT training. The measure used to compare interviewers was completes to first refusals ratio (CFR) and is the number of completed interviews to the number of nonimmediate hang-up first refusals. As an example of what this measures, a CFR of 0.33 is two first refusals to one accept. CFR measures for interviewers in the study ranged from 0.58 to 0.66 over the three experiments. Authors attribute the lack of finding a significant quantitative difference in interview performance between experimental treatment and control to the selection of experienced interviewers only, the call center environment, integrity of the experiment, and the changing nature of refusals. In this call center interviewers receive an ongoing array of training and coaching on an ongoing basis so that elements of the training program may be picked up in the informal environment of the center. For all three of these studies, it would be useful to additionally see the averages for traditional measures for interviewers such as historical calls per hour, refusals per hour, and completes per hour for interviewers as well as study-specific measures for treatment and control group interviewers.

One problem with the widespread practice of using only response rate type metrics to evaluate interviewers as highly productive is the implicit assumption that high cooperation at sample recruitment translates to effectiveness in the interview. Steinkamp (1964) provided empirical evidence that suggests that interviewer productivity should be about more than just sample recruitment. By measuring other metrics (e.g. percent of ambiguous answers, average length of interview, variability in length of interview, item refusals on financial holdings, percent of respondents using records for financial reporting, and interviewer evaluation of questionnaires as complete) his results show that interviewers with high cooperation did not directly translate to a high rate of data capture on variables important for completing the study. He suggests that productivity measures that only evaluate respondent cooperation (response rate measures) is a questionable practice for survey organizations when this is compared with the rate at which interviewers pick up key quantitative study information. His results failed to find a significant relationship between higher response rates and measures of in-interview effectiveness. He suggests that the most likely means of identifying effective interviewers is in analyzing patterns of performance (considering more than one measure) rather than in terms of performance in a single area. He found that there is considerably more interviewer variability in terms of effectiveness by validating survey information that people are sensitive toward. Examples of these variables might be a pickup rate for income, assets, or debts.

17.3.2 Focused Training Designed to Improve Productivity

How do survey researchers build survey skill among their interviewing staff and how do they know their training activities make a difference in survey outcomes? More importantly, how do surveyors identify interviewers who are assets and contributing to production from those they need to quickly identify as liabilities or as having deficiencies? Across survey organizations, it is relatively unknown what particular strategies are consistently used to identify training needs and to increase interview skills.

The kind of training interviewers need depends on the type of respondent role to be addressed. Some respondent interactions require more active participation and thinking by the interviewer. To prolong an interaction, whether at the time of introduction or during the interview at the time of an indication of abandonment, requires the interviewer to constantly stimulate respondents and build rapport. Our observation is that most new inexperienced interviewers do not do this naturally and need training in this area. If a particular survey proves to have difficulty in recruitment, is lengthy, has sensitive subject matter, and/or is otherwise unappealing to respondents, then there is a greater demand on interviewers for having skills and tools to prolong interaction and persuade respondents to stay with the interview. The more interviewers need to stimulate respondents in a given survey, the more training is required.

One method commonly used as a strategy for practicing interviewing and learning the survey content is the role playing by interviewers. But when is role playing most beneficial? Role playing is not deemed to be very effective when (1) interviewers are left on their own to carry out the task without first learning what they are to practice, (2) role playing is unguided, and (3) role playing is not assessed for changes in interviewer behavior or knowledge. It may be difficult for an individual interviewer to accurately recall, discuss, and act out with another interviewer how survey introductions actually play out or to accurately describe how they maintain rapport with respondents. Even more difficult for new interviewers is to fully comprehend the nuances of the respondent–interviewer interaction and to know what they can do or say to respondents. At the lowest experience level, interviewers may only acknowledge that all unsuccessful survey introductions are abject refusals. In a training setting it may be more beneficial and valid for interviewer's to first "see" or "hear" actual interviews and then witness interviewer behaviors and interactions that were successful and unsuccessful in gaining respondent cooperation. Group discussion of the element of interchanges is helpful after interviewers can hear live examples of both successful and unsuccessful introductions. In this way interviewers learn to recognize differences in interview behavior, tie it to outcomes, and start to understand what they themselves do when they are in an interview.

The use of role playing can be very pertinent to training. It is necessary for interviewers to know how they and respondents might act under the stress of a given situation. Respondent stresses can be related to timing, attitudes about surveys or sponsorship, rights as a respondent, and confidentiality to name a few. Interviewer stresses may be related to lack of experience or ability, knowledge, confidence, and expectations about the interaction. To role play effectively requires training tools and training methods—for interviewees (acting as respondents) to test their partner on recognizing the full gamut of respondent themes and for interviewers to recognize and respond appropriately to each theme.

Role playing can add to the demonstration of actual interviewing to teach and test interviewers for respondent theme recognition, practice of listening, and response skills. Role playing gives interviewers an opportunity to practice their skills with stressful and/or negative respondent interactions that can be perceived and acted upon by the interviewer. We believe that many small to medium survey centers rely on ad hoc role playing as a training method. Role playing may be used more as a method to familiarize interviewers with a new questionnaire rather than a way of testing for knowledge or recognition of themes. Ad hoc role playing as part of interviewer training may not constitute thorough enough interaction training and may have a high degree of performance variability between interview partners. It is highly likely in ad hoc role playing that interviewers are practicing only a small set of the types of respondent interactions. More formalized teaching of respondent themes and role playing with instructional materials and formalized objectives for themes has been suggested and shown by some researchers in large survey facilities such as the U.S. Census Bureau (Mayer and O'Brien) as a way to improve interviewer performance.

Role playing during training by an "interviewer" and an "interviewee" partnering also permits observation. Not only can the interviewee assess the quality of the interviewer's interactions and instincts, but they can also immediately provide feedback on correct recognition of respondent themes. Basically, good scene construction for interviewer training materials requires well-articulated respondent problems, practical solutions, and thorough testing. It is best if role-play sessions can occur as part of regular training sessions, remedial trainings, and also as part of project-based training sessions.

Another type of interviewer training that is useful in improving interviewer productivity is sometimes called "rapid response training" to denote a brief and quick training session. This training is designed to address a particular interviewer skill or productivity problem, such as one of the following: (1) refusal aversion; (2) nonspecific respondent concerns; (3) time and burden concerns, (4) government concerns, (5) dealing with hostile respondents, (6) company or sample member "no survey" policy concerns, (7) confidentiality concerns, (8) pass off to another contact in sample unit, (9) active listening, (10) persuasion techniques, (11) taking the easy way out—avoiding self-selection bias, (12) survey content concerns, and (13) selection/sampled concerns.

Shuttles and his colleagues (2002, 2003) provide an example of this kind of training to help telephone interviewers avoid refusals and increase their overall productivity. Their training incorporates many elements of successful training programs carried out by the Bureau of Labor Statistics, the Department of Agriculture, and the Bureau of the Census. This training teaches interviewers to focus their efforts on five specific interviewing skills: (1) recognition: learning the themes of respondents concerns; (2) diagnosing: learning to classify respondent's actual wordings into themes; (3) modification: learning desirable interview behaviors to address concerns; (4) delivery: learning to deliver a set of statements relevant to the respondent's concerns; and (5) rapid response—increasing the speed of response performance through practice.

17.4 RESULTS OF A SURVEY OF TELEPHONE INTERVIEW ORGANIZATIONS

We wanted to know how other survey research centers that conduct telephone interviews measure interviewer productivity and performance and how they use this information to manage interviewers, meet project goals, and staff the telephone facility. To accomplish this, we designed an Internet questionnaire for survey research organizations. The sample consisted of all survey organizations listed in the 2005–2006 American Association for Public Opinion Research (AAPOR) Blue Book supplemented by the listing of organizations in attendance at the 2005 International Field Directors Technology conference, including those with international addresses. Questions for the survey were developed from the questions and issues identified in this paper.

The sample frame consisted of a total of 490 organizations, all of whom were mailed (or e-mailed) a letter inviting them to participate in the Internet survey. Those for whom e-mail addresses were available were also sent an e-mail message requesting their participation in the survey. A total of 161 organizations participated in the Internet survey for a response rate of 33 percent. Of those responding, about 6.8 percent were ineligible because they did not conduct telephone interviews in-house for their organization. This left 150 organizations that completed the entire survey. Responding organizations are characterized in Table 17.1. The majority of organizations, 86 percent, are U.S. based organizations and 13.9 percent are international organizations that are primarily government statistical services. These can be classified further as academic (37.3 percent), government (11.4 percent), nonprofit (9.3 percent), and commercial (44.7 percent). More than two-thirds of the organizations have 50 or less CATI stations at their main interviewing locations. Less than a quarter of the organizations reported having additional CATI locations beyond their main location. Only 19.3 percent have CATI stations at locations beyond their main location.

Most organizations primarily rely on part- time interviewers supplemented with a few full-time interviewers for their telephone interviewing workforce. The average number of part-time interviewers currently on staff was reported as 88 and the number of interviewers hired annually was reported as 169. Over half (54 percent) of the organizations reported having 35 or fewer part-time interviewers currently on staff. Organizations have few full-time interviewers. From the survey, over half of all organizations reported having three or fewer full-time interviewers currently on staff. On average, pay for entry-level interviewers is reported as $8.38 dollars per hour, and the average hourly pay for all interviewers at the organizations responding to the survey was $9.37.

A majority of survey respondents are personally "somewhat or very involved" in the design and administration of telephone surveys (62.5 percent) and a majority are "somewhat or very involved" in the supervision and training of interviewers (63.2 percent). These results suggest that the survey was answered by the appropriate respondents and that survey respondents are well qualified to provide organizational-level interviewer productivity and training information.

While interviewer performance data is important to most organizations (70 percent say it is very important, and 26 percent say it is somewhat important), a handful of organizations (4 percent) think it is unimportant. Over three quarters of organizations are satisfied with the way their organization uses data on interviewer performance, but over 20 percent are dissatisfied.

Table 17.1. Characteristics of Survey Organizations Responding to the Survey (n = 166)

Type of organization	
Academic	37.3%
Government	3.9%
Commercial	47.7%
Nonprofit	11.1%
Location	
International	13.3%
United States	86.7%
CATI locations	
Single main location	80.7%
Other additional locations	19.3%
Number CATI stations at main location	
None	8.6%
1–20	30.7%
21–50	32.1%
51–100	15.0%
100+	13.6%
Number of part-time interviewers currently on staff	
<25	32.1%
26–50	20.0%
50–150	23.6%
>150	24.3%
Number of full-time interviewers	
<3	35.0%
3–10	15.4%
11–30	24.8%
>30	24.8%
Number of telephone interviewers hired in 2005	
Range	0–5000
Mean	166
Median	36

17.4.1 Measuring and Evaluating Interviewer Productivity

The majority of survey organizations that conduct telephone interviews (94.9 percent) said that their organization measures the performance of their telephone interviewers. Many (55.4 percent) indicate that interviewer performance data is available continually, and others collect it by shift (8 percent), per day (14 percent), per week (12.3 percent), or less often (10 percent). Over 80.5 percent of respondents said that interviewer productivity data is available through their CATI system. Despite the widespread availability of performance data on interviewers, only 44.9 percent of respondents said that their organization has standard productivity requirements that

Table 17.2. Percent of Survey Organizations that Produce Interviewer Performance Measures

Performance measure	Percentage of organizations producing
CATI system produces	
Number of call attempts	93.5%
Number of interviews completed	97.6%
Number of refusals	89.3%
Number of ineligibles	73.3%
Number of minutes/hours worked	92.6%
Length of completed interviews	90.0%
Attendance and tardiness	75.2%
Cooperation rate	60.7%
Quality measures collected	
Supervisor ratings of interviewers	81.1%
Monitoring scores	66.4%
Number of questionnaire items with missing values	35.2%
Number of organizations	**122**

interviewers are expected to meet in order to keep their jobs, get promoted, or receive raises. Most response rate productivity measures, (completes, refusals, ineligibles, hours interviewing, length of interviews) are available through organizations' CATI systems for more than 89 percent of organizations. Only attendance and tardiness are not readily available, with only 29.6 percent of organizations saying this measure was available through their CATI system. Table 17.2 displays the percentage of survey organizations that produce and collect each kind of interviewer performance measure.

The majority of survey organizations collect all of these performance measures, but surprisingly about 25 percent do not collect data on attendance or tardiness, and about 40 percent do not collect interviewer's cooperation rates. Survey respondents were asked to indicate the three most important measures of interviewer efficiency in their organization. The most frequently reported measures included calls per hour, completes per hour, rates of refusals, and general survey quality measures such as supervisory evaluations and monitoring scores.

When asked specifically about quality measures collected on interviewers, the most reported measures were supervisor ratings (81 percent) and monitoring scores (66.4 percent). Less than 35.2 percent indicated that they collected information on missing values on questionnaire items as a measure of the quality of interviewing.

Organizations reported that they emphasized and used multiple measures as part of their standards to evaluate interviewers. A few organizations indicated having thresholds for performance, with the lowest 20 percent of interviewers advised to improve their performance or face termination. For the most part, evaluation scorings are comparative. Interviewers are compared to others working on the same survey or survey project and across all studies worked. An assessment of an interviewer can be compared to their own performance over time or on other studies. Another frequently mentioned performance standard is regular monitoring assessments, which are a composite scoring of an individual interviewer's behavior for specific aspects of skill and acceptable interviewing practice including asking questions correctly and as worded, nonbiased probing, appropriate interaction with respondents, refusal avoidance, and an ability to convert previous refusals, demeanor, and voice quality. Some organizations describe this as interviewing to set protocols or standards. Measures that look at basic work performance and workplace ethics were also mentioned, such as number of absences, tardiness, regular availability for work, adhering to schedule, following supervisor instructions, ability to handle technology such as CATI use, time reporting technology, longevity, and tenure as an employee. Some organizations report they have unionized interviewers and must comply with union reporting for layoffs.

A number of organizations report setting productivity measures and establishing benchmarks that are used for decision making. The most commonly mentioned thresholds for individual interviewer statistics called for comparisons to a group mean or average. Some organizations reported using information about the available interviewing budget to set interviewing performance thresholds on key measures. Some organizations indicate that they set quotas, pay incentives, and pay on a per completed interview basis. Decision makers in organizations are actively using productivity measures to establish interviewer levels of pay, to promote and reward interviewers, and to terminate interviewers who are not meeting the organization's set standards for performance. Some of the threshold levels of performance mentioned included the following:

- Lowest 20 percent advised to improve performance
- Minimum of 56 dials/calls per interviewer hour
- Completed interviews per hour for an individual is around 50 percent group mean
- Individual's measures must be at average or above average compared to peers to stay employed
- Individual's measures must be at average or above average compared to peers to receive pay raises
- Individual's average across all projects worked must be at average or above average compared to group mean for 12 months to receive pay increases
- Bell curve rating of employees on production and refusal rates
- Quality control database of 75 variables tracked continuously
- Dial rates set at 90 percent of team rate
- Meet productivity goals for a given project—if goals are not met, individual interviewers monitored more heavily, coached, and retrained
- Productivity bonuses awarded to top 50 percent of productivity performers.

To better understand how organizations measure productivity, the survey asked several questions about how many calls per hour, completes per hour, and refusals

Table 17.3. Telephone Interviewer Productivity Expected—Number of Call Attempts

Number of call attempts	Percent
Less than 20	9.1
21–30	19.0
31–40	19.8
41–50	19.0
51–60	5.8
61–70	3.3
71–80	3.3
More than 80	0
Do not know	20.7
Total	100

per hour, organizations expected for a 10-minute and a 20-minute RDD telephone survey. The results of these questions are summarized in the tables below.

Table 17.3 displays respondents' expectations about the number of call attempts that telephone interviewers should be able to produce hourly for an RDD survey of the general public. The results show a lack of consensus about the number of call attempts to be expected, although the majority of respondents suggest the number should be between 21 and 50 calls per hour. Less than 10 percent think the number is less than 20 calls per hour and over 21 percent do not know. Commercial firms tended to have higher expectations than academic organizations.

Table 17.4 displays respondents' expectations about the number of refusals that would be expected from a telephone interviewer every hour from RDD surveys of the general public, of 10 and 20 minutes in length. Again, there is no clear consensus on what should be expected, with over 38 percent of organizations expecting fewer than three refusals per hour, and 36 percent expecting four or more refusals per hour. Commercial firms tended to expect a higher number of refusals than academic or government organizations.

Table 17.4. Telephone Interviewer Productivity Expected—Number of Refusals

Number of refusals	10-minute RDD survey (%)	20-minute RDD survey (%)
Less than one	4.9	6.6
One	9.0	9.1
Two	12.3	10.7
Three	13.1	10.7
Four	5.7	9.1
Five	10.7	9.1
More than five	18.9	21.5
Do not know	25.4	23.1
Total	100	100

Table 17.5. Telephone Interviewer Productivity Expected—Number of Completions

Number of completed interviews per hour	10-minute RDD survey (%)	20-minute RDD survey (%)
About half an interview	0	8.9
More than half but less than one	2.4	22.8
About one	16.9	33.3
Up to one and a half	21.8	18.7
About two	28.2	4.1
More than two	18.5	0
Do not know	12.1	12.2
Total	100	100

Table 17.5 displays respondents' expectations about the number of completed interviews that would be expected from a telephone interviewer every hour from RDD surveys of the general public, of 10 and 20 minutes in length. About 20 percent of all organizations would expect no more than one 10-minute interview per hour. Another 50 percent of organizations would expect between one and two 10-minute interviews per hour. In comparison, about 65 percent of organizations would expect no more than one 20-minute interview per hour for a 20-minute telephone interview. Commercial firms again tended to expect a higher number of completed interviews than academic or government organizations.

17.4.2 Using Interviewer Performance Information

This section describes how survey research organizations use the data they collect on interviewer performance, including whether they use it to forecast the progress of a survey or interviewing needs, and in particular whether they use it for improving their training of telephone interviewers. The survey results also allow us to describe how survey organizations deal with the issue of productivity in training interviewers.

Most organizations, 70.1 percent, indicate that interviewer performance is very important to their organization but only 26.8 percent of organizations can say they are very satisfied with the way their organization uses interviewer performance data, and this exemplifies the challenge organizations face in trying to change their business operations and to estimate their costs of production. Table 17.6 displays the percentage of organizations that say they use productivity measures in each of several ways. The majority of organizations said they used productivity to accomplish all but one of the goals listed in the table. Almost 96 percent of organizations said they used performance measures to communicate expectations to interviewers, but only 62.4 percent of organizations said they used them to recalibrate their interviewing standards. The least use of performance interview information is in making adjustments to fielded sample. While almost 86 percent of organizations used performance measures to terminate poor or excessively unproductive interviewers, only 68.3 percent used them to

Table 17.6. How Survey Organizations Use Interviewer Performance Measures

Performance measure used for	Percent of organizations
Forecasting the number of interviewers needed	84.0%
Forecasting the number of days required to reach survey goals	85.8%
Rewarding highly productive interviewers	68.3%
Terminating poor or excessively unproductive interviewers	85.7%
Reassigning interviewers based on their productivity	75.6%
Making adjustments in fielded sample replicates	41.5%
Communicating expectations to interviewers	95.7%
Retraining unproductive interviewers	84.9%
Training new interviewers on productivity issues	87.4%
Resetting or calibrating standards of interviewer performance	62.4%
Number of organizations	120

reward highly productive interviewers. Performance measures are important to interviewer training, since almost 85 percent of organizations say they use them to retrain unproductive interviewers, and 87 percent say they use them to train new interviewers.

17.4.3 Training and Retraining Telephone Interviewers

This section describes the telephone interviewing skills that survey organizations indicate are important to be a productive telephone interviewer. We also discuss how survey organizations train and retrain telephone interviewers to improve their performance and productivity.

The basic skills required of interviewers include interacting with a computer through a keyboard and/or mouse, and sufficient typing ability to enter respondent comments and other text into a computer. All of these skills are trainable, although it is generally easier to find people who already have typing skills than to train people in this skill. Average time reported for training the basics of interviewing in the survey was 6.5 hours or less for 64.9 percent of centers. Only about one third of organizations reported more than 6 hours of basic training for interviewers. Longer trainings for basic interviewing was associated with firms that annually hired larger numbers (151–200+ annually) of part-time interviewers.

The vast majority (99.2 percent) of survey research organizations conduct trainings for all new telephone interviewers before they can begin interviewing on a study. More than 95.8 percent reported holding at least two types of regular trainings—interviewing basics and project-level trainings. More than 91 percent of these organizations reported following written standardized basic training procedures for developing the skills of newly hired telephone interviewers (Table 17.7). Supervisors or trainers of more than 86 percent of the organizations follow a standardized basic training system. There is a large variation in the reported average training times per interviewer devoted to specific trainings. For instance, training the basics

Table 17.7. Types of Interviewer Training and Average Number of Training Hours

Types of trainings	N	Percent of organizations		Average number of hours
		No	Yes	
Training the basics of interviewing	121	1.7	98.3	6.6
Specific project training	120	4.2	95.8	3.0
Computer-based individual interviewer training	119	26.9	73.1	3.9
Remedial or performance improvement training	114	26.3	73.7	1.9
Advanced interviewer training	119	43.7	56.3	3.3

of interviewing for organizations ranged from a low of 30 minutes to a high of 30 hours, with the average training time for basic interviewing at 6.6 hours, with 4 hours as the most frequently reported duration of training. Of the types offered, the most intensive trainings (as measured by average training hours per interviewer) were interviewing basics, computer-based trainings done as individuals (73 percent, with an average of 3.9 hours/interviewer), and advanced interviewer trainings (56 percent, with an average of 3.3 hours/interviewer), respectively. Almost three fourths of all organizations (73.7 percent) answering the survey indicated that they conducted remedial or performance improvement trainings. Of the 84 organizations that conducted remedial trainings, 53.7 percent spent one hour or less for this type of training.

Table 17.8 shows the results from the questions that asked organizations to rate the extent to which they included and used each of 22 separate topics in basic interviewer trainings. Organizations reported that they included most of these topics or activities "always" into their basic training with the exception of "pretesting with actual respondents." Only three activities were described as *not* always included by 60 percent or more of organizations, including explanation of survey and interviewer errors, how to tailor survey introductions, and pretesting with actual respondents. The top twelve ranked training topics are essential elements for performing interviewing and understanding how to recruit and talk with respondents and code their responses into a database. The remaining 10 topics and activities are associated with understanding how to do particular aspects of interviewing better. These latter topics are somewhat more complex and theoretical in nature. If we assume that organizations included most of these topics in their interviewer training and we assume that an average length of basic interviewer training was about 7 hours, then, on average, less than 20 minutes can be devoted to each topic or activity in the basic interviewing training.

Advanced interviewer trainings were conducted by about half (55.6 percent) of the organizations with the average advanced training time reported for an individual interviewer as 3.5 hours. There are perhaps overall two reasons for organizations to conduct advanced training for interviewers. The first would be to train experienced

Table 17.8. Topics and Activities Included in Basic Telephone Interviewer Training

		Percentage			
Topics and activities	*N*	Always	Sometimes	Rarely/ never	Do not know
Rules for standardized interviewing	115	94.8	4.3	0.9	0
Explanation and examples of probing and feedback	113	94.7	3.5	1.8	0
Demonstration of how to use CATI system	115	90.4	4.3	5.2	0
Explanation of types of questions (categorical versus open-ended)	113	89.4	6.2	4.4	0
Explanation of respondent reluctance and cooperation	112	89.3	8.9	1.8	0
Proper interaction with respondents	115	87.8	9.6	2.6	0
Addressing respondent concerns	113	86.7	10.6	2.7	0
Explanation of survey introductions	114	85.1	13.2	1.8	0
Practice and role play by interviewers	115	81.7	12.2	6.1	0
Explanation of refusal behavior and interactions	113	84.1	11.5	3.6	0.9
Explanation/use of case disposition codes	112	83.0	9.8	4.5	2.7
Practice and role playing mock interviews	115	86.1	10.4	3.5	0
Data entry errors and ways to make corrections	113	80.5	11.5	7.9	3.5
Explanation of research ethics	113	79.6	13.3	6.2	0.9
How to control making interviewer errors	115	73.9	14.8	8.7	2.7
Expectations of interview performance and how performance is measured	112	73.2	14.3	19.9	2.7
Explanation and practice of dialing	115	72.2	12.2	14.8	0.9
Human subjects research and confidentiality	112	67.9	17.0	12.6	2.7
Special emergency situations and what to do	112	64.3	23.2	11.7	0.9
Scenario testing of questionnaire branching	114	63.2	13.5	19.3	4.4
Explanation of survey and interviewer error	115	56.5	15.7	23.4	4.3
How to tailor survey introductions	114	52.6	21.1	25.3	0
Pretest with actual respondents	112	18.7	35.7	35.7	0.9

Approximately, 49–52 of the 161 respondents to survey did not answer questions in this table. Of the responding organizations, 12 indicated they did not have CATI stations at main location, 11 did not conduct telephone interviews in-house, 5 had no part-time interviewers, 10 hired less than three telephone interviewers in 2005.

interviewers for supervisory and other kinds of positions (e.g., monitoring). The goal of advanced training would then be to prepare them to evaluate other interviewers in their job function and to provide specific feedback to rectify identified problems and to maintain or improve quality of interviewing of the group or the pool of interviewers. The second purpose for advanced training might be to train specific interviewers more in-depth, providing them with more background, tools, and theoretical information with the specific intent to keep them interviewing. This

Table 17.9. Advanced Interviewer Training Topics and Activities

		Percent of organizations			
Topics and activities	N	Always	Sometimes	Rarely/ never	Do not know
Recognition of respondent concerns	107	60.7	18.7	16.8	3.7
Refusal conversion techniques	107	57.0	25.2	13.1	4.7
Diagnosing poor interactions or introductions	108	48.1	25.0	21.3	5.6
Recognition of improper interactions	106	52.8	22.6	19.8	4.7
Monitoring and scoring interviewer performance	106	46.2	24.5	22.7	6.6
Recognition of errors	106	44.3	26.4	22.6	6.6
Expectations of performance measurement system	105	42.9	24.8	25.7	4.7
How to provide feedback on performance	108	39.8	24.1	26.9	9.3

second group of interviewers may be asked to do recontacts of refusals or more difficult interviewing. Interviewers identified for advanced training are counted for high performance interviewing. Table 17.9 provides a list of topics included in advanced interviewer training and shows the frequencies with which organizations cover various topics. For advanced interviewer trainings, the focus of topics is more toward improving an individual interviewer's skill for increasing survey participation and countering nonresponse rather than for training about overall performance issues, performance measurement, or rating other interviewers' performance. Table 17.10 describes the basic methods that survey organizations use to train telephone interviewers, and Table 17.11 describes some of the specialized methods used to train interviewers.

Table 17.10. Methods Organizations Use to Train Interviewers

		Percentage			
	N	Always	Sometimes	Rarely/ never	Do not know
Basic interviewing skills training	115	85.2	11.3	3.4	0
Project-specific training at the beginning of all new telephone projects	114	89.5	9.6	0.9	0
Scenario testing of specific questionnaire skip patterns for projects	114	75.4	14.0	10.6	0
Audio sample of interview	114	8.8	22.8	68.5	0
Interviewer knowledge test to certify onto a project	113	21.2	23.0	54.9	0.9
Practice mock interviews	114	64.0	25.4	9.7	0.9
Pretest with actual respondents	111	21.1	34.2	43.9	0.9

Table 17.11. Use of Specialized Training for Improving Interviewer Performance

		Percentage			
	N	Always	Sometimes	Rarely/ never	Do not know
Actual audio samples of survey introductions	113	6.4	20.4	70.8	2.7
Audio samples of effective interviewing interactions	112	4.4	23.9	69.1	2.7
Audio examples of respondent–interviewer interactions	112	4.5	25.9	67.0	2.7
Practice drills to increase recognition and rapid response to respondent concerns	113	18.6	29.2	47.7	4.4
Practice or role-play mock interviews	111	52.3	25.2	18.9	3.6

17.5 DISCUSSION AND CONCLUSIONS

The common concern of all survey organizations is how to prepare a telephone interviewer workforce for conducting telephone interviews in the most efficient and cost effective manner possible. All survey organizations face the same situation of declining response rates and the expenditure of more effort to recruit respondents and achieve completed interviews. It is useful to know what other survey research centers are doing in this environment to improve the effectiveness of their most important and scarce resource—telephone interviewing hours. Yet, our survey found that less than half of the survey organizations use interview productivity information to make adjustments in their fielded sample for surveys. This suggests that many organizations are not fully utilizing the productivity information available to them to improve their production processes on an ongoing basis. One of our goals in this paper is to distill the experiences of survey organizations into a set of replicable practices that any organization could further develop or adopt to improve its own survey practice. In the survey methodology literature, common themes for countering low response have emerged, but it is relatively unknown to what extent organizations are aware of these ideas and whether they have put them into practice. Inherent to this idea is that the best way of handling the interviewing function is to determine what most organizations do to regularly measure and develop skills related to this most important activity. Organizations have provided the level of interviewer hours on average per interviewer that they schedule to activities and topics. Any organization can evaluate its own interviewer training and performance system against these lists of topics, activities, and averages to know how they must align to be comparable to most organizations. We propose that one way to evaluate the interview function is by comparing the performance rate or yield of interviewing across organizations. To do this, we asked survey centers to provide their standard yield rate for bidding survey costs based on the set average lengths of interviews. We argue that this is a fair measure, as this is often one of the only measurable aspects known at the beginning of a study when a survey firm is asked to provide a cost estimate for a study. If interviewer job performance is lacking, the average number of completed interviews per hour will be lower than the average compared to other survey centers in this study.

Innovation and early adoption of effective interviewing and interviewer management strategies may be the key to survival, let alone the competitiveness of survey centers today. There is momentum behind knowing "good or better practices" as this implies that the most successful survey firms are those that are looking for ways to change that improves their efficiency or financial position. While a large percentage of survey centers (84 percent) regularly collect productivity information on interviewers and it is readily available to them for decision making, only 54 percent use this information as a way to analyze and to make decisions about interviewers. The main differences we find between survey organizations are in the average number of hours per interviewer they devote to trainings, inclusion of training topics, use of specialized training activities, and the way they monitor and evaluate interviewer productivity. We found very few firms consistently using specialized training tools and activities that look at effectiveness of interviewing techniques such as analyzing and demonstrating audio samples of introductions or in-interview interactions, rapid response drills to respondent concerns, and knowledge certification. However, some organizations are beginning to venture into regularly using these tools as a way to improve surveying.

17.5.1 Measuring Interviewer Productivity

Survey organizations that conduct telephone interviews are concerned about productivity, and they collect a variety of productivity data and use these data to identify interviewers to receive further training or to be let go. The majority of organizations use productivity measures based on interview completions, such as the number of completed interviews or completion rate, although many also look at measures such as number of calls per hour, rate of refusals, and simple attendance. As our analysis of some of these measures demonstrates, the absolute levels of these measures change during the course of a survey so that organizations need to take into account at what stage of a survey productivity is measured.

The most important productivity measures for telephone interviewing are call attempts per hour and completes per hour. The majority of survey organizations collect this information routinely, and it is available to most through their CATI system continuously while a survey is in the field. There is not a clear consensus on the number of call attempts, refusals, or the number of completed interviews a telephone interviewer should be able to produce in one hour. The major differences seem to be between commercial survey organizations and the noncommercial organizations. Commercial firms expect a higher number of completed interviews per hour from their telephone interviewers, and they also tolerate a higher number of refusals per hour than the noncommercial organizations. For a 10-minute RDD survey, 69 percent of commercial firms expect two or more completed interviews per hour in comparison to only 36 percent of noncommercial firms that expect this. For a 10-minute RDD survey, 44 percent of commercial firms expect five or more refusals per hour in comparison to only 7 percent of noncommercial firms that expect this. The results for a 20-minute RDD survey similarly reveal differences between the commercial survey firms and the noncommercial survey organizations. Over 77 percent of commercial firms would expect one or more completed interviews per hour in comparison to only 48 percent of noncommercial organizations. Over 60 percent of commercial firms would expect five or more refusals per hour, from a 20-minute RDD survey, in comparison to fewer than 18 percent of noncommercial organizations. So clearly, standards of interviewer productivity differ among survey research organizations and are dependent on the specific goals of each organization.

These differences between commercial and noncommercial survey organizations are also apparent in how productivity information is used. A higher percentage of commercial firms (48 percent) use performance measures to make adjustments in fielded sample than noncommercial firms (36 percent). In addition, a higher percentage of commercial firms (83 percent) than noncommercial firms (68 percent) use performance measures to reassign interviewers based on their productivity and performance.

17.5.2 Training, Evaluating, and Rewarding Telephone Interviewers

This paper presents the results of a survey of telephone interview organizations and their practices with respect to managing and training interviewers, and measuring productivity. From these data, we suggest a number of themes regarding what measures of productivity are most useful and how to use this information to train interviewers.

Evidence from the survey of survey research organizations suggests that best practice for interviewers must be a continual evaluation of their most important activity—interviewing. The majority of organizations depend on systems for developing interviewing skills through ongoing assessment of individuals and providing them with routine feedback on behaviors related to the use of study questionnaires and principles of interviewing. Systems include having written training protocols, trainers, and a supervisory and monitoring force. Organizations offering the most innovation in training are focusing on productivity measures and using this system to trigger training and establish reinforcement for improving interviewer behavior. Other signs of innovation are the use of measures other than response rates that extend productivity to also encompass interviewer effectiveness such as the variability of average interview length, use of audio recordings, analyses of survey content variables for outliers, and interviewer knowledge certification. The fact that interviewer behaviors can be measured over time allows for the assessment of individual change, improvement or declines in real time, and feedback that can result in improved performance and that ultimately affects survey quality and quality of statistics generated from survey data.

We believe that there is much more that survey organizations can do with the productivity data that they routinely collect and that this data will become increasingly important as a way to control survey costs and improve the management and training of telephone interviewers.

CHAPTER 18

Telephone Interviewer Voice Characteristics and the Survey Participation Decision

Robert M. Groves
University of Michigan and Joint Program in Survey Methodology, University of Maryland, USA

Barbara C. O'Hare and Dottye Gould-Smith
Arbitron Inc., USA

José Benkí and Patty Maher
University of Michigan, Michigan, USA

18.1 INTRODUCTION

Early methodological research led to a now well-accepted result that telephone survey response rates often fell between those of comparable mail questionnaire surveys and face-to-face surveys, with face-to-face generally achieving the highest response rates (Hochstim, 1967; Groves and Kahn, 1979; Cannell et al., 1987; Collins et al., 1988; Sebold, 1988). Face-to-face and telephone surveys have in common the use of an interviewer, to seek access to the sample unit, to describe the survey, to answer questions and address concerns of the sample person, and to attempt to persuade them of the value of accepting the survey request. The response rate differences are compatible with the conclusions that interviewers are effective recruiting agents for survey cooperation, relative to written recruitment protocols.

The centralization of telephone survey administration also informed the field about how variable interviewers can be in their ability to elicit the cooperation of

Advances in Telephone Survey Methodology, Edited by James M. Lepkowski, Clyde Tucker, J. Michael Brick, Edith de Leeuw, Lilli Japec, Paul J. Lavrakas, Michael W. Link, and Roberta L. Sangster

the sample. With centralized telephone interviewing facilities, samples were often distributed to interviewers indiscriminately (essentially randomly). Interviewers were assigned relatively similar mixes of cases throughout the data collection. However, variation in response rates across interviewers was quite high. Groves and Fultz (1985) reported large variation in the response rates of interviewers. Lyberg and Lyberg (1991) report a model that, after controlling for assignment area demographic variables, shows residual variability among interviewers of 15–20 percentage points. Interviewers clearly made a difference in achieved telephone survey response rates, and different interviewers achieved widely different response rates.

If telephone interviewers manifest such large differences in response rates, all the stimuli producing those differences must be generated through the audio channel, information gleaned from the words, and vocal properties of the interviewer. This chapter reviews the past literature on interviewer effects on telephone survey response rates (Section 18.2), speculates on properties of interviewer voices that may explain some portion of among-interviewer variability in response rates (Section 18.3), and describes a research project that analyzed the voice characteristics of interviewers as predictors to the decision of respondents to participate in the survey (Section 18.4–18.5).

18.2 RESEARCH ON THE ROLE OF INTERVIEWERS IN TELEPHONE SURVEY RESPONSE RATES

The findings of large variability among telephone interviewers in response rates grew all the more interesting when differences in the nature of the survey introductions between modes were described. Early observations on "cold-call" telephone surveys (those with no prenotification to the sample household) found that refusals to telephone surveys often occurred in the first few seconds of the introduction (see Oksenberg and Cannell, 1988). Decisions over the telephone by respondents were apparently being made quite quickly. Thus, there was less information about the survey on the telephone, relative to the face-to-face mode, that was being used to make the participation decision.

Looking back over the literature, given the finding of large interviewer variation in response rates, it is interesting that during the 1970s and 1980s, the field spent a great amount of energy in attempting to optimize the words that the interviewers used to introduce the study in the first few moments. The fact that telephone interviewers tended to be scripted in their introductions and *still* manifested large variation in response rates led to several sets of experiments on interview introductions. Dillman et al. (1976) found no large effect of having interviewers emphasize the social utility of the survey; O'Neil et al. (1979) in experimental variations found a similar result.

In retrospect, the research energy spent scripting interviewer introductions on the telephone seems largely wasted. Morton-Williams (1993) in the United Kingdom mounted a face-to-face interview experiment in which experienced interviewers were scripted to use the same wording for the introductory statement versus unscripted introductions. The unscripted introductions achieved higher response rates. At about the same time, qualitative work reported by Groves and Couper in 1998 showed that experienced interviewers were self-aware of customizing their introductory behavior to fit different respondent attributes. This led to notions of "maintaining interaction," the effort of interviewers to continue an interaction with respondents, and "tailoring," the creation of individualized presentations of the survey introduction to address any concerns of the respondent perceived by the interviewer. The interviewers attempted to maintain interaction in order to maximize the number of cues from respondents. Later, Campanelli et al. (1997), Dijkstra and Smit (2002), Maynard and Schaeffer (2002a), Couper and Groves (2002), and Houtkoop-Steenstra and van den Bergh (2002) expanded the notions and described how they act on the joint behavior of interviewers and respondents at the recruitment step.

18.2.1 Mode-Specific Features Affecting Interviewer Introductory Behavior

If maintaining interaction and tailoring are common tools of interviewers to discover and address the concerns of interviewers, the telephone mode presents significant challenges to apply them.

First, early studies of the social use of the telephone (Short et al., 1976; Pool, 1977) noted that telephone conversations tended to be more task-oriented and less effective as emotional communication than face-to-face modes. Cannell et al. (1987) noted that telephone survey interactions were shorter.

Second, the telephone mode presents its own set of challenges to accessing sample persons. These include answering machines, caller IDs, telephone butlers, and a growing set of other devices and services that subscribers can use to control what telephone calls the subscriber must deal with immediately. Tuckel and O'Neill in a series of papers (1996, 2002, 2005) have tracked the penetration of answering machines and caller IDs. In face-to-face interview surveys, they have found that significant percentages of U.S. residents use the devices to track call attempts to their residence while they are away from home. Thus far, a minority of residents report using the devices to filter out calls from strangers, but such usage patterns are gradually increasing (Council for Marketing and Opinion Research, 2003). The key interviewer-related issue regarding answering machines is whether or not to leave messages, announcing the fact that the survey organization is attempting contact. The results are mixed (Xu et al., 1993; Koepsell et al., 1996; Arbitron, Inc., 2001; Tuckel and Schulman, 2002). Perhaps the most comprehensive study of this is the multistate Behavioral Risk Factor Surveillance Survey, which over several states concluded no increased response rates from this interviewer behavior (Link and Mokdad, 2005b).

Third, the telephone mode in the United States also became the preferred mode for sales solicitation. One problem encountered by telephone interviewers is a respondent perception that they are telemarketers. The extent of this misperception has not been measured, to our knowledge. However, the extent of the U.S. public's animosity toward telemarketing was well measured in the October 2004 introduction of the "do not call list," which allowed households to request removal from telemarketing lists. With the majority of the U.S. households voluntarily placing themselves on the list, the rate of telemarketing calls declined. Despite this, however, there is a popular interviewer practice in telephone survey introductions—the use of the "I'm not selling anything" announcement at the very beginning of the interaction. This attempt to correct potential misclassification of the call's purpose on the part of the telephone answerer has been studied in some split-half experiments. Pinkleton et al. (1994), for example, find no enhanced effect in a university-based survey. However, de Leeuw and Hox (2004), in a meta-analysis of 29 such experiments, find that the statement does reduce refusals to telephone interviews and recommend the interviewers deliver it early in their introductions.

Fourth, telephone interviewers are generally under more supervisory oversight than are face-to-face interviewers. Early speculation (Groves and Kahn, 1979) was that this would lead to more homogeneity of cooperation rates, as interviewers are consistently exposed to the call behavior of the successful interviewers. Whether this is or is not the case remains uncertain because of the lack of experimental studies on dispersed and centralized interviewing.

Fifth, there appear to be ubiquitous effects of relatively fixed interviewer attributes on cooperation rates. Groves and Couper (1998) replicated by Hox and de Leeuw (2002) show that interviewer experience and a set of attitudinal states appear to predict interviewer-level cooperation rates. They find a presurvey confidence self-report to be indicative of later cooperation rate. The experience effect is an interesting one. The reported empirical data are based on cross-section analysis, comparing interviewers with shorter and longer tenure at a given point in time. The fact that more experienced interviewers have higher cooperation rates could result from less successful interviewers ending their employment earlier or through learning on the job. We suspect that some combination of self-selection and learning is present in the experience effect.

18.3 THE ROLE OF INTERVIEWER VOICE QUALITIES IN TELEPHONE SURVEY PARTICIPATION DECISIONS

Whatever are the attributes of telephone interviewers that produce the large variation in response rates, it is clear that they operate through the voice of the interviewer. All the information about interviewers—who they are, whether they are a threat, whether they are attractive, whether they are trustworthy, whether they are educated, whether they are professional, whether their purpose is desirable—all of these are communicated by the words and aural properties of the voice.

Linguistics and the social psychology of language note that voices can communicate much more than semantic meaning. From studies controlling for word content, it is known that listeners can identify a variety of indexical properties of speakers, including sex (Strand, 1999), height, and weight (Krauss et al., 2002) as well as ethnic, racial (Purnell et al., 1999), and socioeconomic status (Smedley and Bayton, 1978). Thus, although the early research on telephone interviewer variation in response rates attempted to reduce it by standardizing the words used by interviewers, it is possible that other vocal properties are productive of the variation.

The vocal and linguistic properties of speech that convey such indexical properties are diverse. Although the above-cited studies indicate that the way in which speakers produce vowels and consonants are known to convey indexical properties and other information in addition to the semantic content of an utterance, most research on the cues for such paralinguistic information have focused on laryngeal vibration. The rate of laryngeal vibration in speech is the fundamental frequency (f_0), perceived as pitch. Although speakers actively control pitch within a range, an individual's pitch range is in large part physiologically determined by larynx size, such that females and children have higher pitch ranges than males (Peterson and Barney, 1952). The quality of the vibration (voice quality) also systematically differs with sex, with females having breathier voices than males (Klatt and Klatt, 1990).

English does not use either pitch or voice quality to convey semantic meaning. However, pitch is one of the primary cues for the intonational and the consequent pragmatic content of utterances (Liberman and Pierrehumbert, 1984). Because of the role of laryngeal vibration in conveying extra-semantic content such as intonation and indexical properties of the talker, characteristics of laryngeal vibration such as pitch and voice quality appear to be promising candidates for the sources of variation in telephone interviewer response rates.

Some past studies have examined vocal properties of telephone interviewers. The seminal study was done by Oksenberg et al. (1986), which compared telephone interviewers with very high response rates to those with very low response rates. They used raters to code various vocal properties of an introduction from each interviewer and found that "higher pitch, greater variation in pitch, loudness, faster rate of speaking, clearer and more distinct pronunciation, and good speaking skills were all associated with attractiveness (whether determined by the *general attractiveness* measure or by *positive approach* or *competence)* and with higher perceived social class. Interviewers with these vocal and personal characteristics had low refusal rates" (p. 108).

Using acoustic analysis of the same recordings as used by Oksenberg et al., Sharf and Lehman (1984) reported that the average f_0 of an interviewer's introduction is significantly correlated with the historical response rate of the interviewer. The standard deviations of f_0 were also found to be significantly correlated with response rate, but no other acoustic properties that they measured (rate of speaking, mean pause duration, mean intensity, and standard deviation of intensity) were found to correlate with response rate.

In a subsequent study, Oksenberg and Cannell (1988) conducted a similar investigation of listener ratings and acoustic properties of the introductions from three groups of interviewers. The findings with respect to vocal properties generally and fundamental frequency and pitch in particular were inconsistent, with a significant association between response rate and average f_0 in *opposite* directions for two of the groups, and only one group showing a significant association (positive) between standard deviation of f_0 and response rate. Comparisons of listener ratings of the introductions from the high- and low-response-rate interviewers from each group indicate that the high-response-rate interviewer introductions were perceived as "speaking relatively rapidly, loudly, and with a standard American accent, and as sounding more confident and more competent" (p. 265).

Both the Sharf and Lehman (1984) and Oksenberg and Cannell (1988) studies also investigated the intonational pattern on keywords during the interviewer introduction. Sharf and Lehman observed that the three low-response-rate interviewers displayed a rising f_0 pattern for "hello," while the high-response-rate interviewers displayed a falling f_0 pattern. This finding was replicated by Oksenberg and Cannell (1988) for some keywords ("hello," "Bureau," interviewer name) but not others ("older," "survey"). Neither study investigated the intonation pattern of either the entire introduction or even the end of the introduction, the latter of which is known to be particularly important in signaling the pragmatic intent of an utterance (Liberman and Pierrehumbert, 1984).

Although the research has had some impact on the field, it has largely lain unreplicated. Further, these studies have various weaknesses: (a) they were all based on a small number of interviewers, generally 6–12 interviewers per group; (b) they forced uniform scripts on each interviewer; (c) the interviewers were not a representative sample of interviewer performance but of only the tails of the response rate distribution across interviewers; (d) ratings and analyses were made on the basis of a single recording per interviewer; (e) the studies involved mostly bivariate measures of association of interviewer characteristics with refusal rates, with high and low performers pooled into two groups; and (f) voice properties were not examined with respect to other measures of performance of the interviewer.

There remain, therefore, a set of unanswered questions

(a) Are the vocal properties of interviewer introductions predictive of the outcome of a contact with a sample household?
(b) Do electronically measured acoustic properties of interviewer vocal presentations relate in a systematic way to raters' impressions of interviewer voices?
(c) What portion of interviewer variation in cooperation rates is related to acoustic properties of the voice and what to rater impressions of voices?

18.4 RESEARCH DESIGN

The study to address these questions was a collaboration between Arbitron Inc., a radio ratings survey firm, and the University of Michigan Survey Research Center. Probability samples of interviewers from the two different interviewing environments were identified between February and March 2005. The sample consisted of 36 randomly selected Arbitron interviewers and all 22 Survey Research Center interviewers working on the Survey of Consumer Sentiment survey.

18.4.1 Recordings

During the survey period, all introductions of the chosen interviewers were digitally recorded. Following the initial greeting, interviewers informed respondents that the call was being recorded and requested permission to continue recording, following University of Michigan Institutional Review Board (IRB) approved human subjects protocol and Arbitron policy. The UM SRC interviews were digitally recorded at a sampling rate of 22 kHz and stored as mono (single-channel) WAV files using a Dynametric TLP-124A recorder on each interview station. Each UM SRC recording began with the telephone dial tone before the call and ended with the hang-up of the interviewer line or the beginning of the interview, whichever came first. The Arbitron interviews were digitally recorded at 8 kHz in a compressed proprietary format using the TANTACOMM Auditor system and later converted without loss to mono WAV files. Each Arbitron recording began with the interviewer greeting (after the respondent answered the call). The recording ended with a hang-up or the beginning of the interview, whichever came first.

18.4.2 Acoustic Measurements

The Praat speech analysis software package (Boersma and Weenink, 1992–2005) was used for all acoustic analyses and measurements. The interview WAV files were first segmented into conversational turns by trained listeners. Each turn was numbered according to its position in the interview and coded as either interviewer (I), respondent (R), overspeech (O), or telephone (T). *Turns* began whenever the interviewer or respondent began to speak and was not interrupted for at least a single syllable, and ended whenever the other speaker began speaking or with the hang-up. Portions of the interview containing speech from both speakers, or less than a syllable from one speaker and interrupted by the other speaker, were coded as overspeech turns. Telephone dialing or hang-up sounds were coded as telephone turns.

Praat produces spectrograms, or time–frequency visual images of acoustic signals, like that in Fig. 18.1 constructed for the short segment of speech, "hi, my name is." In the spectrogram of Fig. 18.1, the x-axis is time in seconds; the left-hand y-axis is frequency in hertz. Concentrations of acoustic energy at particular frequencies and times are plotted with white representing energy minima, black representing energy maxima, and intermediate shades of gray representing intermediate amounts of energy.

The line in Fig. 18.1 is an f_0 track superimposed on the spectrogram, representing the instantaneous computer-estimated rate of fundamental frequency of laryngeal vibration. The f_0 track has the same x-axis (time) as the spectrogram but has a different frequency scale, plotted on the right-hand side. The peak in the f_0 track at 0.35 seconds is 363 Hz, near the end of the vowel in the word "hi."

In order to assess overall pitch, variation in pitch, final intonation, and speaking rate during each recording, a set of f_0 and speaking rate measurements were carried out on the initial interviewer turns of the selected interviews using Praat's autocorrelation algorithm,[1] Overall pitch was assessed as the median f_0 value of the track. Variation in pitch was assessed by the standard deviation of the f_0 values in

[1]UM SRC WAV files were downsampled to 8 kHz to match the sampling rate of the Arbitron files. The autocorrelation pitch analysis algorithm (2, 1993) used the following parameter settings: pitch floor, 75 Hz; pitch ceiling, 600 Hz; 40 ms Gaussian window.

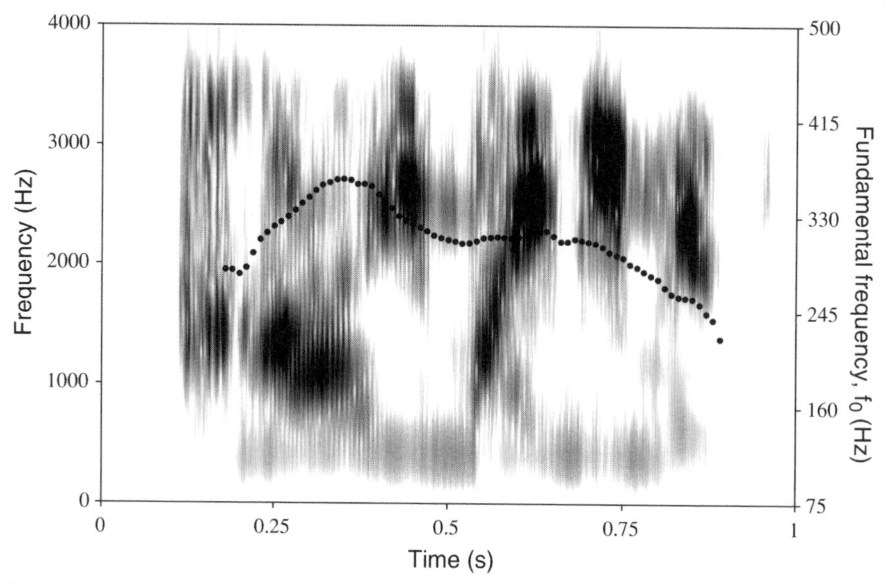

Figure 18.1. A time–frequency plot (spectrogram) of the utterance "hi, my name is" by a female interviewer, with a superimposed f_0 track.

the track. Variation in pitch was also assessed by the low f_0 and high f_0 values of each recording.[2] The pitch pattern at the end of the turn was assessed by final f_0, defined as the median f_0 over the last 50 ms of the turn containing voiced speech (final f_0). Turns ending on a pitch rise have high values for final f_0, while turns ending on a pitch fall have low values for final f_0. Speaking rate in syllables per second was estimated by counting the acoustic energy peaks of the interviewer turn, which roughly correspond to syllable centers.[3]

18.4.3 Ratings of Interviewer Introductory Speech

Eight raters, chosen from the supervisory and monitoring staffs of Arbitron and Survey Research Center, were asked to listen to digital recordings of the first conversation turn of the interviewer. Mulitple raters were chosen in an attempt to achieve some stability in average ratings. The raters were chosen such that the researchers believed that they could not identify the interviewers by voice. The raters were given a computer-assisted training for the rating task, in which the researchers described the purposes of the voice rating. First, the rater judged the likelihood that

[2]Because of occasional errors inherent to the autocorrelation algorithm, 5th and 95th percentile f_0 values of each track were used as proxies for low and high f_0 values, respectively.

[3]An intensity contour was calculated by squaring the amplitude values and convolving with a 40 ms Gaussian analysis window. Each peak in the intensity contour was counted only if it was coincident with periodicity as estimated by the f_0 track, so that high-energy consonant bursts and voiceless strident fricatives would not be counted as syllables.

someone would agree to the survey request Then, on 7-point scales, raters chose values describing the extent to which the interviewer sounded scripted, competent, conversational, confident, enthusiastic, masculine, attractive, easy to understand, fast (rate of speech), high-pitched, breathy, and loud. Raters evaluated all introductions on a single attribute before proceeding to the subsequent attribute. An example introduction was provided for the final four attributes, and raters were instructed to treat the example as the middle of the 7-point scale.

All raters rated all sample interviewer introductions. Raters varied in their ratings of the same introduction; for example, the mean interrater correlation on the rating of likelihood of the contact yielding an "agree" outcome is 0.25. In the analyses that follow, we use for each sample introduction the mean rating across the eight different raters, in order to gain some stability in the ratings, when we use them to predict outcomes of contacts.

18.5 ANALYSIS OF RESULTS

18.5.1 Descriptive Statistics on Interviewer–Householder Contacts

Because of the unique nature of the dataset, we think that mere descriptive statistics are of interest to practicing survey researchers. On average the interviewer group was employed over 2 years—a relatively experienced group of interviewers (see, e.g., Groves and Couper (1998) for interviewer tenure statistics in face-to-face surveys). Most interviewers (79 percent) were female, as is typical of the workforce. The percentage of contacts that led to an agree decision was 41 percent.

The contact-level statistics are interesting. Using the definition of turns that separated overspeech (i.e., moments when both the interviewer and the household informant were talking) from individual interviewer and telephone answerer speech, nonagree contacts contained an average 12 turns and agree contacts, 30 turns during the introduction. The different outcomes do not differ in the average number of seconds of the interviewers' first conversational turn (10 versus 12 seconds), but the total duration of the contact (measured from the opening of the conversation to the termination of the call or beginning of the interview) is twice as long for agree contacts (138 seconds) as nonagree contacts (69 seconds). This compares to the frequently cited statistic of over half of the refusals occurring in the first 30 seconds (Oksenberg and Cannell, 1988) and to recent Arbitron analysis where over one third of all refusals terminate during the first two sentences (Arbitron Inc., 2005 unpublished).

None of the ratings showed large differences between agree contacts and nonagree contacts. Even the mean score for the item about the likelihood of a person agreeing is only slightly higher for the agree contacts than nonagree contacts (4.3 versus 4.1).

18.5.2 Exploratory Analyses

Because we were preparing for multivariate analysis of the acoustic and rater measures predicting contact outcomes, we were curious to measure the intercorrelations of the various items. Table 18.1 presents the correlations for the acoustic measures. That table shows that the median fundamental frequency measure (Median f_0 Hz)

Table 18.1. Correlation Matrix of Acoustic Measures

	Low f_0	High f_0	sd f_0	Last f_0	Length of I'r turn	Syllables per second
Median f_0 Hz	0.83	0.58	−0.02	0.38	−0.13	−0.03
Low f_0 Hz		0.43	−0.24	0.36	−0.13	−0.07
High f_0 Hz			0.66	0.43	−0.07	0.05
SD f_0 Hz				0.19	0.07	0.12
Last f_0 Hz					−0.08	−0.06
Length of I'r turn, seconds						0.10

is highly correlated with the lowest fundamental frequency during the interviewer's speech (r =.83), which means that interviewers have relatively consistent pitch. Median f_0 is also correlated with the highest f_0 measure (High f_0, $r = 0.58$). Similarly, as expected, the higher the "high" fundamental frequency, the higher the standard deviation of the fundamental frequency ($r = 0.66$). In sum, the acoustic measures show expected interitem correlations.

Because full transcriptions of the recordings are not presently feasible, it was not possible to investigate the reported finding of a falling f_0 pattern during certain keywords of the introduction (Sharf and Lehman, 1984; Oksenberg and Cannell, 1988). Instead, as a prelude to investigating the intonational pattern and pragmatic interpretation of the end of the turn, we were able to measure f_0 during the final syllable of the interviewer turn (last f_0) as mentioned above. Given that this has correlations among other measures between 0.35 and 0.45, it appears to be a useful variable to examine.

Finally, the length of the interviewer turn measured in seconds appears relatively uncorrelated with fundamental frequency levels.

18.5.3 Exploratory Multivariate Models of Contact Outcome, Acoustic Predictors

These analyses ask the question: "*can electronic acoustic measurements predict the voice properties that raters hear?*" Since the management question concerns whether machine measurement can replace human observation, this is an appropriate way to ask the question. If the acoustic measurement is not predictive of the raters, we know that there is separable information in the two.

We fit models at the contact level using the acoustic measures, the rater measures, and combinations of both. By any measure of model fit and stability, there was little predictive value of individual ratings and acoustic properties. We believe that this reflects that the models use information *only* from the interviewers' behaviors and include no information about the sample persons' attributes or behaviors. At this level of coding of the contact interaction (just the first interviewer turn), it appears that much heterogeneity of contact situations remain, some of it determinant of whether or not the contact ends in cooperation. Given this outcome, we moved to the interviewer-level analysis.

18.5.4 Exploratory Multivariate Models—Interviewer Level, Mean Historical Response Rate of Interviewers as Dependent Variable

The original work by Oksenberg and colleagues used the interviewer as the unit of analysis. Their work correlated voice attributes with interviewer-level response rates (largely by examining interviewers with high or low response rates and in one instance interviewers with high, middle, and low response rates). This section addresses the question: "*do the typical acoustic properties of an interviewer's voice add anything to the predictive value of the average ratings of the voice, in identifying high performance interviewers from others?*" To address this question we compute a response rate for each interviewer, based on all recordings of the interviewer introductions made during the project, not just the approximately 800 coded and acoustically analyzed. The total number of recordings was over 40,000, and the response rate for contact-level outcome for the 66 interviewers averaged about 52 percent, with substantial variability among them.

Table 18.2 shows a replication of the Oksenberg and Cannell (1988) finding that louder sounding, higher pitch voices are correlated with higher response rates. There is no support for interviewers with faster delivery of introductory speech or those judged more "confident" having higher response rates.

None of the reports by Oksenberg and colleagues fit multivariate models predicting interviewer-level response rates. Table 18.2 shows that the results of multivariate models are dependent on whether controls on the different response rates of the two organizations are made. Without such controls, louder and higher pitch voices appear to have somewhat higher response rates, controlling confidence, and rate of speaking. When organizational controls are entered, the coefficients are driven to statistically insignificant levels. This result partially reflects different distributions for the two organizations on the other independent variables.

We examined many more attributes of interviewer voices in our study than those found important in the Oksenberg et al. work. Thus, we first examine bivariate correlation coefficients between interviewer-level response rates and the means of

Table 18.2. Bivariate Correlations and Multivariate Regression Coefficients for Variables Found Related to Response Rates in Oksenberg et al., Research

	Bivariate correlation	Regression coefficient ignoring organization effect	Regression coefficient controlling on organization effect
Median f_0	0.24*	0.00085*	0.000068
Rating of speech rate	−0.20	−0.065	0.037
Loudness	0.32***	0.10**	−.000068
Confidence	−0.18	0.023	0.0029
Adjusted R^2		0.12	0.64

*$p < 0.10$.
**$p < 0.05$.
***$p < 0.01$.

the rater and acoustic measures. We found that the higher response rate interviewers tended to be judged more conversational ($r = 0.36$), with higher pitched ($r = 0.31$) and breathy ($r = 0.35$) voices. There was a tendency for higher response rate interviewers to be judged more attractive ($r = 0.29$), less masculine ($r = -0.23$), but also (surprisingly) less confident ($r = -0.24$) and less competent ($r = -0.21$). Using the acoustic measures, we found that higher response rate interviewers tended to have introductions with high median f_0 ($r = 0.24$), a high highest f_0 ($r = 0.34$), high variability of f_0 ($r = 0.25$), and a high ending f_0 ($r = 0.27$).

Since the Oksenberg et al. studies, there is increasing evidence that tailoring and customization of interviewer speech during survey requests is productive of higher cooperation rates (Morton-Williams, 1993; Groves and Couper, 1998). We anticipated that variation over contacts in the behavior of interviewers would be related to high response rates. Admittedly, in these data the variation across contacts is likely to be based only on the interviewers' knowledge of the case (e.g., city of residence) or the "hello" of the respondent. Despite the possible weak test of the tailoring hypothesis, we estimate correlations between interviewer-level response rates and the standard deviation of those attributes related to tailoring notions. High-response-rate interviewers show significantly greater variability in their ratings of loudness ($r = 0.41$) and of pitch ($r = 0.28$). Higher-response-rate interviewers show *less* variation over contacts than other interviewers on how masculine they sound ($r = -0.31$). Finally, higher-response-rate interviewers exhibit greater variability in their lowest f_0 measures during their delivery ($r = 0.50$).

Because our purpose was to maximize the prediction of whether interviewers achieve high response rates, we ran a series of stepwise regression models. All models used a dummy variable to remove base differences in response rates between the two organizations. We used means from the ratings and the acoustic measures. We also used as predictors, the standard deviations over contacts within an interviewer. We had two hypotheses about mechanisms by which the standard deviations over contacts affect interviewer-level response rates.

(a) When the predictor attribute was a universally desirable trait (e.g., interviewer attractiveness), variation over contacts in the trait would be negatively related to interviewer-level response rate (controlling on the mean level of the trait). In essence, variation in ratings would measure differences among raters in the judgment of the interviewer.

(b) When the measurement was a trait that might be differentially valued across respondents (e.g., enthusiasm), variation over contacts in the rated trait would be positively related to interviewer-level response rate (controlling the mean level of the trait).

The stepwise regressions were not sensitive to such hypotheses. When both means and standard deviations are introduced as predictors, whichever variates were most powerful were entered into the models. When just means of ratings are used (the first column of numbers in Table 18.3), high response rates are found predicted by low ratings of being scripted ($ß = -0.07$) and low ratings of sounding masculine ($ß = 0.014$). When both means and standard deviations of ratings are candidate predictors, the final stepwise model includes mostly predictors measuring the standard deviations. However, variation predicts higher response rates only for rated enthusiasm and judged age but lower response rates for ratings of being scripted, masculine sounding, and breathy.

When acoustic measures are used as predictors, the mean speed of speech is found related to high response rates, the sole predictor in the stepwise regression using means over contacts. When both means and standard deviations of acoustic measures are candidate predictors, only standard deviations survive as significant predictors. The variation in the speed of speech is negatively related to response rates. (We suspect this might be related to the tendency for low-response-rate interviewers to quickly deliver their introductions when they perceive reluctance on the part of the telephone answerer.) The variation in the low f_0 is positively related to response rates. (We suspect this is an indirect effect of interviewers who are adept at varying the pitch of their voices to fit different circumstances.)

When means and standard deviations from both ratings and acoustic measures are candidate predictors (last column of Table 18.3), new predictors enter the final stepwise model. The new predictors included are means for sounding scripted, confident, and breathy. However, the most important development is the large number

Table 18.3. Regression Coefficients in Forward Stepwise Regressions Predicting Interviewer-level Response Rates, Rater and Acoustic Means and Standard Deviations, after Controlling for Organizational Differences in Mean Response Rates

	Rater variables only		Acoustic variables only		
Predictors	Means only	Means and standard deviations	Means only	Means and standard deviations	All variables
Rater items					
Mean: How scripted?	−0.073**				−0.068**
Mean: How confident?					−0.09***
Mean: How breathy?					−0.12***
Mean: How well could you understand?		0.040*			
Mean: How masculine?	−0.014*				
S. D.: How scripted?		−0.27***			−0.21***
S. D.: How masculine?		−0.36***			−0.31***
S. D.: How enthusiastic?		0.38***			0.35***
S. D.: How breathy?		−0.36**			−0.45***
S. D.: How old?		0.48**			0.36**
Acoustic measures					
Mean: Syllables per second			0.040*		
S. D.: Syllables per second				−0.12***	−0.11***
S. D.: Low f_0				0.0019**	0.0017*
S. D.: Duration					0.008***
Adjusted R^2	0.70	0.80	0.67	0.74	0.86

*$p < 0.10$.
**$p < 0.05$.
***$p < 0.01$.

of entered predictors that measure variation in interviewer behavior over contacts. Variation on some attributes appears to depress response rates; on others, it appears to increase rates. As with all stepwise procedures, however, it is prudent to contrast results with a more conceptually motivated model.

Hence, we also fit a set of models that had specification requirements that the predictors have some track record in the survey methodological literature. For each of these we fit two models, using the means only as predictors and using the means and standard deviations as predictors. Based on prior research findings, we included ratings for scripted (Morton-Williams, 1993), confidence (Groves and Couper, 1998), enthusiasm (Groves and Couper, 1998), and ratings of age (Norris and Hatcher, 1994). Regarding confidence and enthusiasm, we note that in contrast to interviewer perceptions of their own attributes, we have ratings of listener perceptions. Interviewer reports of their own confidence and enthusiasm might be related to higher cooperation, but listener-perceived ratings may behave differently.

Further, we wanted to have measures of interviewer gender (Groves and Fultz, 1985). One was how masculine the interviewer sounded and the other was how "breathy" the voice sounded. Female voices differ from male voices for physiological and stylistic reasons in a number of ways, including some acoustic properties that are perceived as increased breathy voice quality for females (Klatt and Klatt, 1990; Hanson, 1997; Hanson and Chuang, 1999). Listeners can reliably judge breathiness in short utterances (Shrivastav and Sapienza, 2003). We have less of a research record to tap for the acoustic measurements, and we chose to include rate of speech (found important by Oksenberg) and length of introductory conversational turn (found important above).

The first model in Table 18.4 includes only predictors that are means of interviewer behaviors over contacts. Here interviewers who sounded less scripted, less breathy, less confident, who talked faster and longer, tended to have higher response rates. All of these relationships achieve magnitudes detectable at the 0.90 or better confidence level. For the most part, they are consistent with prior results. It is notable that, controlling on impressions of the judged "masculinity" of the voice, breathy voices are connected with lower response rates (this might reflect a perceived lack of professional status). Similarly, the voices consistently judged more confident tend to produce lower response rates. This judgment might be related to an increased social distance evoked by such voices. Since this reverses the result from interviewer self-perceived confidence, some further analysis might be useful.

The findings of the second model in Table 18.4 show large predictive power from variation in interviewer behaviors over contacts. The fit of the model is improved from an adjusted R^2 of 0.72 to 0.81, when predictors reflecting variation in behavior over contacts are added. In contrast to the stepwise results, this result measures the marginal effects of variation, controlling on the mean values of attributes. In short, consideration of variation in interviewer behavior over contacts appears important in understanding cooperation in surveys.

It follows from leverage-salience theory (cf. Groves et al., 2000) that when the specific speech attribute being rated is differentially valued over different types of respondents, then the standard deviations over contacts should predict higher response rates. This would arise from successful tailoring of vocal properties to different respondents' desires thus raising the salience of what the interviewer is saying to the respondent and how it is being said. In contrast, those universally valued atttributes should have negative impact of variability over contacts.

There is some support in Table 18.4 for the logic above. Given the ubiquitous finding that scripting has negative effects on cooperation, the findings that interviewers who vary in "scripting" ratings have lower response rates makes sense. Similarly, monitoring of telephone interviewers often notes that ineffective interviewers often accelerate their rate of speech under stress; Table 18.4 finds that variation in pace over contacts is negatively related to interviewer-level response rates. Further, the standard deviation of the enthusiasm rating has a positive coefficient (0.45, $p = 0.0003$). This is the evidence that some respondents positively value an "enthusiastic" vocal presentation and others do not. Interviewers who show variation over contacts in this vocal attribute achieve higher response rates.

However, other findings in Table 18.4 do not comport with simple explanations. We have no explanation why interviewers who tend to vary in the breathiness or masculine-sounding vocal properties have lower response rates. These puzzles require more study, and probably more data, to resolve.

Table 18.4. Regression Coefficients for Models Using Mean Over Contacts and Standard Deviation Over Contacts as Predictors, Rater Predictors and Acoustic Predictors, After Controlling for Organization Differences in Historical Response Rates

Predictor variable	Means as predictors	Means and standard deviations as predictors
Mean: How scripted?	−0.33***	−0.045
Std. deviation: How scripted?		−0.19**
Mean: How breathy?	−0.14***	−0.081**
Std. deviation: How breathy?		−0.30*
Mean: How masculine did the voice sound?	−0.015	−0.0092
Std. deviation: How masculine?		−0.31**
Mean: How confident?	−0.16**	−0.14***
Std. deviation: How confident?		−0.031
Mean: How enthusiastic?	0.031	−0.010
Std. deviation: How enthusiastic?		0.45***
Mean: How old?	0.043	−0.012
Std. deviation: How old?		0.32
Mean: Syllables per second	0.071**	−0.0012
Std. deviation: Syllables per second		−0.096***
Mean: Duration of interviewer's first turn	0.0061*	0.00091
Std. deviation: Duration of interviewer's first turn		0.0073
Adjusted R^2	0.72	0.81

*$p < 0.10$.
**$p < 0.05$.
***$p < 0.01$.

18.6 CONCLUSIONS

Perhaps the most important summary of this study is that listener ratings and acoustic properties of interviewer's introductory statements predict variation in interviewer-level response rates. They do not, however, successfully predict contact-level cooperation propensities. This contrast in results flows, no doubt, from the large variability in respondent reactions to any given stimulus. (Recall that no analyses in the project incorporated characteristics of the sample persons.) Thus, knowing merely what the interviewers did in their first conversational turn, without knowing the reaction of the respondent, is not very useful in predicting the outcome of the contact.

The memorable findings of Oksenberg et al. on interviewer voices were that higher-response-rate interviewers were rated as louder, higher pitch, apparently more confident, and faster talking. In bivariate analysis we did not find the positive effects of fast speech or of rated confidence of the interviewer. In multivariate models we find that variation over contacts on these attributes is an important part of the impact of vocal characteristics on response rates.

When we add additional predictors to the set used by Oksenberg et al., using stepwise procedures, interviewers whose introductions were judged less scripted, less breathy, and less masculine-sounding had higher response rates. Variability in behaviors was of importance in predicting response rates. It is clear that the acoustic measurements add predictive power for interviewer-level response rates.

We contrasted the stepwise model to a more conceptually justified model. With little loss of model fit or predictive value, those more theoretically motivated models showed less breathy voices related to higher response rates. There were large impacts of variation in interviewer introductions over contacts, most reflecting lower response rates associated with higher cross-contact variation. Indeed, a major finding of the project was the importance of variation in interviewer behavior over contacts in response rates.

What practical value has been learned? We suspect that interviewers who speed up their speech under stress might profit from training interventions. We believe that attempting to dampen consistently the perception of scriptedness is a wise training goal.

Some findings that were not consistent across models deserve practical investigations, in our judgment. One is the finding that mean duration of the introductory turn and/or variation in that duration is related to higher response rates. The length of the turn can be related to informational content of the turn, to the amount of time the respondent is given to judge familiarity of the voice, to understanding the vocal patterns of the interviewer, to perceptions that the interviewer is illegitimately using the time of the respondent, and a host of other possible features of the decision to participate. More study of the length of the interviewer's introductory statements is merited.

In any case, with the ease of electronic recording of interviewer and respondent voices, the exploration of how vocal properties affect the interaction of interviewers and respondents is a valuable opportunity for survey methodology. We hope that other research pursues this opportunity.

CHAPTER 19

Monitoring Telephone Interviewer Performance

Kenneth W. Steve, Anh Thu Burks, Paul J. Lavrakas, Kimberly D. Brown, and J. Brooke Hoover
Nielsen Media Research, USA

19.1 LITERATURE REVIEW

One of the most important advantages that telephone surveys have over other survey modes is that they allow researchers to establish centralized procedures to monitor the quality of the data collection (Lavrakas, 1993, 1997, in press). No other survey data collection mode permits constant, real-time, systematic observation of the data collection process. A telephone survey that does not institute adequate interviewer monitoring procedures misses the great opportunity the mode allows researchers to try to measure and reduce nonresponse, nonresponse error, and measurement error (cf., Groves, 1987). This chapter focuses on some of the ways researchers can deploy monitoring of the work done by telephone interviewers to try to reduce and/or measure possible nonresponse and measurement errors in the telephone surveying.

Obtaining cooperation from all types of respondents, including those indifferent or disinterested, is important for a representative survey sample. Groves (1990) states that the vast majority of refusals take place in the first minute of the phone call. To decrease the amount of refusals, Groves et al. (1992) advocate tailoring the initial delivery of the script to maintain the interaction and thus reduce nonresponse errors; interviewers tailor their responses based on the cues given by the respondent in an attempt to overcome any negative reactions to the request for survey participation. The attractiveness of the survey topic, reciprocation for cooperation, authority of the interviewer's request, and affective states of both the interviewer and the respondent

Advances in Telephone Survey Methodology, Edited by James M. Lepkowski, Clyde Tucker, J. Michael Brick, Edith de Leeuw, Lilli Japec, Paul J. Lavrakas, Michael W. Link, and Roberta L. Sangster

are among the many sociopsychological components that affect the rates of survey response propensity (Groves et al., 1992). If these psychological components are observed and confronted appropriately by the interviewer during the brief initial interaction between the respondent and the interviewer, respondents who initially refuse or express reluctance or indifference may become interested in participating (cf., Groves et al., 2000).

Item and unit nonresponse (i.e., respondent refusals) are phenomena that can easily be recognized and quantified. Measurement errors inherent in the answers respondents provide, however, are often more difficult to detect. Groves (1987) cites the interviewer, the questionnaire, the respondent, and the data collection mode as the four main sources of measurement errors in telephone surveys. Although it is important to try to minimize all sources of error in telephone surveys, this chapter focuses on errors resulting from interviewer behaviors when interacting with respondents. Tucker (1983) states that item variance due to interviewer effects can be a greater source of error than sampling variance. Stokes and Yeh (1988) offered a means for statistically isolating the effects of "discrepant" interviewers. Over the years, in fact, many methods have been proposed for measuring the variability in responses to telephone survey items that can be attributed to interviewer effects (Stock and Hochstim, 1951; Franzen and Williams, 1956; Hansen et al., 1961; Kish, 1962; Freeman and Butler, 1976). Such approaches however, by design, use post hoc methods to account for interviewer errors (i.e., after data collection is complete). As such, they provide little in the way of correcting and reducing such errors in real time, rather than simply accounting for them.

Cannell and Oksenberg (1988) cite a need for quantifiable, objective data on interviewer behavior and performance as a way to measure and reduce interviewer variance, subsequently reducing survey error. Besides identifying sources of measurement error, an interviewer monitoring system (IMS) that incorporates these components can be designed to promote consistent practices by providing feedback to interviewers during the course of the survey. An interviewer monitoring form (IMF) allows researchers to systematically and objectively measure the extent to which interviewer behavior influences nonresponse and thereby may influence nonresponse error (cf., Lavrakas 1987, 1993). Telephone monitoring as a quality control method can support the collection of high-quality representative data for phone survey organizations. A valid and reliable monitoring process requires the establishment of clear, objective evaluation guidelines. Typically, subjective ratings of interviewer quality will not produce consistency among raters and cannot be compared across studies.

Fowler and Mangione (1990) outlined some essential components of a successful monitoring system, including the direct observation of 10 percent of all interviewers by trained staff using a standardized scoring/rating form. They also suggested that entire interviews be observed from obtaining cooperation to case wrap-up and that interviewer feedback be given promptly. Cannell and Oskenberg (1988) highlight a number of additional important attributes of a quality monitoring system that may improve the overall function of measuring and reducing error: testing behavioral measures for reliability, designing objective monitoring criteria around task-related behaviors, presenting feedback in a nonthreatening manner, and emphasizing positively the role of the monitoring system in the collection of quality data. Consistency sessions that train monitoring staff on coding procedures, especially in areas where subjective judgment is involved, further help staff to evaluate the quality of interviewers' work. Holding consistency sessions whenever a new monitor is hired and forms or procedures are updated promotes consistent understanding of coding criteria. Morton-Williams (1979) discussed holding sessions to train monitoring staff on the standards and to ensure the standards are commonly understood by all raters. Cannell and Oskenberg (1988) stated the importance of comparing and discussing the coding of interviews, with training aimed at generating agreement among coders.

A quality monitoring system developed with these items in mind will become part of a comprehensive approach to controlling measurement error and reducing total survey error. Lyberg and Kasprzyk (1991) advocate objective rating systems designed to produce statistics on interviewer errors, productivities, response rates, and quality of interviewing skills. Over the years a number of approaches have been used to try to develop objective method for monitoring interviewer performance. Barioux (1952) offered a method for tracking the number of technical mistakes made by interviewers while delivering a script. His method, however, accounted only for negative behaviors and did not include those behaviors the interviewer performed correctly. Nor did it include behaviors related to the quality of an interviewer's voice, which has been shown to be linked to respondent refusals (Oskenberg et al., 1986; also see Groves et al. (2007), Chapter 18 in this volume).

Sudman (1967) proposed an approach for producing standardized scores of interviewer performance where each error was assigned an error weight, a quantified value based on the severity of the error. An error where the interviewer fails to probe a vague answer given by a respondent might receive a greater error weight compared with the acceptance of partial answers or an unexplained change of code or answers. Individual error scores were then standardized using the mean and the standard deviation for all calls monitored in a given study. Unfortunately, this procedure produces scores that cannot be interpreted relative to specific interviewer behaviors (i.e., the scores are norm referenced). Norm-referenced scores do not allow one to infer what the interviewer is capable of doing, only how well they perform relative to the other interviewers; criterion-referenced scores contain information regarding the interviewers' behavior repertoire (Berk, 1984; Crocker and Algina, 1986; Oosterhof, 1999).

Morton-Williams (1979) adopted the "verbal-interaction coding" frame of Cannell and Robison (1971) for identifying problematic survey questions. By rating a variety of interviewer and respondent behaviors for each question, they were able to objectively monitor and evaluate the performance of the interviewing staff. Both Tucker (1983) and Groves and Magilavy (1986) have focused on interviewer behaviors to evaluate their impact on measurement error. For example, how deviation from verbatim administration of questions, probing, and pacing affects the way in which a respondent answers survey items. Dykema et al. (1997) identified variables to use when creating objective measures for monitoring, such as the verbatim reading of questions, voice skills, pacing, and probing techniques.

Finally, Mudryk et al. (1996) used control charts to track groups of interviewer behaviors such as question delivery (asking, wording, and professionalism/voice), respondent relations (probing/anticipation and judgment), subject matter (definitions and concepts), and data processing (data entry and notes). Again, however, it seems that these efforts appear to be focused primarily on tracking interviewers' mistakes and effectively ignoring all the behaviors that the interviewer performs correctly. It is argued here that an interviewer monitoring system that does not formally and explicitly target "good" behavior is incomplete in that it does not lend itself to the use of tried and true behavior modification principles (e.g., positive reinforcement) during feedback and training.

It is common for a variety of statistics generated from monitoring sessions to be used as part of interviewer performance feedback. Even though monitoring data and feedback can help reinforce skills and behaviors learned in training and ensure interviewers follow rules, such a system will likely not overcome the deleterious effects of inadequate training. Since retraining of interviewers is a necessary and recommended component of maintaining a quality interviewing staff, Fowler (1991) and Lyberg and Kasprzyk (1991) suggest using objective monitoring data to help find areas for individual and group improvement. Carefully designed monitoring systems and feedback practices help interviewers in their attempts to obtain accurate data and gain cooperation from all types of people, reducing nonresponse and measurement errors. When studying how interviewer monitoring as a form of supervision can improve data quality, Fowler and Mangione (1990) found that the type of feedback given to interviewers communicates the values of the organization. When interviewers were given feedback solely on response rates and efficiency, they were less likely to rate data quality as important to their job. When interviewers were given feedback about the quality of their data collecting behaviors, they tended to focus more on accurate data collection. One might easily take this a step further and argue that the way in which the quality feedback is delivered will also have a large impact on the way the interviewers perceive their job. If they only receive feedback when they have done something wrong or need to improve, group morale will surely suffer and individual motivation will likely be stunted.

In summary, it appears that most telephone survey researchers are aware of the importance of monitoring interviewers and that there is some consensus as to what types of things are important to develop a good IMS. There does not, however, appear to be a widely accepted standard IMS approach. This makes it very difficult for practitioners to discuss and compare the impact of interviewer performance and training on the quality of data they collect, not only with each other but also to their clients. This chapter attempts to establish a framework through which such a standard may be achieved. Any such solution must be scalable so that quality assessments can be compared across studies, both within and across different survey organizations. However, prior to discussing this new approach, a review of current state of the art in telephone interview monitoring practices is in order.

19.2 STATE OF THE ART

In the summer of 2005 and the spring of 2006, a web-based survey was conducted to gather information about how survey organizations monitor the quality of work performed by their interviewers. The sampling frame for this survey was constructed using various membership directories, including the American Association for Public Opinion Research (AAPOR), the Council of American Survey Research Organizations (CASRO), the European Society for Opinion and Marketing Research (ESOMAR), and the International Field Directors and Technologies Conference (IFDandTC). The sample was stratified across directories so that all organizations belonging to AAPOR, CASRO, and IFDandTC were selected, while a subset of the organizations on other membership directories was selected. The sampling approach used reflected the fact that AAPOR, CASRO, and IFDandTC lists had fewer organizations in total, yet a higher percentage of organizations that conduct telephone survey data collection. ESOMAR, for example, had many hundreds of organizations, but proportionately fewer that conducted telephone surveys. Thus a sample was drawn from this list rather than include every organization. In total, 767 organizations were selected to participate.

Survey research organizations were recruited via e-mail (54 of which were returned undeliverable). Of the 713 organizations apparently contacted, only four indicated that they were ineligible via e-mail. In the e-mail invitation, the organizations were asked to have the person who was most knowledgeable about their internal monitoring practices complete the survey. In the first phase of sampling, organizations were given approximately 6 weeks to complete the survey (July 26 to September 2, 2005). In the second phase of sampling, organizations were given 3 weeks to complete the survey (April 5 to April 26, 2006). In total, 193 organizations logged on to the survey Web site to provide information. Of these, 187 included valid data that could be included in the data analyses for an AAPOR 1 response rate of 24.4 percent. The survey consisted of 53 questions pertaining to telephone survey research in general, facilities and operational, and telephone monitoring practices.

19.2.1 Survey Participants

Of the organizations that completed the survey, 31 percent were established prior to 1975, 20 percent were established between 1976 and 1985, 25 percent were established between 1987 and 1995, and 24 percent were established between 1996 and 2005. Nearly 67 percent were commercial survey research firms, while a small percentage were from academia (18 percent), not-for-profit sector (9 percent) or government (5 percent). Half of the organizations reported having a relatively small call volume (average less than 10,000 completed interviews in a year—see Table 19.1). Taking into account how many computer-assisted telephone interview (CATI) work stations they had, 36 percent were classified as small (20 stations or less), 29 percent were medium sized (21–50 stations), 20 percent were considered to be large (51–100 stations), and 16 percent were very large (more than 100 stations).

Table 19.1. Typical Call Volumes

Interviews completed per year	*f*	%
Less than 10,000	70	50
10,001–50,000	44	31
Greater than 50,000	26	19

n = 187, 47 missing cases.

While all of these organizations indicated that they conduct telephone survey research, just over half indicated that they conduct their telephone interviews internally. Approximately one-third indicated that they conduct interviews both internally and through an external source, and one tenth completely outsourced all their interviewing. The average pay for research interviewers across research organizations was $8.03 per hour. Nearly half of these research organizations (50 percent) reported having the ability to conduct telephone surveys in a language other than English.

19.2.2 Monitoring Staff and Practices

Of all organizations surveyed, 82 percent reported that they monitor their telephone interviews. Many of the staff who conduct monitoring within these organizations have had previous experience as telephone survey research interviewers (45 percent). In addition to past work experience, most monitoring personnel received some formal training (instructional classes = 30 percent, on-the-job training = 23 percent, managerial instruction = 24 percent). In fact, only 4 percent of organizations indicated that they have no formal training established for their monitoring staff. The average interviewer/monitor ratio within these organizations was 19:1. The ratio of interviewers to monitoring staff ranged widely, from 2:1 to 150:1. Most survey organizations reported having eight or fewer staff members dedicated to monitoring. On average, approximately two people were dedicated to monitoring full time. Among the relatively small organizations, an average of 385 hours is spent monitoring each year (see Table 19.2). As expected, the amount of time spent monitoring increases as the call volume increases. For call centers with relatively small call volume, however, the percentage of call time monitored appears to be greater. When asked about the percentage of contacts monitored, 11 percent of the organizations said they monitor 5 percent or less, 26 percent reported monitoring between 5 and 10 percent, 17 percent

Table 19.2. Hours Monitoring by Organization Size (Call Volume)

Call volume	Completed interviews	Average hours/year	(Hours/year)/call volume
Small	Less than 10,000	385	0.460
Medium	10,001–50,000	1526	0.058
Large	Greater than 50,000	9081	0.042

Table 19.3. Purposes for Monitoring

Reason	*f*	%
Interviewer performance assessment	112	82
Project quality assessment	101	74
Coaching training	106	78
Consistency tracking	94	69
Mandated by clients	68	50
Internal research (R and D)	33	24
Other	18	14

monitor between 10 percent and 20 percent, and 13 percent of the organizations indicated that they monitor more than 20 percent of the interviews they complete. Sixty percent of the organizations indicated that they determine which calls to monitor based on interviewer performance.

These survey organizations offered a number of reasons for monitoring their telephone interviews. Assessing interviewer performance was the most frequent reason given for monitoring (82 percent of those responding—see Table 19.3), followed by interviewer training/coaching (78 percent), evaluating project quality (74 percent), and tracking consistency (69 percent). With respect to the feedback provided to the interviewers, organizations indicated that the majority of the feedback is qualitative in nature rather than quantitative. When feedback is given to interviewers, typically it is provided immediately after the call (73 percent of survey organizations). Given the emphasis on qualitative feedback, it is not surprising that less than half (41 percent) of the organizations indicated that their monitoring process leads to an overall score of interviewer performance.

19.2.3 Interviewer Monitoring Systems

The methods employed to monitor interviewers vary greatly across organizations. A majority (51 percent) of the organizations indicated that they had a standard form used when conducting monitoring sessions. Of these, 35 percent utilize both paper and computer formats, 48 percent utilize a paper only, and 17 percent are strictly computerized. Nearly half of the organizations (46 percent) reported that their interviewers were unaware when they were being monitored. Approximately three fifths (58 percent) of the organizations said they used audio monitoring equipment, while 47 percent used some sort of visual display; 43 percent used both.

Across organizations a variety of behaviors were reported to be monitored, including voice characteristics, reading questions verbatim, script tailoring, probing, respondent participation/reluctance, and data entry errors (see Table 19.4). Most organizations thought that there was "some need" to improve their monitoring processes and procedures. More specifically, organizations indicated that improvements could be accomplished by acquiring better technology, improving on the consistency of monitoring, automating the monitoring process, and by dedicating

Table 19.4. Behaviors Typically Monitored

Behavior	*f*	%
Voice characteristics	118	89
Read questions as worded	133	99
Script tailoring	114	86
Nondirective probing	118	90
Gain respondent cooperation	122	91
Counter respondent reluctance/refusal	118	89
Data entry errors	95	73

more staff to monitoring. When asked on a scale of 0–10 how important telephone monitoring is to their organization, the majority indicated monitoring is "quite" important (mean = 8).

The results of this survey underscore the relative importance of monitoring telephone interviewers. It is clear that most survey research organizations exert a certain rigor in an attempt to reduce interviewer-related measurement error and that new tools to assist these efforts are likely to be embraced. This survey of organizations' monitoring practices also reflects the existing literature in that there does not appear to be a standard approach to monitoring telephone interviewers.

19.3 A NEW APPROACH TO MONITORING TELEPHONE INTERVIEWER BEHAVIOR

What follows is an explanation of how a systemic approach to monitoring telephone interviewers' behaviors was devised at one large survey organization. Although the specific system described below is unique to that organization (Nielsen Media Research), the principles that guided its structure and content should generalize to essentially any survey organization that monitors its telephone interviews. As part of Nielsen Media Research's (NMR) TV Diary surveys, well over 10 million RDD numbers are called each year to recruit households to keep a diary of their TV viewing. In 2001, Nielsen set out to completely reconceptualize and retool the IMF that was being used to assess telephone interviewers' performance on these diary surveys. Concerns with the previous IMF centered on the fact that it produced data that did not reflect interviewer performance differences across interviewers. The new IMS was designed to accomplish four main objectives. First and foremost, the new system was to facilitate coaching efforts to improve the skills level of the interviewers. Second, the system was to direct monitoring attention to several parts of the diary recruitment interview that are annually audited by NMR's client-sponsored auditors. Third, the new system was to be a source of valid information for the human resource department when determining interviewers' merit increases. Finally, the form was to be a source of data for the methodological research department when testing the effectiveness of new call center methodologies.

19.3.1 Developing the New Form

For the new IMF to be an effective coaching tool, it was concluded that the form must contain the *specific verbal behaviors* that interviewers were trained to utilize during diary recruitment. Anchoring the form to observed behaviors was also a way of establishing the construct validity of the new form (cf., Crocker and Algina, 1986; Anastasi, 1988; Oosterhof, 1999). The project began with a series of meetings to identify the range of verbal behaviors that would be measured. This process was one of the collaborations between the methodology research department and the quality and training department within the Nielsen call centers. Several joint sessions were held to operationally define the behaviors and ensure the definitions did not overlap. The goal was to define each behavior using simple, but explicit terms that could be objectively and reliably observed. This process was critical to establishing both construct and content validity of the IMF.

The IMS was comprised of two parts: the actual form used to code the call being monitored and a reference guide. The reference guide was a supporting document containing the operational definitions for each behavior and the corresponding coding options. A list of the behavioral definitions can be found in Appendix 19A. The reference guide also contained instructions for using the form (not shown in the appendix). The form itself was comprised of three basic sections that parallel the normal progression of a TV diary recruitment call (i.e., introduction, data collection, and closing—see Appendix 19B). The quality monitoring staff at the call centers were first required to complete some basic information to assist in tracking and trending the calls being monitored (e.g., interviewer ID, respondent phone number, etc.). This was followed by a brief description of the call outcome or disposition. This was necessary since many calls did not result in a contact with a qualified respondent. The monitors were instructed to use the note areas to keep track of any "problem" behaviors during the call. Their notes could then be used to code the interviewers' behaviors once the call was complete.

As with many surveys that are monitored, the behaviors required of the interviewer during the different parts of the diary recruitment call were not identical. Therefore, the behaviors that were monitored and coached for each part of the call were modulated as necessary (see Table 19.5). Those monitored in more than one section of the call were weighted differently when scored depending on the relative importance of the behavior. Behaviors with a weight of zero in Table 19.5 were not monitored in this part of the call. The speed, modulation, and enunciation with which the interviewer delivered the script were monitored in every part of the call. While these behaviors were coded on a 3-point scale (e.g., speed was "too slow," "proper," or "too fast"), they were scored dichotomously—either on target (OT) or below target (BT). Thus, talking too fast or too slow for the respondent—that is, "below target"—had the same effect on the interviewer's final score.

The interviewer's flexibility, enthusiasm, and confidence were monitored during the introduction and closing sections of the call. All were coded as being "not enough," "correct," or "too much" (see Appendix 19A), where correct was OT and the others were BT. These behaviors were not monitored during the data collection

Table 19.5. Points and Assigned Coding Options to Given Behaviors by Section[a]

	Section/points			Coding options				
Item	A	B	C	SBT	BT	OT	AT	SAT
Speed	5	1	1		X	X		
Modulation	5	1	1		X	X		
Enunciation	5	1	1		X	X		
Flexibility	9	0	6		X	X		
Enthusiasm	9	0	6		X	X		
Confidence	9	0	6		X	X		
Neutrality	5	6	2		X	X		
Accuracy	5	0	2		X	X		
Coding	5	4	0		X	X		
Probing	7	4	0		X	X		
Media employment	0	7	0		X	X		
Other verbatim	0	6	0		X	X		
Address verify	0	7	0		X	X		
Other verify	0	6	0		X	X		
Zip code	0	7	0		X	X		
Resistance	No points: sets the point value of persuaders[b]							
Persuaders	0–30	0–30	0–30	X	X	X	X	X
Unexpected behavior	8	0	0	X	X	X	X	X
Appropriate reaction	8	0	0	X	X	X	X	X
Return importance	0	0	10	X	X	X	X	X
Return appreciation	0	0	10	X	X	X	X	X

[a]X denotes a coding option is available. SBT = significantly below target, BT = below target, OT = on target, AT = above target, SAT = significantly above target.

[b]No resistance = 0, mild resistance = 1, strong resistance = 2, prevent = 3, multiplied by 10 to assign points for persuaders.

portion of the call. It was reasoned that the goal of the interviewer during this portion of the call was to remain neutral and collect accurate data, that is, to engage in standardized survey interviewing (cf., Fowler and Mangione, 1990). Behaviors associated with policies and procedures varied greatly depending on which section of the call was being monitored. Neutrality was monitored throughout every section of the call, whereas accuracy was monitored only during the introduction and closing, and coding and probing were monitored only during the introduction and the data collection. In the data collection section, there were several study-specific behaviors (i.e., media employment, other verbatims, address verify, other verify, and zip code) critical to external audits the call center regularly receives. All policy and procedure-related behaviors were coded dichotomously: the interviewers either get it right or they do not.

The amount of "respondent resistance" and interviewer's use of "persuaders" were also monitored throughout every section of the call (persuaders is a generic term used to refer to the behaviors the interviewer exhibits to overcome respondent resistance). Persuading behaviors were carefully addressed in interviewer training (see Appendix 19A). These were deemed the most critical behaviors in determining whether or not the respondent would agree to keep a TV viewing diary. Respondent resistance included anything that made it difficult for the interviewer to conduct the survey (e.g., questions, interruptions, negative comments, etc.). Respondent resistance was classified into four categories (none, mild, strong, and prevent). If the respondent exhibited no behavior that made it difficult for the interviewer to complete a given section of the call, the person monitoring the call would mark "none" for that section. If the respondent demonstrated a behavior that made it somewhat difficult for the interviewer to complete a given section, the respondent resistance is coded as "mild" for that section, and so on. The final coding option of "prevent" was coded when the interviewer did not need to use reactive persuading techniques as a result of their use of proactive strategies.

The interviewers were trained to use two types of strategies when dealing with respondent resistance: reactive and proactive. Reactive strategies were those commonly associated with persuading respondents after they had express explicit disinterest. Proactive strategies consist of the interviewers' ability to anticipate the concerns of the respondents essentially before they had a chance to voice them. If the interviewer used these proactive strategies effectively, there would be no overt manifestation of resistance on the part of the respondent. In this case, the interviewer's behavior would be marked "significantly above target." If they used these strategies but the respondent still exhibited resistance, their use of persuaders was still significantly above target, assuming the interviewers used the reactive persuading techniques to their full advantage. Prevent would not be coded for resistance in this case (see Appendix 19A).

In administering the introduction, "unique circumstance" and "appropriate reaction" were included to record atypical respondent behaviors (e.g., respondent says someone is at the door and sets the phone down) and whether or not the interviewer handled them appropriately. In the closing section the respondent has already agreed to keep the diary, but that is no guarantee that they will actually return it. At this point in the call, the interviewer is required to let the respondent know how important it is to return the diary and thank the respondent in advance for doing so. Return importance and return appreciation items were added to this section to monitor how well the interviewer accomplishes these two tasks.

Finally, the call summary area at the bottom of the form contained overall information about the call (see Appendix 19B). Some of this information came from the person monitoring the call, while the rest was a product of an automated scoring procedure. The adverse item would be marked whenever an interviewer exhibited a behavior that demanded immediate disciplinary action (e.g., the interviewer used vulgar language). If an adverse behavior was observed, the interviewer would automatically receive a score of 0 for that call. The interviewer impact item was designed to allow the persons monitoring give their impression of the overall effectiveness of the interviewer (i.e., whether or not the interviewer's behavior caused the respondent to hang up or agree to keep a diary). This was used in conjunction with the call outcome, which was recorded by the person monitoring the call. Once all the data were entered into the computer, the final score/rating was computed using the scoring component described below.

19.3.2 Developing the Scoring Component

In keeping with the ideas that guided the design of the monitoring form, the scoring procedure was developed so that scale scores would reflect interviewer performance relative to expected phoning behaviors (i.e., criterion referenced). All calculations were based on the desire to produce a set of final scores that were scaled such that they would range from 0 to 100, where a score of 50 reflected OT behavior. This stemmed from the fact that the survey organization used a coaching and reward program where interviewer performance was categorized as significantly below target (SBT), BT, OT, above target (AT), or significantly above target (SAT). This was accomplished by assigning item weights to the individual behaviors and scaling by awarding partial credit for behaviors that were less than optimal to the success of the call.

19.3.3 Weighting the Behaviors (Items)

Each item was assigned a number between 0 and 10 to reflect its relative importance in making a successful diary recruitment call. These decisions were jointly made by managerial staff at the call centers and senior researchers in the methodological research department. The worth of each point (i.e., point value) was determined by dividing 100 by the total number of points available to the interviewer. The value of 100 was used so that the final scale will range from 0 to 100. Item weights were determined by multiplying the point value by the number of points assigned to that item.

$$W_{\mathrm{I}} = P_{\mathrm{I}}\left(\frac{100}{\Sigma P_{\mathrm{F}}}\right) \tag{19.1}$$

The item weights were then added up to compute the total score. Within this simple equation, the item weight (W_{I}) was equal to the point value for an item (P_{I}) multiplied by the ratio of 100 over the sum of all points available on the form (ΣP_{F}). Unfortunately, this approach produced a score that was inversely related to the probability of the respondent accepting the diary (odds ratio $= 0.999$, $p < 0.001$).

This was due to entering items on the form that assess respondent resistance directly into the calculation of the final score. To correct this problem, respondent resistance items were given a point value of zero and used instead to determine the amount of bonus points assigned to items measuring how well the interviewer used persuading techniques within the section of the call where the resistance occurred. Algebraically, this was accomplished by adding a term "B" to the formula to reflect the bonus points should they be awarded.

$$W_{\mathrm{I}} = (P_{\mathrm{I}} + B_{\mathrm{I}})\left(\frac{100}{\Sigma P_{\mathrm{F}}} + \Sigma B_{\mathrm{F}}\right) \tag{19.2}$$

In this revised equation, the bonus points for a given item (B_{I}) are added to the point value for that item (P_{I}), and the total bonus points for the entire form (ΣB_{F}) are added to the total number of points for the form (ΣP_{F}). While the weight being added to persuading techniques is referred to as "bonus points," conceptually it is more appropriate to think of persuader items as having variable weight depending on how much was needed. This addition to the item weight calculations was a critical part of ensuring the criterion validity of the final score when used to predict diary accept and return rates.

Furthermore, the form was designed so that the introduction (Section A), data collection (Section B), and closing (Section C) portions of the diary recruitment call were monitored separately. If a respondent hung up the phone during the introduction, Sections B and C would be left blank. Thus "incomplete" forms were produced by completed calls and could not be scored. If incomplete forms were not scored, refusal calls would be selectively removed from the final data distribution. This would hinder the ability to use the scores for predicting response and cooperation rates. To allow for the scoring of incomplete forms, the weights had to be calculated for each section individually. The weights were then multiplied by a ratio of the total number of points available for that section over the total number of points available across all completed sections.

$$W_{\mathrm{I}} = \left[(P_{\mathrm{I}} + B_{\mathrm{I}}) \left(\frac{100}{\Sigma P_{\mathrm{S}}} + \Sigma B_{\mathrm{S}} \right) \right] \left(\Sigma P_{\mathrm{S}} + \frac{\Sigma B_{\mathrm{S}}}{\Sigma P_{\mathrm{F}}} + \Sigma B_{\mathrm{F}} \right) \tag{19.3}$$

In this final item weight equation (19.3), the constant of 100 was divided by the total number of points available for the section ($\Sigma P_{\mathrm{S}} + \Sigma B_{\mathrm{S}}$), and the total quantity was multiplied by a ratio defining the relative contribution that section made toward the final score ($\Sigma P_{\mathrm{S}} + \Sigma B_{\mathrm{S}}/\Sigma P_{\mathrm{F}} + \Sigma B_{\mathrm{F}}$). The final score was then determined by adding up all the item weights (ΣW_{I}). When item weights were calculated this way, all monitored calls were scored regardless of where the call ended. The amount each section contributed to the final score depended on the number of sections completed and the number of bonus points awarded to each section (i.e., they were linked to the amount of resistance on the part of the respondent).

19.3.4 Scaling the Final Score

Coding options for each item on the form were identified as being SBT, BT, OT, AT, or SAT. The coding option marked for each item determined the percentage of the weight for the item that was awarded toward the final score (i.e., SBT = 0 percent, BT = 25 percent, OT = 50 percent, AT = 75 percent, and SAT =100 percent). This anchored "on target" behavior at a score of 50. In other words, if all items were coded OT, 50 percent of all the item weights would be awarded and the resulting score would be 50. If all items contained coding options ranging between SBT and SAT, the scores resulting from adding up the item weights would range on a scale from 0 to 100. Given that this was not the case (e.g., speed), a second set of equations was needed to transform the scores above and below 50 to get the scale range back to 0 to 100. Scores below 50 were rescaled into the final score (S_{F}) using Eq. (19.4).

$$\mathrm{S}_{\mathrm{F}} = 50 - \left[(50 - \Sigma W_{\mathrm{I}}) \left(\frac{50}{R_{\mathrm{L}}} \right) \right] \tag{19.4}$$

In this equation (ΣW_I) is the sum of item weights and (R_L) is the lower range of scores (i.e., 50—the lowest possible score). Here, the lowest score possible on the quality monitoring form was allowed to vary as a function of the number of complete sections and the number of bonus points awarded. The same was true for the highest possible score. Scores above 50 were rescaled into the final score (S_F) using Eq. (19.5),

$$S_F = 50 + \left[(\Sigma W_I - 50) \left(\frac{50}{R_U} \right) \right] \tag{19.5}$$

where (ΣW_I) is the sum of item the weights and (R_U) is the upper range of scores (i.e., the highest possible score—50).

19.3.5 Computational Formulas

As one might expect, the values for (R_L) and (R_U) were not easily determined (i.e., the number of combinations of bonus points available and sections complete can be quite large). This task was simplified by taking a more computational approach to calculating the final scale score. For this approach, the scores for each section (ΣW_S) were scaled prior to combining the section scores into the final score. In this case, the ratio determining how much each section contributes to the final score was removed from the item weight calculation.

$$W_I = (P_I + B_I) \left(\frac{100}{\Sigma P_S} + \Sigma B_S \right) \tag{19.6}$$

The section scores were scaled prior to computing the final score using slightly modified versions of the two weight equations.

$$S_S = 50 - \left[(50 - \Sigma W_S) \left(\frac{50}{R_L} \right) \right] \tag{19.7}$$

$$S_S = 50 + \left[(\Sigma W_S - 50) \left(\frac{50}{R_U} \right) \right] \tag{19.8}$$

The sum of the item weights was limited to those for the section being scaled (i.e., the section score). Thus, the values for R_L and R_U had only four possible values for each section depending on the amount of resistance exhibited by the respondent, and the ratio removed from the item weight calculation was then brought back when compiling the scaled section scores into the final scaled score. The final scaled score was now a weighted sum of the three section scores,

$$S_F = [(\Sigma W_A)(R_A)] + [(\Sigma W_B)(R_B)] + [(\Sigma W_C)(R_C)] \tag{19.9}$$

where the contribution of each section was defined by the ratio of the points available for the section over the points available across all completed sections.

$$R_F = \left(\sum P_S + \frac{\sum B_S}{\sum P_F} + \sum B_F \right) \tag{19.10}$$

19.3.6 Evaluating the Resulting Scores

To evaluate the new scoring system, a total of 3180 calls were monitored across 464 interviewers by a group of 56 monitoring staff. Each call was monitored in its entirety. A typical monitoring session was five calls per interviewer. Some interviewers were monitored more than once by more than one quality staff member. An interviewer might have as many as 40 calls monitored. However, most ranged between 5 and 15 monitored calls (1–3 sessions). The number of times a given quality staff member used the new monitoring form during this time was dictated to a large degree by his or her production work schedules. Most staff members were monitored between 20 and 100 calls each. To avoid problems with imputed values when scoring the forms, at least one section on the form had to be complete and no items within a completed section could be left blank. Procedures were put in place at the call center to check for missing data prior to scoring the forms. A total of 610 calls could not be scored due to incomplete data (i.e., the call did not get to the introduction). The resulting dataset was comprised of 2570 calls.

The distribution of scores had good dispersion and were relatively normally distributed. There were, however, an exceptionally large number of interviewers who received a score of 50. It was hypothesized that this was due to a tendency for those monitoring to code interviewer behavior as being OT, especially during early stages when they were first becoming used to using the new form. The minimum observed score was 24.3 when adverse behaviors were ignored (i.e., those that trigger a score of zero). The maximum score observed was 92.1. The final scale scores had a mean of 47.7 and standard deviation of 10.6 and did not appear to vary as a function of the frequency with which the interviewer was monitored. When these scores were placed in their appropriate performance category, 113 calls were SBT (83 adverse), 431 were BT, 1837 were OT, 151 were AT, and 38 were SAT.

19.3.7 Reliability

Fifty calls were monitored by more than one staff member in order to estimate interrater reliabilities. The interrater reliability for each behavior was computed by simply dividing the number of agreements by the number of paired observations (see Table 19.6). Some behaviors produced little variance when monitored, making it difficult to obtain traditional estimates of reliability and criterion validity. Several items had zero variance across observations or near zero interrater reliabilities, indicating that they may not be working as intended. If there is truly zero variance within these behaviors, the question arises as to whether or not they should be retained on the form. These behaviors were kept due to their perceived value to the call center

Table 19.6. Interrater Reliabilities for Individual Behaviors by Section

	Introduction		Data collection		Closing	
Behavior	r	n	r	n	r	n
Speed	0.32	49	0.00	35	0.02	32
Modulation	0.05	50	ZV	35	0.04	32
Enunciation	0.00	50	ZV	35	ZV	31
Flexibility	0.04	50			ZV	32
Enthusiasm	0.20	50			0.53	32
Confidence	0.04	50			0.05	32
Accuracy	ZV	48			ZV	32
Neutrality	ZV	48	0.05	34	ZV	32
Coding	ZV	48	ZV	34		
Probing	0.69	48	0.36	34		
Resistance	0.87	49	0.95	34	0.66	50
Persuaders	0.73	22	0.40	50	ZV	2
Unexpected behavior	0.56	45				
Appropriate reaction	0.29	46				
Media employment			0.89	33		
Other verbatim			0.41	33		
Address verify			ZV	32		
Other verify			0.26	31		
Return import.					0.30	32
Return apprec.					0.49	32

r = number of agreements/number paired observations; ZV = zero variance.

management (i.e., content validity). The prevalence of items with zero variance and/or low reliability may be due to a poorly defined construct, and inadequate training of the monitoring staff. Such cases require a review of the operational definition and corresponding training materials (i.e., construct validity). If a behavior is properly defined and has good content validity, its item weight should be reduced unless a second criterion for success is identified and a more complex measurement model is fit. There were, however, a number of verbal behaviors that had acceptable levels of interrater reliability (e.g., respondent resistance, $r = 0.87$ and the interviewer's use of persuading techniques, $r = 0.73$). These behaviors played a key role in obtaining a successful outcome, and thus more time was devoted to their development. It was hoped that over time other behaviors on the form would be refined to such a degree.

19.3.8 Validity

Although the final scale scores for the prototype form significantly predicted whether or not a respondent agreed to keep the diary (odds ratio = 1.066, $p < 0.001$), the strength of the prediction was not as high as one would like. There was some question about the validity (construct and criterion) of items that produce zero and/or low variance. If an item does not vary, it is not measuring the intended behavior unless the behavior itself does not vary. Regardless of whether or not the item is working, inclusion of these items in calculations of the final score adds measurement error to the validation model, thus lessening the ability to accurately predict accept rates with the final score. If the item does not differentiate between differences in variable behavior, changes to the item or the way it is used should be implemented to validate the item. If the behavior itself truly does not vary, then the item designed to monitor that behavior should be removed from the form.

19.3.9 Expanding the Interviewer Monitoring System

Once the base form was developed, software was written to convert the IMS to an electronic platform so that behaviors could be coded directly into a database with a mouse click using drop-down menus. When a call is finalized, monitoring scores are produced by the computer based on the algorithms described above. All of the data are warehoused on a mainframe computer so that performance reports can be easily run by call center managers. This functionality has proven to be invaluable to the quality and training staff. Being able to run reports on individuals and groups allows the quality and training department to efficiently allocate resources to areas that need improvement the most. The system also gives them a precise framework with which to make these improvements since the information can be tied back to specific behaviors.

The new IMS was also expanded to include all the telephone studies conducted by NMR. At the top of the interface, the person monitoring simply selects the study and the behaviors to be monitored appear on the screen. Since the behaviors for each study are not the same, each study draws from an overall library of behaviors. To adjust the scoring from study to study, behaviors not being monitored are assigned a neutral value (i.e., on target) when the study is selected, and different scaling values are applied to get the final score. As such the resulting scores for all studies are on the same scale, with the same normative interpretations (i.e., SBT, BT, OT, AT, and SAT). No matter what study is being phoned or where a call terminates, scores can also be objectively decomposed into behavioral criteria for individual coaching and identifying areas to improve training efforts.

19.4 DISCUSSION

It is clear that telephone interviewer monitoring is an area of survey research that is central to ensuring the quality of data collected through telephone surveying. The existing literature may, however, leave the practitioner with more questions than answers. When it comes to the details of developing a new interviewer monitoring system, the efforts described here are an attempt to establish a framework for answering some of these questions. While it may not be the final solution to standardizing telephone monitoring practices, it may help to frame the problem in a way that such a goal may be realized.

The IMS described in this chapter is most certainly imperfect. Even so, the impact it has had on the efficiency and the quality of Nielsen's call centers has been immense. Attempts have been made to incorporate behavior modification principles into the coaching process (i.e., reinforcing good behaviors, giving clear examples of target behaviors, etc.). By focusing on clearly defined, objectively observed behaviors the interviewers can see exactly what is expected of them and exactly what they need to do to improve their performance. More importantly, the traditional corresponding friction between the interviewers and monitoring staff has been significantly reduced. This was in large part due to establishing objective scoring criteria. The old form generated highly subjective scores as made by the monitoring staff at that time.

The new form and scoring procedure described above eliminated these concerns because the monitoring staff no longer compute the final score. The new IMS was perceived by the call center staff as being much more objective. By adjusting the focus of feedback sessions based on a formal assessment of positive interviewer behaviors, the morale of the interviewers has increased substantially. Furthermore, the resulting data distribution not only reflects the actual performance differences across interviewers but also captures fluctuations in performance for individual interviewers. The ability to efficiently diagnose weaknesses and monitor the quality of daily production efforts has improved due to the fact that call center managers can quickly and easily run performance reports by study, interviewer, team, quality representation, trainer, and so on.

Although this new approach has had a huge impact on the way the NMR conducts telephone survey research, the bigger question is whether or not it can have an impact on the way the rest of the survey research community does so. The IMS described in the chapter is not intended to be a static form, but more customizable to the setting at hand. Such flexibility is crucial given the variations in purpose for survey research studies. One should not expect that the same set of behaviors would be important across all studies. This being the case, such an approach would require the researcher to report which behaviors were being monitored, how they were operationally defined (if no standard exists), and some indication as to their reliability and validity.

Still, more needs to be done for this approach tool to prove to be an effective tool in standardizing the way research organizations monitor their telephone interviewers. More research is needed to develop a "library" of telephone interviewer behaviors that are both comprehensive and well defined. For example, are these behaviors static across studies, or do they vary depending on study objectives? More specifically, how should a given behavior be operationally defined? It is also still somewhat unclear how behaviors should be weighted once validated. Should the weights be set based on their relative importance (i.e., the implications of a given type of error) or based solely on their ability to predict the outcome? Finally, what are the criteria for assessing the validity of these different types of behaviors? Clearly, the criteria for coding errors are not the same as those for persuading. Although this system may not solve all the problems surrounding telephone interviewer monitoring and attempts to establish a standard practice, the authors believe it to be one step further in that direction.

APPENDIX 19A: OPERATIONAL DEFINITIONS OF INTERVIEWER BEHAVIORS

Voice skills

Speed—How quickly or slowly the script is delivered in relation to the respondent. (Slow—Interviewer spoke too slowly for the respondent. Proper—Interviewer spoke at a proper rate of speed for the respondent. Fast—Interviewer spoke too fast for the respondent.)

Modulation—The natural rise and fall of the voice. (Monotone—The interviewer sounded disinterested, bored, and/or robotic. Natural—The interviewer sounded natural, energetic, and interested in the respondent. False—The interviewer's voice was exaggerated or overly dramatic/false sounding.)

Enunciation—Speaking clearly and distinctly. (Unclear—The interviewer did not speak with enough clarity and distinction, was hard to understand. Clear—The interviewer spoke clearly and distinctly. Exaggerated—Interviewer overemphasized the pronunciation of words.)

Personality

Flexibility—The use of conversational delivery and/or modifying the script. (Not enough—Interviewer did not display a conversational delivery and/or script modification. Correct—Interviewer displayed a conversational delivery and/or modified the script. Too much—Interviewer was too loose with his/her delivery of the script and/or strayed from task.)

Enthusiasm—The interviewer is eager to speak with the respondent (RSPN) and deliver the script. (Not enough—The interviewer did not display enough energy and excitement to generate RSPN interest. Correct—The interviewer displayed the energy and excitement needed to generate RSPN. Too much— The interviewer demonstrated excessive energy and excitement.)

Confidence—Speaking with assurance and certainty. (Not enough—The interviewer sounded uncertain, timid, hesitant, and so on. Correct—Interviewer spoke in a confident, yet respectful manner. Too much—The interviewer sounded arrogant, overbearing, and/or abrupt.)

Policies/procedures

Accuracy—The interviewer delivered information and responded to concerns correctly. (Accurate—The interviewer provided accurate information and addressed questions correctly. Inaccurate—The interviewer provided inaccurate information, left out important details, and/or did not address questions correctly.)

Neutrality—Behavior that does not bias or influence data. (Neutral—The interviewer did not "lead" the respondent or misrepresent NMR or the TV ratings process. Break—The interviewer "led" the respondent or misrepresented NMR or the TV ratings process.)

Coding—The interviewer properly coded all responses provided in the introduction. (Correct—Interviewer properly coded all responses. Incorrect—Interviewer miscoded one or more responses.)
Probing—Additional questions were asked to obtain a clear and complete response. (Correct—The interviewer probed correctly when necessary. Incorrect—The interviewer probed in a way that resulted in an unclear or incomplete response. N/A—Probing was not necessary.)
Media employment—Accurate delivery of the media employment question. (Correct—The interviewer delivered the media employment question verbatim. Incorrect—The interviewer did not deliver the media employment question verbatim or completely failed to deliver the media employment question. N/A—The call ended prior to question.)
Other verbatim—The accurate delivery of verbatim questions. (Correct—The interviewer delivered all data collection questions verbatim. Incorrect—The interviewer failed to deliver one or more questions verbatim.)
Address verify—The interviewer verified mailing address to create an accurate mailing label. (Yes—The interviewer completely and correctly verified mailing address. No—The interviewer did not verify mailing address at all or failed to do so thoroughly. Not applicable—The call ended prior to question.)
Other verify—The interviewer used the operator's alphabet (O/A) when entering new information. (Yes—The interviewer used O/A to verify any new information such as name, address, cable. No—The interviewer failed to use O/A to verify any new information such as name, address, cable. Not applicable—The call ended prior to use.)

Interviewer/respondent exchange

Respondent resistance—Did the respondent demonstrate resistance, for example, interrupt, etc. (None—The respondent demonstrated *no* behavior that made it difficult for the interviewer to complete the survey; the interviewer did not use or effectively use proactive strategies. Mild—The respondent demonstrated behavior that made it *somewhat* difficult for the interviewer to complete the survey. Strong—The respondent demonstrated behavior that made it *very* difficult for the interviewer to complete the survey, may include hang up. Prevent—The respondent demonstrated *no* behavior that made it difficult for the interviewer to complete the survey due to the RI's effective use of proactive strategies.)
Persuaders—Methods for reducing RSPN resistance by (1) identifying RSPN concerns, (2) effectively addressing these concerns, and/or (3) taking a proactive approach to prevent the need to persuade. (Not applicable—There was no need and/or opportunity to persuade. Significantly below target—The interviewer completely failed to use persuaders when needed. Below target—The interviewer attempted to persuade but was unsuccessful. On target—The interviewer used persuaders in an appropriate and timely manner. Above target—The interviewer used persuaders and advanced persuading strategies. Significantly above target—The interviewer used persuaders and advanced strategies. Proactive strategies were used prior to resistance.)

Unique circumstance—The interviewer experienced a unique circumstance or response. Example: Respondent said, "I don't live here, I'm just the babysitter, but I can give you my number and I would be happy to do the survey." (Yes—The interviewer experienced unique circumstance/response. No—The interviewer did not experience unique circumstance/response.)

Appropriate reaction—The interviewer's reaction to the unique circumstance/response. (Yes—The interviewer appropriately reacted to unique circumstance/response; for example, the interviewer said, "I appreciate your interest but unfortunately this number has been scientifically selected and we are unable to add additional numbers at this time". No—The interviewer inappropriately reacted to unique circumstance/response; for example the interviewer says, "It doesn't matter if you're a household member or not, you can participate!" Not applicable—The interviewer did not experience a unique circumstance/response.)

Return importance—The interviewer communicated the importance of returning the diary. (Below target—The interviewer completely failed to mention the importance of returning the diary and/or importance was mentioned but not emphasized. On target—The interviewer properly emphasized the importance of returning the diary. Above target—The interviewer properly emphasized the importance of returning the diary and explained that his/her input is important to his/her local community. It is important to send all the dairies back for his/her household to be included in the TV ratings.)

Return appreciation—The interviewer expressed sincere appreciation to the RSPN for taking part in the survey and returning the diary. (Significantly below target—The interviewer completely failed to express appreciation for taking part in the survey and returning the dairy. The interviewer expressed appreciation but did not sound sincere. On target—The interviewer expressed sincere appreciation to the RSPN for taking part in the survey and returning the diary. Above target—the interviewer expressed sincere appreciation to the RSPN for taking part in the survey and returning the diary, and attempts to personally connect with the respondent to gain and secure a commitment.)

APPENDIX 19B: QUALITY MONITORING FORM

QR ID: ______ DATE: __/__/__ TIME: __:__ AM PM CALL CENTER: WF RC INTERVIEWER: ______

RI ID: ______ BOOTH: ______ WEEK: 1 2 3 4 PHONE # - - OUTCOME: AM/PM AC RF NC OS

DISPOSITION

LANGUAGE	E S A O

SAMPLE	AOS	BOS	HOS

ATTEMPT NUMBER														
1	2	3	4	5	6	7	8	9	10	11	12	13	14	15

ANSWR MACH/PRVCY MNGR	AM	PM	NO
APPROPRIATE MESSAGE LEFT	N/A	YES	NO

PART A: INTRODUCTION

	VOICE SKILLS		
A1: SPEED	SLOW	PROPER	FAST
A2: MODULATION	MONOTONE	NATURAL	FALSE
A3: ENUNCIATION	UNCLEAR	CLEAR	EXAGGERATED

	PERSONALITY		
A4: FLEXIBILITY	NOT ENOUGH	CORRECT	TOO MUCH
A5: ENTHUSIASM	NOT ENOUGH	CORRECT	TOO MUCH
A6: CONFIDENCE	NOT ENOUGH	CORRECT	TOO MUCH

	POLICIES / PROCEDURES	
A7: ACCURACY	ACCURATE	INACCURATE
A8: NEUTRALITY	NEUTRAL	BREAK
A9: CODING	CORRECT	INCORRECT
A10: PROBING	CORRECT	INCORRECT

	INTERVIEWER / RESPONDENT EXCHANGE					
A11: RSPN RESISTANCE	NONE		MILD	STRONG		PREVENT
A12: PERSUADERS	N/A	SBT	BT	OT	AT	SAT
A13: UNIQUE CIRCUMSTANCE		YES			NO	
A14: APPROPRIATE REACTION		YES			NO	

PART B: DATA COLLECTION

	VOICE SKILLS		
B1: SPEED	SLOW	PROPER	FAST
B2: MODULATION	MONOTONE	NATURAL	FALSE
B3: ENUNCIATION	UNCLEAR	CLEAR	EXAGGERATED

	POLICIES / PROCEDURES	
B4: MEDIA EMPLOYMENT	CORRECT	INCORRECT
B5: OTHER VERBATIM	CORRECT	INCORRECT
B6: ADDRESS VERIFY	YES	NO
B7: OTHER VERIFY	YES	NO
B8: ZIP CODE	YES	NO
B9: NEUTRALITY	NEUTRAL	BREAK
B10: CODING	CORRECT	INCORRECT
B11: PROBING	CORRECT	INCORRECT

	INTERVIEWER / RESPONDENT EXCHANGE					
B12: RSPN RESISTANCE	NONE		MILD	STRONG		PREVENT
B13: PERSUADERS	N/A	SBT	BT	OT	AT	SAT

PART C: CLOSING

	VOICE SKILLS		
C1: SPEED	SLOW	PROPER	FAST
C2: MODULATION	MONOTONE	NEUTRAL	FALSE
C3: ENUNCIATION	UNCLEAR	CLEAR	EXAGGERATED

	PERSONALITY		
C4: FLEXIBILITY	NOT ENOUGH	CORRECT	TOO MUCH
C5: ENTHUSIASM	NOT ENOUGH	CORRECT	TOO MUCH
C6: CONFIDENCE	NOT ENOUGH	CORRECT	TOO MUCH

	POLICIES / PROCEDURES	
C7: ACCURACY	ACCURATE	INACCURATE
C8: NEUTRALITY	NEUTRAL	BREAK

	INTERVIEWER / RESPONDENT EXCHANGE					
C9: RSPN RESISTANCE	NONE		MILD	STRONG		PREVENT
C10: PERSUADERS	N/A	SBT	BT	OT	AT	SAT
C11: RTRN IMPORTANCE	SBT	BT		OT	AT	SAT
C12: RTRN APPRECIATION	SBT			OT	AT	

CALL SUMMARY

ADVERSE SITUATION	MISCODE	UNPROF.	LANGUAGE	OTHER
INTERVIEWER IMPACT	NEGATIVE		NEUTRAL	POSITIVE

CALL OUTCOME	AC		RF	CB	OS
FINAL RATING	SBT	BT	OT	AT	SAT

CHAPTER 20

Accommodating New Technologies: Mobile and VoIP Communication

Charlotte Steeh
Independent Consultant

Linda Piekarski
Survey Sampling International

20.1 INTRODUCTION

In the 1970s, survey methodologists believed that the telephone mode would lead to higher response rates in general population surveys than the face-to-face mode. And, for a time, this prediction seemed to be confirmed—at least in a few cases (Groves and Kahn, 1979). Now, 30 to 40 years later, some researchers are predicting the demise of the telephone survey altogether. In this chapter, we examine the proposition that the two newest technologies—wireless communication and voice over Internet protocol (VoIP) may offer a way to rejuvenate the telephone survey. We describe the penetration of these new technologies and how they are evolving and merging. Because studies have been conducted over the past 5 years using mobile phones as the mode of administration, we will also summarize research findings that reveal the major obstacles that future mobile surveys will have to overcome. Since VoIP has not yet been used explicitly in telephone surveys, most of our discussion will center on mobile phones.

20.2 MOBILE PHONES OFFER CHALLENGES AND POTENTIAL

Mobile telephones already outnumber landline telephones in a majority of countries throughout the world (see the statistics for 2005 on the website of the International

Advances in Telephone Survey Methodology, Edited by James M. Lepkowski, Clyde Tucker, J. Michael Brick, Edith de Leeuw, Lilli Japec, Paul J. Lavrakas, Michael W. Link, and Roberta L. Sangster

Telecommunications Union: http://www.itu.int/ITU-D/ict/statistics/ index.html). The popularity of mobile phones and a recent trend toward "wireless-only" or "primarily wireless" households and individuals around the globe pose immediate as well as long-term challenges to telephone sampling and survey methodology. There are many important differences between the mobile environments in the United States and the rest of the world. The Global System of Mobile Communications (GSM), the standard in most countries except the United States, works virtually everywhere and anywhere. In the United States GSM exists side by side with other, sometimes incompatible systems. Many of these other systems do not work outside the United States or in all parts of the United States or even in all parts of a single town. The billing structures are also different. Only in the United States a subscriber's mobile telephone is tied to a single service provider, a single number, and a single rate plan for which he or she pays to receive calls. Outside the United States, a person buys a telephone and may have one or several service providers, telephone numbers, and plans, switching between them by simply changing the SIM (subscriber identity module) cards in their mobile phone.

20.2.1 Penetration of Mobile Phones

Mobile telephone penetration in the United States lags behind that of other countries, primarily due to differences in reliability, technology standards, existing landline infrastructures, and the cost model described above. Mediamark Research Inc. estimated that in the fall of 2005 seventy percent of U.S. households had at least one mobile telephone (Baim, personal communication 2005), a finding corroborated by Tuckel and O'Neill (personal communication 2005) in a personal interview survey based on a much smaller sample size. Individuals and households that have only mobile telephones have the potential to introduce bias to telephone surveys limited to a frame of landline numbers. The number of wireless-only households in the United States has grown to a level that is causing concern for researchers.

Several personal interview studies conducted in 2004 suggested that somewhere between 6 and 7 percent of households in the United States had only mobile phones. Data from Mediamark Research, Inc. (Baim, personal communication 2006) reported that the percent of households with only a mobile telephone has risen to

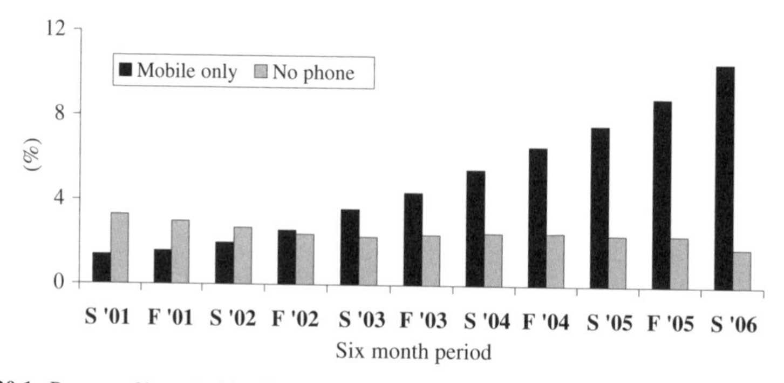

Figure 20.1. Percent of households with only wireless telephone service or no telephone service, 2001–2006. *Data source*: Baim, Julian (Executive Vice President), Mediamark Research Inc., New York. Personal communication. Steeh, C. *Accomondating New Technologies: Mobile and VoIP Communication*; © Mediamark Research, Inc., 2006

10.6 percent (Fig. 20.1). Similar statistics from the 2005 National Health Interview Survey (NHIS) are in Blumberg et al. (2007, Chapter 3 in this volume). In addition, more and more households are primarily mobile and only use their landline for burglar alarm systems, emergency situations, and computer or fax communications. If respondents who only use their landlines for faxes and computers were added to the mobile only, the percentage of respondents without landline access would rise. For example, in the 2003 mobile telephone survey described later in this chapter, it increased from 11 to 19 percent of respondents. This is evidence that we have underestimated the segment of the telephone population that cannot be contacted through a landline. When these three factors are combined with the approximately 2 percent of households, according to the 2000 Census and NHIS data (see Blumberg et al., 2007, Chapter 3 in this volume), that have no telephone service, it is clear that a frame of landline numbers fails to cover more than 10 percent of all households thus increasing the possibility of bias.

As a result of these trends, mobile phones will need to be included in telephone surveys in the not-too-distant future, but sampling them presents numerous challenges. In 2003, a group of telephone methodologists convened a Mobile Telephone Summit meeting to explore the impacts of mobile phones on telephone sampling. After a second summit was held in New York City in February 2005, these challenges and concerns related to sampling mobile phones were defined together with a statement on ethical practices (Lavrakas and Shuttles, 2005b).

20.2.2 Frames for Sampling Mobile Phones

It helps that in most countries mobile numbers can be distinguished from landline numbers. Sometimes the first digits of the number contain the relevant information as in the Netherlands where "06" indicates mobile service (de Leeuw, 2005, personal communication). Because mobile numbers are not included in standard telephone directories in the United States, and the creation of a comprehensive, independent directory of mobile numbers seems unlikely anytime in the near future, list assisted methods cannot be applied when constructing a sample frame of mobile numbers. Exchange-type codes on databases from the Telcordia Routing Administration, the official source for telecommunications industry routing and rating databases, distinguish most mobile telephone numbers from landline numbers. However, some prefixes and 1000-blocks are classified as "shared" by Telcordia and may contain both landline and mobile services.

It is possible to treat landline and mobile numbers as separate but equal frames *and* provide coverage of the mixed exchanges by expanding the mobile frame from a 1000-block to a 100-block frame, similar to the standard list-assisted landline frames available from most sample suppliers. A 100-block organization allows mobile numbers from the mixed or shared exchanges to be included rather than excluded from the frame (as was necessary in the earliest mobile telephone samples) by incorporating those 100-blocks in shared service prefixes with no listed numbers, i.e. those 100-blocks that are not on the list-assisted frame. The file is then sorted according to whatever outside information is available, systematically selecting every *n*th element until the desired sample size is reached, and finally appending a two-digit random number. With a few possible exceptions (see Section 20.2.3), sampling from both frames will ensure complete coverage of all households and individuals with a telephone. Although sampled separately for operational reasons (see Section 20.3), the two frames can be treated as a single virtual frame by applying the same sampling ratio during sample selection.

However, there are two important issues that confound both the comparison of results from these frames and their combination into a single frame. First of all, the mobile telephone accesses individuals rather than households in developed countries. Although it is still not clear to what extent mobile phones are strictly personal devices and landline phones are household devices, it is clear that there will be difficulties in weighting a landline household sample and a mobile individual sample comparably. Secondly, the mobile frame overlaps the landline frame creating multiple probabilities of selection for households and individuals. These two issues are perhaps the most serious challenges to the future of random digit dial (RDD) telephone surveys since determining those probabilities and weighting parameters can be complicated, particularly for households that have both types of telephone service (Brick, 2005b).

20.2.3 Challenges of Calling Mobile Phones

The practicality of using mobile phones to interview respondents depends in large part on the characteristics of the mobile system within a country. Whether or not the called party is charged for the call constitutes one major distinction. In the United States, Canada, and Hong Kong, the called person and the person who initiates the call *both* pay a per minute charge (Lau, 2005; Bacon-Shone and Lau, 2006). In Europe and most other countries, only the caller pays as long as the intended recipient is in the caller's country of residence. The effects of charging the called party have been mitigated in the United States to some extent by the many different plans mobile telephone companies offer. Typically these plans provide a limited number of free minutes during the week days and unlimited free minutes on the weekends and after 9:00 p.m. on week nights.

There are other differences in telephone systems that seem to coincide with national boundaries. In most European countries, it is cheaper to call a mobile telephone from another mobile telephone rather than from a landline telephone, and prepaid service is much more common than in North America (Callegaro and Poggio, 2004). In the United States the legal status of mobile telephone surveys is less favorable than in other countries. For instance, the federal Telephone Consumer Protection Act (TCPA) requires hand dialing of calls to mobile phones thus outlawing the use of autodialers.

Since mobile telephone subscribers in the United States more than likely pay to receive calls, in dollars or minutes, one would expect a higher percentage of refusals and a lower response rate than in a conventional telephone survey. Because the number of possible outcomes of a call attempt is much greater in a mobile telephone survey than that in a landline survey, call disposition codes must also be remapped and response rate formulas modified (Callegaro et al., 2007).

Although mobile phones are truly mobile and, unlike landline phones, can be transported anywhere, they may, nevertheless, be more difficult to contact since the telephone can be turned off, subscribers can be roaming or in an area without service, calls can be dropped, or the network can be busy, just to name a few of the possibilities (Callegaro et al., 2007). In addition, the prefix merely identifies the rate center where the mobile service was purchased and not where the telephone is physically located. A mobile telephone can accompany its owner to college, on a trip to Europe, to a new job location, to restaurants, entertainment venues, in the car, virtually anywhere. This portability means a loosening of the traditional geographic precision of landline telephone numbers.

Other considerations complicate the administration of mobile telephone surveys. Since mobile ownership is unevenly distributed in the U.S. adult population, a sample of mobile numbers cannot be representative of the general population, at least not demographically. Although telephone researchers worry that the landline survey is suffering unacceptable noncoverage bias, the same can be said for mobile telephone surveys. Finally, the omnipresence and easy accessibility of the mobile telephone threaten the quality of the data collected in this manner. Respondents may answer in environments that are not conducive to accurate data collection. Driving a car, eating in a restaurant, or flying in a helicopter are not situations that induce thoughtful responses to interviewer questions. Furthermore, the perceived costs may lead respondents to give the shortest answer possible or to opt out of difficult sequences. Researchers also need to determine how best to protect the respondent's safety.

As mobile phones become more convenient, less expensive, and more popular, they will continue to have an impact on the coverage of landline telephone samples. Given TCPA restrictions on calling mobile phones, sample inefficiencies, and operational considerations in the United States, most researchers do not want mobile telephone numbers in their RDD samples. Nevertheless, despite precautions taken by commercial sampling firms to exclude prefixes and 1000-banks dedicated to mobile service, mobile numbers appear sporadically in list assisted telephone samples. Some get there because mobile numbers are published in white page telephone directories but are not in a 1000-block identified by Telcordia as belonging to a mobile service provider. Some come from rural areas where prefixes or 1000-banks or even 100-banks contain multiple types of service. mobile and unlisted landline numbers in prefixes providing mixed services will not usually be included in a list assisted sample frame unless they are in 100-blocks with directory-listed numbers. Other mobile numbers get in landline samples because a subscriber has ported a landline number to a mobile service provider and it was not successfully scrubbed, that is, removed by sample providers to comply with the regulations of the TCPA. As of December 2005, over 2 million landline subscribers had ported their telephone number to a mobile service. Based on continuing analyses conducted by Survey Sampling International, 89 percent of these ported numbers were in POTS (Plain Old Telephone Service) prefixes, but most of these ported POTS numbers (85 percent) are in exchanges or 100-banks with no directory listed residential numbers suggesting that businesses are the more likely users of landline to mobile porting. The most common reason researchers will encounter a mobile telephone in an RDD sample is call forwarding. Many people and businesses find it very convenient to forward calls received on their home or office telephone to their mobile telephone when they are away from their home or office.

There are some positives associated with including mobile phones in telephone samples. It will of course improve coverage, particularly of young single males. It also has the potential to improve contact rates, particularly with those households or individuals that rarely or never answer their landline telephone. Text messaging may also offer opportunities for researchers. A text message can be a substitute for an advance letter, or may even be used to administer a short survey. Currently text messaging in the United States is generally a premium service for which most subscribers pay extra, and it is unclear whether or not sending a text message now or in the future will violate various state and federal legislation that seeks to ban unsolicited commercial messages.[1] Again the key word is "commercial" and the research community might easily find itself lumped with e-marketing in the same way we have been confused with telemarketers.

20.3 RESEARCH USING MOBILE PHONES IN SURVEYS

Just as penetration rates have generally been higher in Europe than in the United States, so research related to sampling and surveying persons and households through mobile telephony began earlier there.

20.3.1 International Research

The first systematic methodological research focusing on mobile telephones was carried out in Finland. In the mid-1990s Statistics Finland began a program of research that led to the routine inclusion of mobile telephones in its labor force studies. By July 2001, Statistics Finland conducted approximately 50 percent of its CATI interviews on mobile phones (Kuusela and Simpanen, 2002). In 1998, Mediametrie, a French media ratings company, concerned about how a decline in traditional landline subscribers might affect their ratings, also began to conduct mobile telephone interviews and by 2003 had incorporated a mobile telephone only stratum in its standard radio survey (Roy and Vanheuverzwyn, 2002; Vanheuverzwyn and Dudoignon, 2006). In Italy, a major mobile provider, Telecom Italia Mobile, conducted customer satisfaction surveys in 1999 drawing samples from its list of subscribers (Perone et al., 1999).

Experimental studies that compared the results from a survey using a traditional RDD landline frame with the results from a survey based on an RDD mobile telephone frame were also undertaken first outside the United States: in Germany during 1999 and 2002 (Fuchs, 2000a, 2000b, 2002a, 2002b, 2006), Brazil in 2002 and 2003 (Roldao and Callegaro, 2006), Slovenia in 2003 (Vehovar et al., 2004a), and

[1]House Resolution 122 proposes to amend section 227 of the Communications Act of 1934 to prohibit the use of the text, graphic, or image messaging capabilities of mobile telephone systems to transmit unsolicited commercial messages (http://frwebgate.access.gpo.gov/cgi-bin/getdoc.cgi? dbname=108_cong_bills&docid=f:h122ih.txt.pdf)

Hong Kong in late 2003 and early 2004 (Lau, 2005, Bacon-Shone and Lau, 2006). During May 2005, NOP World in the United Kingdom carried out a mobile telephone survey in addition to its traditional landline survey before the British General Election (Moon, 2006). Although the outcomes of these experiments differed to some extent by country, they all confirmed that conducting interviews over mobile phones was not only quite feasible but also desirable.

20.3.2 Research in the United States

The Arbitron Surveys

Skepticism that a survey of the general population could be successfully carried out with a sample of mobile numbers delayed systematic research in the United States. Thus methodologists did not begin to evaluate the feasibility of using mobile phones in telephone surveys until 2002 when Arbitron Inc., worried, like Mediametrie, about the validity of its radio rating system that relied on landline telephone contact, launched a program of mobile telephone research to investigate whether or not the general public—and particularly 18 – 24 year olds—would respond to a request for an interview made over a mobile phone.

Arbitron's extensive program began with a pilot study of 200 interviews in three states and continued through a series of three additional surveys conducted in 2004 and 2005. The results of the 2004 survey were, for the first time, compared to data obtained from a conventional RDD sample on an issue critical for Arbitron: how often respondents would agree to keep a diary of their radio listening habits when contacted by mobile phone. The last survey in the program, conducted during the summer of 2005, tested a method for blending landline and mobile numbers—screening a mobile frame for mobile only households, adding these cases to data collected from a traditional landline sample, and then weighting the combined file to the distribution of the complete telephone owning population (Fleeman, 2005, 2006; Cohen, 2006).

The Arbitron studies made significant contributions to our understanding of how mobile phones will function as a mode of survey administration. They demonstrated that (1) despite pessimistic expectations, some people in the United States will participate when an incentive is offered and the interview is short; (2) a systematic sample of mobile numbers is not overloaded with nonworking, ineligible units; (3) mobile subscribers who agree to participate in a survey are likely to be heavy mobile telephone users; and (4) the geographic match between a respondent's residence and the area code of the mobile telephone number is much less precise, especially at the local level, than in landline surveys (Cohen, 2003, Fleeman and Estersohn, 2006). In addition, they learned that the diary placement rate deteriorated when the mode of contact was a mobile telephone (Fleeman, 2005, 2006). The emphasis on mobile only households in these studies has also been the focus of much of the research conducted to date on mobile phones in the United States.

Although the last Arbitron survey revealed significantly different listening patterns for mobile only respondents, overall estimates of these habits for the blended sample showed almost no differences from estimates obtained using the landline survey alone. Thus, Arbitron concluded that noncoverage of the mobile only population of adults is not currently affecting overall radio listening patterns and that the research program had accomplished its purpose. In 2008, however, Arbitron plans to include mobile only households in some of its standard telephone surveys (Cohen, 2006).

Experimental Studies

The first systematic mobile telephone survey at the national level was conducted in 2003 by Georgia State University as part of a mode comparison study that also included a traditional RDD survey.[2] The surveys were carried out in as identical a manner as possible in order to determine (1) how mobile telephone interviews differ from standard telephone interviews, (2) how well conventional telephone survey methods accommodate the unique characteristics of mobile technology, and (3) whether or not the quality of data collected in this manner is at least equal to the quality of the data gathered in a standard RDD survey. The project was a first step in exploring the hypotheses that adding mobile numbers to telephone samples would improve the coverage and response rates of telephone surveys and thus the quality of the data, issues still unsettled. The questionnaire covered detailed items about telephone use at both the personal and household levels and replicated attitude and behavioral items asked in other well-known general population surveys.

As part of the same project, a second, independent national mobile telephone survey was conducted in early 2004 to determine experimentally whether sending a text message as a prenotification would reduce the noncontact rate. The sample in this study was limited to numbers that had been allocated to the one mobile company, Nextel, that provided the service to all of its customers and also maintained a sophisticated website for sending and tracking text messages.

Also in 2004 the Joint Program in Survey Methodology (JPSM) at the University of Maryland carried out a second pair of matching national surveys, one using a mobile number frame and the other a frame of landline numbers. Although the research design closely resembled the 2003 Mode Comparison Study in both sample sizes and numbers of interviews, the study was conceived from the beginning as a test of dual frame sampling methodology. The emphasis was not so much on detecting differences between the two surveys as on devising a procedure for combining the results to obtain overall estimates of telephone use that would resemble benchmarks from personal interview surveys covering the entire U.S. adult population. Most items on the questionnaire involved the household's telephone service, but a few asked about attitudes. The mobile telephone component also contained experiments that varied monetary incentives and text messaging to determine their effectiveness in gaining cooperation (Brick et al., 2005a, 2005b).

In the spring of 2006 the PEW Research Center conducted a third set of surveys with almost the same parameters as the 2003 and 2004 studies.[3] The project was designed "to assess the feasibility of conducting a telephone survey in a [mobile telephone] sampling frame" (Keeter, 2006). Like the final 2005 Arbitron study, the

---

[2] This research was supported by the National Science Foundation under grant SES-0207843.

[3] The authors thank Scott Keeter, Director of Survey Research at the PEW Research Center, for making available the methodological details of these studies.

Pew effort sought to incorporate mobile telephone interviews into conventional RDD landline samples by focusing on mobile-only respondents. In addition, the study looked at differences between and among respondents grouped by patterns of telephone ownership—landline only, mobile only, and both mobile and landline—as well as by mode of interview. Thus they were able to determine whether the respondents who owned both types of telephone but were interviewed on their landlines differed from the same type of respondent interviewed on their mobile phones. The questionnaire contained a wide array of items on telephone use and a few inquiries about public issues, such as presidential approval, gay marriage, and the Iraq war.

Thus there have been four separate mobile telephone surveys and three landline surveys, all national in scope, from which some initial inferences can be drawn about the differences and similarities of the two data collection methods. In the 2003 surveys, it was assumed that standard telephone methods would apply to mobile telephone surveys. Subsequent studies benefited from the results of this initial effort or chose to test the effectiveness of alternative strategies. Table 20.1 presents the methodological details for each study. Despite the challenges that we have discussed in Section 20.2.3, the results across surveys conducted in the United States are amazingly consistent. Although respondents in conventional surveys overwhelmingly tell interviewers that they would not participate if a survey were conducted by mobile telephone (Tuckel and O'Neill, 2006), actual practice indicates that a fairly stable proportion of adults will agree.

20.3.3 Conducting Mobile Telephone Surveys

Wireless Samples

The samples for the 2003, 2004, and 2006 mobile telephone surveys were drawn using a basic element design. Prefixes were first selected systematically from a frame of all dedicated mobile 1000-blocks sorted by the FIPS State and County of the rate center and within FIPS by the Operating Company Name. As the final step, a three-digit random number was appended to each selected 1000-block.[4] This design excluded mobile numbers in mixed exchanges. Legal restrictions on autodialing and the absence of a reliable directory of mobile numbers suggested that there would be a substantial decrease in the efficiency of mobile samples. Experience in all three studies confirmed this expectation. In 2003, the ratio of hand-dialed mobile numbers to landline numbers, either hand-dialed or autodialed and excluding numbers prescreened as ineligible, was 2.9:1. In 2004 the ratio was 1.6:1, and in 2006 it was 1.5:1. Decreased efficiency of this nature has the effect of either lengthening the field period or requiring additional interviewers to compensate for lower productivity. All of which, in turn, increase the costs of administering the survey.

In the end it would appear from the data in Table 20.1 that the ineligibility rates for mobile surveys are only a little worse, and in one case a bit better, than the rates in traditional landline surveys, supporting the earlier Arbitron finding that mobile samples, unlike the earliest RDD landline samples, are not primarily composed of

[4] The mobile number sample for the 2003 Mode Comparison Study was drawn first from all dedicated mobile 10000-blocks, and a four digit random number was appended to each.

Table 20.1. Methodological Characteristics of Four Survey Projects Involving Mobile Phones

	Mode comparison 2003		Text message, 2004	Dual frame, 2004		PEW Research Center Study, 2006	
Characteristic	Mobile	Land	Mobile	Mobile	Land	Mobile	Land
Response rate[d]	21 (RR2)	33 (RR2)	24 (RR2)[a]	22 (RR3)	34 (RR3)	20 (RR3)	30 (RR3)
Contact rate[d]	75 (CON1)	78 (CON1)	52[a](CON1)	79 (CON1)	79 (CON1)	76 (CON2)	68 (CON2)
Cooperation rate	28 (COOP1)	43 (COOP1)	46[a](COOP1)	27 (COOP1)	37 (COOP1)	28 (COOP3)	50 (COOP3)
Refusal rate[d]	45 (REF1)	38 (REF1)	NA	43(REF2)	38 (REF2)	50 (REF2)	30 (REF2)
Undetermined rate[d]	10	6					
Average length (minutes)	18	17	4	9	9	10[b]	12
Average number of calls (all sample units)	6	10	4	5	6	4.2	3.6
Text message experiment	No	No	Yes	Yes	No	No	No
Incentive offered	$10	$10	No	Experiment $5/$10	No	$10	No
Length of field period in days	181	63	43	58	58	21	21
Original sample size	7999	3821	1890	8000	4488	8414	6662
Valid sample size	3919	1796	1770	3802	1941	4946	2859
RDD Prenotification letters/text message	No	No	Yes for 2 conditions, 3000 sent	Yes, to 86% with TM capability	Yes, to 46% with addresses[c]	No	No
Preidentified as ineligible	None	1095	None	None	1590	None	1025
Pretest	2	1	None	1	1	1	1
R selection	No	Yes	No	No	No	No	Yes
Eligibility based on original sample size	18 + (50%)	18 + (53%)	18 +(59%)	18 +; residential hh (48%)	18 +; residential hh (57%)	18 +(59%)	18 +(43%)

Calling algorithm	Unlimited calls	Unlimited calls	4 calls for first 2 conditions; 7 calls for 3rd	14 if no contact; up to 22 if contact	14 if no contact; up to 22 if contact	10 attempts total; contacts + non contacts	10 attempts total; contacts+non contacts
Voice mail/ answering machine message	On 1st and 3rd calls	On 1st and 3rd calls	1 condition, 1st call only	Yes, if no contact	Yes, if no contact	Yes 1st call if no contact	No
Refusal Conversions	2nd refusal was final	2nd refusal was final	No	2 weeks after initial refusal; 2nd refusal was final for sample of 75% of refusals	2 weeks after initial refusal; 2nd refusal was final—tried to convert all fixed line refusals	Nonhostile refusal conversions attempted; 2nd refusal was final	2nd refusal was final

Note: All rates were calculated using AAPOR formulas as specified.

[a]The average rates for the text message experiment exclude the rates for the text only condition because their inclusion would seriously distort the data.

[b]The average length for the landline survey was based only on the respondents who also had mobile phones. (See Keeter, 2006.)

[c]Of the landline numbers that turned out to be residential 84 percent were sent the advance letter (Brick et al., 2005a).

[d]Based on the total eligible and undetermined sample.

ineligible numbers. However, given present trends, ineligibility is likely to become a problem in the future. With each passing year, more and more children and teenagers possess a mobile telephone for their private use. GfK Technology (2005) reported on the basis of a 2005 youth survey that 44 percent of U.S. respondents 10–18 years owned a mobile phone. Although the percentages of sample units ineligible due to age were very small in the mobile telephone studies examined here (a range from 0.07 to 4.0), subsequent surveys will undoubtedly have a larger rate.

That it is difficult to identify numbers used only for business may help account for the reasonable ineligible rates found in mobile telephone surveys to date. However, assumptions about business use are different than in a landline survey. Since a mobile telephone is more likely to be used for both personal and business purposes than a landline telephone, it was assumed in the early studies that many mobile phones provided by employers would be eligible for the survey and that the percentage of nonresidential numbers would be lower than in a conventional telephone survey. The 2003 data provided corroboration. Some mobile numbers were answered with a business name or by a recording that gave a business name (Toledo Police Department, Bradenton Used Car Sales, e.g.), but, in the end, only 3 percent of the total sample units could be unequivocally classified as business ineligible—substantially lower than the 11 percent in the companion landline sample. This is a characteristic of mobile surveys that will probably not change as mobile telephone penetration increases.

If the ineligibility rate is lower than the one we expected, the percent of numbers with undetermined eligibility is higher. The tendency of many U.S. subscribers to use their mobile phones sporadically exacerbates the problem of determining a number's eligibility status. Prepaid mobile phones, in particular, fall in this category since they are used primarily for outgoing calls and may be turned off most of the time. A second mobile telephone owned by a single individual also presents problems. In addition, imperfections in service cause some nonworking numbers to ring as if they were active numbers without triggering a recorded message. Other mobile phones are still in service but, because they do not have voice mail, appear to be nonworking. Finally, the wording used in many operator intercept messages leaves open the question of whether a number is in service or not. Each mobile provider has its own set of operator messages, and many of these messages are too vague or too poorly worded to accurately describe a number's working status (See Callegaro et al., 2007 for examples of these kinds of messages). In any case the percent of sample units with undetermined eligibility is not only larger than expected but also larger than in a landline survey. When the same standards are used to calculate the undetermined eligibility rate in the two telephone surveys of the 2003 Mode Comparison Study, the difference is statistically significant in the predicted direction (6.2 percent for the landline and 10.0 percent for the mobile telephone survey, $X^2 = 48.383$, $df = 1$, $p < 0.001$). For surveys without the unlimited call attempts and extended field periods of that study, the percentage would be much higher. This assumption is confirmed by the fact that the unknown eligibility rate for the control condition in the text message experiment, to be more fully described below, jumped to 18.5 percent when only four call attempts were allowed (Steeh et al., 2007).

Currently, it is not clear how a mobile number should be treated when a large number of call attempts spread over a long period remain inconclusive. Furthermore, the turnover (churn) in mobile numbers is undoubtedly greater than when sampling is based solely on landline telephones. In sum, higher percentages of numbers with undetermined eligibility drive down response rates when all of these numbers are treated as eligible. Certainly these conditions demand new definitions of eligibility and new methods of calculating "e," a factor that can be applied to the base of the response rate to adjust for the fact that only a portion of these numbers are actually eligible.

As for resolving which analytical unit is more appropriate for a mobile telephone survey, the household or the individual, the four surveys provided a practical answer by generally choosing to abandon the traditional landline respondent selection procedures. In each of the mobile telephone surveys in Table 20.1, the person who answered became the respondent. Nevertheless, two of the studies did not entirely abandon gathering household composition data from mobile respondents. The 2003 Mode Comparison study tried to collect enough data about the respondent's household to allow analysis at either level, and the 2004 Dual Frame study specifically treated the data as relevant only to households. Faced with some evidence from the 2004 Current Population Survey (CPS) Supplement that mobile phones may be shared often enough in the United States for them to be considered household devices, the PEW survey addressed the issue unequivocally by asking, "do any other people age 18 or older regularly *answer* [our italics] your mobile phone or just you?" Approximately 85 percent of the respondents essentially replied "no, just me," confirming that in the United States the mobile telephone is primarily an individual rather than a household device and supporting the 2003 finding that 93 percent of mobile respondents said that the telephone on which they were speaking was used most of the time by them. Seeing the same process operating in Japan, researchers there have referred to this trend as "personalization in communications" (Kohiyama, 2005).

In the future, researchers may be able to eliminate any remaining ambiguity about who should be the respondent by screening for the primary user of the mobile phone. In any case, changing to the individual level will require the redefinition of standard eligibility criteria since we will now have an opportunity to interview groups traditionally excluded from landline surveys, such as homeless individuals, those living in group quarters, and people who have never before had regular telephone service. As a result, the representativeness of telephone surveys will broaden.

Noncoverage. Issues of both undercoverage and multiplicity can be present in a mobile telephone survey. Undercoverage results from the less than universal ownership of mobile phones within a society and will more than likely decrease substantially in magnitude over time. Multiplicity, on the other hand, results from the possibility that individuals may have more than one telephone number. Multiplicity will not only complicate the calculation of selection probabilities as stated in Section 20.2.2 but also increase the percentage of numbers that cannot be identified as either eligible or ineligible, especially when the maximum number of call attempts is low.

The skewed distribution of mobile telephone ownership at present makes undercoverage a serious obstacle to surveys conducted totally via mobile phones. Although we usually think of undercoverage in terms of landline surveys that do not include mobile only users and young adults (Keeter, 2006), undercoverage in the mobile mode is much worse. According to data from the 2003 NHIS and the 2004 Current Population Survey Supplement, approximately 41 to 51 percent of the adult population did not have a mobile telephone in those years. By 2005, 10 percent of the adult population had become mobile only but approximately 37 percent of adults still did not own mobile phones (see Blumberg et al., 2007, Chapter 3 in this volume).

Furthermore, the no-mobile telephone population is extremely biased since it is predominantly made up of people aged 65 and over. This imbalance means that the 65 and over age group is disproportionately missing from mobile sample frames. For example, in 2003, 16 percent of the adult population was 65 or older but only 5 percent of respondents in the 2003 mobile telephone survey fell into that category. By 2006, this discrepancy had narrowed but was still substantial—16 percent versus 8 percent (Blumberg et al., 2006; Keeter, 2006). On the contrary, men seem to be overrepresented when the mode is a mobile telephone in contrast to their usual underrepresentation in traditional telephone surveys. Partly this results from coverage error since there is evidence that slightly more men than women own mobile phones (47 percent of men and 45 percent of women in the 2003 NHIS). It is, however, more likely to reflect either a greater willingness of men to use and answer their mobile phones or the greater accessibility of men to survey interviewers as a result of mobile technology.

Although we do find larger percentages of young adults 18 – 24 year in mobile telephone surveys than exist in the total adult population—21 percent in the 2006 PEW mobile survey versus 13 percent according to the 2005 NHIS (Keeter, 2006; Blumberg et al., 2007, Chapter 3 in this volume), this does not appear to be due to a greater propensity of young adults to own mobile phones. Instead they are less likely to own a mobile telephone than the middle age segments of the adult population (Tucker et al., 2005). Young adults appear to be more cooperative when reached on a mobile telephone for reasons discussed at greater length in the next section. Thus partially because of coverage error, males and young adults are somewhat overrepresented in surveys conducted only by mobile phone, and the elderly are vastly underrepresented. In sum, we conclude on the basis of the experimental studies described in Table 20.1 that a landline sample is still more representative of the general population of 18 years of age and older than a mobile telephone sample despite the loss of many young adult respondents and all of the mobile only population.

Nonresponse. The response rates in Table 20.1 confirm that mobile telephone surveys have a problem. Across the four surveys, response rates ranged from 21 to 24 percent, approximately 10 percentage points lower than the rates for the companion landline surveys. The reasons for such high levels of nonresponse vary. As we have already noted, the characteristics of the telephone system in a country make a huge difference. Thus when response rates across studies are calculated using approximately the same formula, they fall in the 40 percent range for the random digit dial mobile surveys conducted in Germany and Slovenia, but never rise above 30 percent for those conducted in Hong Kong and the United States where the called party is charged for the call (Steeh, 2003, Vehovar et al., 2003, Lau, 2005). In addition, the increase in the percentage of numbers that fall in the unknown eligibility category has a dampening effect upon response rates.

Nonresponse is composed primarily of noncontacts and refusals, and the noncontact rates for the mobile telephone surveys fell within the conventional range. In the three studies that conducted companion mobile and landline surveys, for example, the noncontact rates for the mobile telephone surveys were similar—25, 21, and 24 percent of the respective eligible samples (see Table 20.1) and not consistently higher than the noncontact rates for the matching traditional surveys (Steeh, 2004a; Brick et al., 2007; Keeter, 2006). The exceptionally long field periods and the large numbers of call attempts in each case certainly contributed to these favorable results. In surveys with very short interviewing periods and a limited number of callbacks, however, the noncontact rate, like the unknown eligibility rate, will surely rise. When the number of call attempts was limited to four in the 2004 Text Message study, the combined noncontact rate for the control group and one treatment group approximately doubled to 48 percent.

Furthermore, mobile telephone surveys with short field periods and limited call attempts have a new reason to extend both. Since a potential respondent can answer a mobile telephone at any time in any place, interviewers may need additional call attempts to find an appropriate time to conduct a serious interview. Concerns about respondent safety may also result in rescheduling. Thus mobile surveys will require more call attempts than the two or three that are usual with many public opinion polls. The possible frequency of this outcome begs for an interim disposition code in mobile telephone surveys that designates a respondent who wanted to complete the interview but was in an unfavorable or unsafe environment at the time of the contact.

Despite evidence from mobile surveys in Europe that refusals declined (Fuchs, 2000a, 2000b, 2002a, 2002b; Kuusela, 2003), refusals increased in the mobile components of the U.S. studies. In a country where the called party pays for the call, this is only to be expected. A traditional RDD survey conducted in 2002 in one U.S. state asked respondents who had a mobile telephone whether they would participate in an interview conducted by a research organization over that phone, and why or why not (unpublished data, Georgia State Poll, 2002). Respondents overwhelmingly cited cost as their reason for not wanting to participate. They also frequently attributed their reluctance to feeling that mobile phones are private devices that should only be used for conversations with family and close friends. Refusal rates are also higher because refusal conversions are less effective in mobile telephone surveys than in landline surveys (Steeh, 2004a; Brick et al., 2007). However, contrary to expectations, partial and break-off interviews did not occur in the 2003 mobile survey as often as in the landline survey. Taken together, partial and break-off interviews were almost three times larger for the conventional survey compared to the mobile survey (4.1 percent of the valid sample versus 1.4 percent, respectively), a finding that is difficult to explain and may not be replicated.

Because of these factors, it is no surprise that mobile telephone surveys may contain substantial nonresponse bias depending on the topic being surveyed. Respondents tend to be particularly heavy users of mobile technology and thus are not typical even of the population of mobile owners (Steeh, 2004a). Brick et al. (2006) hypothesized that the introduction to their questionnaire made the topic of the survey salient and thus caused respondents to consent based on their interest in telecommunications. This kind of self-selection process may explain why break-off and partial interviews were relatively rare in the 2003 mobile telephone survey. However, heavy users of mobile phones will be more likely to participate–regardless of the topic or the wording of the introduction–because they carry their mobile phones with them most of the time, have generous calling plans that make them less conscious of the costs of being interviewed, and are most likely without access to a landline telephone (Steeh, 2004a; Brick et al., 2006).

As a result of these nonresponse and coverage biases, the mobile telephone cannot be employed as an independent mode in the United States at the present time if the survey aims to represent the general public. The evidence from European surveys suggests that, although bias due to differential nonresponse may be less severe than in the United States, undercoverage of older populations exists there as well (Kuusela and Simpanen, 2002; Fuchs, 2000a, 2000b, 2002a, 2002b). Of course, the age bias may be beneficial if the target population happens to be young adults.

Changes in methodological procedures. Within the context of these very large and difficult problems, we now examine the day-to-day procedures and best practices that have informed general population telephone surveys in the United States and ask if they continue to be effective when mobile phones are the instruments of communication. Mobile phones do share some features with landline telephones. For example, mobile phones, like landline phones, currently have limited channel capacity. Nonverbal communication is restricted as are visual displays that help respondents deal with complicated questions and response categories. However, the spread of smart phones with cameras and access to videos and the internet will increase channel capacity and give the telephone survey of the future many features of a web survey. In other ways the features of the mobile telephone are different enough to require changes in our conventional methods and procedures.

Calling schedules. What we know about mobile phones leads United States to hypothesize that established calling strategies will not work as well when data collection is carried out over mobile phones. This is especially true in the United States where the cost structure of the mobile industry does not accommodate the usual pattern of calling in the late afternoon and early evening. Experience with the 2003 survey suggests that fewer than five call attempts do not give interviewers sufficient opportunity to contact and convince respondents to cooperate unless the interview is very short. Nor do we want to call respondents at times when they will be most likely to refuse. Fortunately, the U.S. mobile industry seems to be moving toward free incoming calls. When this tendency becomes universal, calling strategies unique to mobile phones may no longer be necessary, and calling during week days may become desirable.

The two major projects conducted in 2003 and 2004 handled the issue of call scheduling differently. Researchers for the 2004 Dual Frame Study reasoned that the tendency of many users to carry their mobile phones with them most of the time would increase productivity during the day time hours. Thus their interviewing schedules routinely included mornings and afternoons as well as evenings and weekends (Yuan et al., 2005). On the contrary, the 2003 Mode Comparison Study ultimately tried to maximize contacts during the periods when incoming calls would be free. For the first replicates, constituting about 5000 sample units, traditional calling procedures were followed. Very few initial calls occurred during daytime hours with most calls being placed during the evening hours from 5:00–9:00 p.m. local time and on weekends. However, the high refusal rate caused a mid-study reevaluation. When the last replicates were released, calls were confined to the weekends in the hope that fewer people would refuse if they knew they were not being charged for the call. Because the two sets of replicates were independent random samples, the different treatments measured how effective weekend-only calling might be as a general practice.

The results from both studies proved inconclusive. Although the 2004 Dual Frame survey found that the contact rate increased with daytime calling, some of their analyses, based only on the first call attempt, showed that completion rates were lowest and refusal rates highest during the week day. Conversely the highest completion rates and the lowest refusal rates occurred on weekend days (Yuan et al., 2005). Analysis of the 2003 data indicated that the response rate rose from 17 percent to 22 percent when interviewing was restricted largely to weekends ($X^2 = 71.94$, df $= 7$, $p < 0.001$). However, there were several uncontrolled factors that may have produced this result. For example, interviewer experience with the questionnaire undoubtedly increased by the time the last replicates were released, and greater experience undoubtedly led to more completions.

The different calling schedules of the 2003 and 2004 surveys may explain why the 2004 study had over three times the number of young, ineligible sample units. During the day, children under age 18 may be more likely to answer a ringing mobile phone. It is also possible that people who are willing to be interviewed during the day, when there is the least possibility of avoiding charges, are likely to be heavy mobile telephone users with generous calling plans and lots of "anytime" minutes thus adding to self-selection bias. On the other hand, calling primarily on weekends extends the field period for the survey considerably and perhaps unacceptably for most public opinion research.

Inducements to participate. The characteristics of the mobile telephone system and its customers, in the United States particularly, would seem to require inducements to gain cooperation. A survey call to a mobile telephone number is the quintessential "cold" contact. The person answering does not expect (or want) a call from a stranger (Steeh, 2003), may be paying for the call, and has no information from an advance letter or even a caller-id screen to gauge the legitimacy of the survey. Without a directory of mobile telephone numbers, prepaid incentives delivered with an advance letter are not an option. However, the national studies carried out in 2003 and 2004 incorporated inducements that might work with mobile phones and designed tests to measure their effectiveness.

Advance notification. Most studies of the effect of sending some kind of advance notification—primarily in the form of a letter mailed several days before the first calls are placed—have confirmed its benefits (Sangster, 2003; Link and Mokdad, 2005a; de Leeuw et al., 2006). Only two substitutes exist for the mobile telephone survey—sending a text message in advance of a call or leaving a scripted message on the telephone's voice mail. Of course, these procedures are imperfect because they apply at the present time only to those sample units equipped with these services, thus increasing the chances of nonresponse bias by differentially encouraging the participation of people already well represented (Link and Mokdad, 2005a). Nevertheless, the mobile surveys utilized both methods to encourage participation.

Prior to 2004, text messages were not a feasible option, and so they were not included in the 2003 survey design. Scripted messages left on phones with voice mail capability were tried as an alternative in both the 2003 Mode Comparison Study and the 2006 PEW Study. The primary purpose of these scripted messages was to alert the mobile telephone user to the postpaid incentive (more on that in the next section), to communicate the sponsors of the survey—a university in one case and a respected survey organization in the other, and to leave a toll-free number that respondents could call to ask questions or participate in an interview. Two different messages were left on specific call attempts in the 2003 surveys. One scripted message was used only in the PEW mobile telephone survey but not in the landline. The process of leaving scripted messages on certain call attempts was apparently very difficult for interviewers to carry out according to the call histories for the 2003 study. Furthermore, neither of these efforts was tested experimentally. To judge from the number of completed interviews resulting from call-ins to the mobile telephone component of the 2003 study, only four percent of respondents, it had limited effectiveness.

By the end of 2003, text messaging had achieved enough penetration to allow a test of its use for advance notification. Thus the 2004 Text Message Survey was designed as an experiment with three randomly selected groups—two treatment groups receiving some form of text message and a control group receiving only the cold call as in the 2003 survey. The Nextel interface used to send the messages confirmed the delivery status of each message through an email addressed to the sender. Although no significant improvement in response rates occurred as a result of sending a text message, the experiment indicated that the refusal and the undetermined rates significantly declined. In addition, it yielded an unexpectedly robust finding. Determining that a message had been delivered served to identify a working number, just as a listing in a telephone book indicates the likely working status of a landline number. This benefit accrues even though none of the respondents may actually have read the text message (Steeh et al., 2007).

The 2004 Dual Frame Survey, on the other hand, included a text message experiment from the beginning but did not make systematic use of scripted voice mail follow-ups (Yuan et al., 2005). Like the results from the 2003 study, no significant difference occurred in response rates between the random half of the sample that received the text message and the random half that did not (Yuan et al., 2005). However, other outcome rates were not compared, and no attempt was made to determine delivery status.

At the present time text messages do not seem to function well as traditional advance notices, but they can provide auxiliary information that helps to weed out nonworking numbers and streamline survey operations (Steeh et al., 2007). As more and more subscribers begin to use the function on a regular basis, text messaging may also become effective as a means of prenotification.

Incentives. Survey methodologists have concluded that, since many owners will have to pay either in minutes or dollars for participating in a mobile telephone survey, reimbursement should be offered (Lavrakas and Shuttles, 2005b). The principal considerations are the amount and type of the incentive and how to present it to prospective respondents. Since the university sponsoring the 2003 survey would not allow payments of cash without asking for social security numbers, an American Express gift check in the amount of $10 was chosen as the incentive. The crucial issue thus became making sure sample units knew about it. The voice mail messages left on the phones of nonresponding sample units (whenever it was possible) were designed to accomplish this purpose.

The 2004 Dual Frame Survey incorporated an incentive experiment only within the mobile telephone component. One random half of the sample received a $5 check and the other half a $10 check. Overall, response rates were substantially higher when $10 was the incentive amount (Yuan et al., 2005). Scripted voice mail messages were not used to communicate this information to nonresponding sample units, but the incentive amount was mentioned in the advance text message. From this experiment we learn that postpaid incentives can have an effect on participation and that a larger incentive is better than a smaller one (Brick et al., 2005). The 2006 PEW study simply offered all potential mobile telephone respondents ten dollars.

We have not yet learned what the cost effective amount of the incentive should be. Since some companies charge as much as 45 cents a minute to prepaid customers or to those who have exceeded their anytime limit, an eighteen minute questionnaire could cost such a respondent $8.10. An incentive amount of ten dollars just barely compensates for these charges. However, an amount higher than $10 may drive the costs of a mobile telephone survey out of range for most survey organizations. In addition, we have not learned what form the incentive should take. Only the PEW study employed the cash payments that the methodological literature recommends most highly, but the survey did not include an experiment that would allow us to judge its effectiveness versus checks or gift certificates.

Type of introduction. The introduction to a mobile telephone survey requires special crafting. It must overcome the surprise that many mobile telephone users have expressed at being called by a stranger, and it must communicate vital information concisely and in a way that does not encourage self-selection. Both the 2003 and 2004 studies used introductions that mentioned telephone usage. In contrast, the PEW survey began with a generic statement, "We are conducting a very short survey of a random sample of Americans." Nevertheless, their response and contact rates were no higher than those in the other two surveys, and the overrepresentation of heavy mobile telephone users continued. Most of the PEW questions, like the items in the 2003 and 2004 studies, involved telephone behavior. This leaves open the extent to which the overrepresentation of heavy mobile telephone users will occur when the survey topic is something other than telephone use.

Presenting the incentive may also involve changes in methodological procedures. Normally, an incentive is labeled in introductions as a token of appreciation rather than as a payment. Of the four surveys examined here, two offered the incentive as a token of appreciation, one as "thank you for your time and participation," and the fourth "for helping us out." However, since mobile telephone respondents may be paying a substantial amount to participate, it seems most reasonable to present the incentive as a reimbursement for any costs the respondent might bear (Lavrakas and Shuttles, 2005b). Needless to say, the incentive should be mentioned as quickly as possible in the introduction.

20.4 VOICE OVER INTERNET PROTOCOL (VoIP) IS THE LATEST CHALLENGE

Another new technology has burst onto the scene over the past several years, VoIP. What VoIP means for telephone sampling is still unclear. Undoubtedly it will present challenges but may also offer great potential for the future. VoIP technology converts analog voice signals into digital packets that are transmitted over high-speed broadband data lines instead of the traditional copper wire. The technology also allows voice communication to be converted to e-mail and vice versa and allows companies to combine their data and voice networks. Transmitting voice over Internet lines is relatively inexpensive and for the moment beyond the reach of most government regulation and taxation.

20.4.1 Penetration of VoIP

Companies such as Vonage and Skype saw VoIP as an untapped market niche and quickly began offering local and long-distance telephone service, in the United States and internationally, at significantly lower rates than local telephone service providers. Cable companies, and more recently local exchange carriers and ISPs, have effectively reacted to the competitive threat from these stand-alone VoIP service providers. By bundling multiple services—broadband, internet access, VoIP, mobile service, and television—into single discounted packages, their one-stop-shopping offerings should be more appealing to subscribers. This could in turn provide them with the long-term competitive edge and encourage wider adoption of VoIP.

VoIP also promises enormous cost savings to survey organizations and call centers. Internet surveys have gained in popularity because they are inexpensive relative to telephone, mail, and in-person interviews. VoIP systems can reduce the costs of conducting telephone surveys to levels that are comparable with internet surveys, thus facilitating the conduct of survey research. However, on June 21, 2006 the U.S. Federal Communications Commission adopted an order changing the Universal Service Fund (USF) contribution methodology to include VoIP providers (Federal Communications Commission, 2006b). The USF helps subsidize telecommunications services in schools, libraries, and high-cost regions of the country. This order will affect reported revenues beginning in August 2006 and may increase overall costs to providers, to subscribers, and ultimately to survey researchers.

VoIP technology is highly flexible. Since it uses internet lines, a subscriber can carry their Internet telephone or adapter when they travel or move and simply plug it in where there is a broadband connection. Companies such as Vonage also allow you to choose a secondary or "virtual" number in any area code in North America or Europe that will ring to your primary Vonage line. VoIP is quickly moving from an exclusively landline environment into the mobile world. Many companies are already marketing WiFi handsets and mobile VoIP services. It is estimated that by 2015 28 percent of voice minutes will be via mobile VoIP (Gardner, 2006). As with all new technologies, penetration numbers are difficult to obtain, but most media reports suggest that the service is quickly gaining in popularity, especially with businesses. Bernstein Research estimated that there will be 2.4 million residential subscribers by the middle of 2005 (The Associated Press, 2005), while TeleGeography estimates that there will be 4.1 million residential subscribers in the United States by the end of 2005 and Digitaltrends reported that 40 percent of household with broadband will subscribe to VoIP by 2010 (Duncan, 2006). Infonetics Research predicts that VoIP revenues will increase by 1431 percent by 2009 (Silverberg, 2005).

20.4.2 Challenges of VoIP

There are still many issues related to VoIP technology that need to be worked out before it can become the telecommunications technology of choice. The most pressing deficiency is the current lack of universal access to Enhanced 911 (E911). When a call is placed to a 911 dispatcher, E911 service displays the caller's telephone number and address (or current location). Since VoIP technology is not tied to a specific computer and can be either landline or wireless, a VoIP telephone number lacks a permanent location and so can be a problem for E911 services. This problem is being successfully addressed for mobile phones by displaying the closest mobile tower location and, in more and more cases, by using Global Positioning System (GPS) technology. As a result most mobile service providers are now in compliance with the Federal Communications Commission (FCC) legislated requirement to provide E911 services to all telecommunications customers (United States Federal Communications Commission, 2001). However, the same is not the case for VoIP services. Subscribers wishing to retain their existing telephone number, when switching to a VoIP service, have frequently discovered that E911 capability is unavailable in their area (The Associated Press, 2005).

Since VoIP requires a high-speed Internet connection, DSL and cable broadband will soon replace dial-up modems and perhaps allow consumers to finally give up their second and third lines. By the end of 2005, 43 million residential end users had a high-speed line (United States Federal Communications Commission, 2006a). Because VoIP uses the Internet, there are issues with security, reliability and voice quality to be worked out. If your Internet Service Provider (ISP) goes down, so does your telephone service. If your house or business loses electricity, you will also lose your telephone service, unless it is the mobile variety. Because it is the Internet, the quality still cannot match that of old fashioned wire. All the Internet threats (communication interception, e-mail bombardment, hackers, spam, viruses, spyware, phishing, and farming) will now also apply to the telephone system. In fact there are already new acronyms germane to these new technologies: SPIT for Spam over Internet Telephony and SPIM for Spam over Instant Messaging (Biever, 2004).

20.4.3 Survey Dilemmas with VoIP

The existence of internet based telephone systems in surveys brings its own set of dilemmas. The existence of "virtual numbers" is chief among these. VoIP companies can now assign a subscriber several different numbers that can be used to decrease the cost of long distance calls for family and friends who still maintain conventional, landline telephones. Thus a number in Florida may actually connect to a telephone in New York. In addition, virtual numbers complicate probabilities of selection at the individual level since the same individual may be reached in multiple ways. It is not known currently how popular these virtual numbers will become since VoIP and mobile phones will gradually replace the landline alternative and make virtual numbers obsolete.

Other dilemmas involve the erosion of the geographical grounding of telephone surveys beyond anything we have yet imagined with mobile phones. As VoIP penetration increases, it may have serious consequences for the geographic precision of telephone samples significantly limiting our ability to do small area sampling. The extent to which a VoIP connection is more likely to be thought of as private is also not clear and may affect cooperation rates. Nor can we assume that the technology will service individuals rather than households. If there is only one computer in a household, a VoIP telephone may well be used by all household members while mobile VoIP is more likely to service individuals.

Currently VoIP numbers are not assigned a special classification by Telcordia, and so landline VoIP numbers can be found in standard list assisted samples and mobile VoIP numbers in mobile samples. Unless Telcordia decides to create a separate classification for VoIP numbers in the future, we do not have to worry about incorporating them into our RDD surveys whether landline or wireless. However, given the challenges we have described, it is not clear how their presence in telephone sample frames may affect survey results. The issue is not critical so long as VoIP penetration is minimal, an unlikely scenario given the current market place focus, but it does seem wise for surveys to begin immediately to collect information about the type of telephone service a respondent has.

Despite all the drawbacks, the future is clearly internet telephony. The industry is developing and marketing an endless variety of new devices and services that mix and match modes of telecommunication. In some ways conducting voice interviews through the internet will resemble the conventional telephone survey, in others it will resemble the mobile telephone study, and in still other ways, it will resemble internet surveys.

20.5 A TEMPORARY ALTERNATIVE TO LANDLINE SURVEYS

Since it is unlikely that any one type of service will dominate in the near future as the landline telephone dominated in the past, it is clear that we need to develop a survey design that will mix mobile and landline services if the telephone mode is to keep its legitimacy as a data collection method. Telephone samples in the foreseeable future must include landline, mobile, and VoIP numbers as a matter of course. The advantages are clear. Covering all types of telephone service will reduce the two errors that threaten to become more and more prevalent in standard landline surveys—nonresponse and noncoverage error. Although researchers hypothesized that we would be able to contact members of disadvantaged groups more easily through mobile phones, results so far have not confirmed this expectation (Steeh, 2004a, b), but, by drawing telephone samples from the largest possible pool, we will have a better chance of contacting groups that have never been part of previous telephone surveys. Thus a mixed sample of mobile, VoIP, and landline numbers will be more representative of the general population and present fewer possibilities for bias.

While there may be general agreement that sampling all kinds of telephone service is the way to rejuvenate and extend the usefulness of telephone surveys, how to accomplish this goal is the major dilemma facing methodologists. Conceptually the sampling frame should be a list of all three types of telephone numbers. Only one set of weights that take into account the number of times an individual appears in the frame would need to be calculated. However, identification of mobile numbers would be required in order to comply with the FCC regulation for hand dialing of mobile numbers. Another option is a multiframe design in which landline and mobile services would be sampled separately, perhaps weighted separately, and then combined into one sample using a mixing parameter.

The first test of the multiframe alternative with data from the 2004 project (Brick et al., 2005b) showed that the combined weighted estimates of telephone use were still biased in comparison to similar estimates from the U.S. Current Population Survey and the U.S. National Health Interview Survey that were carried out face-to-face. The Arbitron and PEW studies explored a partial multiframe design that combines part of a mobile telephone survey (the mobile only segment) with a complete landline survey and then weights the results to estimates of telephone use and demographic characteristics determined for the general population by personal interview surveys (Cohen, 2006; Keeter, 2006; Tompson and Kennedy, 2006). While this may be a feasible, temporary stopgap method, the PEW data, in particular, reveal that on many measures there are noticeable differences in telephone use and attitudes between respondents with landline phones who are interviewed on that telephone and respondents with landlines who are interviewed on their mobile phone. By excluding those people who own a landline telephone but primarily use a mobile phone, we take the chance of leaving out a sizeable group who should be represented in a telephone survey—a group that is much larger than the mobile only group. In yet another approach, Jay and DiCamillo (2006) suggest that some weighting adjustments can be made to landline surveys by identifying "recent mobile phone-only households." These are households that were without landline service for 1 month or more in the past 3 years but reported having a mobile telephone during that time. In sum, we do not know yet quite how to combine the two modes.[5]

---

[5]A session at the 2005 Joint Statistical Meetings was devoted to sample designs and weighting procedures for surveys that included both mobile and landline phones (Lepkowski and Kim, 2005, Brick et al., 2005a, Steeh, 2005c).

Several features of future surveys are clear, however. We are going to have to ask more questions about telephone ownership and use in order to calculate accurate weights. None of the surveys conducted to date have collected sufficient detail. We must determine the number of landline and mobile numbers in the household that belong to each adult and the propensity with which each is answered. We will also need to determine geographic eligibility and ask explicitly about dedicated computer and fax lines since it appears we have been understating the coverage deficiencies of landline surveys. We also must assess the reliability of self-reports and proxy reports of telephone use. Certainly the data will be more accurate when respondents describe their own behavior rather than the behavior of other household members. The challenge will be to keep these questions from consuming the major part of the interview.

20.6 CONCLUSION

Changes in technologies will have a major impact on telephone survey methodology. The future of the telephone as a survey mode of administration hinges on our ability to change our procedures to accommodate new demands. Accommodation depends on our willingness to undertake the kinds of research projects that will begin to give us answers to some of the more difficult statistical and methodological riddles we have described in this chapter. It is a time for us in the survey profession to give free rein to our methodological imaginations. Since it appears that the traditional general population telephone survey is still viable, we have a window of time to experiment and explore alternatives although the window is closing rapidly.

The nascent convergence of internet telephony with mobile technology promises the research community a kaleidoscope of as yet untried possibilities that will enhance the survey experience for respondents and practitioners. Respondents might be contacted by telephone, e-mail, voice mail, text messaging, or some combination of these at any time of day and in almost any location. A news reporter described the future of communications technology by dreaming of "a telephone that roams without a hiccup between mobile and mobile Internet networks, delivering clearer connections indoors and out and paid for with a single bill. Movies, games or business projects started on a desktop computer and finished on a mobile handset. Access to the same personalized digital world of e-mail, messaging and video on any device" (Ho, 2006). Telephone surveys are not dead or dying. They are simply evolving into something new, challenging, and exciting.

PART V

Nonresponse

CHAPTER 21

Privacy, Confidentiality, and Respondent Burden as Factors in Telephone Survey Nonresponse[1]

Eleanor Singer
Survey Research Center, Institute for Social Research, University of Michigan, USA

Stanley Presser
Sociology Department and Joint Program in Survey Methodology, University of Maryland, USA

21.1 INTRODUCTION

Leverage-salience theory (Groves et al., 2000, p. 301) posits that in determining survey cooperation "the achieved influence of a particular feature is a function of how important it is to the potential respondent, whether its influence is positive or negative, and how salient it becomes to the sample person during the presentation of the survey request." Many factors—for example, the survey's topic, sponsor, and purpose—may exert a positive influence for some people but a negative influence for others. One factor, a monetary incentive, is apt to influence everyone positively. Three other factors—privacy, confidentiality, and burden—are likely to influence everyone negatively: they may *deter* participation but not enhance it. Because their effect is exclusively negative, these factors are often invoked to explain declining response rates. However, the degree to which concerns about privacy, confidentiality, and burden do, in fact, exert a negative effect on cooperation is an unsettled question.

[1]We are grateful to Roger Tourangeau for very helpful comments on an earlier version of this chapter.

Advances in Telephone Survey Methodology, Edited by James M. Lepkowski, Clyde Tucker, J. Michael Brick, Edith de Leeuw, Lilli Japec, Paul J. Lavrakas, Michael W. Link, and Roberta L. Sangster

In this chapter, we review what is known about the impact of each of these three factors on survey nonresponse, especially telephone survey nonresponse. Then we examine the issue of whether attitudes toward privacy, confidentiality, and perceptions of burden have contributed to declining response rates by examining how each of them has changed over time. Finally, we attempt to relate the findings about nonresponse rates to nonresponse bias, though the evidence here is virtually nonexistent. We draw on some European and Canadian research, but most of our evidence and conclusions come from studies done in the United States.

By "privacy," we refer primarily to informational privacy: "the claim of individuals, groups, or institutions to determine for themselves when, how, and to what extent information about them is communicated to others" (Westin, 1967, p. 7). However, particularly with interviewer-administered surveys, another aspect of privacy also comes into play, namely the "right to be let alone," first enunciated by Thomas McIntyre Cooley (see Warren and Brandeis, 1890).

By "confidentiality" we refer to "the safeguarding, by a recipient, of information about another individual" (National Research Council, 1979, p. 1). By "burden," survey methodologists often refer to two different, though related, things: an objective property of a survey, typically measured by length (as in the U.S. Office of Management and Budget's burden hours), and a subjective property of what is entailed in answering a survey, usually measured by an indicator of effort, for example, the difficulty of answering (Bradburn, 1977). We will examine both aspects, though there is much more evidence on the former than the latter.

This book is about telephone surveys. But most research on privacy, confidentiality, and burden does not distinguish these concerns by mode. In discussing the findings, we refer to the mode by which they were obtained; where possible, we specify whether the questions asked about telephone surveys, other kinds of surveys, or other kinds of entities altogether.

Aside from ambiguities in how well some of the findings we report apply to telephone surveys, we are hampered by a shortage of evidence altogether. Many questions have been asked over the years about how individuals feel about privacy and related issues, and a smaller number have been asked about the demands made by various kinds of surveys and survey organizations. But the number of experimental studies from which effects can be demonstrated is very small, and in the area of privacy and confidentiality, most of the studies have been sponsored by the U.S. Census Bureau, limiting their generalizability. In what follows, we do our best to draw meaningful generalizations from the available literature without stretching the evidence further than it can comfortably go. We begin by describing changes in the legal landscape that provide part of the context for attitudes about privacy, confidentiality, and burden.

21.2 CHANGES IN THE LEGAL LANDSCAPE

Laws enacted in the past 10 years or so have affected the climate for conducting telephone surveys by strengthening protections for consumer privacy and data confidentiality. We briefly discuss several of these laws: the European Union Directive on Data Protection; legislation in the United States and elsewhere designed to regulate telemarketing, some of which has created Do Not Call lists; and the Confidential Information Protection and Statistical Efficiency Act of 2002 (CIPSEA). We do not, however, have direct evidence for the effect of any of these laws on cooperation with survey requests.

21.2.1 European Union Directive on Data Protection

The European Union (EU) has a very detailed set of regulations designed to protect the confidentiality of information provided by survey respondents, based on the 1995 European Union Directive on Data Protection (http://www. europa. eu. int/comm/internal_market/privacy/index_en.htm). The directive was designed to ensure both a high level of protection for the confidentiality of individual information in all member states and the free movement of personal data within the EU. Because it prevents the transfer of personal data to countries that do not meet EU data protection standards, the directive influences such standards beyond the EU's borders. In addition, because it prohibits certain study protocols—for example, the use of health registers as sampling frames without the consent of participants—some have suggested it may interfere with the ability to carry out needed research (see, e.g., Angus et al., 2003; Coleman et al., 2003). The Council of American Survey Research Organizations has created a special program designed to help research companies meet the privacy requirements for data collection, storage, and dissemination that are imposed by laws like the European Union Directive, and it is also actively lobbying for the creation of one federal "Protecting Personally-identifiable Information" law in the United States.

What role, if any, the passage of the European Directive has played in European trends in privacy and confidentiality attitudes (see, for instance, our discussion of the Eurobarometer surveys below) is unknown.

Regulations Restricting Telephone Calls

One reason often alleged for declining telephone survey response rates in recent years is the flood of telemarketing calls. In response to concerns about fraudulent and misleading telemarketing, the U.S. Congress enacted the Telephone Consumer Protection Act (TCPA) in 1991, and the Federal Trade Commission issued the Telemarketing Sales Rule (TSR) in 1995 (amended in 2003). The TCPA restricts telemarketers from calling numbers placed on a Do Not Call (DNC) registry.[2] The TSR requires telemarketers to disclose clearly and promptly the sales purpose of the call. Both the TCPA and TSR also prohibit the use of autodialers to call cell phones without the prior consent of the called party. Survey research calls are not covered by the TCPA or TSR, except for the ban on autodialing cell phones.

Public response to the 2003 legislation was overwhelming. More than 50 million telephone numbers had been registered for the Do Not Call List by the end of the

[2]The TCPA also limits calls to individuals whose names are not on the list to the hours of 8 am to 9 pm, and it requires telemarketers to allow their telephone number to be identified by caller ID systems.

initial registration period on August 31, 2003, and more than 91 million numbers had been registered 18 months later (Link et al., 2006). In 2004, the 10th Circuit Court of Appeals in Denver ruled that the FTC's Do Not Call Registry was constitutional and did not violate telemarketers' First Amendment rights. In doing so, it overturned a lower court's ruling that the DNC Registry was unconstitutional because it unfairly singled out telemarketing calls.[3]

In principle, these regulations should benefit survey research by reducing "sugging" and "frugging," that is, selling or fundraising under the guise of conducting research and enabling the industry and the public to distinguish more clearly between survey research and telemarketing. In practice, the effect of the new regulations is difficult to assess. Response rates continued to decline after the new rules were enacted. Whether they would have declined even further without this regulatory response to the increased number of unsolicited calls is not known, although survey researchers surveyed by Survey Sampling Inc. in 2004 overwhelmingly (71 percent) agreed that "over time, a reduction in telemarketing calls to consumers will favorably impact telephone survey cooperation rates" (www.surveysampling.com/frame_ detail.php?ID=65¤t=no&issueID=25&yr=2004). As of 2005, however, a study by Link et al. (2006) had found no evidence to this effect.

Similar legislation has been introduced or adopted in other countries. Britain, for example, passed the Communications Act of 2003, which gave the Office of Communications the power to take action against persons or companies who persistently misuse electronic communications networks or services in a way likely to cause unnecessary annoyance, inconvenience, or anxiety. In response, the Market Research Society, together with the British Market Research Association, has produced a "Code of Practice for Automated Dialing Equipment" to guide market researchers' practices in compliance with the Act (http://www.mrs.org.uk/standards/fieldwork.htm#dialling). Comparable legislation was introduced in Canada in December 2004.

The Confidential Information Protection and Statistical Efficiency Act (CIPSEA)
CIPSEA, which was enacted as Title V of the E-Government Act of 2002 (P.L. 107-347), created confidentiality protection for data collected by federal agencies for exclusively statistical purposes under a pledge of confidentiality. The act extended to all federal agencies the level of protection enjoyed previously only by the U.S. Census Bureau and the National Center for Health Statistics: Penalties for willful disclosure are $250,000 in fines and up to 5 years' imprisonment. CIPSEA also protects such information against Freedom of Information Act (FOIA) requests, and the protections extend as well to data collected by *agents* of the federal agencies, who may be organizations or individual researchers, and to administrative data that are linked to data collected for statistical purposes. The U.S. Office of Management and Budget is preparing regulations to implement CIPSEA, which will define more precisely both the reach of protection for confidential statistical records and the opportunities for research access. By strengthening the protections against disclosure, the law and attendant regulations could ease confidentiality concerns.

[3]Calls soliciting charitable or political contributions are exempt from the FTC regulations on the grounds that a significant purpose of the call is to solicit support for a cause, not merely to ask for money.

21.3 PRIVACY AND CONFIDENTIALITY

21.3.1 Effects on Nonresponse

All of the studies investigating the impact of privacy and confidentiality concerns on survey cooperation have demonstrated statistically significant, though sometimes small, effects. The earliest study (National Research Council, 1979) was designed to shed light on how changing the length of time for which census returns were kept confidential might affect the mail return rate. The survey introduction varied the assurance of confidentiality respondents were given. A random fifth of the sample received an assurance of confidentiality "in perpetuity"; another, an assurance for 75 years; a third, for 25 years; a fourth received no mention of confidentiality; and a fifth group was told that their answers might be given to other agencies and to the public. Among those persons who were reached and to whom the introductory statement, including the experimental treatment, was read, refusals varied monotonically from 1.8 percent for those who received the strongest assurance to 2.8 percent for those told their answers might be shared with other agencies and the public; this increase, though small, was statistically significant (Fig. 21.1).

Two studies have examined the relationship between census return rates and concerns about privacy and confidentiality expressed on an attitude survey. The first matched responses to the Survey of Census Participation, a face-to-face attitude survey carried out in the summer of 1990, a few months after the decennial census, with mail returns to the census from the same households.

As shown in Fig. 21.2, both an index of privacy concerns and an index of confidentiality concerns significantly predicted census mail returns, together explaining 1.3 percent of the variance net of demographic characteristics (Singer et al., 1993). In a subsequent analysis, controlling for other presumed deterrents to returning the census forms (e.g., illiteracy, distrust in government), only the confidentiality index continued to show an effect on the return rate (Couper et al., 1998).

The study was repeated for the 2000 census, using respondents to a Gallup random digit dial survey in the summer of 2000 and matching their responses to census

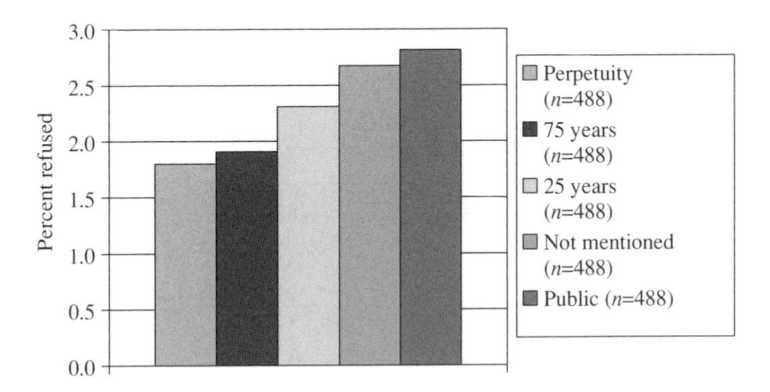

Figure 21.1. Percent refused by length of confidentiality assurance. (National Research Council, 1979).

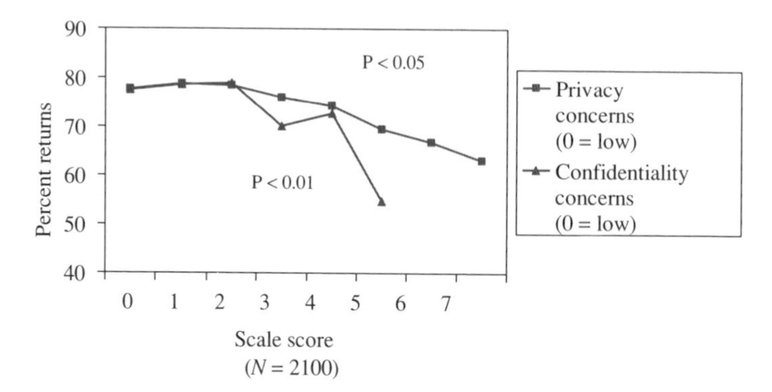

Figure 21.2. Effects of privacy and confidentiality concerns on census mail returns, 1990 (Singer et al., 1993).

returns from the same households (Singer et al., 2003). Attitudes about privacy and confidentiality were estimated to account for 1.2 percent of the variance in census mail returns, almost identical to the 1990 result. The relationships were stronger in one-person households, where the respondent to the survey and the person who returned the census form had a higher probability of being identical than in the sample as a whole. The belief that the census may be misused for law enforcement purposes, as measured by an index of responses to three questions, was a significant negative predictor of census returns (Fig. 21.3). Privacy concerns were also correlated with reduced willingness to provide an address to the Gallup interviewer to permit matching of survey responses to census forms.

Findings from three other studies support the conclusion that privacy concerns reduce participation in the census. Martin (2001, Table 4) shows that respondents who received a long form or were concerned about privacy were more likely to report returning an incomplete census form or failing to return it at all. In addition, an experiment by Junn (2001) reported that respondents who were asked questions

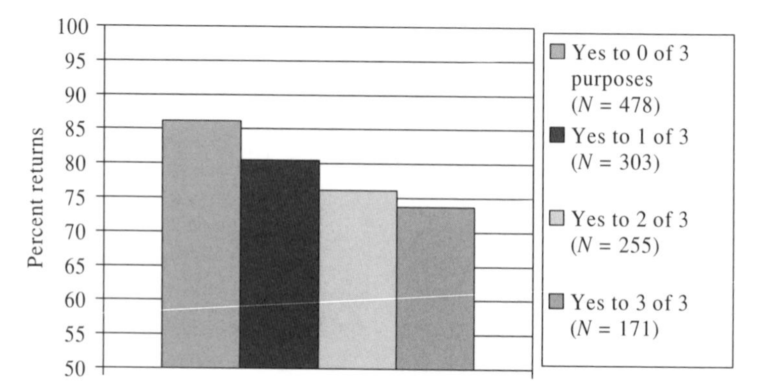

Figure 21.3. Effects of belief that census is used for law enforcement on mail returns, 2000 (Singer et al., 2003).

designed to raise privacy concerns about the census were less likely to respond to long-form questions administered experimentally than respondents who were given reasons for asking intrusive questions, or those in a control group who received neither. Additional evidence is presented in Hillygus et al. (2006).

Census Bureau research currently in progress, which analyzes interviewers' coding of "doorstep concerns" expressed by households selected for the National Health Interview Study, shows that privacy and confidentiality concerns significantly predict interim refusals but not final refusals. This is true of several other concerns expressed by sample persons—for example, being too busy and asking questions of interviewers—suggesting that if interviewers can successfully address such concerns, they need not lead to nonparticipation (Bates et al., 2006). Because these data come from face-to-face surveys, they also suggest that the relation of privacy concerns to final refusals may differ between modes, in part because interviewers may be less able to counter them effectively on the telephone.[4]

A final set of experiments that bears on the relationship between privacy/confidentiality concerns and survey participation involved the impact of asking for a Social Security number (SSN). A 1992 Census Bureau experiment found that requesting SSNs led to a 3.4 percentage point decline in response rates to a mail questionnaire and an additional 17 or so percentage point increase in item nonresponse (Dillman et al., 1993). In experiments conducted in connection with the 2000 census, when SSNs were requested for all members of a household, the mail return rate declined by 2.1 percentage points in high coverage areas, which make up about 81 percent of all addresses, and 2.7 percent in low coverage areas, which contain a large proportion of the country's black and Hispanic populations as well as renter-occupied housing units. Some 15.5 percent of SSNs were missing when a request for the SSN was made for Person 1 only, with increasing percentages missing for Persons 2–6 when SSNs were requested for all members of the household (Guarino et al., 2001, Table 5).

21.3.2 Effects on Reported Willingness to Participate

The experiments discussed so far have been embedded in ongoing surveys or have involved matching survey and census responses; they thus have a great deal of external validity but limit the kind of additional information that can be obtained. From a series of other studies, many of them laboratory experiments, we know that when questions are sensitive, involving sexual behavior, drug use, other stigmatizing behavior, financial information, and the like, stronger assurances of confidentiality elicit higher response rates or better response quality (Berman et al., 1977; Singer et al., 1995). On the other hand, when the topic of the research is innocuous, stronger assurances of confidentiality appear to have the opposite effect, leading to less

[4]A Response Analysis Survey (RAS) of a small number of respondents and nonrespondents to the American Time Use Survey (ATUS), which is conducted by phone, found greater confidentiality concerns among nonrespondents—32 percent, compared with 24 percent among respondents. Because the RAS response rate among nonrespondents to the ATUS was very low (32 percent, compared with 93 percent for respondents), the actual difference between respondents and norespondents is probably larger (O'Neill and Sincavage, 2004).

reported willingness to participate, less actual participation, and greater expressions of suspicion and concern about what will happen to the information requested (Frey, 1986; Singer et al., 1992).

These experiments vary either the assurance of confidentiality given to respondents or the method of protecting confidentiality—for example, randomized response or self-administration versus interviewer administration. The manipulations are assumed to produce an effect either because they address respondents' privacy or confidentiality concerns, or because they evoke a concern that did not previously exist, or both.

Singer (2004) investigated the assumed mechanisms underlying these effects in a study conducted as part of the University of Michigan's RDD Survey of Consumer Attitudes. Interviewers read two hypothetical introductions to respondents describing ongoing studies at the University of Michigan's Survey Research Center, the Health and Retirement Survey (HRS), and the National Survey of Family Growth (NSFG). The descriptions were altered to make them as comparable as possible to each other except for the hypothesized independent variables (financial data for the HRS and sexual behavior for the NSFG); the order in which the studies were described was systematically varied. The descriptions included information about the study's purpose and sponsorship as well as the social and personal benefits of participating. They also indicated the information that would be required (access to "government records" in the case of the HRS and to "medical records" for the NSFG) and included a confidentiality assurance consisting of several sentences.

Immediately following this description, respondents were asked how likely it was that they would be willing to participate in the survey. If they indicated that they were willing, they were asked about their reasons for participating; if they were unwilling, they were asked why they would not be willing. They were then asked how likely they thought it was that a member of their family, their employer, a business, or a law enforcement agency would be able to see their answers and how much they would mind if they did; they were also asked how much they thought various groups would benefit from the results of the study and whether or not they themselves would benefit.

Two findings are of particular interest in the present context: the variables that predicted respondents' expressed willingness to participate in the study and the reasons they gave for participation and nonparticipation. Respondents' judgments of how likely it was that others would have access to their answers, together with their name and address, as well as their rating of how much they would mind such a disclosure—that is, their ratings of the "risks" and "harms" of disclosure—significantly predicted their expressed willingness to participate in the study; those who thought it more likely, and who minded more, were significantly less willing to participate, and the perceived risk/benefit ratio also had a negative relationship to participation. Perceived benefits, on the contrary, whether personal or social, significantly increased willingness to participate.

Analysis of responses to the open-ended questions supported these findings. For example, some 48 percent of the sample said that they would not be willing to take part in the survey described; the most frequent reasons given, by 59 percent of those who said they would not want to participate, were that the surveys were too personal or intrusive, that they objected to giving out financial or medical information, or that they objected to providing access to medical or financial records. Although this finding is clearly context-dependent—that is, different introductions to different surveys might have led to different reasons for nonresponse—it indicates that descriptions of surveys on sensitive topics do arouse concerns about privacy and about the confidentiality of the information requested.

A 2004 survey sponsored by the Administration of Planning and Statistics of the Flemish government to investigate social and cultural issues (Storms and Loosveldt, 2005) also offers evidence that concerns about privacy affect willingness to participate in future surveys (as well as item nonresponse). At the end of the face-to-face interview, respondents were asked to agree or disagree on a 5-point Likert scale with this statement: "Research like this is an invasion of privacy." The survey also measured general attitudes toward privacy, using four of the questions used for this purpose in the U.S. Census Bureau surveys.[5] The correlation between an index based on these questions and the survey-specific question about privacy was a low 0.10. Trust in government and in government institutions was related to general privacy concerns, whereas the survey-specific privacy question was related to evaluations of how pleasant the survey had been.

Storms and Loosveldt found that the survey-specific question, but not the general index, was a significant correlate of (a) refusals to answer the income question on the survey, (b) reported willingness to cooperate again in similar research, and (c) completing the self-administered portion of the questionnaire. This finding may be subject to nonresponse error: Persons who had general privacy concerns might have been more likely to refuse—a conclusion supported by the finding that privacy concerns expressed at the doorstep significantly predicted cooperation. Only 19 percent of those who said they did not want to give personal information ultimately cooperated, compared with 41 percent of those who expressed generally negative attitudes toward research and 38 percent of those who said they were not interested. On the other hand, 94 percent of those who asked questions about confidentiality, such as who gets access to data and whether answers would be anonymous, ultimately cooperated.

Having established that concerns about privacy and confidentiality affect cooperation with survey requests, we turn next to a consideration of whether these attitudes have changed over time. This is central to assessing whether privacy and confidentiality concerns may have contributed to the secular decline in response rates.

21.3.3 Trends in U.S. Attitudes Toward Privacy

One important source of data on trends in U.S. privacy attitudes is a series of surveys conducted by the Harris Organization, many of them for Equifax or other commercial

[5]"In general, how concerned are you about your privacy? The right of privacy is legally well protected; People have no control anymore about the use of their personal information; Because of the right of privacy, the use of computers in handling personal information has to be strictly regulated."

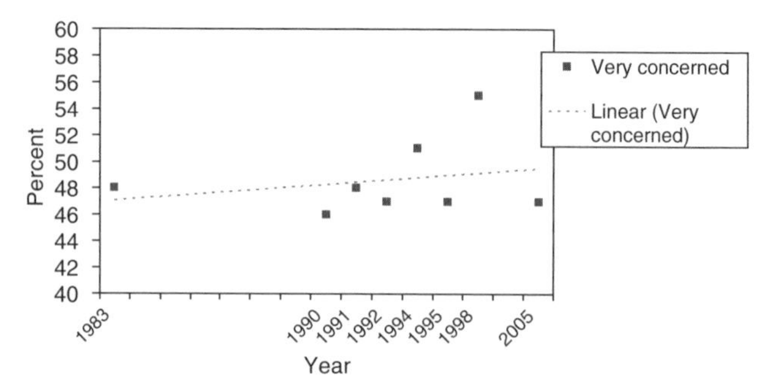

Figure 21.4. Trends in U.S. privacy attitudes (Harris Phone Surveys). "How concerned are you about threats to your personal privacy in America today—very concerned, somewhat concerned, not very concerned, or not concerned at all?" (N~1000 in each year).

sponsors.[6] Answers to a key question asked on these surveys from 1983 to 2005 are shown in Fig. 21.4. Respondents were asked, "How concerned are you about threats to your personal privacy in America today—very concerned, somewhat concerned, not very concerned, or not concerned at all?"[7] Although attitudes have fluctuated somewhat in the past 10 years, generally about half of those surveyed during this period indicated that they were very concerned "about threats to their personal privacy." As we shall see, this is more than double the percentage indicating they were "very worried" about their personal privacy in a series of surveys sponsored by the Census Bureau between 1995 and 2000. This is an instance of Corning and Singer's (2003) observation that privacy and confidentiality attitudes are generally not well informed and may vary significantly depending on question wording, survey organization, and survey sponsor. However, with question wording and sponsor held constant, the trend in privacy concerns has been level or, at most, has shown a slight increase over the past 20 years.

Table 21.1 shows responses to seven other questions asked by Harris in 1994, 2001, 2003, and 2005 (Krane, 2003, and personal communication). Five of the items reveal increasing importance placed on privacy between 1994 and 2001 and then a decline from 2001 to 2003, most likely as a result of the events of 9/11.[8] Since 2003,

---

[6]Trends in attitudes toward privacy and confidentiality are also reviewed by Mayer (2002), who includes qualitative research on privacy and confidentiality.

[7]In all the Harris surveys, this question was placed either at the beginning of the questionnaire or at the beginning of the section on privacy. The largest increase in concern about privacy—from 31 percent to 48 percent—occurred between 1978 and 1983; however, because that change also coincided with a change from personal to telephone interviewing, along with changes in sample design, the 1978 datapoint has not been included in Fig. 21.4.

[8]Harris surveys taken in late September 2001, March 2002, and February 2003 indicate that support for a series of government surveillance measures (e.g., stronger document and security checks for travelers, expanded undercover activities) was extremely high immediately after 9/11 but dropped in each of the two succeeding surveys (Corning and Singer, 2003, Table 10).

Table 21.1. Meaning of Privacy (Harris Poll Data, 1994–2005)

"Privacy means different things to different people. I am going to read you a list of different aspects of privacy. Please tell me how important... "

	Percent extremely important			
	1994	2001	2003	2005
Being in control of who can get information about you	80	84	79	83
Not having someone watch you or listen to you without your permission	79	80	73	75
Controlling what information is collected about you	74	79	69	72
Not being disturbed at home	49	55	62	65
Having individuals in social and work settings not asking you things that are highly personal	49	55	48	52
Being able to go, around in public without always being identified	43	47	48	43
Not being monitored at work	65	40	42	36

N~1,000 for each survey.

these items have shown little change. The other two items produce quite different trends: a decline over the entire time in the importance of freedom from monitoring at work, which Taylor (2003a) attributes to acceptance of telephone monitoring at commercial phone centers, and increasing importance throughout the period of not being disturbed at home, which he attributes to the growth of telemarketing. This interpretation is consistent with the pattern in reports of ever having "personally been the victim of what you felt to be an improper invasion of privacy," which increased from 25 percent to 41 percent between 1995 and 1998 when the phrase "As a consumer. . ." was added to the beginning of the question (Corning and Singer, 2003, pp. 6–7).

In addition, reported protective actions against perceived invasions of privacy, such as refusing to give information to a business or company or asking a company not to pass on personal information, roughly doubled between 1990 and 2001 (Corning and Singer, 2003, p. 7). The available data indicate that concerns about threats to privacy are more frequent among women than among men and that apprehension about the misuse of personal information is higher among older people, African Americans, the less educated, and those with lower incomes (Corning and Singer, 2003).

Because much of the evidence about attitudes toward confidentiality, as distinct from privacy, comes from government surveys, it is useful to compare the perceptions of privacy threats posed by government and business. Corning and Singer (2003, pp. 16–17, Table 9) found consistently more concern about government than about business as a threat to privacy, with the exception of a 2002 Public Agenda Foundation survey.

Responses to seven general questions about privacy asked on surveys done for the U.S. Census Bureau between 1995 and 2000 are shown in Table 21.2. All four surveys

Table 21.2. Trends in U.S. Privacy Attitudes (Surveys Sponsored by the U.S. Census Bureau)

Privacy question	1995 %	1996 %	1999 %	2000 %
How worried about privacy (*very*)	22	—	26	25[a]
Privacy rights well protected (*strongly agree*)	13	9	13	14
People have lost control over personal information (*strongly agree*)	40	44	42	44[a]
Must regulate computers to protect privacy (*strongly agree*)	60	—	59	59
Government knows too much about me (*yes*)	53	—	43	43[a]
Ever victim of privacy invasion? (*yes*)	27	—	29	28
Telephone ever tapped? (*yes*)	10	—	14	17[a]
N	1443	1215	1681	1978

[a]Significant change, 1995 to 2000.

used very similar methods.[9] Two of the questions, including one about "how worried" the respondent is about personal privacy, show a small but statistically significant increase in concern over the 5-year period; one a substantial increase and one a substantial decline. The rest show no change. As pointed out earlier, the concern elicited by the "worried" question is about half that elicited by a somewhat differently worded question in the Harris surveys, but the two items show the same general time trend.

Overall, the evidence suggests that Americans have become more concerned about privacy issues in the last few decades, though many of the indicators show only modest increase and some no change at all.

21.3.4 Trends in Attitudes Toward Confidentiality in the United States[10]

Most of what we know about trends in U.S. attitudes toward confidentiality comes from the four RDD surveys sponsored by the U.S. Census Bureau between 1995 and 2000 that we drew on in the last section. Although these surveys document the trend in attitudes toward the confidentiality of information provided to the U.S. census, they may not be very informative about Americans' concerns regarding the confidentiality of personal information provided to other survey organizations.

[9]The 1995 survey was carried out as part of the University of Maryland's Joint Program in Survey Methodology practicum; the 1996 survey, which used a questionnaire virtually identical to that in 1995, was carried out by Westat for the Census Bureau; the third and fourth surveys, done in July through October 1999, just before the start of the public relations campaign and nationwide field recruiting for Census 2000 and from April to July of 2000 after delivery of census forms, were conducted by The Gallup Organization for the Census Bureau (under a contract with the University of Michigan). All the surveys randomly selected one member of the household aged 18 or over. Response rates in these RDD samples from the contiguous United States were between 60 and 62 percent, and the completed number of interviews ranged from 1215 to 1978.

[10]The discussion of trends in attitudes toward confidentiality is based on Singer (2003, pp. 3–5.).

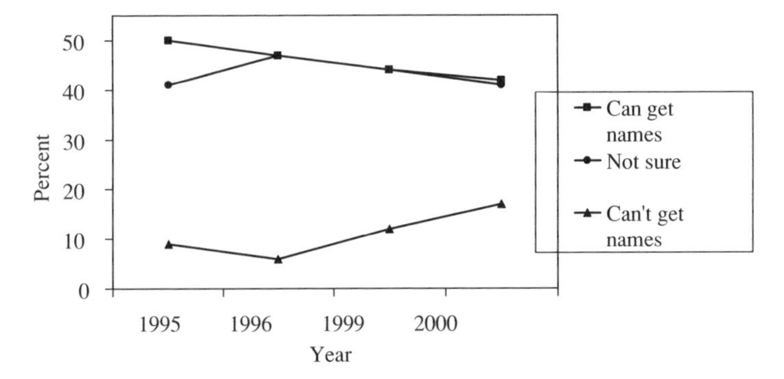

Figure 21.5. Trends in U.S. confidentiality attitudes (Singer et al., 2001). "Do you think other goverment agencies, outside the Census Bureau, can or cannot get people's names and addresses along with their answers to the census, or are you not sure?"

These studies show, first, that people are very poorly informed about the Census Bureau's confidentiality practices—when given the opportunity, almost half say that they are "not sure" whether the Census Bureau protects confidentiality (Fig. 21.5). Second, of those who did offer an opinion, most gave an incorrect response (Fig. 21.5), echoing the finding of a similar study done two decades earlier (National Research Council, 1979). Third, although awareness that the Census Bureau is required by law to keep Census records confidential increased from 51 percent to 76 percent between 1996 and 2000 (Singer et al., 2001, Table 2.19), about one third of those aware of the legal requirement said, in all 3 years in which the question was asked, that they did not trust the Census Bureau to uphold the law. Finally, between 1996 and 2000, the proportion saying it would bother them "a lot" if their answers were not kept confidential increased significantly from 37 percent to 50 percent (ibid, Table 2.17). Thus, whereas concern about privacy in the abstract shows only a modest increase over time, concern about the misuse of confidential personal information provided to the U.S. Census Bureau increased significantly during the 4-year period during which these attitudes were monitored.

21.3.5 Trends in European Attitudes

Any attempt to summarize trends in European attitudes toward privacy and confidentiality risks oversimplification. Nevertheless, the availability of three Eurobarometer surveys, from 1991, 1996, and 2003, which asked identical questions in all the European Union nations in each of the 3 years, has encouraged us to try. Here, we summarize trends in responses to only a few questions; for the complete report, see (http://europa.eu. int/comm/public_opinion/archives/ebs/ebs_196_data_protection.pdf).[11]

[11]Data files are available through ICPSR at the University of Michigan. These face-to-face surveys are carried out in the appropriate language of each country with respondents aged 15 and over by the European Opinion Research Group, a consortium of market and opinion research agencies. The sample consists of random points systematically drawn in each country with probability proportional to population size and density; in each sampling point, a starting address is drawn at random, and further addresses are selected at specified intervals by standard random route procedures. In each household, one respondent is selected at random.

Table 21.3. Trend in Privacy Attitudes (Eurobarometer)

"Different private and public organizations keep personal information about us. It is sometimes said that our privacy must be properly protected? Are you very concerned, fairly concerned, not very concerned, or not at all concerned?"

	Percent very concerned		
Country	1991	1996	2003
Belgium	29	23	23
Denmark	13	13	13
Germany	23	14	19
Greece	51	37	58
Spain	15	12	13
France	54	40	37
Ireland	47	28	36
Italy	47	11	14
Luxembourg	32	23	27
Netherlands	15	11	15
Austria	NA	15	19
Portugal	16	31	13
Finland	NA	6	15
Sweden	NA	59	54
England	44	43	41
EC12[a]	35	NA	NA
EU15[a]	NA	24	25

[a]Weighted by population. *N*'s are approximately 1000 per country, except Luxembourg, where *N*~600.

The report is titled "Data Protection," and that is a clue to an apparent difference between European and American attitudes in this area. Although the term data protection seems to refer primarily to the protection of confidentiality, in fact the questions in these surveys (at least in translation) often use the term privacy. For example, Q. 30 asks, "Different private and public organizations keep personal information about us. It is sometimes said that our privacy must be properly protected. Are you concerned or not that your privacy is being protected?" Responses to this question, by country and averaged across all the EU countries, are shown in Table 21.3. Averaging across the 15 European member countries about one-quarter said they were "very concerned" in both 1996 and 2003. The percentage in 1991 was about one-third but included only the 12 countries who were EU members at that time (i.e., excluding Austria, Finland—both of which have very low levels of concern— and Sweden, which in 2003 expressed the second highest level of concern of all 15 countries). Thus, the conclusion that on average there has been little change in opinion since 1996 conceals a great deal of intercountry and temporal variation. Almost all countries surveyed in both 1991 and 2003, for example, show a decline in concern during this period, perhaps as a result of the Data Protection Law of 1995, whereas 9 of the 15 show an increase in concern between 1996 and 2003.

Table 21.4. Trust in Market and Opinion Research Companies (Eurobarometer)

"The following organizations may keep personal information about us. Do you trust market and opinion research companies to use this information in a way you think acceptable?"

	Trust (%)		Do not trust (%)		Don't know (%)	
Country	1996	2003	1996	2003	1996	2003
Belgium	48	44	25	40	25	16
Denmark	66	56	20	27	13	17
Germany	41	43	29	27	29	30
Greece	47	42	37	47	16	10
Spain	45	37	36	46	19	17
France	46	41	41	44	13	16
Ireland	48	34	24	44	28	22
Italy	54	50	21	27	26	23
Luxembourg	56	57	22	32	22	11
Netherlands	63	47	23	40	15	13
Austria	49	44	24	37	27	19
Portugal	54	50	27	27	18	23
Finland	48	42	38	40	14	18
Sweden	49	46	27	35	24	18
England	46	40	37	36	17	24
EU15[a]	47	43	31	35	21	22

[a]Weighted by population. *N*'s are approximately 1000 per country except Luxembourg, where *N*~600.

The question perhaps most relevant to telephone surveys concerns trust in market and opinion research companies "to use [personal] information in a way you think acceptable." On average, such companies were trusted by 43 percent of respondents in the 15 EU countries in 2003—a 4 percentage point decline since 1996 (see Table 21.4). (Trust in these organizations was not asked in 1991.) Particularly large increases in distrust were shown in Ireland, the Netherlands, and Belgium. Although Denmark showed the highest degree of trust in 2003, it too showed a substantial decline from 1996. As a point of comparison, however, mail order companies were trusted by only 21 percent of the population to use personal information in an acceptable way in 2003, although opinion had become slightly more favorable since 1996.

21.3.6 Trends in Canadian Attitudes

The only Canadian data on attitudes toward privacy that we have been able to locate show fluctuations, but no clear trend. Agreement with the statement "I feel I have less personal privacy in my daily life than I did ten years ago" ranged from 47 to 62 between 1992 and 2005, with the first and last figures 60 and 62 percent, respectively. The 2005 survey also asked, in split-ballot form, a variant question substituting "less protection of personal information" for "less personal privacy"; significantly more respondents (72 percent) agreed with that version.[12] Like their counterparts in the United States, Canadians are not well informed about the issue: only 20 percent say they are "clearly" aware of "any laws that help Canadians deal with privacy and the protection of personal information," and an additional 28 percent are "vaguely" aware. Only about half of those with any awareness could name a law protecting privacy; most of these named the Privacy Act.

21.4 RESPONDENT BURDEN

21.4.1 Effects on Nonresponse

Although respondent burden has two major meanings in survey research—length and difficulty—research on the link between burden and cooperation has focused almost exclusively on length. Moreover, most studies of the relationship of questionnaire length to nonresponse deal with mail surveys. Surprisingly, as Bogen (1996) notes in her review of this literature, the findings have been mixed: many studies of mail surveys show either no length effect or only a weak one. Likewise, Crawford, Couper, and Lamias found no overall effect of announced length (10 minutes versus 20 minutes) on a Web survey's response rate.

For interviewer-administered studies, nonexperimental research has found that interview length does affect cooperation. Both Marton (2004) and Holbrook et al. (Chapter 23 in this volume) report that cooperation rates were inversely associated with questionnaire length for a series of RDD surveys. Dijkstra and Smit (2002) show that initial cooperation rates to a telephone survey were higher when the interviewers departed from the introductory script by not mentioning interview length than when the interviewers conveyed that part of the script. In addition, among those who did read the length part of the script, cooperation was higher when the interviewer later made a mitigating statement that the interview actually would not take that long.

Of course, the absence of experimental control suggests caution in interpreting these results. The surveys drawn on by Marton and Holbrook et al. varied in many ways in addition to length, but Marton was able to control for only two such factors (length of data collection period and proportion of new interviewers), and Holbrook et al. were unable to control for any. Likewise, Dijkstra and Smit could not control for interviewer confidence, which may have been responsible for interviewers' departing from the script and making a mitigating statement as well as for the differences in cooperation.

We know of only three experiments measuring the impact of burden on unit nonresponse in interviewer-administered surveys. The earliest experiment, by Sharp and Frankel (1983), manipulated both interview length and effort required to answer

[12]This was, apparently, the first time Canadians were asked about protection of personal information as distinct from personal privacy. For all the results in this section see www.privcom.gc.ca/information/survey/ekos e.asp

in an in-person survey of 500 suburban Philadelphia, PA residents selected from a multistage area frame. A random subsample of respondents was also recontacted for a follow-up survey 10 months later. For the initial survey, two treatments were crossed: Respondents were told the interview would last either 25 or 75 minutes, and they were asked either to recall expenditures (low effort) or to retrieve the information from records (high effort).

Length did not significantly affect the initial interview unit refusal rates (30 and 27 percent), although the difference was in the expected direction. In a postinterview debriefing of a random 60 percent subsample, however, long-version respondents were almost four times as likely as short ones to say the interview was too long ($p < 0.0001$) and twice as likely to say they would be unwilling to be reinterviewed ($p < 0.01$). But in the subsample that was recontacted (the 40 percent not asked the debriefing questions), the follow-up survey response rate was unrelated to whether respondents had initially received the long or short interview. Length was also unrelated to item nonresponse.

The effort manipulation was not related to any of the dependent variables: item nonresponse, debriefing ratings, and reinterview response rate. However, the majority (61 percent) of respondents in the retrieval condition did not consult records, and almost a quarter of those in the recall condition did. Thus, the effort manipulation was not a powerful one.

The second experiment, by Collins et al. (1988), was conducted with a nationwide sample of British households that had listed telephone numbers. One random subsample was told the phone interview (drawn from the British Social Attitudes Study) would last 40 minutes and the other 20 minutes. The refusal rates were 14 and 9 percent, respectively. Although the authors report neither a significance test nor *N*'s for the two groups, given the total number of cases fielded (429), the response rate difference does not appear to exceed what would be expected by chance.

Two factors may have reduced the likelihood of effects in these experiments. First, in both experiments, the power of the length manipulation was limited to the extent that not all potential respondents may have heard the relevant information. In fact, Sharp and Frankel report that the majority of their refusals came before the respondent was apprised of length, and the same may have occurred in the Collins et al. study. This will often be the case in telephone surveys, especially when no advance letter is sent.

Second, although the difference between the length conditions was relatively large in both experiments (20 minutes in one and 50 minutes in the other), even the short versions (20 and 25 minutes) were still somewhat extended. Similar length increases from a briefer starting point (e.g., expanding a 5 minute interview to 25 minutes) may have a larger impact on response rates. It seems more reasonable to expect length to have a step, as opposed to linear, relation to perceptions of burden.

This expectation is confirmed by the results of the third experiment, which had three length conditions (CMOR, 1996).[13] In a nationwide U.S. RDD telephone

[13]We thank Harry O'Neill, under whose direction this study was done, for making the study available to us.

survey of 1920 adults, respondents were told the interview would last 10, 20, or 30 minutes. The refusal rates were 55, 62, and 61 percent, respectively, with the difference between the 10 minute version and each of the other two versions significant at $p < 0.05$.

Taken together these experiments suggest that the burden associated with particular features of a survey is not likely to be an important determinant of unit nonresponse for most telephone interviews that exceed 15 or 20 minutes. This is partly because when deciding whether to cooperate, potential respondents will often be unaware of a survey's features that may be burdensome: Refusals may occur before the interviewer mentions a feature; the feature may not be included in the introduction; or the feature may be included in the introduction but, as one of many elements, may not be processed by respondents.

However, even when respondents are unaware of burdensome features, the features can affect response rates indirectly via the interviewer. As Bogen (1996) notes, "interviewers' expectations about respondents' negative attitudes toward long interviews may play a more important role than the respondents' negative attitudes themselves." For example, interviewers may sound less confident or exert less effort on more burdensome studies than on less burdensome ones. The importance of this factor is illustrated by the finding from the CMOR (1996) study discussed above that the refusal rate difference between the 10 minute versus 20 and 30 minute versions was almost as large in the experimental condition that did not mention interview length in the introduction as in the condition in which it was mentioned.

Daniel Kahneman and his colleagues (Kahneman et al., 1993) provide a useful perspective in understanding why length may not play an important role in respondent decisions to cooperate with surveys. They find that people tend to ignore duration in evaluating experiences, a phenomenon Kahneman refers to as "duration neglect." In a series of experiments with both pleasant and unpleasant events, people's evaluations were driven mostly by reactions to the most intense part of the experience and to the very end of the experience ("peak value" and "final value"), and very little by how long the experience lasted. Put differently, evaluations are driven by what is salient about an experience, which is rarely its duration. Kahneman views duration neglect "as a special case of a broader family in which categories or sets (e.g., a set of moments) are evaluated by judging a prototype" (Ariely et al., 2000, 525). This suggests the usefulness of thinking about perceptions of burden in terms of reaction to surveys more generally.

An individual's perception of a particular survey's burden may be affected not only by characteristics of the survey (e.g., its length), but also by the individual's cumulative experience with prior surveys (e.g., "I have spent too much time answering surveys in the past") or by attitudes about survey burden in general (e.g., "surveys are too long"). Although the effect of a particular survey's burden is best estimated through experiments that vary the burden, the effects of prior survey experience and of general attitudes toward survey burden may be estimated through correlational studies.

Considering only the impact of burden, one might expect that people who have participated in a recent survey would be more likely to refuse a new survey request than those who had not been respondents in the recent past. In fact, the opposite is true. For example, Frankel (1980) reports that 37 percent of the respondents to the initial Philadelphia survey described above said they had "participated in a survey in the last 12 months" compared to only 11 percent of the refusals who later agreed to a short interview when recontacted by a supervisor ($p < 0.001$). Similarly, refusal rates to household panel surveys are almost always higher at wave one than at later waves (sometimes dramatically so; for instance, the wave one refusal rate on the University of Michigan's Panel Survey of Income Dynamics was over 20 percent compared to barely 2 percent in each of the many waves after the second one). Likewise, in surveys of South Dakota farm operations, a population that receives frequent survey requests from the National Agricultural Statistics Service, Ott (2002) found that "refusal rates do not increase as the number of times an operation is in sample increases. In fact, the rate decreases.... regardless of interview mode."[14]

Thus, although participation in surveys may shape perceptions of survey burden, the causation may run the other way round as well: perceptions of burden independent of actual experience (a la Kahneman's prototype) may influence survey participation. Consider in this regard Stinchcombe et al.'s (1981) study of nonresponse to a U.S. Department of Agriculture survey. Farmers who initially refused were twice as likely as initial cooperators to say they were asked "much too often" to participate in surveys. Yet a check of prior survey records showed that "[t]his subjective reaction is not related to how often farmers are actually asked to participate - it measures instead an irritation which finds being asked once or twice just as bad as being asked ten or a dozen times" (pp. 370–371).

In summary, the existing evidence suggests that perceived burden is much more important than actual burden in determining survey cooperation. Thus we turn next to a consideration of how perceptions of survey burden, as well as attitudes toward surveys more generally, have changed over time.

21.4.2 Trends in Attitudes

The belief that general attitudes about surveys (including their burden) are an important determinant of response rates gave rise to a biennial series of Industry Image Studies in the United States. The studies, conducted by telephone, were first sponsored by Walker Research Inc., from 1978 to 1992 and since then by the Council for Marketing and Opinion Research (CMOR).

Schleifer (1986) reported modest changes in attitudes in the Walker Research surveys from 1978 to 1984. For instance, the judgment among respondents that their last survey experience was too long increased from 14 percent to 23 percent and the feeling that that interview was unpleasant doubled from 6 percent to 12 percent (though these increases were not monotonic). Similarly, agreement with the statement "Answering questions in polls or research surveys is a waste of time" grew from 17 percent to 22

[14]In a review of many nation-wide farm surveys, McCarthy, Beckler, and Qualey (2006:108) conclude that "none of the types of burden we looked at appeared to be systematically related to future survey cooperation."

percent. (Reports of participating in the last year in a "survey" that turned out to be a sales pitch rose from 13 percent in 1980 to 15 percent in 1982 and 17 percent in 1984.)

The more recent CMOR studies have shown a mixed picture of stability and change (CMOR, 2003).[15] Attitudes about negative attributes of surveys changed little. For instance, the proportions saying their last interview experience was too long and unpleasant in 2003 were no different from the corresponding proportions in the mid-1980s. On the other hand, attitudes about preferred mode of data collection shifted considerably. In the past few years, selection of the mail, the leading choice in all years, declined considerably as a result of the rise of the Internet. But preference for the telephone, the second most popular choice in all years, remained essentially unchanged over the entire period.

The most notable shift depicted by the Industry Image Studies involved the proportion of respondents who reported participating in surveys (of any mode) in the preceding year. This grew astronomically from 19 percent in 1978 saying they had participated in a survey at least once in the last year to 82 percent in 2003! However, as several authors, most notably Bickart and Schmittlein (1999), have observed, there are severe problems in relying on surveys that suffer from large nonresponse to estimate the population distribution of characteristics correlated with nonresponse. The Industry Image Surveys (which have always had relatively low response rates) almost certainly underestimate negative views of surveys and overestimate survey requests, and Bickart and Schmittlein's simulations suggest that the biases may be very large. This would be less of a problem for inferences about temporal change if the Industry Image Study response rates had remained stable over time. In fact, the refusal rate was fairly steady from 1978 to 1988 (fluctuating between 42 and 49 percent) but then soared from 53 percent in 1990 to 79 percent in 2003 (CMOR, 2003; total nonresponse in 2003 was 93 percent).[16]

Nonetheless it is hard to resist the conclusion that the volume of survey calls (not to mention sales and fundraising contacts by telephone) has grown over time and is, at least partly, responsible for a decline in survey response rates. A probability sample of survey organizations reporting the number of survey requests they make to individuals or households would provide invaluable evidence on part of this, but as far as we know, no such data exist.

Presser and Kenney (2006) have, however, estimated the change over time in U.S. government surveys that required Office of Management and Budget approval. For data collections covering households and individuals, they report that there were about twice as many burden hours approved in 2004 as in 1984, though the pattern in the intervening years was not constant. The nature of the change over time in other sectors of the survey research industry is unknown.[17]

[15]We thank Harry Heller, Director of Respondent Cooperation at CMOR, for providing portions of this report.

[16]This problem may also compromise some of the privacy trends we reviewed in earlier sections (though not the confidentiality trends, which are based on studies with very similar response rates).

[17]Holbrook et al. (Chapter 23 in this volume) found that nine of the ten private organizations they surveyed reported no change in the length of their interviews between 2000 and 2004, though the remaining one reported longer interviews.

Moreover, as we noted in our discussion of the changing legal landscape, the environment in some nations has changed recently with the institution of Do Not Call lists, which may have reduced the number of calls people on the list receive (though it may also have increased the annoyance such individuals feel at receiving any unsolicited calls, including survey requests).

21.5 PRIVACY, CONFIDENTIALITY, BURDEN, AND NONRESPONSE BIAS

We have reviewed the evidence on the extent to which concerns about privacy, confidentiality, and burden increase survey nonresponse. But nonresponse is often only a proxy for the real concern: nonresponse error or bias. Although some studies have found evidence of increased bias with increased nonresponse (e.g., Baumgartner and Rathbun, 1997; Singer et al., 1999a), even substantial amounts of nonresponse do not necessarily increase bias (Curtin et al., 2000; Keeter et al., 2000; Keeter et al., 2006). Based on a meta-analysis of thirty studies, Groves (2006) concludes that "the nonresponse rate of a survey alone is not a very good predictor of the magnitude of the bias."

Our review has shown that people with privacy or confidentiality concerns are less likely to respond to mail and face-to-face surveys, and probably to telephone surveys as well. There is also evidence that perceptions of survey burden are related to willingness to cooperate. If survey measures are strongly related to attitudes about privacy, confidentiality, or burden, then nonresponse bias is likely. An extreme case of this is the Industry Image Studies' overestimate of the extent to which the population is surveyed and views surveys positively. On the other hand, when the survey measures are unrelated, or only weakly related, to privacy, confidentiality, or burden concerns, the reduced participation of those with such concerns is unlikely to affect the estimates obtained.

21.6 CONCLUSION

Concerns about privacy and confidentiality have increased in the past two decades. There is also evidence that perceptions of the number of survey requests have grown during this time but that evidence is less compelling. At the same time, cooperation with survey requests by government, academic, and commercial research firms has declined. To what extent are the two trends causally related? And, if they are related, how does this affect survey estimates?

Experimental evidence indicates that concerns about privacy and, especially, confidentiality reduce cooperation, but the *net* effects of such attitudes, although statistically significant, are relatively small: Privacy and confidentiality concerns are not the major factors in declining telephone survey response rates.

Based on the available evidence, we conclude that concern about confidentiality seems to be a bigger factor than concerns about privacy in government surveys; concerns about privacy—in the sense of a right to be let alone, which is closely related to perceptions of burden—seem to be a bigger factor in nongovernment surveys. We suspect that the impact on response rates of all three factors—privacy, confidentiality, and perceived burden—will increase in the future because of increasing impatience with the intrusion of telephones into private space and the increasing salience of privacy and confidentiality issues.[18] Although many survey researchers are optimistic about the eventual effect of the Do Not Call Registry, marketers may find ways around the prohibition; fundraising calls for charitable and political organizations are exempt from the regulation; and the public does not draw clear distinctions between marketing calls and legitimate survey requests. A study by Link and his colleagues (2006) finds no evidence that the DNC Registry has so far reversed or even slowed trends in declining response to telephone surveys. Furthermore, although interviewers may be able to counter concerns about confidentiality if they are raised by respondents in face-to-face interviews, the tendency to hang up in the first few seconds of a phone call from a stranger often precludes this strategy in telephone surveys.

When it comes to nonresponse bias, concerns about privacy, confidentiality, and burden will affect survey estimates to the extent these concerns are related to the estimates. The evidence that women, minority, and lower SES groups tend to have greater privacy and confidentiality concerns suggests that estimates of variables associated with these characteristics may be particularly prone to nonresponse bias. We know less about the correlates of burden perceptions, and these may have more sizable consequences for telephone survey nonresponse. Thus, a shift in focus of methodological research from nonresponse levels to nonresponse bias is urgently needed. In the interim, the major consequence of growing concerns about privacy, confidentiality, and burden will likely be to further fuel rising survey costs.

[18]Increases in distrust in government and other institutions might also have this effect. But we note that The American National Election Studies showed a large increase in trust in the federal government during the 1990s, which was a time of decreasing response rates (see www.umich.edu/_nes/nesguide/toptable/tab5a_1. htm)

CHAPTER 22

The Use of Monetary Incentives to Reduce Nonresponse in Random Digit Dial Telephone Surveys

David Cantor
Westat and Joint Program for Survey Methodology, University of Maryland, USA

Barbara C. O'Hare
Arbitron Inc., USA

Kathleen S. O'Connor
Centers for Disease Control and Prevention, National Center for Health Statistics, USA

22.1 INTRODUCTION

Random digit dial (RDD) telephone surveys have been widely used as a cost effective methodology for studying general populations. A major concern with this methodology is that response rates have been declining precipitously over the past 5-years (Curtin et al., 2005). This has raised fears that the methodology is increasingly subject to bias due to nonresponse. For this reason, survey researchers are increasingly looking toward the use of special methods to increase RDD survey response rates.

A special method that can be employed is the payment of a monetary incentive to a respondent in exchange for interview completion. It is relatively common to pay a respondent a monetary incentive, although incentive use varies across the different sectors (commercial, university, government, nonprofit organizational surveys, etc.) and modes used by the survey research industry. There is a large literature on the use

Advances in Telephone Survey Methodology, Edited by James M. Lepkowski, Clyde Tucker, J. Michael Brick, Edith de Leeuw, Lilli Japec, Paul J. Lavrakas, Michael W. Link, and Roberta L. Sangster

of monetary incentives across different interviewing modes (Singer et al., 1999b). However, there has not been a systematic assessment of monetary incentive use in RDD surveys. Incentive use in RDD surveys did not become commonplace until the late 1990s and early 2000s (see also Holbrook et al., 2007, Chapter 23 in this volume for analysis of incentives in the context of RDD surveys).

The application of the monetary incentive technique to RDD surveys poses unique problems. In a telephone survey, the opportunity to offer an incentive before the interview is limited to either mentioning it on the phone at the time of first contact or delivering it in an advance letter. An offer over the phone must be interpreted with limited cues as to the purpose of the call and may be interpreted as a "sales gimmick." Incentive offers in an advance mailing are limited to those phone numbers for which an accurate address can be obtained; however, this address-matched group of households constitutes only a portion of the total sample. Once the advance letter is sent to these households by regular or express mail, there is a chance that the letter will be thrown away before it is read (Couper et al., 1995); thus, the intended recipient may not receive the incentive. Parsons et al. (2002) reported that only about 60 percent of respondents completing their RDD interview report seeing a prenotification letter (see also Link and Mokdad, 2005a).

Additionally, RDD surveys often consist of multiple stages (Fig. 22.1). The *screening* stage is used to establish that the telephone number represents a residential household, and to select an appropriate respondent (e.g., random adult, most knowledgeable adult). Once a respondent is selected, a separate interview is administered. This *extended* interview collects the primary data for the study. For

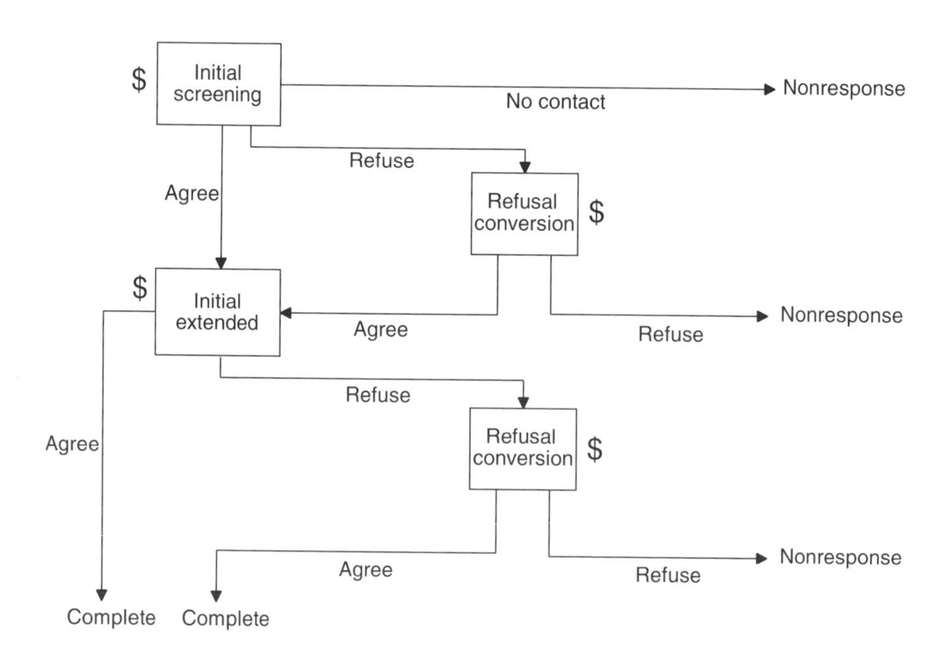

Figure 22.1. Simplified Flow of RDD Survey.

many of the studies reviewed below, there is an attempt to conduct the extended interview immediately after the screening interview is completed. For some surveys, a different mode or methodology might be administered to collect the primary data (e.g., filling out a paper diary, request to do a survey on the Internet). Increasingly, RDD surveys are exerting more effort to convert those that initially refuse the survey request at either the screening or the extended interviewing stage. Incentives can be offered at any point in this process. As seen in Fig. 22.1, this is potentially at four different points (indicated by a dollar sign ($)). The context and logistics associated with offering an incentive at these various points differ. However, there is very little in the research literature that summarizes the performance of incentives in these different contexts.

The purpose of this chapter is to review the evidence testing the effects of monetary incentives in RDD surveys. The discussion in this chapter is restricted to the empirical literature on the effectiveness of incentives when tested in an experimental design. We recognize that paying money to complete surveys may not be universally approved in all situations. Some institutional review boards may view certain types of incentives as "coercive" if they are judged to cloud decisions to participate in potentially harmful protocols (Amdur, 2003). In the United States, for studies conducted by the Federal Government, the Office of Management and Budget (OMB) generally requires a specific rationale to provide incentives and may even require proof that they are worth using (COPAFS, 1992; Kirkendall, 1999; Kulka, 1999). Perhaps most controversial is using different incentives for various populations or incentives that target refusers (Singer et al., 1999a). It is beyond the scope of this chapter to include a meaningful discussion of the ethics related to the use of incentives in these different situations. The goal of this chapter is to review the empirical tests that have been conducted.

In the next section, the theories that have guided incentive use are discussed and applied to the RDD context. Section 22.3 reviews the methodology that we used to collect studies for this review. Section 22.4 reviews the evidence regarding observed incentive effects on response rates and conditioning effects of incentives. Section 22.5 discusses the differential effects of incentives across population groups. Section 22.6 reviews the impact of incentives on estimates and data quality. Section 22.7 explores monetary costs and benefits, by examining net costs and cost savings associated with incentives. The final section summarizes the primary conclusions.

22.2 THEORETICAL LITERATURE AND PREDICTIONS FOR RDD SURVEYS

Several related theories have been used to guide research regarding respondent participation in surveys. Social exchange theory (SET; Dillman, 2000) explains survey participation as a function of the rewards, costs, and trust related to participating (Dillman, 2000). Based on early theories of social exchange (Homans, 1961; Blau, 1964), Dillman applied these concepts to study the decision to cooperate in a survey. The key element in the decision process is whether the respondent believes that future rewards will outweigh the costs. Rewards from participation include nontangible benefits such as gratification, appreciation from the survey sponsors, or enjoyment in talking about the topic. Examples of costs include lost opportunities, loss of privacy, or simple boredom. A fundamental aspect of SET is that token incentives create a sense of obligation for future participation that reinforces the importance of trust in the social exchange (Dillman, 2000, p. 14). This is distinct from the concept of economic rewards, which weigh the monetary returns against the cost of participation.

Under SET, the timing and presentation of monetary incentives are critical. Token payments sent prior to completing the survey establish a trust that the respondent will reciprocate by participating in the survey, and create a future obligation with an unspecified cost (Dillman, 2000, p. 19). In comparison, a promised incentive suggests a bargaining arrangement. The promise requires respondents to trust that participation and the incentive will produce a net reward greater than the burden of the survey.

Leverage-salience theory (LST) is a related perspective that assumes respondents perform a calculus that weighs the positive and negative reasons to participate in a survey (Groves et al., 2000). The factors that are weighed depend on the features of the survey that are made salient at the time the decision is being made (e.g., when the interviewer calls the household). LST expects that the "leverage" of a particular feature varies across respondents and situations. For some respondents, incentives could play a critical role in tipping the decision from declining a request to participate to accepting the invitation. In particular, if a respondent would otherwise decide not to participate, an incentive may increase the perceived positive aspects that may tip the scales to cooperation. Alternatively, there are also respondents who are either too negative, too positive, or simply do not attribute any leverage to an incentive for it to play a critical role in their decision.

The application of these theories leads to the development of several predictions about the effects of incentives. First, SET posits that small incentives that are offered prior to survey participation will be more effective than incentives that are promised to be paid after completion. In an RDD survey, a promised incentive can be offered at least two different times: at the screening stage (without any prior personal contact with the respondent), and/or at the extended stage (upon completion of the screening interview(s)).

We make a second prediction that the incremental size of the effect of a prepaid incentive will decrease as the amount of the incentive increases. According to SET, the amount is not as important as the establishment of a social exchange (and trust). A token incentive of $1 has an effect on the respondent's willingness to participate because it creates an imbalance in the exchange once the respondent accepts the money. Larger amounts of money may engender a higher commitment. But we hypothesize that the increase will not be proportional to the dollars sent. Third, LST predicts that the effect of incentives will vary by individual characteristics, survey characteristics, and situations. Incentives will serve as a motivating influence leading to cooperation within the context of other positive or negative motivators. From a survey design perspective, this implies that the most effective use of an incentive is to target those respondents or situations where it seems to be needed the most. For example, response rates for surveys with a higher level of perceived burden may be more affected by an incentive than those surveys with a lower level of perceived burden. Incentives should perform better in the high perceived burden situation.

Similarly, certain respondents may be known to have less motivation than others (e.g., young males). If it is possible, incentives might be used to target those who are less motivated to participate in a survey.

Thus, LST predicts that the effect of an incentive should increase as the response rate without an incentive decreases. This uses the response rate without an incentive as an aggregate measure of how motivated respondents are to participate. Surveys with lower response rates should then be affected more dramatically by an incentive than surveys with high response rates. There is some support of this hypothesis from other modes, but it has not been examined for RDD surveys (e.g., Singer, 2002).

We integrated these theoretical perspectives and predictions in our review of the results of experimental studies that have tested different types of incentives. Before these results are summarized, Section 22.3 describes the methodology used to review studies that have examined the use of monetary incentives in an RDD context.

22.3 METHODOLOGY

Information was collected on studies that analyzed the use of incentives in RDD surveys. This search involved: (1) an online search of multiple data bases available through the library systems of the University of Maryland and the Centers for Disease Control and Prevention (e.g., Current Index of Statistics; JSTOR, Soc Index; Social Sciences Abstracts; Web of Science), (2) review of proceedings of the annual research meetings of the American Association for Public Opinion Research and the American Statistical Association; and (3) interviews with key informants from a number of commercial, university affiliated, and nonprofit survey research organizations that conduct or have conducted RDD surveys. In addition, we drew upon the in-house experience brought by the three co-authors, who represent a commercial firm (Arbitron, Inc.), a government contractor (Westat), and a government agency (Centers for Disease Control and Prevention, National Center for Health Statistics). Studies from Arbitron are cited as unpublished. For that reason, they are not included in the references at the end of this volume. Details of these studies are shown in Table 22.1.

The goal of the search was to find experimental and nonexperimental studies that have drawn conclusions about the effects of monetary incentives, the consequences of using incentives across different population groups, effects on estimates, and the costs and benefits of using incentives. The review below primarily focuses on those studies that conducted experiments using two or more randomly assigned incentive treatments. Restricting the scope in this way allowed us to focus on studies that tested a variety of treatments, including whether the offer was: (1) prepaid or promised, (2) made at the screener or extended interview, or (3) made at the initial contact or at refusal conversion. The number of studies found is not large enough to conduct a formal meta-analysis (Table 22.1).[1] However, it is possible to provide

[1]Many of the studies were conducted in the United States. This is because the RDD methodology is not used extensively outside of the United States. In an effort to expand the geographic representation, we asked a number of European researchers for references to studies related to RDD surveys. None were revealed from this effort.

Table 22.1. Description of Experimental Studies Used in Literature Review

		Amounts	Type	Sample Size	Year	Refusal Conv?	Timing	Rate (%)	Other
Incentives to Complete Screening Interview									
Lavrakas and Shuttles (2004)	Test 1	0, 1, 3, 5	Pre	1,000	2003–2004	No	Init.	40[a]	Nielson TV diaries
	Test 2	0, 2, 3	Pre	10,000	2003–2004	No	Init.	40	Commercial
									National sample
Singer et al. (2000)	Test 1	0, 5	Pre	200	1996–1997	Yes	Init.	70[a]	Consumer attitudes
	Test 2	0, 10, 20	Pro	200	1996–1997	Yes	Init.	70[a]	University of Michigan
									National sample
Curtin et al. (2005)		0, 5, 10	Pre	450	2004	Yes	Init.	52[a]	see Singer et al. (2000)
Arbitron (1998, unpublished)		0, 1	Pre	5,000	1998	No	Init.	78[b]	Radio diaries
									Commercial
									National sample
Arbitron (2001)	Test 1	0, 2	Pre	2,200	2001	Yes	Ref.	37[a]	Radio diaries
	Test 2	0, 5	Pre	1,500	2001	Yes	Ref.		Commercial
									National sample
Arbitron (2003a)	Test 1	0, 1	Pre	3,000	2003	Yes	Init.	34[a]	Radio diaries
	Test 2	0, 5	Pro	1,000	2003	Yes	Init.		Commercial
									National sample
Arbitron (2003b)	Test 1	0, 1	Pre	1,700	2003	Yes	Init.	34[a]	Radio diaries
	Test 2	0, 5	Pro	450	2003	Yes	Init.		Commercial
									National sample
Arbitron (2003c)	Test 1	0, 1	Pre	7,000	2003	Yes	Init.	34[a]	Radio diaries
	Test 2	0, 5	Pro	2,500	2003	Yes	Init.		Commercial
									National sample

Cantor et al. (1998)	0, 2, 5	Pre	1,600	1997	Yes	Init.	70[a]	Survey on children and health
						Ref.		Private nonprofit corporations Regional sample
Cantor et al. (2003a)	0, 2, 5	Pre	1,800	2002	Yes[c]	Init.	65[a]	Survey on children and health
						Ref.		Private nonprofit corporations National sample
Rizzo et al. (2004)	0, 2	Pre	1,700	2002	Yes	Init.	57[a]	Survey on cancer communications
						Ref.		National Cancer Institute (Federal) National sample
Keeter et al. (2004)	0, 2	Pre	1,000	2003	No	Init.	16[a]	Survey on public opinion Pew Research Center
Brick et al. (2005)	0, 2, 5	Pre	1,000	2003	Yes	Init.	58[a]	Survey on education of children National sample
						Ref.		U.S. Department of Education National sample

(continued)

Table 22.1. ***(Continued)***

		Amounts	Type	Sample Size	Year	Refusal Conv?	Timing	Rate (%)	Other
Kropf et al. (2000)		0, 5[d]	Pro	300	1999	Yes	Init.	50[b]	Omnibus survey University of Maryland National sample
Traub et al. (2005)		0, 2, 5	Pre	2,235	2003	Yes	Ref.		Consumer survey Scarborough research 75 major markets
Strouse and Hall (1997)	Test 1	0, 5, 10	Pro	600	1993, 1994	No	Init.	85[a]	Health care issues Individual states Regional
	Test 2	0, 15, 25, 35	Pro	900	1996, 1997	No	Init.	38[a]	Health care issues Robert Wood Johnson National sample
Cantor et al. (2003b)	Test 1	0, 20, 25[a]	Pro	300–2,000	2002	Yes	Init, Ref	76[a]	Survey on children and health Private nonprofit corporations National sample of caregivers
	Test 2	0, 20, 25[a]	Pro	100–700	2002	Yes	Init, Ref	85[a]	Survey on children and health Private nonprofit corporations National sample of childless adults

Fesco (2001)	Test 1	0, 10, 20, 30	Pro	60	1999	No	Init.	15[b]	Attitudes of science and technology National Science Foundation Calling active cases after survey period National sample
	Test 2	0, 20	Pro	270–530	1999	No	Init.	8	Attitudes of science and technology National Science Foundation Calling active cases after Test 1.
Other Studies Providing Incentives to Complete the Extended Interview									
Abt Associates (2005)		10, 15[e]	Pro	2,800	2004	Yes	Ref.	NA	Adult immunization (≥ 50 years of age) Centers for Disease Control and Prevention National Sample
Ballard-LeFauve et al. (2004)		10, 15[e]	Pro	1,000	2004	Yes	Ref.	NA	Young child immunization survey Centers for Disease Control and Prevention National, state samples

(*continued*)

Table 22.1. (*Continued*)

		Amounts	Type	Sample Size	Year	Refusal Conv?	Timing	Rate (%)	Other
Olson et al. (2004)	Test 1	15, 25[f]	Pro	1,000–3,000	2003, 2004	Yes	Ref.	NA	Survey on children and health National Center for Health Statistics National sample
Currivan (2005)	20	Pro	NA	2003	Yes	Ref.	NA		New York Adult Tobacco Survey New York State
Sokolowski et al. (2004)	20, 50	Pro	NA	2002	Yes	Ref.	NA		Making connections Private nonprofit foundation

Type: Pre—prepaid incentive; Pro—promised incentive.
Sample size: Approximate sample size of each experimental group.
Timing: Init—Initial contact; Ref—refusal conversion; Init/Ref—both initial and refusal conversion.
Rate: Approximate rate for the $0 incentive condition.
[a]Response rate.
[b]co-operation rate.
[c]“0” condition is offer to donate $5 to charity.
[d]Refusal conversion letter was sent using USPS priority mail
[e]Two conditions offered. Both promised $10. One sent $5 prior to refusal conversion call. The balance was promised at time of interview.
[f]Four conditions were offered. Two promised $15, $25. Two sent $5 prior to refusal conversion call. The balance was promised at time of interview.

meaningful summaries of how incentives affect study results with respect to response rates, population groups, survey estimates, and costs.

22.4 INCENTIVES AND RESPONSE RATES

This section summarizes the results of studies that have tested the effects of incentives on response rates.[2] For purposes of discussion, we have divided the results into three types: (1) the use of prepaid incentives to screen households and select respondents (the most common type of experiment found in the literature search); (2) the use of promised incentives; and (3) the effect of offering an incentive at the screener stage on response rates at later stages of telephone data collection.

22.4.1 Prepaid Incentives To Screen Households and Select Respondents

As shown in Fig. 22.1, there are several points at which an incentive can be sent in advance to calling. At the screener, money can be included in the advance letter prior to making initial contact or with a letter sent prior to attempting refusal conversion. It is different for the extended interview because most surveys typically try to interview the extended respondent on the same call as when completing the screener. This eliminates the possibility of sending money ahead of the survey request. It is possible to send money prior to refusal conversion for the extended interview.

Prepaid incentives sent prior to initial contact for the screener. Table 22.2 lists the separate experiments found in the literature that tested prepaid incentives to interview households at the initial screening attempt.[3] This methodology can be applied only to that portion of the sample where an address can be linked to the sampled telephone number. The entries in Table 22.2 represent the percentage point difference in response rates when comparing prepaid incentives of various amounts. For example, the Brick et al. (2005c) study reported a response rate of 57.9 percent when no money was sent and a response rate of 63.4 percent when $2 was sent prior to making the first contact with the household. The difference between these two rates of 5.5 is shown in Table 22.2. Most of the studies listed in Table 22.2 have tested incentives between $1 and $5, with a single study testing $10. These studies consistently found significant effects. The sizes of the effect range from 2.2 to 16 percentage points. Restricting attention to the most common incentive amounts of between $1 and $5, the largest increase is 12.1 percentage points.

---

[2]All monetary amounts are expressed in United States currency, that is, dollars (USD). Reported dollar values are not standardized to a referent year, and thus reflect the value of the currency in the year the study was conducted.

[3]For purposes of this discussion, "screener" incentives are those applied to increase the response by the person who answers the telephone when interviewers are initially calling the household. These incentives are not targeted to any one individual, but are intended to get the respondent to listen and consider the initial request when calling the household.

Table 22.2. Percent Difference in Screener Response Rates for Experiments Testing a Prepaid Incentive Sent Prior to Contacting the Household[a]

	Dollar levels of Incentives that are being compared								
Study	0–1	0–2	0–3	0–5	0–10	1–2	2–3	2–5	5–10
Brick et al. (2005)		5.5[*]		7.4[*]				1.9[*]	
Lavrakas and Shuttles (2004)	4.1[*]	5.9[*]	6.3[*]			1.8[*]	0.4		
Singer et al. (2000)				10.7[*]					
Curtin et al. (2005)				12.1[*]	16[*]				3.9
Cantor et al. (1998)		3.5[*]		3.4[*]				−0.1	
Cantor et al. (2003a)		3.5[*]							
Rizzo et al. (2004b)		2.2[*]							
Arbitron (2003c)[b]	6.4[*]								
Keeter et al. (2004)		12.0[*]							
Mean for column	5.2	5.4	6.3	8.4					

[a]Each entry is the response rate at the higher incentive minus the response rate at the lower incentive noted in the column.

[b]Validated over four separate unpublished studies.

[*]Significantly different from 0 at the 5 percent level (two-sided test).

One important caveat when comparing across studies relates to the use of refusal conversion. The rates for some of these studies reflect one round of conversion attempts, while others are rates before conversion (Lavrakas and Shuttles, 2004; Keeter et al., 2004). As Singer et al. (1999b) notes, differences between incentive conditions have generally been found to decrease as more effort is expended to convert refusals.

A weak relationship appears between the size of the incentive and the size of the effect. Of the studies listed in Table 22.2, every study, with the exception of Cantor et al. (1998), found a higher rate for larger incentives. In several of these cases the difference is statistically significant (Lavrakas and Shuttles, 2004; Brick et al., 2005c). In addition, there is clearly a decreasing effect for each dollar added. For example, Brick et al. (2005) found a significant difference between sending \$2 and \$5. There was a 5.5 percent difference between \$0 and \$2 (2.8 percent/dollar) while the \$5 incentive resulted in a 1.9 percent increase relative to the \$2 (0.6 percent/dollar). Similar effects for amounts ranging from \$3 to \$10 can be seen for the other studies listed in the table.

The wide variation in Table 22.2 is consistent with LST that attributes multiple factors as influencing the decision to participate in a survey. The effect of the incentive must be evaluated in the context of other attributes of the survey. Survey sponsorship, perceived burden, and survey topic are among the factors that contribute to this decision (Groves and Couper, 1998). For example, the perceived authority and credibility of the sponsor varies across government, university, and private sector sponsorship, with highest cooperation rates expected for government and university sponsored surveys. A simple tabulation of the studies above by sponsorship (university/government versus other; see Table 22.1 for characteristics of the studies) shows no consistent relationship of sponsorship and size of the incentive effect. The largest and smallest effects of the incentives are from federal government and university sponsored studies. Rizzo et al. (2004) is federally sponsored and yields one of the smallest effects (2.2), while Curtin et al. (2007) is conducted by a university and yields one of the largest effects (12.1).

A crude test of the LST is to examine whether incentives have their largest effects in situations where the overall propensity to respond is low (Singer et al., 1999b). One way to test this is to look at the effect sizes displayed in Table 22.2 by the level of the response rate when no incentive is used. One would expect larger incentive effects for surveys with lower response rates. There is some support for this hypothesis. The correlation for eight of the studies shown in Table 22.2 is −0.47.[4] One of the largest effects is for the study with the lowest response rate (Keeter et al., 2004). There is a general decline in the effect as the response rate increases. The major exceptions are the two survey of consumer attitudes (SCA) studies, which both have a high response rate and a high effect size.[5]

A more specific test of this hypothesis was carried out by Groves et al. (2004b) who sampled from different lists of respondents that were thought, a priori, to be interested in surveys of particular topics (e.g., teachers and surveys on education). They generally confirmed the LST hypothesis that persons with an interest in the topic were more likely to cooperate on the survey. They also tested whether there was the predicted interaction between interest in the topic and an advance mailing with $5.[6] According to LST, those most interested in the topic should exhibit a smaller effect of the incentive when compared to those with less interest. For two of the three groups that they included in their sample, this hypothesis was confirmed. For the third group the effect was in the opposite direction—those with the most interest in the topic exhibited the largest effect of the incentive.

Prepaid incentives for refusal conversion at the screener. A common procedural variation is to use prepaid incentives at the time of refusal conversion. Once a respondent refuses to be interviewed, the household is sent a letter with a small monetary incentive asking him/her to reconsider the initial decision. This approach is consistent with LST by recognizing that some respondents will be willing to participate without an incentive, while other respondents may be motivated to cooperate only with an incentive.

The offer of an incentive after refusal and before conversion suggests a different exchange than that proposed by SET. The effectiveness of a token prepaid incentive is that it establishes trust with the respondent. This trust fosters motivation to

[4]The correlation was calculated for the eight studies that published a response rate (Table 22.1). For those studies that varied the amount of the incentive, the $2 incentive was used, if that was offered. If it was not offered, the smallest incentive offered in the experiment was used to calculate the correlation.

[5]The above analysis does not consider any sampling error when examining the effect sizes. The differences found by Singer et al. (2000) are based on relatively small samples (n = 350; 95 percent confidence interval = ±6.4 percent). However, these results were replicated (Curtin et al., 2007; n = 450) in a later study. Many of the other studies had relatively large sample sizes (e.g., $n \geq 1000$).

[6]This study is not listed in Table 22.2 because it did not separate the effect of the prenotification letter from the incentive.

reciprocate by participating in the survey. When a respondent refuses without an initial incentive, the perceived costs of participation outweigh the benefits. Offering an incentive before refusal conversion is one way of increasing the benefits to the respondent when making the second request. Although sending money prior to the call may still evoke an obligation to participate, this obligation may not be as strong as when the incentive is sent at the initial contact.

The evidence for RDD surveys is generally consistent with LST. In four of five experiments using prepaid incentives for refusal conversion, a significant effect was found. These studies offered either $2 or $5 that was sent prior to calling to attempt conversion. The increases in response rates were between 3 and 9 percent (Cantor et al., 1998; Arbitron, 2001 (unpublished); Cantor et al., 2003a; Brick et al., 2005c; Traub et al., 2005; for negative evidence see Rizzo et al., 2004). Brick et al. (2005c) and Arbitron (2001, unpublished) are two studies that varied the amount offered. Brick et al. (2005) found that $5 led to a slightly higher, but nonstatistically significant response rate when compared to offering $2 (64.6 percent versus 61.9 percent). Arbitron (2001, unpublished) found a slightly higher refusal conversion rate when comparing $2 and $5 (16.7 percent versus 20.7 percent), although the difference was not statistically significant.

Comparing prepaid at initial contact and refusal conversion for the screener. Two studies directly compared use of incentives at the initial contact with use at refusal conversion. As discussed above, SET would lead to the expectation that sending money at the initial contact would be most effective. Cantor et al. (2003a) directly compared a $2 advance letter incentive with a $5 refusal conversion incentive. They found the two are approximately equivalent with respect to increases in the final response rate. Brick et al. (2005c) compared the same payment levels at each point in the process for $2 and $5. In both cases, the payment at the initial contact produced slightly higher response rates when compared to sending the same amount at refusal conversion, although the difference in response rates were very small and not statistically significant (63.4 percent vs. 61.9 percent for $2 payments; 65.3 percent vs. 64.6 percent for $5 payments).

22.4.2 Promised Incentives

SET predicts that respondents given the promise of an incentive are less likely than those given a prepaid incentive to believe that the gains from participation will outweigh the costs. In this section, we discuss whether this applies for experiments with RDD surveys. Promised incentives have been offered at both the screener and at the extended interview stages.

Promised incentives at the screener. Tests of promised incentives at the screener have ranged from $5 to $25. None of the studies found a statistically significant effect (Table 22.3).

Offering a promised incentive over the phone is difficult. In face-to-face interviews, there are more opportunities to establish rapport than on the telephone.

Table 22.3. Percentage Point Difference in Screener Response Rates for Experiments Testing a Promised Incentive. Cantor, D. *The Use of Monetary Incentives to Reduce Non-response in Random Digit Dial Telephone Surveys*; Lee Giesbrecht, 1996

	Dollar level of Incentives that are being compared[a]			
Study	0–5	0–10	0–20	0–25
Singer et al. (2000)	2.0			
	7.4			
		0.4		
			1.2	
Cantor et al. (1998)	– 7.3			−1.1
				−1.1[b]
Arbitron (2003c)	–.3[c]			
Kropf et al. (2000)	4.9[d]			

[a]Response rate for higher incentive minus response rate for lower incentive.
[b]Offered at refusal conversion.
[c]Validated over three separate studies.
[d]$5 compared to contribution to charity.

For mail surveys, the request and incentive offer can be carefully reviewed. For an RDD screener the telephone interviewer has only a few seconds to establish legitimacy and engage the respondent. Offering an incentive in this context may raise suspicions about the survey request. Interviewer debriefings at both Arbitron and Westat suggest that many interviewers did not find the promised incentive offer at the screening interview to be an effective tool because it was too difficult to make the offer without sounding desperate to gain the respondent's cooperation. Consequently, while promised incentives have been beneficial to mail or field surveys (e.g., Church, 1993; Singer et al., 1999b), they seem to be much less effective at this stage of RDD surveys.

One might think that promises made to households that had received a prenotification letter with information related to the money may have more credibility than a promise made to households that did not receive any prenotification. In most of the experiments discussed above, the offer was made in conjunction with an advance letter. For example, in a series of three experiments by Arbitron a promised incentive did not work for either households that received prenotification or those that did not receive any prenotification. One possible explanation for the ineffectiveness of the letter is the difficulty related to delivering the letter to the person who answers the telephone.

Promised incentives at the extended interview. Interviewers promising money to complete the extended interview may not face the same credibility issues as those at the screener. At the point of the extended interview, someone in the household has completed the screener and has a sense of what the phone call is about. Perhaps for this reason, promising money at the extended interview has met with slightly more success than promising at the screening interview. Strouse and Hall (1997) tested several different payment levels across two different studies.[7] They did not find a significant effect for promised incentives in the $0–$10 range; however, for incentives in the $15–$35 range, they did find a significant effect. Prior to any refusal conversion, for example, they found a $35 incentive led to an increase of 10 percent for the sample that was not listed in the phonebook and 16 percent for the sample that was listed in the phonebook. Cantor et al. (2003b) reported an effect of 9.1 percent when offering $20 compared to offering no money to a sample of caregivers of children age 0–17 years. However, in this same study, no effect was found for a sample of childless adult respondents.

Promises made at refusal conversion are fairly common across survey organizations. A number of studies have reported having success with offering relatively large amounts of money (e.g., $25 or greater) at the end of the field period (e.g., Fesco, 2001; Olson et al., 2004; Currivan, 2005; Curtin et al., 2005, Hanway et al., 2005). The SCA, for example, has offered between $25 and $50 to convert refusals. Several experimental studies have mixed a prepaid incentive with a promise as part of refusal conversion (Cantor et al., 2003b; Olson et al., 2004). These have also found moderate success; for example, Cantor et al. (2003b) found prepaying $5 and promising an additional $15 resulted in a higher response rate than offering nothing to caregivers (6.4 percent), although this strategy did not work for adults without children. The literature contains no clear message on how mixing payment types compares to a simple promise or prepaying a small amount of money. In one study, Olson et al. (2004) found that the prepayment in refusal conversion added to the promise by 3–7 percent (varying by total amount offered).

22.4.3 Conditioning Effects of Incentives

Incentives may also lead to conditioning effects on cooperation at later stages of the survey process. According to SET, providing incentives will establish a social obligation with the respondent. This obligation might carry over to subsequent survey requests. This, of course, assumes that the person who sees the incentive at the initial stages of the survey is the same person who is asked to participate in subsequent stages, or at least communicates something about the incentive to whoever is asked to participate.

A number of different studies have provided incentives at the screening stage and tracked cooperation at later stages in the survey process (Arbitron, 2003a–c, unpublished; Cantor et al., 2003b; Rizzo et al., 2004; Lavrakas and Shuttles, 2004; Brick et al., 2005c). In three of these studies, a significant effect was found. Rizzo et al. (2004) examined response rates for the extended interview by whether the household had received an incentive to complete the screening interview. They found a significant positive effect of 6 percent for a $2 prepaid incentive sent prior to the initial contact with the household. No significant effect was found when the incentive was sent at refusal conversion. Both Arbitron (2003a–c, unpublished) and Lavrakas

[7] Strouse and Hall (1997) mention the incentive to the person who is answering the screening interview by saying that the person who completes the survey will receive the payment.

and Shuttles (2004) found evidence that incentives in the range of $1–$5 provided as part of the RDD recruitment interview increased the return rate of completed diaries between 2 and 6 percent (depending on the study and amount of the incentive).

The evidence that does not support a carry-over effect comes from Brick et al. (2005c), who found no difference in extended interview response rates when providing an incentive to complete the screener. A less direct test is provided by Cantor et al. (2003b), who examined the effects of an incentive provided at the screener stage on different types of incentives provided at the extended interview. A promised incentive at the extended interview was found to be more effective when preceded by a screener incentive for one group (adults without children). However, even when preceded by an incentive at the screener stage, the promised incentive was no better than the no-incentive group. These mixed results may reflect different leverage points that certain types of respondents have regarding incentives, especially as they relate to carry-over effects of these incentives. It may also be that the extended interview respondent has no knowledge of the screener incentive.

22.4.4 Summary and Theoretical Implications of Observed Incentive Effects

A summary of the above results is as follows:

- *Screener*: Prepaid incentives between $1 and $5 increase the response rate from 2 to 12 percentage points.
- *Screener*: Larger prepaid incentives lead to higher response rates at a decreasing rate.
- *Screener*: Prepaid incentives at refusal conversion are about as effective as prepaid incentives sent prior to the initial contact with the household.
- *Screener*: Promised incentives between $5 and $25 at initial contact or refusal conversion do not increase response rates.
- *Extended*: A number of studies have found that promised incentives of $15–$35 increase response rates.
- *Both Stages*: There is mixed evidence that incentives at early stages of the process (screening) have a small positive effect on participation at later stages (e.g., extended interview, completion of mail survey).

These results are generally consistent with those found for other interviewer-mediated surveys, with a few notable exceptions. One exception is the ineffectiveness of promised incentives relative to prepaid incentives. The meta-analysis completed by Singer et al. (1999b) found mixed evidence related to these two types of incentives. The meta-analysis found that prepaid and promised incentives have similar effects. A more detailed review of five studies that directly compared the two found that prepaid had larger effects than promised. The RDD studies reviewed above found almost no effect of promised incentives at the screening stage. The effect for the extended interview was significant for a few studies where at least $15 was offered.

However, even this effect was not uniform across populations (Cantor et al., 2003b). We speculate that these differences are related to the perceived credibility of the promise in the two different contexts. Promised incentives at the RDD screener have very low credibility and are not judged to have an expected net gain in the exchange being proposed. They work better at the extended interview because there is more trust with the interviewer. Even in this case, however, there is a need to offer relatively large amounts of money.

A second exception is that there is a curvilinear effect of incentives for each dollar offered for prepaid incentives on RDD surveys. Additional money added at the screener increases the response rate at a decreasing rate. This differs from other interviewer mediated surveys, which generally show a linear relationship (Singer et al., 1999b).We speculate that this difference is due to the prepaid incentive serving a dual role when motivating participation. One role is to create the obligation (per SET). The second is to engage respondents in the mailings and get them to read, and digest advance materials. No studies have measured the effect of the incentive amount and thoroughness of review of advance mailings. Receiving money in unsolicited mailings may be unusual enough that it generates engagement even at small incentive levels.

Consistent with LST, there is some evidence that the effect of the incentive is negatively related to the propensity to respond to the survey. The less motivated respondents are to participate without an incentive, the more effective the incentive seems to be. This conclusion is supported by a correlation for a small number of studies ($n = 8$), as well as the experimental evidence discussed in Groves et al. (2004). Also consistent with LST, refusal conversion payments at the screening interview are about as effective as payments sent before the initial contact. Correlating the effects of incentives with other survey design characteristics expected to influence response rates is difficult to do because of the small and non-systematic number of experiments available for analysis.

22.5 DIFFERENTIAL EFFECTS OF INCENTIVES ACROSS POPULATION GROUPS

There are at least two reasons why the effect of incentives might vary across the sample population. First, there may be systematic differences in address availability that introduces selectivity into who receives a prepaid incentive, which may cause imbalance in the final response distributions by subgroups. Second, certain individuals may be more influenced by an incentive. LST is based on the framework that individual motivations vary across the population. Based on this, one might expect that certain population characteristics may be correlated with the effects of incentives. For example, a common hypothesis for incentives is that they will be more effective for low income populations (Singer, 2002). For both of those reasons, one might expect to see the effect of incentives in RDD surveys to vary across potential respondents.

22.5.1 Address Availability for Phone Numbers

The first challenge in delivering an incentive to potential respondents in an RDD survey is to obtain an address to which to deliver the advance mailing. Telephone directories are the primary source of addresses for phone numbers, and they generally provide addresses for approximately 30–40 percent of RDD phone numbers (Geisbrecht et al., 1996). Other sources of addresses include private companies that generate address/phone lists compiled from sources such as credit records, warranty cards, auto registrations, and similar data. These sources are used by some survey organizations to supplement address matching for numbers not listed in phone directories. These "unlisted" numbers are converted to "mailable" status and can be sent prenotification mailings. These sources of addresses typically add addresses to another 10–15 percent of sample (Arbitron, Westat experience), bringing the total of selected RDD phone numbers with addresses to approximately 45–50 percent. A majority of these without addresses are nonworking or nonresidential numbers. Experience at Arbitron is that addresses can be obtained for all except approximately 15–18 percent of residential phone numbers before contact with the household.

Table 22.4. Percentage of Telephone Households that are Listed in the White Pages by Selected Household Characteristics

	Percent listed		Percent listed
Region		Household income	
Northeast	63.3	Less than $5,000	47.5
Midwest	63.6	$5,000–$7,499	52.3
South	61.1	$7,500–$9,999	55.1
West	52.6	$10,000–$14,999	57.1
		$15,000–$24,999	54.8
Race[a]		$25,000–$34,999	55.3
White	63.3	$35,000–$49,999	58.2
Black	41.5	$50,000–$74,999	60.3
American Indian, Aleut, Eskimo	48.0	$75,000 or more	63.3
Asian/Pacific Islander	48.6		
Other	36.6	Length of residence	
		Less than 1 month	19.3
Hispanic origin[a]		1–6 months	19.3
Hispanic	35.6	7–11 months	29.1
Non-Hispanic	62.2	1–2 years	55.1
		3–4 years	62.2
Presence of children under 3		5 years or more	70.2
No child	61.6	Refused/don't know	55.1
One child	48.8		
Two children	45.8		
Three or more children	37.7		

[a]The characteristic is that of the reference person.

Source: Giesbrecht et al. (1996), based on Current Population Survey (CPS) data.

Address availability varies by key sociodemographic characteristics. As seen in Table 22.4, unlisted households are more likely to be non-white, be Hispanic, have at least one child living at home, have lower income, and have length of residence of less than a year (Geisbrecht et al., 1996).

As noted above, commercial suppliers can also match to phone numbers that are not listed in the telephone directory. A profile of a sample of phone numbers with addresses from both phone directories and a commercial supplier provides information compatible with Table 22.4. In a logistic regression analysis predicting "no address" and using household characteristics available from the commercial supplier, Arbitron has found that a phone number is more likely not to have a matched address when it is known to be a multifamily dwelling unit, the length of residence is less than 3 years, or when length of residence is missing. The odds of not having an address are significantly reduced if a person 65 or older is resident (Arbitron, 2000, unpublished). These two studies indicate that phone numbers for which an address is unobtainable are more likely to be associated with persons who are non-white, are Hispanic, are in large households, have lower income, are more mobile, and are more urban. This result is consistent with studies that have looked at address availability among interviewed samples (Keeter et al., 2004; Curtin et al., 2005).

Demographic differences in households with and without a matched address suggest that there would be differences in call outcomes as a result. These differences may be related to address accuracy, as well as differences in household likelihood of contact and cooperation. With the relatively recent success in matching addresses to RDD phone numbers not listed in telephone directories, there is limited documentation of how this segment of the sample may respond differently than other listed sample or other unlisted sample. Analysis at Arbitron has shown that numbers that are not listed in the telephone book but do have a mailable address have rates of known residential status, no answer, and known business that are between those listed and those unlisted and unmailable. In addition to the concern of correctly identifying a residential address through this matching, there is also the concern that the address provided for the phone number does not match that provided by the respondent. Together, addresses that are not residential or are not associated with the sample phone number are a survey cost without any benefit.

22.5.2 Are Incentives More Effective for Certain Population Groups?

A second consideration is whether the propensity to respond to an incentive varies by population groups. Prepaid incentives may raise response rates among those groups that already have a higher propensity to respond, thus actually making the sample *less* representative of the total population (Curtin et al., 2007). These differences would be of concern if the survey estimates of interest are correlated with response propensity in a way that is not compensated for by weighting.

A limitation in evaluating the effect of incentives by population groups in an RDD survey is the lack of known characteristics of the households before contact. Studies have relied on two primary approaches to assess the presence of differential incentive effects by population groups. One is to look for changes in the demographic composition of cooperating respondents, with and without incentives. A second method is to examine response rate differences by area characteristics such as geodemographic census data.

The evidence from studies using the first method suggests that offering an incentive, either prepaid or promised, does not affect the demographic composition of the interviewed sample (Singer et al., 2000; Arbitron 2003a, 2003b, unpublished; Cantor et al., 2003a; Brick et al., 2005c).

The evidence related to using external data is mixed. In the National Survey of America's Families (NSAF), the effect of prepaid incentives at the screening interview was examined by analyzing census data linked to the household. Many of the characteristics tested were not significant. It was found that an incentive worked better in areas with more African Americans, lower employment, higher migration, and a shorter commute time to work. The incentive also performed better in states with low response rates (Cantor et al., 2003a). These differences, however, were relatively small.

An Arbitron study examined the effect of a $5 promised incentive in conjunction with a $1 prepaid incentive on survey cooperation rates, and found no difference in the effect in high density African American or Hispanic areas as defined by ZIP codes, compared to nonethnic areas (Arbitron, 2003a, unpublished).

While studies support geodemographic indicators as good predictors of area response rates (Harris-Kojetin and Fricker, 1999), these variables have not been found to be good predictors of survey participation at the individual respondent level (Arbitron, 1998b, unpublished; Burks et al., 2005). Given this, it may be difficult to isolate variation in incentive effects using these ecological data.

One potential benefit of identifying populations by geography would be in targeting advance notice and incentives to population groups that are underrepresented or have lower propensities to respond. While this raises issues regarding the fairness of targeting incentives to selected segments of the population, it may be that certain population groups increase their response propensity in parity with response rates of other groups with the incentive offer.

A practical concern with the use of incentives and advance mailers is the cost. Ecological characteristics offer some opportunity to know something about a household before sending the incentive, and may be used to cost effectively target the incentives to increase response propensity in low response areas.

In summary, the evidence indicates that incentives used to increase cooperation in an RDD survey do not vary greatly across populations. This is somewhat different than research on incentives in other interview modes (e.g., Singer, 2002). For the research on RDD studies, the different methodologies used to test the effects of incentives on the composition of the sample are not highly specific and are restricted to a relatively small number of studies. Consequently, this is an area deserving further study, to better understand differences across populations in sensitivity to incentive amount. At a minimum, this information is important to efficiently allocate incentive dollars. More broadly, it will provide survey practitioners with better information on the consequences of using incentives in their sample.

22.6 DO INCENTIVES IMPACT ESTIMATES AND DATA QUALITY?

Incentives can impact estimates in several different ways. First, by only providing incentives to those with addresses, there might be disproportionate influence of these groups on the final results. Second, if selected segments of the population are more likely to respond to incentives than others and these segments differ on key survey measures, then estimates could be affected by the use of an incentive. With the differential availability of addresses, then at least for prepaid incentives, this is a real possibility. Third, if the incentive changes the processes used to answer questions, estimates may also be affected by the incentive. Respondents may become more attentive to the survey, as a result of the incentive, or conversely may feel an obligation to respond, but put forth minimal effort.

22.6.1 Effect of Incentives on Estimates

The correlation between address availability and demographics suggests that prepaid incentives may introduce disproportionate influence on survey response distributions. Two studies that looked specifically at the effect of address availability on survey estimates found no significant effect. Curtin et al. (2007) found no significant effect of address availability in a logistic model predicting the Index of Consumer Sentiment (ICS). Models estimated with and without demographic controls yielded the same results. There was a suggestion that having a listed address did interact with demographics when predicting the ICS. Arbitron (2002b, unpublished) examined the exclusion of unlisted sample without an address on final radio audience estimates. Comparison of audience levels by time of day, station format, race/ethnicity, household size, and other key indicators showed no meaningful differences with and without the unlisted non-mailable sample. When the sample was disaggregated differences suggested that those without an address available may listen to radio differently than those with addresses.

The most common approach to evaluate the effects of incentives on estimates is to make comparisons across different incentive conditions (Singer et al., 2000; Sokolowski et al., 2004; Rizzo et al., 2004; Curtin et al., 2005). These studies primarily focus on the final estimates for households with addresses receiving incentives. Generally, these studies do not find significant differences between estimates with an incentive and without. One exception is Singer et al. (2000), who found that 8 of the 18 variables tested indicated an effect of either a prepaid or a promised incentive on response distributions. In five of these cases, the responses tended to be more optimistic, in support of theoretical work by Schwartz and Clore (1996). In a replication of this experiment, Curtin et al. (2007) did not find a direct effect of an incentive on the Index of Consumer Sentiment, but did find that the incentive had a significant interaction with race. The result suggested that non-whites provided more optimistic answers under an incentive condition than whites.

22.6.2 Effect of Incentives on Data Quality

Another dynamic that may occur as a result of the incentive is the inclusion of respondents who view the response task differently than those not motivated by the incentive. It is possible that they may put more effort into providing correct answers because of the incentive, which may be indicated through fewer "don't know" responses and less missing data. Alternatively, respondents may be more likely to provide answers, but this could take the form of putting less thought into their answers due to the lack of intrinsic motivation. These hypotheses have been tested indirectly by looking at different measures of data quality.

Across the studies, there are mixed results (Singer et al., 2000; Cantor et al., 2003a,b; Curtin et al., 2005). In some cases, there was no difference in data quality; however, other studies indicated improvements in data quality, while still other research findings demonstrated a decline in data quality. This suggests that the effect may vary by other survey attributes such as survey topic and size of incentive. Further research will help clarify this relationship of incentives and data quality. At a minimum, this suggests that the possible effect of the incentive on data quality should be considered when designing a study, developing an analytic plan, and analyzing the data.

In summary, studies reveal no clear pattern of effects of incentives on survey estimates and data quality in RDD surveys. As discussed in a previous section, the increase in response rates from an incentive is relatively small. This may explain why analyses to date have not found big effects of incentives on sample composition or estimates. More research is needed to better understand when incentives might affect estimates and the sensitivity to dollar amounts.

22.7 MONETARY COSTS AND BENEFITS

Over time, telephone survey interviewers have had to expend an increasing amount of effort to contact households and complete interviews. This has serious cost implications for RDD surveys (Curtin et al., 2000., Keeter et al., 2004, Curtin et al., 2007). Incentives can reduce the amount of effort to make contact and, therefore, have a positive effect on the overall cost of the survey. Data generated from other modes of data collection (mail, in-person, and mixed mode) suggest that the use of monetary incentives can be cost effective by reducing the number of hours per interview, contacts, interviewer costs, follow-up field visits, and mailings (Berlin et al., 1992; Church, 1993; Shettle and Mooney, 1999; Link et al., 2001). It is reasonable, therefore, to hypothesize that incentive use in an RDD survey may reduce costs in a similar way.

Sparse data are generally available across RDD surveys to measure cost components (effort expenditure, net costs, cost savings, and provision of incentives), including government surveys funded and conducted with taxpayer dollars (COPAFS, 1992; Singer et al., 1999b). Cross-survey comparison of cost data is frequently hampered for two reasons: Many of these cost components are not detailed in peer-reviewed publications for proprietary reasons, and it can be difficult to directly compare these metrics across surveys due to the myriad of conditions that can be implemented. However, a number of RDD studies have discussed the impact of incentives on the level of effort and the net cost of using an incentive.

22.7.1 Incentives and Cost Savings

Prepaid incentives do reduce the number of calls to complete an interview, regardless of whether they are for the screening (Singer et al., 2000; Arbitron, 2003b, unpublished; Lavrakas and Shuttles, 2004; Curtin et al., 2005; Brick et al., 2005) or extended interview (Strouse and Hall, 1997; Link et al., 2001; Olson et al., 2004). Prepaid incentives do not appear to be related to a reduction in the number of calls to initially contact the household, or to the outcome of the first call (Singer et al., 2000). This is logical, since respondents have no sure way of knowing who is calling until they answer the telephone. The reduction is attributed to two primary sources: once respondents have been on the telephone, they require fewer callbacks, and less refusal conversion is required because respondents are initially more cooperative (Singer et al., 2000; Arbitron, 2003b, unpublished; Brick et al., 2005). For example, a prepaid incentive reduced the percentage of first refusals for the SCA from 45.9 percent to 28.2 percent (Singer et al., 2000). The initial refusal rate was also reduced from 42.1 percent to 36.0 percent in the Arbitron radio survey with a letter and $1 incentive (Arbitron, 2003b, unpublished).

Another reason for a reduction of calls is that incentives can increase the use of a toll-free telephone number for respondents to call to complete the survey. Data from the National Postsecondary Student Aid Study (NPSAS) found that a targeted $20 refusal conversion incentive resulted in 21.3 percent of incentive cases calling the project's toll-free telephone number to complete the interview, as compared to 10 percent of cases who were not offered an incentive (Link et al., 2001). Of the incentive cases who called the toll-free number, 86.9 percent completed an interview compared to 66.7 percent of control cases (Link et al., 2001). In the National Survey of Children's Health (NSCH) refusal conversion study, the refusal conversion letter alerted respondents to a toll-free telephone number that could be dialed if they wished to participate immediately. Eighteen percent of completed interviews in the prepaid incentive group and 14 percent of completed interviews in the promised-incentive group were conducted using this toll-free telephone number (Olson et al., 2004). Overall, 16 percent of the incentive case interviews were completed with respondents who dialed the toll-free number (Blumberg et al., 2005b). Arbitron found that a promised incentive of $5, in addition to $1 in advance, increased the number of call-in agree cases from 0.53 percent to 1.7 percent compared to a postcard advance notice. However, second stage diary completion rates were 10 percent lower in the incentive group (Arbitron, 2003a).

Regardless of the reason, the size of the reduction in the number of calls is not particularly large. Different studies have used slightly dissimilar methods to measure this reduction. For example, Lavrakas and Shuttles (2004) examined the mean total number of dialing attempts per number to complete the screening interview, and found a reduction of 0.1 dialing attempts for each $1 increase in the incentive (mean of 5.6 calls for $0, 5.5 calls for $1, 5.3 calls for $2, and 5.2 calls for $3). Brick et al. (2005) reported a range of calls between 6.7 and 7.3 across the 10 different screener incentive treatments. At the extended level, Link et al. (2001) found it necessary to make, on average, 16.1 call attempts for incentive cases and 20.5 attempts for nonincentive cases. Strouse and Hall (1997) reported similar variation across their incentive conditions at the extended stage.

22.7.2 Net Costs: The Cost Effectiveness of Using a Monetary Incentive

The net cost of an incentive is a function of the monies spent to provide monetary incentives and the cost savings resulting from a more cooperative sample requiring less survey effort. When computing the net costs of incentives, researchers must consider the baseline survey parameters. Cost parameters are, to some degree, specific to the survey and organization. Findings cannot necessarily be generalized across all possible incentive conditions, organizational sponsors, or other parameters (Singer et al., 1998, Traub et al., 2005). The savings achieved from an incentive are highly affected by the size of the response rate increase. An incentive that dramatically increases the initial cooperation rate will save more money because it reduces the level of effort more than an incentive that does not have as dramatic an effect on the cooperation rate.

With one exception (Singer et al., 2000), studies have concluded that the savings realized by a decrease in the level of effort from an incentive do not offset the increased costs associated with providing one (Strouse and Hall, 1997; Link et al., 2001; Gelman et al., 2003; Curtin et al., 2005; Brick et al., 2005c). Curtin et al. (2007) report, for example, that "the decision to offer prepaid incentives will often involve a tradeoff between higher costs and higher response rates" (p. 10). Brick et al. (2005) report that incentives add to the costs of the survey, even though it cost approximately twice as much to complete an interview if the respondent ever refused ($12 if never refused versus $25 if refused at least once).

The net cost of implementing an incentive is illustrated in a paper by Gelman et al. (2003) that quantifies the costs and benefits of incentives to help survey decision makers to determine if incentives should be used. Data from the Singer et al., meta-analysis were used to create Bayesian regression models to more precisely examine the impact of incentive timing, value, form (gift or cash), as well as survey mode on overall response rates. The model estimates were then applied to the cost structure of the Social Indicators Survey (SIS) with a response rate of 50 percent.

In this analysis, Gelman estimated the average noninterview cost to be $0.78 per call. This figure was multiplied by the reduction in the number of calls with an incentive to estimate the amount saved due to fewer calls. It was estimated that a $5 prepaid incentive could reduce the number of calls made for the SIS caregiver survey by 8041 calls, resulting in a savings of $6272 (see Table 22.5). The net cost of the incentive is the total cost of the incentive minus the estimated amount saved due to fewer calls, or $56,686 USD in this example using a $5 prepaid incentive. The $5 postpaid incentive not only was projected to save less money and cost less but it also was projected to achieve a smaller increase in the response rate (Gelman et al., 2003).

One important exception to the above conclusion occurs when incentives allow the survey to avoid expensive procedures. Link et al. (2001) examined cost savings associated with a reduction in the number of in-person interviews in a mixed mode

Table 22.5. Estimated Impact (Using Bayesian Regression Modeling) of a $5 Prepaid Incentive and $5 Postpaid Incentive Based on 1500 Completed Interviews for the Caregiver Survey of the Social Indicators Survey (SIS)

Metric	Impact of $5 prepaid incentive (compared to $0 incentive)	Impact of $5 postpaid incentive (compared to $0 incentive)
Total cost of the incentive	$62,958	$9,375
Reduction in the number of calls	8,041	3,376
Amount saved due to fewer calls	$6,272	$2,634
Net cost of the incentive	$56,686	$6,741
Assumed increase in response rate (percent)	5.3	1.5

Source: Gelman et al. (2003).

national longitudinal study of college students. A $5 prepaid incentive was sent to noncontacts, hard-to-reach persons, and initial refusers; upon interview completion over the telephone a $15 check was mailed. The incentives reduced by 6.5 percent the cases requiring in-person interviews. This cost savings more than paid for the incentives that were used.

Brick et al. (2005) discuss the possibility of applying an incentive to a subsample of those who refuse to complete the interview. This approach is advantageous when the cost of finding respondents through screening is high relative to the cost of actually completing the extended interview. Brick et al. (2005) find that for the National Household Education Survey, one of the most cost-efficient methods is to sample 60 percent of the screener refusals and prepay $5 prior to calling the household. This has a number of advantages, including limiting the amount of money that is needed to pay for the incentive while achieving a relatively high response rate.

22.8 CONCLUSION

The results of our review of studies demonstrate a range of response rate effects, depending on the timing, type and amount of the incentive, as seen in Table 22.6. Overall, five major conclusions can be drawn from this chapter:

(1) Small prepaid incentives ($1–$5) consistently produce modest increases (2–12 percents) in the screener response rate. Increasing incentives to more than $1 generally lead to marginally larger gains in the response rate. However, as the amount of money is increased, the improvement in the response rate slows.

(2) Small prepaid refusal conversion payments to complete the screener are about as effective as similar payments provided before the initial survey request.

Table 22.6. Summary of Response Rate Effects by Timing and Type of Incentive Offer

	Screening interview		Extended interview	
	Prepaid	Promised	Prepaid	Promised
Initial offer	Consistent significant effects for $1–$5 (2.2–12.1 percent gain)	Generally no significant effects for $5–$20	No opportunity to offer, if continuing on from screening interview	Mixed evidence No benefit from incentives less than $10 Significant effects for incentives of $15–$35
Refusal conversion	Either $2 or $5 refusal conversion incentive increases the response rate by 3–9 percents About as effective as a prepaid incentive at the initial survey request	No significant effects for $25	Evidence of significant effect of pre-paid and promised together	Limited mixed evidence of $25–$50 incentives, usually at end of field period Smaller amounts have not been tested

"Significant" effects are those identified by the authors as being statistically significant.

(3) Promised incentives ($5–$35) do not increase the response rate at the screening stage. There is evidence that promised incentives of at least $15 are effective in increasing the response rate for the extended interview.

(4) Very little evidence was found that incentives substantially change survey results or data quality.

(5) Incentives reduce the level of effort to complete an interview. The reduction in effort does not generally compensate for the cost of providing incentives. The exception is when incentives save the cost of very expensive follow-up (e.g., in-person interviewing).

Several of these conclusions are consistent with SET and LST. SET predicts greater effectiveness of prepaid over promised incentives. LST predicts that using incentives at refusal conversion will be as (or almost as) effective as sending it before the initial contact.

Very little evidence has been found that incentives have different effects across predefined demographic subgroups. This is not consistent with LST, which predicts that respondents will treat incentives differently when deciding on survey cooperation. The one exception to this is the finding that respondents with intrinsically low-response propensity are more affected by an incentive. The analyses that have been conducted to date have not had highly specific measures of respondent characteristics, especially as they relate to decisions on survey cooperation. Nonetheless, when examining response rates by ecological characteristics or data from the sample frame, very few interactions have been found with incentive treatments. These analyses were based on a relatively small number of experimental studies, and could not control for other differences between the studies (e.g., sponsorship, burden, and saliency).

More generally, these findings suggest that unless incentives can be targeted to underrepresented groups they will not significantly affect total survey error (e.g., by reducing nonresponse bias). This is not entirely surprising given that the gains in response rates from an incentive are not large. Groves et al. (2004) calculate maximum reductions in nonresponse bias from treatments that raise the response rate by varying amounts. They conclude that a treatment that increases response rates by 5–10 percentage points, like many of the incentive experiments discussed above, will have relatively little impact on nonresponse error of estimated proportions. Larger gains in the response rate, coupled with a large difference between who is brought into the survey by an incentive, would be required to significantly affect estimates and total survey error.

Taken within a larger context, incentives may still be seen as significantly adding to the quality of the survey. Incentives can serve an important function for studies that cannot institute extensive follow-up or other methods proven to increase response rates. In other cases, they are a proven method of meeting targeted completion goals for particular segments of the population. For example, in a large commercial survey, factors that are considered when deciding whether or not to implement an incentive include the response rate gain, the relative cost, potential benefits to data quality, and client satisfaction. Given that incentives do increase response rates, they are one way to achieve an optimal mix of these different factors. Incentives may also increase the response rate enough to maintain face validity of the survey, especially if it can make the difference of crossing a critical milestone (e.g., 47 percent vs. 52 percent).

The use of incentives for any particular study has to be considered according to the goals and specific survey design. More research is needed regarding the optimal situation and timing of incentive use to maximize effectiveness, as well as the effect on data quality, and the specific conditions under which incentives operate for RDD surveys. As this chapter is being written, it is becoming even more difficult to complete RDD surveys. Response rates are declining and the quality of the frame is coming into question. Incentives will have some role in the evolution of RDD methods. For this role to be better defined, it is important to further assess the impact on data quality, as well as determine optimal guidelines and approaches for when incentives seem appropriate.

CHAPTER 23

The Causes and Consequences of Response Rates in Surveys by the News Media and Government Contractor Survey Research Firms[1]

Allyson L. Holbrook

Survey Research Laboratory and Graduate Program in Public Administration, University of Illinois at Chicago, USA

Jon A. Krosnick

Departments of Communication, Political Science, and Psychology, Stanford University, USA

Alison Pfent

Department of Psychology, The Ohio State University, USA

[1]The authors thank ABC News, Abt Associates, CBS News, The New York Times, the Gallup Organization, the Kaiser Family Foundation (KFF), the Los Angeles Times, Mathematica Policy Research, Inc., the Pew Research Center for the People and the Press, the RAND Corporation, Research Triangle Institute (RTI), Schulman, Ronca, Bucuvalas, Inc. (SRBI), the Washington Post, and Westat for their willingness to share the data presented in this paper. Jon Krosnick is University Fellow at Resources for the Future. Correspondence regarding this chapter should be sent to Allyson Holbrook, Survey Research Laboratory (MC336), 412 S. Peoria St., Chicago, IL 60607 (e-mail: allyson@uic.edu) or to Jon A. Krosnick, 434 McClatchy Hall, 450 Serra Mall, Stanford University, Stanford, California, 94305 (e-mail: krosnick@stanford.edu).

Advances in Telephone Survey Methodology, Edited by James M. Lepkowski, Clyde Tucker, J. Michael Brick, Edith de Leeuw, Lilli Japec, Paul J. Lavrakas, Michael W. Link, and Roberta L. Sangster

On October 13, 1998, columnist Arianna Huffington wrote: "It's no wonder that the mushrooming number of opinion polls, coupled with the outrageous growth of telemarketing calls, have led to a soaring refuse-to-answer rate among people polled" (The New York Post, p. 27). And Huffington has not been alone in expressing this view: numerous survey researchers have shared her sense that response rates have been dropping in recent years, supported by solid data documenting this trend (e.g., de Heer, 1999; Steeh et al., 2001; Tortora, 2004; Curtin et al., 2005). As a result, researchers have been increasingly inclined to implement data collection strategies to combat this trend, including longer field periods, increased numbers of call attempts, sending advance letters, offering incentives, attempting refusal conversions, and more (de Heer, 1999; Curtin et al., 2000, 2005).

These efforts have been inspired by a concern about the quality of survey data, because conventional wisdom presumes that higher response rates assure more accurate results (Backstrom and Hursh, 1963; Babbie, 1990; Aday, 1996; Rea and Parker, 1997), and response rates are often used to evaluate survey data quality (Atrostic et al., 2001; Biemer and Lyberg, 2003). Generalizing the results of a survey to the population of interest is based on the assumption that the respondents who provided data are a representative sample of the population. If survey nonresponse (i.e., failure to contact or elicit participation from eligible respondents) creates nonresponse *error* (because respondents differ from nonrespondents), survey estimates of means, proportions, and other population parameters will be biased (Caetano, 2001).

But in fact, it is not necessarily so that lower response rates produce more nonresponse error. Lower response rates will only affect survey estimates if nonresponse is related to substantive responses in a survey. In other words, nonresponse bias will occur if respondents and nonrespondents differ on the dimensions or variables that are of interest to the researchers. But it is quite possible that nonrespondents are sometimes essentially a random subset of a full survey sample, at least random with respect to the variables being measured (if nonresponse is caused by other factors that are uncorrelated with the variables of interest).

When nonresponse produces no bias, strategies to increase response rates may needlessly increase the expense of a survey without increasing data quality. Furthermore, the interviews yielded by many call attempts or by converting refusals may actually produce lower quality reports contaminated by more measurement error, for example, by increasing item nonresponse (Retzer et al., 2004). Therefore, in order to decide how many resources to devote to increasing response rates, it is useful to understand the impact of nonresponse on survey results.

23.1 THE CURRENT INVESTIGATION

The research we describe here was designed to improve understanding of response rates in several ways. First, we surveyed leading survey organizations to explore whether the survey administration procedures being used (e.g., number of call attempts, use of refusal conversions, advance letters, and offering incentives) have changed over time in recent years, perhaps in response to concerns about response rates. Second, we used an extensive set of more than 100 random digit dialing (RDD) telephone studies conducted over a 10 year period (between 1996 and 2005) by leading survey organizations to get an overall picture of response rates in recent years. Third, we used a subset of these studies that involved the same topic, same interview length, same sponsor and conducting organization, and same methodology to assess whether response rates have changed between 1996 and 2003.

Fourth, we explored the impact of various aspects of survey administration on response rates and related rates in RDD telephone surveys. To complement past studies of the impact of individual survey administration strategies (e.g., refusal conversions, increased call attempts, incentives, and advance letters) one at a time in experiments (e.g., Singer et al., 1999b), we explored whether the use of particular survey administration procedures affects response rates and other rates in a multivariate, correlational, observational (nonexperimental) statistical analysis.

Finally, we gauged the extent to which response rates affect survey data accuracy. Specifically, we assessed whether lower response rates are associated with less demographic representativeness of a sample.

We begin below by defining response, contact, cooperation, and refusal rates, on which our analyses will focus. Then we review the findings of past studies examining telephone surveys on the issues we will explore. Next, we describe the data we collected to assess the effects of survey administration procedures and changes in these procedures over time, and the consequences of response rates for demographic representativeness. We then describe the results of our analyses, discuss their limitations, and discuss the implications of our findings for survey research practice.

23.2 DEFINITIONS

The response rate for an RDD survey is the proportion of eligible households with whom interviews are completed (we used AAPOR's response rate 3). Response rates are a function of two different aspects of the interaction with respondents: *contacting* respondents and gaining their *cooperation*. The processes of contacting respondents and gaining their cooperation involve very different strategies. As such, researchers are often interested in separating the influence of contact and cooperation, and separate contact and cooperation rates can be calculated. For an RDD survey, the contact rate is defined as the proportion of eligible households in which a housing unit member was reached (we used AAPOR's contact rate 2). The cooperation rate is the proportion of successfully contacted households from which an interview is obtained (we used AAPOR's cooperation rate 1).

These separate rates help researchers interested in increasing response rates (or those concerned about low response rates) to determine the extent to which contact and cooperation each contribute to response rates and to tailor strategies to increase response rates that target contact (e.g., increased number of call attempts) or cooperation (e.g., offering an incentive). Response rates are also decreased when potential respondents refuse to participate in surveys, and strategies such as refusal conversions target this particular problem. The refusal rate for an RDD survey is the proportion of eligible households that refuse to participate (we used AAPOR's refusal rate 2). Although one might imagine that the refusal rate is 100 percent minus the cooperation rate, the refusal rate is in fact the proportion of *all eligible households* in which a refusal occurred, whereas the cooperation rate is the proportion of *all contacted households* that yielded an interview.

23.3 ANTECEDENTS AND CONSEQUENCES OF RESPONSE RATES

23.3.1 Survey Administration Procedures and Response Rates

As a survey is constructed and conducted, researchers must make many decisions about how to conduct the survey. These decisions include

(1) the purpose of the survey (e.g., whether it is for news media release or not),
(2) whom to interview (e.g., whether the sample will be from the nation as a whole or from a single state or region, whether list-assisted sampling will be used to generate telephone numbers, the method for choosing a household member to interview, and interviewing in languages other than English),
(3) whether to attempt to provide information about the study to respondents prior to initial interviewer contact with the respondent (e.g., via advance letters or messages left on answering machines),
(4) the amount of effort to be made to contact respondents (e.g., the field period length and maximum number of contact attempts),
(5) whether to attempt to persuade respondents to participate (e.g., by offering incentives or attempting to convert refusals),
(6) procedural aspects that affect respondent burden (e.g., the length of the survey, allowing respondents to make appointments and to initiate contact to be interviewed).

A researcher's decisions on these issues are usually driven by the purpose of the survey and the resources available to conduct it.

Many researchers have explored how survey administration procedures affect RDD telephone survey response rates. This research has been used, in part, to identify procedures that effectively maximize response rates (e.g., Frankovic 2003; Brick, et al., 2003b). We offer a brief, partial review of this literature next, along with our hypotheses about the potential impact of various design features.

Whom to Interview

List-assisted samples. To increase efficiency and maximize calls to working residential numbers, many telephone surveys today use list-assisted samples in which calls are made only in 100-banks of numbers with at least one residential listing (called "1+ banks"; Casady and Lepkowski, 1993; Tucker et al., 2002), two or more listed residential numbers ("2+ banks"), or three or more listed residential numbers ("3+ banks"). However, using list-assisted samples may have costs for sample representativeness, because numbers from banks that do not meet the requirement (i.e., banks with very few or no listed telephone numbers) are not included in the sample. If the characteristics of households in these banks differ from those included in the sample, the use of list-assisted sampling could bias the representativeness of the survey sample (Giesbrecht et al., 1996) while increasing the response rate and increasing administration efficiency.

Within-household respondent selection. When conducting an RDD telephone survey, researchers are usually interested in obtaining a random sample of the population of *people* rather than a random sample of households. In order to do this, interviewers select one household member using one of various techniques (see Rizzo et al., 2004a; Gaziano, 2005, for reviews). Acquiring a roster of all eligible members of the household permits randomly selecting one person to be interviewed, yielding equal probability of selection. Less invasive quasi-probability and nonprobability techniques are also sometimes used to select an adult from all those in the household. For example, some techniques involve asking for the adult in the household who had the most recent birthday. Still other techniques involve asking for the person *at home* with the next or last birthday or asking first for the youngest male at home and then for the oldest female at home if no male is available. An even less burdensome procedure involves interviewing any knowledgeable adult. Although some studies have found significantly higher cooperation rates or completion rates (i.e., the number of completes divided by the number of completes plus refusals) when using less intrusive quasi-probability and nonprobability selection methods than when using more intrusive probability methods (e.g., O'Rourke and Blair, 1983; Tarnai et al., 1987), others have found no significant differences in cooperation or completion rates between these respondent selection techniques (e.g., Oldendick et al., 1988; Binson et al., 2000).

Spanish interviewing. The Latino population is one of the fastest growing ethnic groups in the United States, making it increasingly tempting for survey researchers to translate survey interviews into Spanish and to have bilingual interviewers. Having bilingual interviewers who can conduct the interview in Spanish may increase response rates because they minimize the number of eligible respondents who cannot be interviewed due to language barriers. Spanish interviewing may also reduce the perceived burden of responding for respondents who are bilingual but have difficulty with English.

Attempts to Provide Additional Information

Advance letters. Researchers sometimes send advance letters without incentives to tell respondents about the survey sponsor, topic, and purpose. In RDD telephone surveys, this cannot be done for the entire sample, because (1) researchers cannot typically get mailing addresses for all the RDD telephone numbers,[2] (2) only

[2]The proportion of listed RDD sample telephone numbers varies greatly in published reports, from less than 40 percent to more than 70 percent (e.g., Traugott et al., 1987; Brick et al., 2003b), and may vary based on factors such as the geographic area being surveyed, the extent to which the sample has been cleaned to eliminate nonworking or disconnected numbers, and the recency with which the sample has been updated by the company that provided it.

a portion of the people who receive the advance letter read it, and (3) the household member who reads the advance letter may not be the same person who answers the telephone. For example, in studies involving lists of respondents for whom addresses were known, only about three quarters of respondents reported that they had received an advance letter (Traugott et al., 1987). Experimental studies suggest that people who receive advance letters are more likely to participate in a survey and less likely to refuse than those who do not (Dillman et al., 1976; Traugott et al., 1987; Camburn et al., 1995; Smith et al., 1995; Hembroff et al., 2005; Link and Mokdad, 2005a).

Messages on answering machines. Now that answering machines and voicemail are ubiquitous (see Roth et al., 2001), interviewers can choose to leave messages on answering machines, or they may forego this opportunity. An answering machine message may act as a form of an advance letter to give potential respondents information about the survey and to increase the perceived legitimacy of the project. However, answering machine messages may not be effective if respondents do not remember them at the time of later contact by an interviewer, and repeated answering machine messages may be irritating to potential respondents, thus reducing participation.

Experimental tests of the effects of answering machine messages have produced mixed results. Some evidence suggests that answering machine messages increase reported willingness to participate (Roth et al., 2001) and participation (Xu et al., 1993), particularly if repeat messages are not left (Tuckel and Shukers, 1997). Other researchers have found no effect of leaving answering machine messages on participation (Tuckel and Schulman, 2000; Link and Mokdad, 2005b). Messages explaining that the interviewer is not selling anything may be especially effective (Tuckel and Shukers, 1997), but providing information about university sponsorship, the importance of the research, a monetary incentive, or a number respondents can call to complete the survey may not increase response rates more than a basic introductory message without such information (Xu et al., 1993; Tuckel and Schulman, 2000).

General Contact Effort

Field period length. The length of the field period is the number of days during which interviewing is conducted. Longer field periods may increase the probability of contact, because respondents are less likely to never be available (e.g., be out of town or ill) during a longer field period. Some studies indicate that longer field periods are associated with higher response rates (e.g., Groves and Lyberg, 1988; Keeter et al., 2000).

Maximum number of call attempts. One aspect of survey administration is the maximum number of times that interviewers attempt to reach each household, after which the telephone number is retired from the active sample. Higher maximum numbers of call attempts have been found to be associated with higher response rates in some studies (e.g., O'Neil, 1979; Massey et al., 1981; Traugott, 1987; Merkle et al.,

1993). This effect is not linear; each additional call attempt increases response rates less than the previous attempt does.[3]

Direct Efforts to Persuade and Gain Compliance

Incentives. Many studies have shown that offering respondents material incentives for participation increases response rates (e.g., Yu and Cooper, 1983; Singer et al., 1999b; Singer et al., 2000). Typically, cash incentives have been more effective than other material gifts, and prepaid incentives (provided before respondents complete the interview) are usually more effective than promised incentives (to be provided after an interview is completed; Singer et al., 1999b). Prepaid incentives may be particularly effective because they invoke the norm of reciprocity (Dillman, 1978; Groves et al., 1992).

Refusal conversion. If a potential respondent initially refuses to be interviewed, a "refusal conversion" interviewer can call back sometime later to attempt to convince the individual to complete the survey. If refusal conversion interviewers are at least sometimes successful at obtaining completed interviews, they will increase a survey's response and cooperation rates, and recent evidence suggests that response rates in studies would be substantially lowered (5–15 percentage points) if refusal conversions were not done (Curtin et al., 2000; Montaquila et al., Chapter 25 in this volume) and that 7–14 percent of refusals are successfully converted to completed interviews when refusal conversions are attempted (e.g., Brick, et al., 2003b; Retzer et al., 2004).

Convenience and Respondent Burden

Interview length. Conventional wisdom suggests that people are less likely to agree to participate in a survey that is longer because of the increased burden. Most potential respondents do not know how long a survey will be at its start, which presumably minimizes any impact of interview length on participation, but interviewers may subtly communicate the length of the survey even if it is not mentioned. In one study that manipulated the stated length of a survey, respondents told the interview would be 40 minutes were more likely to refuse to participate than those told the interview would be only 20 minutes (Collins et al., 1988).

Appointments and respondent-initiated contact. Organizations sometimes allow interviewers to make appointments with respondents to be interviewed at a later time, and some organizations allow respondents to call in to make an appointment or to complete an interview. These procedures allow the survey organization to use resources more efficiently to contact respondents more easily and allow greater convenience for respondents, and may therefore increase response rates (e.g., Collins et al., 1988).

[3]The timing of calls (across time of day and days of the week) may also influence their success (e.g., Cunningham et al., 2003).

23.3.2 Effects of Response Rates on the Accuracy of Survey Results

Methods for assessing effects of response rates on accuracy. A great deal of research has explored the impact of nonresponse on telephone survey results by assessing whether respondents and nonrespondents differ from one another (see Groves and Couper, 1998 for a review). This has been done by (1) conducting a follow-up survey to interview people who did not respond to the initial survey (e.g., Massey et al., 1981), (2) comparing the wave-one characteristics of respondents who were and were not lost at follow-up waves of interviewing in panel studies (e.g., Schejbal and Lavrakas, 1995), (3) comparing early versus late responders to survey requests (under the assumption that late responders are more similar to nonresponders than early responders; e.g., Merkle et al., 1993), (4) comparing people who refuse an initial survey request to those who never refuse (e.g., O'Neil, 1979; Retzer et al., 2004), (5) using archival records to compare the personal and/or community characteristics of households that do and do not respond to survey requests (e.g., Groves and Couper, 1998), and (6) comparing the characteristics of respondents in an RDD survey sample to those of the population as a whole (e.g., Mulry-Liggan, 1983; Keeter et al., 2000).

Many of these studies have focused on the relation of nonresponse to the demographic characteristics of the samples, and some have tested whether nonresponse is related to substantive survey responses. However, there are reasons to hesitate about generalizing evidence from some of these approaches to nonresponse in a cross-sectional survey. For example, nonresponse in panel studies after the first wave is not the same phenomenon as nonresponse in the initial wave of such a survey. Similarly, reluctant respondents and late responders may not be the same as nonrespondents.

Some of the most direct evidence about nonresponse bias comes from research comparing responses from similar surveys that achieved different response rates (e.g., Traugott et al., 1987; Keeter et al., 2000; Groves et al., 2004b). For example, Keeter et al. (2000) varied the amount of effort put into obtaining high response rates in two surveys with identical survey questionnaires by manipulating the field period length, extent of refusal conversion attempts, and number of call attempts. As a result, one survey had a much higher response rate than the other. Demographic representativeness and substantive survey responses could then be compared to assess the effects of response rates on them.

Findings regarding demographic characteristics. Some past studies indicate that respondents and nonrespondents had different demographic characteristics, so the survey samples were unrepresentative of the population. But in every case, the body of evidence is actually quite mixed.

For example, some evidence indicates that women were overrepresented in RDD surveys relative to the population (Chang and Krosnick, 2001). Consistent with this, researchers have found that males were more difficult to contact than females (Traugott, 1987; Shaiko et al., 1991; Merkle et al., 1993) and that males were more difficult to find for later waves of a panel survey (Schejbal and Lavrakas, 1995). However, Keeter et al. (2000) found that the proportion of men and women did not differ between survey samples with different response rates. Similarly, Mulry-Liggan (1983) found no difference in the proportion of men and women in an RDD survey sample relative to that in the population. And Retzer et al. (2004) found no significant difference in the rate of refusal conversions among male and female respondents.

Some evidence also suggests that respondents and nonrespondents sometimes differ in terms of income. One study found that an RDD survey sample included more high-income respondents and fewer low-income respondents than the population (Chang and Krosnick, 2001). Consistent with this, panel surveys suggest that lower income respondents may be more difficult to locate for later waves (e.g., Schejbal and Lavrakas, 1995). Some panel survey follow-up studies have found that lower income respondents were more likely to refuse telephone survey requests (e.g., O'Neil, 1979). However, other researchers have found no differences in the income levels of respondents interviewed via refusal conversions and those who did not initially refuse (e.g., Retzer et al., 2004). And in a comparison of surveys with different response rates, the survey with the higher response rate underrepresented low-income respondents more than the survey with the lower response rate (e.g., Keeter et al., 2000).

Respondents of different races may also respond at different rates to telephone surveys. For example, some evidence suggests that RDD survey samples may underrepresent racial minorities, particularly African American respondents (Chang and Krosnick, 2001), although there is some evidence that other racial minority groups may be overrepresented (Mulry-Liggan, 1983; Chang and Krosnick, 2001). White respondents have been underrepresented in some surveys (e.g., Keeter et al., 2000; Chang and Krosnick, 2001), overrepresented in others (e.g., Green et al., 2001), and accurately represented in others (e.g., Mulry-Liggan, 1983). In a comparison of surveys with different response rates, the one with the higher response rate resulted in less underrepresentation of white respondents than the one with a lower response rate (Keeter et al., 2000). However, evidence from studies examining difficult to reach respondents suggests that nonwhites may be more difficult to contact than whites (Merkle et al., 1993; Traugott, 1987) and more difficult to find for later waves of a panel survey (Schejbal and Lavrakas, 1995). Other studies found no significant racial differences between respondents who were interviewed as a result of refusal conversions and those who did not initially refuse (e.g., Retzer et al., 2004).

Education was also found to be related to likelihood of responding in some telephone surveys. Some studies documented underrepresentation of low education respondents and overrepresentation of high education respondents (e.g., Mulry-Liggan, 1983; Chang and Krosnick, 2001). Likewise, some researchers have found that more educated people are easier to locate for later waves of a panel survey (Schejbal and Lavrakas, 1995) and less likely to be interviewed as a result of a refusal conversion (O'Neil, 1979; Retzer et al., 2004). However, other studies have found that more educated people require more call attempts (Merkle et al., 1993), and that surveys with higher response rates may overrepresent high education respondents *more* than surveys with lower response rates (Keeter et al., 2000).

Compared to the population, RDD studies have sometimes underrepresented the youngest (Chang and Krosnick, 2001) and oldest adults (Mulry-Liggan, 1983; Chang and Krosnick, 2001). Older adults (those 65 and older) are easier to contact (Traugott,

1987; Shaiko et al., 1991; Merkle et al., 1993), perhaps because they are less likely to work and therefore more likely to be at home. Older people are also easier to locate for later waves of panel surveys (Schejbal and Lavrakas, 1995), perhaps because they are more tied to the community and less likely to move between waves of panel surveys. However, considerable evidence also suggests that older people may be more likely to refuse to be interviewed and may make up a larger proportion of respondents who require a refusal conversion than respondents who do not (O'Neil, 1979; Massey et al., 1981; Struebbe et al., 1986; Retzer et al., 2004).

Findings regarding responses to substantive questions. Nearly all research focused on substantive variables has concluded that response rates are unrelated to or only very weakly related to the distributions of substantive responses (e.g., O'Neil, 1979; Smith, 1984; Merkle et al., 1993; Curtin et al., 2000; Keeter et al., 2000; Groves et al., 2004; Curtin et al., 2005). For example, comparing two similar surveys with different response rates, Keeter et al. (2000) found statistically significant differences for only 14 of 91 items they compared. Although this is larger than the proportion that would be expected by chance alone, the 14 differences were all small in magnitude. Other surveys have found comparably small effects of response rates on substantive responses (O'Neil, 1979; Smith, 1984; Merkle et al., 1993; Curtin et al., 2000; Groves et al., 2004; Curtin et al., 2005).

23.4 METHODS

Thus, the evidence accumulated provides little support for the idea that response rates in telephone surveys are associated with the distributions of substantive survey responses and mixed evidence as to whether low response rates are associated with reduced demographic representativeness. To further explore the causes and effects of nonresponse, we contacted 14 major survey data collection organizations who agreed to provide information about their RDD telephone procedures and information about specific surveys: ABC News, Abt Associates, CBS News, the New York Times, the Gallup Organization, the Kaiser Family Foundation, the Los Angeles Times, Mathematica Policy Research, Inc., the Pew Research Center for the People and the Press, the RAND Corporation, Research Triangle Institute, Schulman, Ronca, Bucuvalas, Inc., the Washington Post, and Westat. These organizations come from two broad classes: ones that primarily conduct surveys with short data collection periods for news media release, and ones that primarily conduct surveys with much longer data collection field periods that are often sponsored by government agencies. All surveys we examined involved data collected by or for one or more of these organizations. In some cases, organizations cosponsored a survey, or one organization designed and directed the research and subcontracted data collection to another organization.

23.4.1 Changes in Survey Administration Procedures Over Time

From each organization, we requested the name of their field director or a person at their organization who could tell us about changes in survey administration procedures in recent years. For organizations that did not collect their own data, we obtained contact information for a person who could answer questions about changes in survey administration procedures over time at the organization that did their data collection. Because some surveys involved the collaboration of several organizations and because the same subcontractor did the data collection for multiple organizations, there is not a one-to-one association between these individuals and the 14 organizations initially contacted. We identified 12 such people, to whom we sent a questionnaire asking about differences in survey administration procedures between 2000 and 2004. Ten respondents provided data to us. The other two organizations did not conduct any RDD surveys in one of these years and therefore could not answer our questions.

23.4.2 RDD Study Methodologies and Response Rates

From each organization, we requested information about recent general population RDD surveys they had conducted. We requested three types of information about each survey: (1) frequencies for final disposition codes, (2) unweighted demographic distributions for all people who responded to the questionnaire, and (3) information about the survey administration procedures.

Information about surveys was collected in three phases. In January of 2003, we contacted 12 organizations and requested information about survey administration procedures for up to five national and five state-level RDD general population surveys that were in the field for at least 3 days. For organizations that conducted a large number of these surveys, we requested that they send us information about the five surveys conducted nearest the beginning of the last five quarters (starting 1/1/02, 4/1/02, 7/1/02, 10/1/02, and 1/1/03). These requests yielded disposition code frequencies and survey administration information for 49 surveys and unweighted demographic distributions for 27 of these surveys (starting 1/1/04, 4/1/04, 7/1/04, 10/1/04, and 1/1/05). In February of 2005, we contacted six organizations (five of the organizations contacted in January, 2003, and one new organization) and requested disposition code frequencies, information about survey administration procedures, and unweighted demographic frequencies for recent general population RDD telephone surveys that were in the field for at least 3 days. We asked especially for national surveys but said we would accept general population surveys of state and regional samples as well. This request yielded disposition code frequencies, and survey administration procedure information for an additional 22 surveys and unweighted demographic frequencies for 18 of these surveys. One additional organization was contacted and asked for disposition code frequencies, unweighted demographic frequencies, and information about survey administration processes for all general population RDD surveys for which the organization had records. This request yielded disposition code frequencies and survey administration information about an additional 43 surveys and unweighted demographic frequencies for 36 of these surveys.

In total, we received usable disposition code frequencies for 114 RDD surveys conducted between 1996 and 2005, which we used to gauge the relations between survey administration procedures and response, contact, cooperation, and refusal rates. Of these, 90 were national surveys (either all 50 states or the contiguous United States), 19 were surveys of samples within a single state, and five involved some other sort of more limited geographic area (e.g., city, county, or metropolitan area). We included surveys of a state or region in these analyses to maximize sample size and because doing so reduced the extent to which survey administration procedures were confounded with one another. Our analyses examining the effect of survey administration variables controlled for mean differences in rates between national and nonnational surveys, and we also repeated all analyses with only the national surveys to see whether the results differed notably.

Of the 90 national surveys, unweighted demographic frequencies were provided for 81 of them. For three surveys, the disposition codes were for RDD screeners for surveys that dealt with special populations of respondents. In these cases, demographic information was not collected from all screened respondents, and could therefore not be used in our research. For the remaining 6 surveys, demographics were not provided.

Among the 90 national surveys, 26 were surveys conducted on the same topic by the same organization using the same methodology between 1996 and 2003. We used these 26 surveys to more directly assess changes in response, contact, cooperation, and refusal rates over this time period.

Study characteristics. For each survey, we asked the organization about its administration procedures, including whether the survey involved a national, state, or regional sample, the type of sample used (e.g., all working blocks versus all blocks with at least two listed residential numbers), the respondent selection technique used, the languages in which the interviewing was conducted, whether advance letters were sent, whether answering machine messages were left, the field period length, the maximum number of call attempts, the use of incentives and refusal conversions, and procedures for making appointments and allowing respondents to contact the survey organization (see Tables 23.1 and 23.2 for a list of these variables and descriptive statistics). We used this information to assess the impact of survey administration procedures on response rates.

We also created a variable indicating which organization contributed the data. Because surveys from the same organizations are not fully independent of one another (and therefore may be more similar than surveys from different organizations), it would be ideal to control for organization in our analyses, if possible. But doing so meaningfully requires sufficient independence of survey organization from the implementation procedure variables. We conducted all analyses with and without controlling for survey organization to assess the plausibility of that approach.

23.4.3 Calculating Response, Cooperation, Contact, and Refusal Rates

Final disposition code frequencies were used to estimate AAPOR response rate 3, contact rate 2, cooperation rate 1, and refusal rate 2 using the AAPOR response rate calculator available online (www.aapor.org). The response, contact, and refusal rates we used include a portion of unknown eligibility cases with those known to be eligible, using the proportion *e*. We chose to use the CASRO method of estimating *e*

Table 23.1. Descriptive Statistics for Continuous Variables

Variable	Mean	Standard deviation	Minimum	Maximum	N
Rates					
Response rate	0.30	0.13	0.04	0.70	114
Contact rate	0.67	0.13	0.33	0.92	114
Cooperation rate	0.44	0.15	0.09	0.84	114
Refusal rate	0.29	0.09	0.04	0.55	114
e	0.55	0.13	0.26	0.84	114
Demographic discrepancies (in %)					
Gender[a]	2.58	3.17	0.02	17.33	57
Income	3.00	1.88	0.60	10.93	74
Race	2.99	3.08	0.40	24.90	80
Education	5.68	2.21	1.10	11.08	81
Age	2.20	1.26	0.37	5.90	81
Continuous survey administration procedures					
General efforts to contact					
Field period length (in days)	35.76	64.41	2.00	399.00	113
Maximum number of call attempts[b]	8.46	4.03	3.00	20.00	69
Convenience and respondent burden					
Interview length (in minutes)	16.91	7.04	3.50	34.00	89
Survey organization variables					
Number of surveys per organization	10.31	10.93	1	43	114

[a]Twenty-four surveys (from four organizations) that used gender quotas were excluded from estimates of gender discrepancies.
[b]For seven surveys, no maximum number of calls was set; these are excluded from the statistics in the last row of this table.

because it could be applied equivalently across studies and did not require additional methodological information about the studies that we did not have. This method results in a conservative estimate of e (i.e., this procedure likely overestimates e and underestimates response rates). The estimates of e for the surveys we examined are likely to be relatively high, because the bulk of the surveys were for media release and involved fairly short field periods, which leads to high estimates of e (Smith, 2003b). Whenever we were uncertain about the correspondence between the disposition codes used by an organization and the AAPOR codes, we worked with the organization to assign cases the most appropriate AAPOR codes.

23.4.4 Demographic Representativeness

Unweighted demographic data for age, race, gender, income, and education were compared to data from the Current Population Survey March Demographic Supplement

Table 23.2. Descriptive Statistics for Categorical Survey Administration Procedures

Procedure	Value	Number of surveys	Percent
Purpose of survey			
News media release	For media release	92	80.70
	Not for media release	22	19.30
	Missing	0	0.00
Whom to interview			
Geographic area	National (all 50 states or lower 48)	90	78.95
	Not national (state or region)	24	21.05
	Missing	0	0.00
List-assisted sampling	All blocks	6	5.26
	All working blocks (1+ listed)	17	14.91
	All working blocks (2+ listed)	20	17.54
	All working blocks (3+ listed)	56	49.12
	Missing	15	13.16
Respondent selection technique	Any adult at home	3	2.63
	Youngest male/oldest female at home	52	45.61
	Last/next birthday at home	17	14.91
	Last/next birthday all adult residents	21	18.42
	Modified Kish	8	7.02
	Other	12	10.53
	Missing	1	0.88
Spanish interviewing	Yes	28	24.56
	No	76	66.67
	Missing	10	8.77
Attempts to provide additional information			
Sent advance letters	Yes	14	12.28
	No	95	83.33
	Missing	5	4.39
Answering machine messages	Never leave messages	81	71.05
	Sometimes leave messages	14	12.28
	Always leave messages	1	0.88
	Missing	18	15.79
Direct efforts to persuade and gain compliance			
Incentives offered	Yes	8	7.02
	No	101	88.60
	Missing	5	4.39
Refusal conversions	Yes	95	83.33
	No	14	12.28
	Missing	5	4.39

Table 23.2. ***(Continued)***

Procedure	Value	Number of surveys	Percent
Convenience and respondent burden			
Made appointments with any household member	Yes	34	29.82
	No	68	59.65
	Missing	12	10.53
Respondent could call to make appointment	Yes	14	12.28
	No	88	77.19
	Missing	12	10.53
Respondent could call to complete interview	Yes	13	11.40
	No	88	77.19
	Missing	13	11.40

from the year in which the target survey was conducted. For each demographic variable, the demographic discrepancy was the average of the absolute value of the discrepancies between the survey data proportion and the current population survey (CPS) proportion for all the response categories of that variable.[4]

23.5 RESULTS

23.5.1 Changes in Survey Administration Procedures 2000–2004

Whom to Interview

Sampling. Five organizations reported no changes in their sampling procedures. One organization reported a greater use of listed numbers (rather than RDD) in 2004 than in 2000, and two organizations reported more cleaning or screening of numbers in 2004 than in 2000.

Within-household respondent selection. Nine organizations reported no changes in respondent selection techniques. One organization reported changing from oldest male/youngest female at home in 2000 to last birthday by 2004.

Spanish interviewing Two organizations did not interview in Spanish. Three others that did so reported no change in Spanish interviewing between 2000 and 2004. Four organizations reported that they conducted more interviews in Spanish in 2004 than in 2000. One organization reported conducting Spanish interviews in fewer surveys in 2004 than in 2000, but conducting the same proportion of interviews in Spanish in those surveys in 2000 and 2004.

[4]Coding of demographic variables was done as consistently as possible across surveys. Gender was coded male and female. Race was coded white, black, and other races. Education was coded less than high school education, high school education (or GED), some college, and 4-year college degree or more. The original coding of age and income varied widely, so it was impossible to code them identically across all surveys. Age was always coded into six categories, but the specific categories varied across the surveys. Income was coded into 4 or 5 categories and was coded as similarly as possible.

Attempts to Provide Additional Information

Advance letters. Seven organizations did not use advance letters. The other three organizations all reported that they sent advance letters in more studies in 2004 than in 2000. One organization reported that when advance letters were sent, addresses were available for a greater proportion of sample in 2004 than in 2000. No other changes in the use of advance letters were reported.

Answering machine messages. No organizations reported any changes in their procedures regarding leaving messages on answering machines.

General Contact Effort

Field period length. One of the 10 organizations reported longer field periods in 2004 than in 2000. All others reported no change.

Maximum number of call attempts. Four organizations reported changes in their call attempts between 2000 and 2004. Two reported that the average and maximum number of call attempts was greater in 2004 than in 2000, and two reported that the average (but not maximum) number of call attempts was greater in 2004 than in 2000. No organizations reported making fewer call attempts in 2004 than in 2000.

Direct Efforts to Persuade and Gain Compliance

Incentives. Seven organizations did not offer incentives. Of the remaining three, one reported no change, one reported using incentives in more studies in 2004 than in 2000, but no change in the amount of incentives offered between 2000 and 2004, and the last reported using incentives in more studies in 2004 than in 2000 and incentives of larger size in 2004 than in 2000.

Refusal conversions. Two organizations did not do refusal conversions. Of the remaining eight, five reported no change in the procedures for refusal conversions or the proportions of refusals that were followed up by conversion attempts. One organization reported that the number of refusals for which conversions were attempted was higher in 2004 than in 2000. Another organization reported that the number of conversion attempts for each refusal was greater in 2004 than in 2000, and the final organization reported attempting refusal conversions with a larger proportion of refusals in 2004 than in 2000, and making more refusal conversion attempts per refusal in 2004 than in 2000.

Convenience and Respondent Burden

Interview length. One of the ten organizations reported longer interviews in 2004 than in 2000. All others reported no change.

Appointments and respondent-initiated contact. No organizations reported any changes in their procedures regarding making appointments with respondents or other household members, or their procedures regarding allowing respondents to call the survey organization to make an appointment or to complete an interview.

Summary
Overall, few changes in survey administration procedures were made by these organizations. The changes that were made involved more use of techniques to increase response rates (e.g., increasing number of call attempts, refusal conversions, more, and larger incentives) in 2004 than in 2000.

23.5.2 Response, Contact, Cooperation, and Refusal Rates

Descriptive statistics for response, contact, cooperation, and refusal rates and the value of e are shown in the top panel of Table 23.1. Response rates varied from 4 percent to 70 percent and averaged 30 percent. Contact rates ranged from 33 percent to 92 percent and averaged 67 percent. Cooperation rates ranged from 9 percent to 84 percent and averaged 44 percent. Refusal rates ranged from 4 percent to 55 percent and averaged 29 percent. The estimate of e varied from 0.26 to 0.84 and averaged 0.55, which are similar to those reported in previous work (e.g., Smith, 2003b).

Contact and cooperation rates were related to response rates as one would expect. Cooperation rates were highly significantly and positively correlated with response rates ($r = 0.89, p < 0.001, N = 114$). Contact rates were also highly significantly and positively correlated with response rates, but more weakly ($r = 0.62, p < 0.001, N = 114$). So the variation in response rates across studies is attributable mostly to cooperation rates and less to contact rates. Refusal rates were negatively associated with cooperation ($r = -0.45, p < 0.01, N = 114$) and response rates ($r = -0.20, p < 0.05, N = 114$), and positively associated with contact rates ($r = 0.52, p < 0.001, N = 114$). This is consistent with existing evidence that increased contact provides greater opportunities for respondents to refuse (Brick et al., 2003b; Sangster and Meekins, 2004).

23.5.3 Changes in Response Rates Over Time 1996–2003

The correlation between the year (1996–2005) in which a survey went into the field and its response rate was negative and highly significant ($r = -0.38, p < 0.001, N = 113$), indicating that later surveys had lower response rates. Some of this trend could be attributable to changes in survey administration procedures over time (although this seems unlikely given the changes in survey administration procedures described in the previous section), or to the fact that some organizations gave us data from surveys conducted only during certain segments of the time interval and not others (thus confounding organization with time). We therefore examined 26 surveys conducted, between December 1996 and October 2003, by a single organization identical in terms of length and topic with very consistent survey administration procedures. Although the length of the field period in these surveys varied from 3 to 14 days, field period length was not related to response, contact, cooperation, or refusal rates. Additionally, none of the results regarding over-time changes were affected by controlling for length of field period. Date of survey was coded as the number of months after January 1996 that the survey went into the field.

Date of the survey and response rate were strongly negatively correlated ($r = -0.70, p < 0.001, N = 26$). This suggests that response rate decreased by two tenths of a percentage point each month during this time period. Although this is a small change, over the 82 month time period in which these studies were conducted, this regression suggests response rates dropped 16.4 percentage points, a dramatic decrease.

We also explored how contact rates, cooperation rates, and refusal rates changed over time. Contact rates ($r = -0.61$, $p < 0.001$, $N = 26$) and cooperation rates ($r = -0.47$, $p < 0.001$, $N = 26$) dropped significantly during this time period. Interestingly, refusal rates were not significantly associated with date of survey ($r = -0.07$, not significant, $N = 26$), consistent with evidence that refusal rates may not have been changing over time where cooperation and contact rates were (e.g., Brick et al., 2003b).

23.5.4 Survey Administration and Response Rates

Bivariate Analyses

Purpose of the survey. Surveys that were intended for news media release had response rates 6 percent lower on average than those that were not intended for news media release ($b = -0.06$, $p < 0.05$; see row 1 of column 1 in Table 23.3). This difference was a function of lower contact rates ($b = -0.12$, $p < 0.01$) and higher refusal rates ($b = -0.04$, $p < 0.10$), but not lower cooperation rates ($b = -0.03$, not significant; see row 1 of columns 2, 3, and 4 in Table 23.3).

Whom to interview. Response rates for national surveys were significantly higher than those for state or regional surveys ($b = 0.09$, $p < 0.01$; see row 2 of column 1 in Table 23.3). This effect was primarily due to greater cooperation in national surveys ($b = 0.11$, $p < 0.01$), and not to differences in contact rates ($b = 0.03$, not significant) or refusal rates ($b = -0.002$, not significant; see row 2 of columns 2, 3, and 4 in Table 23.3).

Using list-assisted samples (instead of pure RDD samples) was associated with higher response rates ($b = 0.14$, $p < 0.01$), contact rates ($b = 0.10$, $p < 0.01$), and cooperation rates ($b = 0.19$, $p < 0.01$), and weakly associated with higher refusal rates ($b = 0.05$, $p < 0.10$; see row 3 of columns 1–4 in Table 23.3).[5]

Response rates were lower when the respondent selection technique used limited the number of eligible respondents more severely. Interviewing any knowledgeable adult (the baseline group in the regression analysis shown) yielded the highest response rates. Respondent selection techniques that involved selecting an adult from those at home at the time of contact (either using the next/last birthday method or youngest male-oldest female method) had somewhat lower response rates ($b = -0.20$, $p < 0.01$; $b = -0.25$, $p < 0.01$; see rows 4 and 5 of column 1 in Table 23.3), and techniques that involved selecting a respondent from among all adult residents of the

[5]List-assisted sampling was coded 0 if no list assisted sampling was done, 0.33 for surveys using only 1+ listed blocks, 0.67 for surveys using only 2+ listed blocks, and 1 for surveys using only 3+ listed blocks for all regression analyses. The effect of list assisted sampling on response rate was stable across different ways of coding this variable (e.g., a series of dummy variables or a binary variable distinguishing surveys that used list-assisted sampling from those that did not).

Table 23.3. Bivariate Associations of Survey Administration Procedures with Response, Contact, Cooperation, and Refusal Rates (Standard Error in Parentheses)

	Unstandardized OLS regression coefficients				
Procedure	Response rate	Contact rate	Cooperation rate	Refusal rate	N
Purpose of survey					
For news media release	−0.06* (0.03)	−0.12** (0.03)	−0.03 (0.04)	−0.04+ (0.02)	114
Whom to interview					
National survey	0.09** (0.03)	0.03 (0.03)	0.11** (0.03)	−0.002 (0.02)	114
List-assisted sampling[a]	0.14** (0.04)	0.10** (0.04)	0.19** (0.04)	0.05+ (0.03)	99
Respondent selection technique[b]					113
Youngest male/oldest female at home	−0.20** (0.05)	−0.08 (0.06)	−0.18** (0.07)	0.08+ (0.04)	
Last/next birthday at home	−0.25** (0.06)	0.17** (0.06)	−0.19** (0.08)	0.05 (0.05)	
Last/next birthday all adult residents	−0.34** (0.05)	−0.23** (0.06)	−0.31** (0.07)	0.03 (0.05)	
Modified Kish	−0.35** (0.06)	−0.16* (0.07)	−0.35** (0.08)	0.17** (0.05)	
Other	−0.28** (0.06)	−0.02 (0.06)	−0.30** (0.08)	0.16** (0.05)	
Spanish interviewing	0.05+ (0.03)	0.03 (0.03)	0.03 (0.03)	−0.05* (0.02)	104
Attempts to provide additional information					
Sent advance letters	0.04 (0.04)	0.17** (0.03)	−0.05 (0.04)	0.09** (0.03)	109
Answering machine messages left[c]	0.11+ (0.07)	0.24** (0.06)	0.04 (0.08)	0.07 (0.05)	96
General contact effort					
Field period length (in days)	0.001** (0.0002)	0.001** (0.0002)	0.0004* (0.0002)	0.001 (0.0001)	113
Maximum number of call attempts[d]	0.005+ (0.003)	0.02** (0.002)	0.001 (0.003)	0.008** (0.002)	76
Direct efforts to persuade and gain compliance					
Incentives offered	0.12* (0.05)	0.11* (0.05)	0.08 (0.06)	0.02 (0.04)	109
Refusal conversions	0.03 (0.04)	0.10** (0.04)	−0.03 (0.04)	0.05+ (0.03)	109
Convenience and respondent burden					
Interview length (in minutes)	−0.006** (0.002)	−0.001 (0.002)	−0.008** (0.002)	0.0005 (0.001)	89

(continued)

Table 23.3. (*Continued*)

	Unstandardized OLS regression coefficients				
Procedure	Response rate	Contact rate	Cooperation rate	Refusal rate	N
Made appointments with any household member	−0.04 (0.03)	0.01 (0.03)	−0.04 (0.03)	0.05* (0.03)	102
Respondent could call to make appointment	0.04 (0.04)	0.14** (0.04)	−0.02 (0.05)	0.06** (0.03)	102
Respondent could call to complete interview	0.02 (0.04)	0.16** (0.04)	−0.05 (0.05)	0.09** (0.03)	101

[a]List-assisted sampling was coded 0 if no list-assisted sampling was done, 0.33 for surveys using only 1+ listed blocks, 0.67 for surveys using only 2+ listed blocks, and 1 for surveys using only 3+ listed blocks for all regression analyses.
[b]Any adult at home was used as the comparison group.
[c]Answering machine messages were coded 0 if no answering machine messages were left, 0.5 if they were left some of the time, and 1 if they were always left for all regression analyses.
[d]For regressions involving maximum number of call attempts, surveys that made unlimited calls were given a value of 20, the highest maximum number of calls reported.
$^{**}p < 0.01$.
$^{*}p < 0.05$.
$^{+}p < 0.10$.

household (either using the next/last birthday method or a modified Kish method) had the lowest response rates ($b = -0.34, p < 0.01$; $b = -0.35, p < 0.01$; see rows 6 and 7 of column 1 in Table 23.3). Surveys that used other respondent selection techniques had response rates somewhere between these latter two types ($b = -0.28, p < 0.01$; see row 8 of column 1 in Table 23.3). The effects of respondent selection technique on response rates were not only primarily due to lower cooperation rates (see rows 4–8 of column 3 in Table 23.3), but were also somewhat the result of reduced contact rates and higher refusal rates (see rows 4–8 of columns 2 and 4 in Table 23.3).

Surveys that involved interviewing in Spanish had marginally significantly higher response rates than those that did not ($b = 0.05, p < 0.10$; see row 9 of column 1 in Table 23.3). Spanish interviewing increased response rates one half of a percentage point on average. Interviewing in Spanish was not associated with contact rates ($b = 0.03$, not significant) or cooperation rates ($b = 0.03$, not significant; see row 9 of columns 2 and 3 in Table 23.3), but did lower refusal rates ($b = -0.05, p < 0.05$; see row 9 of column 4 in Table 3), suggesting that the ability of an interviewer to speak Spanish may have been important for avoiding refusals.

Attempts to provide additional information. Surprisingly, sending advance letters did not enhance overall response rates ($b = 0.04$, not significant; see row 10 of column 1 in Table 23.3). Interestingly, advance letters increased contact rates ($b = 0.17$, $p < 0.01$) and refusal rates ($b = 0.09, p < 0.01$), and did not affect cooperation rates ($b = -0.05$, not significant; see row 10 of columns 2, 3, and 4 in Table 23.3). Thus, people for whom an address was available who read the letter may have been easier to contact, perhaps because they were less likely to avoid talking with an interviewer (e.g., by screening calls using an answering machine or caller ID). But among potential respondents who were successfully contacted, advance letters may have provided the opportunity to prepare reasons to refuse to participate.

Leaving messages more frequently was marginally significantly associated with higher response rates ($b = 0.11$, $p<10$; see row 11 of column 1 of Table 23.3).[6] Leaving messages on answering machines appears to have increased response rates primarily by increasing contact ($b = 0.\ 24$, $p < 0.01$; see row 11 of column 2), and was not associated with either cooperation rates ($b = 0.04$, not significant) or refusal rates ($b = 0.07$, not significant; see row 11 of columns 3 and 4 in Table 23.3). Leaving an answering machine message may have increased contact by reducing the extent to which respondents screened calls from the survey organization to avoid contact.

General contact effort. Longer field periods were associated with higher response rates ($b = 0.001$, $p < 0.01$; see row 12 of column 1 in Table 23.3); one extra day of calling yielded one tenth of a percentage point increase in response rates. Longer field periods appear to have increased response rates primarily by causing more contact ($b = 0.001$, $p < 0.01$; see row 12 of column 2 in Table 23.3). Field period length was more weakly related to cooperation rate ($b = 0.0004$, $p < 0.05$; see row 12 of column 3 in Table 23.3) and was unrelated to refusal rate ($b = 0.001$, not significant; see row 12 of column 4 in Table 23.3).[7]

A higher maximum number of call attempts was marginally significantly associated with higher response rates ($b = 0.005$, $p < 0.10$; see row 13 of column 1 in Table 23.3). Increasing the maximum number of call attempts by one increased response rates by one half of a percentage point.[8] A higher maximum number of call attempts was also associated with higher contact rates ($b = 0.02$, $p < 0.01$; see row 13 of column 2 in Table 23.3) and refusal rates ($b = 0.008$, $p < 0.01$; see row 13 of column 4 in Table 23.3), but not greater cooperation ($b = 0.001$, not significant; see row 13 of column 3 in Table 23.3). A higher maximum number of call attempts may have been associated with higher levels of refusals both because greater contact provided more opportunity for respondents to refuse and because researchers may have increased their the maximum number of call attempts when the refusal rate for a survey was high.

[6]Answering machine messages were coded 0 if no answering machine messages were left, 0.5 if they were left some of the time, and 1 if they were always left. The effect of leaving answering machine messages on response rate was significant if leaving messages was coded as a series of dummy variables, but it was not significant if leaving messages was coded as a binary variable distinguishing surveys that left messages some or all of the time from those that never left messages.

[7]The refusal rate was calculated using calls that were coded as refusals by interviewers and did not include callback and answering machine disposition codes possibility indicating soft refusals, which occur more frequently in surveys with longer field periods (Sangster and Meekins 2004).

[8]In analyses examining the relation of number of callbacks and response rates, surveys with an unlimited number of callbacks were given a maximum number of calls value of 20, the highest maximum number of calls reported.

Direct efforts to persuade and gain compliance. As expected, surveys that offered incentives had significantly higher response rates than those that did not ($b = 0.12$, $p < 0.05$; see row 14 of column 1 in Table 23.3), a difference of 12 percentage points on average. Surprisingly, though, incentives were associated with increased contact ($b = 0.11$, $p < 0.05$) rather than increased cooperation ($b = 0.08$, not significant) or decreased refusal ($b = 0.02$, not significant; see row 14 of columns 2–4 in Table 23.3). This could have occurred because interviewers communicated the incentive to household members other than the respondent, and this information decreased immediate hang-ups but did not increase respondent willingness to participate.

Attempting refusal conversions was not associated with increased response rates ($b = 0.03$, not significant) or cooperation rates ($b = -0.03$, not significant; see row 15 of columns 1 and 3 in Table 23.3). Surprisingly, attempting refusal conversions was associated with higher contact rates ($b = 0.10$, $p < 0.01$) and slightly *higher* refusal rates ($b = 0.05$, $p < 0.10$; see row 15 of columns 2 and 4 in Table 23.3).

Convenience and respondent burden. As expected, response rates were lower for longer interviews ($b = -0.006$, $p < 0.01$; see row 16 of column 1 in Table 23.3). A 1 minute increase in survey length reduced the response rate by sixth tenths of a percentage point. This suggests that a 15 minute increase in the length of an interview would result in a 9 percentage point decrease in response rates. Longer interview length appears to have decreased response rates primarily by decreasing cooperation, because this rate was strongly correlated with interview length ($b = -0.008$, $p<0.01$; see row 16 of column 3 in Table 23.3). Interview length was unrelated to contact rate or refusal rate ($b = -0.001$ and $b = 0.0005$, respectively, not significant; see row 16 of columns 2 and 4 in Table 23.3).

Procedures designed to increase the convenience of scheduling the interview did not increase response rates (see row 17–19 of column 1 in Table 23.3). Being able to make appointments with any household member was associated with significantly higher refusal rates ($b = 0.05$, $p < 0.05$; see row 17 of column 4 in Table 23.3), and allowing a respondent to call and make an appointment or complete the interview were associated with increased contact rates ($b = 0.14$, $p < 0.01$ and $b = 0.16$, $p < 0.01$, respectively; see rows 18 and 19 of column 2 in Table 23.3), and increased refusal rates ($b = 0.06$, $p < 0.01$ and $b = 0.09$, $p < 0.01$, respectively; see rows 18 and 19 of column 4 in Table 23.3)

Multivariate Analyses

These survey administration procedures were implemented in correlated ways, so it is interesting to separate their effects on rates by conducting multivariate regressions predicting the rates using the procedural variables in Table 23.3. One of the difficulties with doing so is the amount of missing data for some of the survey administration procedures (see Tables 23.1 and 23.2). In order to maintain a reasonable sample size to detect effects, we included only independent variables that were significant predictors in the bivariate analyses shown in Table 23.3 and for which we had data for at least 80 percent of the surveys. The latter requirement led to exclusion of the average length of the interview and the maximum number of call attempts. In the regressions shown in Table 23.4, response, cooperation, contact, and refusal rates were regressed on variables representing: news media release surveys (versus those that were not for news media release), national surveys (versus state or regional), list-assisted sampling, respondent selection techniques (using a series of dummy variables for various techniques with interviews of any knowledgeable adult as the comparison group), whether interviews were conducted in Spanish, leaving answering machine messages, the length of the field period length, and the use of incentives.

These analyses' results were primarily consistent with those of the bivariate analyses, although some of the bivariate effects changed. Taken together, these survey administration procedures explained 80 percent of the variance in response rates.

Purpose of interview. The response, contact, cooperation, and refusal rates for news media release surveys were not different than those for surveys not for news media release ($b = 0.03$, not significant; $b = 0.03$, not significant; $b = -0.05$, not significant; and $b = -0.02$, not significant, respectively; see row 1 in Table 23.4). The effect of survey purpose on response rates, which was significant in the bivariate analyses, presumably became nonsignificant in the multivariate analyses because the many methodological differences between media and nonmedia surveys accounted for the apparent effect of survey purpose.

Whom to interview. National surveys had significantly higher response ($b = 0.10$, $p < 0.05$) and cooperation rates ($b = 0.16, p < 0.01$) than those conducted in smaller geographic areas (see row 2 of columns 1 and 3 in Table 23.4). Using list-assisted sampling was associated with higher response ($b = 0.12$, $p < 0.01$), contact ($b = 0.17, p < 0.01$), cooperation ($b = 0.11, p < 0.01$) and refusal rates ($b = 0.14, p < 0.01$; see row 3 of Table 23.4).

Response rates also varied by the method used to select the respondent to interview (see rows 4–8 of Table 23.4). Consistent with the bivariate analysis, interviewing any knowledgeable adult had the highest response rates; techniques that selected an adult from those at home had somewhat lower response rates ($b = -0.35$, $p < 0.01$ and $b = -0.33$, $p < 0.01$; see rows 4 and 5 of column 1 in Table 23.4); and techniques that selected a respondent from all adult household members had the lowest response rates ($b = -0.46, p < 0.01$ and $b = -0.43, p < 0.01$; see rows 6 and 7 of column 1 in Table 23.4). The multivariate analyses showed more clearly that these differences are primarily due to differences in cooperation and refusal rates (see rows 4–8 of columns 3 and 4 in Table 23.4) rather than to differences in contact rates (see rows 4-8 of column 2 in Table 23.4).

Interviewing in Spanish increased response rates ($b = 0.05$, $p < 0.05$; see row 9 of column 1 in Table 23.4) but was not associated with contact, cooperation, or refusal rates ($b = 0.03$, not significant; $b = 0.05$, not significant; $b = -0.03$, not significant, respectively; see row 9 of columns 2–4 in Table 23.4).

Attempts to provide additional information. Leaving answering machine messages was not associated with response rates, as was the case in the bivariate analyses ($b = -0.13$, not significant; see row 10 of column 1 in Table 23.4). The multivariate

Table 23.4. Unstandardized OLS Regression Coefficients Predicting Response, Contact, Cooperation, and Refusal Rates with Survey Administration Procedures (Standard Error in Parentheses)

Predictor	Response rate	Contact rate	Cooperation rate	Refusal rate
Purpose of interview				
For news media release	0.03	0.03	−0.05	−0.02
	(0.05)	(0.08)	(0.07)	(0.06)
Whom to interview				
National survey	0.10*	0.003	0.16**	−0.04
	(0.04)	(0.05)	(0.05)	(0.04)
List-assisted sampling	0.12**	0.17**	0.11*	0.14**
	(0.04)	(0.05)	(0.05)	(0.04)
Respondent selection technique[a]				
Youngest male/oldest female at home	−0.35**	0.01	0.45**	0.19**
	(0.06)	(0.09)	(0.08)	(0.07)
Last/next birthday at home	−0.33**	−0.01	0.39**	0.19**
	(0.06)	(0.09)	(0.08)	(0.07)
Last/next birthday all adult residents	−0.46**	−0.12	−0.57**	0.18**
	(0.05)	(0.08)	(0.07)	(0.06)
Modified Kish	−0.43**	0.004	−0.56**	0.31**
	(0.06)	(0.09)	(0.09)	(0.07)
Other	−0.28**	−0.05	−0.26**	0.15*
	(0.05)	(0.08)	(0.07)	(0.06)
*Spanish interviewing	0.05*	0.03	0.05	−0.03
	(0.02)	(0.03)	(0.03)	(0.02)
Attempts to provide additional information				
Answering machine messages left	−0.13	0.60**	−0.60**	0.42**
	(0.11)	(0.16)	(0.15)	(0.12)
General contact effort				
Field period length (in days)	0.001**	0.0002	0.001*	0.00001
	(0.0002)	(0.0003)	(0.0003)	(0.0002)
Direct attempts to persuade and gain compliance				
Incentives offered	0.10*	−0.07	0.16**	−0.10*
	(0.04)	(0.06)	(0.05)	(0.04)
R^2	0.80	0.63	0.75	0.59
N	87	87	87	87

Note: Only variables that were significantly or marginally significantly associated with response rates in the bivariate analyses shown in Table 23.3 and that had valid data for at least 80 percent of the 114 cases were included as predictors in these analyses. The latter criterion led to the exclusion of the length of the survey interview, which was missing for 22 percent of the 114 surveys and maximum number of call attempts, which was missing for 33 percent of the 114 surveys.

[a]Any knowledgeable adult at home was used as the comparison group.

**$p < 0.01$.

*$p < 0.05$.

analyses suggest, however, that answering machine messages increased contact rates ($b = 0.60, p < 0.01$), reduced cooperation rates ($b = -0.60, p < 0.01$), and increased refusal rates ($b = 0.42, p < 0.01$; see row 10 of columns 2–4 in Table 23.4). These opposing effects may have canceled one another out, leading to no overall effect of leaving answering machine messages on response rates.

General contact effort. Field period length was significantly and positively associated with response rates ($b = 0.001, p < 0.05$; see row 11 of column 1 in Table 23.4). Surprisingly, field period length was not associated with contact rate ($b = 0.0002$, not significant), but was associated with a higher cooperation rate ($b = 0.001, p < 0.05$; see row 11 of columns 2 and 3 in Table 23.4). Field period length was unassociated with refusal rate ($b = 0.00$, not significant; see row 11 of column 4 in Table 23.4). Thus, it appears that a longer field period may have led to more willingness to cooperate by people who initially provided soft refusals.

Direct attempts to persuade and gain compliance. Offering incentives was significantly and positively related to response rates ($b = 0.10, p < 0.05$; see row 12 of column 1 in Table 23.4). This was due to increased cooperation rates ($b = 0.16, p < 0.01$) and reduced refusal rates ($b = -0.10, p < 0.05$; see row 12 of columns 3 and 4 in Table 23.4) but not to changes in contact rates ($b = -0.07$, not significant; see row 12 of column 2 in Table 23.4).

Analyses Using Only National Surveys

The analyses of survey administration procedures reported thus far used both national and nonnational surveys. Multivariate analyses like those in Table 23.4 conducted with only national surveys yielded results similar to those produced using the full set of surveys, but there were some differences. Some effects that were significant when using all surveys became nonsignificant using only the national surveys. For example, respondent selection technique was no longer significantly associated with refusal rates; leaving answering machine messages was no longer significantly associated with either contact or refusal rates; and using incentives was no longer significantly associated with any of the rates. And some effects became significant that had not been before. For example, national news media surveys yielded significantly higher cooperation rates and lower refusal rates than national nonnews media release surveys ($b = 0.24, p < 0.01$; $b = -0.19, p < 0.01$; respectively). But in general, the results were closely comparable to those shown in the Tables.

Controlling for survey organization. We also conducted the analyses shown in Table 23.4 controlling for nonindependence among surveys conducted by the same organization (by specifying organization as the "psu" using "svy" commands in Stata). The results of these analyses were very similar to those reported in Table 23.4, with just a few exceptions. The effect of Spanish interviewing on refusal rates became significant ($b = -0.03, p < 0.05$); the effect of answering machine messages on cooperation became nonsignificant ($b = -0.60$, not significant); and using incentives became significantly negatively associated with contact ($b = -0.07, p < 0.05$).

When we estimated these parameters controlling for survey organization and using just the national surveys, the results of analyses predicting response and contact rates did not change notably. There was little variation in the administration procedures within survey organization, so the effects of different procedures on cooperation and refusal rates could only be disentangled from survey organization by including nonnational surveys.

23.5.5 Response Rates and Demographic Representativeness

Finally, we explored the relation of response rates to the demographic representativeness of the interviewed samples using a subset of 81 national surveys for which we had unweighted demographic frequencies. Response rates in these surveys varied from 0.05 to 0.54. We examined the absolute value of the discrepancies between unweighted survey estimates and population estimates from the Census Bureau's March Supplement of the CPS from the year in which the survey was conducted for gender, income, race, education, and age (see rows 6–10 of Table 23.1).

The average discrepancies for these variables ranged from 2.20 percent for age to 5.68 percent for education. In order to gauge the association of demographic discrepancy with response rate, we regressed each of the demographic discrepancies on (1) response rate only, and (2) response rate and response rate squared to test for nonlinearity.

For gender, income, and race, the association was significantly negative and linear (see rows 1–3 of Table 23.5); response rate accounted for less than 10 percent of the variance in demographic discrepancy in each case. The demographic discrepancies for gender, income, and race predicted by response rates in the regression equations

Table 23.5. Effect of Response Rates on Demographic Representativeness

		Unstandardized regression coefficients (standard errors in parentheses)		
Demographic discrepancy	N	Linear effect of response rate	Squared effect of response rate	R^2
Gender[a]	57	−9.59** (3.59)	—	0.10
Income	74	−4.61* (2.22)	—	0.06
Race	80	−8.55* (3.55)	—	0.07
Education	81	−29.63** (10.54)	37.79* (16.43)	0.13
Age	81	−20.67** (5.51)	23.97** (8.59)	0.26

[a] Twenty-four surveys (from four organizations) that used gender quotas were excluded from this estimate.

**$p < 0.001$.

*$p < 0.05$.

in rows 1–3 of Table 23.5 are shown in Fig. 23.1(*a*)–(*c*), respectively. For example, a 45 percentage point increase in response rate was associated with in a decrease in demographic discrepancy of 3 percentage points for age and 4 percentage points for education and race.

For education and age, the effect of response rates on demographic discrepancies was nonlinear. In both cases, the linear effect of response rates was negative and significant, and the squared effect of response rates was positive and significant, indicating decreased marginal returns (see rows 4 and 5 of columns 2 and 3 in Table 23.5). These effects are illustrated by the demographic discrepancies predicted from response rates using the regressions in rows 4 and 5 of Table 23.5 shown in Fig. 23.1(*d*)–(*e*). Education and age demographic discrepancies were greater as response rate dropped below 0.30, but increases in response

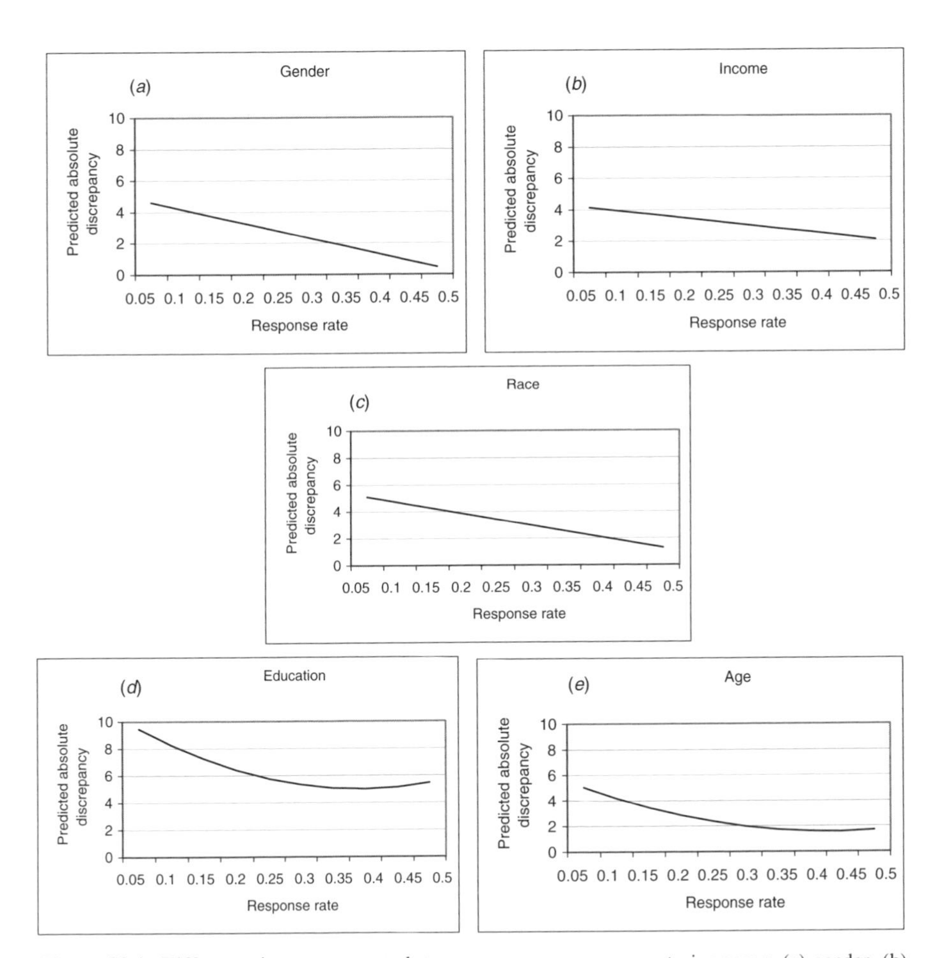

Figure 23.1. Difference in response rate between groups as reponse rate increases: (a) gender, (b) income, (c) race, (d) education, and (e) age.

rates above 0.30 were not associated with further reductions in education and age discrepancies.

When controlling for survey organization, the associations between response rates and demographic discrepancies involving gender, income, and race became nonsignificant. The nonlinear relations of response rate with education and age discrepancies were unchanged by controlling for survey organization.

23.6 DISCUSSION

23.6.1 Summary of Findings and Implications

This research examined response rates in the largest set of surveys to date. The response rates in the surveys we examined were somewhat lower than those reported in other large reviews of RDD response rates (e.g., Massey et al., 1997 found an average response rate of 62 percent among 39 RDD surveys), but our surveys were conducted later and included more surveys that were designed for news media release than previous reviews of response rates in the field.

Consistent with a growing body of evidence, we found that response rates have decreased in recent years (see also Battaglia et al., 2007, Chapter 24 in this volume) and that when organizations have changed their survey administration procedures in recent years, they have done so in ways designed to increase response rates. These findings suggest that trends observed regarding previous time periods (de Heer, 1999; Steeh et al., 2001; Tortora, 2004; Curtin et al., 2005) have continued in recent years.

We also explored the antecedents and consequences of response rates, and much of our evidence dovetails nicely with findings reported by other investigators. Like others, we found that surveys with longer field periods (Groves and Lyberg, 1988; Keeter et al., 2000), shorter interviews (Collins et al., 1988; Singer and Presser, 2007, Chapter 21 in this volume), more call attempts (O'Neil, 1979; Traugott, 1987; Massey et al., 1981; Merkle et al., 1993), incentive offers (Yu and Cooper, 1983, Chapter 22 in this volume; Singer et al., 1999b; Singer et al., 2000; Cantor et al., 2007, Chapter 22 in this volume), and less invasive, easier to implement respondent selection techniques (O'Rourke and Blair, 1983; Tarnai et al., 1987) yielded higher response rates. As expected, surveys that used list-assisted samples and Spanish interviewing had higher response rates as well.

Although leaving messages on answering machines was weakly associated with response rates in a bivariate analysis (as found by Xu et al., 1993), this effect disappeared in our multivariate analysis, suggesting that the bivariate association was due to other methodological differences across studies.

Sending advance letters was also not associated with increased response rates. Although advance letters may enhance the perceived legitimacy of survey interviewers, obtaining this benefit requires that the advance letter be read by the same person who answers the telephone when the interviewer calls and that the letter be associated by the potential respondent with the telephone call. And these conditions may only rarely occur. Furthermore, advance letters may also forewarn potential respondents who want to avoid being interviewed, allowing them to think of excuses not to participate or to screen their calls to avoid interviewers. Obtaining addresses and mailing letters uses valuable resources, so implementing this procedure may not have enhanced response rates because the funds used to pay for it could have been more effectively used to increase response rates in other ways.

Implementing refusal conversion attempts did not yield higher response rates. Although refusal conversions in a single study may increase its response rate (Curtin et al., 2000; Brick et al., 2003b; Retzer et al., 2004), there are at least two reasons why refusal conversions may not be associated with response rates across studies. First, the time and financial resources devoted to refusal conversion attempts would have been devoted instead to other, equally successful strategies if the refusal conversions had not been attempted (i.e., simply more calling to other, nonrefusal households). Second, the decision to attempt refusal conversions may have been made partway through the field period of a study, and researchers may have been particularly likely to attempt refusal conversions when the response rate was low.

Making appointments and allowing respondent-initiated contact (for appointments and interviews) were not associated with increased response rates. Perhaps the people who took advantage of these opportunities would have been easy to contact and elicit cooperation from regardless.

Response rates were positively associated with demographic representativeness, but only very weakly. This conclusion is consistent with much past evidence showing that efforts to increase response rates may have minimal effects on demographic data quality (e.g., Keeter et al., 2000; Brick et al., 2003b; Frankovic, 2003). In general population RDD telephone surveys, lower response rates do not notably reduce the quality of survey demographic estimates. So devoting substantial effort and material resources to increasing response rates may have no measurable effect on the demographic accuracy of a survey sample.

23.6.2 Limitations and Future Research

This research has a number of limitations. First, we have examined correlations of response rates and other rates with administration procedures. Because the procedures that we examined were not manipulated experimentally and because many of them were correlated with one another in implementation, we cannot be certain that we have effectively disentangled their impact on response rates.

Second, our findings regarding the weak associations between response rates and demographic representativeness apply only to the specific range of response rates we examined and should not be generalized beyond that range. Furthermore, although the response rates we examined varied from 4 percent to 70 percent, 77 percent of the surveys in our sample had response rates between 20 percent and 50 percent, and our findings seem most likely to apply to this range of surveys. Our findings might not generalize to improvements in response rates above 50 percent or to declines in response rates below 20 percent. Our findings should be generalized to studies other than general population RDD surveys with caution as well.

23.7 CONCLUSION

Response rates in RDD surveys continue to decrease over time, and lower response rates do decrease demographic representativeness within the range we examined, though not much. This evidence challenges the assumptions that response rates are a key indicator of survey data quality and that efforts to increase response rates will necessarily be worth the expense.

CHAPTER 24

Response Rates: How Have They Changed and Where Are They Headed?[1]

Michael P. Battaglia
Abt Associates Inc.

Meena Khare
Centers for Disease Control and Prevention, National Center for Health Statistics

Martin R. Frankel
Baruch College, City University of New York and Abt Associates Inc.

Mary Cay Murray
Abt Associates Inc.

Paul Buckley
buckley.sphor, llc

Saralyn Peritz
Abt Associates Inc.

[1]The findings and conclusions in this chapter are those of the authors and do not necessarily represent the views of the Centers for Disease Control and Prevention.

Advances in Telephone Survey Methodology, Edited by James M. Lepkowski, Clyde Tucker, J. Michael Brick, Edith de Leeuw, Lilli Japec, Paul J. Lavrakas, Michael W. Link, and Roberta L. Sangster

24.1 INTRODUCTION

Response rates are generally considered an important indicator of survey quality. It is often assumed that the higher the response rate, the more likely the survey reflects the characteristics of the target population, thereby giving analysts and consumers of the data greater confidence in estimates and analyses based on the survey data (Groves and Lyberg, 1988). The Office of Management and Budget (2006) emphasizes the importance of high response rates for surveys collecting "influential" information. Considerable research has been carried out on issues related to survey quality, including the topic of nonresponse (Groves, 1989, and Groves and Couper, 1998). Many other issues contribute to survey quality, for example, measurement error as impacted by questionnaire design and mode of data collection (Biemer, 2001).

The discussion in this chapter focuses on unit nonresponse as it has been experienced in random digit dialing (RDD) telephone surveys. Unit nonresponse occurs when it is not possible to determine if a sample telephone number is a residential number or when an interview is not completed with a residential sample telephone number. Massey et al. (1997) previously found little consistency in how response rates were calculated in government-sponsored telephone surveys. In an effort to standardize reporting of response rates, the American Association for Public Opinion Research (2004) released guidelines for calculating response rates.

For some time, response rates in household surveys, whether they are conducted in person or by telephone, have been declining, and the effort associated with collecting high-quality data has been increasing (Brick et al., 2003b). O'Rourke and Johnson (1999) reported a range of 22.2 percent to 70.1 percent in the response rate for 35 RDD surveys in a 1998 study on an inquiry of RDD survey response rates. Recently, Curtin et al. (2005) showed that the overall response rate in the Survey of Consumer Attitudes (SCA) declined considerably from 1997 to 2003 at the average annual rate of 1.5 percentage points to 48.0 percent. The National Household Education Survey (2004) reported a decline in the response rate from 72.5 percent in 1999 to 62.4 percent in 2003, an annual rate of decline of 2.5 percentage points. The Behavioral Risk Factor Surveillance System (2003) reported a decline in the median response rate for the 50 states from 68.4 percent in 1995 to 53.2 percent in 2003 (an average decline of 1.9 percentage points per year). The RDD component of the National Survey of America's Families (2003) reported a decline in the overall response rate from 65.1 percent in 1997 to 62.4 percent in 1999 and to 55.1 percent in 2002 among the surveys of children, and 61.8 percent, 59.4 percent, and 51.9 percent among the adult surveys in 1997, 1999, and 2002, respectively. Atrostic et al. (2001) reported on six in-person surveys conducted by the U.S. Census Bureau. All six showed a decrease in the response rate between 1990 and 1999. Refusal rates also increased in all the six surveys across this time frame.

24.2 THE NATIONAL IMMUNIZATION SURVEY

In 1992 the Childhood Immunization Initiative (CII) (CDC, 1994) was established to (1) improve the delivery of vaccines to children; (2) reduce the cost of vaccines for parents; (3) enhance awareness, partnerships, and community participation; (4) improve vaccinations and their use; and (5) monitor vaccination coverage and occurrences of disease. To fulfill the CII mandate of monitoring vaccination coverage, the National Immunization Survey (NIS) was implemented by the National Immunization Program (NIP) and the National Center for Health Statistics (NCHS) of the Centers for Disease Control and Prevention (CDC). Beginning with the second quarter of 1994, the NIS data collection effort has conducted independent quarterly surveys in each of the 78 Immunization Action Plan (IAP) areas consisting of 28 urban areas including the District of Columbia, 31 states, and 19 rest of state (i.e., balance of state) areas (Smith et al., 2001a). The data collection for the NIS (1994–2004) was conducted by Abt Associates Inc.

The target population for the NIS is children aged from 19 to 35 months living in households at the time of the interview. Samples of telephone numbers were drawn independently, for each calendar quarter, within 78 Immunization Action Plan (IAP) areas. Also, these 78 independent surveys shared a common sample design that employed list-assisted RDD (Lepkowski, 1988; Tucker et al., 1993). This method selects a random sample of telephone numbers from "banks" of 100 consecutive telephone numbers (e.g., 617-495-0000 to 617-495-0099) that contain one or more directory-listed residential telephone numbers. The sampling frame of telephone numbers was updated each quarter to reflect new telephone exchanges and area codes.

To meet the IAP area target of completed interviews for age-eligible children per quarter, a quarterly sample of telephone numbers was drawn for each IAP area and divided into random subsamples called replicates (Buckley et al., 1998). Before a replicate was loaded into the computer-assisted telephone interviewing (CATI) system, several processes removed business and nonworking numbers by matching sampled numbers with databases of listed residential and business telephone numbers and then processed the unmatched sample numbers through an autodialer system, which removed a portion of the nonworking numbers (Battaglia et al., 2005). The overall percentage of telephone numbers eliminated from the initial sample for 2004 was 42.8 percent. Numbers that were not identified as business or nonworking, along with those identified as residential, were then loaded into the CATI system. These telephone numbers were also matched with a commercial database of residential addresses, and advance letters were mailed for the telephone numbers with a match (Camburn et al., 1995).

The retained telephone numbers were called and screened to determine the existence of a household and the presence of age-eligible children. The screening interview was brief for households that did not contain any age-eligible children. When an eligible child was located, a household interview was conducted with the adult who is the most knowledgeable about the child's vaccination history. This adult was often the mother of the child. The child's vaccination provider was then contacted to obtain a vaccination history. The provider-reported vaccination histories were used to estimate vaccination rates (Smith et al., 2005). This chapter just focuses on the RDD component of the NIS.

Since its inception, the NIS sampling and data collection methodology has undergone a number of changes listed below. Several enhancements were implemented to improve the response rate. Some changes were made after examining trade-offs between survey costs and nonresponse bias. For example, reductions were made in the maximum number of call attempts to noncontact telephone numbers due to the relatively small number of interviews obtained and the costs associated with making those call attempts.

- Calling local telephone company business offices (TCBOs) to determine residential status of noncontact telephone numbers was discontinued after 1996.
- Used Genesys-ID from 1994 to 2000 and Genesys-ID*plus* starting in the fourth quarter of 2000 to eliminate business and nonworking numbers, and also to eliminate noncontact telephone numbers identified as nonworking numbers at the end of the quarter.
- Number of call attempts to noncontact telephone numbers was reduced from 24 to 15 at the end of 1999 and from 15 to 10 call attempts in the first quarter of 2002.
- Calls to answering machine cases was reduced from 24 to 15 in 2000.
- Changed reverse-address matching vendor with a more comprehensive database in the fourth quarter of 1999 to mail advance letters; match rate increased from 40 percent to 67 percent.
- As a follow-up to an NCHS IRB decision, number of refusal conversion attempts was reduced from two to one attempt in 2002.
- Test conducted in 2004 on using monetary incentives to improve participation among households with refusals or partial interviews.

24.3 DEFINITION OF RESPONSE RATES FOR RDD SAMPLES

CASRO (Frankel, 1983) and AAPOR (2004) have published guidelines for calculating response rates for RDD surveys. The formulas make assumptions about specific unknown rates and leave it to the survey statistician to specify those rates in the calculation of the response rate. Some RDD surveys sample one adult per household, while other RDD surveys first screen for households containing eligible individuals before sampling one eligible individual from the eligible sample households. The NIS falls in the latter category except that an interview is conducted for each age-eligible child. The CASRO task force on completion rates (Frankel, 1983) classified RDD surveys in which all or virtually all households are eligible as Type I (single stage, no screening). This would include surveys that sample one adult per household because almost all households contain one or more persons aged 18 years or older. Surveys that screen for household containing one or more eligible individuals were classified as Type II (single stage, screening required). For either type of survey, the response rate can be expressed by completed interviews divided by eligible reporting units.

The numerator for Type I or Type II surveys is the number of completed interviews. The definition of the denominator is more complicated. For a Type I RDD survey, all households in the sample are eligible. For a Type II survey, only eligible households are included in the denominator. For both Type I and Type II surveys, one must decide how to handle unresolved (undetermined) telephone numbers. These are sample numbers where it is never determined if the sample number is residential, business, or nonworking. The main category of unresolved telephone numbers is noncontact numbers—numbers that are called repeatedly and the telephone is never answered and no messaging device is ever encountered. For Type I and Type II RDD surveys, some of the unresolved numbers are eligible households and should be included in the denominator. For a Type II survey, one must also account for known households where the screening is not completed and the eligibility of the household is unknown. Some of these telephone numbers are eligible households and should be included in the denominator.

The CASRO guidelines for Type II surveys lead to a response rate commonly called the CASRO response rate. It can be expressed as the product of three component rates (resolution rate, screener completion rate, and interview completion rate):

$$\frac{\text{Resolved numbers}}{\text{Total sample}} \times \frac{\text{Screened households}}{\text{Known households}} \times \frac{\text{Completed interviews}}{\text{Eligible households}}$$

The CASRO response rate was used as the primary response rate in the NIS from 1994 to 2004 and is widely used in other RDD surveys. It can be shown to be equivalent to AAPOR response rate 3 applied to a Type II survey (AAPOR, 2004). To show its application to the NIS, it is first necessary to list the final sample disposition categories used in the NIS (see Table 24.1).

Table 24.1. Categories of Final Dispositions for Telephone Numbers in the NIS Household Survey, 1995–2004

Label	Final Disposition Category
Total	Total sample in all released replicates
F	Telephone numbers found to be nonworking or nonresidential through list-assisted and pre-CATI autodialing procedures
Total to CATI	All sampled telephone numbers initially loaded into the CATI system
D	Nonworking, out-of-scope
E	Nonresidential
G	Noncontact
I	Answering machine or service containing a message that does not identify whether the number is a residential or business number (working number, household status unknown)
S2	Potential household (e.g., hang-up during the introduction) but it is never determined if the number is actually residential
S1	Known household, screening for eligible child incomplete
J	Screened household, no eligible children
K	Screened household, eligible children
L	Completed household interview

For a given quarter or calendar year, the CASRO response rate was calculated for each individual IAP area and for the overall sample.[2] The number of completed interviews comes from a single category, L. Several categories, however, contribute to the denominator, the number of eligible households in the sample (Ezzati-Rice et al., 2000). Households screened as eligible (category K) form one component of the denominator. Categories G (noncontact), I (answering machine or service, household status unknown), S1 (known household, screening for eligible child incomplete), and S2 (potential household) may also contain eligible households, whose numbers should be estimated and included in the number of eligible reporting units. Inclusion of these categories involves three assumptions that underlie the CASRO response rate.

(1) The proportion of households among unresolved numbers (p) is equal to the proportion of households found among resolved numbers.
(2) The proportion of eligible households in the unresolved group (r) is the same as the proportion of eligible households found among households for which screening was completed.
(3) The proportion of eligible households among the known but unscreened households (q) is equal to the proportion of eligible households found among screened households.

Before applying these assumptions, the CASRO response rate in the NIS can be written as

$$\frac{L}{K + p \times r \times (G + I + S2) + q \times S1}$$

The three assumptions imply that

$$p = \frac{J + K + S1}{J + K \times S1 + D + E + F} \quad \text{and} \quad r = q = \frac{K}{J + K}$$

It means that only information from the sample is used to estimate p, r, and q. P is estimated from the observed household residential working number rate, and r and q are assumed to be equal and are estimated from the household eligibility rate among screened households in the sample. Under the three assumptions it can be shown that the CASRO response rate is equivalent to the product of the resolution rate, the screening completion rate, and the interview completion rate shown above. For 2004 the overall CASRO response rate was 73.2 percent. The resolution rate was 83.9 percent, the screener completion rate was 94.9 percent, and the interview completion rate was 92.0 percent. The estimated percentage of residential numbers among the unresolved numbers, p, was 31.6 percent. The observed household eligibility rate, r, was 3.29 percent.

[2] For the overall sample one can weight each sample telephone number by the reciprocal of the probability of selection of the telephone number, which varies by IAP area and for a given IAP area by a small amount from quarter to quarter, and calculate a weighted CASRO response rate. Several time periods were examined, and for the NIS the weighted overall response rate is within 0.2 percentage points of the unweighted response rate. Unweighted response rates are therefore reported in the discussion of response rates in the NIS.

The observed eligibility rate found among households screened in the NIS is lower than the eligibility rate among telephone households estimated from other sources (Shapiro et al., 1996). This difference suggests that the eligibility rate among unscreened households may be higher than the rate found among screened households. But the number of unscreened households is not large enough, relative to the number of screened households, to account for all the additional eligible children. Thus, the most likely explanation for the difference is that some screened households have eligible children but do not mention them, because the screening questions are not properly understood, indicating that no age-eligible children are present is an easy way to end the interview process, or the parent does not feel comfortable indicating that a young child is present in the household. Massey (1995), Ezzati-Rice et al. (2000), and Smith et al. (2005) suggest estimating the denominator of the response rate, the number of eligible households in the NIS sample, using sources other than the survey itself.

24.4 NIS RESPONSE RATES, 1995–2004

Although the CASRO response rate remained high in the NIS, it experienced a gradual decline over 10 years. Table 24.2 and Figs. 24.1–24.4 show trends in the NIS response rate and its components from 1995 to 2004. Figure 24.1 shows the CASRO response rate for each quarter and for each calendar year. For the first two years (1995 and 1996), the CASRO response rate (Row 11, Table 24.2) was 85.8 percent or higher.

From 1997 to 1999 it declined from 84.6 percent to 80.3 percent. Further decline occurred from 78.7 percent in 2000 to 73.1 percent in 2004 with the lowest response rate in 2003 at 69.8 percent. Overall from 1995 to 2004, the CASRO response rate declined by 17.3 percentage points, an average of 1.9 percentage points per year. To understand the decline in the response rate over time, it is necessary to examine the three components that comprise the overall CASRO response rate—the resolution rate, the screening completion rate, and the interview completion rate.

24.4.1 Resolution Rate

The resolution rate (Row 5, Table 24.2, and Fig. 24.2) has shown a substantial decline over time and is the primary contributor to the decline in the CASRO response rate. The resolution rate was higher than 94 percent in the first 2 years of the NIS. This high resolution rate was in part due to the practice of submitting the unresolved telephone numbers to the local telephone company business offices (TCBO) to determine their residential status (Shapiro et al., 1995, Frankel et al., 2003a). This practice was discontinued after 1996, because it ultimately resulted in few completed interviews. From 1997 to 1999 the resolution rate declined from 92.1 percent to 88.6 percent. The resolution rate held at 88.1 percent in 2000. This was most likely due to the introduction in the fourth quarter of 2000 of a new automated method, Genesys-ID*plus* (Battaglia et al., 1995b, Battaglia et al., 2005) of prescreening the RDD sample to remove a larger portion of the nonworking and business telephone

Table 24.2. Response Rates and Key Monitoring Statistics, National Immunization Survey, 1995–2004

Row	Key indicator	1995	1996	1997	1998	1999	2000	2001	2002	2003	2004
1	Total selected telephone numbers in released replicates	1,917,474	2,021,133	2,118,796	2,239,721	2,533,608	2,662,722	3,042,911	3,361,396	3,744,489	3,607,627
2	Phone numbers resolved before CATI	407,259	397,276	395,488	407,496	483,903	671,215	1,055,376	1,306,025	1,534,473	1,545,789
	by GENESYS-IDplus (Row 2/ Row 1)	(21.2%)	(19.7%)	(18.7%)	(18.2%)	(19.1%)	(25.2%)	(34.7%)	(38.9%)	(41.0%)	(42.8%)
3	Total telephone numbers called	1,510,215	1,623,857	1,723,308	1,832,225	2,049,705	1,991,507	1,987,535	2,055,371	2,210,016	2,061,838
4	Advance letters mailed	565,194	537,322	573,748	589,944	746,824	1,146,845	1,191,713	1,285,751	1,420,131	1,332,878
	(Row 4/ Row 3)	(37.4%)	(33.1%)	(33.3%)	(32.2%)	(36.4%)	(57.6%)	(60.0%)	(62.6%)	(64.3%)	(64.6%)
5	Resolved telephone Numbers[1] –	1,851,274	1,905,956	1,950,500	2,024,343	2,243,904	2,345,183	2,641,723	2,849,329	3,131,074	3,023,174
	Resolution Rate (Row 5/ Row 1)	(96.5%)	(94.3%)	(92.1%)	(90.4%)	(88.6%)	(88.1%)	(86.8%)	(84.8%)	(83.6%)	(83.8%)
6	Households Identified	885,069	929,066	943,834	945,122	1,009,543	1,014,714	1,054,561	1,056,429	1,070,351	959,422
	(Row 6/ Row 5)	(47.8%)	(48.7%)	(48.4%)	(46.7%)	(45.0%)	(43.3)	(39.9%)	(37.1%)	(34.2%)	(31.7%)
7	Households successfully screened for presence of age-eligible children – *Screening Completion Rate*	853,536	899,549	924,328	923,970	979,606	973,784	1,014,363	1,020,404	1,006,499	909,866
	(Row 7/ Row 6)	(96.4%)	(96.8%)	(97.9%)	(97.8%)	(97.0%)	(96.0%)	(96.2%)	(96.6%)	(94.0%)	(94.8%)

8	Households with no age-eligible	819,825	864,528	889,758	889,489	943,268	937,824	978,378	986,203	972,532	877,228
	children (Row 8/ Row 7)	(96.1%)	(96.1%)	(96.3%)	(96.3%)	(96.3%)	(96.3%)	(96.5%)	(96.6%)	(96.6%)	(96.4%)
9	Households with age-eligible children	33,711	35,021	34,570	34,481	36,338	35,960	35,985	34,201	33,967	32,638
	– *Eligibility Rate* (Row 9/ Row 7)	(3.9%)	(3.9%)	(3.7%)	(3.7%)	(3.7%)	(3.7%)	(3.5%)	(3.4%)	(3.4%)	(3.59%)
10	Households with age-eligible children with completed RDD interviews										
	– *Interview Completion Rate*	31,520	32,911	32,434	32,271	33,932	33,477	32,796	30,974	30,134	30,019
	(row 10/ row 9)	(93.5%)	(94.0%)	(93.8%)	(93.6%)	(93.4%)	(93.1%)	(91.1%)	(90.6%)	(88.7%)	(92.0%)
11	CASRO response rate[2] (Row 5 * Row 7 * Row 10)	87.1%	85.8%	84.6%	82.7%	80.3%	78.7%	76.1%	74.2%	69.8%	73.1%
12	Age-eligible children with completed RDD Interviews	31,997	33,305	32,742	32,511	34,442	34,087	33,437	31,693	30,930	30,987

[1] Includes telephone numbers resolved before CATI (Row 2).

[2] CASRO: Council of American Survey Research Organizations.

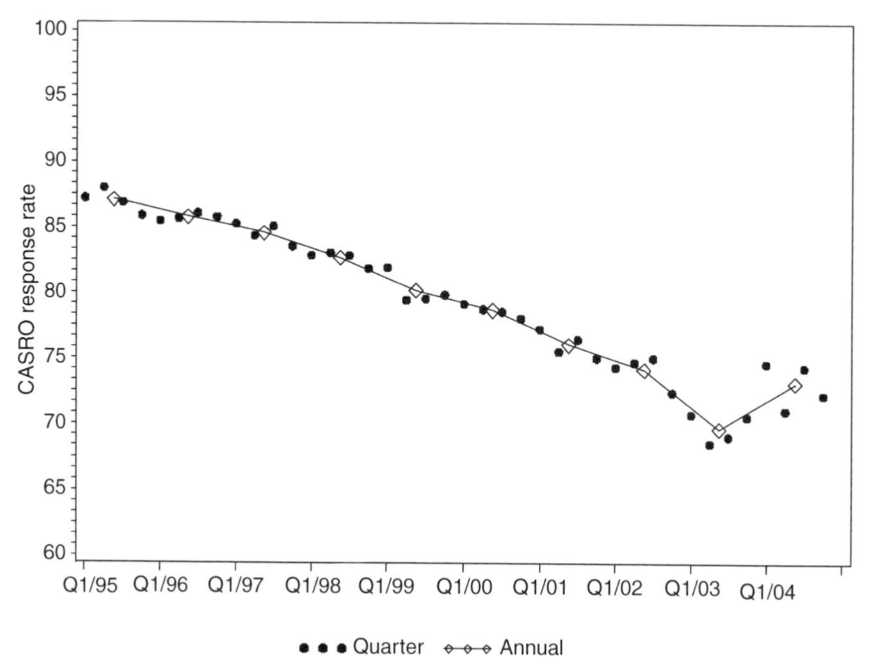

Figure 24.1. CASRO response rate by quarter and year, NIS.

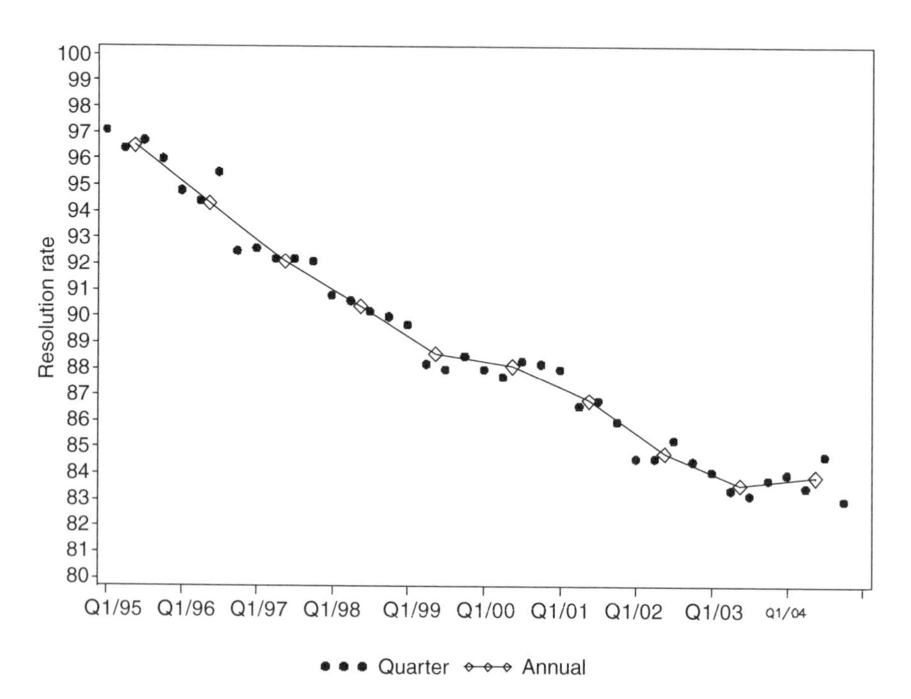

Figure 24.2. Resolution rate by quarter and year, NIS.

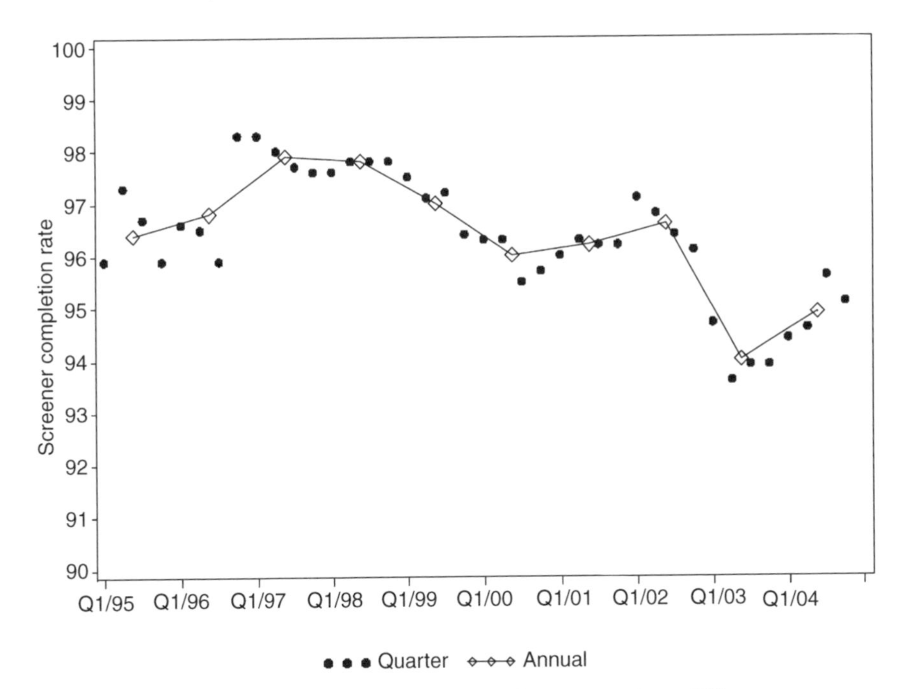

Figure 24.3. Screener completion rate by quarter and year, NIS.

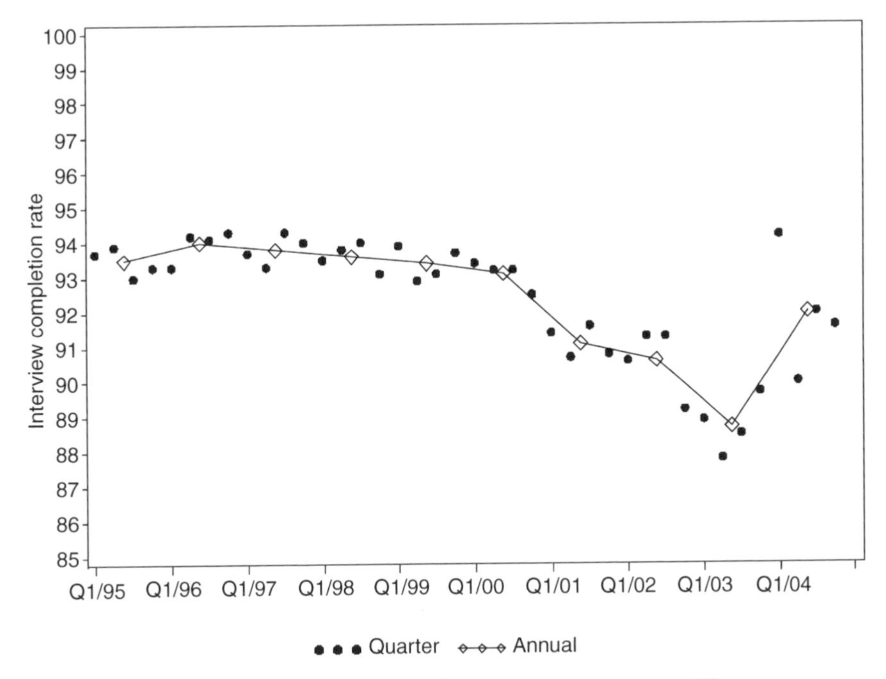

Figure 24.4. Interview completion rate by quarter and year, NIS.

Table 24.3. Trend in Types of Unresolved Telephone Numbers, National Immunization Survey, 1995–2004

Year of survey	1995	1996	1997	1998	1999	2000	2001	2002	2003	2004
Unresolved numbers	3.4	5.7	7.9	9.6	11.4	11.9	13.2	15.2	16.4	16.2
Ring-no-answer, or busy	1.4	3.2	5.1	5.5	6.9	7.3	7.3	8.0	8.5	8.3
Answering machines	0.6	0.8	1.1	1.1	1.3	1.4	1.5	2.5	2.5	3.1
Potential households	1.4	1.6	1.7	2.9	3.2	3.2	4.4	4.7	5.4	4.8

numbers. The prescreening process can essentially be viewed as a first call attempt that reduces survey costs but that most of the additional out-of-scope numbers purged by the new method would have eventually been identified as out-of-scope during the interviewer dialing of the sample. In 2001, the resolution rate further declined to 86.8 percent and held at around 84 percent from 2002 to 2004.

The decline in resolution rate means more sample telephone numbers end up as unresolved (e.g., ring-no-answer to all call attempts). Table 24.3 shows trends in unresolved telephone numbers by type. The percentage of telephone numbers unresolved increased from 3.4 percent in 1995 to 11.4 percent in 1999 and increased to 16.2 percent in 2004. This increase was primarily due to noncontacts after repeated calls. In 2004 the 16.2 percent unresolved sample consisted of 8.3 percent noncontacts, 3.1 percent answering machines and other voice-mail devices, and 4.8 percent potential households with unknown residential status (e.g., HUDIs). The noncontact component of the sample increased from 1.4 percent in 1995 to 6.9 percent in 1999 and to 8.3 percent in 2004. Some of the noncontact telephone numbers are actually unassigned numbers. Tucker et al. (2002) report that the number of telephone numbers in the United States increased from 435 million to 770 million from 1990 to 1999 and that this increase was much larger than the increase in residential telephone lines. The number of call attempts to noncontact numbers was reduced from 24 to 15 at the end of 1999 and again from 15 to 10 calls at the start of 2002. This contributed to a gradual decrease in the resolution rates from 2000 to 2003. Sangster (2002) found that making a large number of call attempts to noncontact telephone numbers can add to the cost of a survey while yielding only a small number of additionally resolved telephone numbers.

24.4.2 Screener Completion Rate

The NIS screening interview determined the presence of age-eligible children and is therefore a brief interview for most households. The screening interview represents the first contact with a household, and one might expect to see a decline in the screener completion rate over time. The screener completion rate (row 7,

Table 24.2, and Fig. 24.3), however, remained almost unchanged (96–98 percent) over the first 8 years of the NIS (1995-2002); it then declined slightly in 2003 to 94.0 percent but increased to 96.4 percent in 2004. The decline in 2003 was most likely due to an NCHS IRB decision to only make one refusal conversion attempt on refusal cases. The increase in 2004 was most likely caused by the use of a $15 cash incentive to households determined to contain a child under 3 years of age where it had not been determined if the child was in the 19–35 month eligible age range (Olson et al., 2004).

24.4.3 Interview Completion Rate

The decline in the interview completion rate in recent years is a secondary contributor to the decline in the CASRO response rate. The interview completion rate (Row 10, Table 24.2, and Fig. 24.4) declined by a small amount (0.9 percentage points) between 1996 and 2000; it further declined by an additional 2.5 percentage points between 2000 and 2002 and by an overall 5.3 percentage points from 1995 to 2003. The interview completion rate was 90.6 percent in 2002 but declined to 88.7 percent in 2003. The decline in 2003 was most likely due to an NCHS IRB decision to only make one refusal conversion attempt on refusal cases. The interview completion rate slightly increased to 92.0 percent in 2004. This increase in 2004 was almost certainly due to the use of a $15 incentive to interview refusals and partial interviews (Olson et al., 2004).

Respondents in 2 to 3 percent of known households refuse the NIS screener interview or never complete the age eligibility screening questions. Some respondents in age-eligible households who complete the screener interview refuse the household interview or complete only a part of the interview (break off/partial interview). The interviewing staff made every effort to reduce the number of refusals or partial-interview cases among eligible households. Figure 24.5 shows the trends in interview refusal and partial-interview cases among all age-eligible households. During 1995–1998, the interview refusal rate was around 5 percent; it then increased to nearly 6 percent by 2000 and continued increasing to 7 percent in 2001, almost 8 percent in 2002 and 9 percent in 2003 among the screened age-eligible households. This decision to only make one conversion attempt produced a slight increase in the refusal rate and probably accounts for much of the decline in the interview completion rate in 2003. In 2004 the NIS interview refusal rate declined to 5.5 percent. This was most likely due to a test of the use of incentives to interview refusals and partial interviews.

The NIS made multiple call attempts to complete partial interviews. The final dispositions of partial interview cases are (1) converted to a completed interview, (2) remained incomplete with a nonrefusal, and (3) remained incomplete with a refusal. The percentage of partial interviews, excluding the converted cases, among all age-eligible households was around 6 percent in 1995 and 1996 (Fig. 24.5). From 1997 to 2000, this rate increased to nearly 7 percent, and it continued increasing to nearly 9 percent in 2001 and 2002 and to 11 percent in 2003. The above-mentioned NCHS IRB decision also affected the number of conversion attempts to complete

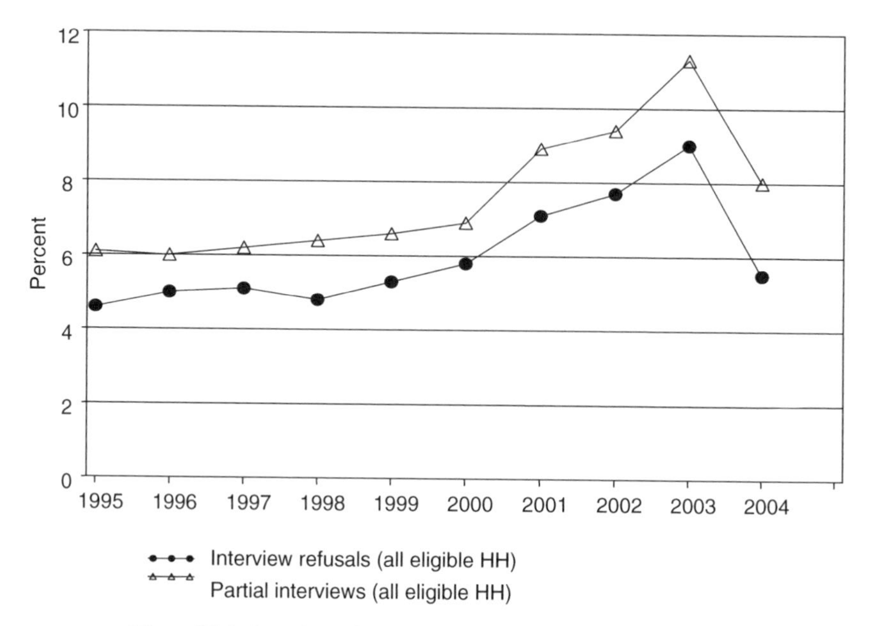

Figure 24.5. Interview refusal and partial interview rates by year, NIS.

partial interviews and resulted in a slight increase in the percentage of households with partial interviews that were not converted and probably accounts for much of the decline in the interview completion rate seen in 2003. In 2004 the use of incentives probably accounts for the reduction in the percentage of partial interviews to 8.0 percent.

The NIS sample design consists of 78 geographic strata representing IAP areas for which estimates of vaccination coverage are produced. Part of the analysis of response rates therefore also looked at trends in the CASRO response rates by IAP area. Three time periods are used—1995, 1999, and 2004. For all 78 IAP areas, the median CASRO response rate declined by 14.3 percentage points from 1995 to 2004 (87.9 to 73.6 percent). The decline from 1995 to 1999 and the decline from 1999 to 2004 are of the same magnitude (7.2 percentage points). In 2004 the IAP areas with the highest response rates were all rural—Montana, Nebraska, North Dakota, South Dakota, and Wyoming. The IAP areas with the lowest response rates were all urban, and all but one was on the east coast of the United States – Orleans Parish, LA, Maryland Rest of State, New Jersey Rest of State, Newark, NJ, and New York City.

The NIS sample was classified into the nine census divisions as defined by the U.S. Census Bureau. In 2004 the lowest CASRO response rates were seen in the Middle Atlantic, New England, and East South Central Census Divisions—all covering the eastern United States (see Table 24.4). The largest decline from 1995 to 2004 also occurred in those three census divisions. Only in the East South Central Census Division there was the difference in the response rates from 1999 to 2004, noticeably

Table 24.4. CASRO Response Rate and Change in CASRO Response Rate for the Nine Census Divisions and Four Categories of the Percent of Households Located in Metropolitan Statistical Areas, National Immunization Survey 1995, 1999, and 2004

	1995	1999	2004	1999–1995	2004–1999	2004–1995
Census division						
New England	87.9%	78.6%	70.6%	−9.3%	−8.0%	−17.3%
Middle Atlantic	84.5	75.9	68.1	−8.6	−7.8	−16.4
East North Central	89.0	80.7	74.4	−8.3	−6.3	−14.6
West North Central	90.6	86.1	78.2	−4.5	−7.9	−12.4
South Atlantic	86.1	78.3	70.3	−7.8	−8.0	−15.8
East South Central	88.3	82.5	72.2	−5.8	−10.3	−16.1
West South Central	86.4	81.2	71.9	−5.2	−9.3	−14.5
Mountain	88.6	82.8	76.8	−5.8	−6.0	−11.8
Pacific	84.2	78.6	76.3	−5.6	−2.3	−7.9
Percent of households in IAP area located in a metropolitan statistical area in 2000						
0.0–49.9%	90.8	84.7	80.5	−6.1	−4.2	−10.3
50.0–74.9%	88.9	82.7	75.3	−6.2	−7.4	−13.6
75.0–89.9	85.9	81.7	74.1	−4.2	−7.6	−11.8
90.0–100.0	86.7	78.2	70.7	−8.5	−7.5	−16.0

larger than for 1995 to 1999. The IAP areas were also divided into four categories based on the percentage of households in metropolitan statistical areas based on 2000 census data (see Table 24.4). For all 3 years there is a monotonic relationship between the percentage of households located in an MSA and the response rate—the higher the percent in MSAs the lower the response rate. For IAP areas with 90 to 100 percent of their households in an MSA, the decline in the response rate from 1995 to 2004 was largest—16.0 percentage points.

Given the decline in the CASRO response rate over time, two methods were used to assess nonresponse bias in the NIS (1) assessing nonresponse bias using ecological variables and (2) assessing nonresponse bias using level of effort.

24.5 ASSESSING NONRESPONSE BIAS USING ECOLOGICAL VARIABLES

Logistic regression models based on 2001–2002 NIS sample data were used to predict (1) resolved number versus unresolved number, (2) known household—screener completed versus not completed, and (3) interview completed versus not completed. For the first two levels of nonresponse, separate logistic regression models were fit within each IAP area for directory-listed sample telephone numbers versus sample numbers that are not directory listed. For the third level of nonresponse—interview completed versus not completed—the stepwise logistic models were developed for each of the nine census divisions by residential-directory-listed status of the telephone number because the sample size was too small to develop separate IAP area models.

No strong theoretical basis existed for including specific ecological exchange variables in the models. It was decided to give consideration to virtually all of the exchange variables that Genesys Sampling Systems (Marketing Systems Group, 1993) makes available. These included variables related to the age, race/ethnicity, and education distribution of the population of the exchange. Other variables related to housing/household characteristics such as household income, owned versus rented units, average rent, and median home value. Finally, categorical variables related to the population size of the county and the degree of urbanicity were also used.

Forward stepwise logistic regression with an inclusion criterion of variables entering at the 0.10 significance level was used. To determine the final model, a stopping rule used the Schwarz criterion (Schwarz, 1978). If, however, the final model contained only one predictor variable and a predictor variable entered the model at the second step, it was also included. The area under the receiver operating characteristic (ROC) curve provides a measure of how well a model discriminates between response versus nonresponse. Using the general rule provided by Hosmer and Lemeshow (2000), the ecological predictors in the models examined can be characterized as having weak (generally the resolution models) to acceptable (generally the screener completion and interview completion models) discrimination.

Seventy-six percent of the 78 resolution rate models for residential-directory-listed numbers contained an age distribution of the population variable. Forty-five percent of the models included a household income distribution variable, 33 percent included the median home value variable, 29 percent included a variable related to the minority population distribution, 19 percent included mean rent, 17 percent included a variable related to the population size of the county, 13 percent included the percent of households that are owner-occupied, 10 percent included the percent of the adult population that are college graduates, and 10 percent included median years of education of the adult population. For telephone numbers that are not residential-directory-listed, a somewhat similar set of results were observed: 82 percent of the models included age distribution of the population, 62 percent included household income distribution, 58 percent included minority population distribution, 45 percent included median home value, 45 percent included percent owner occupied, 27 percent included percent college graduate, 23 percent included median years of education, 21 percent included MSA status, and 15 percent included mean rent.

The 78 models for screener completion among residential-directory-listed numbers included fewer predictors than the corresponding models for resolved numbers. Seventy-one percent of the models included an age distribution variable, 36 percent included a household income distribution variable, 29 percent included a minority distribution variable, and 12 percent included an MSA status variable. Variables such as percent college graduate, county size, average rent, median home value, and median years of education entered 10 percent or less of the 78 models. For numbers that are not residential-directory-listed, the age distribution variable was included in 59 percent of the 78 models. Similar to the directory-listed models, the household income distribution variable entered 33 percent of the models, and the minority population distribution variable entered 27 percent of the models. Variables like percent owner occupied, median household income, median years of education, median home value, average rent, percent college graduate, MSA status, and county size entered 10 percent or fewer of the 78 models. The largest difference between the screener-completed models and the resolved models was the inclusion of median home value and/or average rent in a substantial percentage of the resolved models.

Finally, the interview-completed numbers had nine census division models for residential-directory-listed numbers. The minority population distribution variable entered five of the nine models. The age distribution of the population variable entered three of the nine models, as did the household income distribution variable, and MSA status was included in two models. Variables like median household income, median home value, percent owner occupied, average rent, and county size each entered only one model. For numbers that are not residential-directory-listed, age distribution of the population entered six of the nine models. Household income distribution entered four models; minority population distribution entered two models, as did average rent. MSA status, median home value, percent owner occupied, and median years of education each entered only one model. The most consistent pattern among the three levels of nonresponse was the importance of the age distribution of the population variable.

The models predicting screener response represents the initial contact with a household in an attempt to determine if the household contains any age-eligible children. A closer focus on the results for those 156 (78 IAP areas by residential-directory-listed status) logistic regression models points to some patterns related to the probability of completing the screening to determine age eligibility (see Table 24.5). For 29 percent of the models, the higher the percentage of the population aged 0–17 years in the telephone exchange, the lower the probability of completing the screener. Recall that the NIS screens for households with young children, and this finding may account for some of the difference between the expected and observed household eligibility rates. Also, for 12 percent of the models, the higher the percentage of the population aged 45–54 or 55–64 years in the telephone exchange, the higher the probability of completing the screener. Households with persons in those age ranges may be less likely to have young children; the screening interview is brief for sample household without any age-eligible children.

For 13 percent of the models, the higher the percent black population or the higher the percent Hispanic population in the telephone exchange, the lower the probability of the screener being completed. It was also the case that the higher the percent Asian population in the telephone exchange, the higher the probability of completing the screener. For 10 percent of the models, the higher the average rent or the higher the median home value in the telephone exchange, the lower the probability of completing the screener. Also, for 11 percent of the models, predictor variables related to metropolitan statistical area or size of county almost always indicated that the probability of completing the screener is lower in MSA central city or suburban telephone exchanges compared to non-MSA telephone exchanges and in larger counties compared to smaller counties.

Table 24.5. Ecological Predictors of Screener Nonresponse for 156 Logistic Regression Models (78 IAP Areas by Residential-Directory-Listed Status of the Telephone Number)

IAP Area	Predictor #1	Predictor #2	Predictor #3	Predictor #4
		Directory-Listed Numbers		
Connecticut	Percent age 18–24 years			
MA-Rest of State	Percent age 0–17 years	Percent population non-Hispanic Asian		
MA-City of Boston	Percent age 0–17 years	Center city of MSA	Center city county of MSA	
Maine	Percent HHs with income $50,000–74,999	Percent population Hispanic		
New Hampshire	Percent population non-Hispanic Asian	Percent age 35–44 years		
Rhode Island	Percent age 0–17 years	Percent HHs with income $75,000–99,999		
Vermont	Percent age 35–44 years			
NJ-Rest of State	Percent age 0–17 years	Percent age 55–64 years		
NJ-City of Newark	Percent college graduate for population age 25+	Percent HHs with income $25,000–34,999		
NY-Rest of State	County is in largest 21 MSAs	County is in MSA with > 85,000 HHs	County with > 20,000 HHs or in an MSA with > 20,000 HHs	Percent age 25–34 years
NY-NYC	Percent age 0–17 years	Percent population Hispanic		
District of Columbia	Percent age 35–44 years	Percent HHs with income $25,000–34,999		
Delaware	Percent age 0–17 years	Percent age 35–44 years		
MD-Rest of State	Percent population non-Hispanic black	Percent age 35–44 years		
MD-Baltimore City	Percent HHs with income $25,000–34,999	Percent HHs with income $35,000–49,999		

PA-Rest of State	Percent HHs with income < $10,000	Mean rent	
PA-Philadelphia	Percent age 0–17 years	Percent population Hispanic	
Virginia	Percent age 0–17 years	Percent population non-Hispanic Asian	
West Virginia	Percent population Hispanic	Percent age 35–44 years	
AL-Rest of State	Percent age 55–64 years	Percent HHs with income $25,000–34,999	
AL-Jefferson County	Percent age 0–17 years	Percent population non-Hispanic Asian	
FL-Rest of State	Percent age 25–34 years	Percent age 45–54 years	
FL-Duval County	Percent population non-Hispanic Asian	Percent HHs owner occupied	
FL-Miami-Dade County	Percent population non-Hispanic Asian	Percent HHs owner occupied	
GA-Rest of State	Percent HHs with income $10,000–14,999	Percent age 35–44 years	
GA-Fulton/DeKalb Counties	Percent age 0–17 years	Median home value	Percent HHs with income $35,000–49,999
Kentucky	Percent age 0–17 years	Percent HHs with income < $10,000	
Mississippi	Percent population non-Hispanic black	Mean rent	
North Carolina	Percent population non-Hispanic Asian	Percent age 55–64 years	

(*continued*)

Table 24.5. *(Continued)*

IAP Area	Predictor #1	Predictor #2	Predictor #3	Predictor #4
South Carolina	Percent age 0–17 years	Percent population Hispanic		
TN-Rest of State	Percent population Hispanic	Percent age 55–64 years		
TN-Shelby County	Percent HHs owner occupied	Percent age 25–34 years		
TN-Davidson County	Percent age 0–17 years	Percent HHs with income $35,000–49,999		
IL-Rest of State	Percent HHs with income $50,000–74,999	County is in largest 21 MSAs	County is in MSA with > 85,000 HHs	County with > 20,000 HHs or in an MSA with > 20,000 HHs
IL-City Chicago	Center city of MSA	Median years education for population age 25+		
IN-Rest of State	Center city of MSA	Center city county of MSA	Percent age 25–34 years	Other county of MSA
IN-Marion County	Percent age 0–17 years	Percent age 25–34 years		
MI-Rest of State	Percent population non-Hispanic black	Percent age 25–34 years		
MI-Detroit	Percent age 35–44 years	Percent age 55–64 years		
Minnesota	Percent age 18–24 years	Percent age 0–17 years		
OH-Rest of State	Percent population non-Hispanic Asian			
OH-Cuyahoga County	Percent HHs with income < $10,000			
OH-Franklin County	Percent age 0–17 years	Center city of MSA		
WI-Rest of State	Percent HHs with income $50,000–74,999	Percent HHs with income $35,000–49,999		
WI-Milwaukee County	Percent population non-Hispanic black	Percent age 35–44 years		

Arkansas	Percent age 0–17 years	Percent HHs owner occupied		
LA-Rest of State	Percent age 55–64 years	County is in MSA with > 85,000 HHs	County with > 20,000 HHs or in an MSA with > 20,000 HHs	
LA-Orleans Parish	Percent age 0–17 years	Percent HHs with income $35,000–49,999		
New Mexico	Percent age 0–17 years	Median home value		
Oklahoma	Percent college graduate for population age 25+	Mean rent		
TX-Rest of State	Percent age 0–17 years	Percent HHs with income $15,000–24,999		
TX-Dallas County	Center city of MSA	Center city county of MSA	Percent age 55–64 years	
TX-El Paso County	Percent HHs owner occupied	Mean rent		
TX-City of Houston	Percent age 0–17 years	Center city of MSA	Center city county of MSA	
TX-Bexar County	Percent age 18–24 years	Center city of MSA		
Iowa	Percent age 35–44 years	Percent age 55–64 years		
Kansas	Percent HHs with income < $10,000	Percent HHs with income $15,000–24,999		
Missouri	Percent age 55–64 years	County is in largest 21 MSAs	County is in MSA with > 85,000 HHs	County with > 20,000 HHs or in an MSA with > 20,000 HHs
Nebraska	Percent population non-Hispanic black	Percent HHs with income < $10,000		
Colorado	Mean rent	Percent age 45–54 years		
Montana	Percent age 55–64 years	Percent HHs with income $25,000–34,999		

(*continued*)

Table 24.5. (*Continued*)

IAP Area	Predictor #1	Predictor #2	Predictor #3	Predictor #4
North Dakota	Center city of MSA	Center city county of MSA	Percent age 25–34 years	Other county of MSA
South Dakota	Percent age 55–64 years	Percent HHs with income $25,000–34,999		
Utah	Percent HHs with income $50,000–74,999	Percent age 55–64 years		
Wyoming	Percent population non-Hispanic black	Percent HHs with income $50,000–74,999		
AZ-Rest of State	Percent age 0–17 years	Center city of MSA	Center city county of MSA	Other county of MSA
AZ-Maricopa County	Percent age 18–24 years	Percent age 0–17 years		
CA-Rest of State	Percent age 0–17 years	Mean rent		
CA-Los Angeles County	Percent age 0–17 years	Percent population non-Hispanic black		
CA-Santa Clara	Median years education for population age 25+	Mean rent		
CA-San Diego County	Percent age 55–64 years	Percent HHs with income $75,000–99,999		
Hawaii	Percent population Hispanic	Percent age 25–34 years		
Nevada	Percent age 18–24 years	County is in MSA with > 85,000 HHs		
Alaska	Percent HHs with income $25,000–34,999	Median years education for population age 25+		
Idaho	Percent HHs with income $10,000–14,999	Percent population Hispanic		
Oregon	Percent HHs with income $50,000–74,999	Percent age 55–64 years		

WA-Rest of State	Percent college graduate for population age 25+	Percent HHs with income \$25,000–34,999	
WA-King County	Percent age 0–17 years	Percent HHs with income < \$10,000	
		Nondirectory-Listed Numbers	
Connecticut	Percent age 18–24 years		
MA-Rest of State	Percent HHs with income \$10,000–14,999		
MA-City of Boston	Percent population non-Hispanic black	Percent HHs owner occupied	
Maine	Median HH income		
New Hampshire	Percent age 35–44 years	Percent age 55–64 years	
Rhode Island	Percent population Hispanic		
Vermont	Percent age 35–44 years	Percent age 55–64 years	
NJ-Rest of State	Percent age 0–17 years		
NJ-City of Newark	Percent age 18–24 years	Median years education for population age 25+	
NY-Rest of State	Percent population non-Hispanic black	Median home value	
NY-NYC	Percent population non-Hispanic black	Percent age 25–34 years	
District of Columbia	Percent age 25–34 years	Percent HHs with income < \$10,000	
Delaware	Percent age 18–24 years	Center city of MSA	Center city county of MSA
MD-Rest of State	Percent college graduate for population age 25+	Mean rent	

(continued)

Table 24.5. ***(Continued)***

IAP Area	Predictor #1	Predictor #2	Predictor #3	Predictor #4
MD-Baltimore City	Percent age 0–17 years	Percent HHs with income $35,000–49,999		
PA-Rest of State	County is in largest 21 MSAs	County is in MSA with > 85,000 HHs	County with > 20,000 HHs or in an MSA with > 20,000 HHs	Percent age 45–54 years
PA-Philadelphia	Percent population Hispanic	Percent college graduate for population age 25+		
Virginia	Percent age 0–17 years	Percent HHs with income $15,000–24,999		
West Virginia	Median HH income	Percent age 35–44 years		
AL-Rest of State	Percent population non-Hispanic black	County is in MSA with > 85,000 HHs	County with > 20,000 HHs or in an MSA with > 20,000 HHs	
AL-Jefferson County	Percent population non-Hispanic black	Percent HHs with income $50,000–74,999		
FL-Rest of State	Percent age 0–17 years	Median home value		
FL-Duval County	Percent HHs owner occupied	Percent HHs with income $35,000–49,999		
FL-Miami-Dade County	Percent population non-Hispanic black	Percent HHs owner occupied		
GA-Rest of State	Percent HHs with income $50,000–74,999			
GA-Fulton/DeKalb Counties	Percent age 0–17 years	Percent HHs with income $50,000–74,999		
Kentucky	Percent age 35–44 years			
Mississippi	Percent age 18–24 years	Percent HHs with income $25,000–34,999		

North Carolina	Percent population non-Hispanic Asian	Percent HHs with income $25,000–34,999	
South Carolina	County is in MSA with > 85,000 HHs	County with > 20,000 HHs or in an MSA with > 20,000 HHs	Percent HHs with income $35,000–49,999
TN-Rest of State	Percent HHs with income $15,000–24,999		
TN-Shelby County	Percent age 0–17 years	Percent age 35–44 years	
TN-Davidson County	Percent age 0–17 years	Percent HHs with income $50,000–74,999	
IL-Rest of State	Percent age 0–17 years	Percent population non-Hispanic black	
IL-City Chicago	Percent age 35–44 years	Median home value	
IN-Rest of State	Percent age 0–17 years	Percent HHs with income $75,000–99,999	
IN-Marion County	Percent population non-Hispanic Asian	Percent HHs with income $35,000–49,999	
MI-Rest of State	Percent age 0–17 years	Percent age 25–34 years	
MI-Detroit	Percent age 18–24 years		
Minnesota	Percent age 0–17 years	Percent HHs with income $50,000–74,999	
OH-Rest of State	Percent age 35–44 years	Percent HHs with income $75,000–99,999	
OH-Cuyahoga County	Percent population non-Hispanic Asian	Percent college graduate for population age 25+	
OH-Franklin County	Percent population non-Hispanic Asian	Mean rent	
WI-Rest of State	Percent HHs with income $75,000–99,999	Percent HHs with income $15,000–24,999	

(*continued*)

Table 24.5. ***(Continued)***

IAP Area	Predictor #1	Predictor #2	Predictor #3	Predictor #4
WI-Milwaukee County	Percent population non-Hispanic Asian			
Arkansas	Percent age 18–24 years	Percent age 35–44 years		
LA-Rest of State	Percent population non-Hispanic Asian	Percent age 25–34 years		
LA-Orleans Parish	Percent HHs with income < $10,000	Percent HHs with income $35,000–49,999		
New Mexico	Percent age 0–17 years			
Oklahoma	Percent age 0–17 years	County is in MSA with > 85,000 HHs	County with > 20,000 HHs or in an MSA with > 20,000 HHs	
TX-Rest of State	Percent age 0–17 years			
TX-Dallas County	Percent population non-Hispanic Asian	Percent HHs owner occupied		
TX-El Paso County	Percent age 35–44 years			
TX-City of Houston	Percent age 0–17 years	Percent population non-Hispanic Asian		
TX-Bexar County	Percent HHs owner occupied	Percent age 55–64 years		
Iowa	Percent population Hispanic			
Kansas	Percent population non-Hispanic black			
Missouri	Mean rent			
Nebraska	Percent age 0–17 years	Percent HHs with income $25,000–34,999		
Colorado	Percent age 0–17 years	Percent age 35–44 years		

Montana	Percent college graduate for population age 25+			
North Dakota	Percent age 0–17 years	Percent age 35–44 years		
South Dakota	Percent age 55–64 years	Mean rent		
Utah	Median years education for population age 25+	Percent age 25–34 years		
Wyoming	Percent HHs with income \$10,000–14,999	Percent age 45–54 years		
AZ-Rest of State	Percent population Hispanic	Percent HHs with income \$25,000–34,999		
AZ-Maricopa County	Percent HHs owner occupied	Percent population Hispanic		
CA-Rest of State	Percent age 55–64 years	Mean rent		
CA-Los Angeles County	Percent age 0–17 years	Percent HHs with income \$10,000–14,999		
CA-Santa Clara	Percent age 0–17 years	Percent age 55–64 years		
CA-San Diego County	Percent HHs with income \$10,000–14,999	Mean rent		
Hawaii	Percent age 0–17 years	Percent age 25–34 years		
Nevada	Percent HHs with income \$10,000–14,999	Percent population non-Hispanic black		
Alaska	Percent age 18–24 years	Percent age 25–34 years		
Idaho	Center city of MSA	Center city county of MSA	Percent age 45–54 years	Other county of MSA
Oregon	Percent population non-Hispanic black	Percent age 25–34 years		
WA-Rest of State	Mean rent	Percent HHs with income \$35,000–49,999		
WA-King County	Center city of MSA	Percent age 35–44 years		

24.6 ASSESSING NONRESPONSE BIAS USING LEVEL OF EFFORT

The usual formula for the amount of nonresponse bias in the sample mean, $\bar{y}_1$, is $E(\bar{y}_1)-\bar{Y} = W_2(\bar{Y}_1-\bar{Y}_2)$, where W_2 is the proportion of nonresponse in the population and $\bar{Y}_2$ is the unknown population mean for the nonrespondent population. Lacking any information on the value of $\bar{Y}_2$, one can say that as the response rate declines, the potential for nonresponse bias increases; however, the magnitude of that bias is unknown. The Office of Management and Budget (2006) has provided guidance on the design of nonresponse bias studies for statistical surveys with response rates below 80 percent. Recent studies that have examined the relationship between level of effort (e.g., number of call attempts) and nonresponse bias have found that RDD surveys with low response rates may not be subject to substantial nonresponse bias with respect to means and proportions (Keeter et al., 2000). One limitation of this approach to assessing the magnitude of nonresponse bias is that the basis of comparison is a survey that itself does not have a high response rate (e.g., 60 percent). This leaves open the possibility that substantial nonresponse bias exists as one moves from a 60 percent response rate to 100 percent, but that bias does not appear as one moves from a 40 percent response rate to a 60 percent response rate.

In 1999 the NIS achieved a high CASRO response rate of 80.3 percent. This response rate was achieved by making up to 24 call attempts to noncontact telephone numbers and a large number of additional call attempts to known households in an effort to complete the eligibility screening and the interview, if the household was eligible. Refusal conversion procedures were also extensively used. Using the record of call data files for the four quarters in 1999, the maximum number of call attempts for all sample telephone numbers was limited to 1, 5, 10, 15, and 20 in order to simulate lower response rates resulting from a lower level of calling effort in the NIS. In other words, a new final disposition was assigned to each sample telephone number based on the call attempt results that occurred up to and including the maximum number of call attempts being considered. These lower maximum call attempt levels resulted in CASRO response rates of 23.8 percent, 53.4 percent, 64.5 percent, 70.0 percent, and 72.7 percent.

The numbers of completed interviews for age-eligible children were 6,371, 21,484, 27,620, 30,216, and 31,424 compared to the full 1999 sample of 34,442 completed child interviews. For each maximum call attempt level, a final RDD weight was calculated using the weighting methodology employed for the full sample. This weighting methodology was carried out at the IAP area level and included separate unit nonresponse adjustments for unresolved telephone numbers, known households where the screener was not completed, and interview nonresponse among eligible households. This was followed by an interruption-in-telephone-service adjustment to compensate for the exclusion of children in nontelephone households (Frankel et al., 2003b), and poststratification to Vital Statistics control totals based on age group, maternal education, and maternal race/ethnicity (Hoaglin and Battaglia, 2006).

Sixteen variables from the RDD survey were identified for the nonresponse bias assessment. These variables are shown in Table 24.6. They include demographic variables not included in the poststratification and substantive variables related to

Table 24.6. Estimates of Nonresponse Bias by Maximum Number of Call Attempts

Variable	Estimates by maximum number of call attempts						Absolute value of estimated bias				
	1 (23.8 percent response rate)	5 (53.4 percent response rate)	10 (64.5 percent response rate)	15 (70.0 percent response rate)	20 (72.7 percent response rate)	Full sample (80.3 percent response rate)	1−Full sample	5−Full sample	10−Full sample	15−Full sample	20−Full sample
Vaccination record used during interview	54.34%	55.19%	52.96%	52.22%	51.60%	49.73%	4.61%	5.46%	3.23%	2.49%	1.87%
Child up-to-date on four key vaccines	42.24	41.13	40.28	39.99	39.77	39.13	3.11	2.00	1.15	0.86	0.64
Child up-to-date on varicella	53.69	54.57	54.44	54.59	54.58	54.51	0.82	0.06	0.07	0.08	0.07
Child up-to-date on hepatitis B	66.11	66.29	65.62	65.46	65.11	64.41	1.70	1.88	1.21	1.05	0.70
Maternal age less than or equal to 19 years	3.80	3.69	3.68	3.73	3.76	3.76	0.04	0.07	0.08	0.03	0.00
Located in central city of an MSA	34.99	36.81	36.46	36.14	36.07	36.06	1.07	0.75	0.40	0.08	0.01
Moved from a different state	9.52	9.88	9.66	9.51	9.51	10.01	0.49	0.13	0.35	0.50	0.50
Child ever participated in WIC program	53.89	54.46	54.21	54.09	53.98	54.01	0.12	0.45	0.20	0.08	0.03
Child currently participating in WIC program	29.07	29.34	29.01	28.68	28.43	28.05	1.02	1.29	0.96	0.63	0.38
Household income is above $75,000	12.03	11.46	11.80	11.86	11.94	11.70	0.33	0.24	0.10	0.16	0.24
Child is not the first born	59.47	59.97	59.89	60.07	60.01	59.77	0.30	0.20	0.12	0.30	0.24
Only one child in the household	26.32	26.97	27.10	27.25	27.39	28.06	1.74	1.09	0.96	0.81	0.67
Mother was the respondent to survey	86.76	86.46	86.34	86.31	86.14	85.82	0.94	0.64	0.52	0.49	0.32
Child has 1 vaccination provider	63.59	65.24	65.56	65.99	66.09	64.27	0.68	0.97	1.29	1.72	1.82
Child has provider-reported vaccination data	67.03	67.25	66.96	66.95	66.74	64.18	2.85	3.07	2.78	2.77	2.56
Mean difference							1.25	1.16	0.85	0.76	0.64

the parental report of the child's vaccination status, the number of vaccination providers, and whether the child has vaccination data reported by their provider(s). In Table 24.6 we show the estimates for the five maximum call attempt levels, the estimates based on the full sample, and the absolute value the difference between each estimate and the full sample estimate. For the percentage of respondents using a vaccination record using the interview and the percentage of respondents reporting that the child is up-to-date on their vaccinations, the absolute value of the estimated bias is large in magnitude at a maximum of one or five call attempts. The first five call attempts followed a set of optimal calling rules developed for the NIS (Dennis et al., 1999), with the first call attempt generally made during afternoon hours to eliminate business numbers. Therefore, the one-attempt bias estimates should be viewed with some caution.

The mean absolute value of the difference declines in a monotonic fashion as the response rate increases (based on a higher number of call attempts). Because the estimates for the 16 variables vary in magnitude, it is useful to examine the relative absolute value of the difference between each estimate and the full sample estimate. These results are presented in Table 24.7. At a 23.8 percent response rate, the mean relative bias is 3.3 percent. As the response rate increases, the mean relative bias decreases to 1.6 percent by the time a 72.7 percent response rate is reached. This represents a 51.5 percent decline in the mean relative bias. The results of simulating a lower CASRO response rate indicate that for response rates around 65 percent and higher, there does not appear to be any substantial bias. The mean relative bias is somewhat larger as the response rate declines to around 50 percent and grows considerably larger as the response rate approaches 20 percent.

24.7 CONCLUSIONS

The NIS achieved high response rates compared to other RDD survey over the 10-year period examined. This is partially due to the nature of the eligible target population, the brief screener interview for ineligible households, and the topic of the questionnaire for age-eligible households. The NIS also employed a large number of techniques to help attain a high response rate but at the same time made cost versus bias trade-offs in reducing efforts in areas that yielded few completed interviews with eligible households. This was accomplished using experiments embedded in the survey and by carrying out simulations using the NIS survey data (e.g., see Srinath et al., 2001). From 1995 to 2004 the CASRO response rate declined in a gradual and steady fashion. Most of the decline was caused by the decline in the resolution rate; it became more difficult to determine if a sample telephone number was residential. A secondary contributor in more recent years was a decline in the interview completion rate; it became somewhat more difficult to complete the interview with age-eligible households after 2000.

Two methods were used to assess nonresponse bias in the NIS. Focusing on the examination of ecological correlates of screener nonresponse, age of the population, and race/ethnicity of the population were often found to be statistically significant

Table 24.7. Estimates of Relative Bias by Maximum Number of Call Attempts

Variable	1 versus Full sample	5 versus Full sample	10 versus Full sample	15 versus Full sample	20 versus Full sample
Vaccination record used during interview	9.3%	11.0%	6.5%	5.0%	3.8%
Child up-to-date on four key vaccines	7.9	5.1	2.9	2.2	1.6
Child up-to-date on varicella	1.5	0.1	0.1	0.1	0.1
Child up-to-date on hepatitis B	2.6	2.9	1.9	1.6	1.1
Maternal age less than or equal to 19 years	1.1	1.9	2.1	0.8	0.0
Located in central city of an MSA	3.0	2.1	1.1	0.2	0.0
Moved from a different state	4.9	1.3	3.5	5.0	5.0
Mother is divorced or separated	1.9	2.2	2.7	1.2	1.2
Child ever participated in WIC program	0.2	0.8	0.4	0.1	0.1
Child currently participating in WIC program	3.6	4.6	3.4	2.2	1.4
Household income is above $75,000	2.8	2.1	0.9	1.4	2.1
Child is not the first born	0.5	0.3	0.2	0.5	0.4
Only one child in the household	6.2	3.9	3.4	2.9	2.4
Mother was the respondent to survey	1.1	0.7	0.6	0.6	0.4
Child has 1 vaccination provider	1.1	1.5	2.0	2.7	2.8
Child has provider-reported vaccination data	4.4	4.8	4.3	4.3	4.0
Mean relative bias	3.3	2.8	2.3	1.9	1.6

predictors. The level of effort nonresponse bias analysis found that bias appeared to be small in magnitude at response rates around 65 percent and above but increased in magnitude at response rates in the 25 percent to 50 percent range.

Over the past 10 years, there has been a growth of impediments to the continued viability of RDD surveys (Frankel, 2004). Key factors that have lead to difficulties in conducting RDD surveys include telemarketing, the large volume of telephone surveys, more time spent outside the household, and the use of answering machine and caller ID devices. In the coming years telephone surveys based on RDD methods will certainly change. Here are some of the factors that may influence the nature of telephone surveys.

- *Federal do not call list*—The FTC do not call list has resulted in a decrease in the number of unwanted telephone calls received by many U.S. households.
- *A decline in the number of telephone surveys*—There will likely continue to be a decline in telephone surveys conducted by market, opinion, and political survey researchers.
- *In some surveys respondents will be paid for their participation*—In earlier years a large majority of the population was willing to participate in surveys without receiving any monetary benefits. Going forward, potential respondents who do not find the "nonmonetary" benefits of survey participation sufficient will most likely not be persuaded by offers of $1 or even $5. It will probably be necessary to provide higher compensation for surveys involving substantial respondent burden.
- *Better methods will be developed to invite and achieve survey participation*—RDD telephone surveys will likely pay more attention to the survey recruitment process. The positive impact of the "advance letter" is well documented, but the simple advance letter may not be enough. The growth and success of the direct mail business has led to modification in mail-opening behavior. It may be necessary to introduce a survey and request participation via some type of premium mail or delivery method.
- *Multiple modes of participation will be offered to respondents*—In order to maximize respondent cooperation, it will probably be necessary to offer potential respondents multiple modes by which they may respond to survey questions. It will likely be necessary to allow a respondent to respond over the web, by mail, and even by calling a toll-free number at the respondent's convenience.
- *Under represented groups*—There will be a focus on increasing the participation of groups, such as young adults, that are currently under represented in RDD surveys.
- *Mobile telephones*—It will almost certainly become necessary to include mobile-only telephone households in RDD samples. Research on methods for accomplishing this is already underway.

Over the next years, response rates will most likely continue to decline at a gradual rate, and at the same time telephone surveys will become more expensive. Even with this higher cost, RDD sampling will remain in use until an equally valid, but lower cost, probability sampling approach is developed. With all of their limitations, telephone surveys offer the only scientifically sound alternative to more costly in-person surveys based on area or list-based probability sampling. Given this lack of options for probability sample designs, RDD surveys may decline in use, but they will most likely survive.

CHAPTER 25

Aspects of Nonresponse Bias in RDD Telephone Surveys

Jill M. Montaquila and J. Michael Brick
Westat and University of Michigan, USA

Mary C. Hagedorn
Westat, USA

Courtney Kennedy and Scott Keeter
The Pew Research Center for the People & the Press

25.1 INTRODUCTION

The monograph from the first Telephone Survey Methodology (TSM) conference held in 1987 included a series of papers on telephone survey nonresponse. Groves and Lyberg (1988) gave an overview of issues related to nonresponse in telephone surveys, emphasizing the need for more research into approaches for combating potential declines in telephone survey response rates and reducing the potential for nonresponse bias. All of the remaining papers on nonresponse in the monograph focused on data collection methods aimed at attaining high response rates, and none dealt specifically with nonresponse bias. Since then, it has become increasingly difficult to attain high response rates in surveys, particularly telephone surveys. Thus, while survey practitioners have continued to study and use various approaches aimed at eliciting respondents' cooperation, some research has turned to examinations of the potential for nonresponse bias.

Estimation strategies aimed at reducing biases are used in virtually every sample survey. Propensity-based and cell-weighting adjustments (also referred to as weighting class adjustments) for nonresponse (Kalton and Flores-Cervantes 2003) are specifically aimed at reducing unit nonresponse bias, and calibration adjustments such

Advances in Telephone Survey Methodology, Edited by James M. Lepkowski, Clyde Tucker, J. Michael Brick, Edith de Leeuw, Lilli Japec, Paul J. Lavrakas, Michael W. Link, and Roberta L. Sangster

as poststratification and raking may also serve that purpose if the response propensities are associated with the calibration variables. But survey organizations also are making a greater effort to reduce nonresponse—effort that requires resources that could be devoted to reducing error in other aspects of the survey process. Is this effort worthwhile? This paper compares the bias that would result from different levels of effort in the survey, assuming weighting adjustments are used for each level of effort. For each level of effort, the sample is reweighted so that the estimates represent what would have been obtained had the particular level of effort been used.

We begin with an overview of issues associated with declining response rates and nonresponse bias, and then discuss various models of nonresponse and provide a theoretical framework for the relationship between response rates and nonresponse bias. The results of the empirical evaluations for two quite different surveys are then presented. The last section addresses the implications of this evaluation for telephone surveys.

25.1.1 Declining Response Rates in RDD Surveys

For the past several decades, the response rate has been one of the most important and widely used indicators of survey quality. It has been viewed by many as a fundamental measure of the potential for nonresponse bias. Achieving high response rates in sample surveys has become increasingly difficult in recent years and is particularly problematic for random digit dial (RDD) telephone surveys (de Leeuw et al., 2002; Curtin et al., 2005; Battaglia et al., 2007, Chapter 24 in this volume; Holbrook et al., 2007, Chapter 23 in this volume). This difficulty with attaining high response rates persists despite the introduction of additional procedures aimed at increasing response, as discussed later.

A number of studies have examined the causes and mechanisms of nonresponse. Steeh et al. (2001) argued that a decline in response rates observed over the 1960s and 1970s leveled off during the 1980s and 1990s but that the composition of nonrespondents changed over that period; refusal rates declined while noncontact rates increased. In their discussion of the leverage-salience theory of response, Groves et al. (2004) made the point that items directly related to the survey topic are most susceptible to nonresponse bias and that people who are interested in or have characteristics related to the survey topic may be more likely to cooperate. To the extent that this is true, comparisons of easier and harder to interview respondents might confirm it, and some studies have attempted to do so. For example, Voigt et al. (2003) found "intermediate and late responders" to be slightly younger, more likely to be non-white, and less educated than "early responders." In that same study, the authors found initial refusers to be older and more likely to be non-white, less likely to be currently married, and less educated than early responders.

25.1.2 Methods to Increase Response Rates and Reduce Nonresponse Bias

In recognition of the potential for differences among initial cooperators, refusers, and reluctant (or "late" or "resistant") respondents, a number of data collection and estimation strategies have been used in RDD surveys in an effort to attain high response rates and minimize the effects of nonresponse bias (Holbrook et al., 2007, Chapter 23, in this volume). Some strategies have been around since the early days of RDD sampling, while others have been introduced more recently.

Efficient calling strategies that result in multiple call attempts, on varying days of the week at various times of day, are an important part of this effort. The field period must be long enough to allow sufficient numbers of call attempts (e.g., Weeks, 1988; Greenberg and Stokes, 1990). Good questionnaire design principles (e.g., well-written introductions and clear, well-executed skip patterns), limits on the length of the interview, and the use of well-trained interviewers contribute to effective data collection efforts. Translation of questionnaires into multiple languages (or the use of interpreters) and using multiple modes of administration are relatively new strategies in the ongoing attempt to counter the decline in response rates. Tactics such as refusal conversion, mailings (both advance letters and mailings to follow up with nonrespondents), incentives (Curtin et al., 2007; Cantor et al., 2007, Chapter 22 in this volume), and answering machine messages have been widely used in attempts to elicit cooperation.

Despite these efforts, RDD response rates have continued a downward trend. Thus, estimation strategies aimed at reducing the effects of nonresponse bias are increasingly important. For example, Biemer (2001) provides a theoretical framework for assessing nonresponse bias using a nonresponse follow-up survey. Nonresponse follow-up is an important tool in understanding and assessing nonresponse bias. By determining the characteristics associated with nonresponse, auxiliary variables can be more effectively used in estimation to reduce the effects of nonresponse bias. One such application is described by Teitler et al. (2003), who used a mixed-mode study to examine how characteristics and costs per case vary by mode. They used proxy data available from an early stage of interviewing to examine nonresponse and found large differences between respondents and nonrespondents in education and race. By using education and race in a cell-weighting adjustment for nonresponse or in a poststratification adjustment (Kalton and Flores-Cervantes, 2003), the bias associated with these differences between respondents and nonrespondents may be substantially reduced.

25.2 MODELS OF NONRESPONSE IN RDD SURVEYS AND STATISTICAL ADJUSTMENTS

In this section, theory relating nonresponse rates and nonresponse biases are explored. Simple relationships are established first and then used to provide methods for bounding the effect on the bias due to different levels of effort in data collection. To implement the proposed bounds, the potential sources of nonresponse bias are considered. The theory is extended to include the effect of weighting adjustments on nonresponse bias. Two models that have been proposed in the literature are examined and are shown to lead to weighting adjustments that reduce nonresponse bias when the model assumptions hold.

25.2.1 Relationships between Response Rates and Bias

Two different perspectives on nonresponse in surveys view response as either deterministic or stochastic. The deterministic view is that the population can be partitioned into respondent and nonrespondent strata (Cochran, 1977: 361–362). The bias due to nonresponse of an estimated mean then depends on the relative sizes of the strata and the differences in the characteristics in the two strata. This approach is very descriptive, but does not provide a simple mechanism to assess what the bias might be if the survey procedures and response rates were different. For this reason, the deterministic perspective is not considered further.

The stochastic perspective assumes that for a given survey each unit has a positive response propensity ϕ_i, and nonresponse is treated much like a second phase of sampling. An important difference is that response propensities, unlike sampling probabilities, are not controlled and typically are unknown. Under this model, the respondent mean is unbiased if there is no correlation between the response propensity and the characteristics being estimated (ϕ and y, respectively).

Response propensity models can be used to develop estimators to reduce nonresponse bias (Little, 1986; Bethlehem, 1988; Kalton and Maligalig, 1991; Lee and Valliant, 2007, Chapter 8 in this volume). Response propensity models also provide a method for estimating bias, if the relationship between the response propensities and the characteristics of interest is known or can be estimated. Colombo (2000) models response propensity distributions, but does not address the key to assessing the magnitude of bias for an estimate—the correlation between the specific characteristics and the response propensity. Groves et al. (2004) use response propensity models for bounding nonresponse bias, directly incorporating the correlation.

Our goal is to develop bounds on estimates of differences in bias for estimates of a proportion (P) when the estimates are the result of different levels of effort in data collection. The first step is to express the bias of the estimate and corresponding bounds on the bias using response propensities. Assume the average response propensities for the units with and without the characteristic are ϕ_1 and ϕ_2, respectively. The estimator based on respondent reports without any auxiliary data is $\hat{P}_r = \sum_r w_i y_i / N$, where $y_i = 1$ if unit i has the characteristics and 0 otherwise, w_i is the inverse of the probability of selecting unit i, and the sum is over respondents. The bias of the estimator of a mean in general is $N^{-1}\bar{\phi}\sum(\phi_i - \bar{\phi})(y_i - \bar{y})$. For a proportion, $bias(\hat{P}_r) = P(1-P)(1-\lambda)\{P + (1-P)\,\lambda\}^{-1}$, where $\lambda = \phi_2\,\phi_1^{-1}$, and the bias depends only on the ratio of the response propensities, λ.

If the average response propensities are the same for those with and without the characteristics ($\lambda = 1$), then the estimate is unbiased. The bias is negative when $\lambda > 1$ and positive when $\lambda < 1$. Table 25.1 gives the relative bias of an estimate (bias divided by P) for different values of λ. The table shows that if ϕ_2 is less than 1.5 times ϕ_1 (e.g., response rates of 30 percent and 45 percent for those with and without the characteristics), then the relative bias in the respondent estimate for a 50 percent characteristic is less than -20 percent. For example, suppose the statistic to be estimated is the percent of adults who are not registered to vote. (Note that since "not registered to vote" is the characteristic of interest in this example, ϕ_1 is the average response propensity for those not registered to vote, and ϕ_2 is the average response

Table 25.1. Relative Bias (Bias Divided by P) of the Respondent Proportion for Different Ratios of Response Propensities ($\lambda = \phi_2\phi_1^{-1}$)

	λ_1			
P (%)	1.1 (%)	1.5 (%)	2.0 (%)	2.5 (%)
1	−9	−33	−50	−60
5	−9	−32	−49	−59
10	−8	−31	−47	−57
25	−7	−27	−43	−53
50	−5	−20	−33	−43
75	−2	−11	−20	−27
90	−1	−5	−9	−13
95	0	−2	−5	−7
99	0	0	−1	−1

NOTE: Response propensities for units with the characteristic is ϕ_1; those without the characteristic is ϕ_2.

propensity for those registered to vote.) Assume those who are registered have a response rate of 65 percent and those who are not registered have a response rate of 50 percent. Thus, $\lambda = 0.65/0.50 = 1.3$. If $P = 25$ percent of adults are not registered, then the bias is just under 5 percent.

Effect of level of effort on nonresponse bias. Next, consider the effect on the bias from increasing the response rate by using higher levels of effort. The percent reduction in the bias of the estimate is

$$\kappa = 100 - 100\frac{\text{bias}(\hat{P}_{r2})}{\text{bias}(\hat{P}_{r1})} = 100 - \frac{100(1-\lambda_2)(P+\lambda_1(1-\lambda_1)^{-1})}{P(1+\lambda_2)+\lambda_2} \tag{25.1}$$

where $\phi_{1,1}$ is the average response propensity for those with the characteristic at effort level 1, $\phi_{1,2}$ is the average response propensity for those with the characteristic at effort level 2, $\phi_{2,1}$ and $\phi_{2,2}$ are defined similarly for those without the characteristic, and $\lambda_1 = \phi_{2,1}/\phi_{1,1}$ and $\lambda_2 = \phi_{2,2}/\phi_{1,2}$. The expression is undefined if the bias $(\hat{P}_{r1}) = 0$.

If the extra data collection effort changes response rates but does not change the ratios of the average response propensities ($\lambda_1 = \lambda_2$), then there is no reduction in the bias ($\kappa = 0$ percent). At the other extreme, suppose the extra effort results in $\lambda_2 = 1$ and there is no bias in the estimate from the higher response rate survey, then $\kappa = 100$ percent. Table 25.2 gives the percent reduction in bias for different values of the population percentage (P) for $\lambda_2 = 1.1$ (e.g., a 40 percent response rate in the high level of effort survey for those with the characteristic and 44 percent for those without) and varying values of λ_1. The percent reduction in bias is not very sensitive to P. In addition, the percent reduction is greatest when the low-level effort survey estimate has a large value of $\lambda_1 = \phi_{2,1}/\phi_{1,1}$, as would be expected. It is worth noting that the bias due to nonresponse increases with higher response rates if $\lambda_2 > \lambda_1$ and $\kappa < 0$. Merkle et al., (1998) describe a case in which incentives led to higher response rate and greater bias.

Table 25.2. Percent Reduction in Bias of Estimate due to Higher Level of Effort (κ), with λ_2 = 1.1 and Different Values of λ_1

	λ_1			
P %	1.2 (%)	1.5 (%)	2.0 (%)	2.5 (%)
1	45	73	82	85
5	46	73	82	85
10	46	73	83	86
25	47	74	84	87
50	48	76	86	89
75	49	78	88	91
90	50	79	89	92
95	50	80	90	93
99	50	80	90	93

NOTE: $\lambda_1 = \phi_{2,1}/\phi_{1,1}$ and $\lambda_2 = \phi_{2,2}/\phi_{1,2}$, where $\phi_{1,1}$ and $\phi_{2,1}$ are the response propensities for those with and without the characteristic, respectively, at effort level 1, and $\phi_{1,2}$ and $\phi_{2,2}$ are the response propensities for those with and without the characteristic, respectively, at effort level 2.

Suppose in the registered voter example extra call attempts and refusal conversion are added to increase the response rate for those who are registered from 65 percent to 70 percent and those who are not registered from 50 percent to 60 percent. Thus, $\lambda_1 = 1.3$, $\lambda_2 = 0.7/0.6 = 1.17$, and the percent reduction in bias is $\kappa = 40.5$ percent. In this case, the increased effort results in the bias decreasing from 5 percent to 2 percent.

Causes of nonresponse bias. One way to form bounds on the reduction in the bias in the estimates is to assign values for λ_1 and λ_2. Informed choices of these values depend on understanding the causes of nonresponse, especially causes that would yield response propensities that differ for those with and without the characteristic. The two primary causes of nonresponse in RDD surveys are inaccessibility (inability to contact sampled units) and unamenability (unwillingness of those contacted to respond). Within the framework described above, mechanisms likely to give rise to different response propensities for those with and without the characteristic are either a direct cause of the nonresponse or a variable highly correlated with a direct cause. For example, different response propensities due to inability to contact respondents in telephone surveys might be reasonable in surveys estimating statistics such as time use, travel, or the use of technology. Under this theory other characteristic, such as being in a single person household, might also be highly correlated to differential propensities due to inaccessibility.

With respect to amenability, the topic and sponsorship of the survey are features that sometimes cause differential response propensities for those with and without the characteristic. This hypothesis is consistent with leverage-saliency theory (Groves et al., 2000). For telephone surveys, advance letters and introductions that identify the topic might result in persons being more or less amenable to participating based on whether they have a specific characteristic. For example, habitual drug users may be more likely to refuse a drug survey, persons who have not participated in education activities may be less likely to respond to an education survey, and those who do not fish might be less likely to respond to a survey about fishing and wildlife recreation. Groves et al. (2004) find some support for this relationship, but the differences in response propensities are not very large.

25.2.2 Statistical Adjustments for Nonresponse

The expressions given above reflect the nonresponse bias in estimates without the use of adjustments aimed at reducing that bias. These expressions give indications of the *potential* for bias due to differential response propensities, but do not account for reductions in bias due to adjustments used in estimation. Brick and Kalton (1996) review weighting methods used to reduce bias including cell weighting adjustments. In the cell weighting method, the sampled units are partitioned into cells or classes based on data known for both respondents and nonrespondents. The weights for the respondents in a cell are then increased to account for the nonrespondents in the same cell. The adjusted estimator is $\hat{p}_a = \Sigma_c \hat{p}_{cr} A_c$, where the sum is over the adjustment cells, $A_c = \Sigma_s w_i / \Sigma_r w_i$, and $\hat{p}_{cr}$ is the unadjusted estimator in cell c. This method includes response propensity stratification (Little, 1986), where the cells are formed based on estimated response propensities to maximize variation in the observed rates across cells.

Brick and Kalton (1996) express the bias of the cell weighting estimator of a general mean as $N^{-1}\Sigma_c \bar{\phi}_c^{-1} \Sigma_i (\phi_{ci} - \bar{\phi})(y_{ci} - \bar{y}_c)$. For a proportion, this can be written in terms of $\lambda_c = \phi_{c2}\phi_{c1}^{-1}$ (the ratio of the response propensities within cell c) as

$$N^{-1}\Sigma_c N_c \frac{(1-\lambda_c)(1-P_c)P_c}{(1-\lambda_c)P_c + \lambda_c} \tag{25.2}$$

where N_c is the number of units in the population in cell c and P_c is the population proportion of units in cell c with the characteristic.

The cell weighting adjustment eliminates any bias if either the units within a cell have the same response propensity or they all either have or do not have the characteristics. This is consistent with Little (1986) who noted that the bias is eliminated if the units are missing at random within the cells. Using the stochastic view, the bias is also equal to zero if the correlation between the response propensity and characteristic within a cell is zero. Even if these conditions do not hold exactly, the cell weighting estimator may have less bias than the unadjusted estimator if weighting cells are formed using variables that are correlated with either the response propensities or the characteristic being estimated.

As noted above, response propensity models can also be used (as an alternative to cell weighting adjustments) to develop estimators to reduce nonresponse bias. (See, for example, Lepkowski et al., 1989.) These models are logistic regression models that use auxiliary data that are available for both respondents and nonrespondents to predict each unit's probability of response. The adjusted estimator is then computed using the respondents' adjusted weights $w_i^A = w_i\hat{\phi}_i$.

Calibration methods, including raking (also referred to as "sample balancing"), can also be used to reduce biases, including nonresponse bias. The raking adjustment (See, e.g., Kalton and Maligalig, 1991) involves creating a k-dimensional table using variables that are known for respondents, and iteratively adjusting weighted totals for each margin to external population totals until the adjusted weighted marginal totals converge to the population totals for all k dimensions simultaneously. When raking is used to adjust for nonresponse, the underlying response model is that for a given cell in the k-dimensional table, the response propensities of units in the cell are constant and are the product of marginal effects for each dimension; specifically, response propensities depend only on main effects of the variables used to form the table and not on interaction effects.

Although the set of statistical methods used to adjust for nonresponse that are described in this section is not exhaustive, it includes the most common adjustments, and the preceding discussion illustrates the importance of the link between the model of nonresponse and the choice of statistical adjustment methods. The following section describes two common models used to characterize nonresponse.

25.2.3 Models of Nonresponse

The classes model. A model of survey participation described by Lin and Schaeffer (1995) is the "classes model." The classes model assumes that nonrespondents are heterogeneous, and that there are groups of nonrespondents whose responses can be estimated from the observed data of similar groups of respondents. The researcher must specify the groups or classes. A common approach is to use respondents who refused the survey request at least once, but later participated, as one group and assume that the remaining refusers who never complete the survey are in the same class (e.g., Benson et al., 1951; Robins, 1963; O'Neil, 1979; DeMaio, 1980; Smith, 1984). More textured classes models (Stinchcombe et al., 1981; Lin and Schaeffer, 1995; Voigt et al., 2003) incorporate both refusals and noncontracts (i.e., amenability and accessibility). In this latter set, accessibility has been measured as either the number of call attempts required for completion or the time duration between the initial attempt and interview completion. In these models, one class includes "temporary refusers" and final refusers, and another class has "difficult to contact" and the final noncontact cases. If the assumptions of the classes model hold, then it is simple to implement a cell weighting adjustment to reduce the bias of the estimates by defining the weighting cells to be identical to the groups or classes in the model.

Continuum of resistance model. A second model frequently cited in the literature on nonresponse bias is what Lin and Schaeffer (1995) refer to as the "continuum of resistance." For earlier uses of this approach, see Filion (1976) and Fitzgerald and Fuller (1982). The rationale for this model is that with less effort, the harder to complete cases (e.g., those requiring refusal conversion or a larger number of call attempts) would have been nonrespondents. The model extends that logic and assumes that the remaining nonrespondents are homogeneous and similar to those units that were difficult to complete. If the assumptions of the model hold, then weighting adjustments could be used to reduce the bias of the estimates as compared to the simple unadjusted estimates. Dunkelberg and Day (1973) proposed one weighting method to accomplish this. A simple cell weighting adjustment also could be used, where the weights of the difficult-to-complete cases are increased to account for the nonrespondents. Unfortunately, there is little evidence that the assumptions of the continuum of resistance model hold in RDD surveys. See, for example, Lin and Schaeffer (1995), Groves and Wissoker (1999), Curtin et al. (2000), Keeter et al. (2000), and Teitler et al. (2003).

25.3 EMPIRICAL EVALUATION

We conducted an empirical evaluation to investigate nonresponse bias, using two RDD surveys that are very different from each other in terms of topic, field period, and sponsorship. We selected a set of alternative scenarios to evaluate, including different levels of effort and response rates. In a departure from many of the other nonresponse bias studies, we reweighted for each scenario. In addition to studying estimates of population proportions and rates, we also examined multivariate analyses. This section provides background information on the two studies used in the evaluation, describes the evaluation design and methods used, and gives key findings from this evaluation. In the discussion of each of the studies and the evaluation, we describe the statistical adjustments used to reduce nonresponse bias and relate our findings to the theoretical discussion given in the previous section.

25.3.1 Study Background

This evaluation uses a study conducted by the Pew Research Center for the People & the Press, and the Adult Education and Lifelong Learning Survey of the 2001 National Household Education Surveys Program (AELL-NHES: 2001). The overall response rate for the Pew study was 44 percent (American Association for Public Opinion Research (AAPOR) *RR3* formula; see AAPOR, 2004), and the overall response rate for the AELL survey was 53 percent. As discussed later, this analysis enables us to examine the differential bias that would be expected as a result of curtailed data collection efforts; it does not examine the nonresponse bias due to nonresponse to the full-scale effort.

Overview of Pew studies. The Pew Research Center for People & the Press conducts surveys examining a variety of political and social phenomena. In 2003, a typical 5-day survey conducted by the Pew Research Center obtained interviews in roughly one quarter (26 percent) of the sampled households, a decrease of about 10 percent from a similar survey conducted in 1997. The data used in this analysis come from a methodological experiment conducted in 2003, which was a replication and extension of another study on nonresponse conducted by Pew Research Center in 1997. (See Keeter et al., 2000.) The approach of the experiment was to compare the responses from a sample of people obtained through Pew's usual methodology to a sample obtained with a more rigorous survey effort over a much longer field period. To do this, the two surveys used identical survey questionnaires addressing political and social opinions, electoral behavior, media use, knowledge items, social integration, crime, attitudes about polling, and demographic characteristics.

The "standard" study was designed to complete 1000 interviews in 5 days. The sample was list-assisted, and respondents within households were selected using a procedure that interviewed the youngest male/oldest female currently at home. A minimum of ten call attempts were made for each sampled telephone number. Interviews were conducted in English only. The standard survey was conducted from June 4 through 8, 2003, but data collection on the sample continued until October 30 in an effort to gain additional information about the sample. After the first 5 days, households that refused, and for which an address was available, were sent a letter, and efforts were made to persuade most refusals (from both listed and unlisted households) to consent to an interview. A total of 1370 interviews were conducted over the course of the standard 5 day field period and the extended calling period.

The "rigorous" study began at the same time as the standard survey and was conducted until October 30. It employed a more extensive effort to locate and interview reluctant and difficult-to reach-individuals, featuring advance letters (a random half of that included a \$2 incentive), refusal conversion letters, letters with \$2 incentives to households where no contact had been made as of July 24, and answering machine messages. No maximum number was set for the number of call attempts. The response rate was 51 percent, 1089 interviews were completed.[1]

The Pew Research Center findings reported in this analysis are based on the *combined* dataset for the 2003 standard and rigorous samples.[2] Weighting was done using an iterative process (sample balancing; see Section 25.2.2) to match national parameters for sex, age, education, race, Hispanic origin, and region obtained from the latest available March demographic supplement of the Current Population Survey (CPS). Weights were trimmed and varied from 1.0 to 3.1.

The National Household Education Surveys Program. The National Household Education Surveys Program (NHES) is an RDD survey system developed by the National Center for Education Statistics (NCES) in the U.S. Department of Education. It is designed to collect information on a range of educational issues through CATI surveys of households. Each NHES administration includes a brief screening interview (Screener) and two or three topical interviews. NHES, which has been conducted by Westat eight times from 1991 through 2005, also has experienced declining response rates, especially at the initial household screening stage (Brick and Broene, 1997). Screener unit response rates in 1991 and 1993 were greater than

[1]Questions for which a change in aggregate opinion may have occurred during the long field period (such as President Bush's approval rating) were excluded from the analysis.

[2]We were able to merge the cases from these samples because the extended calling period (conducted on the standard sample after the first 5 days) brought the total level of effort expended on the standard sample in line with the effort expended on the rigorous sample. Ultimately, both samples were called from June 4 – October 30 and featured multiple refusal conversion attempts and an essentially unlimited number of call attempts.

80 percent, but fell to 64 percent in 2001 and 65 percent in 2003 (Brick et al., 2005c), despite increased effort in the form of extended calling protocols, increased refusal conversion efforts, and the use of monetary incentives.

The Adult Education and Lifelong Learning survey (AELL) of NHES:2001 is the survey used in this research. The AELL survey addressed participation in a range of educational activities over a 12 month period. The sample for NHES:2001 included 179,211 telephone numbers selected using the list-assisted procedure. An advance letter was mailed to about 57 percent of the households with matched addresses. The NHES data collection protocol included up to 20 call attempts to complete a Screener. Difficult-to-complete households with matched addresses received a letter encouraging response. If a household refused, the case was held for about 2 weeks before attempting a refusal conversion call, a refusal conversion letter and study brochure were sent by FedEx during this period to those cases with matched addresses. If a case refused a second time, it was again held for a 2-week period and another conversion attempt was made. At the screening stage, household members were enumerated as needed to identify eligible household members and apply the sampling algorithm. The NHES interviews were conducted in English and Spanish. Screeners were completed with 48,385 households, and the Screener response rate was 65 percent.

Once a household member was selected for a topical survey, up to 20 additional call attempts were made to complete the interview. If the selected adult refused, the case was held for 2 weeks and was then released to interviewers for refusal conversion. If a second refusal was received, the case was again held for 2 weeks and another refusal conversion attempt was made after sending a letter by FedEx. Letters were also sent to cases with large numbers of calls to encourage response. Interviews were completed with 10,873 sampled adults, for an AELL response rate of 77 percent. Taking into account the Screener response rate, the net response rate for the AELL survey was 53 percent.

Survey weights were developed to produce national estimates. Base weights accounted for differential sampling of telephone numbers and the selection of adults within households. A cell weighting adjustment (as described in Section 25.2.2) for Screener nonresponse was applied using cells formed by variables associated with Screener response propensities, specifically the mailable status of the telephone number, an indicator of whether any answering machine message was left, region, listed status of the telephone number, and select characteristics of the telephone exchange (percent White, median home value, percent renters, and percent owners). The nonresponse-adjusted household weights were poststratified to national totals using data from the March 2001 CPS. A cell weighting adjustment was used to adjust for AELL survey nonresponse, with cells formed by the cross classification of whether the sampled adult was also the Screener respondent, region, whether the sampled adult participated in adult education in the previous 12 months, and sex. The nonresponse-adjusted weights were raked (which is essentially equivalent to the sample balancing method used in the Pew surveys) to national totals from the CPS. Additional detail on the sampling, data collection, and estimation methods used in NHES:2001 is given in Nolin et al. (2004).

25.3.2 Evaluation Design

As noted above, a variety of procedures have been used to increase response rates in RDD surveys and most of those were used in the Pew study and in the AELL survey. The effects of such procedures on response rates and nonresponse bias can be measured by building experiments into the RDD survey. (See, for example, Brick et al., 2005c.) However, if procedures (e.g., refusal conversion) are known to be effective in increasing response rates, then not using the procedures on an experimental group for the sake of the experiment will result in lower response rates for the survey as a whole. The Pew methodological studies undertaken in 1997 and 2003 (see Keeter et al., 2000 and Keeter et al., 2006, respectively) examined nonresponse bias by incorporating methods that could not be built into the regular survey protocol, such as extending data collection beyond the standard 5-day period that is needed for their fast response surveys.

The approach taken in this evaluation was not to alter the procedures used in data collection but rather to manipulate the data after data collection to evaluate various hypothetical scenarios. Only certain procedures can be manipulated retrospectively to create a dataset with a lower response rate. Thus, the scenarios that can be evaluated after data collection are somewhat limited, but include some of the most effective methods for increasing response rates.

Evaluation scenarios. The three procedures at the screener level that were manipulated in this evaluation are the total number of call attempts, the number of refusal conversion attempts, and the number of call attempts before being able to classify a telephone number as residential or not. The number of call attempts before classifying a number's residential status is not a standard measure of effort, but because a large number of attempts are often made in RDD surveys in an attempt to resolve residential status and because this may be viewed as a measure of accessibility, this measure was investigated.

To evaluate potential scenarios, we computed response rates for four different procedures: not making more than eight Screener call attempts, not making more than 16 Screener call attempts, no Screener refusal conversion attempts,[3] and not more than one Screener refusal conversion attempt. After examining these response rates, we arrived at four alternative scenarios to consider (in addition to the full effort, labeled F0, which imposed no restriction on the numbers of refusal conversion attempts or call attempts).

- S1—No screener refusal conversion (no restriction on the number of call attempts)
- S2—Only one screener refusal conversion attempt and no more than 8 call attempts
- S3—Only one screener refusal conversion attempt and no more than 16 call attempts
- S4—No screener refusal conversion and no more than 8 call attempts

[3]Throughout this paper, the term "refusal conversion attempt" refers to the process of attempting to convert a refusal case. This often involves multiple call attempts.

In addition to the Screener-level procedures, procedures at the extended AELL interview level were investigated for NHES:2001. The most natural one was the elimination of completed extended interviews that were obtained by doing refusal conversions of the sampled adult (labeled E2). Another, more radical, option was the elimination of all completed interviews with adult males between the ages of 25 and 35 (labeled E1). This is not an operational procedure like the others, but was considered because of its likely effect on the bias of the estimates. Adult education participation, a key characteristic derived from the survey, is correlated with age and sex (Kim, 2004). Treating all sampled males in an age group as nonrespondents could help better understand the effect of large, differential nonresponse rates in specific groups.

Methods. For each of the six AELL study scenarios and each of the four Pew study scenarios, the sample was reweighted to create a new dataset for analysis. Although additional nonresponse bias might be introduced by curtailing the effort, if the weighting adjustments completely remove the additional bias then the net effect is no additional bias. Through the reweighting, interviews completed outside of the limits of the scenario were treated as nonrespondents. For example, for scenario S1, all cases completed following refusal conversion were treated as nonrespondents. Each dataset was then processed as if it were the completed dataset and the full set of weighting adjustments was made separately for each dataset. The weighting steps used in this reweighting included the same adjustments, done exactly as they had been done for the full effort.[4] (See Section 25.3.1.)

The analysis focused on computing various key estimates from each scenario-specific dataset, and contrasting the estimates to those from the full effort dataset (F0). Although estimates from F0 may themselves have nonresponse bias, those biases cannot be directly measured or estimated in this study, and it is assumed that any such biases would also be present in estimates from the alternative scenarios. The purpose of this study is to assess the incremental bias introduced by reducing the level of effort below the full effort. Later we discuss extrapolations to the units that never responded.

For both the AELL and Pew studies, a variety of demographic characteristics for which a reliable parameter exists were considered. For the AELL survey, the analysis also included estimates of the adult education participation rate, a key statistic from that survey, computed for a variety of demographic and other domains that are typically used in reports on adult education. Since the estimates from a scenario will be highly correlated with F0, the standard errors of the estimated differences between a scenario and F0 were computed to account for this correlation. With such highly correlated estimates, it was likely that even small differences would be statistically significant. Thus, in determining which statistically significant differences are important, practical

[4]In a few cases, some of the original weighting classes were too sparse under a specific scenario and some collapsing of cells was required. Since the goal was to compare the estimates from the different datasets after they have been subjected to the same weighting procedures, every effort was made to minimize additional collapsing.

importance must also play a role (e.g., differences of less than 3 percent in the estimates might be deemed to be practically insignificant). In addition to these univariate comparisons, the effects of the curtailed efforts on multivariate relationships were examined by fitting logistic regression models for each scenario.

25.3.3 Findings

Effect of weighting on nonresponse bias. While it is generally difficult to account for weighting adjustments theoretically, their effects on bias may be important especially if there are large nonresponse biases. The potential effect of the weighting, especially the adjustment to control totals using raking, was the main rationale for going through the effort to reweight each of the scenarios before computing estimated biases. As reported later in this section, the bias estimates in our evaluation were small and not significant for nearly all of the scenarios and characteristics considered. This result might be due to the inherent lack of relationship between the response propensities and the characteristics, or it might be due to the effectiveness of the weighting methods, or some combination of the two. Thus, before evaluating the effect of the level of effort, we directly evaluate the effect of the weighting on the estimated bias. As we report later in this section, the results of our evaluation show that the weighting adjustments do have a significant effect on bias.

Two methods of evaluating the effectiveness of the adjustments for nonresponse and raking are presented using the AELL scenario S3 for illustration. The first, and primary, method is the *base weighting scheme.* It compares the estimates from the fully adjusted weights for scenario S3 to estimates from weights for scenario S3 that contain no nonresponse or raking adjustments (i.e., only base weights). While the S3 base weights only account for the probabilities of selection, they do account for all household and adult level sampling stages. The comparison of these two sets of estimates gives a direct account of the overall effectiveness of the weighting in reducing bias.

We began by examining differences between the base weighted and fully weighted scenario S3 estimates and computing the *t*-test statistic value (*t value*) for the differences. All of the characteristics examined[5] have differences with statistically significant *t*-values and over half of the characteristics examined—most notably, the overall adult education participation rate, the key estimate from the AELL survey—had differences that were deemed to be substantively important (at least 3 percent). These findings show that the weighting adjustments in this case are effective in reducing the nonresponse biases arising from the induced S3 nonresponse scheme, and that using the unadjusted estimates would lead to much larger biases.

The second method is the *abridged weighting scheme*, which compares the estimates from the fully adjusted weights for scenario S3 to estimates computed using an abridged weighting approach. The abridged weighting approach uses the scenario

[5]The characteristics considered in this comparison were overall adult education participation rate, participation in credential/vocational programs, and adult education participation by gender and by age; region; mailable status of the telephone number; percent white in the telephone exchange; presence of children in the household; sex; age; educational attainment; and marital status.

F0 weights, applied to only the subset of units that are respondents for scenario S3. It is a useful approach since most other studies that have used data on level of effort to estimate nonresponse bias (e.g., Curtin et al., 2000; Keeter et al., 2000) used an abridged weighting approach in their analyses. (Blumberg et al. (2005a) applied a separate poststratification adjustment to the full-effort weights of respondents in each of their evaluation scenarios, but did not fully reweight each scenario.) To compare the abridged weighting scheme and the fully weighted scheme, we computed the difference between estimates from the fully weighted S3 scenario and estimates from the S3 scheme computed using abridged weights. This comparison revealed statistically significant differences in the percentage of adults with mailable telephone numbers (i.e., telephone numbers that could be matched to a mailing address by vendors who maintain databases with address-telephone number linkages); the percentage of adults in telephone exchanges that are 20 to 50 percent white; the percentage of adults ages 25 to 34; the percentage of adults with some college or an associate's degree; the percentage of adults who are married; and the percentage of adults who are separated, divorced, or widowed. However, since the standard error of the difference is small due to the positive correlation between the fully weighted estimates and those computed using the abridged weighting scheme, even substantively unimportant differences may be statistically significant.

Among the characteristics considered, none have estimates computed with abridged weights that differed by at least 3 percentage points from the fully weighted S3 estimates. This suggests that although the fully adjusted weights may reduce nonresponse bias, the marginal effects using even abridged weights are relatively small. In many practical situations where the adjustments are generally based on correlates of nonresponse rather than the direct cause, the abridged weights should be adequate for estimating the bias due to different levels of effort.

Response rates. Table 25.3 shows the response rates that would have been achieved if the level of effort had been restricted according to the given rule used to define each of the evaluation scenarios. For both studies, response rates and sample yield vary considerably among the scenarios. Table 25.3 also shows the marginal effect each restriction would have had on the response rate as a proportion of the achieved response rate. For example, if all the Screeners completed as a result of refusal conversion in NHES:2001 were treated as nonresponse, then the response rate would only be 38.0 percent (71 percent of the NHES:2001 F0 rate) and the AELL interview yield would have been 73 percent of the actual (F0) yield.

If data collection procedures are modified in an effort to increase response propensities, the modification will affect bias only if the change in response propensities differs between the proportion of the population having the characteristic of interest and the proportion that does not have the characteristic (i.e., in the theory given earlier this corresponds to stating that $\kappa = 0$ unless $\lambda_1 \neq \lambda_2$). Thus, one way to evaluate the effect of the different levels of effort in the scenarios is to examine changes in the response rates for subgroups. Keeter et al. (2004) contains an extensive discussion of the effect of the level of effort on response propensities for subgroups, based on comparisons of the 2003 Pew standard and rigorous samples. These comparisons revealed

Table 25.3. Effect of Procedures on Response Rate

	NHES:2001			Pew 2003		
		Marginal effect (%)**			Marginal effect (%)	
Scenario	Response rate (%)*	Response rate	Sample yield	Response rate (%)	Response rate	Sample yield
F0: Full effort	53.4	100	100	51.4	100	100
S1: No refusal conversion	38.0	71	73	37.4	73	79
S2: No 2nd refusal conversion and no more than 8 screener call attempts	43.3	81	85	32.9	64	68
S3: No 2nd refusal conversion and no more than 16 screener call attempts	49.3	92	94	37.6	73	81
S4: No refusal conversion and no more than 8 screener call attempts	34.0	64	67	27.6	54	60
E1: No adult males ages 25 to 35	48.3	90	91	†	†	†
E2: No AELL interview refusal conversion	47.8	89	90	†	†	†

SOURCE: U.S. Department of Education, National Center for Education Statistics, National Household Education Surveys Program, 2001; Pew Research Center for People & the Press study conducted in 2003.

*Response rate given is the AELL-NHES:2001 overall unit response rate (which reflects both screener nonresponse and nonresponse to the AELL extended interview).

**The marginal effect is the procedure response rate divided by the full effort response rate.

†Not applicable.

few differences between estimates from the standard and rigorous samples. To examine this for the AELL study, we computed response rates for each scenario by characteristics of the telephone exchange or household. Although the overall magnitudes of the response rates vary across scenarios, the patterns among these general household- and area-level characteristics remain the same (e.g., in the AELL study, for each scenario, the Northeast is the region with the lowest response rate and the Midwest is the region with the highest response rate; response rates increase as the percent white in the exchange increases). The characteristic that fits the pattern least is mailable addresses. For example, using scenario S2 the values of the change in response rate (compared to the full effort) for mailable addresses are $\lambda_1 = 37.9/59.3 = 0.64$, $\lambda_2 = 54.6/72.9 = 0.75$, and the percent reduction in bias computed using κ is about 33 percent if the percent with matched addresses is 75 percent. While these computations indicate there can be reasonable reduction in the percent of bias due to using the more extensive F0 data collection efforts, the percent reduction is not large in absolute terms and may not be substantively important (Brick et al., 2005d).

Variation in characteristics of level of effort. A second way to examine the effect of effort on bias is by studying how characteristics vary according to the level of effort. For sake of illustration, we focus the discussion here on estimates of one particular characteristic—the percentage of adults in one-person households—although several key substantive and demographic characteristics were considered. For the AELL survey, the key outcome measure is participation in adult education; for this characteristic, both the overall estimate and estimates for subgroups were considered. The subgroups used in this evaluation were defined by age, sex, age by sex, race/ethnicity, educational attainment, and household income. In addition to overall adult education participation, overall estimates of participation in two particular types of courses—credential/vocational and formal courses—were considered. The demographic characteristics whose distributions were examined under each scenario were age, sex, age by sex, race/ethnicity, educational attainment, household income, household size, marital status, employment status over the past 12 months, nativity, the first language the adult learned to speak, long-term disability status, whether the family participated in the Child Health Insurance Program (CHIP), whether the family received food stamps, whether the family received Medicaid, and whether the family received Women, Infants, and Children (WIC) assistance. For the Pew study, the characteristics considered in the evaluation were sex, educational attainment, household size, income, marital status, political party affiliation, age, and parenthood.

Figure 25.1 shows the effect of refusal conversion on the AELL estimate of the proportion of adults in one-person households. Figure 25.2 shows the effect on the same characteristic by the number of call attempts required to obtain a completed screening interview. (Although each type of figure is shown for only one of the studies, similar patterns were observed in both studies.) These two figures provide glimpses into the potential nonresponse bias due to refusals and noncontacts, although they are limited because no data on the final refusals and noncontacts are available. Nevertheless, under the response theories described earlier, it might be speculated that the final refusals are similar in characteristics to the ones that required refusal conversion, and the final noncontacts are similar to the cases that required many calls to complete the Screener.

Figure 25.1 can be viewed as a depiction of amenability. Adults in households with no refusals are more likely to live in one-person households. Additional analyses of the AELL and Pew results revealed that adults in households with no refusals are more likely to be adult education participants and to have an educational attainment of a bachelor's degree or higher. Notice that for the refusal conversion graph, the estimated percent of adults by category is not highly variable. This implies that for this statistic, the change in nonresponse bias associated with refusal conversion may not be very large.

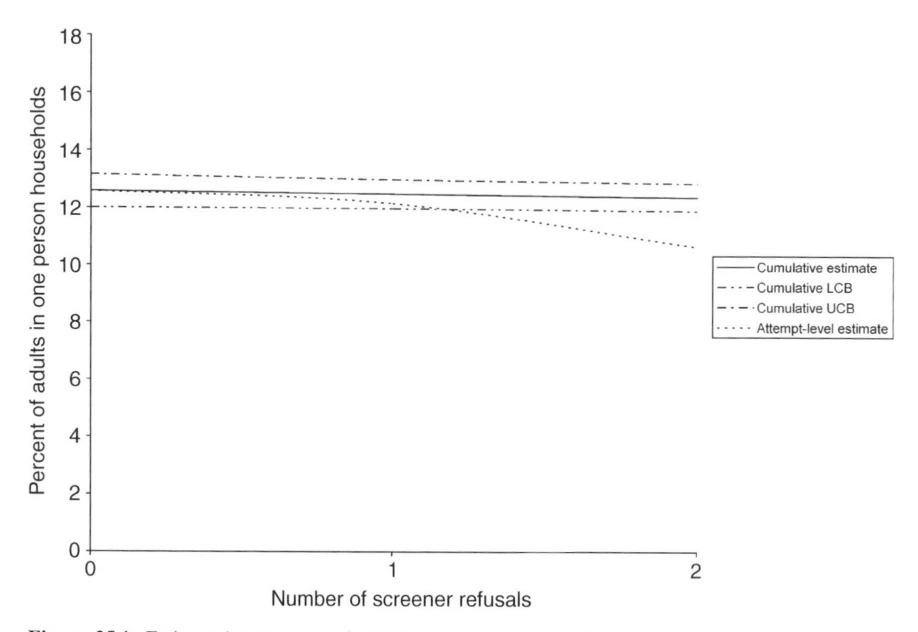

Figure 25.1. Estimated percentage of adults in one-person households by number of call attempts resulting in refusals, AELL. *NOTE*: LCB and UCB are the *lower confidence bound* and *upper confidence bound*, respectively, for a 95 percent confidence interval. (*SOURCE*: U.S. Department of Education, National Center for Education Statistics, National Household Education Surveys Program, 2001.)

Figure 25.2 shows that adults in households requiring fewer call attempts are more likely to live in households with more than one household member. Although not statistically significant, further analysis of the AELL data suggests that adults in households requiring fewer call attempts may be more likely to be adult education nonparticipants. These findings are consistent with the conventional wisdom that one-person households are harder to contact because it is less likely that someone is at home when the survey calls are placed. The same hypothesis is reasonable for adult education participants.

Although not shown here, the estimates from each evaluation scenario of characteristics of adults were also examined by the number of call attempts required to resolve residency status (Brick et al., 2005d). Since residency status is typically resolved on the first call attempt on which contact is made, this can be viewed as primarily an examination of accessibility. The AELL estimates showed that adults in households requiring fewer attempts (0–4 calls) to resolve residency status are more likely to be adult education nonparticipants, are more likely to be married or to have been married, and are less likely to be college graduates. It should be noted that because the majority of the telephone numbers were resolved in eight or fewer calls in the AELL study, there is more variability in the estimates for the categories involving larger numbers of calls.

Estimated biases of estimates under each scenario considered. A third way to evaluate the effect of the different levels of effort is to compute differences between

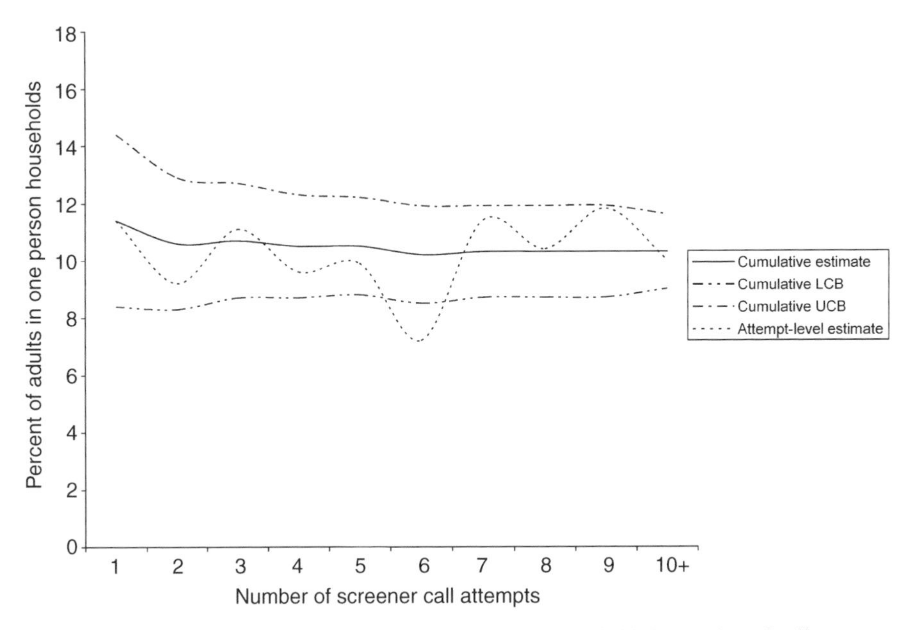

Figure 25.2. Estimated percentage of adults in one-person households by number of call attempts, Pew 2003 study. *Note*: LCB and UCB are the *lower confidence bound* and *upper confidence bound*, respectively, for a 95 percent confidence interval. (*SOURCE*: Pew Research Center for People & the Press study conducted in 2003.)

estimates computed under each particular scenario and the corresponding estimates computed from the full effort. These differences could be viewed as measures of the additional bias, relative to the full effort, incurred by curtailing the effort. As mentioned previously, estimates of a variety of characteristics (listed above) were computed for each scenario and compared to the full effort estimates. Table 25.4 gives, for each scenario, the median absolute relative difference and the maximum absolute relative difference.[6] In terms of the relative difference measures shown in Table 25.4, these findings suggest that the bias incurred by the curtailed efforts of the evaluation scenarios is small.

The AELL analysis showed that although estimates of a few characteristics (whether the family received food stamps, whether the family received Medicaid, and whether the family received WIC) differ between scenario F0 and the other Screener scenarios studied, these differences are small; only a few differences exceed 1 percentage point, and no difference is as large as 2 percentage point. (See Brick et al., 2005d, for details.) The results for the AELL interview scenarios, E1 and E2, are similar; although a few characteristics differ between scenario F0 and each of the AELL interview scenarios, the only differences exceeding 2 percentage points are for estimates of age and sex involving scenario E1.

[6]The absolute relative difference is $\left|\hat{\theta}_{Sn} - \hat{\theta}_{F0}\right| / \hat{\theta}_{F0}$, where $\hat{\theta}_{F0}$ is the full effort estimate and $\hat{\theta}_{Sn}$ is the estimate for the particular scenario.

Table 25.4. Median Absolute Relative Difference and Maximum Absolute Relative Difference Among a Set of Estimates for the Scenarios Considered, AELL and Pew Study

	Median absolute relative difference (%)		Maximum absolute relative difference (%)	
Scenario	AELL	Pew study	AELL	Pew study
S1: No refusal conversion	1.9	0.7	9.6	3.9
S2: No 2nd refusal conversion and no more than 8 screener call attempts	0.7	2.4	6.4	3.9
S3: No 2nd refusal conversion and no more than 16 screener call attempts	0.5	1.0	3.2	4.5
S4: No refusal conversion and no more than 8 screener call attempts	1.7	1.4	12.1	3.8
E1: No adult males ages 25 to 35[a]	0.9	†	12.3	†
E2: No AELL interview refusal conversion	0.8	†	5.8	†

NOTE: Estimates of less than 10 percent or above 90 percent were excluded from these calculations. For the AELL study, 71 estimates were used in these calculations. For the Pew study, 8 estimates were used in these calculations.

SOURCE: U.S. Department of Education, National Center for Education Statistics, National Household Education Surveys Program, 2001; Pew Research Center for People & the Press study conducted in 2003.

[a]The calculation of maximum absolute relative difference for scenario E1 excludes estimates involving age, since very large differences in these estimates were induced by the design of this scenario.

†Not applicable.

The only differences in adult education participation rates between the full effort and the scenarios studied that are of both statistical and substantive (a t value of at least 2 and at least 3 percentage point difference) significance are for scenarios S1 and S4, the two scenarios with the greatest reduction in response rates. These differences are for estimates of adult education participation rates for the "other" race/ethnicity category (i.e., non-Hispanics other than whites and blacks), and for scenario S4 for adult education participation rates for males aged 35–44. Scenarios S1 and S4 are the two scenarios in which no Screener refusal conversion was used. Even scenario E1, which excluded an entire population subgroup, did not give any indication of substantial differential nonresponse bias in estimates of adult education participation.

The 2003 Pew study analysis showed very small, substantively unimportant differences between the estimates from the scenarios and the full effort estimates. None of the characteristics considered differed by as much as 2 percentage points.

Using the AELL survey, we considered the two sources of nonresponse bias—unamenability and inaccessibility—for estimating the percent of adults who participated in adult education. The percent of adults participating in adult education is 48 percent with no refusal conversion, and decreases to 43 percent when two or more conversions are required. This result is consistent with theories that suggest that those with an interest in a topic (e.g., adult education participants are more likely to be interested in the AELL survey topic than nonparticipants) are likely to be more amenable to the survey. On the contrary, we found that as more Screener call attempts are made, the proportion of respondents who are adult education participants increases. That result is also not unexpected, since persons participating in adult education activities might be more difficult to contact because of these activities. The result is that while nonresponse due to refusals might be expected to bias the estimates of adult education participation upward, the bias due to noncontact appears to be in the opposite direction. The biases may be partially offsetting. This notion of counteracting effects on amenability and accessibility on bias is consistent with the results reported by O'Neil (1979).

We can estimate the bias expected based on the results given above by applying the theory developed earlier. Suppose the only source of nonresponse was refusal to participate, and that the survey did not do any refusal conversion. Further, assume that the adults in refusing households are like those adults in households with two or more refusals who completed the survey, and have an adult education participation rate of 43 percent. Under these assumptions, in the half of the households that respond, the adult education participation rate is 48 percent, and in the other half that do not respond, the participation rate is 43 percent. This results in a nonresponse bias of 2.5 percentage points ($= 48 - (48+43)/2$).

Using the expressions given earlier for the bias in terms of λ, the ratio of the response propensity for those not participating in adult education (ϕ_2) to the response propensity of those who did participate in adult education (ϕ_1) under the assumptions given above is $\lambda = 0.9$. The values of λ for the four studies discussed by Groves et al. (2004) are 0.80, 0.87, 0.90, and 1.0. In the Groves et al. study, the frames and topics were chosen to maximize the differences in response rates, and the values of λ they observed were not that different from that used to estimate the bias for AELL.

Multivariate relationships. To examine whether the reduced effort under each scenario has an effect on multivariate relationships, logistic regression models were fit for each scenario. For AELL, these models examined the relationship between adult education participation (the dependent variable) and predictor variables including age, sex, race/ethnicity, educational attainment, household income, employment status, and occupation. For the Pew study, these models examined the relationship between political party affiliation and predictor variables including various demographics (race, age, income, marital status) and political views (Presidential approval, views on homosexuality, and ideology). For this evaluation, the logistic regression models were run for F0 and the other scenarios. The coefficients of the regression parameters from the scenarios were contrasted with the F0 coefficients and the statistical significance of the differences was evaluated. We found no significant differences between the full-effort parameter estimates and the study scenario parameter estimates. This finding shows that for the multivariate analysis given here the lack of differential nonresponse bias is consistent with the previous univariates analysis.

The argument could be made that there are no significant differences in this analysis because all the important characteristics are included as predictors. To examine this, the model of adult education participation from AELL was refit for scenarios F0 and E1, excluding age. Again, the findings suggested no apparent differential nonresponse bias in the parameter estimates. These results demonstrate that even when the entire subgroup of males, aged from 25 to 35 years, are omitted and an important covariate such as age is excluded from the model, there is no evidence of nonresponse bias in the parameter estimates.

Another form of multivariate analysis was examined using the Pew study. For this analysis, a 15-item index of ideology was computed for the 494 cases "hardest-to-complete" cases, those that had refused the interview at least twice before complying and/or required 21 or more calls to complete, and for the other easier to complete cases. The index of ideology is based on variables such as tolerance of homosexuality, views on censorship, and favorability toward various racial/ethnic groups. The Cronbach's alpha for the complete dataset was 0.71. For the hardest-to-complete cases, Cronbach's alpha was 0.71, and for the easy-to-complete cases it was 0.68. There is little substantive importance to a difference of this size.

In summary, when data from the AELL and Pew studies are used together with the concepts of nonresponse theory presented earlier, the possibility of large nonresponse bias in the estimates seems remote within the ranges of response rates considered. Two features of nonresponse in these studies, and in many RDD surveys, appear to mitigate the potential for large biases. The first is that there is evidence that the two main sources of nonresponse, refusals and noncontacts, tend to have at least partially offsetting biases. The second is that the differences in response propensities among subgroups of analytic interest are not large enough to cause large biases in the estimates. The behavior of the other characteristics examined in this evaluation are either similar to those discussed here or show even less evidence of differences in response propensities. Given that there does not seem to be a strong direct link between the source of nonresponse bias and the statistics being examined from the survey, the theory and data suggest that biases due to nonresponse should be relatively small.

25.4 DISCUSSION

In recent years, several methodological experiments have attempted to inform researchers about the nature of nonresponse, with particular emphasis on the circumstances under which nonresponse can judiciously be ignored in practice. Studies by Curtin et al. (2000), Keeter et al. (2000), and Blumberg et al. (2007, Chapter 3 in this volume) have led some to conclude that nonresponse in RDD telephone surveys does not pose a serious threat to the validity of results. All three studies suggest that increased effort results in a relatively low reduction in the bias of an estimate. Thus, there is little evidence that call designs with a high number of callbacks yield estimates that are substantially less biased than those with a more modest level of effort. (For an exception, see Traugott, 1987.) The scope of this conclusion, however, is limited to nonresponse among sampled persons who require extra effort but ultimately participate.

The results of these analyses add to the body of evidence that for many RDD telephone surveys, response propensity may not have a strong correlation with many estimates of interest. Consequently, the classes and continuum of resistance models of nonresponse bias have at best narrow applicability to such surveys. According to the theory presented here, biases in estimates of proportions only arise when the average response propensities for those with the characteristic are different from those without the characteristic. The evaluation showed that, while the extra data collection efforts used in the two studies increased the response rates, they did not result in different response propensities for those with and without the characteristic over a wide range of estimates. Thus, the increases in the response rates due to the additional efforts did not reduce nonresponse bias. The following section discusses these findings in the context of the causes of nonresponse and nonresponse bias.

25.4.1 Mechanisms of Nonresponse That Have the Potential for Increasing Nonresponse Bias

When nonresponse is random, it is essentially ignorable and only affects sample size and costs of data collection. In such cases, response rates are completely unrelated to the potential for nonresponse bias (e.g., those with and without a particular characteristic have the same response propensities). There is a large body of research showing that nonresponse is generally not random. Typically, females are more likely to respond than males, those living in rural areas respond at higher rates than those in urban areas, and older persons are more likely to respond than the younger ones (see Groves and Couper, 1998 for other characteristics related to response rates). The evidence is convincing that there are characteristics of persons or households that are correlated to their probability of responding to a survey. These differences in response rates are present even in surveys with relatively high response rates. The idea of completely random nonresponse is simply inconsistent with experience in RDD surveys.

Most studies recognize that there will be differential nonresponse rates associated with the characteristics of the households or persons in RDD surveys and that nonresponse biases may result unless some methods are used to reduce the biases. The most common approach to reducing biases in RDD surveys is to adjust the survey weights. The effectiveness of the weight adjustment procedures depends on the correlation between the characteristics estimated in the survey and the auxiliary data or the response propensities. For example, educational attainment is an auxiliary variable used in AELL because it is a correlate of participation in adult education activities and is also related to response propensities. In general, weighting adjustments that include as many auxiliary variables as possible, without causing large variations in the weights that increase the variance of the estimates, have the potential to reduce nonresponse biases substantially.

Other potential sources of nonrandom response rate differences may not be reduced by the typical demographic weighting adjustments. The two sources discussed earlier are those related to amenability and accessibility. Amenability may be associated with differential salience of the survey topic and may not be highly correlated with demographic characteristics. For example, in a survey of technology use, it may be that those who are active computer and Internet users may find the survey topic of greater interest and be more likely to respond than others. The resulting nonresponse bias could yield inflated estimates of technology use. A survey on a sensitive topic, such as drug use or domestic violence, may experience lower participation among those who engage in these behaviors, due to a resistance to report negative or criminal behavior.

A correlation between accessibility and survey content is another potential source of nonresponse bias that might not be addressed adequately by weighting adjustments using demographic auxiliary data. A survey focusing on travel or community participation may suffer from nonresponse biases because those who travel often or have high levels of community involvement may be more difficult to contact, especially if the survey has a relatively short data collection period. To return to the example of a technology use survey, it might be that those who are heavy technology users are more difficult to contact. In these examples, biases would be reflected in lower estimates of the measures of interest, e.g., travel episodes or technology use.

The interaction of amenability and accessibility also may play an important role in the magnitude of the nonresponse bias. For some surveys, amenability and accessibility effects may be in the same direction, that is, those who are amenable may also be accessible or vice versa. For example, a telephone survey focusing on effectiveness of the national "Do Not Call" registry would likely miss many of those who are registered. Participation in this registry may be indicative of a lack of amenability to receive unsolicited calls, and may also be a barrier to accessibility (particularly if many registry participants also use other mechanisms to prevent unsolicited contact, such as answering machines or call screening devices). In other cases, accessibility and amenability may be inversely related. In the travel survey example, those who travel extensively might be more willing to respond to a travel survey, but they may be more difficult to contact due to their absences from home.

25.4.2 Limitations of the Analysis

An important limitation of the type of analysis presented here is that only biases within the observed levels of response can be investigated, and biases associated with units that never responded cannot be directly evaluated. Without knowing the characteristics of those who did not respond to the surveys at all, it is not possible to ascertain these biases in this type of study. As a somewhat extreme example, suppose some of the sample units were persistent nonrespondents, defined as a group or set of sampled units who would not respond to the survey under any reasonable survey conditions. If these persistent nonrespondents were very different from the other sampled units, then the type of nonresponse bias analyses presented here would not be useful at identifying this source of bias.

This limitation is not as severe as it might seem in this situation because other independent assessments of the bias in survey estimates have been conducted for these studies. These other assessments have not suggested large nonresponse biases. In NHES (including the AELL survey), the estimates from the survey are compared to data from other surveys with high response rates, notably, the CPS, to identify potential biases. In general, these analyses have revealed few significant differences between NHES and CPS estimates of common data elements, and those differences that have been observed are generally small in magnitude (Nolin et al., 2004). While this type of comparison confounds nonresponse bias with differences due to other sources of error, the absence of large differences is consistent with the analyses presented here. For the Pew studies, Keeter et al. (2000) and Keeter et al. (2006) compared the demographic characteristics of responders with population estimates from the CPS and found few differences and no large substantive differences. Comparisons for other behavioral or attitudinal measures with other high response rate surveys are not available. Nevertheless, for the Pew studies, there is other convincing evidence of the absence of bias. The final Pew preelection survey, which was conducted the weekend before the 2004 general election and used procedures very similar to other survey organizations, nearly exactly matched the election outcome (51 percent Bush, 48 percent Kerry) even though there were many factors other than nonresponse that might have caused differences between the survey estimates and the actual vote. These results indicate that it is very unlikely that nonresponse bias could be a large contributor to error in the Pew surveys.[7]

The consistency of the findings between the Pew and AELL surveys and those of other studies (e.g., Curtin et al., 2000; Merkle and Edelman 2002; Groves et al., 2004) suggests that the Pew and AELL results are not merely case studies, and some generalization to other studies is not unreasonable. However, it would be unwise to assume that these results apply to all RDD surveys. Nonresponse can and does result in biases in survey estimates in many RDD surveys, especially when amenability and accessibility are associated with the measures of interest. Further research, especially from studies with varying sponsors, topics, protocols, and response rates would help consolidate understanding of the potential for nonresponse due to different sources.

25.4.3 Implications for Data Collection and Estimation Methods

Historically, response rates have been viewed as an important indicator of survey quality. For example, some scholarly journals refuse to publish articles that have response rates that are below a specific rate. The presumption underlying this

[7]The average of several major national surveys taken in the days leading up to the election showed President George W. Bush with a 1.5 percentage point advantage over Senator John Kerry, a slight underestimation of Bush's actual margin of victory (2.4 percent). Polling summaries can be found at http://www.realclearpolitics.com/bush_vs_kerry.html, accessed on April 24, 2006.

perspective is that lower response rates are associated with greater potential for nonresponse bias. However, the findings of this and other recent research have provided little evidence of a correlation between overall response rates and bias. Both theoretically and empirically, it is clear that nonresponse bias is a function of the interaction between differential nonresponse and the survey characteristics. A logical conclusion is that it may be more effective to focus efforts on design features and data collection strategies to minimize the potential for bias rather than simply increasing the overall response rate. The judicious selection of strategies is especially important in that most surveys are conducted under temporal and financial resource constraints.

One way to reduce the potential for bias is to carefully evaluate and test those factors that might induce differential response rates. For example, in RDD surveys the goal of the initial introduction is communicate the legitimacy of the survey and its content in a way as to promote interest as broadly as possible among the sample members. In designing surveys with topics that may be strongly associated with amenability, especially those that involve sensitive content, introductions that have polarizing effects should be avoided. Cognitive testing and pretesting introductions specifically for this type of effect may be very useful. Another technique that could be used is to seed the sample with persons who have known interest in the survey to evaluate whether these persons respond at the same rate as the overall population (see Groves et al., 2004 for examples).

Similarly, some methods exist to avoid or reduce nonresponse bias due to inaccessibility. One approach is to focus data collection strategies and efforts on the hard-to-reach respondents. In an RDD survey in which the survey content is likely to be correlated to availability, a data collection protocol that includes a relatively large number of call attempts on different days and at different times, over a relatively long period may be appropriate. Another possibility is to allow for proxy respondents, such as other adults in the household responding for those that are away on travel. Weighting adjustments might also be devised to reduce this bias. For example, if key characteristics are related to the size of the household and one-person households are harder to contact, then adjustments that account for household size might reduce bias.

This research furthers the idea that the focus in RDD surveys should be on identifying potential sources of bias, concentrating efforts in data collection on these sources even if it means further reducing the overall response rate, and using auxiliary data to the greatest extent possible in weighting to further reduce biases. As technologies continue to evolve and mobile phones and the Internet become even more important, some of these relationships may also change. The theory shows that the source of the bias, not the absolute magnitude, is the most important variable that must be monitored.

CHAPTER 26

Evaluating and Modeling Early Cooperator Effects in RDD Surveys

Paul P. Biemer

RTI International and University of North Carolina, USA

Michael W. Link

Centers for Disease Control and Prevention, GA, USA

26.1 INTRODUCTION

During recent years, response rates for random digit dial (RDD) telephone surveys have continued to decline (Groves and Couper, 1998; de Leeuw and de Heer, 2002). For example, in the behavioral risk factor surveillance system (BRFSS), the median state-level response rate dropped from approximately 65 percent in 1996 to just over 50 percent in 2004 (Centers for Disease Control, 2005a).[1] Similarly, response rates for the survey of consumer attitudes and behavior (SCA) have declined roughly 1 percent per year for the past quarter century, with the rate of decline being greater in more recent years (Curtin et al., 2005).[2] Traditionally, low response rates are regarded as an indication of high nonresponse bias and poor data quality. Several recently published studies, however, report that for response rates within the range of 40–70 percent, nonresponse bias is constant (Curtin et al., 2000; Keeter et al., 2000;

[1]The behavioral risk factor surveillance system (BRFSS) is one of the world's largest, on-going RDD telephone surveys conducting more than 300,000 surveys annually, tracking health conditions and risks behaviors in the United States (Mokdad et al., 2003).

[2]The survey of consumer attitudes and behavior (SCA) measures changes in consumer attitudes and expectations in an effort to understand why these changes occur; to evaluate how they relate to consumer decisions to save, borrow, or make discretionary purchases; and to forecast changes in aggregate consumer behavior (Curtin et al., 2005).

Advances in Telephone Survey Methodology, Edited by James M. Lepkowski, Clyde Tucker, J. Michael Brick, Edith de Leeuw, Lilli Japec, Paul J. Lavrakas, Michael W. Link, and Roberta L. Sangster

Carlson and Strouse, 2005). This finding may indicate that persons who cooperate early in a survey do not differ appreciably on measures of interest from persons who ultimately respond to the survey but require more contact attempts to be reached or more persuasive invitations to participate. This finding would further explain why a survey with a 40 percent response rate, representing mostly early cooperators, would show no greater bias due to nonresponse than the same survey pushed to a 70 percent response rate. Such a conclusion has important implications, because if the bias due to nonresponse is minimal for key measures of interest, then survey researchers would have little motivation or reason to continue to expend resources trying to achieve a higher response rate by pursuing reluctant or difficult to reach respondents.

Lacking sufficient theories capable of predicting when nonresponse rates imply nonresponse bias and when they do not, survey researchers have taken different approaches to handling potential nonresponse bias. One approach is to maximize response rates assuming that as they increase the potential for nonresponse decreases, possibly as a result of either reducing the size of the nonresponding group or from interviewing a more diverse and representative sample of respondents (Keeter et al., 2000). In either case, the implication is that as response rates increase, the survey estimates will likely be less biased, thereby making the improvement of response rates an essential goal for minimizing nonresponse bias.

A second perspective assumes that key characteristics of the survey design and administration such as topic, burden, sponsorship, and interviewer behavior directly affect survey participation and that potential respondents who find one or more of these characteristics compelling will tend to participate, whereas other people will remain nonrespondents. "Thus," as noted by Keeter and colleagues (2000: 127), "as the response rate increases, the respondent pool is being increased by more of those persons located and contacted for whom the survey conditions are minimally acceptable." This statement implies that survey distributions on variables of interest are likely to remain relatively fixed as the response rate increases and that exhaustive attempts to improve rates may actually exacerbate nonresponse bias by increasing the magnitude of the differences between respondents and nonrespondents. A better understanding of the relationships among early responders, later responders, and nonrespondents may shed valuable light on which of these two perspectives on nonresponse are more applicable as well as suggest new statistical approaches for handling nonresponse bias.

26.1.1 Early Cooperator Effect

In this paper, a *cooperator* will refer to a sample member that eventually will respond at some callback or request to participate. Cooperators that respond to initial calls (say, five or fewer callbacks) will be called *early cooperators* and those that respond beyond these initial calls (say, six or more) will be called *later cooperators*. The *early cooperator effect* (ECE) is the expected difference between two estimates of a population mean: one based upon early cooperators and the other based upon all cooperators (i.e., early plus later cooperators). Research on ECE has a long history in survey research stretching back to the late 1940s and 1950s (Politz and Simmons, 1949, 1950; Simmons, 1954). Yet despite the relatively large body of literature on this subject, there is little consensus on exactly when ECE may become problematic and how such effects might be effectively dealt with without the expenditure of considerable resources.

A number of studies have reported differences in the demographic distribution of early and later responders (Lavrakas et al., 1992; Triplett et al., 1996). Some studies found that reluctant or difficult-to-reach respondents tended to be older and less educated than respondents who cooperated earlier; however, differences with respect to income, occupation, race, and marital status were less consistent (O'Neil, 1979; Groves, 1989; Lavrakas et al., 1992; Kristal et al., 1993; Triplett et al., 1996). Other studies found education and race to be the more important distinguishing characteristics, with age showing inconsistent effects across early and later responders and initial refusers (Voigt et al., 2003). Similar findings have led some researchers to conclude that no clear demographic patterns exist that are predictive of ECE across different populations and study topics (Fitzgerald and Fuller, 1982; Lin and Schaeffer, 1995; Voigt et al., 2003).

A central issue to the ECE debate is the potential biasing effects on survey estimates.[3] Again, the research in this area, although broad in scope, is narrow in practical guidance. A number of studies have shown that prevalence rates on measures of interest often differ among early responders, later responders, and nonrespondents because these subgroups represent different segments of the study base in regard to the measure of interest (Oakes et al., 1973; Siemiatycki and Campbell, 1984; Cottler et al., 1987; Macera et al., 1990; Decouflé et al., 1991; Paganini-Hill et al., 1993; Helasoja et al., 2002; Voigt et al., 2003; and Stang and Jöckel, 2004). Other analyses, however, have drawn very different conclusions, reporting that although the characteristics of early and later responders may differ, the inclusion of data from reluctant or hard-to-reach respondents has little effect on the final results obtained (Siemiatycki and Campbell, 1984; Lavrakas et al., 1992; Kristal et al., 1993; Brøgger et al., 2003). As Rogers and colleagues (2004, 89) noted, "... the implications of these findings regarding potential response bias remains complex and seemingly situational."

Most ECE research has implicitly or explicitly used a "continuum of resistance" model, assuming that nonparticipants are both homogeneous, and similar to survey participants who require considerable effort to be interviewed (Filion, 1976;

[3]In addition to the potential problem of ECE, reluctant responders may also provide less accurate data than early cooperators (Cannell and Fowler 1963). Studies have shown that converted refusers have high item nonresponse due to "refusals to answer" or "don't knows" (Blair and Chun 1992; Triplett et al 1996). Lower motivation to participate in surveys is hypothesized to lead to "satisficing", that is, applying minimal cognitive effort in answering a question (Krosnick and Alwin 1987). Interviewer behavior may inadvertently encourage satisficing by treating reluctant respondents more gingerly, with less probing, particularly late in the field period (Triplett et al 1996). Reluctant responders may also tend to respond thoughtlessly resulting in higher rates of misclassification error later in a study (Stang and Jöckel 2004). These considerations call into question the practice of single-mindedly pursuing high response rates (Voigt et al., 2005).

Fitzgerald and Fuller, 1982; Lin and Schaeffer, 1995). A large number of studies across a wide range of topics have used level of effort (LOE) or similar measures as a proxy for nonresponse to evaluate the potential for bias. These studies have reported mixed results. Some of them found strong effects on final estimates when the responses from reluctant responders were removed from the final estimates (Hill et al., 1997; Parker and Dewey, 2000), whereas for other studies the effects were small (Sheikh and Mattingly, 1981, Siemiatycki and Campbell, 1984; Paganini-Hill et al., 1993; Boersma et al., 1997; Etter and Perneger, 1997; Gasquet et al., 2001; Holt et al., 1997; McCarthy and MacDonald, 1997; Carlson and Strouse, 2005; Battaglia et al., 2007, Chapter 24, in this volume).

An alternative approach uses a "class model" and assumes greater heterogeneity among nonrespondents by grouping them into classes based on their behavior during the course of attempts to contact them. By identifying groups of respondents whose behavior before participation appears to have been similar to that of nonrespondents, these participant groups can be used to estimate characteristics of nonparticipants (O'Neil, 1979; Stinchcombe et al., 1981; Smith, 1984; Lin and Schaeffer, 1995). For example, Lin and Schaeffer (1995) defined seven groups of participants and nonparticipants: (1) final refusals, (2) hard refusals who are converted, (3) soft refusals who are converted, (4) interviewed, easy to locate (fewer than four attempts), (5) interviewed, hard to locate (five or more calls), (6) unlocated (have telephone number), and (7) unlocated (no number found). After examining both the class and the continuum of resistance models, however, Lin and Schaeffer (1995) concluded that neither approaches provided a dependable framework for predicting the effects of nonresponse on estimates derived from a given sample.

26.1.2 Objectives of Study

For RDD surveys, limited data are available on nonrespondents that can be effectively used for traditional nonresponse adjustment approaches. However, a widely used method for reducing the nonresponse bias is the poststratification adjustment (PSA) (see, for example, Massey and Botman, 1988; Carlson and Strouse, 2005). This method requires the joint distribution of the control variables to be used in the PSA be known for the telephone population (for example, counts of persons in telephone households by age, race, and sex). These data are usually drawn from the latest census or, if census data are not current, from a large-scale current survey such as the U.S. Current Population Survey.

The PSA has several disadvantages in adjusting for nonresponse. First, for state estimates and smaller areas, the sample sizes for the external data may be quite small and unsuitable for poststratification purposes. Second, the variables available for the PSA may be confined to demographic characteristics that, as Groves and Couper (1998) discuss, may not be the best predictors of response propensity. Thus, the adjustment may not be effective at reducing nonresponse. Third, often the population of interest for the RDD survey may be such that external data for the PSA are not available. For example, the survey may be restricted to vehicle owners or cigarette smokers for whom reliable post-stratifying data are not available at any level.

In this paper, we discuss an approach for adjusting for nonresponse bias that can be used for RDD surveys and for other surveys when data on the LOE associated with sample units are available either in combination with or in lieu of a PSA. Our proposed methodology is an extension of the method originally proposed by Drew and Fuller (1980, 1981) and Drew and Ray (1984), all stemming from early ideas of Politz and Simmons (1949, 1950). Using the call history data captured during data collection, response propensity is modeled as a function of LOE for a sample unit and one or more grouping variables may be known only for the respondents in the sample. The model produces an estimate of the joint distribution of the grouping variables that can then be used as weighting adjustments to compensate for nonresponse.

Drew and Fuller's model also allows for the possibility that *hard-core nonrespondents* (HCNRs) exist for some survey designs. HCNRs might also be called *noncooperators* in our terminology since they are sample members that have 0 probability of responding to any callback under the survey design. Drew and Fuller define total nonresponse bias as the combination of the biases due to the nonresponding cooperators and the HCNRs.

We propose two models for adjusting for nonresponse bias in an RDD survey: one that is aimed at reducing the ECE, or equivalently, the bias associated with the exclusion of later cooperators and another that attempts to adjust for total nonresponse bias, i.e., both the ECE and the bias due to HCNRs.

Considerable research has gone into determining the LOE (typically in terms of numbers of calls or contacts) that will optimize response, while minimizing bias and costs (Rao, 1982; Croner et al., 1985; Fitch, 1985; Groves, 1989; Kalsbeek et al., 1994). If successful, our approach has important implications for saving survey costs and would also allow researchers to have a higher degree of confidence in estimates while reducing the LOE and costs associated with obtaining those results.

In the following section, we consider the evidence for ECE from a major RDD survey, the BRFSS. This section demonstrates that for some characteristics and population subgroups, the magnitude of ECE can be problematic. Section 26.3 develops the model for adjusting for ECE and nonresponse bias and various elaborations of the basic model for practical use in RDD surveys. Section 26.4 presents results that demonstrate the proposed method's performance for BRFSS. Section 26.5 provides a summary of the findings and conclusions regarding the utility of the methodology for general RDD surveys.

26.2 EARLY COOPERATOR EFFECTS IN BRFSS

26.2.1 Description of BRFSS Dataset

In 1984, the Centers for Disease Control and Prevention (CDC) initiated the state-based BRFSS to collect prevalence data on risk behaviors and preventive health practices that affect health status. As one of the largest RDD telephone health surveys in the world, BRFSS is a monthly, state-based, cross-sectional survey administered by the 50 state health departments and health departments in the District of Columbia, Puerto Rico, Guam, and the Virgin Islands with support from CDC. Interviews are conducted with the noninstitutionalized adult population aged 18 years or older, focusing on health practices and risk behaviors linked to chronic diseases, injuries, and preventable infectious diseases. One adult within a household is randomly selected for interview. Interviewing is conducted by different data collection organizations for each state (although some states use common contractors) using a standardized questionnaire and protocols to determine the distribution of risk behaviors and health practices among adults. Responses are forwarded to CDC, where the monthly data are aggregated for each state, returned with standard tabulations, and published by each state at year's end. The BRFSS questionnaire was developed jointly by CDC and the states. Data derived from the questionnaire provide health departments, public health officials, and policy-makers with necessary public health information. When combined with mortality and morbidity statistics, these data enable public health officials to establish policies and priorities and to initiate and assess health promotion strategies. The content and size of the BRFSS make it an ideal vehicle for investigating ECE.

26.2.2 Key Variables for the Analysis

The data examined here come from the 2004 BRFSS, which contains survey data as well as call center paradata for more than 1.9 million telephone numbers dialed in an effort to obtain almost 300,000 interviews across the 3143 counties in the United States. Early and later cooperators were defined based on the overall number of call attempts made to obtain a completed interview. The BRFSS standard protocol requires that a minimum of 15 call attempts be made to reach a final disposition for a telephone number (i.e., to assign a code of complete, refused, language barrier, incapable, unavailable, or ineligible). However, states may choose to make more than 15 call attempts. For the purposes of the present analysis, we defined an "early cooperator" as a sample person who was interviewed with five or fewer call attempts. All other respondents (i.e., those requiring more than five attempts) were labeled as "later cooperators."

Information on the demographic characteristics and health-related issues derived from either responses to the 2004 BRFSS questionnaire or information on the sample file. The eight demographic characteristics used in our analyses included sex, race, age, education, marital status, household size, number of adults in the household, and type of area in which a person lives based on the metropolitan statistical area (MSA) code associated with the telephone number (e.g., city center, outside of center, in county, in an MSA with no center, and outside of an MSA). In the analyses, we looked at more than 20 different health-related survey responses, of which we focus on seven in this chapter: access to health care coverage, general health status, asthma, diabetes, having a limiting physical, mental, or emotional problem, alcohol consumption, and receiving an influenza vaccination (see Table 26.1 for exact wording of these items).

Table 26.1. Health-related Variables Used in Analysis

Variable	Question wording
Health coverage	Do you have any kind of health care coverage, including health insurance, prepaid plans such as HMOs, or government plans such as medicare?
General health	Would you say that in general your health is excellent, very good, good, fair, or poor?
Asthma	Have you ever been told by a doctor, nurse or other health professional that you had asthma?
Diabetes	Have you ever been told by a doctor you have diabetes?
Drink alcohol	A drink of alcohol is 1 can or bottle of beer, 1 glass of wine, 1 can or bottle of wine cooler, 1 cocktail, or 1 shot of liquor. During the past 30 days, how many days per week or per month did you have at least one drink of alcoholic beverages?
Physical, mental or emotional problem	Are you limited in any activities because of physical, mental, or emotional problems?
Influenza shot	During the past 12 months, have you had a flu shot?

To observe the effects of early cooperators on the estimates, only the base weight associated with each interviewed unit was used in the analysis. This weight took into account the probabilities of selection at the household level as well as the within-household random selection. Postsurvey adjusted weights were not used because their function is to attenuate the effects of nonresponse and frame undercoverage of the telephone population.

26.2.3 Demographic Differences between Early and Later Cooperators

Replicating earlier studies, we determined whether early and later cooperators differed significantly on demographic characteristics. One school of thought is that people who respond, or cooperate, later in a survey should have different characteristics to the extent that response propensities are related to those characteristics. A counter viewpoint holds, however, that respondents interviewed later should not differ appreciably from cooperators interviewed earlier, because both sets of participants are responding to similar stimuli meant to encourage participation; i.e., they are both responders. Table 26.2 provides a summary of the characteristics of early and later cooperators.

Slightly more than two thirds (68.3 percent) of the respondents were early cooperators, completing a survey within five call attempts, whereas 31.7 percent were later cooperators. Looking first at bivariate relationships, early and later responders differed significantly on all eight demographic variables examined. The largest differences were noted across age and racial/ethnic groups. Slightly more than 80 percent of adults aged 65 or older were early cooperators, compared with 61.4 percent of adults aged 18–34 years. In terms of race/ethnicity, a higher

Table 26.2. Demographic Characteristics and Size of Early and Later Cooperator Groups

		Early cooperators		Later cooperators		
Characteristics	(*n*)	Percent	95 percent CI	Percent	95 percent CI	AOR
Total	293,881	68.3	68.2,68.5	31.7	31.5,31.8	—
Sex						
Male	113,759	65.0	64.7,65.3	35.0	34.7,35.3	1.00
Female	180,066	70.4	70.2,70.7	29.6	29.3,29.8	1.28
Race-ethnicity						
White, non-Hispanic	234,501	70.0	69.8,70.1	30.0	29.9,30.2	1.00
Black, non-Hispanic	23,502	63.4	62.8,64.0	36.6	36.0,37.2	0.80
Hispanic	18,158	56.9	56.1,57.6	43.1	42.4,43.9	0.65
Age						
18–34 years	60,813	61.4	61.0,61.8	38.6	38.2,39.0	1.00
35–49 years	85,391	62.8	62.5,63.1	37.2	36.9,37.5	1.04
50–64 years	79,007	69.3	68.9,69.6	30.7	30.4,31.1	1.35
65 or more years	66,626	80.9	80.6,81.2	19.1	18.8,19.4	2.47
Education						
< High school	30,751	71.1	70.6,71.6	28.9	28.4,29.4	1.00
High school	89,797	69.8	69.5,70.1	30.2	29.9,30.5	0.95
> High school	172,628	67.1	66.9,67.3	32.9	32.7,33.1	0.91
Marital status						
Single	41,441	62.3	61.8,62.8	37.7	37.2,38.2	1.00
Married	167,808	68.0	67.8,68.2	32.0	31.8,32.2	1.10
Divorced/separated	48,445	67.2	66.8,67.6	32.8	32.4,33.2	1.00
Widowed	34,145	80.6	80.6,80.1	19.4	19.0,19.9	1.21
Number of adults						
One	97,949	70.9	70.6,71.2	29.1	28.8,29.4	1.00
Two	159,286	67.8	67.5,68.0	32.2	32.0,32.5	1.00
Three or more	36,636	64.0	63.6,64.5	36.0	35.5,36.4	0.96
Number of children in HH						
None	189,110	71.1	70.9,71.3	28.9	28.7,29.1	1.00
One or more	102,727	63.6	63.0,63.7	36.4	36.3,36.8	0.98
MSA code						
City center	80,850	66.5	66.2,66.8	33.5	33.2,33.8	1.00
Outside center	41,058	65.1	64.7,65.6	34.9	34.4,35.3	0.91
In county	56,614	66.2	65.8,66.6	33.8	33.4,34.2	0.95
In MSA with no center	7,567	57.4	56.3,58.5	42.6	41.5,43.7	0.65
Outside of MSA	107,792	72.8	72.6,73.1	27.2	26.9,27.4	1.25

AOR, adjusted odds ratio; CI, confidence interval.

percentage of non-Hispanic whites (70.0 percent) were early cooperators than were non-Hispanic blacks (63.4 percent) and Hispanics (56.9 percent).

Using logistic regression to model, the odds of a respondent being an early cooperator, we obtained a clearer picture of the background characteristics associated with early cooperation in the BRFSS. Age showed the strongest effect: the odds of respondents aged 65 years and older being early cooperators were nearly two-and-a-half times the odds for respondents aged 18–34 years. Respondents aged 50–64 years were 35 percent more likely to be early cooperators than respondents aged 18–34 years, whereas the odds for cooperators aged 35–49 years were only 4 percent higher. Across racial/ethnic groups, the odds of non-Hispanic blacks being early cooperators were almost 20 percent lower than whites, whereas the odds for Hispanics were 35 percent lower. Additionally, the odds of a woman being an early cooperator were 28 percent higher than for men. Several other factors (i.e., education, marital status, number of adults in household, and type of area in which person lives) were significantly related to the likelihood of a respondent being an early cooperator; however, the impact of these variables tended to be more modest.

26.2.4 Differences between Early and Later Cooperators in Selected Health-Related Variables

In Table 26.3, we examine the differences between early and later cooperators across 20 different health-related variables, focusing here on seven of them for illustrative purposes. These differences characterize the two groups of cooperators and suggest the potential bias with including only early cooperators in the survey. Even at the aggregate level, considerable variation in the magnitude of the differences across the variables was observed. For instance, for asthma the overall difference was quite small at 1 percent. In contrast, the difference between the two groups on the question of having received an influenza vaccination was considerably larger at 7.6 percent, indicating that early cooperators were much more likely than later cooperators to state that they received vaccination. The reverse relationship, however, was found with alcohol consumption: the percentage of early cooperators who stated that they had a drink of alcohol within the past 30 days was 6.4 percent lower than the percentage of later cooperators who said they drank.

Even greater differences were noted when the overall estimates were disaggregated. For example, on the asthma question, little variation was noted across the various subgroups. In contrast, for the alcohol question significant differences were noted for nearly every subgroup, with the differences between early and later cooperators being highest for respondents who were non-Hispanic whites, aged 65 years or older, had only a high school education or less, lived in a one adult household, had no children living in the household, and lived in less populated areas. The most persistent differences across variables were found among respondents who had less than a high school education, lived in households with only one adult, and lived in households with no children. Differences were significantly higher on most of the health-related items for these three characteristics. Although lower levels of education have been shown in previous studies to be related to early and later cooperator difference, the other two characteristics have not. Other factors identified in previous studies as having an important impact, in particular race and age, had more periodic, less consistent effects here.

But while early and later cooperators do vary considerably on a number of these survey items, does the exclusion of later responders bias the final estimates? Table 26.4,

Table 26.3. Difference Between Early and Later Cooperators in Health and Risk Measures

Demographic characteristics	Health conditions and risk factors													
	Health care coverage		General health		Asthma		Diabetes		Drink alcohol		Physical, mental, emotional problem		Influenza shot	
	% Diff	95 % CI	% Diff	95 % CI	% Diff	95 % CI	% Diff	95 % CI	% Diff	95 % CI	% Diff	95 % CI	% Diff	95 % CI
Total	2.4	1.9,2.9	3.6	3.1,4.1	1.0	0.5,1.4	2.2	1.8,2.5	−6.4	−7.1,−5.7	6.5	6.0,7.0	7.6	6.9,8.2
Sex														
Male	2.3	1.5,3.1	3.4	2.7,4.2	0.2	−0.2,0.6	2.5	1.9,3.0	−4.5	−5.6,−3.5	6.5	5.8,7.3	7.9	6.9,8.9
Female	2.3	1.6,2.9	3.6	2.9,4.2	1.2	0.6,1.8	1.9	1.4,2.4	−6.2	−7.1,−5.3	6.3	5.6,6.9	7.1	6.2,7.9
Race/ethnicity														
White, non-Hispanic	0.7	0.2,1.2	5.1	4.7,5.6	0.4	−0.1,0.9	2.5	2.1,2.8	−8.7	−9.4,−7.9	6.3	5.8,6.9	7.7	7.0,8.4
Black, non-Hispanic	−0.1	−1.9,1.7	6.5	4.8,8.2	1.4	−0.1,3.0	3.8	2.5,5.1	−5.7	−7.8,−3.5	6.5	5.0,8.1	2.9	1.1,4.8
Hispanic	2.5	−0.2,5.2	−1.6	−4.1,0.9	3.2	1.6,4.9	0.4	−1.1,1.9	−1.9	−4.7,0.8	3.1	1.5,4.7	2.9	0.7,5.1
Age, years														
18–34	−0.1	−1.4,1.2	−0.5	−1.3,0.4	1.0	−0.1,2.0	0.3	−0.2,0.7	−4.0	−5.4,−2.6	2.1	1.4,2.9	0.8	−0.4,1.9
35–49	0.3	−0.6,1.2	1.1	0.2,2.0	1.6	0.8,2.4	0.3	−0.3,0.9	−3.4	−4.6,−2.2	3.9	3.0,4.7	2.0	1.0,3.0
50–64	0.4	−0.5,1.2	3.2	2.1,4.3	1.7	0.8,2.6	1.6	0.7,2.5	−4.7	−6.1,−3.4	6.7	5.6,7.8	3.1	1.8,4.4
65 or more	0.8	0.2,1.3	2.7	1.2,4.1	0.8	−0.2,1.8	0.2	−1.1,1.5	−7.1	−8.8,−5.5	6.2	4.8,7.7	3.2	1.6,4.8
Education														
< High school	11.3	9.0,13.5	3.2	0.9,5.5	4.0	2.6,5.5	2.8	1.3,4.4	−7.7	−9.9,−5.7	10.4	8.6,12.1	8.4	6.4,10.5
High school	2.4	1.4,3.4	4.4	3.4,5.3	0.3	−0.6,1.1	2.7	2.1,3.4	−7.9	−9.1,−6.6	5.9	5.0,6.8	8.6	7.5,9.7
> High school	1.0	0.4,1.5	3.0	2.5,3.5	0.8	0.2,1.4	1.7	1.2,2.1	−4.9	−5.4,−4.1	5.9	5.3,6.5	6.9	6.1,7.8

Marital status														
Single	0.6	−0.3,1.3	3.0	2.3,3.6	0.6	0.1,1.3	1.9	1.4,2.3	−5.1	−6.1,24.1	4.8	4.1,5.6	3.5	2.6,4.4
Married	0.7	0.4,1.0	3.4	3.1,3.8	0.8	0.4,1.1	1.9	1.6,2.2	−6.0	−6.6,25.5	5.3	4.9,5.7	6.8	6.2,7.3
Other status	3.9	3.3,4.5	6.6	6.0,7.1	−0.3	−0.7,0.3	3.4	3.0,3.9	−10.2	−11.0,29.5	7.8	7.1,8.4	10.3	9.6,11.0
Number of adults in household														
One	2.4	1.6,3.1	6.9	6.0,7.7	0.3	−0.4,1.1	3.6	3.0,4.3	−9.7	−10.8,28.6	8.6	7.7,9.5	10.2	9.1,11.2
Two	1.7	1.2,2.3	3.4	2.8,4.0	0.5	−0.1,1.1	1.8	1.4,2.3	−6.1	−6.9,25.3	5.6	5.0,6.2	7.6	6.8,8.4
Three or more	2.5	0.9,4.0	1.4	0.1,2.8	2.5	1.3,3.6	1.6	0.7,2.6	−4.8	−6.6,23.0	5.9	4.7,7.1	3.7	2.2,5.3
Number of children in household														
None	2.3	2.0,2.6	6.1	5.6,6.5	0.5	0.2,0.9	2.7	2.4,3.3	−9.3	−9.8,28.8	7.3	6.8,7.7	9.2	8.7,9.7
One or more	−0.1	−0.6,0.4	1.1	0.7,1.5	0.8	0.4,1.3	0.9	0.5,1.2	−4.2	−5.0,23.7	3.1	2.7,3.5	2.0	1.4,2.5
Metropolitan statistical code														
City center	3.4	2.4,4.4	2.6	1.6,3.5	1.4	0.5,2.2	1.6	0.9,2.3	−4.1	−5.3,22.9	6.5	5.6,7.3	7.8	6.6,8.9
Outside center	2.8	1.5,4.2	1.7	0.4,0.3	0.7	−0.5,1.8	1.7	0.7,2.6	−3.3	−5.1,21.5	5.9	4.6,7.1	7.8	6.2,9.4
In county	0.9	−0.1,1.8	4.5	3.6,5.4	0.8	−0.2,1.7	2.4	1.7,3.1	−7.7	−9.1,26.4	5.8	4.8,6.8	6.8	5.5,8.1
Outside of MSA	2.5	1.6,3.4	5.2	4.3,6.0	1.3	0.6,2.1	3.0	2.3,3.6	−8.2	−9.3,27.0	6.7	5.8,7.6	7.2	6.1,8.3

% Diff, percent age for early cooperator (LOE 1–5)—percent age for later responder (LOE 5–15); 95 % CI, 95 percent confidence interval.

Table 26.4. Difference between Early Cooperators and Overall Estimates in Health and Risk Measures

	Health conditions and risk factors													
	Health care coverage		General health		Asthma		Diabetes		Drink alcohol		Physical, mental, emotional problem		Influenza shot	
Demographic characteristics	% Diff	95 % CI	% Diff	95 % CI	% Diff	95 % CI	% Diff	95 % CI	% Diff	95 % CI	% Diff	95 % CI	% Diff	95 % CI
Total	0.8	0.6,1.0	1.2	1.1,1.4	0.3	0.2,0.5	0.7	0.6,0.9	−2.2	−2.4,−1.9	2.1	2.0,2.3	2.6	2.3,2.8
Sex														
Male	0.9	0.5,1.2	1.3	1.0,1.6	0.1	−0.2,0.3	0.9	0.7,1.1	−1.7	−2.1,−1.3	2.4	2.1,2.7	2.9	2.6,3.3
Female	0.7	0.5,0.9	1.1	0.9,1.3	0.4	0.2,0.6	0.6	0.4,0.8	−2.0	−2.3,−1.7	1.9	1.7,2.1	2.2	2.0,2.5
Race/ethnicity														
White, non-Hispanic	0.2	0.1,0.4	1.6	1.5,1.8	0.1	−0.1,0.3	0.8	0.7,0.9	−2.7	−2.9,−2.5	1.9	1.8,2.1	2.4	2.2,2.6
Black, non-Hispanic	−0.1	−0.7,0.6	2.5	1.9,3.1	0.6	−0.1,1.1	1.5	1.0,2.0	−2.2	−3.1,−1.4	2.4	1.8,3.0	1.1	0.4,1.8
Hispanic	1.2	−0.1,2.4	−0.7	−1.9,0.4	1.5	0.7,2.3	0.2	−0.5,0.9	−0.9	−2.2,0.4	1.4	0.7,2.2	1.4	0.3,2.4

Age, years																
18–34	−0.1	−0.6,0.5	−0.2	−0.5,0.2	0.4	−0.1,0.8	0.1	−0.1,0.3	−1.6	−2.2,−1.1	0.9	0.6,1.2	0.3	−0.2,0.8
35–49	0.1	−0.2,0.5	0.4	0.1,0.7	0.6	0.3,0.9	0.1	−0.1,0.3	−1.3	−1.7,−0.8	1.5	1.1,1.8	0.8	0.4,1.2
50–64	0.1	−0.2,0.4	1.0	0.7,1.4	0.5	0.3,0.8	0.5	0.2,0.8	−1.5	−1.9,21.1	2.1	1.8,2.4	1.0	0.6,1.4
65 or more	0.2	0.1,0.3	0.5	0.3,0.8	0.2	−0.1,0.4	0.1	−0.2,0.3	−1.5	−1.8,−1.1	1.2	1.0,1.5	0.7	0.3,1.0
Education														
< High school	3.9	3.1,4.6	1.1	0.3,1.9	1.4	0.9,1.9	1.0	0.4,1.5	−2.6	−3.4,−1.9	3.5	2.9,4.1	2.9	2.2,3.6
High school	0.8	0.5,1.1	1.4	1.1,1.8	0.1	−0.2,0.4	0.9	0.7.1.1	−2.5	−2.9,−2.1	1.8	1.6,2.1	2.7	2.4,3.1
> High school	0.3	0.2,0.5	1.0	0.9,1.2	0.3	0.1,0.5	0.6	0.4,0.7	−1.7	−2.0,−1.4	2.0	1.8,2.2	2.4	2.1,2.7
Marital status														
Single	0.1	−0.2,0.5	1.2	1.0,1.5	0.3	0.1,0.6	0.7	0.5,0.9	−2.0	−2.4,21.6	1.9	1.6,2.1	1.4	1.1,1.7
Married	0.3	0.1,0.4	1.1	0.9,1.2	0.4	0.2,0.5	0.5	0.4,0.6	−1.9	−2.1,21.7	1.6	1.5,1.7	2.1	1.9,2.3
Other status	1.1	0.9,1.2	1.9	1.8,2.1	−0.1	−0.3,0.1	1.0	0.8,1.1	−2.9	−3.2,22.7	2.2	2.0,2.4	3.0	2.8,3.2
Number of adults in household														
One	0.7	0.5,0.9	2.1	1.8,2.3	0.1	−0.1,0.3	1.1	0.9,1.3	−2.9	−3.2,22.6	2.5	2.3,2.8	3.1	2.7,3.4
Two	0.6	0.4,0.8	1.1	0.9,1.3	0.2	−0.1,0.4	0.6	0.5,0.8	−2.0	−2.3,21.8	1.8	1.7,2.0	2.5	2.3,2.8
Three or more	0.9	0.4,1.5	0.5	0.1,1.1	0.9	0.5,1.4	0.6	0.3,1.0	−1.8	−2.5,21.1	2.2	1.7,2.7	1.4	0.8,2.0

which shows the estimated ECE for the characteristics in our analysis, attempts to address this question. In this table, the ECE was estimated as the difference between the mean based on just early cooperators and the corresponding mean based on the entire sample. We found that even for variables in which the difference in responses between early and later cooperators was relatively high, ECE estimates were minimal. For instance, on the alcohol and influenza vaccination items where the difference between the two groups was approximately 6–7 percent, the estimates of ECE were less than 3 percent (−2.2 percent for the alcohol consumption item and 2.6 percent for the influenza vaccination item). Moreover, even at the subgroup level the ECEs were relatively minor. For the alcohol item, none of the subgroup ECE estimates exceeded 3 percent. For the vaccination question, the estimated ECEs were 3.1 percent for the respondents with only one adult in the household and lower for all other subgroups.

Although these differences are relatively small, they may have practical importance for some applications. Moreover, although differences at the national level may be small, the same characteristics may exhibit much larger differences at the state or local levels. Extending our analysis to states and smaller geographic areas was beyond the scope of this paper. In the following section, we consider how the ECE can be reduced using a callback model for nonrespondents.

26.3 NONRESPONSE ADJUSTMENT USING NUMBER OF CALLBACKS

The idea of using callback data to adjust for nonresponse predates the paper by Drew and Fuller by many years. We noted the famous work of Politz and Simmons (1949, 1950) and Simmons (1954). In addition, various models were explored by Deming (1953), Dunkelberg and Day (1973), Proctor (1977), Potthoff et al. (1993), and Anido and Valdés (2000). In this section, we propose an approach that extends the ideas in Drew and Fuller (1980, 1981). In their paper, Drew and Fuller modeled nonresponse as a function of a single grouping variable, G, and the LOE required to obtain a response. They applied their method to a mail survey with two follow-up mailings (i.e., three total attempts). An RDD telephone survey will typically have many more follow-up attempts to obtain a response. In addition, unlike a mail survey, where the outcome of a mailing is either a returned questionnaire or no response, a telephone call attempt can have many different types of outcomes. These and other differences between the two modes will be considered as we extend Drew and Fuller's ideas to RDD surveys.

Many RDD surveys specify a rule for the minimum number of call attempts that should be made to each number to reach a final disposition of interviewed or refused. As previously noted the BRFSS requires that each telephone number be called at least 15 times, although data collectors are allowed to exceed this number and often do. Our motivation for considering callback models for nonresponse was twofold. First, considerable resources could be saved in RDD surveys if a smaller, fixed number of call attempts, say 5 or 10, were made to reach a final disposition. Whereas some variables may incur a relatively small ECE by limiting the call attempts, for other variables the ECE could be important as in the case of the BRFSS. Secondly, even when a maximum call attempts strategy is used, a callback model may still be useful for reducing the overall nonresponse bias in the final estimates. The next two sections examine these issues in the context of the BRFSS.

The key idea in Drew and Fuller's approach was to use the information on the outcome of each follow-up attempt to estimate response propensities for population subgroups and then adjust the base weights to compensate for nonresponse using the adjustment factors from the callback model. Another innovation was to incorporate a dichotomous latent variable representing two classes in the population: (1) people who will respond to some number of follow-up attempts and (2) people who have zero probability of responding at each attempt. Drew and Fuller referred to the latter group as the hard-core nonrespondents (HCNR). In our study, we applied a similar model to adjust for total nonresponse bias and to evaluate its effectiveness for situations in which the nonresponse bias was extreme.

26.3.1 Model

To illustrate the ideas and simplify the development of the model, assume that a simple random sample (SRS) of size n households is selected from a population of size N households. Initially we consider the case where only units of known eligibility are included in the analysis. Extensions for units with unknown eligibility (for example, never answered numbers) as well as unequal probability sampling are considered in Section 26.3.6.

At the conclusion of the survey, the eligible sample can be classified into three mutually exclusive categories or final dispositions: interviews, noninterviews, and noncontacts. For our purposes, partial interviews will be classified as "noninterviews," although in practice they could constitute a fourth category. *Noninterviews* will include refusals, language barriers, unavailable or infirm sample members, and other types of nonresponse obtained upon contact with the household. "Noncontacts" will include telephone numbers where an eligible sample member was never successfully contacted, although someone else in the household may have been.

The LOE variable is generally the maximum amount of effort applied to a case to obtain a final disposition and can be defined in various ways. For example, the LOE can be defined simply as the number of call attempts that were made for a case or, it can be a range of call attempts combined with other activities aimed at obtaining a response from a unit. The only requirement is that the LOE variable increases as the amount of effort expended towards contacting and interviewing the sample members increases. In addition, the maximum number of levels, L, should be less than 10 in order to avoid an overly complex model. When the number of call attempts is large, they can be combined into ranges such as 1, 2–3, 4–5,...,16+.

26.3.2 Notation and Assumptions

Let s denote the state of a unit at some point in time, where $s = 1$ if interviewed; $s = 2$ if noninterviewed; and $s = 3$ if not contacted. Note that, by definition, all units begin in state $s = 3$ and either remain there or will move to the other two states as the LOE increases. States 1 and 2 are assumed to be "absorbing" states, that is, once a case is interviewed or noninterviewed, the unit is no longer worked. Thus, we need to only consider the probabilities of transitions from state 3 to 1 and 3 to 2 because all other transitions have zero probability of occurrence.

Denote the LOE variable by $l = 1, \ldots, L$, where L is the maximum number possible LOE. At the conclusion of the survey, a maximum call attempt required to reach a final disposition can be assigned to every case denoted by l^*. In our application, l^* for a contacted case is the call attempt associated with the final contact with the unit. For noncontacts, $l^* = L$, however, in general it is the final attempt made to contact a unit. Cases that were not contacted by the end of the survey and that had not yet reached the LOE L (so-called *right-censored cases*) were excluded from our analysis in order to further simplify the exposition of the methodology. However, one possible extension of our model to include right-censored cases will be described in Section 26.3.6.

Thus each sample unit i can be associated with an $L \times 1$ outcome vector $\mathbf{O}_i = [o_{il}]'$ where o_{il} ($l = 1, \ldots, L$) is the state the unit occupied at LOE l. For units that reach state $s = 1$ or 2 at LOE l^*, we set $o_{il} = s$ for $l \geqslant l^*$. Thus, there are only $2L + 1$ possible outcome patterns. For example, the pattern $\mathbf{O}_i = [1, \ldots, 1]'$ denotes a case that was completed at $l^* = 1$ and $[2, \ldots, 2]'$ denotes a case that was noninterviewed at $l^* = 1$. Likewise, $[3, 1 \ldots 1]'$ denotes a case that was contacted and interviewed at $l^* = 2$. To simplify the notation, we drop the subscript i when it is clear we are referring to a single unit.

Denote the outcome pattern for a unit that reached absorbing state, s, at LOE l^* by $\mathbf{O}_{l^*,s}$. Note that $\mathbf{O}_{L,3}$ denotes outcome pattern $[3, \ldots, 3]'$ for never contacted units. Let $\pi_{l,s} = P(o_l = s)$ denote the probability that a unit is in state s at LOE l. Let $\pi_{l,s|l-1,s'} = P(o_l = s | o_{l-1} = s')$ for $l > 1$ denote the probability of being in state s at LOE l given the unit occupied state s' at LOE l-1. Then the probability of observing a case with outcome pattern $\mathbf{O}_{l^*,s}$ is

$$P(\mathbf{O}_{l^*,s}) = \pi_{l^*,s} = \pi_{o_1=3}\pi_{o_2=3|o_1=3} \ldots \pi_{o_{l^*}=s|o_{l^*-1}=3} \tag{26.1}$$

Drew and Fuller's approach assumes that the process is *stationary*, that is, $P(S_l)$ is the same for all l. In our framework, this means $\pi_{o_l = s | o_{l-1} = 3} = \pi_s$, say, for all $l > 1$ and (26.1) simplifies to

$$P(\mathbf{O}_{l^*,s}) = (\pi_3)^{l^*-1}\pi_s \tag{26.2}$$

which is the trichotomous geometric distribution. For our approach, stationarity is not required and Drew and Fuller's assumption will be relaxed.

Population heterogeneity of response propensities is modeled by introducing a grouping variable, G, and specifying that the π-probabilities depend on G. This may be a single variable or a combination of variables. We assume that G is observed for all interviewed units and is missing for all noninterviewed and noncontacts; however, the model can be easily extended to also include variables that are known for all sample members (such as state or region). In addition, the missing data mechanism is likely to be nonignorable, that is, the probability that the final disposition is noninterview or noncontact depends upon G. We next consider how these notions are represented in the model.

26.3.3 Grouping Variables

Consider a grouping variable G with categories $g = 1, \ldots, K$ that are known only for units that are ultimately interviewed (i.e., G is known when $o_{l^*} = 1$ for some l^* and unknown for all other patterns). Assume the π-probabilities defined above depend upon G and, thus, missingness of G also depends upon G. The focus of our analysis is to obtain an estimate of the population distribution of G that compensates for this missing information. Using maximum likelihood estimation of the incomplete data likelihood, an estimate of the distribution G will be obtained. If G is chosen to be highly correlated with the π-probabilities, weighting the sample by the estimate distribution of G will be effective for reducing nonresponse bias as will be shown later.

Let $n(l^*,s)$ denote the number of units in the sample having outcome pattern $\mathbf{O}_{l^*,s}$ for $l^* = 1, \ldots, L$ and let $n(l^*,s,g)$ denote the number of units in group g having this pattern. It can be shown that the incomplete data likelihood is given by

$$\log \mathcal{L}(\pi) = \sum_{g} \sum_{l^*} n(l^*,1,g) \log \pi_g \pi_{l^*,1|g} + \sum_{l^*} n(l^*,2) \log\left(\sum_{g} \pi_g \pi_{l^*,2|g} \right) + n(L,3) \log\left(\sum_{g} \pi_g \pi_{L,3|g} \right) \quad (26.3)$$

where from (26.1), $\pi_{l^*,s|g} = P(O_{l^*,s} \mid g) = \pi_{o_1=3|g} \pi_{o_2=3|g,o_1=3} \ldots \pi_{o_{l^*}=s|g,o_{l^*-1}=3}$. Note that the first summation in (26.3) corresponds to the interview cases in which the full table G is observed; for the second and third summation, corresponding to noninterviews and noncontacts, respectively, G is not observed. Note also that by allowing the patterns $\mathrm{O}_{l^*,s}$ to depend upon G in all states, nonresponse is assumed to be nonignorable in the sense of Little and Rubin (1987).

26.3.4 Modeling Hard-Core Nonrespondents

The model in (26.3) assumes group homogeneity, that is, within each group defined by G and for a given LOE, units have the same probability of being interviewed, noninterviewed, and not contacted. Even if G is chosen to be highly correlated with call outcomes, this assumption can only be approximated in practice. One reason for this is the existence of the so-called hard-core nonrespondents (HCNRs). These are sample members who, under the current survey design protocol, have 0 probability of responding regardless of the value of L, the maximum allowable number of call attempts.[4] If the HCNR group is sizable and not well predicted by G, the model in (26.3) will not hold.

[4]Note that we do not assume that HCNRs are nonrespondents under any survey protocol. In fact, the HCNR group may be highly dependent upon the survey design. As an example, some HCNRs under the current telephone survey protocol might respond if mode of data collection were changed to mail/self-administered or face-to-face.

To represent the HCNRs, Drew and Fuller (1980) defined a dichotomous latent variable which we denote by X, where $X = 1$ for a HCNR and $X = 2$ otherwise. Let $\pi_{x=1}$ denote the proportion of the population who are HCNRs under the survey design protocol, that is, units having 0 probabilities of being interviewed or $P(o_l = 1|X = 1) = 0$ for all l. Extending the likelihood in (26.3), the joint distribution of $\mathbf{O}_{l^*,S}$ and G can be rewritten in terms of X to obtain the following likelihood

$$\log \mathcal{L}(\pi) = \sum_{g}\sum_{l^*} n(l^*,1,g)\log \pi_g \pi_{x=2}\pi_{l^*,l|g,x=2} + \sum_{l^*} n(l^*,2)\log\left(\pi_{x=1} + \pi_{x=2}\sum_{g}\pi_g \pi_{l^*,2|g,x=2}\right)$$
$$+ n(L,3)\log\left(\pi_{x=1} + \pi_{x=2}\sum_{g}\pi_g \pi_{l,3|g,x=2}\right) \tag{26.4}$$

where $\pi_{l^*,s|g,x} = P(O_{l^*,s} \mid G = g, X = x)$. In this likelihood, X is not allowed to depend on G. Extensions of (26.4) to allow X to vary across the levels of G were also considered in our analysis and those results will be discussed later.

26.3.5 Estimation

The likelihood equations in (26.3) and (26.4) are identifiable and can be maximized by the EM algorithm (Dempster et al., 1977) to yield estimates of the π parameters. The maximum likelihood estimates (MLEs) of the population proportions denoted by $\hat{\pi}_g$ from (26.3) or (26.4) can be used to adjust the sample mean for nonresponse.

Under SRS, an estimate of the expected response rate for units in group g is

$$r_g = \frac{n_{rg}}{n_g} \tag{26.5}$$

where nrg is the number of completed interviews in group g and ng is the sample size including nonrespondents. Let Y denote a variable of interest in the survey, let yi denote the value of Y for the ith respondent and let $\hat{r}_g$ denote (26.5) after substituting $n\hat{\pi}_g$ for ng. Let $\bar{y}_{rg}$ denote the mean of the respondents in group g that is assumed to be unbiased for the subgroup mean, $\bar{Y}_g$. This corresponds to the missing at random (MAR) assumption of Little and Rubin (1987). Then an unbiased estimate of the population mean, $\bar{Y}$, is the weighting class estimator (see, for example, Särndal et al., 1991, p. 580) given by

$$\bar{y}_{wc} = \frac{1}{n}\sum_{g=1}^{K}\sum_{i=1}^{n_{rg}}\frac{y_{gi}}{r_g} = \sum_{g=1}^{K}\pi_g \bar{y}_{rg} \tag{26.6}$$

In general, n_g is not known but can be estimated by $n\hat{\pi}_g$, where $\hat{\pi}_g$ is an estimate of π_g, the proportion of the population in group g. Substituting for π_g in (26.6) produces the following estimator of $\overline{Y}$

$$\tilde{y} = \sum_{g=1}^{K} \hat{\pi}_g \overline{y}_{rg} \tag{26.7}$$

An approximate expression for the variance of $\tilde{y}$ can be obtained by the delta method. Let $\mathbf{\Omega}_{\hat{\pi}}$ denote the variance–covariance matrix for the vector $\hat{\boldsymbol{\pi}} = \left[\hat{\pi}_g\right]$ and let $\overline{\mathbf{y}}_r = \left[\overline{y}_{rg}\right]$, where $\overline{y}_{rg} = \sum_{i=1}^{n_{rg}} y_{gi} / n_{rg}$ is the simple expansion mean of all observations in group g, $g = 1,\ldots, K$. The variance of $\tilde{y}$ is approximately

$$V(\tilde{y}) \approx \sum_{g=1}^{K} \pi_g^2 V(\overline{y}_{rg}) + E(\overline{\mathbf{y}}_r)\mathbf{\Omega}_{\pi} E(\overline{\mathbf{y}}_r') \tag{26.8}$$

which is estimated by substituting the appropriate estimators for the parameters in (26.8) by

$$v(\tilde{y}) = \sum_{g=1}^{K} \hat{\pi}_g^2 v(\tilde{y}_g) + \overline{\mathbf{y}}\hat{\mathbf{\Omega}}_{\boldsymbol{\pi}}\overline{\mathbf{y}}' \tag{26.9}$$

26.3.6 Extensions of the Model

In this section, we consider several extensions of the model that may be needed for some applications. First, we consider a modification of the model that partially accounts for the unequal probabilities of selection of individuals in RDD surveys. Second, we discuss the problem of unknown eligibility and a simple modification to account for such cases. Finally, the model in the preceding section is extended to incorporate right-censored observations.

Unequal weighting. For most telephone household surveys, telephone numbers are selected from a frame of working numbers, a contact with the household is made and then a single person is selected from the eligible household members using some method of random sampling. Even in RDD surveys where households are selected with equal probabilities, the within-household selection usually results in an unequal probability sample of individuals. In addition, many telephone surveys incorporate stratification and oversampling of certain types of households, which also results in differential probabilities of selection.

If the unequal selection probabilities are not incorporated in the analysis, a substantial bias in the parameter estimates may arise (Patterson et al., 2002). Skinner (1989) developed the pseudomaximum likelihood (PML) method, which under unequal probability sampling can be used to estimate any statistical model including the latent variable callback models discussed here. A few statistical packages for latent class modeling have incorporated the PML methodology, including Mplus (Asparouhov, 2005) and Latent Gold (Vermunt and Magidson, 2005). Unfortunately, as we complete our analysis, these packages are still unable to handle the complexities of the model in (26.4). Instead we used the ℓEM software

(Vermunt, 1997b) which implements the EM algorithm for MLE to fit all the models considered in the chapter using a weighted and rescaled input table approach as described below.

The data required to obtain the estimates of π_{lg} in (26.6) is the $(2L+1) \times K$ cells of cross-classification table formed by the l outcome variables $O_1, \ldots, O_l$ and G after omitting the cells that are structurally 0. We replaced the sample counts in each cell of this table with a weighted count that reflects the distribution of the variables in the population as follows. First, for each cell of the $G \times \mathbf{O}$ table, the sample count was replaced by the sum of the base weights of the units in the cell. Then these cell totals were multiplied by a scaling factor equal to the original sample size in the cell divided by the sum of the base weights for all cells in the table. For cases in disposition states $S = 2$ and $S = 3$, person-level weights were not available since the number of eligible adults within the household is unknown. As an approximation, base weights that reflected within household selection probabilities were imputed by multiplying the household base weight by the average number of adults in all responding households.

This weighting and rescaling approach ensures that the cell counts reflect the population distribution while retaining the original sample size. The analysis then proceeded using this weighted table as though the data were selected using SRS. As noted in Clogg and Eliason (1987) and Patterson et al., (2002), the weighting and rescaling cell count approach produces model unbiased parameter estimates; however, the standard errors of the estimates will usually be understated. We can compensate for this to some extend by multiplying the sampling variances from the analysis by the so-called unequal weighting effect (UWE) factor.[5]

Unknown eligibility. Following any RDD survey, a substantial fraction of the numbers dialed will remain noncontacted, and their eligibility for the survey is not known with certainty (referred to as "ring-no-answer numbers," or RNAs). For example, after excluding right-censored numbers and known ineligibles, approximately 20 percent of the remaining telephone numbers were RNAs. Prior research (see, for example, Frankel et al., 2003a) has shown that about 20 percent of these RNAs are actually eligible households. All the eligible households in the RNA group have response pattern $\mathbf{O}_{L,3}$ and hence excluding them will reduce the count of cases with this response pattern.

One strategy for compensating for these missing eligible units is to simply increase the count of cases with the $\mathbf{O}_{L,3}$ pattern by about $0.20n_{\text{RNA}}$, where n_{RNA} is the number of RNA's in the sample after applying LOE L. For the BRFSS analysis, this remedy had very little effect on the estimates. Another strategy is to extend the model by adding a fourth disposition state for ineligible units. Then cases for which eligibility is not determined would be treated as right-censored observations. In this manner, the eligibility rate among RNAs would be estimated simultaneously with the other parameters of the model, and this information would be reflected in the model estimates of π_{lg}. This approach was not attempted in our analysis.

[5]The UWE is $1 + cv^2$, where cv denotes the coefficient variation of the sample weights (Kish, 1965).

Modeling right-censored observations The previous development assumed that all noncontact units received the maximum LOE associated with a particular model. Thus, for a model that assumes a maximum of L call attempts, noncontact cases that receive fewer than L call attempts are excluded from the analysis. For the BRFSS data, this amounted to about 12 percent of all noncontact cases for models with $L = 5$, about 25 percent for $L = 10$ attempts, and about 30 percent for $L = 15+$ call attempts. For smaller datasets, removing such cases from the analysis could have a substantial impact on the precision of the estimates. In addition, even though these cases provide only partial information, deleting them from the analysis could substantially bias the estimates of π_{lg} if the right censoring is nonignorable relative to G and $\mathbf{O}$. In this section, we consider a natural extension of the callback model to handle right censoring so that these cases can be retained in the analysis.

The method we propose is derived from the approach described by Vermunt (1997a, p. 283) for partially observed event history data based on the work of Fay (1986, 1989). An important advantage of Fay's method for modeling missing data is that it easily allows for nonignorable nonresponse or, in our case, censoring mechanisms that depend upon grouping variables in the model that are not observed for the censored units.

As before, let $\mathbf{O}' = [o_l,\, l = 1,\ldots, L]$ denote the outcome vector associated with a unit in the sample. For any case in the sample, o_l has some probability of being either missing or observed before applying LOE l. Define the response vector $\mathbf{R}' = [r_1,\ldots,r_L]$, where $r_l = 1$ if o_l is observed; otherwise $r_l = 2$. Next we consider the joint likelihood of G, $\mathbf{R}$, and $\mathbf{O}$.

The joint likelihood can be partitioned into four parts: interviewed units for which G and $\mathbf{O}$ are both observed and $\mathbf{R} = \mathbf{1}$; noninterviewed units for which $\mathbf{O}$ is observed, G is not observed, and $\mathbf{R} = \mathbf{1}$; uncensored noncontacts for which $\mathbf{O} = \mathbf{3}$ and $\mathbf{R} = \mathbf{1}$; and censored noncontacts for which $\mathbf{O}$ is only partially observed and $\mathbf{R} \neq \mathbf{1}$. We consider each of these four cases separately and then combine them to obtain the joint likelihood.

Consider the likelihood for cases that are interviewed at some LOE, say $l^* \leq L$. For such a case, the first l^*-1 elements of $\mathbf{O}$ are 3, and the remaining elements are 1. For example, for $l^*=3$ and $L=5$, $\mathbf{O}' = [3,3,1,1,1]$ and $\mathbf{R}'=[1,1,1,1,1]$. The probability of observing these patterns for a unit in group g is

$$\begin{aligned}\pi_{l^*=3,s=1,\mathbf{R}|g} &= \pi_{o_1=3|g}\pi_{r_2=1|o_1=3,g}\pi_{o_2=3|o_1=3,g}\pi_{r_3=1|o_2=3,g}\pi_{o_3=1|o_2=3,g} \\ &= \pi_{l^*=3,s=1}\pi_{\mathbf{R},l^*=3|g}\end{aligned} \tag{26.10}$$

where $\pi_{l^*,s|g}$ was previously defined and $\pi_{\mathbf{R},l^*=3|g} = \pi_{r_2=1|o_1=3,g}\pi_{r_3=1|o_2=3,g}$. Note that $\pi_{\mathbf{R},l^*=3|g}$ does not depend on the final status s. Therefore, for any l^* and s we can write

$$\pi_{l^*,s,\mathbf{R}|g} = \pi_{l^*,s|g}\pi_{\mathbf{R},l^*|g} \tag{26.11}$$

where $\pi_{\mathbf{R},l^*|g} = \pi_{r_2=1|o_1=3,g}\cdots\pi_{r_{l^*}=1|o_{l^*-1}=3,g}$.

Now the component of the log-likelihood corresponding to interviewed cases can be written as

$$\sum_{g,l^*} n(l^*,1,g) \log \pi_g \pi_{l^*,1|g} \pi_{R,l^*|g} \tag{26.12}$$

By the same reasoning, we can write the component corresponding to noninterviewed cases as

$$\sum_{l^*} n(l^*,2) \log\left(\sum_g \pi_g \pi_{l^*,2|g} \pi_{R,l^*|g} \right) \tag{26.13}$$

For uncensored noncontact cases, the log-likelihood component is

$$n(L,3) \log\left(\sum_g \pi_g \pi_{L,3|g} \pi_{\mathbf{R},L|g} \right) \tag{26.14}$$

Finally, we consider the probability, a noncontact case is censored at LOE l^*+1, that is, the state is observed at l^* but not observed for all $l > l^*$. Write

$$\pi_{l^*+1,3,R|g} = \pi_{l^*,3|g} \pi_{R,l^*|g} \pi_{r_{L^*+1}=2|g} \tag{26.15}$$

which implies that the likelihood component for censored cases is

$$\sum_{l^*<L} n(l^*,3) \log\left(\sum_g \pi_g \pi_{l^*,3|g} \pi_{R,l^*|g} \pi_{r_{L^*+1}=2|g} \right) \tag{26.16}$$

Combining (26.12–26.14) and (26.16), the joint likelihood is therefore

$$\begin{aligned} \log \mathcal{L}(\pi) = {} & \sum_{g,l^*} n(l^*,1,g) \log \pi_g \pi_{l^*,1|g} \pi_{R,l^*|g} + \sum_{l^*} n(l^*,2) \log\left(\sum_g \pi_g \pi_{l^*,2|g} \pi_{R,l^*|g} \right) \\ & + n(L,3) \log\left(\sum_g \pi_g \pi_{L,3|g} \pi_{\mathbf{R},L|g} \right) + \sum_{l^*<L} n(l^*,3) \log\left(\sum_g \pi_g \pi_{l^*,3|g} \pi_{R,l^*|g} \pi_{r_{L^*+1}=2|g} \right) \end{aligned} \tag{26.17}$$

The corresponding model for HCNRs is a straightforward extension of (26.17) following the approach used to extend the likelihood in (26.3) to the one in (26.4) and will not be shown here.

It can be shown that (26.17) differs from (26.3) only by an additive constant if we assume that **R** is independent of G. This is the case where the censoring process is ignorable with respect to the variables in the model and, thus, excluding the right-censored units in the analysis will have no effect on the estimates.[6] The model in (26.17) has not been explored in the current paper since the ignorable censoring assumption seems reasonable for these data. Nevertheless, such models can be readily fit with the ℓEM software as shown in Vermunt (1997b).

[6]As an example, the censoring process would be ignorable if truncation is independent of the number of times a case has been called.

26.4 APPLICATION TO THE BRFSS DATA

An important goal of this application is to evaluate the ability of the estimator in (26.6) to reduce the ECE associated with cutoff rules below the current rule of 15 used by CDC. To examine the utility of the ECE adjustment model, we used data from the 2004 BRFSS, which consists of survey data as well as call record data for almost 1 million telephone numbers that were dialed in an effort to obtain almost 300,000 completed interviews across the 3143 counties in the United States. Only telephone numbers that are known with reasonable certainty to reach U.S. households were retained in the analysis. For example, telephone numbers that resulted in an immediate hang-up during the introduction to the survey (so-called HUDIs) were included and assumed to reach households. Numbers that were never answered (so-called ring-no-answers or RNAs) were not included in the analysis. Preliminary investigations to incorporate RNAs in the analysis using the methods of Section 26.3.6, however, suggest that RNAs effects on the results are small.

The original dataset used for this analysis consisted of 749,198 household telephone numbers. Using the definitions of the three categories defined above, approximately 39 percent of these numbers were completed interviews, 36 percent were noninterviews, and 25 percent were noncontacts. The number of call attempts ranged from 1 to more than 60, with an average of approximately 9 calls. About 53 percent of the noncontact cases (or 13 percent of all cases) were terminated prematurely (i.e., fewer than 15 call attempts were made) and were excluded from the analysis. As shown in Section 3.6, including these right-censored cases in the analysis would have added considerable complexity to the model. Given the survey design protocol of the BRFSS, however, right-censoring was likely to be ignorable with respect to our choice of grouping variable G.[7] Therefore, including them is not likely to make an appreciable difference in the main results. The final dataset used in the analysis consisted of 648,223 cases.

26.4.1 Model Specification

To reduce the dimensionality of the data table, the call attempts were recoded into 10 levels of the LOE variable. These were defined as $l = 1,\ldots,5$ denoting call attempts 1–5, respectively; $l = 6$ denoting 6 or 7 call attempts; $l = 7$ denoting 8 or 9 attempts; $l = 8$ denoting 10 or 11 call attempts; $l = 9$ denoting 12, 13 or 14 attempts; and, $l = 10$ denoting 15 or more attempts. A number of variables were considered for use in the analysis as grouping variables, including age, race/ethnicity, sex, education, marital status, household size, employment status, MSA/non-MSA, census region, and census division. Some combination of these variables will be used to form the

[7]In the BRFSS, most of the right-censored cases are from replicate samples that were released late in the fielding period and not worked to completion. Some states will release replicate samples to try to meet completion goals, but do so at the risk of having more partially worked noncontact numbers. Since the replicates are random samples, their characteristics should be similar to cases that received the requisite number of callbacks.

variable G in model (26.4) that defines the weighting classes for the estimator in (26.6). Therefore, the ideal variables are those that have the greatest correlation with RDD response rates. Our preliminary analysis suggested that a combination of the three following variables performed quite well: age (18–34, 35–49, 50–64, and 65+), race/ethnicity (white, non-Hispanic, black non-Hispanic, and Hispanic), and number of adults in the household (1, 2, 3, and 4 or more persons). This was not an exhaustive search for the optimal combination of variables, and it is possible that other combinations could perform better.

Using the general likelihood in (26.4), a large number of model specifications can be considered by altering the assumptions for the model parameters. However, only two basic types of models were explored in the analysis. One type of model assumes no latent variable (i.e., no HCNRs) as in Equation (26.3) (referred to as model A), and the other type of model incorporates a latent variable to represent the HCNR group, as in Equation (26.4) (referred to as model B). For each model type, the most parsimonious model that provided an adequate fit to the data was used to generate the results. For model A, the best model among those tested assumed (i) $\pi_{X=1} = 0$ and (ii) $\pi_{l,s|g} = \pi_{l',s|g}$ for $l \neq l'$, where $l,l' = 3,\ldots,9$. That is, model A assumed no HCNRs and set the response probabilities equal for LOE 3,..., 9. The probabilities for LOEs 1 and 2 were not restricted, nor was the LOE 10 in models where $L = 10$. Model B assumptions were similar except assumption (i) was replaced by (i′) $\pi_{X=1} \geqslant 0$ to represent the HCNR group in the population.

26.4.2 Estimators of the ECE

Let $\bar{y}_L$ denote the mean of all completed cases in the survey for a given callback cutoff L. For example, $\bar{y}_5$ denotes the mean of all units interviewed at or before LOE 5 (i.e., 5 call attempts and $\bar{y}_{10}$ denotes the unadjusted estimator based on all the interviews (i.e., 15 or more call attempts). An estimate of the ECE associated with a cutoff rule of L is

$$ECB_L = \bar{y}_L - \bar{y}_{10} \tag{26.18}$$

for $L = 1,\ldots, 9$. In words, the estimated ECE for the sample with cutoff L is the difference between the mean of cases that were completed with LOE L or less and the mean of all cases completed in the sample.

Let $\hat{\pi}_{ALg}$ denote the estimate of π_g from model A for $g = 1,\ldots, K$. Substituting this estimate into (26.7) yields the following estimate of $\bar{Y}$:

$$\tilde{y}_{AL} = \sum_{g=1}^{K} \hat{\pi}_{ALg} \bar{y}_{Lg} \tag{26.19}$$

where $\bar{y}_{Lg}$ is the mean of all cases in group g that were completed with LOE L or less. Similarly, the estimator under model B, denoted $\bar{y}_{BL}$, is (26.19) after replacing $\hat{\pi}_{ALg}$ by $\hat{\pi}_{BLg}$ that is the estimate of π_g obtained from model B. The difference

$D_{AL} = \tilde{y}_{AL} - \bar{y}_{10}$ is an estimate of the ECE after the model A adjustment. We hypothesize that $|D_{AL}| \leq |ECE_L|$ for the cutoff rules of interest.

Note that $\bar{y}_{10}$ is subject to both early cooperator bias and hard-core nonresponse bias, since the nonrespondents to the survey represent both groups. Since $\tilde{y}_{B10}$ is adjusted for both biases, it is expected to smaller nonresponse bias and can be regarded as a type of "gold standard" or "preferred estimator" of $\bar{Y}$ for purposes of gauging the magnitude of the nonresponse bias in $\bar{y}_{10}$.

Let $D_B = \bar{y}_{10} - \tilde{y}_{B10}$, where D_B is an indicator of the nonresponse bias in $\bar{y}_{10}$. In addition to evaluating D_B for the survey characteristics of interest, we will also compare $\tilde{y}_{BL}$ and $\tilde{y}_{B10}$ for $L = 1, \ldots, 9$. These comparisons will provide an indicator of the ability of model B to adjust not only for early cooperator bias but also for hard-core nonresponse bias.

To obtain the estimates of π_{Lg} under each model, an input table was formed by crossing the L outcome variables $O_1, \ldots, O_L$ and G to form the $(2L + 1) \times K$ table with cells corresponding to the $2L + 1$ possible patterns of $\mathbf{O}$ for each level of G. This table was weighted by the base weights only and rescaled as described in Section 26.3.6. The analysis proceeded with this weighted table under the SRS assumption. The process was repeated for each cutoff value of interest and for both models. The ℓEM software (Vermunt, 1997b) was used to fit the models via the EM algorithm.

26.4.3 Results for ECE Analysis

With regard to the comparisons of $|D_{AL}|$ and $|ECB_L|$, Table 26.5 presents the results for the seven key variables shown in Table 26.1. Column 2 is $\bar{y}_{10}$ corresponding to response category shown in column 1. Column 3 is ECB_5 and column 4 is D_{A5}. As

Table 26.5. Comparison of the Ece for a Maximum Five Callbacks Strategy Before and After (Model A) Adjustment

		ECE for five callbacks	
Category	All data	Unadjusted	Adjusted
Health coverage	85.96	0.81	−1.29
General health			
Excellent	20.72	−0.92	−0.55
Very good	33.13	−0.35	−0.4
Good	29.58	0.06	0.11
Asthma	13.43	0.33	0.54
Diabetes	8.8	0.72	0.25
Drink alcohol	52.84	−2.18	−1.84
Physical, mental, or emotional problem	18.73	2.16	0.9
Influenza shot	35.77	2.54	−0.77

ECE, early cooperator effect.

noted in Section 26.4.2, only three items exhibit an ECE of any importance (viz., ALCOHOL, FLUSHOT, and PHYMO). In all three cases, $|D_{A5}|$ is substantially smaller than $|ECB_5|$. When $|ECB_5|$ is small, the performance of the adjusted estimator is mixed.

In Table 26.6, we summarize the results for $\bar{y}_L$, $\tilde{y}_{AL}$, and $\tilde{y}_{BL}$ for three possible cutoff values: 5 call attempts ($L = 5$), 9 call attempts ($L = 7$), and 11 call attempts ($L = 8$) as well as for the full sample ($L = 10$). The estimates in the row labeled $\tilde{y}_{BL}$ may be interpreted as estimates of $\bar{Y}$ that have been adjusted for total nonresponse bias and, thus, the $\bar{y}_{10} - \tilde{y}_{B10}$ can be interpreted as an estimate of the nonresponse bias in the full sample estimate, $\bar{y}_{10}$. These bias estimates suggest little nonresponse bias for the full sample estimate. The largest bias is 1.16 for FLUSHOT.

The difference $ECE_L = \bar{y}_L - \bar{y}_{10}$ is the early cooperator bias for a cutoff strategy corresponding to LOE L. Likewise $D_{AL} = \tilde{y}_{AL} - \bar{y}_{10}$ reflects the effect of the model A adjustment on ECB_L. In the table, $\bar{y}_{10}$ has been bolded to facilitate these comparisons. As noted in the discussion for Table 26.4, there seems to be little to choose between $\bar{y}_L$, $\tilde{y}_{AL}$, and $\tilde{y}_{BL}$ when the ECE is small for all values of L. However, when the ECE is large, $\tilde{y}_{AL}$ tends to be closer to $\bar{y}_{10}$ than the other two estimators.

26.4.4 Comparisons with a PSA Estimator

In this section, we compare the model-based estimators of $\bar{Y}$ with the corresponding PSA estimates using PSA adjustment factors that were derived from the Current Population Survey (CPS).[8] To facilitate the interpretation of the comparisons, grouping variables used in the model-based estimators also formed the PSA adjustment cells for the PSA estimator. Since the CPS provides very good estimates of the joint distributions of these variables, the PSA estimates will be regarded as the "gold standard" for these comparisons.

Let $\hat{\pi}_{\mathrm{CPS},g}$ denote the estimates of π_g from the CPS and let $\bar{y}_{\mathrm{CPS},L}$ denote (26.7) with $\hat{\pi}_{\mathrm{CPS},g}$ in place of $\hat{\pi}_g$. Our criterion for the comparison will be based upon the relative difference

$$RD_L = \frac{\hat{\bar{Y}}_L - \bar{Y}_{\mathrm{CPS},L}}{\bar{Y}_{\mathrm{CPS},L}} \tag{26.20}$$

where $\hat{\bar{Y}}_L$ is any estimator of $\bar{Y}$ based upon maximum LOE L. Under the assumption that $\bar{y}_{\mathrm{CPS},L}$ is unbiased, RD_L is an indicator of the relative bias of $\hat{\bar{Y}}_L$. Let RD_{UL}, RD_{AL}, and RD_{BL} correspond to (26.20) after replacing $\hat{\bar{Y}}_L$ by $\bar{y}_L$, $\tilde{y}_{AL}$, and $\tilde{y}_{BL}$, respectively.

[8]The current population survey (CPS) is a monthly survey of about 50,000 households conducted by the U.S. Bureau of the Census for the Bureau of Labor Statistics. The CPS is the primary source of information on the labor force characteristics of the U.S. population, including demographic information. The sample is scientifically selected to represent the civilian, noninstitutional population. Respondents are interviewed to obtain information about the employment status of each member of the household 15 years of age and older. The sample provides estimates for the nation as a whole and serves as part of model-based estimates for individual states and other geographic areas.

Table 26.6. Comparison of Three Estimators of $\overline{Y}$ for Key Brfss Variables and Alternative Cutoffs (L)

Category	Estimator	Number of call attempts				Category	Estimator	Number of call attempts			
		5	9	11	15+			5	9	11	15+
General health		$L=5$	$L=7$	$L=8$	$L=10$			$L=5$	$L=7$	$L=8$	$L=10$
Excellent	$\overline{y}_L$	19.80	20.28	20.44	**20.72**	Diabetes	$\overline{y}_L$	9.52	9.15	9.03	**8.80**
	$\tilde{y}_{AL}$	20.17	20.33	20.47	20.73		$\tilde{y}_{AL}$	9.05	9.05	8.96	8.76
	$\tilde{y}_{BL}$	19.24	20.54	20.67	20.84		$\tilde{y}_{BL}$	9.20	8.83	8.71	8.67
Very good	$\overline{y}_L$	32.78	33.07	33.11	**33.13**	Influenza shot	$\overline{y}_L$	38.31	36.9	36.51	**35.77**
	$\tilde{y}_{AL}$	32.73	32.87	32.89	32.89		$\tilde{y}_{AL}$	35.00	35.39	35.07	34.42
	$\tilde{y}_{BL}$	31.14	33.06	33.11	33.03		$\tilde{y}_{BL}$	34.07	34.72	34.60	34.61
Good	$\overline{y}_L$	29.64	29.56	29.52	**29.58**	Health coverage	$\overline{y}_L$	86.77	86.41	86.30	**85.96**
	$\tilde{y}_{AL}$	29.69	29.64	29.60	29.69		$\tilde{y}_{AL}$	84.67	85.31	85.24	84.94
	$\tilde{y}_{BL}$	30.69	29.56	29.54	29.62		$\tilde{y}_{BL}$	81.54	85.04	85.10	85.12
Drink alcohol	$\overline{y}_L$	50.66	13.59	52.13	52.84	Physical, mental, or emotional problems	$\overline{y}_L$	20.89	19.71	19.35	**18.73**
	$\tilde{y}_{AL}$	51.00	51.70	51.99	**52.69**		$\tilde{y}_{AL}$	19.63	19.13	18.82	18.23
	$\tilde{y}_{BL}$	50.41	52.02	52.32	52.84		$\tilde{y}_{BL}$	18.74	18.85	18.60	18.22
Asthma	$\overline{y}_L$	13.76	13.59	13.51	13.43						
	$\tilde{y}_{AL}$	13.97	13.67	13.58	**13.49**						
	$\tilde{y}_{BL}$	14.12	13.70	13.59	13.51						

For this analysis, the characteristic, Y, is one based upon a "worst case scenario" regarding the bias in $\hat{\bar{Y}}$. It is well known that nonresponse bias is extreme when the Y_i are proportional to the response propensities, p_i (see, for example, Bethlehem, 2002). Therefore, we consider a set of characteristics with the property: $Y_i = \alpha p_i$ for some constant α. We further assume that the p_i are constant for all units i in group g (i.e., the MAR assumption). Let p_g denote the group g propensity, which is estimated by r_g, the group g response rate. We compute r_g as in (26.5) by replacing n_g with an estimate of this number from the CPS (viz., $n\hat{\pi}_{\text{CPS},g}$). Thus, the response rate is $\hat{r}_{\text{CPS},g} = n_{rg}(n\hat{\pi}_{\text{CPS},g})^{-1}$ for $g = 1,\ldots, K$.

As an example, consider RD_{BL} and let $\bar{y}_{rg} = \hat{r}_{\text{CPS},g}$ in (26.7). This yields

$$\tilde{y}_{\text{BL}} = \frac{n_r}{n} \sum_{g=1}^{K} \frac{\hat{\pi}_{Lg}\hat{\pi}_{\text{BL}g}}{\hat{\pi}_{\text{CPS},g}} \tag{26.21}$$

where $\hat{\pi}_{\text{BL}g}$ and $\hat{\pi}_{Lg}$ are the adjusted and unadjusted estimates, respectively, of π_g using just the completed cases after L callbacks. Further,

$$\begin{aligned} RD_{\text{BL}} &= \sum_{g=1}^{K} \hat{\pi}_{Lg} \left(\frac{\hat{\pi}_{\text{BL}g} - \hat{\pi}_{\text{CPS}g}}{\hat{\pi}_{\text{CPS}g}} \right) \\ &= \sum_{g=1}^{K} \frac{\hat{\pi}_{Lg}\hat{\pi}_{\text{BL}g}}{\hat{\pi}_{\text{CPS}g}} - 1 \end{aligned} \tag{26.22}$$

This measure of relative bias reflects the most extreme case possible given the set of response propensities from the BRFSS. Note that RD_{BL} is intuitively appealing since RD_{BL} is 0 when $\hat{\pi}_{\text{BL}g} = \hat{\pi}_{\text{CPS}g}$. The ratios $\hat{\pi}_{\text{BL}g}/\hat{\pi}_{\text{CPS}g}$ are weighted by the observed group proportions, $\hat{\pi}_{Lg}$, so the larger the group, the greater its contribution to relative bias.

We hypothesize that $|RD_{\text{BL}}| < |RD_{\text{UL}}|$ for every L cutoff since, if this were true, inference could be improved by using model B instead of the unadjusted estimator. Furthermore, a small value of $|RD_{\text{BL}}|$ implies that $\tilde{y}_{\text{BL}}$ and $\bar{y}_{\text{CPS},L}$ produce very similar estimates and, therefore, $\tilde{y}_{\text{BL}}$ can be used when CPS or other gold-standard estimates of π_g are unavailable.

The estimates $\hat{\pi}_{Lg}, \hat{\pi}_{\text{AL}g}$, and $\hat{\pi}_{\text{BL}g}$ from the analysis in Section 4.2 were used to compute the measures of relative bias. The estimates $\hat{\pi}_{\text{CPS},g}$ were obtained from the 2004 March supplement of the CPS for adults aged 18 years old and older living in households with telephones. Table 26.7 shows the values of $|RD_{\text{AL}}|$, $|RD_{\text{UL}}|$, and $|RD_{\text{BL}}|$ for $L = 5,\ldots,10$ corresponding to 5, 7, 9, 11, 14, and 15+ call attempts, respectively.

The results in Table 26.7 clearly show much greater agreement with the CPS PSA estimator using either model A or model B than with the unadjusted estimator, particularly for smaller values of L. Further, model B agrees more closely with the CPS than model A for $L<14$. If the CPS PSA estimate is regarded as unbiased, then the relative differences in the table can be interpreted as estimates of the maximum relative bias given the distribution of BRFSS response propensities. The table shows

Table 26.7. Absolute Relative Differences Comparing Adjusted and Unadjusted Estimates with CPS PSA Estimate as a Function of *L*

L	Number of attempts	$\lvert RD_{UL}\rvert \times 100$ %	$\lvert RD_{AL}\rvert \times 100$ %	$\lvert RD_{BL}\rvert \times 100$%
5	5	8.76	5.22	1.35
6	7	6.87	3.99	2.46
7	9	5.83	3.41	2.88
8	11	8.76	3.41	2.88
9	14	4.50	2.62	2.77
10	15	3.95	2.29	2.43

that the relative bias in the unadjusted estimator is almost 9 percent at $L = 5$ and only 1.4 percent for the model B estimator. At $L = 10$ (the maximum callback strategy), the unadjusted relative bias is 4 percent compared with the adjusted relative bias of about 2.4 percent.

26.5 SUMMARY AND CONCLUSIONS

This analysis demonstrates that early cooperators are different from later cooperators on a wide range of characteristics. At the aggregate level, we found considerable variation in the magnitude of the differences, and even greater differences were noted when the estimates were disaggregated. However, even for items showing the greater differential bias, the ECE was still in an acceptable range due to the small size of the later cooperator group.

These results suggest that unlimited call attempts to achieve higher response rates in RDD surveys may not lead to large gains in estimator accuracy. A better strategy for survey resource allocation would be to direct resources to areas of the survey design that would show greater gains. For example, the money saved by limiting the number of callback attempts to a maximum of from 5 to 10 could be used to reduce measurement errors through improved questionnaire design, more extensive pretesting or enhanced interviewer training.

Determining whether to reduce the maximum number of callbacks and accept a lower response rate is still a difficult decision because of the risk that for some important estimates from the survey, the nonresponse bias may increase to unacceptable levels. Although poststratification offers some protection against this bias, good poststratifying variables are rarely available for RDD surveys. Besides, a nonresponse adjustment combined with a PSA ought to work better than a PSA alone and offer even more bias protection.

To avoid the risk of bias for limited callback protocols, we proposed a weight adjustment based upon a callback model that can be used either in combination with or in lieu of a PSA. The general model, which can be applied for as few as five callbacks, incorporates a latent indicator variable for HCNRs. Using information on the level of effort expended in finalizing a unit as well as other information available only on interviewed units, the model provides weight adjustment factors that can be used to correct the estimates for the nonresponse bias.

Two variations of the general model were explored. Model A excluded the latent variable and aimed at reducing the ECE between the limited callback estimate and the full effort estimate. Model B incorporated a latent variable to represent the HCNRs in the sample. Since these nonrespondents are indistinguishable from nonrespondents who would eventually respond to a future callback, HCNR status is unobservable.

The BRFSS data were used to simulate a range of limited callback protocols from five to the maximum number available in the dataset. For reducing ECE, model A performed better than model B. This was expected since model A is designed to reduce the ECE while model B is designed to compensate for HCNRs that do not contribute to the ECE. We further found that the model A adjusted estimates were not always better than the unadjusted estimates. We speculated that the effectiveness of the model A adjustment depends upon the size of the ECE. In our analysis, model A substantially reduced the ECE when the ECE was large; however, when the ECE was small or negligible, there was little or no improvement in using model A.

In a second analysis, model B was employed to adjust for total nonresponse bias using a somewhat different approach. We considered the situation in which the nonresponse bias is quite large. This occurs when a characteristic of interest is highly correlated with response propensity. Under this scenario, a relative difference measure was derived for comparing both the model A and model B adjusted estimates with the traditional PSA estimator, where the PSA factors were derived from CPS. Using this metric, the model B adjusted estimate was clearly superior to the corresponding unadjusted estimate for all the limited callback protocols considered. The model A adjustment also performed better than the unadjusted estimate though reduction in bias was less than for model B for lower callback maximums. These results suggest that model B is effective at reducing both ECE and the HCNR bias in situations where nonresponse bias is important. The data required to apply the model are fairly minimal. Most RDD surveys maintain information on the number of call attempts as well as the outcome of each call that can be used to construct the LOE variables.

Our experience suggests that good results can be obtained with as few as five callbacks to all noncontacts. However, there is still much to learn about many aspects of the proposed approach before it can be recommended as a safeguard against ECE for RDD surveys of early cooperators. Some issues that remain include the following.

What grouping variables are most effective for reducing ECE? The present study only considered variables that were available from the CPS to allow the comparisons in Section 26.4.4. However, virtually any combination of variables in the survey is a candidate for G in the model. The most effective choices are variables that are most highly correlated with response propensity.

How does the callback modeling approach perform for small samples? The BRFSS dataset was extremely large, which allowed for a very precise estimation of the model parameters. Sparse cells posed no problem for the current analysis even though the fitted models contained as many as 13 categorical manifest variables each with at least three categories. Applications to smaller datasets may encounter convergence problems with this many variables. In addition, the standard errors of the adjustment factors may be so large with small samples that the unadjusted estimator may have smaller MSE except in cases of extreme bias. The standard error of the estimator in (26.7) has been largely unexplored.

How should LOEs be defined to make the most effective use of paradata on callback attempts and outcomes in RDD surveys? In considering only the number of callback attempts to obtain a final disposition, our application did not make use of other information in the call history record that could be useful in modeling nonresponse, including refusal conversion attempts, appointments, whether contact was made with someone in the household other than an eligible adult. Incorporating this information into the model could improve the model's performance.

Should the callback model adjustment for ECE and nonresponse be used in combination with the traditional PSA adjustment? In field surveys, nonresponse adjustments are used in conjunction with PSAs in a two-step approach. The same could be done for RDD surveys in which the model B adjustments would be followed by the PSA adjustments. In this way, control variables that are not available externally for the entire telephone population could still be incorporated into the adjustment process. This approach has not been evaluated here but could lead to greater reductions in nonresponse bias as well as greater protection against the risks of ECE.

References

Abt Associates. (2005). *The National Immunization Survey (NIS) 2004 Annual and Final Methodology Report.* Cambridge, MA: Abt Associates.

Aday, L. (1996). *Designing and Conducting Health Surveys.* San Francisco: Jossey-Bass. Agans, R. P. & Kalsbeek, W. D. (2005). *Nationwide Study of Female Sexual Health.* Chapel Hill, NC: University of North Carolina Survey Research Unit.

Agans, R.P. & Kalsbeek, W.D. (2005). *National Study of Female Health*, Chapel Hill, NC: University of North Carolina Survey Research Unit.

Agrifoglio, M. (2004). Sight translation and interpreting: a comparative analysis of constraints and failures. *Interpreting, 6*, 43–67.

Ahmed, W. M. & Kalsbeek, W. D. (1998). An analysis of telephone call history data from the behavioral risk factor surveillance system. *Proceedings of the Survey Research Section*, American Statistical Association, 224–229.

Alexander, C. H. (1988). Cutoff rules for secondary calling in a random digit dialing survey. In R. M. Groves, P. P. Biemer, L. E. Lyberg, J. T. Massey, W. L. Nicholls, & J. Waksberg (Eds.), *Telephone Survey Methodology* (pp. 113–126). New York: John Wiley & Sons Inc.

Alexander, C. H., Jr. & Wetrogan, S. (2000). Integrating the American Community Survey and the Intercensal Demographic Estimates Program. *Proceedings of the American Statistical Association*, 295–300.

Amdur, R. J. (2003). *The Institutional Review Board Member Handbook.* Boston: Jones and Bartlett Publishers.

American Association for Public Opinion Research. (2004). Standard definitions: final disposition of case codes and outcome rates for surveys. Retrieved November 30, 2006, from http://www.aapor.org/pdfs/standarddefs_3. 1.pdf

American National Standards Institute (2001). Common industry format for usability test reports. Retrieved December 19, 2006, from http://www.idemployee.id.tue.nl/g.w.m.rauterberg/lecturenotes/Common-Industry-Format.pdf

Anastasi, A. (1988). *Psychological Testing.* Old Tappan, NJ: Prentice Hall.

Advances in Telephone Survey Methodology, Edited by James M. Lepkowski, Clyde Tucker, J. Michael Brick, Edith de Leeuw, Lilli Japec, Paul J. Lavrakas, Michael W. Link, and Roberta L. Sangster

Anda, R. F., Dodson, D. L., Williamson, D. F., & Remington, P. L. (1989). Health promotion data for state health departments: telephone versus in-person survey estimates of smoking and alcohol use. *American Journal of Health Promotion, 4*, 32–36.

Anderson, A.H. (2008). Video mediated interactions and surveys. In F.G. Conrad and M.F. Schober (Eds.). *Envisioning the Survey Interview of the Future.* New York: John Wiley & Sons, Inc.

Anderson, J. E., Nelson, D. E., & Wilson, R. W. (1998). Telephone coverage and measurement of health risk indicators: data from the National Health Interview Survey. *American Journal of Public Health, 88*, 1392–1395.

Angus, V. C., Entwistle, V. A., Emslie, M. J., Walker, K. A., & Andrew, J. E. (2003). The requirement for prior consent to participate on survey response rates. *BMC Health Services Research, 3*, 21.

Anido, C. & Valdés, T. (2000). An iterative estimating procedure for probit-type nonresponse models in surveys with call backs. *Sociedad de Estadística e Investigación Operativa, 9*, 233–253.

Anton, J. (1997). *Call Center Management*, West Lafayette, IN: Purdue University Press.

Aquilino, W. S. (1992). Telephone versus face-to-face interviewing for household drug use surveys. *International Journal of Addiction, 27*, 71–91.

Aquilino, W. S. (1994). Interview mode effects in surveys of drug and alcohol use: a field experiment. *Public Opinion Quarterly, 58*, 210–240.

Aquilino, W. S. & Losciuto, L. A. (1990). Effects of interview mode on self-reported drug use. *Public Opinion Quarterly, 54*, 362–395.

Aquilino, W. S. & Wright, D. L. (1996). Substance use estimates from RDD and area probability samples. *Public Opinion Quarterly, 60*, 563–573.

Arday, D. R., Tomar S. L., Nelson, D. E., Merritt, R. K., Schooley, M. W., & Mowery, P. (1997). State smoking prevalence estimates: a comparison between the behavioral risk factor surveillance system and current population surveys. *American Journal of Public Health, 87*, 1665–1669.

Ariely, D., Kahneman, D., & Loewenstein, G. (2000). Joint comment on "When does duration matter in judgment and decision making?" *Journal of Experimental Psychology: General, 129*, 524–529.

Arthur, A. (2004 September). Are landline phones losing ground? *The Source.* Retrieved July 26, 2005, from http://www.mediamark.com/mri/thesource/sorc2004_09.htm.

Asparouhov, T. (2005). Sampling weights in latent variable modeling. *Structural Equation Modeling, 12*, 411–434.

Associated Press. (2005, September 29). Some net phone users can't re-use numbers. Retrieved October 22, 2006, from http://www.msnbc.msn.com/id/9536184/

Atrostic, B. K., Bates, N., Burt, G., & Silberstein, A. (2001). Nonresponse in U.S. government household surveys: consistent measures, recent trends, and new insights. *Journal of Official Statistics, 17*, 209–226.

Axelrod, M. & Cannell, C. F. (1959). A research note on an attempt to predict interviewer effectiveness. *Public Opinion Quarterly, 23*, 571–76.

Babbie, E. R. (1990). *Survey Research Methods.* Belmont, CA: Wadsworth.

Backstrom, C. H. & Hursh, G. (1963). *Survey Research.* Evanston, IL: Northwestern University Press.

Bacon-Shone, J. & Lau, L. (2006, January). Mobile vs. fixed-line surveys in Hong Kong. Paper presented at the *Second International Conference on Telephone Survey Methodology*, Miami, FL.

Ballard-LeFauve, K., Cardoni, J., Murray, M. C. & Olson, L. (2004, May). Is it worth it? Using the mail to enhance RDD surveys. Paper presented at the *Annual Meeting of the American Association for Public Opinion Research*, Phoenix, AZ.

Bankier, M. D. (1986). Estimators based on several stratified samples with applications to multiple frame surveys. *Journal of the American Statistical Association, 81*, 1074–1079.

Barioux, M. (1952). A method for the selection, training and evaluation of interviewers. *Public Opinion Quarterly, 16*, 128–130.

Baker, R. P. (1998). The CASIC future. In R. P. Baker, M. P. Couper, J. Bethlehem, C. Z. F. Clark, J. Martin, W. L. Nicholls, II, & M. O'Reilly (Eds.), *Computer Assisted Survey Information Collection* (pp. 583–604). New York: John Wiley & Sons, Inc.

Baker, R. P. & Lefes, W. L. (1988). The design of CATI systems: a review of current practice. In R. M. Groves, P. P. Biemer, L. E. Lyberg, J. T. Massey, W. L. Nicholls, & J. Waksberg (Eds.), *Telephone Survey Methodology* (pp. 387–402). New York: John Wiley & Sons, Inc.

Barr, D. J. (2003). Paralinguistic correlates of conceptual structure. *Psychonomic Bulletin & Review, 10*, 462–467.

Bartlett, D. L., Ezzati-Rice, T. M., Stokley, S., & Zhao, Z. (2001). Comparison of NIS and NHIS/NIPRCS vaccination coverage estimates. *American Journal of Preventive Medicine, 20* (Suppl.), 25–27.

Bass, R. T. & Tortora, R. D. (1988). A comparison of centralised CATI facilities for an agricultural labor survey. In R. M. Groves, P. P. Biemer, L. E. Lyberg, J. T. Massey, W. L. Nicholls, & J. Waksberg (Eds.), *Telephone Survey Methodology*. New York: John Wiley & Sons, Inc.

Bates, N. (2003). Contact histories in personal visit surveys: The Survey of Income and Program Participation (SIPP) methods panel. *Proceedings of the Joint Statistical Meetings*, American Statistical Association, 7–14.

Bates, N. A., Dahlhammer, J., & Singer, E. (2006, May). Privacy concerns, too busy, or just not interested: using doorstep concerns to predict survey nonresponse. Poster presented at the *Annual Conference of the American Association of Public Opinion Research*, Montreal, Quebec, Canada.

Battaglia, M. P., Malec, D. J., Spencer, B., Hoaglin, D. C., & Sedransk, J. (1995a). Adjusting for noncoverage of nontelephone households in the National Immunization Survey. *Proceedings of the Section on Survey Research Methods*, American Statistical Association, 678–683.

Battaglia, M. P., Ryan, M., & Cynamon, M. (2005). Purging out-of-scope and cellular telephone numbers from RDD samples. *Proceedings of the Annual Meeting of the American Statistical Association* [CD-ROM] (pp. 3794-3801), Alexandria, VA: American Statistical Association.

Battaglia, M. P., Starer, A., Oberkofler, J., & Zell, E. R. (1995b). Pre-identification of nonworking and business telephone numbers in list-assisted random-digit-dialing samples. *Proceedings of the Section on Survey Research Methods*, American Statistical Association, 957–962.

Baumgartner, R. M. (1990, May). Telephone answering machine messages and completion rates for telephone surveys. Paper presented at the *Annual Meeting of the American Association for Public Opinion Research*, Lancaster, PA.

Beatty, P. & Hermann, D. (2002). To answer or not to answer: decision processes related to survey item nonresponse. In R. M. Groves, D. A. Dillman, J. L. Eltinge, & R. J. A. Little (Eds.), *Survey Nonresponse* (pp. 71–85). New York: John Wiley & Sons, Inc.

Beerten, R. & Martin, J. (1999). Household ownership of telephones and other communication links: implications for telephone surveys. *Survey Methodology Bulletin, 44*, 1–7.

Behavioral Risk Factor Surveillance System (2003). Technical information and data: Summary data quality reports. http://www.cdc.gov/brfss/technical-infodata/pdf/2003SummaryDataQualityReport.pdf.

Béland, Y., Bailie, L., Catlin, G., & Singh, M. P. (2000). CCHS and NPHS—an improved health survey program at Statistics Canada. *Proceedings of the Survey Research Methods Section*, American Statistical Association, 677–682.

Béland, Y., Dufour, J., & Hamel, M. (2001). Preventing nonresponse in the Canadian Community Health Survey. *Proceedings of the 2001 Statistics Canada International Symposium* [CD-ROM].

Bennett, D. J. & Steel, D. (2000). An evaluation of a large-scale CATI household survey using random digit dialing. *Australian & New Zealand Journal of Statistics, 42*, 255–270.

Benson, S., Booman, W. P., & Clark, K. E. (1951). Study of interviewer refusals, *Journal of Applied Psychology, 35*, 116–119.

Berk, R. A. (1984). *A Guide to Criterion Referenced Test Construction.* Baltimore: Johns Hopkins University Press.

Berlin, M., Mohadjer, L., Waksberg, J., Kolstad, A., Kirsch, I., Rock, D., & Yamamoto, K. (1992). An experiment in monetary incentives. *Proceedings of the American Statistical Association*, Survey Research Section, 393–398.

Berman, J., McCombs, H., & Boruch, R. F. (1977). Notes on the contamination method: two small experiments in assuring confidentiality of response. *Sociological Methods & Research, 6*, 45–63.

Berry, C. C., Flatt, S. W., & Pierce, J. P. (1996). Correcting unit nonresponse via response modeling and raking in the California tobacco survey. *Journal of Official Statistics, 12*, 349–363.

Berry, S. H. & O'Rourke, D. (1988). Administrative designs for centralized telephone survey centers: implications of the transition to CATI. In R. M. Groves, P. Biemer, L. Lyberg, J. Massey, W. L. Nicholls, & J. Waksberg (Eds.), *Telephone Survey Methodology.* New York: John Wiley & Sons, Inc.

Bethlehem, J. G. (1998). Reduction of non response bias through regression estimation. *Journal of Official Statistics, 4*, 251–260.

Bethlehem, J. G. (2002). Weighting nonresponse adjustments based on auxiliary information. In R. Groves, D. Dillman, J. Eltinge, & R. Little (Eds.), *Survey Nonresponse* (pp. 275–288). New York: John Wiley & Sons, Inc.

Bickart, B. & Schmittlein, D. (1999). The distribution of survey contact and participation in the United States: constructing a survey-based estimate. *Journal of Marketing Research 36*, 286–294.

Biemer, P. P. (1983). Optimal dual frame sample design: results of a simulation study. *Proceedings of the Survey Research Methods Section of the American Statistical Association*, 630–635.

Biemer, P. (1988). Measuring data quality. In R. M. Groves, P. Biemer, L. Lyberg, J. Massey, W. Nicholls, & J. Waksberg (Eds.), *Telephone Survey Methodology* (pp. 321–340). New York: John Wiley & Sons, Inc.

Biemer, P. P. (2001). Nonresponse and measurement bias in a comparison of face-to-face and telephone interviewing. *Journal of Official Statistics, 17*, 295–320.

Biemer, P. P., Herget, D., Morton, J., & Willis, G. (2000). The feasibility of monitoring field interview performance using computer audio recorded interviewing (CARI). *Proceedings of the Survey Research Methods Section*, American Statistical Association, 1068–1073.

Biemer, P. P. & Link, M. W. (2003). The impact of response rates on survey accuracy: When is too little too much? Paper presented at the *Annual Conference of the American Association for Public Opinion Research*, Nashville, TN.

Biemer P. P. & Lyberg, L. (2003). *Introduction to Survey Quality*. New York: John Wiley & Sons, Inc.

Biever, C. (2004, September 24). Move over spam, make way for "spit." *NewScientist.com News Service.* Retrieved October 21, 2006, from http://www.newscientist.com/article.ns?id=dn6445

Binson, D., Canchola, J. A., & Catania, J. A. (2000). Random selection in a national telephone survey: a comparison of the Kish, next birthday, and last-birthday methods. *Journal of Official Statistics, 16*, 53–59.

Bishop, G. F., Hippler, H.-J., Schwarz, N., & Strack, F. (1988). A comparison of response effects in self-administered and telephone surveys. In R. M. Groves, P. P. Biemer, L. E. Lyberg, J. T. Massey, W. L. Nicholls II, & J. Waksberg (Eds.), *Telephone Survey Methodology* (pp. 321–340). New York: John Wiley and Sons, Inc.

Blair, J. & Chun, Y. I. (1992, May). Quality of data from converted refusers in telephone surveys. Presented at the *Annual Conference of the American Association for Public Opinion Research*, St. Petersburg, FL.

Blair, J. & Czaja, R. (1982). Locating a special population using random digit dialing. *Public Opinion Quarterly, 46*, 585–590.

Blair, J., Mennon, G., & Bickart, B. (1991). Measurement effects in self vs. proxy responses: an information processing perspective. In P. P. Biemer, R. M. Groves, L. E. Lyberg, N. A. Mathiowetz, & S. A. Sudman (Eds.), *Measurement Errors in Surveys* (pp. 145–166). New York: John Wiley & Sons, Inc.

Blankenship, A. (1977). *Professional Telephone Surveys*. New York: McGraw-Hill.

Blau, P. M. (1964). *Exchange and Power in Social Life*. New York: John Wiley and Sons, Inc.

Bloom, B., & Dey, A. N. (2006). Summary health statistics for US children: National Health Interview Survey, 2004. *National Center for Health Statistics: Vital Health Statistics, 10*, 227.

Bloom, J. E. (1999). Linguistic markers of respondent uncertainty during computer-administered survey interviews. Unpublished doctoral dissertation, New School for Social Research.

Blumberg, S. (2005). NHIS estimates of cell only population size and characteristics. Presentation to the Cell Phone Summit II Meeting, New York, NY, February.

Blumberg, S., Luke, J., & Cynamon, M. (2005a). The prevalence and impact of wireless substitution: updated data from the 2004 National Health Interview Survey. Paper presented at the *Conference of the American Association of Public Opinion Research*, Miami, FL.

Blumberg, S. J., Luke, J. V., & Cynamon, M. L. (2006). Telephone coverage and health survey estimates: evaluating the need for concern about wireless substitution. *American Journal of Public Health, 96*, 926–931.

Blumberg, S. J., Olson, L., Frankel, M. R., Osborn, L., Becker, C. J., Srinath, K. P., & Giambo, P. (2003). Design and operation of the National Survey of Children with Special Health Care Needs, 2001. *Vital and Health Statistics, Series 1*, No. 41.

Blumberg, S. J., Olson, L., Frankel, M. R., Osborn, L., Srinath, K. P., & Giambo, P. (2005b). Design and operation of the National Survey of Children's Health, 2003. *Vital and Health Statistics, Series 1*, No. 43.

Blumberg, S. J., Olson, L., Osborn, L., Srinath, K. P., & Harrison, H. (2002). Design and operation of the National Survey of Early Childhood Health, 2000. *Vital and Health Statistics, Series 1*, No. 40.

Bocklund, L. & Bengtson, D. (2002). *Call Center Technology Demystified: The No-Nonsense Guide to Bridging Customer Contact Technology, Operations and Strategy.* Annapolis, MD: Call Center Press.

Boersma, F., Eefsting, J. A., & van den Brink, W. (1997). Characteristics of non-responders and the impact of nonresponse on prevalence estimates of dementia. *International Journal of Epidemiology, 26*, 1055–1062.

Boersma, P. (1993). Accurate short-term analysis of the fundamental frequency and the harmonics-to-noise ratio of a sampled sound. *Proceedings of the Institute of the Phonetic Sciences, University of Amsterdam, 17*, 97–110.

Boersma, P. & Weenink, D. (2005). Praat: doing phonetics by computer (Version 4.3.12) [Computer software]. Amsterdam: Institute of Phonetic Sciences. Retrieved December 14, 2006, from http://www.praat.org

Bogen, K. (1996). The effect of questionnaire length on response rates: a review of the literature. *Proceedings of the Survey Research Methods Section*, American Statistical Association, 1020–1025.

Bortfeld, H., Leon, S. D., Bloom, J. E., Schober, M. F., & Brennan, S. E. (2001). Disfluency rates in conversation: effects of age, relationship, topic, role, and gender. *Language and Speech, 44*, 123–149.

Botman, S. L., Moore, T. F., Moriarity, C. L., & Parsons, V. L. (2000). Design and estimation for the National Health Interview Survey, 1995–2004. *Vital Health Statistics, 2*, 130.

Boyle, W. R. & Kalsbeek, W. D. (2005). Extensions to the two-stratum model for sampling rare subgroups in telephone surveys. *Proceedings of the Section on Survey Research Methods*, American Statistical Association, 2783–2790.

Bradburn, N. (1977). Respondent burden. *Proceedings of the Health Survey Research Methods Second Biennial Conference.* Washington, D.C: U.S. Department of Health, Education and Welfare.

Bradburn, N. (1978). Respondent burden. *Health Survey Research Methods.* (DHEW Publication No. 79-3207, pp. 49–53). Washington, DC: U.S. Department of Health, Education, and Welfare.

Braun, M. & Harkness, J. A. (2005). Text and context: challenges to comparability in survey questions. In J. H. P. Hoffmeyer-Zlotnik & J. A. Harkness (Eds.), *Zuma-Nachrichten Spezial No. 11. Methodological Aspects in Cross-National Research.* Mannheim, Germany: ZUMA.

Brennan, S. E. & Schober, M. F. (2001). How listeners compensate for disfluencies in spontaneous speech. *Journal of Memory and Language, 44*, 274–296.

Brennan, S. E. & Williams, M. (1995). The feeling of another's knowing: prosody and filled pauses as cues to listeners about the metacognitive states of speakers. *Journal of Memory and Language, 34*, 383–398.

Brick, J. M., Allen, B., Cunningham, P., & Maklan, D. (1996a). Outcomes of a calling protocol in a telephone survey. *Proceedings of the Survey Research Methods Section.* Alexandria, VA: American Statistical Association.

Brick, J. M., Brick, P. D., Dipko, S., Presser, S., Tucker, C., & Yuan, Y. (2007). Cell phone survey feasibility in the U.S.: Sampling and calling cell numbers versus landline numbers. *Public Opinion Quarterly, 71*, 23–39.

Brick, J. M. & Broene, P. (1997). Unit and item response rates, weighting and imputation procedures in the 1995 National Household Education Survey. (NCES 97-06). Washington, DC: U. S. Department of Education, Office of Educational Research and Improvement.

Brick, J., Burke, J., & West, J. (1992). Telephone undercoverage bias of 14-to 21-year-olds and 3-to 5-year-olds. *National Household Education Survey Technical Report No. 2.* (NCES 92-101). Washington, DC: U. S. Department of Education, Office of Educational Research and Improvement.

Brick, J. M., Cahalan, M., Gray, L., & Severynse, J. (1994). A study of selected nonsampling errors in the 1991 Survey of Recent College Graduates (NCES 95-640). Washington, DC: U.S. Department of Education, Office of Educational Research and Improvement.

Brick, J. M., Dipko, S., Presser, S., Tucker, C., & Yuan, Y. (2005a). Estimation issues in dual frame sample of cell and landline numbers. Unpublished manuscript.

Brick, J. M., Dipko, S., Presser, S., Tucker, C., & Yuan, Y. (2005b). Surveying households on cell and landline phones. Paper presented at the *Joint Statistical Meetings*, Minneapolis, Minnesota.

Brick, J. M., Dipko, S., Presser, S., Tucker, C., & Yuan, Y. (2006). Nonresponse bias in a dual frame sample of cell and landline numbers. *Public Opinion Quarterly, 70*, 780–793.

Brick, J. M., Ferraro, D., Strickler, T., & Liu, B. (2003a). *2002 NSAF Sample Design: NSAF Methodology Report No. 2.* Washington, DC: The Urban Institute.

Brick, J. M., Judkins, D., Montaquila, J., & Morganstern, D. (2002). Two-phase list-assisted RDD sampling. *Journal of Official Statistics, 18*, 203–215.

Brick, J. M., Judkins, D., Montaquila, J., Morganstein, D., & Shapiro, G. (2000). Evaluating secondary data sources for random digit dialing samples. *Proceedings of the Survey Research Methods Section of the American Statistical Association*, 142–150.

Brick, J.M., & Kalton, G. (1996). Handling missing data in survey research. *Statistical Methods in Medical Research, 5*, 215–238.

Brick, J. M., Martin, D., Warren, P., & Wivagg, J. (2003b). Increased efforts in RDD surveys. *2003 Proceedings of the Section on Survey Research Methods*, Alexandria, VA: American Statistical Association, 26–31.

Brick, J. M., McGuinness, R., Lapham, S. J., Cahalan, M., Owens, D., & Gray, L. (1995). Interviewer variance in two telephone surveys. *Proceedings of the Section on Survey Research Methods*, American Statistical Association, 447–452.

Brick, J. M., Montaquila, J., Hagedorn, M. C., Roth, S. B., & Chapman, C. (2005c). Implications for RDD design from an incentive experiment. *Journal of Official Statistics, 21*, 571–589.

Brick, J. M., Montaquila, J., Hagedorn, M. C., Li, L., and Seastrom, M. (2005d). An Analysis of Nonresponse Bias in the 2001 National Household Education Surveys Program (NHES: 2001). Unpublished technical report.

Brick, J. M., Montaquila, J., & Scheuren, F. (2002). Estimating residency rates for undetermined telephone numbers. *Public Opinion Quarterly, 66*, 18–39.

Brick, J. M., Waksberg, J., & Keeter, S. (1996b). Using data on interruptions in telephone service as coverage adjustments. *Survey Methodology, 22*, 185–197.

Brick, J., & West, J. (1992). Multiplicity Sampling for Dropouts in the NHES Field Test. National Household Education Survey Technical Report No 3. (Report No. NCES 92-102). Washington, DC: U.S. Department of Education, Office of Educational Research and Improvement.

Brøgger, J., Bakke, P., Geir, E. E., & Gulsvik, A. (2003). Contribution of follow-up of non-responders to prevalence and risk estimates: a Norwegian respiratory health survey. *American Journal of Epidemiology, 157*, 558–566.

Buckley, P., Dennis, J. M., Saulsberry, C., Coronado, V. G., Ezzati-Rice, T., Maes, E., Roden, A., & Wright, R. A. (1998). Managing 78 simultaneous RDD samples. *Proceedings of the Section on Survey Research Methods*, American Statistical Association, 957–961.

Burke, J., Mohadjer, L., Green, J., Waksberg, J., Kirsch, I., & Kolstad, A. (1994). Composite estimation in national and state surveys. *Proceedings of the Survey Research Methods Section of the American Statistical Association*, 873–878.

Burke. J., Morganstein, D., & Schwartz, S. (1981). Toward the design of an optimal telephone sample. *Proceedings of the Survey Research Methods Section*, American Statistical Association, 448–453.

Burkheimer, G. J. & Levinsohn, J.R. (1988). Implementing the Mitofsky–Waksberg sampling design with accelerated sequential replacement. In R. M. Groves, P. P. Biemer, L. E. Lyberg, J. T. Massey, W. L. Nicholls, & J. Waksberg (Eds.), *Telephone Survey Methodology* (pp. 99–112). New York: John Wiley & Sons, Inc.

Burks, A. T., Lavrakas, P. J., & Bennett, M., (2005, May). Predicting a sampled respondent's likelihood to cooperate: Stage III research. Paper presented at the *Annual Meeting of the American Association for Public Opinion Research*, Miami, FL.

Burstiner, I. (1974). Improving the productivity of a telephone sales force. *Management Review, 63*, 26–33.

Buskirk, T. D. & Callegaro, M. (2002, December). Surveying U.S. mobile phone subscribers. When, where, or if ever. Paper presented at the *Annual Conference of the Pacific Chapter of the American Association for Public Opinion Research*, Monterey, CA.

Butler, D. (2004). *Bottom-Line Call Center Management: Creating a Culture of Accountability and Excellent Customer Service.* Burlington, MA: Elsevier Butterworth-Heinemann.

Buzzi, C., Callegaro, M., Schizzerotto, A., & Vezzoni, C. (2000). *Studio o Lavoro? Le Scelte Formative ed Occupazionali Dei Maturi Trentini. [Study or work? The Occupational and Educational Choices of Graduated High School Students of the Province of Trento].* Trento: Edizioni del Rettorato.

Caetano, R. (2001). Non-response in alcohol and drug surveys: a research topic in need of further attention. *Addiction, 96*, 1541–1545.

California Health Interview Survey (2005). *CHIS 2003 Methodology Series: Report 1 — Sample Design.* Los Angeles, CA: UCLA Center for Health Policy Research.

Callegaro, M. & Poggio, T. (2004). Espansione della telefonia mobile ed errore di copertura nelle inchieste telefoniche [Mobile telephone growth and coverage error in telephone surveys] [Electronic version]. *Polis, 18*, 477–506.

Callegaro, M., Steeh, C., Buskirk, T., Kuusela, V., Vehovar, V., & Piekarski, L. (2007). Fitting disposition codes to mobile phone surveys: experiences from studies in Finland, Slovenia, and the United States. *Journal of the Royal Statistical Society Series A (Statistics in Society), 170*, 647–670.

Camburn, D. P., Lavrakas, P. J., Battaglia, M. P., Massey, J. T., & Wright, R. A. (1995). Using advance letters in random-digit-dialing telephone surveys. *Proceedings of the Section on Survey Research Methods, American Statistical Association*, 969–974.

Camburn, D. P. & Wright, R. A. (1996, May). Predicting eligibility rates for rare population in RDD screening surveys. Paper presented at the *AAPOR Annual Conference*, Salt Lake City, UT. Retrieved May 16, 2006, from http://www.cdc.gov/nis/papers_pubs.htm

Campanelli, P., Sturgis, P., & Purdon, S. (1997). *Can You Hear Me Knocking: An Investigation Into the Impact of Interviewers on Survey Response Rates*, London: Social and Community Planning Research.

Cannell, C. F. & Fowler, F. J. (1963). Comparison of a self-enumerative procedure and a personal interview: A validity study. *Public Opinion Quarterly, 37*, 250–264.

Cannell, C., Groves, R., Magilavy, L., Mathiowetz, N., & Miller, P. (1987). An experimental comparison of telephone and personal health surveys, *Vital and Health Statistics*, Series 2, No. 106, Hyattsville, MD: National Center for Health Statistics.

Cannell, C. F., Lawson, S. A., & Hausser, D. L. (1975). A technique for evaluating interviewer performance: a manual for coding and analyzing interviewer behavior from tape recordings of household interviews. Ann Arbor: University of Michigan Press.

Cannell, C., Miller, P., & Oksenberg, L. (1981). Research on interviewing techniques. In S. Leinhard (Ed.), *Sociological Methodology* (pp. 389–437). San Francisco: Jossey-Bass.

Cannell, C. & Oksenberg, L. (1988). Observation of behaviors in telephone interviews. In R. M. Groves, P. P. Biemer, L. E. Lyberg, J. T. Massey, W. L. Nicholls, & J. Waksberg (Eds.), *Telephone Survey Methodology* (pp. 475–495). New York: John Wiley & Sons, Inc.

Cannell, C. & Robison, S. (1971). Analysis of individual questions. In J. B. Lansing, S. B. Witney, & A. C. Wolfe (Eds.), *Working Papers on Survey Research in Poverty Areas.* Ann Arbor: University of Michigan, Institute for Social Research, Survey Research Center.

Cantor, D. & Phipps, P. (1999). Adapting cognitive techniques to establishments surveys. In M. Sirken, T. Jabine, G. Willis, E. Martin, & C. Tucker (Eds.), *A New Agenda for Interdisciplinary Survey Research Methods* (pp. 74–78). Hyattsville, MD: National Center for Health Statistics.

Cantor, D., Cunningham, P., & Giambo, P. (1998, May). Testing the effects of a pre-paid incentive and express delivery to increase response rates on a random digit dial telephone survey. Paper presented at the *Annual Meeting of the American Association for Public Opinion Research*, St. Louis, MO.

Cantor, D., Cunningham, P., Triplett, T., & Steinbach, R. (2003a, May). Comparing incentives at initial and refusal conversion stages on a screening interview for a random digit dial survey. Paper presented at the *Annual Meeting of the American Association for Public Opinion Research*, Nashville, TN.

Cantor, D., Wang, K., & Abi-Habibm, N. (2003b, May). Comparing promised and prepaid incentives for an extended interview on a random digit dial survey. Paper presented at the *Annual Conference of the American Association of Public Opinion Research*, Nashville, TN.

Carlson, B. L. & Strouse, R. (2005). The value of increasing effort to maintain high response rates in telephone surveys. *Proceedings of the American Statistical Association*, Survey Methodology Section [CD-ROM].

Casady, R. J. & Lepkowski, J. M. (1991). Optimal allocation for stratified telephone survey designs. *Proceedings of the Section on Survey Research Methods*, American Statistical Association, 111–116.

Casady, R. J. & Lepkowski, J. M. (1993). Stratified telephone survey designs. *Survey Methodology, 19*, 103–113.

Casady, R. J. & Lepkowski, J. M. (1999). Telephone sampling. In P. S. Levy & S. Lemeshow (Eds.), *Sampling of Populations: Methods and Applications* (3rd ed.) (pp. 455–480). New York: John Wiley & Sons, Inc.

Casady, R. J., Snowden, C. B., & Sirken, M. G. (1981). A study of dual frame estimators for the National Health Interview Survey. *Proceedings of the Survey Research Methods Section of the American Statistical Association*, 444–453.

Catlin, G. & Ingram, S. (1988). The effects of CATI on cost and data quality. In R.M. Groves, P. P. Beimer, L. E. Lyberg, & J. T. Massey (Eds.), *Telephone Survey Methodology*, New York: John Wiley & Sons.

Cattell, J. (2001, February). The mobile internet revolution and its implications for research. Paper presented at *Net Effects* 4, Barcelona, Spain.

Cellular Telecommunications Industry Association—The Wireless Association (2006). CTIA's Wireless Industry Indices: 1985–2005. Washington, DC: CTIA. Retrieved July 26, 2006, from http://files.ctia.org/pdf/CTIAEndYear2005Survey.pdf

Centers for Disease Control and Prevention (1994). Reported vaccine-preventable diseases - United States, 1993, and the Childhood Immunization Initiative. *Morbidity and Mortality Weekly Report*, 43, 57–60.

Chafe, W. & Tannen, D. (1987). The relation between written and spoken language. *Annual Review of Anthropology, 16*, 383–407.

Chang, L. & Krosnick, J. A., (2001, May). The representativeness of national samples: comparisons of an RDD telephone survey with matched Internet surveys by Harris Interactive and Knowledge Networks. Paper presented at the *Conference of the American Association for Public Opinion Research.*

Chapman, D. W. & Weinstein, R. B. (1990). Sampling Design for a Monitoring Plan for CATI Interviewing. *Journal of Official Statistics*, 6, 205–211.

Chingos, P. T. (2002). *Paying for Performance: A Guide to Compensation Management* (2nd ed.). New York: John Wiley & Sons, Inc.

Christian, L. M. (2003). The influence of visual layout on scalar questions in web surveys. Unpublished masters' thesis, Washington State University, Pullman.

Christian, L. M. & Dillman, D. A. (2004). The influence of symbolic and graphical language manipulations on answers to paper self-administered questionnaires. *Public Opinion Quarterly, 68*, 57–80.

Church, A. H. (1993). Estimating the effect of incentives on mail survey response rates: a meta-analysis. *Public Opinion Quarterly, 57*, 62–79.

Clark, H. H. (1996). *Using Language*. Cambridge: Cambridge University Press.

Clark, H. H. & Brennan, S. E. (1991). Grounding in communication. In L. B. Resnick, J. M. Levine, & S. D. Teasley (Eds.), *Perspectives on Socially Shared Cognition* (pp. 127–149). Washington, D. C.: APA Books.

Clark, H. H. & Schaefer, E. F. (1989). Contributing to discourse. *Cognitive Science, 13*, 259–294.

Clark, H. H. & Schober, M. F. (1992). Asking questions and influencing answers. In J. M. Tanur (Ed.), *Questions About Questions: Inquiries Into the Cognitive Bases of Surveys* (pp. 15–48). New York: Russell Sage Foundation.

Clark, H. H. & Wilkes-Gibbs, D. (1986). Referring as a collaborative process. *Cognition, 22*, 1–39.

Clayton, R. L. & Werking, G. S. (1998). Business surveys of the future: the world wide web as a data collection methodology. In M. P. Couper, R. P. Baker, J. Bethlehem, C. Z. F. Clark, J. Martin, W. L. Nichols II, & J. O'Reilly (Eds.), *Computer Assisted Survey Information Collection* (pp. 543–562). New York: John Wiley & Sons, Inc.

Clayton, R. L. & Winter, D. L. S. (1992). Speech data entry: results of a test of voice recognition for survey data collection. *Journal of Official Statistics, 8*, 377–388.

Cleveland, B. & Hash, S. (2004a). *Call Center Agent Turnover and Retention: The Best of Call Center Management Review* (2nd ed.). Annapolis, MD: Call Center Press.

Cleveland, B. & Hash, S. (2004b). *Call Center Recruiting and New-Hire Training: The Best of Call Center Management Review* (2nd ed.). Annapolis, MD: Call Center Press.

Cleveland, B. & Mayben, J. (2002). *Call Center Management on Fast Forward.* Annapolis, MD: Call Center Press.

Clogg, C. C. & Eliason, S. R. (1987). Some common problems in loglinear analysis. *Sociological Methods and Research, 16*, 8–14.

Cochran, W. G. (1968). The effectiveness of adjustment by subclassification in removing bias in observational studies. *Biometrics, 24*, 295–313.

Cochran, W. G. (1977). *Sampling Techniques.* New York: John Wiley & Sons, Inc.

Cohen, E. (2003). An exploratory study of cellphone users. Paper presented at the *Cell Phone Sampling Summit*, New York.

Cohen, E. (2006). Arbitron's cell phone research and future plans. Retrieved October 21, 2006, from http://www.arbitron.com/downloads/cellphone021606.pdf.

Coleman, M. P., Evans, B. G., & Barrett, G. (2003). Confidentiality and the public interest in medical research: will we ever get it right? *Clinical Medicine, 3*, 219–228.

Collins, M. (1999). Editorial; sampling for UK telephone surveys. *Journal of the Royal Statistical Society Series A (Statistics in Society), 162*, 1–4.

Collins, M. & Sykes, W. (1987). The problems of non-coverage and unlisted numbers in telephone surveys in Britain. *Journal of the Royal Statistical Society, A, 150*, 241–253.

Collins, M., Sykes W., Wilson, P., & Blackshaw, N. (1988). Nonresponse: The U.K. experience. In R. M. Groves, P. B. Biemer, L. E. Lyberg, J. T. Massey, W. L. Nicholls II, & J. Waksberg (Eds.), *Telephone Survey Methodology* (pp. 213–232). New York: John Wiley & Sons, Inc.

Colombo, R. (2001). A model for diagnosing and reducing nonresponse bias. *Journal of Advertising Research*, 40, 85–93.

Conrad, F., Couper, M. P., Tourangeau. R., & Galesic, M. (2005). Interactive feedback can improve the quality of responses in web surveys. Paper presented at the *Conference of the American Association for Public Opinion Research*, Miami Beach, FL.

Conrad, F. G. & Schober, M. F. (2000). Clarifying question meaning in a household telephone survey. *Public Opinion Quarterly, 64*, 1–28.

Conrad, F. G. & Schober, M. F. (2005). Promoting uniform question understanding in today's and tomorrow's surveys. *Journal of Official Statistics, 21*, 215–231.

Conrad, F. G. & Schober, M. F. (2008). *Envisioning the Survey Interview of the Future.* New York: John Wiley and Sons, Inc.

Conrad, F., Schober, M., & Dijkstra, W. (2004). Non-verbal cues of respondents' need for clarification in survey interviews. *Proceedings of the American Statistical Association*, Section on Survey Methods Research, 4943–4950.

Cooke, M., Nielsen, A., & Strong, C. (2003). The use of SMS as a research tool. In R. Banks, J. Currall, J. Francis, L. Gerrard, R. Khan, T. Macer, M. Rigg, E. Ress, S. Taylor, & A. Westlake (Eds.), *ASC 2003. The Impact of Technology on the Survey Process* (pp. 267–276). Chesham, UK: Association for Survey Computing.

Cooper, S. L. (1964). Random sampling by telephone: an improved method. *Journal of Marketing Research, 1*, 45–48.

Corning, A. & Singer, E. (2003). *Survey of U.S. Privacy Attitudes.* Washington, D. C.: Computer Science and Telecommunications Board, Division on Engineering and Physical Sciences, National Research Council, National Academy of Sciences.

Costabile, M. & Addis, M. (2002). Comunicazione mobile in Italia; Passato, presente e futuro. [Mobile communication in Italy: Past, present and future]. In M. Costabile & M. Addis (Eds.), *Mobile Communication. Successi di Marketing Nelle Telecomunicazioni Mobili in Italia [Mobile communication. Marketing successes in Italian mobile Telecommunications.]* (pp. 23–76). Milano: Il Sole 24 Ore.

Cottler, L. B., Zipp, J. F., & Robbins, L. N. (1987). Difficult-to-recruit respondents and their effect on prevalance estimates in an epidemiologic survey. *American Journal of Epidemiology, 125*, 329–339.

Council for Marketing and Opinion Research. (1996). *Respondent Cooperation and Industry Image Study.* Glastonbury, CT: Author.

Council for Marketing and Opinion Research. (2003). *Respondent Cooperation and Industry Image Study.* Glastonbury, CT: Author.

Council of Professional Associations on Federal Statistics (COPAFS). (1992, October). Providing incentives to survey respondents: Final report. Paper resented at the *Providing Incentives to Survey Respondents Symposium*, Cambridge, MA.

Couper, M. P. (1998a). Measuring survey quality in a CASIC environment. *Proceedings of the Section on Survey Research Methods, American Statistical Association,* 41–49.

Couper, M. P. (1998b, August). Measuring survey quality in a CASIC environment. Paper presented at the *Joint Statistical Meetings of the American Statistical Association*, Dallas, TX.

Couper, M. P. (2000a). Usability evaluation of computer-assisted survey instruments. *Social Science Computer Review, 18*, 384–396.

Couper, M. P. (2000b). Web surveys: a review of issues and approaches. *Public Opinion Quarterly, 64*, 464–494.

Couper, M. P. & Groves, R. M. (1992). The role of interviewer in survey participation. *Survey Methodology, 18*, 263–277.

Couper, M. P. & Groves, R. (2002). Introductory interactions in telephone surveys and nonresponse. In D. Maynard, H. Houtkoop-Steenstra, N. Schaeffer, & J. vander Zouwen (Eds.), *Standardization and Tacit Knowledge: Interaction and Practice in the Survey Interview* (pp. 161–178), New York: John Wiley & Sons, Inc.

Couper, M. P. & Hansen, S. E. (2001). Computer assisted interviewing. In J. F. Gubrium, & J. A. Holstein (Eds.), *Handbook of Interviewing* (pp. 557–575). Thousand Oaks, CA: Sage.

Couper, M., Hansen, S., & Sadosky, S. A. (1997). Evaluating interviewer use of CAPI technology. In L. Lyberg, P. Biemer, M. Collins, E. De Leeuw, C. Dippo, N. Schwarz, & D. Trewin (Eds.), *Survey Measurement and Process Quality*, New York: John Wiley and Sons, Inc.

Couper, M. P., Holland, L., & Groves, R. M. (1992). Developing systematic procedures for monitoring in a centralized telephone facility. *Journal of Official Statistics, 8*, 63–76.

Couper, M. P. & Lyberg, L. (2005, April). The use of paradata in survey research. Paper presented at the *54th Session of the International Statistical Institute*, Sydney, Australia.

Couper, M. P., Mathiowetz, N. A., & Singer, E. (1995). Related households, mail handling and returns to the 1990 census. *International Journal of Public Opinion Research, 7*, 172–177.

Couper, M. P., Peytchev, A., Little, R. J. A, Strecher, V. J., & Rother, K. (2005). Combining information from multiple modes to reduce nonresponse bias. *Proceedings of the Survey Research Methods Section of the American Statistical Association*, 2910–2917.

Couper, M. P., Singer, E., & Kulka, R. A. (1998). Participation in the 1990 decennial census. *American Politics Quarterly, 26*, 59–80.

Couper, M. P., Singer, E., & Tourangeau, R. (2003). Understanding the effects of audio-CASI on self-reports of sensitive behavior. *Public Opinion Quarterly, 67*, 385–395.

Couper, M. P., Traugott, M., & Lamias, M. J. (2001). Web survey design and administration. *Public Opinion Quarterly, 65*, 230–254.

Crawford, S. D., Couper, M. P., & Lamias, M.J. (2001). Web surveys: perceptions of burden. *Social Science Computer Review, 19*, 146–162.

Crocker, L. & Algina, J. (1986). *An Introduction to Classical and Modern Test Theory.* Orlando, FL: Harcourt Brace Jovanovich.

Croner, C. M., Williams, P. D., & Hsiung, S. (1985). Callback response in the National Health Interview survey. *Proceedings of the Survey Research Methods Section*, American Statistical Association, 164–169.

Cummings, K. M. (1979). Random digit dialing: a sampling technique for telephone surveys. *Public Opinion Quarterly, 43*, 233–244.

Cunningham, P., Martin, D., & Brick, J. M. (2003). An experiment in call scheduling. *Proceedings of the Survey Research Methods Section*, American Statistical Association, 59–66.

Cunningham, M., Meader, J., Molloy, K., Moore, D., & Pajunen, S. (1997). Using cellular telephones to interview nontelephone households. *Proceedings of the Survey Research Methods Section*, American Statistical Association, 250–254.

Currivan, D. (2005, May). The impact of providing incentives to initial telephone survey refusers on sample composition and data quality. Paper presented at the *Annual Meeting of the American Association of Public Opinion Research*, Miami Beach, FL.

Currivan, D. B., Nyman, A. L., Turner, C. F., & Biener, L. (2004). Does telephone audio computer-assisted self-interviewing improve the accuracy of prevalence estimates of youth smoking? *Public Opinion Quarterly, 68*, 542–564.

Curtin, R., Presser, S., & Singer, E. (2000). The effects of response rate changes on the index of consumer sentiment. *Public Opinion Quarterly, 64*, 413–28.

Curtin, R., Presser, S., & Singer, E. (2005). Changes in telephone survey nonresponse over the past quarter century. *Public Opinion Quarterly, 69*, 87–98.

Curtin, R., Singer E., & Presser, S. (2007). Incentives in telephone surveys: a replication and extension. *Journal of Official Statistics, 23*, 91–105.

Czaja, R. F., Snowden, C. B., & Casady, R. J. (1986). Reporting bias and sampling errors in a survey of a rare population using multiplicity counting rules. *Journal of the American Statistical Association, 81*, 411–419.

Czajka, J. L., Hirabayashi, S. M., Little, R. J. A., & Rubin, D. B. (1992). Projecting from advance data using propensity modeling: an application to income and tax statistics. *Journal of Business and Economic Statistics, 10*, 117–132.

D'Agostino, R. B. Jr. (1998). Propensity score methods for bias reduction for the comparison of a treatment to a non-randomized control group. *Statistics in Medicine, 17*, 2265–2281.

Dahlbäck, N., Jönsson, A., & Ahrenberg, L. (1993). Wizard of Oz studies—why and how. *Knowledge-Based Systems, 6*, 258–266.

Danbury, T. (1975, November). Alternative sampling models for random digit dialing surveys. Paper presented at the *Twenty-Fifth Annual Conference of the Advertising Research Foundation*, New York.

Dashen, M. & Fricker, S. (2001). Understanding the cognitive processes of open-ended categorical questions and their effect on data quality. *Journal of Official Statistics, 17*, 457–477.

Dautch, B. (2005, February). Update on current restrictions for survey research calls to cell phones. Paper presented at the *Cell Phone Sampling Summit II*, New York.

Davern, M., Lepkowski, J., Call, K. T., Arnold, N., Johnson, T. L., Goldsteen, K., Todd-Malmlov, A., & Blewett, L. A. (2004). Telephone service interruption weighting adjustments for state health insurance surveys. *Inquiry, 41*, 280–290.

Davis, M. H. (2005). A "constituent" approach to the study of perspective taking: what are its fundamental elements? In B. F. Malle & S. D. Hodges (Eds.), *Other Minds: How Humans Bridge the Divide Between Self and Others* (pp. 44–55). New York: Guilford.

Dawson, K. (2004). *The Call Center Handbook: The Complete Guide to Starting, Running, and Improving Your Call Center.* San Francisco: CMP Books.

de Heer, W. (1999). International response trends: results of an international survey. *Journal of Official Statistics 15*, 129–142.

de Leeuw, E. D. (1992). *Data Quality in Mail, Telephone, and Face to Face Surveys.* Amsterdam: TT Publications.

de Leeuw, E. D. (1999). How do successful and less successful interviewers differ in tactics for combating survey nonresponse? *Bulletin de Methodologie Sociologique, 62*, 29–42.

de Leeuw E. D. (2005). To mix or not to mix: data collection modes in surveys. *Journal of Official Statistics, 21*, 233–255.

de Leeuw, E. & Collins, M. (1997). Data collection method and data quality: an overview. In L. E. Lyberg, P. P. Biemer, M. Collins, E. De Leeuw, C. Dippo, N. Schwarz, & D. Trewin (Eds.), *Survey Measurement and Process Quality* (pp. 199–221). New York: John Wiley & Sons, Inc.

de Leeuw, E. & de Heer, W. (2002). Trends in household survey nonresponse: a longitudinal and international comparison. In R. M. Groves, D. A. Dillman, J. L. Eltinge (Eds.), *Survey Nonresponse* (pp. 41–54). New York: John Wiley & Sons, Inc.

de Leeuw, E. & Hox, J. (2004). I am not selling anything: twenty-nine experiments in telephone introductions. *International Journal of Public Opinion Research, 16*, 464–473.

de Leeuw, E., Hox, J., Korendijk, E., & Lensvelt-Mulders, G. (2004, August). The influence of advance contacts on response in telephone surveys: a meta-analysis. Paper presented at the *Sixth Conference on Logic and Methodology*, Amsterdam, the Netherlands.

de Leeuw, E., Hox, J., Korendijk, E., Lensvelt-Mulders, G., & Callegaro, M. (2006, January). The influence of advance letters on response in telephone surveys: a metaanalysis. Paper presented at the *Second International Conference on Telephone Survey Methodology*, Miami, FL.

de Leeuw, E. D. & Nicholls, W., II (1996). Technological innovations in data collection: acceptance, data quality and costs. *Sociological Research Online, 1.* Retrieved December 14, 2006, from http://www.socresonline.org.uk/1/4/LeeuwNicholls.html

de Leeuw, E. D., Nicholls, W. L., II, Andrews, S. H., & Mesenbourg, T. L. (2000). The use of old and new data collection methods in establishment surveys. In A. Christiansen, A.-M. Flygare, O. Frank, J. Johansson, P. Lundquist, A. Modig, B. Swensson, & L. E. Öller (Eds.), *Methodological Issues in Official Statistics.* Retrieved December 19, 2006, from the Statistics Sweden (SCB) Web site: http://www.scb.se/Grupp/Omscb/_Dokument/Deleeuw.pdf

de Leeuw, E. D. & van der Zouwen, J. (1988). Data quality in telephone and face to face surveys: a comparative meta-analysis. In R. M. Groves, P. P. Biemer, L. E. Lyberg, J. T. Massey, W. L. Nicholls, & J. Waksberg (Eds.), *Telephone Survey Methodology* (pp. 283–299). New York: John Wiley & Sons, Inc.

Decouflé, P., Holmgreen, P., & Calle, E. E. (1991). Nonresponse and intensity of follow-up in an epidemiologic study of Vietnam veterans. *American Journal of Epidemiology, 133*, 83–95.

DeMaio, T. (1980). Refusals: who, where, and why. *Public Opinion Quarterly, 44*, 223–233.

Deming, W. E. (1950). *Some Theory of Sampling.* New York: John Wiley & Sons, Inc.

Deming, W. (1953). On a probability mechanism to attain an economic balance between resultant error of response and the bias of nonresponse. *Journal of the American Statistical Association, 48*, 743–772.

Deming, W. E. (1977). An essay on screening, or on two-phase sampling, applied to surveys of a community. *International Statistical Review, 45*, 29–37.

Deming, W. E. & Stephan, F. F. (1940). On the least squares adjustment of a sample frequency table when the expected marginal total are known. *Annals of Mathematical Statistics, 11*, 427–444.

Dempster, A., Laird, N., & Rubin, D. (1977). Maximum likelihood from incomplete data via the EM algorithm. *Journal of the Royal Statistical Society (Series B), 39*, 1–38.

Dennis, J. M., Coronado, V. G., Frankel, M., Rodén, A., Saulsberry, C., Speizer, H., & Wright, R. A. (1998). Use of an intranet to manage a telephone survey. *Proceedings of the Survey Research Section*, American Statistical Association, 946–950.

Dennis, J. M., Saulsberry, C., Battaglia, M. P., Rodén, A.-S., Hoaglin, D. C., Frankel, M. et al. (1999, October). Analysis of call patterns in a large random-digit dialing survey: the National Immunization Survey. Paper presented at the *International Conference on Survey Nonresponse*, Portland, OR.

Deville, J. C. & Sarndal, C. E. (1992). Calibration estimators in survey sampling. *Journal of American Statistics Association, 87*, 376–382.

Dijkstra, W. & Smit, J. (2002). Persuading reluctant recipients in telephone surveys. In R. Groves, D. Dillman, J. Eltinge, & R. Little (Eds.), *Survey Nonresponse* (pp. 135—148). New York: John Wiley & Sons, Inc.

Dijkstra, W. & van der Zouwen, J. (1987). Styles of interviewing and the social context of the survey interview. In H. Hippler, N. Schwarz, & S. Sudman (Eds.), *Social Information Processing and Survey Methodology* (pp. 200–211). New York: Springer Verlag.

Dillman, D. A. (1978). *Mail and Telephone Surveys: The Total Design Method.* New York: John Wiley & Sons, Inc.

Dillman, D. A. (2000). *Mail and Internet Surveys: The Tailored Design Method.* New York: John Wiley & Sons, Inc.

Dillman, D. A. (2007). *Mail and Internet Surveys: The Tailored Design Method* (2nd Ed.). New York: John Wiley & Sons, Inc.

Dillman, D. A., Brown, T., Carlson, J., Carpenter, E., Lorenz, F., Mason, R., Saltiel, J., & Sangster, R. L. (1995). Effects of category order on answers to mail and telephone surveys, *Rural Sociology, 60*, 674–687.

Dillman, D. A. & Christian, L. M. (2005). Survey mode as a source of instability in responses across surveys. *Field Methods, 17*, 30–52.

Dillman, D. A., Gallegos, J. G., & Frey, J. H. (1976). Reducing refusal rates for telephone interviews. *Public Opinion Quarterly 40*, 66–78.

Dillman, D. A. & Mason, R. G. (1984, May). The influence of survey method on question response. Paper presented at the *Conference of the American Association for Public Opinion Research*, Delavan, WI.

Dillman, D. A. & Tarnai, J. (1991). Mode effects of cognitively designed recall questions: a comparison of answers to telephone and mail surveys. In P. P. Biemer, R. M. Groves, L. E. Lyberg, N. A. Mathiowetz, & S. Sudman (Eds.), *Measurement Errors in Surveys* (pp. 73–93). New York: John Wiley & Sons, Inc.

Dillman, D. A., Eltinge, J. L., Groves, R. M., & Little, R. J. A. (2002). Survey nonresponse in design, data collection, and analysis. In R. M. Groves, D. A. Dillman, J. A. Eltinge, & R. J. A. Little (Eds.), *Survey Nonresponse* (pp. 3–26). New York: John Wiley & Sons, Inc.

Dillman, D. A., Phelps, G., Tortora, R., Swift, K., Kohrell, J., & Berck, J. (2001). Response rate and measurement differences in mixed mode surveys: Using mail, telephone, interactive voice response and the internet. Unpublished manuscript. Retrieved June 6, 2006, from http://www.sesrc.wsu.edu/dillman/papers.htm

Dillman, D. A., Sangster, R. L., Tarnai, J., & Rockwood, T. H. (1996). Understanding differences in people's answers to telephone and mail surveys. In M. T. Braverman, J. K. Slater (Eds.), *New Directions for Evaluation Series: Advances in Survey Research* (pp. 46–52). San Francisco: Jossey-Bass.

Dillman, D. A., Sinclair, M. D., & Clark, J. R. (1993). Effects of questionnaire length, respondent friendly design, and a difficult question on response rates for occupant-addressed census mail surveys. *Public Opinion Quarterly 57*, 289–304.

Dillman, D. A. & Tarnai, J. (1988). Administrative issues in mixed-mode surveys. In R. M. Groves, P. Biemer, L. Lyberg, J. Massey, W. Nicholls, & J. Waksberg (Eds.), *Telephone Survey Methodology* (pp. 509–528). New York: John Wiley & Sons, Inc.

Dillman, D., West, K., & Clark, J. (1994). Influence of an invitation to answer by telephone on response to census questionnaires, *Public Opinion Quarterly, 58*, 557–568.

Dixon, J. & Figueroa, E. (2003). Using call records to study sources of nonresponse bias in consumer expenditure surveys. *Proceedings of the Joint Statistical Meetings*, American Statistical Association, 78–85.

Dobbs, J., Gibbins, C., Martin, J., Davies, P., & Dodd, T. (1998). Reporting on data quality and process quality. *Proceedings of the Survey Research Methods Section*, American Statistical Association, 32–40.

Doherty-Sneddon, G., Bruce, V., Bonner, L., Longbotham, S., & Doyle, C. (2002). Development of gaze aversion as disengagement from visual information. *Developmental Psychology, 38*, 438–445.

Dovidio, J. F. & Fazio, R.H. (1992). New technologies for the direct and indirect assessment of attitudes. In J. M. Tanur (Ed.), *Questions about Questions: Inquiries into the Cognitive Bases of Surveys* (pp. 204–237). New York: Russell Sage Foundation.

Down, J. & Duke, S. (2003). SMS polling. A methodological review. In R. Banks, J. Currall, J. Francis, L. Gerrard, R. Khan, T. Macer, et al. (Eds.), *ASC 2003. The Impact of Technology on the Survey Process* (pp. 277–286). Chesham, UK: Association for Survey Computing.

Doxa. (2005). *Junior 2004. Indagine sui comportamenti dei ragazzi 5-13 anni [Junior 2004. Survey on children's behaviours, 5 to 13 years old].* Milan: Doxa.

Drake, C. (1993). Effects of misspecification of the propensity score on estimators of treatment effect. *Biometrics, 49*, 1231–1236.

Drew, J. D., Choudhry, G. H., & Hunter, L. A. (1988). Nonresponse issues in government telephone surveys. In R. M. Groves, P. P. Biemer, L. E. Lyberg, J. T. Massey, W. L. Nicholls, & J. Waksberg (Eds.), *Telephone Survey Methodology* (pp. 233–246). New York: John Wiley & Sons, Inc.

Drew, J. H. & Fuller, W. A. (1980). Modeling nonresponse in surveys with callbacks. *Proceedings of the Survey Research Methods Section*, American Statistical Association, 639–642.

Drew, J. H. & Fuller, W. A. (1981). Nonresponse in complex multiphase surveys. *Proceedings of the Survey Research Methods Section*, American Statistical Association, 623–628.

Drew, J. H. & Ray, G. (1984). Some empirical investigations of nonresponse in surveys with callbacks. *Proceedings of the Survey Research Methods Section*, American Statistical Association, 560–565.

Drew, P. (1984). Speakers' reportings in invitation sequences. In J. M. Atkinson & J. Heritage (Eds.), *Structures of Social Action: Studies in Conversation Analysis* (pp. 129–151). New York: Cambridge University Press.

Duncan, Geoff. (2006, April 18). VoIP in 32.6 million U.S. homes by 2010? *Digital Trends.* Retrieved October 21, 2006, from http://news.digitaltrends.com/article10129.html.

Duncan, K. B. & Stasny, E. A. (2001). Using propensity scores to control coverage bias in telephone surveys. *Survey Methodology, 27*, 121–130.

Dunkelberg, W. C. & Day, G. S. (1973). Nonresponse bias and callbacks in sample surveys. *Journal ofMarketing Research, 10*, 160–68.

Durbin, J. & Stuart, A. (1951). Differences in response rates of experienced and inexperienced interviewers. *Journal of the Royal Statistical Society, 114*, 163–206.

Dykema, J., Lepkowski, J. M., & Blixt, S. (1997). The effect of interviewer and respondent behavior on data quality: analysis of interaction coding in a validation study. In L. E. Lyberg, P. P. Biemer, M. Collins, E. De Leeuw, C. Dippo, N. Schwarz, & D. Trewin (Eds.), *Survey Measurement and Process Quality* (pp. 287–310). New York: John Wiley & Sons, Inc.

Edwards, S. & Cantor, D. (1991). Toward a response model in establishment surveys. In P. Biemer, R. Groves, L. Lyberg, N. Mathiowetz, & S. Sudman (Eds.), *Measurement Errors in Surveys* (pp. 211–233). New York: John Wiley & Sons, Inc.

Edwards, S., Martin, D., DiSogra, C., & Grant, D. (2004, August). Altering the hold period for refusal conversion cases in an RDD survey. Paper presented at the *Joint Statistical Meetings*, American Statistical Association, Toronto, Ontario, Canada.

Edwards, T. P., Suresh, R., & Weeks, M. F. (1998). Automated call scheduling: Current systems and practices. In M. P. Couper, R. P. Baker, J. Bethlehem, C. Z. F. Clark, J. Martin, W. L. Nicholls II, & M. O'Reilly (Eds.), *Computer Assisted Survey Information Collection* (pp. 285–306). New York: John Wiley & Sons, Inc.

Edwards, W. S., Brick, J. M., Flores-Cervantes, I., DiSogra, C., & Yen, W. (2002). Sampling race and ethnic groups in RDD surveys. *Proceedings of the Survey Research Methods Section of the American Statistical Association*, 893–898.

Ehlen, P., Schober, M. F., & Conrad, F. G. (2005). Modeling speech disfluency to predict conceptual misalignment in speech survey interfaces. *Discourse Processes,* 44.

Elinson, J. (1992). Methodology issues. In P. B. Sheatsley & W. J. Mitofsky (Eds.), *A Meeting Place* (pp. 105–116). Lenexa, KS: American Association for Public Opinion Research.

Elliot, M. R., Little, R. J. A., & Lewitsky, S. (2000). Subsampling callbacks to improve survey efficiency. *Journal of the American Statistical Association, 95*, 730–738.

Eltinge, J. L. & Yansaneh, I. B. (1997). Diagnostics for formation of nonresponse adjustment cells, with an application for income nonresponse in the U.S. Consumer Expenditure Survey. *Survey Methodology, 23*, 33–40.

Escobedo, L. G., Landen, M. G., Axtell, C. D., & Kaigh, W. D. (2002). Usefulness of telephone risk factor surveys in the New Mexico border region. *American Journal of Preventive Medicine, 23*, 22–27.

Etter, J. F. & Perneger, T. V. (1997). Analysis of non-response bias in a mailed health survey. *Journal of Clinical Epidemiology, 50*, 1123–1128.

Ezzati-Rice, T. M., Cynamon, M., Blumberg, S. J., & Madans, J. H. (1995a). Use of an existing sampling frame to collect broad-based health and health-related data at the state and local level. *Proceedings of the 1999 Federal Committee on Statistical Methodology Research Conference*. Retrieved April 26, 2006, from http://www.cdc.gov/nchs/about/major/slaits/Publications_and_Presentations.htm

Ezzati-Rice, T. M., Frankel, M. R., Hoaglin, D. C., Loft, J. D., Coronado, G., & Wright, R. A (2000). An alternative measure of response rate in random-digit-dialing surveys that screen for eligible subpopulations. *Journal of Economic and Social Measurement, 26*, 99–109.

Ezzati-Rice, T. M., Zell, E. R., Battaglia, M. P., Ching, P. L. Y. H., & Wright, R. A. (1995b). The design of the National Immunization Survey. *Proceedings of the Section on Survey Research Methods of the American Statistical Association*, 668–672.

Fadia, A. (2001). Transform your call center using remote home agents. *Customer Inter@ction Solutions, 20*, 60–63.

Fay, R. E. (1986). Causal models for patterns of nonresponse. *Journal of the American Statistical Association, 81*, 354–365.

Fay, R. E. (1989). Estimating nonignorable nonresponse in longitudinal surveys through causal modeling. In D. Kasprzyk, G. Duncan, G. Kalton, & M. P. Singh (Eds.), *Panel Surveys* (pp. 375–399). New York: John Wiley & Sons, Inc.

Ferrari, P. W., Storm, R. R., & F. D. Tolson, F. D. Computer-assisted telephone interviewing. *Proceedings of the Survey Research Methods Section*, American Statistical Association. Retrieved December 15, 2006, from http://www.amstat.org/sections/SRMS/Proceedings/papers/1984_113.pdf

Fesco, R. S. (2001). Experiences with incentives in a public attitudes survey using RDD. Paper presented at the *Federal Committee on Statistical Methodology Research Conference*, Washington, DC. Retrieved December 15, 2006, from http://www.fcsm.gov/01papers/Fecso.pdf

Filion, F. L. (1976). Exploring and correcting for nonresponse bias using follow-ups of non-respondents. *Pacific Sociological Review, 19*, 401–408.

Fischbacher, C., Chappel, D., Edwards, R., & Summerton, N. (1999, September). The use and abuse of the internet for survey research. Paper presented at the *Third ASC International Conference*, Edinburgh, Scotland.

Fisher, S. & Kydoniefs, L. (2001, September). Using a theoretical model of respondent burden to identify sources of burden in surveys. Paper presented at the *12th International Workshop on Household Survey Nonresponse*, Oslo, Norway.

Fitch, D. J. (1985). Sample size versus call-back efforts: an empirical approach. *Proceedings of the Survey Research Methods Section*, American Statistical Association, 522–524.

Fitzgerald, R. & Fuller, L. (1982). I hear you knocking but you can't come in. *Sociological Methods Research, 2*, 3–32.

Fleeman, A. (2005, May). Will they talk to us? Survey performance rates of a cell phone sample. Paper presented at the *Annual Conference of the American Association for Public Opinion Research*, Miami Beach, FL.

Fleeman, A. (2006, January). Merging cellular and landline RDD sample frames: a series of three cellphone studies. Paper presented at the *Second International Conference on Telephone Survey Methodology*, Miami, FL.

Fleeman, A. & Estersohn, D. (2006, May). Geographic controls in a cell phone sample. Paper presented at the 61st *Annual Conference of the American Association for Public Opinion Research*, Montreal, Quebec, Canada.

Flemming, G. & Sonner, M. (1999, May). Can internet polling work? Strategies for conducting public opinion surveys online. Paper presented at the *Annual Meeting of the American Association for Public Opinion Research*, St. Pete Beach, FL.

Flores Cervantes, I., Shapiro, G., & Brock-Roth S. P. (1998). Effect of oversampling by poverty status in an RDD survey. *Proceedings of the Section on Survey Research Methods of the American Statistical Association*, 469–474.

Ford, E. S. (1998). Characteristics of survey participants with and without a telephone: findings from the third national health and nutrition examination survey. *Journal of Clinical Epidemiology, 51*, 55–60.

Foreman, J. & Collins, M. (1991). The viability of random digit dialing in the U.K. *Journal of the Market Research Society, 33*, 219–228.

Forsman, G. & Danielsson, S. (1997). Can plus digit sampling generate a probability sample? *Proceedings of the Section on Survey Research Methods*, American Statistical Association, 958–963.

Fowler, F. J., Jr. (1984). *Survey Research Methods* (Applied Social Research Methods Series, Volume 1). Newbury Park, CA: Sage Publications.

Fowler, F. J. (1991). Reducing interviewer-related error through interviewer training, supervision, and other means. In P. P. Beimer, R. M. Groves, L. E. Lyberg, N. Mathiowetz, & S. Sudman (Eds.), *Measurement Errors in Surveys* (pp. 259–278). New York: John Wiley & Sons, Inc.

Fowler, F., Jr. (1992). How unclear terms affect survey data. *Public Opinion Quarterly, 56,* 218–231.

Fowler, F. J. & Mangione T. W. (1990). *Standardized Survey Interviewing: Minimizing Interviewer-Related Error.* Newbury Park, CA: Sage Publications.

Fowler, F. J., Jr., Gallagher, P. M., Stringfellow, V. L., Zalavasky, A. M., Thompson, J. W., & Cleary, P. D. (2002). Using telephone interviews to reduce nonresponse bias to mail surveys of health plan members. *Medical Care, 40,* 190–200.

Fowler, F. J., Roman, A. M., & Di, Z. X. (1998). Mode effects in a survey of medicare prostate surgery patients. *Public Opinion Quarterly, 62,* 29–46.

Fox Tree, J. E. (1995). The effects of false starts and repetitions on the processing of subsequent words in spontaneous speech. *Journal of Memory and Language, 34,* 709–738.

Fox Tree, J. E. & Clark, H. H. (1997). Pronouncing "the" as "thee" to signal problems in speaking. *Cognition, 62,* 151–167.

Frankel, J. (1980). *Measurement of Respondent Burden: Study Design and Early Findings.* Washington, DC: Bureau of Social Science Research.

Frankel, L. R. (1983). The report of the CASRO task force on response rates. In Wiseman, F. (Ed.), *Improving Data Quality in Sample Surveys* (pp. 1–11). Cambridge, MA: Marketing Science Institute.

Frankel, M. R. (2004, July). RDD surveys: past and future. In Cohen, S. B. & Lepkowski, J. M. (Eds.), *Eighth Conference on Health Survey Research Methods.* Hyattsville, MD: National Center for Health Statistics.

Frankel, M. R., Battaglia, M. P., Kulp, D. W., Hoaglin, D. C., Khare, M., & Cardoni, J. (2003a, May). The impact of ring-no-answer telephone numbers on response rates in random-digit-dialing surveys. Paper presented at the *Annual Conference of the American Association for Public Opinion Research*, Phoeniz, AZ.

Frankel, M. R., Srinath, K. P., Hoaglin, D. C., Battaglia, M. P., Smith, P. J., Wright, R. A., & Khare, M. (2003b). Adjustments for non-telephone bias in random-digit-dialing surveys. *Statistics in Medicine, 22,* 1611–1626.

Frankovic, K. A. (1992). AAPOR and the polls. In P. B. Sheatsley & W. J. Mitofsky (Eds.), *A Meeting Place* (pp. 117–154). Lenexa, KS: American Association for Public Opinion Research.

Frankovic, K. A. (2003, February). Improving respondent cooperation in the polling environment. Paper presented at the *Second Annual CMOR Respondent Cooperation Workshop on Protecting Our Assets*, Orlando, FL.

Franzen, R. & Williams, R. (1956). A method for measuring error due to variance among interviewers. *Public Opinion Quarterly, 20,* 587–592.

Freeman, J. & Butler, E. W. (1976). Some sources of interviewer variance in surveys. *Public Opinion Quarterly, 40,* 79–91.

Frey, J. H. (1983). *Survey Research by Telephone.* (Sage Library of Social Research, Volume 150). Beverly Hills, CA: Sage.

Frey, J. H. (1986). An experiment with a confidentiality reminder in a telephone survey. *Public Opinion Quarterly, 50,* 267–269.

Fricker, S., Galesic, M., Tourangeau, R., & Yan, T. (2005). An experimental comparison of web and telephone surveys. *Public Opinion Quarterly, 69*, 370–392.

Fromkin, V. A. (Ed.). (1973). *Speech Errors as Linguistic Evidence.* The Hague: Mouton Publishers.

Fromkin, V. A. (1980). *Errors in Linguistic Performance: Slips of the Tongue, Ear, Pen and Hand.* New York: Academic Press.

Fuchs, M. (2000a, August). Conducting a CATI Survey in the cellular phone population experiences from a field experiment. Paper presented at the *American Statistical Association*, Joint Statistical Meetings, Indianapolis, IN.

Fuchs, M. (2000b, October). Non-response in a cellular phone survey. Paper presented at the *International Sociological Association Conference, Section RC33*, Cologne, Germany.

Fuchs, M. (2002a). Eine CATI-umfrage unter handy-nutzern. Methodische erfahrungen aus einem vergleich mit einer festnetz-stichprobe. In Gabler, S., & Häder, S. (Eds.), *Methodische probleme bei telefonstichprobenziehung und-realisierung* (pp. 121–137). Münster: Waxman.

Fuchs, M. (2002b). Kann man umfragen per handy durchführen? Ausschöpfung, interview-dauer und item-nonresponse im vergleich mit einer festnetz-stichprobe. *Planung & Analyse, 2*, 57–63.

Fuchs, M. (2006). Non-response and measurement error in mobile phone surveys. Paper presented at the *Second International Conference on Telephone Survey Methodology*, Miami, FL.

Fuchs, M. (2008). Mobile web surveys. In F. G. Conrad and M. F. Schober (Eds.), *Envisioning the Survey Interview of the Future.* New York: John Wiley & Sons, Inc.

Fuller, W. A. & Burmeister, L. F. (1972). Estimators for samples selected from two overlapping frames. *Proceedings of the Social Statistics Section of the American Statistical Association*, 245–249.

Futterman, M. (1988). CATI instrument logical structures: an analysis with applications. *Journal of Official Statistics, 4*, 333–348.

Gabler, S. & Häder, S. (2002). Idiosyncrasies in telephone sampling: the case of Germany. *International Journal of Public Opinion Research, 14*, 339–345.

Gallup Europe. (2005). *Technical Report on Mobile Phones*, Brussels: Author.

Gardner, W. D. (2006). Mobile VoIP set to surpass fixed net telephony. Retrieved September 21, 2006, from http://www.informationweek.com/story/show Article.jhtml?articleID=193003067.

Garren, S. T. & Chang, T. C. (2002). Improved ratio estimation in telephone surveys adjusting for noncoverage. *Survey Methodology, 28*, 63–76.

Gasquet, I., Falissard, B., & Ravaud, P. (2001). Impact of reminders and method of questionnaire distribution on patient response to mail-back satisfaction survey. *Journal of Clinical Epidemiology, 54*, 1174–1180.

Gaziano, C. (2005). Comparative analysis of within-household respondent selection techniques. *Public Opinion Quarterly, 69*, 124–157.

Gelman, A. & Carlin, J. B. (2002). Poststratification and weighting adjustments. In R. M. Groves, D. A. Dillman, J. A. Eltinge, & R. J. A. Little (Eds.), *Survey Nonresponse* (pp. 289–302). New York: John Wiley & Sons, Inc.

Gelman, A., Stevens, M., & Chan, V. (2003). Regression modeling and meta-analysis for decision making: a cost-benefit analysis of incentives in telephone surveys. *Journal of Business and Economic Statistics, 21*, 213–225.

GfK Technology. (2005, March 9). Backpacks, lunch boxes and cells?...Nearly half of US teens and tweens have cell phones, according to GfK NOP mKids study. Retrieved July 16, 2006, from http://www.gfkamerica.com/news/mkidspressrelease.htm.

Giesbrecht, L. H., Kulp, D. W., & Starer, A. W. (1996). Estimating coverage bias in RDD samples with current population survey data. *Proceedings of the American Statistical Association*, Section on Survey Research Methods, 503–508.

Gile, D. (1997). Conference interpreting as a cognitive management problem. In J. H. Danks, G. M. Shreve, S. B. Fountain, & M. K. McBeath (Eds.), *Cognitive Process in Translation and Interpreting*. Thousand Oaks, CA: Sage.

Glasser, G. J. & Metzger, G. D. (1972). Random digit dialing as a method of telephone sampling. *Journal of Marketing Research, 9*, 59–64.

Glasser, G. J. & Metzger, G. D. (1975). National estimates of nonlisted telephone households and their characteristics. *Journal of Marketing Research, 12*, 359–361.

Glenberg, A. M., Schroeder, J. L., & Robinson, D. A. (1998). Averting the gaze disengages the environment and facilitates remembering. *Memory & Cognition, 26*, 651–658.

Göksel, H., Judkins, D. R., & Mosher, W. D. (1991). Nonresponse adjustments for a telephone follow-up to a national in-person survey. *Proceedings of the Section on Survey Research Methods*, American Statistical Association, 581–586.

Goldman-Eisler, F. (1958). Speech production and the predictability of words in context. *Quarterly Journal of Experimental Psychology, 10*, 96.

Goldstein, K. M. & Jennings, K. J. (2002). Effect of advance letters. *Public Opinion Quarterly, 66*, 608–617.

Goodwin, C. (2000). Practices of seeing: visual analysis: an ethnomethodological approach. In T. van Leeuwen & C. Jewitt (Eds.), *Handbook of Visual Analysis* (pp. 157–182). London: Sage.

Green, M. C., Krosnick, J. A., & Holbrook, A. L. (2001). The survey response process in telephone and face-to-face surveys: Differences in respondent satisficing and social desirability response bias. Retrieved December 14, 2006, from http://scholar.google.com/scholar?hl=en&lr=&q=cache:aqR8DcULFrcJ:www.psy.ohio-state.edu/social/tch62a.pdf

Greenberg, B. S. & Stokes, L. S. (1990). Developing an optimal call scheduling strategy for a telephone survey. *Journal of Official Statistics, 6*, 421–435.

Greenfield, T. K., Midank, L. T., & Rogers, J. D. (2000). Effects of telephone versus face-to-face interview modes on reports of alcohol consumption. *Addiction, 95*, 277–284.

Grembowski, D. & Phillips, D. (2005). Linking mother and child access to dental care: a multimode survey. (Data report no. 05–023). Pullman, WA: Washington State University, Social and Economic Sciences Research Center.

Grote, D. (2005). *Forced Ranking*. Boston: Harvard Business School Press.

Groves, R. M. (1979). Actors and questions in telephone and personal interview surveys. *Public Opinion Quarterly, 43*, 190–205.

Groves, R. M. (1987). Research on survey data quality. *Public Opinion Quarterly, 51* (Suppl. 50th Anniversary), S156–S172.

Groves, R. M. (1989). *Survey Errors and Survey Costs*. New York: John Wiley and Sons, Inc.

Groves, R. M. (1990). Theories and methods of telephone surveys. *Annual Review of Sociology, 16*, 221–240.

Groves, R. M. (2006). Nonresponse rates and nonresponse error in household surveys. *Public Opinion Quarterly, 70*, 646–675.

Groves, R. M., Berry, M., & Mathiowetz, N. (1980). Some impacts of computer assisted telephone interviewing on survey methods. *Proceedings of the American Statistical Association*, Section on Survey Research Methods, 519–524.

Groves, R. M., Biemer, P. P., Lyberg, L., E., Massey, J. T., Nicholls II, W. L., & Waksberg, J. (1988). *Telephone Survey Methodology.* New York: John Wiley & Sons, Inc.

Groves, R. M., Cialdini, R. B., & Couper, M. P. (1992). Understanding the decision to participate in a survey. *Public Opinion Quarterly, 56*, 475–495.

Groves, R. M. & Couper, M. (1998). *Nonresponse in Household Interview Surveys.* New York: John Wiley & Sons, Inc.

Groves, R. M., Fowler, F. J., Couper, M. P., Lepkowski, J. M, Singer, E., & Tourangeau, R. (2004). *Survey Methodology.* New York: John Wiley & Sons, Inc.

Groves, R. M. & Fultz, N. H. (1985). Gender effects among telephone interviews in a survey of economic attitudes. *Sociological Methods and Research, 14*, 31–52.

Groves, R. M. & Heeringa, S. G. (2006). Responsive design for household surveys: tools for actively controlling survey errors and costs. *Journal of the Royal Statistical Society: Series A (Statistics in Society), 169*, 439–457.

Groves, R. M. & Kahn, R. L. (1979). *Surveys by Telephone: A National Comparison with Personal Interviews.* New York: Academic Press.

Groves R. M. & Lepkowski, J. M. (1985). Dual frame, mixed mode survey designs. *Journal of Official Statistics, 1*, 263–286.

Groves, R. M. & Lepkowski, J. M. (1986). An experimental implementation of a dual frame telephone sample design. *Proceedings of the Section on Survey Research Methods*, American Statistical Association, 340–345.

Groves, R. M. & Lyberg, L. E. (1988). An overview of nonresponse issues in telephone surveys. In R. M. Groves, P. Biemer, L. E. Lyberg, J. T. Massey, W. L. Nicholls II, & J. Waksberg, *Telephone Survey Methodology* (pp. 191–212). New York: John Wiley & Sons, Inc.

Groves, R. M. & Magilavy, L. (1986). Measuring and explaining interviewer effects in centralized telephone surveys. *Public Opinion Quarterly, 50*, 251–266.

Groves, R. M. & McGonagle, K. A. (2001). A theory-guided interviewer training protocol regarding survey participation. *Journal of Official Statistics, 17*, 249–265.

Groves, R. M., Presser, S., & Dipko, S. (2004). The role of topic interest in survey participation decisions. *Public Opinion Quarterly, 68*, 2–31.

Groves, R. M., Singer E., & Corning A. (2000). A leverage-saliency theory of survey participation: description and illustration. *Public Opinion Quarterly, 64*, 299–308.

Groves, R. & Wissoker, D. (1999). 1997 early response studies. In *National Survey of America's Families (NSAF) Methodology Series, 7.* Washington, DC: Urban Institute.

Guarino, J., Hill, J. M., & Woltman, H. F. (2001). *Analysis of the Social Security Number-Notification Component of the Social Security Number, Privacy Attitudes, and Notification Experiment.* Washington, DC: U.S. Census Bureau.

Haggerty, C., Nicholls, II, W. L., & Dull, V. (1999). Monitoring practice for telephone surveys. Unpublished manuscript, NORC, University of Chicago.

Hak, T. (2002). How interviewers make coding decisions. In D. W. Maynard, H. Houtkoop-Steenstra, N. C. Schaeffer, & J. van der Zouwen (Eds.), *Standardization and Tacit Knowledge* (pp. 449–470). New York: John Wiley & Sons, Inc.

Hak, T. & Bernts, T. (1996). Coder training: theoretical training or practical socialization? *Qualitative Sociology, 19*, 235–257.

Hall, J. & Bernieri, F. (Eds.). (2001). *Interpersonal Sensitivity.* Mahwah, NJ: LEA.

Hall, J. & Carlson, B. (2002). Sampling and weighting issues over three waves of the community tracking study household survey: using a partial overlap to increase efficiency. *Proceedings of the Survey Research Methods Section of the American Statistical Association*, 1319–1322.

Hansen, M. H. & Hurwitz, W. N. (1946). The problem of nonresponse in sample surveys. *Journal of the American Statistical Association, 41*, 517–529.

Hansen, M. H., Hurwitz, W. N., & Madow, W. G. (1953). *Sample Survey Methods and Theory.* (Volume 1: Methods and Applications). New York: John Wiley & Sons, Inc.

Hansen, M. H., Hurwitz, W. N., & Burshad, M. A. (1961). Measurement errors in census surveys. *Bulletin of the ISI, 38*, 351–374.

Hansen, S. E. (2000). Asking questions with computers: interaction in the computer-assisted standardized survey interview. Unpublished doctoral dissertation, University of Michigan, Ann Arbor.

Hansen, S. E. & Couper, M. (2004). Usability testing to evaluate computer-assisted instruments. In S. Presser, J. M. Rothgeb, M. P. Couper, J. T. Lessler, E. Martin, J. Martin, & E. Singer (Eds.), *Methods for Testing and Evaluating Survey Questionnaires* (pp. 337–360). New York: John Wiley & Sons, Inc.

Hansen, S. E., Fuchs, M., & Couper, M. (1997). CAI instrument usability testing. *Proceedings of the Section on Survey Research Methods*, American Statistical Association, 1023–1028.

Hanson, H. M. (1997). Glottal characteristics of female speakers: acoustic correlates. *Journal of the Acoustical Society of America, 101*, 466–481.

Hanson, H. M. & Chuang, E. S. (1999). Glottal characteristics of male speakers: acoustic correlates and comparison with female data. *Journal of the Acoustical Society of America 106*, 1064–1077.

Hanson, R. H. & Marks, E. S. (1958). Influence of the interviewer on accuracy of survey results. *Journal of the American Statistical Association, 53*, 635–655.

Hanway, S., Steiger, D. M., Chattopadhyay, M., & Chapman, D. (2005). The impact of an extended call design on RDD survey estimates. Unpublished paper, Gallup Organization.

Haraldsen, G. (2002, November). Identifying and reducing response burden in internet business surveys. Paper presented at the *International Conference on Questionnaire Development, Evaluation, and Testing Methods (QDET)*, Charleston, South Carolina.

Haraldsen, G. (2004). Identifying and reducing response burden in internet business surveys. *Journal of Official Statistics, 20*, 393–410.

Harkness, J. A. (1995, May). ISSP methodology translation work group report. Report presented to the *General Assembly of the International Social Survey Program*, Cologne, Germany.

Harkness, J. A. (2003). Questionnaire translation. In J. A. Harkness, F. J. R. v. d. Vijver, & P. P. Mohler (Eds.), *Cross-Cultural Survey Methods* (pp. 35–56). New York: John Wiley & Sons, Inc.

Harkness, J. A. (2004). Problems in establishing conceptually equivalent health definitions across multiple cultural groups. In S. B. Cohen & J. M. Lepkowski (Eds.), *Eighth Conference on Health Survey Research Methods* (pp. 85–90). Hyattsville, MD: National Center for Health Statistics.

Harkness, J. A. (2007). Comparative survey research: goals and challenges. In J. Hox, E. D. de Leeuw, & D. Dillman (Eds.), *International Handbook of Survey Methodology*. Mahwah, NJ: Lawrence Erlbaum.

Harkness, J. A., Pennell, B. E., & Schoua-Glusberg, A. (2004). Survey questionnaire translation and assessment. In S. Presser, J. M. Rothgeb, M. P. Couper, J. T. Lessler, E. Martin, J. Martin, & E. Singer (Eds.), *Methods for Testing and Evaluating Survey Questionnaires* (pp. 435–473). New York: John Wiley & Sons, Inc.

Harkness, J. A., van de Vijver, F. J. R., & Johnson, T. (2003). Questionnaire design in comparative research. In J. A. Harkness, F. J. R. v. d. Vijver, & P. P. Mohler (Eds.), *Cross-Cultural Survey Methods* (pp. 19–34). New York: John Wiley & Sons, Inc.

Harpuder, B. E. & Stec, J. A. (1999). Achieving an optimum number of callback attempts: cost-savings vs. non-response error due to non-contacts in RDD surveys. *Proceedings of the Survey Research Methods Section*, American Statistical Association, 913–918.

Harris Interactive Inc. (2005, June). Nearly one in ten US adults use wireless phones exclusively and landline displacement expected to grow (news release). Rochester, NY: Harris Interactive Inc. Retrieved July 26, 2005, from http://www.harrisinteractive.com/news/allnewsby date.asp?NewsID=943

Harris-Kojetin, B. A. & Fricker, S. S. (1999, October). The influence of environmental characteristics on survey cooperation: a comparison of metropolitan areas. Paper presented at the International Conference on Survey Nonresponse, Portland, OR.

Harris-Kojetin, B. A. & Mathiowetz, N. A. (1998, May). The effects of self and proxy response status on the reporting of race and ethnicity. Paper presented at the *Annual Conference of the American Association of Public Opinion Research*, St. Louis, MO.

Harris-Kojetin, B. & Tucker, C. (1999). Exploring the relation of economic and political conditions with refusal rates in a government survey. *Journal of Official Statistics, 15*, 167–184.

Harris Poll Online. (2006). Harris poll history: Why does polling matter? Retrieved December 14, 2006, from http://www.harrispollonline.com/uk/history.asp

Hartley, H. O. (1962). Multiple frame surveys. *Proceedings of the Social Statistics Section*, American Statistical Association, 203–206.

Hartley, H. O. (1974). Multiple frame methodology and selected application. *Sankhyā, 36*, 99–118.

Hauck, M. & Cox, M. (1974). Locating a sample by random digit dialing. *Public Opinion Quarterly, 38*, 253–260.

Hawkins, D. F. (1975). Estimation of nonresponse bias. *Sociological Methods and Research, 3*, 461–485.

Heckathorn, D. (1997). Respondent-driven sampling: a new approach to the study of hidden populations. *Social Problems, 44*, 174–199.

Heckman, J. J. (1979). Sample selection bias as a specification error. *Econometrica, 47*, 153–161.

Heeringa, S. G. & Groves, R. M. (2007). Responsive design for household surveys. *Proceedings of the section on Survey Research Methods*. American Statistical Association, 3644–3651.

Heerwegh, D. (2005). Effects of personal salutation in e-mail invitations to participate in a web survey. *Public Opinion Quarterly, 69*, 588–598.

Helasoja, V., Prättälä, R., & Dregval, L. (2002). Late response and item nonresponse in the finbalt health monitor survey. *European Journal of Public Health, 12*, 117–123.

Hembroff, L. A., Rusz, R., Raferty, A., McGee, M., & Ehrlich, N. (2005). The cost-effectiveness of alternative advance mailings in a telephone survey. *Public Opinion Quarterly, 69*, 232–245.

Hill, A., Roberts, J., & Ewings, P. (1997). Non-response bias in a lifestyle survey. *Journal of Public Health Medicine, 19*, 203–207.

Hillygus, D. S., Nie, N. H., Prewitt, K., & Pals, H. (2006). *The Hard Count*. New York: Russell sage.

Hippler, H. & Schwarz, N. (1988). Response effects in surveys. In R. M. Groves, P. Biemer, L. Lyberg, J. Massey, W. Nicholls, & J. Waksberg (Eds.), *Telephone Survey Methodology* (pp. 102–122). New York: John Wiley & Sons, Inc.

Ho, D. (2006, July 22). Emerging technology framework puts various gadgets on same page. *Atlanta Journal Constitution*, p. F1.

Hoaglin, D. C. & Battaglia, M. P. (1996). A comparison of two methods of adjusting for non-coverage of nontelephone households in a telephone survey. *Proceedings of the Section on Survey Research Methods*, American Statistical Association, 497–502.

Hoaglin, D. C. & Battaglia, M. P. (2006). Assigning weights in RDD surveys that focus on children. Manuscript submitted for publication.

Hochstim, J. (1967). A critical comparison of three strategies of collecting data from households. *Journal of the American Statistical Association, 62*, 976–989.

Holbrook, A. L., Green, M. C., & Krosnick, J. A. (2003). Telephone versus face-to-face interviewing of national probability samples with long questionnaires: comparisons of respondent satisficing and social desirability response bias. *Public Opinion Quarterly, 67*, 79–125.

Holt, V. L., Martin, D. P., & LoGerfo, J. P. (1997). Correlates and effect of nonresponse in a postpartum survey of obstetrical care quality. *Journal of Clinical Epidemiology, 50*, 117–122.

Holt, D. & Smith, T. M. (1979). Post stratification. *Journal of the Royal Statistical Society, Series A: General, 142*, 33–46.

Homans, G. (1961). *Social Behavior: Its Elementary Forms*. New York: Harcourt, Brace, & World.

Hope, S. (2005). Scottish crime and victimization survey: calibration exercise: a comparison of survey methodologies. Research Report for the Scottish Executive, Glasgow, Scotland. Retrieved December 14, 2006, from http://www.scotland.gov.uk/Publications/2005/12/22132936/29366

Horrigan, M., Moore, W., Pedlow, S., & Wolter, K. (1999). Undercoverage in a large national screening survey for youths. *Proceedings of the Section on Survey Research Methods of the American Statistical Association*, 570–575.

Horvitz, D. G. & Thompson, D. J. (1952). A generalization of sampling without replacement from a finite universe. *Journal of the American Statistical Association, 47*, 663–685.

Hosmer, D. W. & Lemeshow, S. (2000). *Applied Logistic Regression* (2nd ed.). New York: John Wiley & Sons, Inc.

House, C. C. & Nicholls, W. L. II. (1988). Questionnaire design for CATI: design objectives and methods. In R. M. Groves, P. P. Biemer, L. Lyberg, J. T. Massey, W. L. Nicholls II, & J. Waksberg (Eds.), *Telephone Survey Methodology* (pp. 421–436), New York: John Wiley & Sons, Inc.

Houtkoop-Steenstra, H. (2000). *Interaction and the Standardized Survey Interview: The Living Questionnaire.* Cambridge, UK: Cambridge University Press.

Houtkoop-Steenstra, H. (2002). Questioning turn format and turn-taking problems in standardized interviews. In D. W. Maynard, H. Houtkoop-Steenstra, N. C. Schaeffer, & J. van der Zouwen (Eds.), *Standardization and Tacit Knowledge: Interaction and Practice in the Survey Interview* (pp. 243–260). New York: John Wiley & Sons, Inc.

Houtkoop-Steenstra, H. & van den Bergh, H. (2002). Effects of introduction in large-scale telephone survey interviews in Maynard. In H. Houtkoop-Steenstra, N. Schaeffer, & J. van der Zouwen (Eds.), *Standardization and Tacit Knowledge: Interaction and Practice in the Survey Interview* (pp. 205–218). New York: John Wiley & Sons, Inc.

Hox, J. & de Leeuw, E. (2002). The influence of interviewers' attitude and behavior on household survey nonresponse: an international comparison. In R. M. Groves, D. A. Dillman, J. L. Eltinge, & R. J. A. Little (Eds.), *Survey Nonresponse* (pp. 103–120). New York: John Wiley & Sons, Inc.

Huffington, A. (1998, October 13). Questionning the pollsters. *The New York Post*, p. 27.

Huggins, V., Dennis, M., & Seryakova, K. (2002). An evaluation of nonresponse bias in internet surveys conducted using the knowledge networks panel. *Proceedings of the Survey Research Methods Section*, American Statistical Association, 1525–1530.

Hyman, H. (1954). *Interviewing in Social Research.* Chicago: University of Chicago Press.

Iannacchione, V. G. (2003). Sequential weight adjustments for the location and cooperation propensity for the 1995 National Survey of Family Growth. *Journal of Official Statistics, 19*, 31–43.

Iannacchione, V. G., Milne, J. G., & Folsom, R. E. (1991). Response probability weight adjustment using logistic regression. *Proceedings of the Section on Survey Research Methods*, American Statistical Association, 637–642.

Inglis, K., Groves, R. M., & Heeringa, S. (1987). Telephone sample design for the U. S. Black household population. *Survey Methodology, 13*, 1–14.

International Standards Organization (2005). *Market, Opinion and Social Research—Service Requirements* (ISO/DIS 20252). Secretariat: AENOR: Spain.

International Telecommunication Union. (2006). *World Telecommunication Indicators Database* (9th ed.) Geneva, Switzerland: ITU.

Ipsos-Inra. (2004). EU telecom services indicators. Produced for the European Commission, DG Information Society. Retrieved June 15, 2005, from http://europa.eu.int/information_society/topics/ecomm/doc/useful_information/librar y/studies_ext_consult/inra_year2004/report_telecom_2004_final_reduced.pdf

ISTAT (L'Istituto Nazionale di Statistica). (2003). *Famiglie, abitazioni e sicurezza dei cittadini. Indagine multiscopo sulle famiglie "Aspetti della vita quotidiana" Dicembre 2001-Marzo 2002 [Households, dwellings and citizens' safety. Household multipurpose survey "Aspects of every day life " December 2001–March 2001].* Roma: Istituto Nazionale di Statistica.

Jabine, T., Straf, M., Tanur, J., & Tourangeau, R (Eds.). (1984). *Cognitive Aspects of Survey Methodology: Building a Bridge Between Disciplines.* Washington, DC: National Academy Press.

Jackson, C. P. & Boyle, J. M. (1991). Mail response rate improvement in a mixed-mode survey. *Proceedings of the American Statistical Association*, 295–300.

Japec, L. (1995, March). *Issues in Mixed-Mode Survey Design*. Stockholm: Statistics Sweden.

Japec, L. (2002). The interview process and the concept of interviewer burden. *Proceedings of the Section on Survey Research Methods*, American Statistical Association, 1620–1625.

Japec, L. (2003). Interviewer work at Statistics Sweden: Two studies on strategies and work environment. Memorandum, Statistics Sweden (in Swedish).

Japec, L. (2004). Interviewer burden, behaviour, attitudes, and attributes-effects on data quality in the European Social Survey. Unpublished manuscript.

Japec, L. (2005). Quality issues in interview surveys: some contributions. Unpublished doctoral dissertation, Stockholm University.

Japec, L. & Lundqvist, P. (1999). Interviewer strategies and attitudes. Paper presented at the *International Conference on Survey Nonresponse*, Portland, OR.

Jay, E. D. & DiCamillo, M. (2005). Improving the representativeness ofRDD telephone surveys by accounting for "recent cell phone only households." Paper presented at the *Annual Conference of the Pacific Chapter of the American Association for Public Opinion Research*, San Francisco, CA.

Jay, E. D. & DiCamillo, M. (2006). Improving the representativeness of RDD telephone surveys by accounting for recent cell phone-only households. Paper presented at the *Second International Conference on Telephone Survey Methodology*, Miami, FL.

Jordan, L. A., Marcus, A. C., & Reeder, L. G. (1980). Response styles in telephone and household interviewing: a field experiment. *Public Opinion Quarterly, 44*, 210–222.

Judkins, D., Brick, J. M., Broene, P., Ferraro, D., & Strickler, T. (2001). *1999 NSAF Sample Design* (NSAF Methodology Report No. 2). Washington, DC: The Urban Institute.

Judkins, D., DiGaetano, R., Chu, A., & Shapiro, G. (1999). Coverage in screening surveys. *Proceedings of the Section on Survey Research Methods of the American Statistical Association*, 581–586.

Judkins, D., Shapiro, G., Brick, M., Flores Cervantes, I., Ferraro, D., Strickler, T., & Waksberg, J. (1999). *1997 NSAF Sample Design*. (NSAF Methodology Report No. 2). Washington DC: The Urban Institute.

Junn, J. (2001, April). The influence of negative political rhethoric: An experimental manipulation of Census 2000 participation. Paper presented at the *Midwest Political Science Association Meeting*, Chicago.

Kahneman, D., Fredrickson, B., Schrieber, C., & Redelmeier, D. (1993). When more pain is preferred to less: adding a better ending. *Psychological Science, 4*, 401–405.

Kalina, S. (1998). *Strategische Prozesse beim Dolmetschen: Theoretische Grundlagen, empirische Fallstudien, didaktische Konsequenzen*. Tübingen, Germany: Gunter Narr.

Kalsbeek, W. D. (2003). Sampling minority groups in health surveys. *Statistics in Medicine, 22*, 1527–1549.

Kalsbeek, W. D., Botman, S. L., Massey, J. T., & Liu, P. W. (1994). Cost-efficiency and the number of allowable call attempts in the national health interview survey. *Journal of Official Statistics, 10*, 133–152.

Kalsbeek, W. D., Boyle, W. R., Agans, R., & White, J. E. (2007). Disproportionate sampling for population subgroups in telephone surveys. *Statistics in Medicine, 26*, 1657–1674.

Kalsbeek, W. D., Morris, C., & Vaughn, B. (2001). Effects of nonresponse on the mean squared error of estimates from a longitudinal study. *Proceedings of the Section on Survey Research Methods*, American Statistical Association. Retrieved December 21, 2006, from http://www.amstat.org/sections/srms/Proceedings/

Kalton, G. (1983). *Compensating for Missing Survey Data.* Ann Arbor: University of Michigan Press.

Kalton, G. (1993). *Sampling Rare and Elusive Populations.* National Household Survey Capability Programme. New York: United Nations Department of Economic and Social Information and Policy Analysis Statistics Division. Retrieved December 3, 2005, from http://unstats.un.unsd/publications/UNFPR_UN_INT_92_P80_16E.pdf

Kalton, G. (2003). Practical methods for sampling rare and mobile populations. *Statistics in Transition, 6,* 491–501.

Kalton, G. & Anderson, D. W. (1986). Sampling rare populations. *Journal of the Royal Statistical Society, 149*, 65–82.

Kalton, G. & Flores-Cervantes, I. (2003). Weighting methods. *Journal of Official Statistics, 19,* 81–97.

Kalton, G. & Maligalig, D. S. (1991). A comparison of methods of weighting adjustment for nonresponse. *1991 Annual Research Conference Proceedings*, U.S. Department of Commerce, Economics and Statistics Administration, Bureau of the Census, Washington, DC, 409–428.

Katz, J. E. (2003). *Machines That Become Us: The Social Context of Personal Communication Technology.* New Brunswick, NJ: Transaction.

Keeter, S. (1995). Estimating telephone noncoverage bias with telephone survey. *Public Opinion Quarterly, 59*, 196–217.

Keeter, S. (2006). The impact of cell phone noncoverage bias on polling in the 2004 presidential election. *Public Opinion Quarterly, 70*, 88–98.

Keeter, S. (2006, May 15). The cell phone challenge to survey research. Retrieved October 22, 2006, from http://people-press.org/reports/display.php3?ReportID=276

Keeter, S., Best, J., Dimock, M., & Craighill, P. (2004, May). The pew research center study of survey nonresponse: implications for practice. *Paper presented at the Annual Meeting of the American Association for Public Opinion Research*, Phoenix, AZ.

Keeter, S., Kennedy, C., Dimock, M., Best, J., & Craighill, P. (2006). Gauging the impact of growing nonresponse on estimates from a national RDD survey. *Public Opinion Quarterly*, *70*, 759–779.

Keeter, S. & Miller, C. (1998, May). Consequences of reducing telephone survey nonresponse bias or what can you do in eight weeks that you can't do in five days? Paper presented at the *Annual Meeting of the American Association for Public Opinion Research Association*, St. Louis, MO.

Keeter, S., Miller, C., Kohut, A., Groves, R., & Presser, S. (2000). Consequences of reducing nonresponse in a large national telephone survey. *Public Opinion Quarterly 64*, 125–148.

Khare, M., Chowdhury, S., Wolter, K., Wooten, K., & Blumberg, S. J. (2006, July). An evaluation of methods to compensate for noncoverage of phoneless households using information on interruptions in telephone service and presence of wireless phones. Paper presented at the *Joint Statistical Meetings*, American Statistical Association, Seattle, WA.

Khursid, A. & Sahai, H. (1995). A bibliography of telephone survey methodology. *Journal of Official Statistics, 11*, 325–367.

Kim, J. (2004). Remote agents: the challenges of virtual and distributed contact centers. *Customer Inter@ction Solutions, 23*, 54–57.

Kim, S. W. & Lepkowski, J. (2002, May). Telephone household non-coverage and mobile telephones. Paper presented at the *Annual Meeting of the American Association of Public Opinion Research*, St. Petersburg, FL.

Kiraly, D. C. (1995). *Pathways to Translation*. Kent, Ohio: Kent State University Press.

Kirkendall, N. J. (1999, June). Incentives—Perspective of OMB. Paper presented at a *Washington Statistical Society Seminar*, Washington, DC.

Kish, L. (1962). Studies of interviewer variance for attitudinal variables. *Journal of the American Statistical Association, 57*, 92–115.

Kish, L. (1965). *Survey Sampling*. New York: John Wiley & Sons, Inc.

Kjellmer, G. (2003). Hesitation: in defense of ER and ERM. *English Studies, 84*, 170–198.

Klatt, D. H. & Klatt, L. C. (1990). Analysis, synthesis, and perception of voice quality variations among female and male talkers. *Journal of the Acoustical Society of America, 87*, 820–857.

Koepsell, T. D., McGuire, V., Longstreth, Jr., W. T., Nelson, L. M., & van Belle, G. (1996). Randomized trial of leaving messages on telephone answering machines for control recruitment in an epidemiologic study. *American Journal of Epidemiology, 144*, 704–706.

Kohiyama, K. (2005). A decade in the development of mobile communications in Japan (1993–2002). In M. Ito, D. Okabe, & M. Matsuda (Eds.), *Personal, Portable, Pedestrian: Mobile Phones in Japanese Life*. Cambridge, MA: MIT Press.

Kormendi, E. (1988). The quality of income information in telephone and face to face surveys. In R. M. Groves, P. P. Biemer, L. E. Lyberg, J. T. Massey, W. L. Nicholls, & J. Waksberg (Eds.), *Telephone Survey Methodology* (pp. 341–356). New York: John Wiley & Sons, Inc.

Krane, D. (2003, April). Update of privacy research. Paper presented at a *Meeting of the New York Chapter of the American Association for Public Opinion Research*.

Krauss, R. M., Freyberg, R., & Morsella, E. (2002). Inferring speakers' physical attributes from their voices. *Journal of Experimental Social Psychology, 38*, 618–625.

Krings, H. P. (1986). *Was in Den Köpfen Von übersetzern Vorgeht* [What Goes on in the Minds of Translators]. Tübingen, Germany: Gunter Narr.

Kristal, A., White, R., Davis, J. R., Corycell, G., Raghunathan, T. E., Kinne, S., & Lin, T. K. (1993). Effects of enhanced calling efforts on response rates, estimates of health behavior, and costs in a telephone health survey using random-digit dialing. *Public Health Reports, 108*, 372–379.

Kropf, M. E., Scheib, J., & Blair, J. (2000). The effect of alternative incentives on cooperation and refusal conversion in a telephone survey. *Proceedings of the American Statistical Association*, Survey Research Section, 1081–1085.

Krosnick, J. A. (1991). Response strategies for coping with the cognitive demands of attitude measures in surveys. *Applied Cognitive Psychology, 5*, 213–236.

Krosnick, J. A. (2002). The causes of no-opinion responses to attitude measures in surveys: they are rarely what they appear to be. In R. M. Groves, D. A. Dillman, J. L. Eltinge, & R. J. A. Little (Eds.), *Survey Nonresponse* (pp. 87–100). New York, John Wiley & Sons, Inc.

Krosnick, J., Allyson, A., Holbrook, L., & Pfent, A. (2003, May). Response rates in recent surveys conducted by non-profits and commercial survey agencies and the news media. Paper presented at the *Annual Meeting of the American Association for Public Opinion Research*, Nashville, TN.

Krosnick, J. A. & Alwin, D. F. (1987). An evaluation of a cognitive theory of response-order effects in survey measurement. *Public Opinion Quarterly, 51*, 201–219.

Krosnick, J. A. & Berent, M. K. (1993). Comparisons of party identification and policy preferences: the impact of survey question format. *American Journal of Political Science, 37*, 941–964.

Krosnick, J. A. & Fabrigar, L. R. (1997). Designing rating scales for effective measurement in surveys. In L. Lyberg, P. Biemer, M. Collins, L. Decker, E. de Leeuw, C. Dippo, N. Schwarz, & D. Trewin (Eds.), *Survey Measurement and Process Quality* (pp. 141–164). New York: John Wiley & Sons, Inc.

Krug, S. (2005). *Don't Make Me Think! A Common Sense Approach to Web Usability.* Indianapolis, IN: New Riders.

Krysan, M. & Couper, M. (2004). Race of interviewer effects: what happens on the web? Paper presented at the *Annual Meeting of the American Association for Public Opinion Research Conference*, Phoenix, AZ.

Kulka, R. A. (1999, June). Providing respondent incentives in federal statistical surveys: The advance of the real "phantom menance"? Paper presented at a *Meeting of the Washington Statistical Society*, Washington, DC.

Kulka, R. A. & Weeks, M. F. (1988). Toward the development of optimal calling protocols for telephone surveys: a conditional probability approach. *Journal of Official Statistics 4*, 319–332.

Kussmaul, P. (1995). *Training the Translator.* Amsterdam: John Benjamins.

Kuusela, V. (1997). *Puhelinpeittävyys japuhelimella tavoitettavuus Suomessa [Telephone coverage and accessibility by telephone in Finland].* Helsinki, Finland: Tilastokeskus.

Kuusela, V. (2000). *Puhelinpeittävyyden muutos Suomessa [Change of telephone coverage in Finland].* Helsinki, Finland: Tilastokeskus.

Kuusela, V. (2003). Mobile phones and telephone survey methods. In R. Banks, J. Currall, J. Francis, L. Gerrard, R. Kahn, T. Macer, M. Rigg, E. Ross, S. Taylor, & A. Westlake (Eds.). *Proceedings of the fourth Association of Survey Computing International Conference: The Impact of* 33 *New Technology on the Survey Process* (pp. 317–327). Chesham Bucks, UK: Association for Survey Computing.

Kuusela, V. & Notkola, V. (1999, October). Survey quality and mobile phones. Paper presented at the *International Conference on Survey Nonresponse*, Portland, OR.

Kuusela, V. & Simpanen, M. (2002, August). Effects of mobile phones on phone survey practices and results. Paper presented at the *International Conference on Improving Surveys*, Copenhagen, Denmark.

Kuusela, V. & Vikki, K. (1999, October). Change of telephone coverage due to mobile phones. Paper presented at the *International Conference on Survey Nonresponse*, Portland, OR.

Laaksonen, S. & Chambers, R. (2006) Survey estimation under informative non-response with follow-up. *Journal of Official Statistics, 22*, 81–95.

Landon, E. L., Jr. & Banks, S. K. (1977). Relative efficiency and bias of plus-one telephone sampling. *Journal of Marketing Research, 14*, 294–299.

Lau, L. K. P. (2004). Mobile phone surveys in Hong Kong: Methodological issues and comparisons with conventional phone surveys. Paper presented at the *RC33 Sixth International Conference on Social Science Methodology*, Amsterdam, August.

Lau, L. K. P. (2005). Mobile phone surveys in Hong Kong: methodological issues and comparisons with conventional phone surveys. Updated version of a paper presented at the *International Sociological Association's August 2004 Meeting, RC33*, Amsterdam.

Lavrakas, P. J. (1987). *Telephone Survey Methods: Sampling, Selection, and Supervision.* Newbury Park, CA: Sage Publications.

Lavrakas, P. J. (1993). *Telephone Survey Methods: Sampling, Selection, and Supervision* (2nd ed.). Newbury Park, CA: Sage Publications.

Lavrakas, P. J. (1997). Methods for sampling and interviewing in telephone surveys. In L. Bickman, & D. Rog (Eds.), *Applied Social Research Methods Handbook* (pp. 429–472). Newbury Park, CA: Sage Publications.

Lavrakas, P. J. (2004, May). Will a "perfect storm" of cellular-linked forces sink RDD? Paper presented at the *Annual Meeting of the American Association for Public Opinion Research*, Phoenix, AZ.

Lavrakas, P. J. (2007). Surveys by telephone. In W. Donsbach, & M. W. Traugott (Eds.), *Handbook of Public Opinion Research.* Sage Publications.

Lavrakas, P. J., Bauman, S. L., & Merkle, D.M. (1992). Refusal report forms (RRFs), refusal conversions, and nonresponse bias. Paper presented at the *Annual Conference of the American Association for Public Opinion Research*, St. Petersburg, FL.

Lavrakas, P. J. & Shepard, J. (2003, May). CMOR's national survey to help build an advertising campaign to motivate survey response. Paper presented at the *Annual Conference of the American Association for Public Opinion Research*, Nashville, TN.

Lavrakas, P. J. & Shuttles, C. D. (2004, August). Two advance letter experiments to raise survey response rates in a two-stage mixed mode survey. Paper presented at the *Joint Statistical Meetings, American Statistical Association*, Toronto, Canada.

Lavrakas, P. J. & Shuttles, C. D. (2005a). Cell phone sampling, RDD surveys, and marketing research implications. *Alert!, 43*, 4–5.

Lavrakas, P. J. & Shuttles, C. D. (2005b, February). Cell phone sampling summit II statements on "accounting for cell phones in telephone survey research in the U.S." Retrieved July 26, 2005, from Nielsen Media Research website http://www.nielsenmedia.com/cellphonesummit/statements.html

LeBailly, R. K. & Lavrakas, P. J. (1981). Generating a random digit dialing sample for telephone surveys. Paper presented at *ISSUE '81: Annual SPSS Convention*, San Francisco.

Lee, S. (2006). Propensity score adjustment as a weighting scheme for volunteer panel web surveys. *Journal of Official Statistics, 22*, 329–349.

Lehtonen, R. (1996). Interviewer attitudes and unit nonresponse in two different interviewing schemes. In S. Laaksonen (Ed.), *International Perspectives on Nonresponse. Proceedings from the Sixth International Workshop on Household Survey Nonresponse*, Helsinki: Statistics Finland.

Lepkowski, J. M. (1988). Telephone sampling methods in the United States. In R. M. Groves, P. P. Biemer, L. E. Lyberg, J. T. Massey, W. L. N., II, & J. Waksberg (Eds.), *Telephone Survey Methodology* (pp. 73–98). New York: John Wiley & Sons, Inc.

Lepkowski, J. M. & Groves, R. M. (1986). A mean squared error model for dual frame, mixed mode survey design. *Journal of the American Statistical Association, 81*, 930–937.

Lepkowski, J. M., Kalton, G., & Kasprzyk, D. (1989). Weighting adjustments for partial nonresponse in the 1984 SIPP panel. *Proceedings of the Section on Survey Research Methods*, American Statistical Association, 296–301.

Lepkowski, J. M. & Kim, S.-W. (2005, August). Dual-frame landline/cellular telephone survey design. Paper presented at the 2005 *Joint Statistical Meetings*, American Statistical Association, Minneapolis, MN.

Lessler, J. T. & Kalsbeek, W. D. (1992). *Nonsampling Error in Surveys.* New York: John Wiley & Sons, Inc.

Levelt, W. J. M. (1989). *Speaking: From Intention to Articulation.* Cambridge, MA: MIT Press.

Levine, O. S., Farley, M., Harrison L. H., Lefkowitz, L., McGeer, A., Schwartz, B., & The Active Bacterial Core Surveillance Team (1999). Risk factors for invasive pneumococcal disease in children: A population based case-control study in North America, *Pediatrics, 103*, 28.

Liberman, M. & Pierrehumbert, J. B. (1984). Intonational invariants under changes in pitch range and length. In Runoff, M. & Ochre, R. (Eds.), *Language Sound Structure.* Cambridge, MA: MIT Press.

Lin, I. F. & Schaeffer, N. C. (1995). Using survey participants to estimate the impact of nonparticipation. *Public Opinion Quarterly, 59*, 236–258.

Lind, K., Johnson, T., Parker, V., & Gillespie, S. (1998). Telephone non-response: a factorial experiment of techniques to improve telephone response rates. *Proceedings of the Survey Research Methods Section*, American Statistical Association, 848–850.

Lind, K., Link, M., & Oldendick, R. (2000). A comparison of the accuracy of the last birthday versus the next birthday methods for random selection of household respondents. *Proceedings of the Survey Research Methods Section*, American Statistical Association, 887–889.

Link, M. (2002a, January). From call centers to multi-channel contact centers: The evolution of call center technologies. Paper presented at the *Council for Marketing and Opinion Research*, Respondent Cooperation Workshop, New York.

Link, M. (2002b, March). Voice-over internet protocol (VoIP) and other advances in call center technologies. Paper presented at the *Federal Computer-Assisted Survey Information Collection Conference*, Washington, DC.

Link, M. W. (2003, February). BRFSS response rates: Which way is up? Paper presented at the *Behavioral Risk Factor Surveillance System Annual Conference*, St. Louis, MO.

Link, M. (2006). Predicting persistence and performance among newly recruited telephone interviewers. *Field Methods, 18*, 305–320.

Link, M., Armsby, P., Hubal, R., & Guinn, C. (2004). Accessibility and acceptance of responsive virtual human technology as a telephone interviewer training tool. *Computers in Human Behavior, 22*, 412–426.

Link, M. W., Malizio, A. G., & Curtin, T. R. (2001, May). Use of targeted monetary incentives to reduce nonresponse in longitudinal surveys. Paper presented at the *Annual Meeting of the American Association for Public Opinion Research*, Montreal, Quebec, Canada.

Link, M. W. & Mokdad, A. (2004, August). Moving the behavioral risk factor surveillance system from RDD to multimode: a web/mail/telephone experiment. Paper presented at the *Joint Statistical Meetings,* American Statistical Association, Toronto, Ontario, Canada. [CD-ROM].

Link, M. W. & Mokdad, A. (2005a). Advance letters as a means of improving respondent cooperation in RDD studies: a multi-state experiment. *Public Opinion Quarterly, 69*, 572–587.

Link, M. W. & Mokdad, A. (2005b). Leaving answering machine messages: do they increase response rates for the behavioral risk factor surveillance system? *International Journal of Public Opinion Research, 17*, 239–250.

Link, M. W. & Mokdad, A. (2005c). Advance letters as a means of improving respondent cooperation in random digit dial studies: a multistate experiment. *Public Opinion Quarterly 69*, 572–587.

Link, M. W. & Mokdad, A. (2006). Can web and mail survey modes improve participation in an RDD-based national health surveillance? *Journal of Official Statistics, 22*, 293–312.

Link, M. W., Mokdad, A., Jiles, R., Weiner, J., & Roe, D. (2004, May). Augmenting the BRFSS RDD design with mail and web modes: Results from a multi-state experiment. Paper presented at the *Annual Conference of the American Association for Public Opinion Research*, Phoenix, AZ.

Link, M. W., Mokdad, A., Kulp, D., & Hyon, A. (2006). Has the national "do not call" registry helped or hurt state-level response rates? *Public Opinion Quarterly, 70*, 794–809.

Link, M. W. & Oldendick, R. W. (1999). Call screening: is it really a problem for survey research? *Public Opinion Quarterly, 63*, 577–589.

Little, R. J. A. (1986). Survey nonresponse adjustments for estimates of means. *International Statistical Review, 54*, 139–157.

Little, R. J. A. & Rubin, D. (1987). *Statistical Analysis with Missing Data*. New York: John Wiley & Sons, Inc.

Little, R. J. A. (1993). Post-stratification: a modeler's perspective. *Journal of the American Statistical Association, 88*, 1001–1012.

Little, R. J. A. & Vartivarian, S. (2003). On weighting the rates in non-response weights. *Statistics in Medicine, 22*, 1589–1599.

Little, R. J. A. & Vartivarian, S. (2005). Does weighting for nonresponse increase the variance of survey means? *Survey Methodology, 31*, 161–168.

Lohr, S. L. (1999). *Sampling: Design and Analysis*. North Scituate, MA: Duxbury Press.

Lohr, S. L. & Rao, J. N. K. (2000). Inference from dual frame surveys. *Journal of the American Statistical Association, 95*, 271–280.

Lohr, S. L. & Rao, J. N. K. (2006). Estimation in multiple-frame surveys. *Journal of the American Statistical Association, 101*, 1019–1030.

Luke, J., Blumber, S., & Cynamon, M. (2004, August). The prevalence of wireless telephone substitution. Paper presented at the *Joint Statistical Meetings*, American Statistical Association, Toronto, Ontario, Canada.

Lund, R. E. (1968). Estimators in multiple frame surveys. *Proceedings of the Social Science Statistics Section*, American Statistical Association, 282–288.

Lyberg, L., Japec, L., & Biemer, P. P. (1998). Quality improvements in surveys: a process perpective. *Proceedings of the Survey Research Methods Section*, American Statistical Association, 23–31.

Lyberg, L. & Kasprzyk, D. (1991). Data collection methods and measurement error: an overview. In P. Beimer, R. Groves, L. Lyberg, N. Mathiowetz, & S. Sudman (Eds.), *Measurement Errors in Surveys* (pp. 237–257). New York: John Wiley & Sons, Inc.

Lyberg, I. & Lyberg, L. (1991). Nonresponse research at statistics sweden. *Proceedings of the American Statistical Association*, Survey Research Methods Section, 78–87.

Lynch, M. (2002). The living text. written instructions and situated actions in telephone surveys. In D. W. Maynard, H. Houtkoop-Steenstra, N. C. Schaeffer, & J. vander Zouwen (Eds.), *Standardization and Tacit Knowledge: Interaction and Practice in the Survey Interview* (pp. 125–150). New York: John Wiley & Sons, Inc.

Macera, C. A., Jackson, K. L., & Davis, D. R. (1990). Patterns of nonresponse to a mail survey. *Journal of Clinical Epidemiology, 43*, 1427–1430.

Maklan, D. & Waksberg, J. (1988). Within-household coverage in RDD surveys. In R. M. Groves, P. P. Biemer, L. E. Lyberg, J. T. Massey, W. L. Nicholls, & J. Waksberg (Eds.), *Telephone Survey Methodology* (pp. 51–69). New York: John Wiley & Sons, Inc.

Malle, B. F. & Hodges, S. D. (2005). *Other Minds: How Humans Bridge the Divide Between Self and Others*. New York: Guilford.

Manfreda, K. L. & Vehovar, V. (2002). Do mail and web surveys provide the same results? In A. Ferligoj & A. Mrvar (Eds.), *Metodološki zvezki* [Developments in Social Science Methodology], *18*, 149–169. Retrieved December 14, 2006, from http://mrvar.fdv.uni-lj.si/pub/mz/mz18/lozar1.pdf

Manfreda, K. L., Biffignandi, S., Pratesi, M., & Vehovar, V. (2002). Participation in telephone pre-recruited web surveys. *Proceedings of the Survey Research Methods Section*, American Statistical Association, 2178–2183.

Marketing Systems Group. (1993). *GENESYS Sampling System: Methodology*. Fort Washington, PA: GENESYS Sampling Systems.

Martin, E. A. (2001). Privacy concerns and the census long form: some evidence from census 2000. *Proceedings of the American Statistical Association*, Section on Survey Methods. Retrieved January 23, 2007, from http://www.amstat.org/sections/srms/Proceedings/

Martin, J. & Beerten, R. (1999, October). The effect of interviewer characteristics on survey response rates. Paper presented at the *International Conference on Survey Nonresponse*, Portland, OR.

Marton, K. (2004). Effects of questionnaire and field work characteristics on call outcome rates and data quality in a monthly CATI survey. Unpublished doctoral dissertation. Ohio State University, Columbus.

Mason, R. E. & Immerman, F. W. (1988). Minimum cost sample allocation for Mitofsky–Waksberg random digit dialing. In R. M. Groves, P. P. Biemer, L. E. Lyberg, J. T. Massey, W. L. Nicholls, & J. Waksberg (Eds.), *Telephone Survey Methodology* (pp. 127–142). New York: John Wiley & Sons, Inc.

Massey, J. T. (1995). Estimating the response rate in a telephone survey with screening. *Proceedings of the Section on Survey Research Methods*, American Statistical Association, 673–677.

Massey, J. T., Barker, P. R., & Hsiung, S. (1981). An investigation of response in a telephone survey. *Proceedings of the American Statistical Association*, Section on Survey Research Methods, 426–431.

Massey, J. T. & Botman, S. L. (1988). Weighting adjustments for random digit dialed surveys. In R. Groves, P. Biemer, L. Lyberg, J. Massey, W. Nichols II, & J. Waksberg (Eds.), *Telephone Survey Methodology* (pp. 143–160). New York: John Wiley & Sons, Inc.

Massey, J. T., O'Connor, D. J., & Krotki, K. (1997). Response rates in random digit dialing (RDD) telephone surveys. *Proceedings of the American Statistical Association*, Section on Survey Research Methods, 707–712.

Massey, J. T., Wolter, C. Wan, S. C., & Liu, K. (1996). Optimum calling patterns for random digit dialed telephone surveys. *Proceedings of the Survey Research Section*, American Statistical Association, 485–490.

Matthews, R., Bennett, G., & Down, J. (2000, October). Using WAP phones to conduct market research. A report on the preliminary trials. Paper presented at the *Conference Telecommunications: Migration Into a New eService World*? Berlin, Germany.

Mayer, T. S. (2002). Privacy and confidentiality research and the U.S. census bureau: recommendations based on a review of the literature. *Research Report Series, Survey Methodology #2002–01.* Washington, DC: Statistical Research Division, U.S. Bureau of the Census.

Mayer, T. S. & O'Brien, E. (2001, August). Interviewer refusal aversion training to increase survey participation. Paper presented at the *Joint Statistical Meetings of the American Statistical Association*, Atlanta, GA.

Maynard, D. & Schaeffer, N. C. (2002a). Opening and closing the gate: the work of optimism in recruiting survey respondents. In D. Maynard, H. Houtkoop-Steenstra, N. Schaeffer, & J. van der Zouwen (Eds.), *Standardization and Tacit Knowledge: Interaction and Practice in the Survey Interview* (pp. 179–204). New York: John Wiley & Sons, Inc.

Maynard, D. W. & Schaeffer, N. C. (2002b). Standardization and its discontents. In D. Maynard, H. Houtkoop, N. C. Schaeffer, & J. van der Zouwen (Eds.), *Standardization and Tacit Knowledge: Interaction and Practice in the Survey Interview* (pp. 3–45). New York, John Wiley & Sons, Inc.

McAuliffe, W. E., LaBrie, R., Woodworth, R., & Zhang, C. (2002). Estimates of potential bias in telephone substance abuse surveys due to exclusion of households without telephones. *Journal of Drug Issues, 32*, 1139–1153.

McCarthy, G. M. & MacDonald, J. K. (1997). Nonresponse bias in a national study of dentists' infection control practices and attitudes related to HIV. *Community Dental and Oral Epidemiology, 25*, 319–323.

McCarthy, J. S., Beckler, D., & Qualey, S. (2006). An analysis of the relationship between survey burden and nonresponse: if we bother them more are they less cooperative? *Journal of Official Statistics, 22*, 97–112.

McCarty, C. (2003). Differences in response rates using most recent versus final dispositions in telephone surveys. *Public Opinion Quarterly, 67*, 396–406.

McCullagh, P. & Nelder, J. A. (1991). *Generalized Linear Models.* New York: Chapman and Hall.

McKay, R. B., Robison, E. L., & Malik, A. B. (1994). Touch-tone data entry for household surveys: findings and possible applications. *Proceedings of the Survey Research Methods Section*, American Statistical Association, 509–511.

Merkle, D. & Edelman, M. (2002). Nonresponse in exit polls: a comprehensive analysis. In R. M. Groves, D. Dillman, J. Eltinge, & R. Little (Eds.), *Survey Nonresponse* (pp. 243–259). New York: John Wiley & Sons, Inc.

Merkle, D., Edelman, M., Dykeman, K., & Brogan, C. (1998). An experimental study of ways to increase exit poll response rates and reduce survey error. Paper presented at the *Annual Conference of the American Association for Public Opinion Research*, St. Louis, MO.

Merkle, D. M., Bauman, S. L., & Lavrakas, P. J. (1993). The impact of callbacks on survey estimates in an annual RDD survey. *Proceedings of the American Statistical Association*, Section on Survey Research Methods, 1070–1075.

Mitchell, I. (2001). Call center consolidation—does it still make sense? *Business Communications Review, 12*, 24–28.

Mitofsky, W. (1970). Sampling of telephone households, unpublished CBS memorandum.

Mitofsky, W., Bloom, J., Lensko, J., Dingman, S., & J. Agiesta. (2005). A dual frame RDD/registration-based sample design/lessons from Oregon's 2004 National Election Pool survey. *Proceedings of the Survey Research Section*, American Statistical Association, 3929–3936.

Mockovak, B. & Fox, J. (2002, November). Approaches for incorporating user-centered design into CAI development. Paper presented at the *International Conference On Questionnaire Development, Evaluation, and Testing Methods*, Charleston, SC.

Mohadjer, L. (1988). Stratification of prefix areas for sampling rare populations. In R. M. Groves, P. P. Biemer, L. E. Lyberg, J. T. Massey, W. L. Nicholls, & J. Waksberg (Eds.), *Telephone Survey Methodology* (pp. 161–174). New York: John Wiley & Sons, Inc.

Mohadjer, L. & West, J. (1992). *Effectiveness of Oversampling Blacks and Hispanics in the NHES Field Test.* National Household Education Survey Technical Report. Washington, DC: Department of Education.

Mokdad, A., Stroup, D., & Giles, H. W. (2003). Public health surveillance for behavioral risk factors in a changing environment: recommendations from the behavioral risk factor surveillance team. *Morbidity and Mortality Recommendations and Reports, 52*, 1–12.

Moon, N. (2006, January). Can opinion polls be conducted using cell phones? Paper presented at the *Second International Conference on Telephone Survey Methodology*, Miami, FL.

Moore, J. C. (1988). Self/proxy response status and survey response quality. *Journal of Official Statistics, 4*, 155–172.

Moore, R. J. & Maynard, D. W. (2002). Achieving understanding in the standardized survey interview: repair sequences. In D. W. Maynard, H. Houtkoop-Steenstra, N. C. Schaeffer, & J. van der Zouwen (Eds.), *Standardization and Tacit Knowledge: Interaction and Practice in the Survey Interview* (pp. 281–311). New York: John Wiley & Sons, Inc.

Morganstein, D. & Marker, D. A. (1997). Continuous quality improvement in statistical agencies. In L. Lyberg, P. Biemer, M. Collins, E. de Leeuw, C. Dippo, N. Schwarz, & D. Trewin (Eds.), *Survey Measurement and Process Quality* (pp. 475–500). New York: John Wiley & Sons, Inc.

Morganstein, D., Tucker, C., Brick, M., & Espisito, J. (2004). Household telephone service and usage patterns in the United States in 2004: a demographic profile. *Proceedings of the Survey Research Methods Section*, American Statistical Association.

Morton-Williams, J. (1979). The use of verbal interactional coding for evaluating a questionnaire. *Quality and Quantity, 13*, 59–75.

Morton-Williams, J. (1993). *Interviewer Approaches.* Dartmouth, UK: Aldershot.

Moskowitz, J. M. (2004). Assessment of cigarette smoking and smoking susceptibility among youth: Telephone computer-assisted self-interviews versus computer-assisted telephone interviews. *Public Opinion Quarterly, 68*, 565–587.

Mudryk, W., Burgess M. J., & Xiao, P. (1996). Quality control of CATI operations in statistics Canada. *Proceedings of the Survey Research Methods Section*, American Statistical Association, 150–159.

Mulry, M. & Spencer, B. (1990, March). Total error in post enumeration survey (PES) estimates of population: the dress rehearsal census of 1988. *Proceedings of the Bureau of the Census Annual Research Conference*, 326–361.

Mulry-Liggan, M. H. (1983). A comparison of a random digit dialing survey and the current population survey. *Proceedings of the American Statistical Association*, Section on Survey Research Methods, 214–219.

Murphy, W., O'Muircheartaigh, C., Emmons, C.-A., Pedlow, S., & Harter, R. (2003). Optimizing call strategies in RDD: Differential nonresponse bias and costs in REACH 2010. *Proceedings of the Survey Research Methods Section*, American Statistical Association.

Murray, M. C., Battaglia, M., & Cardoni, J. (2004). Enhancing data collection from "other language" households. In S. B. Cohen & J. M. Lepkowski (Eds.), *Eighth Conference on Health Survey Research Methods* (pp. 119–114). Hyattsville, MD: National Center for Health Statistics.

Nathan, G. (2001). Telesurvey methodologies for household surveys: a review and some thought for the future. *Survey Methodology, 27*, 7–31.

Nathan, G. & Eliav, T. (1988). Comparison of measurement errors for telephone interviewing and home visits by misclassification models. *Journal of Official Statistics, 4*, 363–374.

National Center for Health Statistics. (2006). *2004 National Immunization Survey Methodology Report.* (Unpublished internal research report.) Hyattsville, MD: National Center for Health Statistics.

National Household Education Survey. (2004). *National Household Education Surveys Program: 2001 Methodology Report.* (NCES Publication No. 2005071). Retrieved December 14, 2006, from http://nces.ed.gov/pubsearch/pubsinfo.asp?pubid=2005071

National Research Council. (1979). *Privacy and Confidentiality as Factors in Survey Response.* Washington DC: National Academy Press.

National Survey of America's Families. (2003). 2002 NSAF response rates. (Publication No. 8). Retrieved December 14, 2006, from http://www.urban.org/UploadedPDF/900692_2002_Methodology_8.pdf

Nebot, M., Celentano, D. D., Burwell, L., Davis, A., Davis, M., Polacsek, M., & Santelli, J. (1994). AIDS and behavioural risk factors in women in inner city Baltimore: a comparison of telephone and face to face surveys. *Journal of Epidemiology and Community Health, 48*, 412–418.

Neilsen, J. (1999). *Designing Web Usability: The Practice of Simplicity.* Indianapolis: New Riders.

Nielsen Media (2005). Accounting for cell phones in telephone survey research in the U.S. Retrieved October 21, 2006, from http://www.nielsenmedia.com/cellphonesummit/statements.html.

Nelson, D. E., Powell-Griner, E., Town, M., & Kovar, M. G. (2003). A comparison of national estimates from the National Health Interview Survey and the Behavioral Risk Factor Surveillance System. *American Journal of Public Health, 93*, 1335–1341.

Nicolaas, G. & Lynn, P. (2002). Random-digit dialing in the U.K.: viability revisited. *Journal of the Royal Statistical Society: Series A (Statistics in Society), 165*, 297–316.

Noble, I., Moon, N., & McVey, D. (1998). Bringing it all back home: using RDD telephone methods for large scale social policy and opinion research in the U.K. *Journal of the Market Research Society, 40*, 93–120.

Nolin, M. J., Montaquila, J., Nicchitta, P., Collins-Hagedorn, M., & Chapman, C. (2004). *National Household Education Surveys Program 2001: Methodology Report* (NCES 2005-071). Washington, DC: U.S. Department of Education, National Center for Education Statistics.

Norman, K. L. (1991). The psychology of menu selection: designing cognitive control of the human/computer interface. Norwood, NJ: Ablex Publishing.

Norris, D. & Hatcher, J. (1994). The impact of interviewer characteristics on response in a national survey of women, presented at the *Joint Statistical Meetings*, Toronto, Canada.

Nurmela, J., Melkas, T., Sirkiä, T., Ylitalo, M., & Mustonen, L. (2004). Finnish people's communication capabilities in interactive society of the 2000s. *Bulletin of Statistics, 4*, 34.

Nusser, S. & Thompson, D. (1998). Web-based survey tools. *Proceedings of the Survey Research Methods Section*, American Statistical Association, 951–956.

Oakes, T. W., Friedman, G. D., & Seltzer, C. C. (1973). Mail survey response by health status of smokers, nonsmokers, and ex-smokers. *American Journal of Epidemiology, 98*, 50–55.

O'Brien, E. M., Mayer, T. S., Groves, R. M., & O'Neill, G. E. (2002). Interviewer training to increase survey participation. *Proceedings of the Survey Research Methods Section*, American Statistical Association, 5202–5207.

Odom, D. M. & Kalsbeek, W. D. (1999). Further analysis of telephone call history data from the behavioral risk factor surveillance system. *Proceedings of the Survey Research Methods Section*, American Statistical Association, 398–403.

Ofcom. (2005). *The Communications Market 2005: Telecommunications.* London: Ofcom.

Office for National Statistics (2001). *Labour Force Survey Users Guide–Volume 1.* Retrieved December 14, 2006, from the United Kingdom's Office of National Statistics website: http://www.statistics.gov.uk/downloads/theme_labour/LFSUGvol1.pdf

Office of Management and Budget. (2006, January). Questions and answers when designing surveys for information collections. Washington, DC: Office of Information and Regulatory Affairs. Retrieved December 14, 2006, from http://www.whitehouse.gov/omb/inforeg/pmc_survey _guidance_2006.pdf

Oksenberg, L. & Cannell, C. (1988). Effects of interviewer vocal characteristics on nonresponse. In R. M. Groves, P. P. Biemer, L. E. Lyberg, J. T. Massey, W. L. Nicholls II, & J. Waksberg (Eds.), *Telephone Survey Methodology* (pp. 257–269). New York: John Wiley & Sons, Inc.

Oksenberg, L., Coleman, L., & Cannell, C. F. (1986). Interviewers' voices and refusal rates in telephone surveys. *Public Opinion Quarterly, 50*, 97–111.

Oldendick, R. W., Bishop, G. F., Sorenson, S. B., & Tuchfarber, A. J. (1988). A comparison of the Kish and last birthday methods of respondent selection in telephone surveys. *Journal of Official Statistics, 4*, 307–318.

Oldendick, R. W. & Link, M. W. (1994). The answering machine generation. *Public Opinion Quarterly, 58*, 264–273.

Olson, L., Frankel, M., O'Connor, K. S., Blumberg, S. J., Kogan, M., & Rodkin, S. (2004, May). A promise or a partial payment: the successful use of incentives in an RDD survey. Paper presented at the *Annual Conference of the American Association for Public Opinion Research*, Phoenix AZ.

O'Muircheartaigh, C. & Campanelli, P. (1999). A multilevel exploration of the role of interviewers in survey nonresponse. *Journal of the Royal Statistical Society, Series A, 162*, 437–446.

O'Muircheartaigh, C. & Pedlow, S. (2000). Combining samples vs. cumulating cases: a comparison of two weighting strategies in NLSY97. *Proceedings of the Section on Survey Research Methods of the American Statistical Association*, 319–324.

O'Neil, M. J. (1979). Estimating the nonresponse bias due to refusals in telephone surveys. *Public Opinion Quarterly, 43*, 218–232.

O'Neil, M., Groves, R. M., & Cannell, C. C. (1979). Telephone interview introduction and refusal rates: experiment in increasing respondent cooperation. *Proceedings of the Survey Research Methods Section of the American Statistical Association*, 252–255.

O'Neill, G. E. & Sincavage, J. R. (2004). Response analysis survey: a qualitative look at response and nonresponse in the American Time Use Survey. Retrieved December 14, 2006, from http://www.bls.gov/ore/pdf/st040140.pdf. Washington, DC: U.S. Department of Labor, Bureau of Labor Statistics.

Oosterhof, A. (1999). *Developing and Using Classroom Assessments* (2nd ed.). Old Tappan, NJ: Prentice-Hall.

Oosterveld, P. & Willems, P. (2003, January). Two modalities, one answer? Combining internet and CATI surveys effectively in market research. Paper presented at the *ESOMAR Technovate Conference*, Cannes, France.

O'Rourke, D. & Blair, J. (1983). Improving random respondent selection in telephone surveys. *Journal of Marketing Research, 20*, 428–432.

O'Rourke, D. & Johnson, T. (1999). An inquiry into declining RDD response rates, Part III: a multivariate review. *Survey Research: Newsletter from the University of Illinois at Chicago Survey Research Laboratory, 30(2-3)*, 1–2. Retrieved December 14, 2006, from http://www.srl.uic.edu/publist/Newsletter/1999/99v30n2.pdf

Ott, K. (2002). Does a familiar face increase response? Using consistent interviewer assignments over multiple survey contacts. *Proceedings of the Survey Research Methods Section*, American Statistical Association, 2586–2591.

Paganini-Hill, A., Hsu, G., & Chao, A. (1993). Comparison of early and late responders to a postal health survey questionnaire. *Epidemiology, 4*, 375–379.

Pannekoek, J. (1988). Interviewer variance in a telephone survey, *Journal of Official Statistics*, 4, 375–384.

Parackal, M. (2003, July). Internet-based and mail survey: a hybrid probabilistic survey approach. Paper presented at the *Australian Wide Web Conference*, Gold Coast, Australia.

Parker, C. & Dewey, M. (2000). Assessing research outcomes by postal questionnaire with telephone follow-up. *International Journal of Epidemiology, 29*, 1065–1069.

Parsons, J., Owens, L., & Skogan, W. (2002). Using advance letters in RDD surveys: results of two experiments. *Survey Research: Newsletter from the University of Illinois at Chicago Survey Research Laboratory, 33(1)*, 1–2. Retrieved December 14, 2006, from http://www.srl.uic.edu/Publist/Srvrsch/2002/02v33n1.PDF

Patterson, B. H., Dayton, C. M., & Graubard, B. (2002). Latent class analysis of complex sample survey data: application to dietary data. *Journal of the American Statistical Association, 97*, 721–729.

Paxson, M. C., Dillman, D. A., & Tarnai, J. (1995). Improving response to business mail surveys. *Business Survey Methods*. New York: John Wiley & Sons, Inc.

Perone, C., Matrundola, G., & Soverini, M. (1999, September). A quality control approach to mobile phone surveys: the experience of Telecom Italia Mobile. In *Proceedings of the*

Association of Survey Computing International Conference, September 1999. Retrieved October 22, 2006, from http://www.asc.org.uk/Events/Sep99/Pres/perone.ppt

Perron, S., Berthelot, J. M., & Blakeney, R. D. (1991). New technologies in data collection for business surveys. *Proceedings of the American Statistical Association*, Section on Survey Research Methods, 707–712.

Peterson, G. E. & Barney, H. L. (1952). Control methods used in a study of vowels. *Journal of the Acoustical Society of America, 24*, 174–184.

Petrie, R. S., Moore, D., & Dillman, D. (1997). Establishment surveys: the effect of multi-mode sequence of response rate. *Proceedings of the Survey Research Methods Section,* American Statistical Association, 981–987.

Pew Research Center for the People and the Press (2006, May 15). The cell phone challenge to survey research: national polls not undermined by growing cell-only population. Washington, DC: Pew Research Center. Retrieved July 31, 2006, from http://people-press.org/reports/display.php3?ReportID=276

Piazza, T. (1993). Meeting the challenge of answering machines. *Public Opinion Quarterly, 57,* 219–231.

Piekarski, L. (1996). A brief history of telephone sampling. *World Opinion.* Retrieved April 1, 2006, from http://www.worldopinion.com/reference.taf?f=refi&id=1252

Piekarski, L. (2004, February). The changing telephone environment: implications for surveys. Paper preseted at the *Federal CASIC Workshop.*

Piekarski, L. (2005). Wireless Challenge: what we know and don't know. Presentation at the *Cell Phone Summit II*, New York, NY, February.

Piekarski, L., Kaplan, G., & Prestegaard, J. (1999). Telephony and telephone sampling: the dynamics of change, Presented at the *Annual Conference of the American Association of Public Opinion Research*, St. Petersburg, FL.

Pierre, F. & Y. Béland (2002). Étude sur quelques erreurs de réponse dans le cadre de I Enquéte sur la santé dans les collectivités canadiennes. *Proceedings of the Survey Methods Section*, Statistical Society of Canada, Ottawa, 69–75.

Pierzchala, M. & Manners, T. (2001). Revolutionary paradigms of Blaise. In *Proceedings of the 7th International Blaise Users Conference* (pp. 8–26). Washington, DC: Westat.

Pineau, V. & Slotwiner, D. (2003). Probability samples vs. volunteer respondents in internet research: defining potential effects on data and decision-making in marketing applications. Menlo Park, CA.: Knowledge Networks, Inc. Retrieved January 8, 2007, from http://www.knowledgenetworks.com/insights/docs/Volunteer%20white%20paper%20 11-19-03.pdf

Pinkleton, B., Reagan, J., Aaronson, D., & Ramo, E. (1994). Does "I'm not selling anything" affect response rates in telephone surveys? *Proceedings of the American Statistical Association*, 1242–1247.

Poe, G. S., Seeman, I., McLaughlin, J. K., Mehl, E. S., & Dietz, M. S. (1990). Certified versus first-class mail in a mixed-mode survey of next-of-kin respondents. *Journal of Official Statistics, 6*, 157–164.

Politz, A. & Simmons, W. (1949). An attempt to get "not at homes" into the sample without callbacks. *Journal of the American Statistical Association, 44*, 9–31.

Politz, A. & W. Simmons. (1950). A note on "An attempt to get 'not at homes' into the sample without callbacks." *Journal of the American Statistical Association, 45*, 136–137.

Pool, I. (1977). *The Social Impact of the Telephone.* Cambridge: MIT Press.

Potter, F. J. (1988). Survey of procedures to control extreme sampling weights. *Proceedings of the Section on Survey Research Methods*, American Statistical Association, 453–458.

Potter, F. J. (1990). A study of procedures to identify and trim extreme sampling weights. *Proceedings of the Section on Survey Research Methods*, American Statistical Association, 225–230.

Potter, F. J. (1993). The effect of weight trimming on nonlinear survey estimates. *Proceedings of the Section on Survey Research Methods*, American Statistical Association, 758–763.

Potter, F. J., McNeill, J. J., Williams, S. R., & Waitman, M. A. (1991). List-assisted RDD telephone surveys. *Proceedings of the Section on Survey Research Methods*, American Statistical Association, 117–122.

Pothoff, R., Manton, K. G., & Woodbury, M. A. (1993). Correcting for nonavailability bias in surveys by weighting based on number of callbacks. *Journal of the American Statistical Association, 88*, 1197–1207.

Poynter, R. (2000, September 28). We've got five years. Talk presented at a *Meeting of the Association for Survey Computing Meeting on Survey Research on the Internet*, London. Retrieved January 8, 2007, from www.asc.org.uk/Events/Sep00/Poynter.ppt

Presser, S. (1990). Can changes in context reduce vote overreporting in surveys? *Public Opinion Quarterly, 54*, 586–593.

Presser, S. & Blair, J. (1994). Survey pretesting: do different methods produce different results? *Survey Methodology, 24*, 73–104.

Presser, S. & Kenney, S. (2006). U.S. government survey research: 1984–2004. Unpublished manuscript.

Presser, S., Rothgeb, J. M., Couper, M. P., Lessler, J. T., Martin, E., Martin, J., & Singer, E. (Eds.). (2004). *Methods for Testing and Evaluating Survey Questionnaires.* New York: John Wiley & Sons, Inc.

Prinzo, O. (1998). *An Acoustic Analysis of Air Traffic Control Communication.* Washington, DC: Federal Aviation Administration.

Proctor, C. (1977). Two direct approaches to survey nonresponse: estimating a proportion with callbacks and allocating effort to raise the response rate. *Proceedings of the Social Statistics Section, American Statistical Association*, 284–290.

Purdon, S., Campanelli, P., & Sturgis, P. (1999). Interviewers' calling strategies on face-to-face interview surveys. *Journal of Official Statistics, 15*, 199–216.

Purnell, T., Idsardi, W., & Baugh, J. (1999). Perceptual and phonetic experiments on American English dialect identification. *Journal of Language and Social Psychology, 18*, 10–30.

Ramirez, C. (1997). Effects of precontacting on response and cost in self-administered establishment surveys. *Proceedings of the Survey Research Methods Section*, American Statistical Association, 1000–1005.

Ranta-aho, M. & Leppinen, A. (1997). Matching telecommunication services with user communication needs. In K. Nordby & L. Grafisk (Eds.), *Proceedings of the International Symposium on Human Factors in Telecommunications*, 401–408.

Rao, P. S. (1982). Callbacks: an update. *Proceedings of the Survey Research Methods Section*, American Statistical Association, 85–87.

Rea, L. M. & Parker, R. A. (1997). *Designing and Conducting Survey Research: A Comprehensive Guide.* San Francisco: Jossey–Bass.

Redline, C. & Dillman, D. A. (2002). The influence of alternative visual designs on respondents' performance with branching instructions in self-administered questionnaires. In R. Groves, D. Dillman, J. Eltinge, & R. Little (Eds.), *Survey Nonresponse* (pp. 179–195). New York: John Wiley & Sons, Inc.

Redline, C., Dillman, D., Carley-Baxter, L., & Creecy, R. (2005). Factors that influence reading and comprehension of branching instructions in self-administered questionnaires. *Allgemeines Statistisches Archiv* [Journal of German Statistical Society], 89, 21–38.

Reed, S. J. & Reed., J. H. (1997). The use of statistical quality control charts in monitoring interviewers. *Proceedings of the Section on Survey Research Methods*, American Statistical Association, 893–898.

Reedman, L. & Robinson, M. (1997). An improved call-scheduling method using call history and frame information. *Proceedings of the Section on Survey Research Methods*, American Statistical Association, 730–735.

Retzer, K. F., Schipani, D, & Cho, Y. I. (2004). Refusal conversion: monitoring the trends. *Proceedings of the Section on Survey Research Method, American Statistical Association*, 426–431.

Rheingold, H. (2002). *Smart Mobs: The Next Social Revolution.* Cambridge, MA: Perseus Publishing.

Rivers, D. (2000, May). Probability-based web surveying: an overview. Paper presented at the *Annual Conference of the American Association for Public Opinion Research*, Portland, OR.

Rizzo, L., Brick, M. J., & Park, I. (2004a). A minimally intrusive method for sampling persons in random digital surveys. *Public Opinion Quarterly, 68*, 267–274.

Rizzo, L., DiGaetano, R., Cadell, D., & Kulp, D. W. (1995). the women's CARE study: the results of an experiment in new sampling designs for random digit dialing surveys. *Proceedings of the Survey Research Methods Section*, American Statistical Association, 644–649.

Rizzo, L., Park, I., Hesse, B., & Willis, G. (2004b, May). Effect of incentives on survey response and survey quality: a designed experiment within the HINTS I RDD sample. Paper presented at the *2004 Annual Meeting of the American Association for Public Opinion Research*, Phoenix, AZ.

Robins, L. (1963). The reluctant respondent. *Public Opinion Quarterly*, 27, 276–286.

Rogers, A., Murtaugh, M. A., Edwards, S., & Slattery, M. L. (2004). Are we working harder for similar response rates, and does it make a difference? *American Journal of Epidemiology, 160*, 85–90.

Rogers, T. F. (1976). Interviews by telephone and in person: quality of responses and field performance. *Public Opinion Quarterly, 40*, 51–65.

Roldao, L. & Callegaro, M. (2006, January). Conducting a branding survey of cell phone users and nonusers: the Vox Populi experience in Brazil. Paper presented at the *Second International Conference on Telephone Survey Methodology*, Miami, FL.

Rosen, C. (2004). Our cell phones, ourselves. *The New Atlantis, 6*, 26–45.

Rosen, R. J. & O'Connell, D. (1997). Developing an integrated system for mixed mode data collection in a large monthly establishment survey. *Proceedings of the Survey Research Methods Section*, American Statistical Association, 198–203.

Rosenbaum, P. R. (1984a). From association to causation in observational studies: the role of tests of strongly ignorable treatment assignment. *Journal of the American Statistical Association, 79*, 41–48.

Rosenbaum, P. R. (1984b). The consequences of adjustment for a concomitant variable that has been affected by the treatment. *Journal of the Royal Statistical Society, Series A, 147*, 656–666.

Rosenbaum, P. R. & Rubin, D. B. (1983a). Assessing sensitivity to an unobserved binary covariate in an observational study with binary outcome. *Journal of the Royal Statistical Society, Series B, 45*, 212–218.

Rosenbaum, P. R. & Rubin, D. B. (1983b). The central role of the propensity score in observational studies for causal effects. *Biometrika, 70*, 41–55.

Rosenbaum, P. R. & Rubin, D. B. (1984). Reducing bias in observational studies using subclassification on the propensity score. *Journal of the American Statistical Association, 79*, 516–524.

Roth, S. B., Montaquila, J., & Brick, J. M. (2001, May). Effects of telephone technologies and call screening devices on sampling, weighting and cooperation in a random digit dialing (RDD) survey. Paper presented at the *Annual Conference of the American Association for Public Opinion Research*, Montreal, Quebec, Canada.

Rothbart, G. S., Fine, M., & Sudman, S. (1982). On finding and interviewing the needles in a haystack: the use of multiplicity sampling. *Public Opinion Quarterly, 46*, 408–421.

Roy, G. & Vanheuverzwyn, A. (2002, August). Mobile phones in sample surveys. Paper presented at the *International Conference on Improving Surveys*, Copenhagen, Denmark.

Rubin, D. B. & Thomas, N. (1992). Characterizing the effect of matching using linear propensity score methods with normal distributions. *Biometrika, 79*, 797–809.

Rubin, D. B. & Thomas, N. (1996). Matching using estimated propensity scores: relating theory to practice. *Biometrics, 52*, 249–264.

Rust, K. F. & Rao, J. N. K (1996). Variance estimation for complex surveys using replication techniques. *Statistical Methods in Medical Research, 5*, 281–310.

Sangster, R. L. (2002). Calling efforts and nonresponse for telephone panel surveys. *Proceedings of the Annual Meeting of the American Statistical Association* [CD-ROM].

Sangster, R. L. (2003). Can we improve our methods to reduce nonresponse bias in RDD surveys? *Proceedings of the Section on Survey Research Methods*, American Statistical Association, 3642–3649.

Sangster, R. L. & Meekins, B. J. (2004). Modeling the likelihood of interviews and refusals: using call history data to improve efficiency of effort in a national RDD survey. *Proceedings of the Survey Research Methods Section*, American Statistical Association, 4311–4317.

Saris, W. E. (1998). Ten years of interviewing without interviewers: the telepanel. In M. P. Couper, R. P. Baker, J. Bethlehem, C. Z. F. Clark, J. Martin, W. L. Nicholls II, & J. M. O'Reilly (Eds.), *Computer Assisted Survey Information Collection* (pp. 409–429). New York: John Wiley & Sons, Inc.

Saris, W. & Krosnick, J. A. (2000, May). The damaging effect of acquiescence response bias on answers to agree/disagree questions. Paper presented at the *Conference of the American Association for Public Opinion Research*, Portland, OR.

Särndal, C., Swensson, B., & Wretman, J. (1992). *Model Assisted Survey Sampling.* New York: Springer-Verlag.

Schaeffer, N. C. (1991). Conversation with a purpose—or conversation? Interaction in the standardized interview. In P. P. Biemer, R. M. Groves, L.E. Lyberg, N. A. Mathiowetz, & S. Sudman (Eds.), *Measurement Errors in Surveys* (pp. 367–391). New York: John Wiley & Sons, Inc.

Schaeffer, N. C. (2002). Conversation with a purpose—or conversation? Interaction in the standardized interview. In D. Maynard, H. Houtkoop-Steenstra, N. C. Schaeffer, & J. van der Zouwen (Eds.), *Standardization and Tacit Knowledge: Interaction and Practice in the Survey Interview* (pp. 95–124). New York: John Wiley & Sons, Inc.

Schaeffer, N. C. & Maynard, D. W. (1996). From paradigm to prototype and back again: interactive aspects of cognitive processing in standardized survey interviews. In N. Schwarz & S. Sudman (Eds.), *Answering Questions: Methodology for Determining Cognitive and Communicative Processes in Survey Research*, (pp. 65–88). San Francisco: Jossey-Bass Publishers.

Schejbal, J. A. & Lavrakas, P. J. (1995). Panel attrition in a dual-frame local area telephone survey. *Proceedings of the American Statistical Association*, Section on Survey Research Methods, 1035–1039.

Scherpenzeel, A. (2001). Mode effects in panel surveys: a comparison of CAPI and CATI. Bases statistiques et vues d'ensemble. Neuchâtel: Bundesamt, für Statistik, Office Fédéral de la Statistique. Retrieved December 14, 2006, from http://72.14.205.104/search?q=cache:hx3LrJ43LFoJ:www.swisspanel.ch/file/methods/capi_cati.pdf+%22Scherpenzeel%22+%22Mode+effects+*+panel%22&hl=en&gl=us&ct=clnk&cd=2

Schiffrin, D. (1987). *Discourse Markers.* Cambridge, UK: Cambridge University Press.

Schiller, J. S., Martinez, M., & Barnes, P. (2006). Early release of selected estimates based on data from the 2005 National Health Interview Survey. Hyattsville, MD: National Center for Health Statistics. Retrieved July 26, 2006, from http://www.cdc.gov/nchs/about/major/nhis/released200606.htm

Schleifer, S. (1986). Trends in attitudes toward participation in survey research. *Public Opinion Quarterly, 50*, 17–26.

Schneider, S. J., Cantor, D., Malakhoff, L., Arieira, C., Segel, P., Nguyen, K. L., & Tancreto, J. G. (2005). Telephone, internet, and paper data collection modes for the Census 2000 short form. *Journal of Official Statistics, 21*, 89–101.

Schober, M. F. (1999). Making sense of questions: an interactional approach. In M. G. Sirken, D. J. Herrmann, S. Schechter, N. Schwarz, J. M. Tanur, & R. Tourangeau (Eds.), *Cognition and Survey Research* (pp. 77–93). New York: John Wiley & Sons, Inc.

Schober, M. F. & Bloom, J. E. (2004). Discourse cues that respondents have misunderstood survey questions. *Discourse Processes, 38*, 287–308.

Schober, M. F. & Brennan, S. E. (2003). Processes of interactive spoken discourse: the role of the partner. In A. C. Graesser, M. A. Gernsbacher, & S. R. Goldman (Eds.), *Handbook of Discourse Processes* (pp. 123–164). Mahwah, NJ: Lawrence Erlbaum Associates.

Schober, M. F. & Clark, H. H. (1989). Understanding by addressees and overhearers. *Cognitive Psychology, 21*, 211–232.

Schober, M. F. & Conrad, F. G. (1997). Does conversational interviewing reduce measurement error? *Public Opinion Quarterly, 61*, 576–602.

Schober, M. F. & Conrad, F. G. (2002). A collaborative view of standardized survey interviews. In D. Maynard, H. Houtkoop-Steenstra, N. C. Schaeffer, & J. van der Zouwen (Eds.), *Standardization and Tacit Knowledge: Interaction and Practice in the Survey Interview* (pp. 67–94). New York: John Wiley & Sons, Inc.

Schober, M. F., & Conrad, F. G. (2008) Survey interviews and new communications technologies. In F. G. Conrad & M. F. Schober (Eds). *Envisioning the Survey Interview of the Future*, New York: John Wiley & Sons, Inc.

Schober, M. F., Conrad, F. G., & Bloom, J. E. (2000). Clarifying word meaning in computer-administered survey interviews. *Proceedings of the Twenty-second Annual Conference of the Cognitive Science Society* (pp. 447–452). Mahwah, NJ: Lawrence Erlbaum Associates, Publishers.

Schober, M. F., Conrad, F. G., & Fricker, S. S. (2004). Misunderstanding standardized language in research interviews. *Applied Cognitive Psychology, 18*, 169–188.

Schwarz, G. (1978). Estimating the dimension of a model. *Annals of Statistics, 6*, 461–464.

Schwartz, N. & Clore, G. L. (1996). Feelings and phenomenal experiences. In E. T. Higgins & A. Kruglanski (Eds). *Social Psychology: Handbook of Basic Principles* (pp. 433–465). New York: Guilford.

Schwarz, N., Grayson, C. E., & Knäuper B. (1998). Formal features of rating scales and the interpretation of question meaning. *International Journal of Public Opinion Research, 10*, 177–183.

Schwarz, N. & Hippler, H., (1991). Response alternatives: the impact of their choice and presentation order. In P. Biemer, R. Groves, L. Lyberg, N. Mathiowetz, & S. Sudman (Eds.), *Measurement Errors in Surveys* (pp. 41–56). New York: John Wiley & Sons, Inc.

Schwarz, N., Knäuper, B., Hippler, H., Noelle-Neumann, E., & Clark, F. (1991a). Rating scales: numeric values may change the meaning of scale labels. *Public Opinion Quarterly, 55*, 570–582.

Schwarz, N., Strack, F., Hippler, H., & Bishop, G. (1991b). The impact of administration mode on response effects in survey measurement. *Applied Cognitive Psychology, 5*, 193–212.

Schwarz, N. & Wellens, T. (1997). Cognitive dynamics of proxy responding: the diverging perspectives of actors and observers. *Journal of Official Statistics, 13*, 159–179.

Sebold, J. (1988). Survey period length, unanswered numbers, and nonresponse in telephone surveys. In R. M. Groves, P. B. Biemer, L. E. Lyberg, J. T. Massey, W. L. Nicholls II, & J. Waksberg, (Eds.), *Telephone Survey Methodology* (pp. 257–271). New York: John Wiley & Sons, Inc.

Shaiko, R. G., Dwyre, D., O'Gorman, M., Stonecash, J. M., & Vike, J. 1991. Pre-election political polling and the non-response bias issue. *International Journal of Public Opinion Research, 3*, 86–99.

Shapiro, G., Battaglia, M. P., Camburn, D. P., Massey, J. T., & Tompkins, L. I. (1995). Calling local telephone company business offices to determine the residential status of a wide class of unresolved telephone numbers in a random-digit-dialing sample. *Proceedings of the Survey Research Methods Section*, American Statistical Association, 975–980.

Shapiro, G. M., Battaglia, M. P., Hoaglin, D. C., Buckley, P., & Massey, J. T. (1996). Geographical variation in within-household coverage of households with telephones in an RDD survey. *Proceedings of the Section on Survey Research Methods*, American Statistical Association, 491–496.

Sharf, D. J. & Lehman, M. E. (1984). Relationship between the speech characteristics and effectiveness of telephone interviewers. *Journal of Phonetics, 12*, 219–228.

Sharp, D. (2003). *Call Center Operation: Design, Operation, and Maintenance*. Burlington, MA: Elsevier.

Sharp, H. & Palit, C. (1988). Sample administration with CATI: The Wisconsin Survey Research Laboratory's system. *Journal of Official Statistics, 4*, 401–413.

Sharp, L. & Frankel, J. (1983). Respondent burden: a test of some common assumptions. *Public Opinion Quarterly 47*, 36–53.

Sheikh, K. & Mattingly, S. (1981). Investigating non-response bias in mail surveys. *Journal of Epidemiology and Community Health, 35*, 293–296.

Shettle, C. & Mooney, G. (1999). Monetary incentives in U.S. Government surveys. *Journal of Official Statistics 15*, 231–250.

Shneiderman, B. (1997). *Designing the User Interface: Strategies for Effective Human-Computer Interaction*. Reading, MA: Addison Wesley.

Short, J., Williams, E. & Christie, B. (1976). *The Social Psychology of Telecommunications*. London: John Wiley & Sons, Inc.

Shrivastav, R. & Sapienza, C. (2003). Objective measures of breathy voice quality obtained using an auditory model. *Journal of the Acoustical Society of America, 114*, 2217–2224.

Shuttles, C. D., Welch, J. S., Hoover, J. B., & Lavrakas, P. J. (2002, May). The development and experimental testing of an innovative approach to training telephone interviewers to avoid refusals. Paper presented at the *Annual Conference of the American Association for Public Opinion Research*, St. Petersburg Beach, FL.

Shuttles, C. D., Welch, J. S., Hoover, J. B., & Lavrakas, P. J. (2003, May). Countering nonresponse through interviewer: avoiding refusals training, Paper presented at the *Annual Conference of the American Association for Public Opinion Research*, Nashville, TN.

Siemiatycki, J. & Campbell, S. (1984). Nonresponse bias and early versus all responders in mail and telephone surveys. *American Journal of Epidemiology, 120*, 291–301.

Silverberg, D. (2005, September 2). Everything you need to know about voice over internet protocol (VoIP). Retrieved October 22, 2006, from http://www.digitaljournal.com/article/35641/Everything_You_Need_to_Know_About_Voice_over_Internet_Protocol_VoIP

Simmons, W. (1954). A plan to account for "not at homes" by combining weighting and callbacks. *Journal of Marketing, 19*, 42–53.

Singer, E. (2002). The use of incentives to reduce nonresponse in household surveys. In R. Groves, D. Dillman, J. Eltinge, & R. Little (Eds.), *Survey Nonresponse* (pp. 163–177). New York: John Wiley & Sons, Inc.

Singer, E. (2003). Census 2000 testing, experimentation, and evaluation program topic report No. 1, TR-1. *Privacy Research in Census 2000*. Washington DC: U.S. Census Bureau.

Singer, E. (2004). Confidentiality, risk perception, and survey participation. *Chance, 17*, 30–34.

Singer, E., Frankel, M. R., & Glassman, M. B. (1983). The effect of interviewer characteristics and expectations on response. *Public Opinion Quarterly, 47*, 68–83.

Singer, E., Groves, R. M., & Corning, A. (1999a). Differential incentives: beliefs about practices, perceptions of equity, and effects on survey participation. *Public Opinion Quarterly, 63*, 251–260.

Singer, E., Hippler, H-J., & Schwarz, N. (1992). Confidentiality assurances in surveys: reassurance or threat. *International Journal of Public Opinion Research, 4*, 256–268.

Singer, E., Mathiowetz, N., & Couper, M. P. (1993). The impact of privacy and confidentiality concerns on census participation. *Public Opinion Quarterly*, 57, 465–482.

Singer, E. vanHoewyk, J., Gebler, N., Raghunathan, T., & McGonagle, K. (1999b). The effect of incentives on response rates in interviewer-mediated surveys. *Journal of Official Statistics, 15*, 217–230.

Singer, E., Van Hoewyk, J., & Maher, M. P. (1998). Does the payment of incentives create expectation effects? *Public Opinion Quarterly, 62*, 152–164.

Singer, E., Van Hoewyk, J., & Maher, M. P. (2000). Experiments with incentives in telephone surveys. *Public Opinion Quarterly, 64*, 171–188.

Singer, E., Van Hoewyk, J., & Neugebauer, R. (2003). Attitudes and behavior: the impact of privacy and confidentiality concerns on participation in the 2000 Census. *Public Opinion Quarterly, 65*, 368–384.

Singer, E., Van Hoewyk, J., & Tourangeau, R. (2001). Final report on the 1999–2000 surveys of privacy attitudes. (Contract No. 50-YABC-7-66019, Task Order No. 46-YABC-9=00002). Washington, DC: U.S. Census Bureau.

Singer, E., Von Thurn, D. R., & Miller, E. R. (1995). Confidentiality assurances and response. *Public Opinion Quarterly, 59*, 66–77.

Sirken, M. G. (1970). Household surveys with multiplicity. *Journal of the American Statistical Association, 65*, 257–266.

Sirken, M. G. (1997). Network sampling. In P. Armitage & T. Colton (Eds.), *Encyclopedia of Biostatistics* (Vol. 4, pp. 2977–2986). New York: John Wiley & Sons, Inc.

Sirken, M. G. & Casady, R. J. (1988). Sampling variance and nonresponse rates in dual frame, mixed mode surveys. In R. M. Groves, P. P. Biemer, L. E. Lyberg, J. T. Massey, W. L. Nicholls, & J. Waksberg (Eds.), *Telephone Survey Methodology* (pp. 175–188). New York: John Wiley & Sons, Inc.

Sirken, M., Jabine, T., Willis, G., Martin, E., & Tucker, C. (2000). A new agenda for inter disciplinary survey research methods. *Proceedings of the CASM II Seminar.* Hyattsville, Md: National Center for Health Statistics.

Skinner, C. J. (1989). Domain means, regression and multivariate analysis. In C. J. Skinner, D. Holt, & T. M. F. Smith (Eds.), *Analysis of Complex Surveys* (pp. 59–87), Chichester, UK: John Wiley & Sons, Inc.

Skinner, C. J. (1991). On the efficiency of raking ratio estimation for multiple frame surveys. *Journal of the American Statistical Association, 86*, 779–784.

Skinner, C. J. & Rao, J. N. K. (1996). Estimation in dual frame surveys with complex designs. *Journal of the American Statistical Association, 91*, 349–356.

Smedley, J. W. & Bayton, J. A. (1978). Evaluative race-class stereotypes by race and perceived class of subject. *Journal of Personality and Social Psychology, 36*, 530–535.

Smith, P. J., Battaglia, M. P., Huggins, V. J., Hoaglin, D. C., Rodén, A.-S., Khare, M., Ezzati-Rice, M., & Wright, R. A. (2001a). Overview of the sampling design and statistical methods used in the National Immunization Survey. *American Journal of Preventive Medicine, 20*, 17–24.

Smith, P. J., Hoaglin, D. C., Battaglia, M. P., Khare, M., & Barker, L. E. (2005). Statistical methodology of the national immunization survey: 1994–2002. National Center for Health Statistics. *Vital Health Statistics, 2*, 138.

Smith, P. J., Rao, J. N. K., Battaglia, M. P., Daniels, D., & Ezzati-Rice, T. (2001b). Compensating for provider nonresponse using propensities to form adjustment cells: the national immunization survey. *Vital and Health Statistics*, Series 2, No. 133. (DHHS Publication No. PHS 2001–1333).

Smith, T. W. (1984). Estimating nonresponse bias with temporary refusals. *Sociological Perspectives, 27*, 473–489.

Smith, T. W. (1995, May). Little things matter: a sample of how differences in questionnaire format can affect survey responses. Paper presented at the *Annual Conference of the American Association for Public Opinion Research*. Ft. Lauderdale, FL.

Smith, T. W. (2003a). Developing comparable questions in cross-national surveys. In J. A. Harkness, F. J. R. van de Vijver, & P. P. Mohler (Eds.), *Cross-Cultural Survey Methods* (pp. 69–91). Hoboken, New Jersey: John Wiley & Sons, Inc.

Smith, T. W. (2003b). A review of methods to estimate the status of cases with unknown eligibility. Report presented to the *Standard Definitions Committee of the American Association for Public Opinion at the Annual Conference*. Phoenix, AZ.

Smith, T. W. (2004). Developing and evaluating cross-national survey instruments. In S. Presser, J. M. Rothgeb, M. P. Couper, J. T. Lessler, E. Martin, J. Martin, & E. Singer (Eds.), *Methods for Testing and Evaluating Survey Questionnaires* (pp. 431–452). Hoboken, NJ: John Wiley & Sons, Inc.

Smith, V. L. & Clark, H. H. (1993). On the course of answering questions. *Journal of Memory and Language, 32*, 25–38.

Smith, W., Chey, T., Jalaludin, B., Salkeld, G., & Capon, T. (1995). Increasing response rates in telephone surveys: a randomized trial. *Journal of Public Health Medicine 17*, 33–38.

Snijkers, G. (2002). Cognitive laboratory experiences on pretesting computerized questionnaires and data quality. Unpublished doctoral dissertation, University of Utrecht, the Netherlands.

Sokolowski, J., Eckman, S., Haggerty, C., Sagar, A., & Carr, C. (2004, May). Respondent incentives: do they affect your data? Data comparability in an RDD survey. Paper presented at the *Annual Meeting of the American Association for Public Opinion Research*, Phoenix, AZ.

Sonquist, J. A., Baker, E. L., & Morgan, J. N. (1973). Searching for structure (revised edition). Unpublished manuscript, University of Michigan, Institute for Social Research.

Sperry, S., Edwards, B., Dulaney, R., & Potter, D. (1998). Evaluating interviewer use of CAPI navigation features. In M. Couper, R. Baker, J. Bethlehem, C. Clark, J. Martin, W. L. Nicholls, II, & J. M. O'Reilly (Eds.), *Computer Assisted Survey Information Collection* (pp. 351–365). New York: John Wiley & Sons, Inc.

Srinath, K. P., Battaglia, M. P., Cardoni, J., Crawford, C., Snyder, R., & Wright, R. A. (2001, May). Balancing cost and mean squared error in RDD telephone survey: the national immunization survey. Paper presented at the *Annual Meeting of the American Association for Public Opionion Research*, Montreal, Quebec, Canada.

Srinath, K. P., Battaglia, M. P., & Khare, M. (2004). A dual frame sampling design for an RDD survey that screens for a rare population. *Proceedings of the Survey Research Section of the American Statistical Association*, 4424–4429.

Stanford, V., Altvater, D., & Ziesing, C. (2001, May). Programming techniques for complex surveys in Blaise. Paper presented at the *Meeting of the International BLAISE Users Conference*, Washington, DC.

Stang, A. & Jöckel, K. H. (2004). Low response rate studies may be less biased than high response rate studies. *American Journal of Epidemiology, 159*, 204–210.

Statistics Canada. (2003). *CCHS Cycle 1.1 2000–2001 Public Use Microdata Files*. Catalogue No. 82M0013GPE. Ottawa, Ontario, Canada.

Stec, J. A., Lavrakas, P. J., & Shuttles, C. W. (2004). Gaining efficiencies in scheduling callbacks in large RDD national surveys. *Proceedings of the Survey Research Methods Section*, American Statistical Association, 4430–4437.

Stec, J. A., Lavrakas, P. J., & Stasny, E. A. (1999). Investigating unit non-response in an RDD survey. *Proceedings of the Survey Research Methods Section*, American Statistical Assocation, 919–924.

Steeh, C. (2002). Georgia state poll. Retrieved October 22, 2006, from Georgia State University, Andrew Young School of Policy Studies Web site: http://aysps.gsu.edu/srp/georgiastatepoll/index.htm

Steeh, C. (2003, May). Surveys using cellular telephones: a feasibility study. Paper presented at the *Annual Conference of the American Association for Public Opinion Research*, Nashville, TN.

Steeh, C. (2004a). A new era for telephone surveys. Paper presented at the *Annual Conference of the American Association for Public Opinion Research*, Phoenix, AZ.

Steeh C. (2004b). Is it the young and the restless who only use cellular phones? Paper presented at the *Annual Conference of the American Association for Public Opinion Research*, Phoenix, AZ.

Steeh, C. (2005a). Data quality in cell phone interviewing. Paper presented at the *Cell Phone Summit II*, New York.

Steeh, C. (2005b). Quality assessed: cellular phone surveys versus traditional telephone surveys. Paper presented at the *Annual Conference of the American Association for Public Opinion Research*, Miami, FL.

Steeh, C. (2005c). Single frame versus multiple frame designs for telephone samples. Paper presented at the *Joint Statistical Meetings*, American Statistical Association, Minneapolis, MN.

Steeh, C., Buskirk, T. D., & Callegaro, M. (2007). Using text messages in U.S. mobile phone surveys. *Field Methods, 19*, 59–75.

Steeh, C., Kirgis, N., Cannon, B., & DeWitt, J. (2001). Are they really as bad as they seem? Non-response rates at the end of the twentieth century. *Journal of Official Statistics, 17*, 227–247.

Steel, D. & Boal, P. (1988). Accessibility by telephone in Australia: implications for telephone surveys. *Journal of Official Statistics, 4*, 285–297.

Steinkamp, S. (1964). The identification of effective interviewers. *Journal of the American Statistical Association, 59*, 1165–1174.

Stephan, F. F. (1948). History of the uses of modern sampling procedures. *Journal of the American Statistical Association, 43*, 12–39.

Stinchcombe, A., Jones, C., & Sheatsley, O. (1981). Nonresponse bias for attitude questions. *Public Opinion Quarterly, 45*, 359–375.

Stock, J. S. & Hochstim, J. R. (1951). A method of measuring interviewer variability. *Public Opinion Quarterly, 15*, 322–334.

Stokes, L. & Yeh, M. (1988). Searching for causes of interviewer effects in telephone surveys. In R. M. Groves, P. P. Biemer, L. E. Lyberg, J. T. Massey, W. L. Nicholls, & J. Waksberg

(Eds.), *Telephone Survey Methodology* (pp. 357–373). New York: John Wiley & Sons, Inc.

Stokes, S. L. & Greenberg, B. S. (1990). A priority system to improve callback success in telephone surveys. *Proceedings of the Survey Research Methods Section American Statistical Association*, 742–747.

Storms, V. & Loosveldt, G. (2005, July). Who is concerned about their privacy? Socioeconomic background and consequences of the concern about privacy for cooperation with surveys. Paper presented at the *Conference of the European Association for Survey Research*, Barcelona, Spain.

Strand, E. (1999). Uncovering the role of gender stereotypes in speech perception. *Journal of Language and Social Psychology, 18*, 86–99.

Strouse, R. C. & Hall, J. W. (1997). Incentives in population based health surveys. *Proceedings of the American Statistical Association*, Survey Research Section, 952–957.

Struebbe, J. M., Kernan, J. B., & Grogan, T. J. (1986). The refusal problem in telephone surveys. *Journal of Advertising Research, 26*, 29–38.

Suchman, L. & Jordan, B. (1990). Interactional troubles in face-to-face survey interviews. *Journal of the American Statistical Association, 85*, 232–253.

Sudman, S. (1967). Quantifying interviewer quality. *Public Opinion Quarterly, 30*, 664–667.

Sudman, S. (1972). On sampling of very rare human populations. *Journal of the American Statistical Association, 67*, 335–339.

Sudman, S. (1976). *Applied Sampling.* New York: Academic Press.

Sudman, S. & Bradburn, N. (1974). *Response Effects in Surveys: A Review and Synthesis.* Chicago: Aldine Publishing Co.

Sudman, S., Bradburn, N., & Schwarz, N. (1996). *Thinking About Answers: The Application of Cognitive Process to Survey Methods.* San Francisco: Jossey-Bass.

Sudman, S. & Kalton, G. (1986). New developments in the sampling of special populations. *Annual Review of Sociology, 12*, 401–429.

Sudman, S., Sirken, M. G., & Cowan, C. D. (1988). Sampling rare and elusive populations, *Science*, *240*, 991–996.

Sudman, S. & Wansink, B. (2002). *Consumer Panels* (2nd ed.). Chicago: American Marketing Association.

Suessbrick, A., Schober, M. F., & Conrad, F. G. (2005). When do respondent misconceptions lead to survey response error? *Proceedings of the American Statistical Association*, Section on Survey Research Methods, 3929–3936.

Sun, W. (1996). Using control charts in surveys. *Proceedings of the Survey Research Methods Section*, American Statistical Association, 178–183.

Sykes, W. & Collins, M. (1988). Effects of mode of interview: experiments in the U.K. In R. M. Groves, P. P. Biemer, L. E. Lyberg, J. T. Massey, W. L. Nicholls, & J. Waksberg (Eds.), *Telephone Survey Methodology* (pp. 301–320). New York: John Wiley & Sons, Inc.

Tanzer, N. K. (2005). Developing tests for use in multiple languages and cultures: a plea for simultaneous development. In R. K. Hambleton, P. F. Merenda, & C. D. Spielberger (Eds.), *Adapting Educational and Psychological Tests for Cross-Cultural Assessment*, Mahwah, NJ: Erlbaum.

Tarnai, J. & Dillman, D. A. (1992). Questionnaire context as a source of response differences in mail versus telephone surveys. In N. Schwarz & S. Sudman (Eds.), *Context Effects in Social and Psychological Research* (pp. 115–129). New York: Springer-Verlag.

Tarnai, J. & Moore, D. L. (2004). Methods for testing and evaluating computer-assisted questionnaires. In S. Presser, J. M. Rothgeb, M. P. Couper, J. T. Lessler, E. Martin, J. Martin, & E. Singer (Eds.), *Methods for Testing and Evaluating Survey Questionnaires* (pp. 319–335). New York: John Wiley & Sons, Inc.

Tarnai, J., Rosa, E. A., & Scott, L. P. (1987, May). An empirical comparison of the Kish and the Most Recent Birthday Method for selecting a random household respondent in telephone surveys. Paper presented at the *Annual Meeting of the American Association for Public Opinion Research*, Hershey, PA.

Taylor, H. (2003a). Most people are "privacy pragmatists," who, while concerned about privacy, will sometimes trade it off for other benefits. *The Harris Poll*, 17. Retrieved December 14, 2006, from http://www.harrisinteractive.com/harris_poll/index.asp?PID=365

Taylor, S. (2003b). Telephone surveying for household social surveys: the good, the bad and the ugly. *Social Survey Methodology, 52*, 10–21.

Tedesco, H., Zuckerberg, R. L., & Nichols, E. (1999, September). Designing surveys for the next millennium: web-based questionnaire design issues. Paper presented at the *Third ASC International Conference*, Edinburgh, Scotland.

Teitler, J., Reichman, N., & Sprachman, S. (2003). Costs and benefits of improving response rates for a hard-to-reach population. *Public Opinion Quarterly*, 67, 126–138.

Terhanian, G. (2000, May). How to produce credible, trustworthy information through internet-based survey research. Paper presented at the *Annual Meeting of the American Association for the Public Opinion Research*, Portland, OR.

Terhanian, G. & Bremer, J. (2000). *Confronting the Selection-Bias and Learning Effects of Problems Associated with Internet Research*. Rochester, NY: Harris Interactive.

Terhanian, G., Bremer, J., Smith, R., & Thomas, R. (2000). *Correcting Data from Online Survey for the Effects of Nonrandom Selection and Nonrandom Assignment*. Rochester, NY: Harris Interactive.

Thornberry, O. T. & Massey, J. T. (1988). Trends in United States telephone coverage across time and subgroups. In R. M. Groves, P. P. Biemer, L. E. Lyberg, J. T. Massey, W. L. Nicholls, & J. Waksberg (Eds.), *Telephone Survey Methodology* (pp. 25–49). New York: John Wiley & Sons, Inc.

Thorpe, L. E., Gwynn, R. C., Mandel-Ricci, J., Roberts, S., Tsoi, B., Berman, L., Porter, K., Ootchega, Y., Curtain, L. R., Montaquila, J., Mohadjer, L., & Frieden, T. R. (2006, July). Study design and participation rates of the New York City Health And Nutrition Examination Survey, 2004. *Preventing Chronic Disease: Public Health Research, Practice, and Policy*, 3 Retrieved, December 18, 2006, from http://www.cdc.gov/pcd/issues/2006/jul/05_0177.htm

Thurkow, N. M., Bailey, J. S., & Stamper, M. R. (2000). The effects of group and individual monetary incentives on productivity of telephone interviewers. *Journal of Organizational Behavior Management, 20*, 3–26.

Tjøstein, I., Thalberg, S., Nordlund, B., & Vestgården, J. V. (2004, February). Are the mobile phone users ready for MCASI—mobile computer assisted self-interviewing. Paper presented at the *Technovate 2: Worldwide Market Research Technology and Innovation*, Barcelona, Spain.

Thompson, S. K. & Seber, G. A. F. (1996). *Adaptive Sampling.* New York: John Wiley & Sons, Inc.

Tompson, T. & Kennedy, C. (2006, June). A dual frame RDD study of cell phone usage and political attitudes. Presentation to the *New York Chapter of the American Association for Public Opinion Research*, New York.

Tortora, R. D. 2004. Response trends in a national random digit dial survey. *Metodološki zvezki, 1*, 21–32.

Totten, J. W., Lipscomb, T. J., Cook, R. A., & Lesch, W. (2005). General patterns of cell phone usage among college students: a four-state study. *Services Marketing Quarterly, 26*, 13–39.

Tourangeau, R. (1984). Cognitive science and survey methods. In T. Jabine, M. Straf, J. Tanur, & R. Tourangeau (Eds.), *Cognitive Aspects of Survey Methodology: Building a Bridge Between Disciplines* (pp. 73–100). Washington, DC: The National Academy of Sciences.

Tourangeau, R. (2004). Survey research and societal change. *Annual Review of Psychology, 55*, 775–801.

Tourangeau. R., Couper, M. P., & Conrad, F. (2004). Spacing, position, and order: interpretive heuristics for visual features of survey questions. *Public Opinion Quarterly, 68*, 368–393.

Tourangeau, R., Rips, L. J., & Rasinski, K. (2000). *The Psychology of Survey Response.* Cambridge, UK: Cambridge University Press.

Tourangeau, R., Rasinski, K., Jobe, J., Smith, T., & Pratt, W. (1997). Sources of error in a survey of sexual behavior. *Journal of Official Statistics, 13*, 341–365.

Tourangeau, R. & Smith, T. W. (1996). Asking sensitive questions: the impact of data collection mode, question format, and question context. *Public Opinion Quarterly, 60*, 275–304.

Traub, J., Pilhuj, K., & Mallett, D. T. (2005, May). You don't have to accept low survey response rates: how we achieved the highest survey cooperation rates in company history. Poster session presented at the *Annual Meeting of the American Association of Public Opinion Research*, Miami Beach, FL.

Traugott, M. W. (1987). The importance of persistence in respondent selection for preelection surveys. *Public Opinion Quarterly, 51*, 48–57.

Traugott, M. W., Groves, R. M., & Lepkowski, J. M. (1987). Using dual frame designs to reduce nonresponse in telephone surveys. *Public Opinion Quarterly, 51*, 522–539.

Traugott, M. W. & Joo, S.-H. (2003, May). Differences in the political attitudes and behavior of cell and land line telephone users. Paper presented at the *Annual Conference of the American Association for Public Opinion Research*, Nashville, TN.

Trewin, D. & Lee, G. (1988). International comparisons of telephone coverage. In R. M. Groves, P. P. Biemer, L. E. Lyberg, J. T. Massey, W. L. Nicholls, II, & J. Waksberg (Eds.), *Telephone Survey Methodology.* New York: John Wiley and Sons, Inc.

Triplett, T., Blair, J., Hamilton, T., & Kang, Y. C. (1996). Initial cooperators vs. Converted refusers: are there response behavior differences? *Proceedings of the Survey Research Methods Section*, American Statistical Association, 1038–1041.

Triplett, T., Scheib, J., & Blair, J. (2001, May). How long should you wait before attempting to convert a telephone refusal? Paper presented at the *Annual Conference of the Amercian Association of Public Opinion Research*, Montreal, Quebec, Canada.

Troldahl, V. C. & Carter, R. E. (1964). Random selection of respondents within households in phone surveys. *Journal of Marketing Research, 1*, 71–76.

Tuckel, P., Daniels, S., & Feinberg, G. (2006, May). Ownership and usage patterns of cell phones: 2000–2006. Paper presented at the *Annual Conference of the American Association for Public Opinion Research*, Montreal, Quebec, Canada.

Tuckel, P. & Feinberg, B. (1991). The answering machine poses many questions for telephone survey researchers. *Public Opinion Quarterly, 55*, 200–217.

Tuckel, P. & O'Neill, H. (1996). Screened out. *Marketing Research, 8*, 34–43.

Tuckel, P. & O'Neill, H. (2002). The vanishing respondent in telephone surveys. *Journal of Advertising Research, 42*, 26–48.

Tuckel, P. & O'Neill, H. (2005, May). Ownership and usage patterns of cell phones: 2000–2005. Paper presented at the *Annual Conference of the American Association for Public Opinion Research*, Miami Beach, FL.

Tuckel, P. & Schulman, M. (2000). The impact of leaving different answering machine messages on nonresponse rates in a nationwide RDD survey. *Proceeding of the Section on Survey Research Methods*, American Statistical Association, 901–906.

Tuckel, P., & Shukers, T. (1997). The answering machine dilemma. *Marketing Research, 9*, 4–10.

Tucker, C. (1983). Interviewer effects in telephone surveys. *Public Opinion Quarterly, 47*, 84–95.

Tucker, C. (1992). The estimation of instrument effects on data quality in the consumer expenditure diary survey. *Journal of Official Statistics, 8*, 41–61.

Tucker C., Brick, J. M., Meekins, B., & Morganstein, D. (2004). Household telephone service and usage patterns in the U.S. in 2004. *Proceedings of the Section on Survey Research Methods, American Statistical Association* [CD-ROM].

Tucker C., Brick, J. M., & Meekins, B. (2007). Household telephone service usage patterns in the United States in 2004: implications for telephone samples. *Public Opinion Quarterly*, 71, 3–22.

Tucker, C., Casady, R., & Lepkowski, J. (1991). An evaluation of the 1988 current point-of-purchase CATI feasibility test. *Proceedings of the Survey Research Methods Section*, American Statistical Association, 508–513.

Tucker, C., Casady, R. J., & Lepkowski, J. (1993). A hierarchy of list-assisted stratified sample design options. *Proceedings of the Section on Survey Research Methods*, American Statistical Association, 982–987.

Tucker, C., Dixon, J., Downey, K., & Phipps, P. (2005, April). Evaluation of nonresponse bias in the Current Employment Statistics Program. Paper presented at the Fifty-Fifth Session of the International Statistical Institute, Sydney, Australia.

Tucker, C., Lepkowski, J. M., & Piekarski, L. (2001, May). List-assisted sampling: the effect of telephone system changes on design. Paper presented at the *Annual Conference of the American Association of Public Opinion Research*, Montreal, Quebec, Canada.

Tucker, C., Lepkowski, J. M., & Piekarski, L. (2002). The current efficiency of list-assisted telephone sampling designs. *Public Opinion Quarterly, 66*, 321–338.

Turner, C. F., Forsyth, B. H., O'Reilly, J. M., Cooley, P. C., Smith, T. M., Rogers, S. M., & Miller, G. H. (1998). Automated self-interviewing and the survey measurement of sensitive behaviors. In M. P. Couper, R. P. Baker, J. Bethlehem, C. Z. F. Clark, J. Martin, W. L. Nicholls II, & J. M. O'Reilly (Eds.), *Computer Assisted Survey Information Collection* (pp. 455–473). New York: John Wiley & Sons, Inc.

United States Bureau of Labor Statistics and United States Bureau of the Census (2002). *Current Population Survey Technical Paper 63 (revised): Design and Methodology.* Washington, DC: U.S. Government Printing Office.

United States Bureau of the Census. (1989). In *Historic Statistics of the United States, Colonial Times to 1970* (Vol. II, pp. 775–810). White Plains, NY: Kraus International.

United States Bureau of the Census. (2005b). Statistical abstract. Table No. 1149: households with computers and internet access by selected characteristic: 2003. Retrieved December 14, 2006, from http://www.census.gov/compendia/statab/information_communications/

United States Bureau of the Census. (2006). Design and methodology: American Community Survey. Retrieved December 14, 2006, from http://www.census.gov/acs/www/Downloads/tp67.pdf

United States Centers for Disease Control and Prevention. (1994). Reported vaccine-preventable diseases—United States, 1993, and the childhood immunization initiative. *Morbidity and Mortality Weekly Report, 43*, 57–60.

United States Centers for Disease Control and Prevention. (2003). 2003 Behavioral risk factor surveillance system: Summary data quality reports. Retrieved December 14, 2006, from http://www.cdc.gov/brfss/technical_infodata/pdf/2003SummaryDataQualityReport.pdf

United States Centers for Disease Control and Prevention. (2006a). Behavioral risk factor surveillance system operational and user's guide, version 3.0. Retrieved December 21, 2006, from http://ftp.cdc.gov/pub/Data/Brfss/userguide.pdf

United States Department of Transportation, Bureau of Transportation Statistics. (2003). *Transportation Availability and Use Study for Persons with Disabilities, 2002.* Retrieved November 24, 2005, from http://www.bts.gov/programs/omnibus_surveys/targeted_survey/2002_national_transportation_availability_and_use_survey/public_use_data_files/

United States Federal Communications Commission. (2001). In the matter of implementation of 911 act. (FCC 01-351). Retrieved October 8, 2006, from http://wireless.fcc.gov/releases/fcc01-351.pdf

United States Federal Communications Commission. (2003). Implementation of section 6002(b) of the Omnibus Budget Reconciliation Act of 1993: annual report and analysis of competitive market conditions with respect to commercial mobile services, eighth report. Washington, DC: Wireless Telecommunications Bureau, Federal Communications Commission.

United States Federal Communications Commission. (2004). Implementation of section 6002(b) of the Omnibus Budget Reconciliation Act of 1993: annual report and analysis of competitive market conditions with respect to commercial mobile services, ninth report. Washington, DC: Wireless Telecommunications Bureau, Federal Communications Commission.

United States Federal Communications Commission. (2005). Trends in telephone service, Retrieved December 14, 2006, from http://www.fcc.gov/Bureaus/Common_Carrier/Reports/FCC-State_Link/IAD/trend605. pdf

United States Federal Communications Commission. (2006). High-speed services for internet access: status as of December 31, 2005. Retrieved October 23, 2006, from http://hraunfoss.fcc.gov/edocs_public/attachmatch/DOC-266596A1.pdf

United States Federal Communications Commission. (2006). In the matter of universal service contribution methodology. (FCC 06-94). Retrieved October 7, 2006, from http://www.fcc.gov/omd/pra/docs/3060-0855/3060-0855=08.pdf

United States National Center for Health Statistics (2006). State and Local Area Integrated Telephone Survey: National Asthma Survey. Retrieved April 30, 2006, from http://www.cdc.gov/nchs/about/major/slaits/nsa.htm

U.S. Department of Commerce (2004). A N Nation Online: Entering the Broadband Age. Retrieved September 10, 2007, from http://www.ntia.doc.gov/reports/anol/NationOnlineBroadband04.pdf

van de Vijver, F. J. R. & Poortinga, Y. H. (2005). Conceptual and methodological issues in adapting tests. In R. K. Hambleton, P. F. Merenda, & C. D. Spielberger (Eds.), *Adapting Educational and Psychological Tests for Cross-Cultural Assessment* (pp. 39–63). Mahwah, NJ: Erlbaum.

van der Zouwen, J. (2002). Why study interaction in the survey interview? Response from a survey researcher. In D. W. Maynard, H. Houtkoop-Steenstra, N. C. Schaeffer, & J. v. d. Zouwen (Eds.), *Standardization and Tacit Knowledge: Interaction and Practice in the Survey Interview* (pp. 47–65). New York: John Wiley & Sons, Inc.

van der Zouwen, J., Dijkstra, W., & Smit, J. H. (1991). Studying respondent-interviewer interaction: the relationship between interviewing style, interviewing behavior, and response behavior. In P. P. Biemer, R. M. Groves, L. E. Lyberg, N. A. Mathiowetz, & S. Sudman (Eds.), *Measurement Errors in Surveys* (pp. 419–437), New York: John Wiley & Sons, Inc.

Van Goor, H. & Rispens, S. (2004). A middle class image of society: A study of undercoverage and nonresponse bias in a telephone survey. *Quality & Quantity, 38*, 35–49.

Van Liere, K. D., Baumgartner, R. M., Rathbun, P. P., & Tannenbaum, B. (1991, May). Factors affecting response rates in surveys of businesses and organizations. Paper presented at the *Annual Conference of the American Association for Public Opinion Research*, Phoenix, AZ.

Vanheuverzwyn, A. & Dudoignon, L. (2006, January). Coverage optimization of telephone surveys thanks to the inclusion of cellular phone-only stratum. Paper presented at the *Second International Conference on Telephone Survey Methodology*, Miami, FL.

Varedian, M. & Försman, G. (2002, May). Comparing propensity score weighting with other weighting methods: a case study on web data. Paper presented at the *Annual Meeting of the American Association for Public Opinion Research*, St. Petersburg Beach, FL.

Vartivarian, S. & Little, R. (2002). On the formation of weighting adjustment cells for unit nonresponse. *Proceedings of the Survey Research Methods Section*, American Statistical Association, 637–642.

Vartivarian, S. & Little, R. (2003, November). Weighting adjustments for unit nonresponse with multiple outcome variables. *The University of Michigan Department of Biostatistics Working Paper Series*, Working Paper 21. Retrieved December 14, 2006, from http://www.bepress.com/umichbiostat/paper21

Vehovar, V., Belak, E., Batagelj, Z., & Cikic, S. (2004a). Mobile phone surveys: The Slovenian case study. *Metodološki zvezki: Advances in Methodology and Statistics, 1*, 1–19.

Vehovar, V., Manfreda, L. K., Koren, G. P., & Dolnicar, V. (2004b, May). Mobile phones as a threat to the survey industry: a typical example from Europe—the case of Slovenia. Paper presented at the *Annual Conference of the American Association for Public Opinion Research*, Phoenix, AZ.

Vermunt, J. K. (1997a). Log-linear models for event histories. *Advanced Quantitative Techniques in the Social Sciences Series,* Vol. 8, Thousand Oaks, CA: Sage Publications.

Vermunt, J. K. (1997b), ℓEM: a general program for the analysis of categorical data. Unpublished manuscript, Tilburg University, the Netherlands.

Vermunt, J. K. & Magidson, J. (2005). *Latent GOLD 4.0 User's Guide.* Belmont, MA: Statistical Innovations, Inc.

Vigderhous, G. (1981). Scheduling telephone interviews: a study of seasonal patterns. *Public Opinion Quarterly, 45,* 250–259.

Virtanen, V., Sirkiä, T., & Nurmela, J. (2005, July). Reducing nonresponse by SMS reminders in three sample surveys. Paper presented at the *First Conference of the European Association for Survey Research,* Barcelona, Spain.

Viterna, J. S. & Maynard, D. W. (2002). How uniform is standardization? Variation within and across survey research centers regarding protocols for interviewing. In D. Maynard, H. Houtkoop-Steenstra, N. C. Schaeffer, & J. van der Zouwen (Eds.), *Standardization and Tacit Knowledge: Interaction and Practice in the Survey Interview* (pp. 365–397). New York: John Wiley & Sons, Inc.

Voigt, L. F., Boudreau, D. M., & Weiss, N. S. (2005, Feburary 15). Letter in response to "Studies with low response proportions may be less biased than studies with high response proportions" [Letter to the Editor]. *American Journal of Epidemiology, 161,* pp. 401–402.

Voigt, L. F., Koepsell, T. D., & Daling, J. R. (2003). Characteristics of telephone survey respondents according to willingness to participate. *American Journal of Epidemiology, 157,* 66–73.

Voogt, R. & Saris, W. (2005). Mixed mode designs: finding the balance between nonresponse bias and mode effects. *Journal of Official Statistics, 21,* 367–387.

Waite, A. (2002). *A Practical Guide to Call Center Technology.* New York: CMP Books.

Waksberg, J. (1973). The effect of stratification with differential sampling rates on attributes of subsets of the population. *Proceedings of the Section on Survey Research Methods, American Statistical Association,* 429–434.

Waksberg, J. (1978). Sampling methods for random digit dialing. *Journal of the American Statistical Association, 73,* 40–46.

Waksberg, J. (1983). A note on "Locating a special population using random digit dialing." *Public Opinion Quarterly, 47,* 576–578.

Waksberg, J., Brick, J. M., Shapiro, G., Fores-Cervantes, I., & Bell, B. (1997). Dual-frame RDD and area sample for household survey with particular focus on low-income population. *Proceedings of the Survey Research Methods Section,* American Statistical Association, 713–718.

Waksberg, J., Brick, J. M., Shapiro, G., Flores-Cervantes, I., Bell, B., & Ferraro, D. (1998). Nonresponse and coverage adjustment for a dual-frame survey. *Proceedings of Symposium 97,* New Directions in Surveys and Censuses, 193–198.

Waksberg, J., Judkins, D., & Massey, J. (1997). Geographic-based oversampling in demographic surveys of the U. S. *Survey Methodology, 23,* 61–71.

Warren, S. & Brandeis, L. (1890). The right to privacy. *Harvard Law Review, 4,*193–220.

Weeks, M. F. (1988). Call scheduling with CATI: current capabilities and methods. In R. M. Groves, P. P. Biemer, L. E. Lyberg, J. T. Massey, W. L. Nicholls, & J. Waksberg

(Eds.), *Telephone Survey Methodology* (pp. 403–420). New York: John Wiley & Sons, Inc.

Weeks, M. F. (1992). Computer-assisted survey information collection: a review of CASIC methods and their implications for survey operations. *Journal of Official Statistics, 8*, 445–465.

Weeks, M. F., Jones, B. L., Folsom, Jr., R. E., & Benrud, C. H. (1980). Optimal times to contact sample households. *Public Opinion Quarterly, 44*, 101–114.

Weeks, M.F., Kulka, R. A., & Pierson, S. A. (1987). Optimal call scheduling for a telephone survey. *Public Opinion Quarterly, 51*, 540–549.

Weinberg, E. (1983). Data collection: planning and management. In P. H. Rossi, J. D. Wright, A. B. Anderson (Eds.), *Handbook of Survey Research* (pp. 329–358). New York: Academic Press.

Werking, G., Tupek, A. R., & Clayton, R. L. (1988). CATI andtouchtone self-response applications for establishment surveys. *Journal of Official Statistics, 4*, 349–362.

Westat. (2000). *Blaise Design Guidelines: A Report From the Blaise Usability Committee.* Rockville, MD: Westat.

Westin, A. F. (1967). *Privacy and Freedom.* New York: Athenaeum.

Whitmore, R. W., Folsom, R. E., Burkheimer, G. J., & Wheeless, S. C. (1988). Within-household sampling of multiple target groups in computer-assisted telephone surveys. *Journal of Official Statistics, 4*, 299–305.

Whitmore, R. W., Mason, R. E., & Hartwell, T. D. (1985). Use of geographically classified telephone directory lists in multi-mode surveys. *Journal of the American Statistical Association, 80*, 842–844.

Whittaker, S. (2003). Mediated communication. In A. C. Graesser, M. A. Gernsbacher, & S. R. Goldman (Eds.), *Handbook of Discourse Processes* (pp. 243–286). Mahwah, NJ: Erlbaum.

Widman, L. & Vogelius, L. (2002, June). Daily reach using SMS. measuring and reporting reach in real time. Paper presented at the *WAM—Week of Audience Measurement*, Cannes, France.

Williams, S., Lu, R., & Hall, J. (2004, August). Response models in RDD surveys: utilizing Genesys telephone-exchange data. Paper presented at the *Joint Statistical Meetings*, Amercian Statistical Association, Toronto, Ontario, Canada.

Willimack, D. & Nichols, E. (2001, May). Building an alternative response process model for business surveys. Paper presented at the *Annual Meeting of the American Association for Public Opinion Research*, Montreal, Quebec, Canada.

Willis, G. (2005). *Cognitive Interviewing: A Tool for Improving Questionnaire Design.* Thousand Oaks, CA: Sage Publications.

Wilson, P., Blackshaw, N., & Norris, P. (1988). An evaluation of telephone interviewing on the British Labour Force Survey. *Journal of Official Statistics, 4*, 385–400.

Wilson, D. H., Starr, G. J., Taylor, A. W., & Dal Grande, E. (1999). Random digit dialling and electronic white pages samples compared: demographic profiles and health estimates. *Australian and New Zealand Journal of Public Health, 23*, 627–633.

Wolter, K. M. (1985). *Introduction to Variance Estimation.* New York: Springer-Verlag.

Wright, R. D. (Ed.). (1998). *Visual Attention.* New York: Oxford University Press.

Xu, M., Bates, B. J., & Schweitzer, J. C. (1993). The impact of messages on survey participation in answering machine households. *Public Opinion Quarterly 57*, 232–237.

Yansaneh, I. S. & Eltinge, J. L. (1993). Construction of adjustment cells based on surrogate items or estimated response propensities. *Proceedings of the Survey Research Methods Section*, American Statistical Association, 538–543.

Yu, J. & Cooper, H. (1983). A quantitative review of research design effects on response rates to questionnaires. *Journal of Marketing Research 20*, 36–44.

Yuan, Y. A., Allen, B., Brick, J. M., Dipko, S., Presser, S., Tucker, C., Han, D., Burns, L., & Galesic, M. (2005, May). Surveying households on cell phones—results and lessons. Paper presented at the *Annual Conference of the American Association for Public Opinion Research*, Miami Beach, FL.

Zell, E. R., Ezzati-Rice, T. M., Battaglia, M. P., & Wright, R. A. (2000). National Immunization Survey: the methodology of a vaccination surveillance system. *Public Health Reports, 115*, 65–77.

Index

Acoustic measurement, 391–392
Advance letters, 500–504, 510, 514, 518–519, 526, 531–532, 536, 563, 566, 570
Answering machines, 6–8, 15, 214, 319–320, 346–348, 504, 512, 514, 519, 540
Automated telephone answering systems, 222–223
Automatic dialing, 319, 330

Behavior coding, 286–287, 290–292
Bias. *See also* Error; Nonresponse
  estimates of 77–81, 163–165, 564–566, 575–581, 589, 593–600, 611

Call centers
  characteristics of, 25, 317–323, 326–330, 332–340, 347, 362, 369, 406, 408, 417–418
Call forwarding, 428
Call scheduling, *see* CATI: call scheduling
CASI, 11, 110, 152, 207, 299–300
CATI
  call scheduling, 21, 196, 332, 341, 345–348, 350, 352, 354, 357–358, 364, 438
  Canadian Community Health Survey, 300–314
  CAPI comparisons, 299–300
  costs, 5, 17, 104, 234, 280, 298, 319, 322, 340, 348, 357
  interviewers and organizational issues, 235, 236, 294–296
  oral translation, 233–234, 239
  questionnaire design, 19, 276–279, 281–284
  software, 321
  usability testing, 19
  web surveys, 15
Cell telephones, *see* Mobile telephones
Commercial list frames, 135, 161
Commercial or private surveys, 508
Confidentiality, *see* Privacy
Costs, 4, 24, 32, 55, 115, 123, 156, 234, 320
  balancing with survey errors, 150–152
  call centers, 330–332
  CATI, *see* CATI: costs
  incentives, 441, 473–475, 481, 484, 495–496
  interviewer, 359, 361
  Mitofsky–Waksberg design, 17, 130
  mixed mode, 151, 297, 344, 493, 563
  mobile phone charges, 442
  multipurpose surveys, 131–132
  sample management, 353
  stratified sampling, 118
  two phase sampling, 123–125

Coverage
mobile only households, 43, 88, 95–98, 100–102, 429, 430
noncoverage, 49, 60, 86, 114–117, 124, 129, 132, 135, 149, 151–153, 156–161, 163, 164, 168
nontelephone households, 17, 18, 116, 124, 129, 136, 151, 152, 160, 556
overcoverage, 435
undercoverage, 17, 54, 56, 57, 72, 74, 77, 80, 81, 83–86, 130, 435, 436, 438, 593
Cues
clarification, 218–220
in conversation, 214–218
comprehension, 220–222

Directory frame sampling, 33, 36, 53, 54, 430
Disaster recovery, 328
Discourse markers, 218
Disfluencies, 212–214, 218–219
Do not call lists, *see* Privacy: do not call lists
Dual frame designs
list assisted sampling, 36
mixed mode designs, 151–152, 168–169
mobile and landline, 103, 111, 430, 432, 435, 439, 440, 441
principles, 157–162

Error
interviewer bias, 194
interviewer error, 158, 187, 313, 349, 379, 380, 402, 403
interviewer variance, 164, 189, 191, 193, 197, 206, 234, 402
noncoverage bias, 114, 116, 149, 160, 169, 170, 176, 181, 183, 427
response bias, 74, 265, 267, 312, 313, 361, 589
response error, 24, 116, 170
sampling bias, 163
sampling error, 30, 32, 119, 151, 152, 156, 158, 159, 163, 168, 169, 304
sampling variance, 131, 165, 166, 402, 606

Follow-up studies
mail surveys, 343
multiplicity sampling, 129, 138, 141
nonresponse, 163
telephone use, 4

Government surveys, 16, 469–470, 493

Hedges, 218–219
Hesitations, 290
Household roster, 40. *See also* Respondent selection

Incentives
monetary, 20, 25, 139, 362, 430, 441, 449, 471, 473–475, 483, 493, 495, 504, 505, 532, 541, 560, 571
nonmonetary, 495, 505
prepaid, 20, 439, 495–497, 505
Interactive voice response (IVR), 11, 138, 153, 168, 222, 259, 317, 329
Interviewers
burden, 196–207
burden model, 197–198
interpenetration of assignments, 44, 214, 395, 492, 569
interviewer effects, 154, 189, 190, 234, 246, 312, 386, 402
training, 201, 204, 246, 327, 335, 336, 360, 362, 367, 370, 371, 378–382, 407, 410, 615
voice characteristics, 25, 385, 386, 407, 408
Interviewing, process model, 191–196
Internet. *See also* VoIP
access, 10, 98, 153, 257, 438, 442
call center design, 319, 320
multimode surveys, 15, 586
panels, 11
surveys, 18, 138, 297, 372, 442, 444, 473
Intraclass correlation, 15, 22, 32, 39, 103, 105, 388, 568, 589, 603

Kish method, *see* Respondent selection: objective respondent selection

Landline telephones, 53, 56–60, 66, 72, 77, 80, 86–88, 91, 92, 95–100, 102–108, 435
Length of interview, 106, 197, 206, 347, 369, 374
Leverage-salience theory, 398, 449, 474, 562
List-assisted sampling, *see* Sampling: list assisted

Missing data, *see* Nonresponse: item nonresponse
Mitofsky–Waksberg design. *See also* Sampling
  costs, 4, 5, 15–17, 34, 114, 131, 135, 341, 344
  first-stage identification and replacement, 35, 36
  sample allocation, 36, 342
  secondary calling rules, 130, 342
  weighting, 35
Mixed mode surveys, 18, 25, 54, 148, 155, 341, 343, 344, 357, 493, 495, 563
Mobile telephones, 29–31, 43, 53, 56, 57, 64, 80, 84, 86–112, 153, 160, 182, 423, 428, 435, 451
Mode effects
  background, 153–155
  telephone vs. face to face, 74, 164, 297–299, 300–314
  telephone vs. self-administered, 18, 55, 139, 149, 152–153, 155–156, 161–162, 257–258, 297–299, 340, 343, 345, 457, 464, 484
  telephone vs. web, 153, 251, 256–259, 260
Monitoring
  facility design, 323, 324, 326, 327, 332–333
  performance evaluation, 235, 374–375, 384, 401–411, 415–417
  quality control, 204
  sample management, 349–350, 354
Multiple frame designs, *see* Dual frame designs
Multiplicity sampling, *see* Sampling; Weighting adjustments

Noncontacts, 21, 22
  adjustment, 601–603, 607, 609, 616
  call scheduling, 354
  level of effort, 577
  mobile telephones, 437–438
  multiplicity sampling, 146
  trends, 540
Noncoverage, *see* Coverage
Noncoverage bias, *see* Error
Nonresponse
  adjustment, *see* Weighting adjustment
  bias, 163, 543–555, 556–558, 564–566, 584–585, 589
  continuum of resistance, 589–590
  don't know, 144, 145, 195, 199, 206, 463, 489, 493
  early cooperator, 26, 587, 588, 591–598, 600, 611, 615–616
  "hard core," 603–604
  item nonresponse, 14, 137, 188, 193, 206, 209, 307–309, 415, 455, 457, 465, 493, 500, 520, 603, 607
  noncontacts, *see* Noncontacts
  rates, *see* Response rates
  refusals, *see* Refusals
Nontelephone households, *see* Coverage

Open-ended questions, 6, 156, 196, 456

Poststratification, *see* Weighting adjustment
Privacy
  attitude trends, 467–469
  attitudes toward, 12, 23 457–464
  do not call lists, 14, 320, 331, 387, 451, 469, 470, 560, 584
  CIPSEA, 452
  European Union Directive on Data Protection, 451–452
  Federal Communications Commission 14, 30, 32, 58, 442, 443
  multiplicity sampling, 143
  nonresponse, 453–457
  Telephone Consumer Protection Act, 14, 426, 451
  training interviewers, 371, 380
Proxy reporting, 41, 42, 57, 150, 446, 469, 563, 586, 590

Question wording, *see* CATI: questionnaire design; Open-ended questions; Response scales; Sensitive questions
Questionnaire design
  CATI usability, 280
  cognitive aspects of survey methodology (CASM), 19–20
  respondent processes, 188–189, 201
  scalar questions, 251
  screen complexity, 279
  screen design standards, 281, 295–296
  visual elements, 276–278

Rare populations
background, 113–115
disproportionate sampling, 117–123
multiple frames, 126–128, 160, 161
multiplicity sampling, 128–130, 134–136
RDD modifications, 130–131
screening, 115–117
two phase sampling, 123–126
RDD. *See also* Mitofsky–Waksberg design
boundary problems, 37–38
estimation, 44–52
result codes—*see* result codes
sampling, 32–37, 42–44
sampling frame, 30–32, 42–44
Refusals
adjustment, 601–603, 607, 609, 616
call scheduling, 354
conversion attempts, 20, 204, 345, 381, 475, 500, 505, 514, 520, 523, 532, 563, 578–582
level of effort, 577
mobile telephones, 437–438, 439–440
multiplicity sampling, 142–146
rates, 510, 515
trends, 540,
Respondent burden, 464–467
Respondent selection. *See also* Household roster
methods used, 503, 512–513, 517–518, 521–522
mobile telephones, 435
nearest, last, most recent birthday, 17, 40, 41, 89, 503, 512, 516
objective respondent selection, 17, 39–41
Trodahl–Carter, 40, 41
Rizzo modification, 17, 41
Response bias, *see* Error
Response rates
AAPOR standards, 20, 146, 405, 501, 510, 511, 532, 533, 569
adjustments, 47
and bias, 562, 564–567
definition, 532–535, 535–540,
interview completion, 541–543
interviewer evaluation, 368–369, 403
mode differences, 298
models, 500–501, 516–524, 575–577
multiple mode surveys, 149
propensity scores, 179
resolution rates, 535–540
result codes, 346
screener completion, 540–541
trends, 4, 515–516, 530,
Response scales, 242, 245, 251, 260, 264, 271

Sample management systems, 340–358
Sampling
disproportionate stratified, 117–123
dual frame, 37, 126–128, 156–162
frames, 29–32, 42–44, 134, 135
list assisted, 35–37, 502, 503, 517, 518, 521, 522
list frame, 33, 34
Mitofsky–Waksberg, 34, 35, 130, 131
multiple frame, 37, 126–128, 156–162
multiplicity, 25, 53, 114, 128–130, 135–139, 147
RDD, 34
stratification, 44, 117–122, 126, 127, 131, 179, 181, 567, 605
two-phase, 123–126
within households, *see* Respondent selection
Scales, *see* Response scales
Screening
caller ID, 6, 8, 15, 21, 320, 347, 387, 519, 560
rare groups, 115–117, 128–132
tritone, 6
Sensitive questions, 155, 193, 201, 203, 206, 207, 213, 230, 260, 298
Social desirability, 41, 152, 193, 257, 272, 273, 298–300, 312, 313
Stratification, *see* Sampling
STS messaging, *see* Text messaging
Survey period, 41, 352, 390

Touch tone data entry (TDE), 11. *See also* IVR
Telephone households, *see* Nontelephone households
Text messaging, 21, 108, 428, 430, 440, 446
Training, 336–337, 367–371, 378–381, 384, 400, 411
Translation
oral, 231–235
oral vs. written, 245–247
respondent comprehension, 240–241
response scales, 242–243

Spanish interviewing, 503
written, 231–232

Undercoverage, *see* Coverage

Voice characteristics of interviewers, *see* Interviewers: voice characteristics
Voice mail, 8, 15, 108, 320, 433, 434, 440, 441, 446
Voice over Internet Protocol (VoIP), 10, 23, 25, 319, 323, 328, 330, 357, 423, 442–445

Waksberg, *see* Mitofsky–Waksberg design
Web surveys, 10, 15, 18, 55, 110. *See also* Internet: surveys
Weighting adjustment
base, 45, 46, 574
calibration, 45, 51, 52, 54, 172
disproportionate allocation, 122
noncoverage, 45, 48, 49, 181
nonresponse, 45, 46–48, 600–601, 605–606
phone service interruption, 176
poststratification, 141, 142, 145, 174, 179, 181, 304, 556, 561, 563, 575
propensity, 148, 173–181
within household selection, 606
Wireless broadband, 10, 23, 442, 443
Wireless telephones, *see* Mobile telephones

WILEY SERIES IN SURVEY METHODOLOGY

Established in Part by WALTER A. SHEWHART AND SAMUEL S. WILKS

The ***Wiley Series in Survey Methodology*** covers topics of current research and practical interests in survey methodology and sampling. While the emphasis is on application, theoretical discussion is encouraged when it supports a broader understanding of the subject matter.

The authors are leading academics and researchers in survey methodology and sampling. The readership includes professionals in, and students of, the fields of applied statistics, bio-statistics, public policy, and government and corporate enterprises.

ALWIN · Margins of Error: A Study of Reliability in Survey Measurement
*BIEMER, GROVES, LYBERG, MATHIOWETZ, and SUDMAN · Measurement Errors in Surveys
BIEMER and LYBERG · Introduction to Survey Quality
BRADBURN, SUDMAN, and WANSINK · Asking Questions: The Definitive Guide to Questionnaire Design—For Market Research, Political Polls, and Social Health Questionnaires, *Revised Edition*
BRAVERMAN and SLATER · Advances in Survey Research: New Directions for Evaluation, No. 70
CHAMBERS and SKINNER (editors · Analysis of Survey Data
COCHRAN · Sampling Techniques, *Third Edition*
CONRAD and SCHOBER · Envisioning the Survey Interview of the Future
COUPER, BAKER, BETHLEHEM, CLARK, MARTIN, NICHOLLS, and O'REILLY (editors) · Computer Assisted Survey Information Collection
COX, BINDER, CHINNAPPA, CHRISTIANSON, COLLEDGE, and KOTT (editors) · Business Survey Methods
*DEMING · Sample Design in Business Research
DILLMAN · Mail and Internet Surveys: The Tailored Design Method
GROVES and COUPER · Nonresponse in Household Interview Surveys
GROVES · Survey Errors and Survey Costs
GROVES, DILLMAN, ELTINGE, and LITTLE · Survey Nonresponse
GROVES, BIEMER, LYBERG, MASSEY, NICHOLLS, and WAKSBERG · Telephone Survey Methodology
GROVES, FOWLER, COUPER, LEPKOWSKI, SINGER, and TOURANGEAU · Survey Methodology
*HANSEN, HURWITZ, and MADOW · Sample Survey Methods and Theory, Volume 1: Methods and Applications
*HANSEN, HURWITZ, and MADOW · Sample Survey Methods and Theory, Volume II: Theory
HARKNESS, VAN DE VIJVER, and MOHLER · Cross-Cultural Survey Methods
KALTON and HEERINGA · Leslie Kish Selected Papers
KISH · Statistical Design for Research
*KISH · Survey Sampling
KORN and GRAUBARD · Analysis of Health Surveys
LEPKOWSKI, TUCKER, BRICK, DE LEEUW, JAPEC, LAVRAKAS, LINK, and SANGSTER (editors) · Advances in Telephone Survey Methodology
LESSLER and KALSBEEK · Nonsampling Error in Surveys
LEVY and LEMESHOW · Sampling of Populations: Methods and Applications, *Third Edition*
LYBERG, BIEMER, COLLINS, de LEEUW, DIPPO, SCHWARZ, TREWIN (editors) · Survey Measurement and Process Quality

*Now available in a lower priced paperback edition in the Wiley Classics Library.

MAYNARD, HOUTKOOP-STEENSTRA, SCHAEFFER, VAN DER ZOUWEN · Standardization and Tacit Knowledge: Interaction and Practice in the Survey Interview
PORTER (editor) · Overcoming Survey Research Problems: New Directions for Institutional Research, No. 121
PRESSER, ROTHGEB, COUPER, LESSLER, MARTIN, MARTIN, and SINGER (editors) · Methods for Testing and Evaluating Survey Questionnaires
RAO · Small Area Estimation
REA and PARKER · Designing and Conducting Survey Research: A Comprehensive Guide, *Third Edition*
SARIS and GALLHOFER · Design, Evaluation, and Analysis of Questionnaires for Survey Research
SÄRNDAL and LUNDSTRÖM · Estimation in Surveys with Nonresponse
SCHWARZ and SUDMAN (editors) · Answering Questions: Methodology for Determining Cognitive and Communicative Processes in Survey Research
SIRKEN, HERRMANN, SCHECHTER, SCHWARZ, TANUR, and TOURANGEAU (editors) · Cognition and Survey Research
SUDMAN, BRADBURN, and SCHWARZ · Thinking about Answers: The Application of Cognitive Processes to Survey Methodology
UMBACH (editor) · Survey Research Emerging Issues: New Directions for Institutional Research No. 127
VALLIANT, DORFMAN, and ROYALL · Finite Population Sampling and Inference: A Prediction Approach